PRENTICE-HALL BIOLOGICAL SCIENCE SERIES

William D. McElroy and Carl P. Swanson, *Editors*

BIOLOGICAL PHOSPHORYLATIONS
DEVELOPMENT OF CONCEPTS

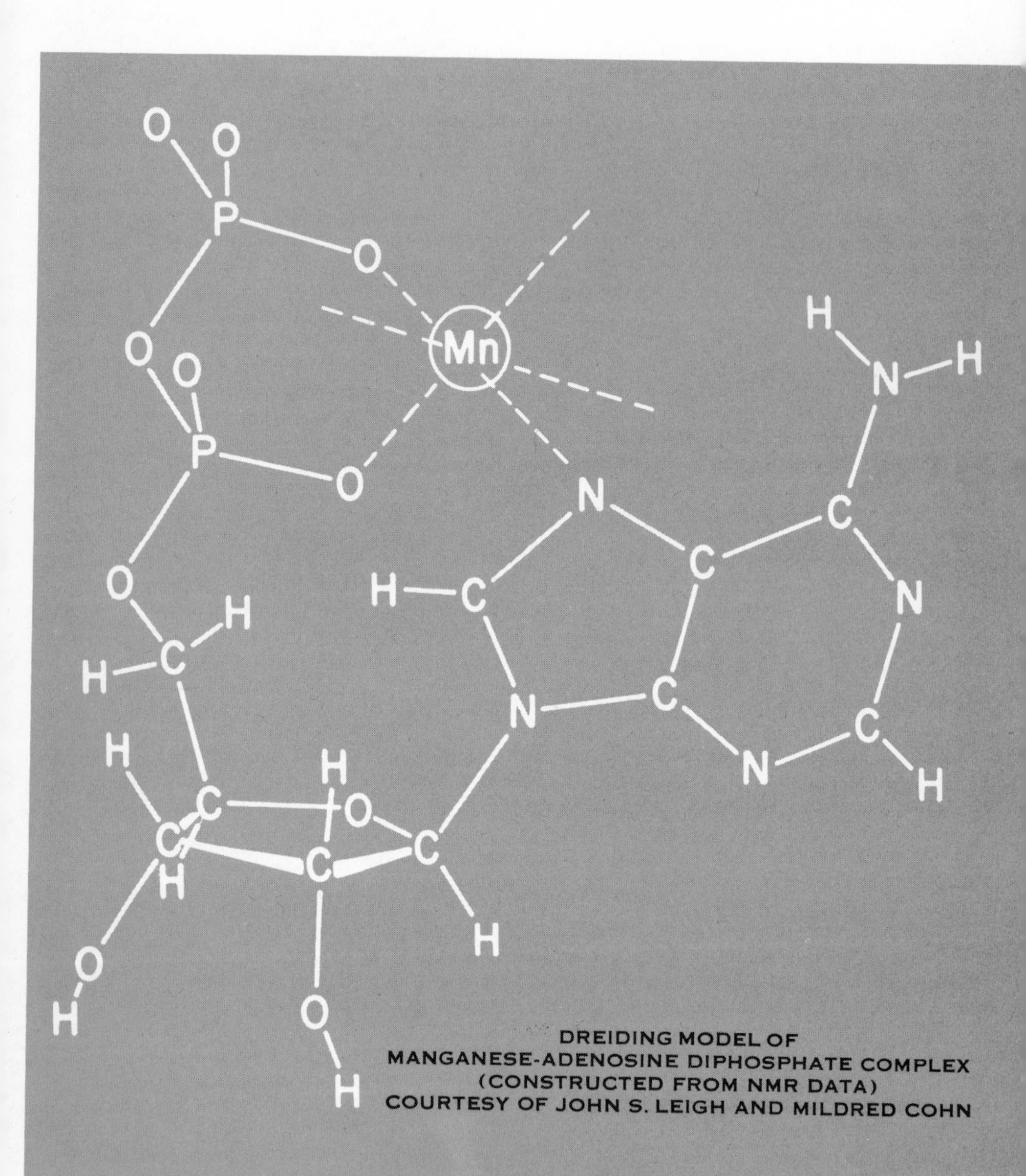

DREIDING MODEL OF
MANGANESE-ADENOSINE DIPHOSPHATE COMPLEX
(CONSTRUCTED FROM NMR DATA)
COURTESY OF JOHN S. LEIGH AND MILDRED COHN

BIOLOGICAL PHOSPHORYLATIONS

DEVELOPMENT OF CONCEPTS

HERMAN M. KALCKAR

Professor of Biological Chemistry
Harvard Medical School

Henry S. Wellcome Research Biochemist
Massachusetts General Hospital

Prentice-Hall, Inc., Englewood Cliffs, New Jersey

Current Printing (last digit): 10 9 8 7 6 5 4 3 2 1

13-076745-X

Library of Congress Catalog Card Number 70-77223

Printed in the United States of America

PRENTICE-HALL INTERNATIONAL, INC., *London*
PRENTICE-HALL OF AUSTRALIA, PTY. LTD., *Sydney*
PRENTICE-HALL OF CANADA, LTD., *Toronto*
PRENTICE-HALL OF INDIA PRIVATE LTD., *New Delhi*
PRENTICE-HALL OF JAPAN, INC., *Tokyo*

To remember
Gerty Therese Cori,
Kaj Ulrik Linderstrøm-Lang,
Einar Lundsgaard, and
Louis Rapkine

FOREWORD

Life on this planet seems to be committed to phosphorylations as the molecular coinage in bioenergetics. This is evident from studies of cells from microorganisms to man. Pasteur and Liebig had some intense arguments about life and fermentations. Pasteur emphasized the intact living cell as the smallest unit necessary for fermentation ("la fermentation c'est la vie sans air"). Later it was disclosed that cell juice — even crystalline pure fermentation enzymes — can catalyze not only fermentation but also coupled phosphorylations. This focused attention on protein structure even more since the crystalline proteins in many cases can respond to simple regulator molecules. However, the importance of more complicated structures — subcellular structures in the phosphate cycle — soon became evident. Oxidative phosphorylation depends on mitochondrial structure. ATP splitting in muscle is tied to myosin or myo-granules.

These discoveries sparked an immense interest for elucidating the structural proteins of muscle and for an understanding of cell membrane function as a basis for bioenergetics differentiation, motility, transport, perception and communication in general. Moreover, the understanding of phosphate bond energy turned out to provide a basis for obtaining insight into DNA, RNA and protein biosynthesis, and thus for the hectic development of present day molecular biology.

The present monograph highlights many of these great developments. For these reasons alone this monograph is a most welcome addition to the literature of the history of biological sciences.

Dr. Kalckar's imagination and historical perspective make this book a sparkling adventure into the subject of phosphorylation and bioenergetics. The complexity of the monograph stems from the very fact that Dr. Kalckar in this ambitious task wants to portray a highly dynamic and wide field which encompasses scores of fundamental disciplines like thermodynamics, physical chemistry, organic chemistry, biochemistry and cell biology.

On the background of this avalanche of events, it is unbelievable that the author could manage to get so many representative articles about exciting discoveries into one volume. Evidently he has succeeded as a detective excavating monumental yet condensed papers in which the new ideas are presented in a concise form. A large variety of international journals, proceedings and transactions had to be scanned. A considerable number of letters and short communications bear further witness to the hectic development in the exciting fields of biochemistry and molecular biology. Very alive indeed are the artistic translations which in my opinion will contribute greatly to enrich American and English scientific literature. With artistic sense Dr. Kalckar has transmitted the human slang, jargon, and local flavor of these restless scientists faced with their new situations including Dr. Kalckar with his oxidative phosphorylation which we managed to get added to the monograph. The unusually original and lively introductions also deserve special mention, not the least of which are the disclosures of forgotten fundamental observations and interpretations now turning out to have surprising relevance and actuality. This book should capture the interest and imagination of students as well as of advanced scholars. What strikes me about the book is not only the great amount of work, energy and lively inventiveness invested, but also the understanding of the importance of human qualities in the development of ideas. The human elements become a dominant theme in the spirited prologue and epilogue, in which interesting philosophical and dialectic problems are discussed, and which closes with a crisp saying by Niels Bohr — "What we try in science is to evolve a way of speaking unambiguously about our experiences."

Dr. Kalckar has made a really worth-while attempt in this direction. This is fundamental biochemistry at its best, and the students of biology have much to learn from this important history.

W. D. McElroy
The Johns Hopkins University
Baltimore, Maryland

ACKNOWLEDGMENTS

Over the last seven years, I have accumulated not only an immense amount of material, but also a great debt to numerous generous colleagues both inside and outside of the formal discipline of biochemistry. For their generous help and reassuring confidence, I am deeply grateful. Their patience, their unselfish and scholarly interest in the introductions and in the translated texts (whether one of the first ten editions or one of the last ten) and their helpful and valuable suggestions encouraged me to finish the huge task. Kind friends and colleagues gladly sent me their precious snapshots of distinguished scientists taken in informal surroundings. Thanks are also due to those publishers who so graciously granted permission to reprint the original texts and illustrations of these classical articles.

Herman M. Kalckar
Boston, Massachusetts

CONTENTS

PART II

OXIDATIVE PHOSPHORYLATION 169

PART VII

RELAXATION AND CONTRACTION 525

PART VIII

PRECURSORS OF POLYMERS 623

DEVELOPMENT OF CONCEPTS

PROLOGUE

In this collection of essays incorporated with reprinted papers in fundamental biochemistry, I have made an attempt to step aside from the more conventional approaches used in textbooks and in most lecture series in biochemistry and molecular biology. I believe that the present more ambitious experiment, to try to recreate a more individualistic style in biology, may be of general human interest. The book tries to portray some important endeavors in the development of biochemistry and cell physiology. I have not concealed, but rather have gone out of my way to offer illustrations of the haphazard and unpredictable advance of these two most dynamic branches of the life sciences, biochemistry and cell physiology. This somewhat different approach may turn out to be less convenient, yet, perhaps, of more lasting interest to students and scholars in biology.

Muddling Through to New Concepts

We should be presumptuous indeed if we were to pretend that a systematic and logical application of knowledge represents the only or even the main factor in the growth of the life sciences. The surprises, pleasant and unpleasant, which workers may encounter in their pursuit of ideas constitute a truly dramatic chapter in the odyssey of natural science, and especially in the development of ideas in biology. Scientists are, in fact, not unlike artists. For both are bound to experience the conflict between the careful form which their craft demands and life itself. A searching scientist, especially a biologist, can scarcely escape the kind of perturbation confronting many an artist. To quote E. B. White: "A paragraph under construction cracks and turns into a snicker—what a man does with this uninvited snicker (which may resemble a sob, at that) decides his destiny"[(1)].

The natural sciences offer striking examples of such fortuitous incidents. For instance, the early recognition of isotopes and their application as tracers in chemistry and biology stemmed from the discerning attitude of a great chemist, George Hevesy. This story begins around 1915, when Lord Rutherford received the Vienna Academy's magnificent donation of radioactive pitchblende. The great physicist hired a number of bright young chemists in order to do justice to the important "pitchblende project." Yet, this apparently feasible project, the separation of lead from radium D, defied all the skills of Rutherford's chemists. Hevesy was one of these brilliant chemists, but even he was not able to solve the riddle of lead and radium. Hevesy accepted the disintegration of the project and turned the problem upside down. If separating radium D from lead defies all known chemical methods, he reasoned, why not exploit this hard, irreducible fact? Hevesy decided to use the highly radioactive element radium D as a tracer for lead. This rather fantastic trick was put to test with enormous success, first in purely inorganic systems and later in the intact living organism. Hevesy was the sole architect of both enterprises.

The behavior of the open, yet prepared mind is illustrated repeatedly in the biological sciences. The discovery of phosphorylation is a good example. Arthur Harden of the Lister Institute originally wanted to make antiserum against Buchner's yeast zymase. Unexpectedly, the antiserum did not exert any inhibition; rather, the immune serum, as well as the normal serum, greatly boosted the action of zymase. It turned out that at least two low molecular activators of alcoholic fermentation were present in serum. One of them was inorganic phosphate. The other was cozymase (DPN). Thus, Sir Arthur Harden received his Nobel prize, not for studies of antizymase, but for his recognition of two "uninvited snickers": cozymase and phosphate as partners in alcoholic fermentation.

Meyerhof believed that he had finally shown that lactic acid formation is an obligatory step in muscle contraction. He was ready, however, to leave the established ideas behind him in view of Lundsgaard's demonstration of the alactacid muscle contraction.

These days we are inclined to believe that studies of the molecular biology of the gene and of gene transcription and translation have given us the key to a rational and fruitful bioengineering science. The future will show that this may well be the case. In any event, the subtleties of biochemical experimentation remain a source of wonder to molecular biologists as well.

In My Beginning Is My End

This may appear to be a rather erratic collection of classical and modern papers, but rather than apologize for either the choices or the omissions, I shall try to explain what may have motivated my selection. The whole idea of this collection originated almost as a joke. On several occasions I could not help but observe that many bright graduate students had failed to develop an awareness of the origin of concepts which they used daily—such as "high-energy phosphate," "Embden-Meyerhof pathway," "PeP," or oxidative phosphorylation.

Many of these concepts stem from observations made during the "golden age" of German cellular physiology, of muscle physiology in particular. I have therefore tried to do some justice to this era. Rather than limit the book to lengthy, elaborate articles, I have compromised and included brief essays, too, and some excerpts from books or extensive articles.

I have stressed originality rather than clarity. Fortunately, in many cases they go together. Although the pioneering work by Embden and by Parnas and his school is largely forgotten, it deserves to be "excavated," as does Lundsgaard's great work. The concept of "high-energy phosphate" originated, as Lipmann emphasized in his 1941 review, from the work of Meyerhof and Lundsgaard in the late twenties. I have tried to find some of the condensed yet original articles which they wrote in this period. In fact, I started my translations here, intending them only for a couple of good graduate students. Then, my colleague Dr. William McElroy stepped in and encouraged me to compile a book about phosphorylation and muscle function. I also included a few pages on mitochondrial responses and phosphorylation, since that offers many analogies to the contraction-relaxation processes in muscle.

Demonstration of the existence of respiration-dependent phosphorylation, so-called oxidative phosphorylation, did lead to various interesting developments. Belitzer in Russia and Ochoa in this country introduced the novel concept of phosphorylation linked to electron transport systems. Since Belitzer's great work has been largely unavailable to American students, I have made a special effort to present his detailed treatise on this subject, and I am able to do so, thanks to Mrs. Rose Ernsberger's translation from the Russian and the good advice of Dr. Simon Black.

The pioneer work of Carl and Gerty Cori on the glycogen phosphorylase stimulated the important development of regulatory mechanisms.

More than thirty years ago, Edsall and von Muralt and Hans Weber laid the foundation of macromolecular biology with respect to muscular contraction. Their work formed the basis of the idea of "mechanochemistry" introduced by Engelhardt and Szent-Györgyi about ten years later.

Since this diverse collection centers around the development of molecular cell biology, and especially that of molecular muscle biology, it needed, for proper direction, at least a few essays describing recent concepts of the structure of muscle and of muscle contraction. The Huxley-Hansen model of the "interdigital creep" has been a challenge to biochemists for the last decade.

It would have been tempting to enlarge on the topic and to discuss phosphorylation in general or the concept of "high-energy" nucleotidyl or pyrophosphoryl compounds, just as it would have been interesting to enlarge the topic of contraction to embrace cellular motion in general; especially since the work of Raaflaub, Lehninger, and Racker has demonstrated the possible relation of mitochondrial motion to that of muscle. This would, however, have required several volumes.

Since the available space is limited I have had to omit many relevant articles.

I have been reluctant to include my own work in this collection. In response to an earnest request by Dr. William McElroy, I have consented to reprint two or three pages of my 1938 dissertation on the discovery of oxidative phosphorylation and the first (1941) report on myokinase by Sidney Colowick and myself.

The material presented here is arranged topically, but the papers in each main section—chemistry, energetics (thermodynamics), regulation, contractile protein, etc.—appear in more or less chronological order.

The Problem of Translation and Fidelity

Most of the translations from the German are my own; the translations from Russian and French were done by others (see acknowledgments). I have tried to preserve the local and personal flavor of the originals and yet make them intelligible to students by translating into simple modern English. At this point, I would like to comment on the labor involved in a faithful and imaginative translation. In most cases I have played the role of the translator right from the beginning, in some cases only from the second to the twelfth or thirteenth version.

A truthful translation should preserve the local flavor, or at least try to re-create it. Since the subject originated largely in Germany, most of the papers reprinted here are translated from German articles. It is not for me to judge whether the original style was good or bad. Obviously, it has been easier to translate papers whose authors took pains with their formulation and wrote a clear style. But that, too, faces the translator with a rather serious responsibility, to try to do justice to such incisive descriptions as those in the monumental papers by Warburg and his coworkers, by Lundsgaard, and some by Meyerhof. Fortunately, Meyerhof wrote several of his own distinguished essays about the early development of muscle biochemistry in English. Some of his German publications (like that on hexokinase) may have been written more hurriedly, but they recall, in a fascinating way, exciting moments in a busy scientist's life. In such cases, I have tried only to make the text intelligible to students, and have abstained from smoothing over the rough descriptions. The same applies to some of Embden's and Parnas' exciting papers.

Much work, and good advice from others, have gone into my effort to preserve the local flavor and humor of colorful slang. I like to recall this example from Parnas, Ostern, and Mann: "dann geht die grosse Ammoniakbildung los." First I tried "unleashed" for "geht los," but since I was afraid of exaggerating and yet wanted to come as close as possible to the German slang of old Galicia, I did not commit myself until Dr. Carl Cori told me that the expression "geht los" was used in old Bohemia and Galicia to describe the rising of the curtain in the theater—the play is "on." Likewise, the ammonia formation is "on"!

I have translated words attached to graphs, as well as ordinates and abscissae, but I did not go all the way. For instance, in the papers from Warburg's laboratory, I purposely preserved expressions like "ln" (instead of "log"); or io/i and cm^2/mol as a dimension for the absorption coefficient β instead of the conventional "OD." Not only would translation to the more conventional "log" and "OD" involve recalculations, but it would in a way become less accurate. In this connection, remember that Otto Warburg grew up in a milieu of physicists and photochemists. In physics and photochemistry, "absorbency" or "OD" is derived from a cross-sectional term, *the molar absorption coefficient,* the dimension of which is cm^2/mol (or $cm^2/mass$). It is good for the student to search for the origin of the parameters he uses routinely. Spectrophotometry has become perhaps the most important tool in present-day biochemistry. Yet students use names like "OD" or "340" with only a faint notion, if any, about the basis or origin of these important parameters. Nevertheless, a number of expressions may need a brief explanatory comment. These will be found occasionally throughout the book, especially in the translated articles.* In rare cases where deviations from English or ambiguities in the terminology of the text in the non-translated articles, especially those published in non-English journals, may obscure the meaning, I have taken the liberty of adding explanatory comments. Otherwise, I have stayed away from correcting idiomatic deviations from standard English which may appear in journals edited outside an English-speaking country. I have corrected simple misprints.

In a number of extensive articles from which I have selected excerpts, the tables and figures as well as the bibliography have been renumbered according to their succession in the new context. However, in order not to lose the local flavor, I have religiously tried to preserve not only the names of institutions and acknowledgements to foundations but also other seemingly less important features. These include titles of authors or names of academicians who may have communicated articles, if such items appeared in the original publications. In the same line, it is, for instance, not without cultural interest to know that Keilin made his great discovery as a Fellow, a Beit Memorial Fellow. I feel that such an approach tends to be in harmony with the spirit of genuine fidelity.

*Whenever brief editorial explanatory comments in the text of translated papers appear, they will be enclosed by brackets. In this way, a clear distinction is made between the translation of the text and explanatory comments. The captions may be a simple listing of names, whimsical quips, or direct quotations whenever the opportunity presents itself. In order not to overburden the brief quotation, I have usually not felt it necessary to add bibliographic references.

Informal photographs of a few of the authors have been included. When I present photographs of scientists who are no longer living, I have added some biographical comments to the photograph captions since these people may not be listed in such books as *American Men of Science* or the *International Who's Who.*

PART I

CHEMICAL BASIS OF CELLULAR PHOSPHORYLATIONS

RECOGNITION THAT inorganic orthophosphate had a specific biological function in the energy metabolism of the cell stems from Harden and Young's discovery [*q.v.*]* that fermentation of sugar in cell-free yeast juice has an absolute requirement for phosphate and that the latter is consumed by esterification with hexose. One should note that, before Pasteur, Lavoisier's thesis stated that combustion, i.e., complete oxidation of organic material by means of oxygen, was the only type of energy-yielding reaction. Pasteur changed this thesis in his famous expression *la fermentation . . . c'est la vie sans air* . . . . Pasteur believed, however, that energy metabolism depended on the intact organization of the living cell. Hence, it was of fundamental interest when Buchner discovered sugar fermentation in cell-free yeast juice. Harden's observation of phosphate esterification coupled with fermentation represented the beginning of a new development, the study of energetic coupling in enzyme systems. Of course, Harden did not express it in this fashion because so little was then known about the pathway of fermentation and hence about the steps which may be coupled with phosphate uptake.

Harden's discovery of the specific effect of phosphate on yeast juice sugar fermentation was, as usually happens in biology, a freak observation in a control tube of a series of experiments designed for another purpose. He was trying to prepare antizymase by injecting yeast juice prepared according to Buchner into animals and testing the sera. It turned out that normal serum greatly stimulated the fermentation of yeast juice. This effect was traced back to inorganic phosphate. Harden also discovered a cofactor necessary for fermentation, later called *cozymase,* and eventually identified by Warburg and von Euler as a pyridine nucleotide.

The oxidation-reduction mechanism in various fermentations was formulated in a more explicit way by Neuberg and by Kluyver and Donker(1)† in the middle twenties. The *unity in biochemistry,* as Kluyver called it, was well demonstrated by the observations which Embden, Meyerhof, and Parnas made on muscle metabolism shortly after Harden and Young's yeast studies. In 1914 Embden and his coworkers had resumed their studies on lactic acid and phosphoric acid formation in extracts of skeletal muscle. Even before the turn of the century, investigators were talking about the existence of "meat phosphoric acid" (*Phosphofleischsäure*). Fletcher and Hopkins' observations on lactic acid formation in muscle [*q.v.*] encouraged Embden and his coworkers(2) to look for a simultaneous formation of lactic acid and phosphoric acid in muscle extracts in order to see whether both might come from the same precursor. They found good indications for such a correlation.

Hexosemonophosphates and Glycogen

Embden and later Meyerhof studied the nature of the lactic acid precursor. Various types of phosphoric acid compounds (including nucleic acids, phospholipids, etc.) were added to muscle extracts but no release of phosphate took place. Embden and Meyerhof then recalled the discoveries

*See under author's name for reference to the reprinted material.

†Figures in superscript parentheses refer to items in list of references at end of introductory material.

by Harden and Young, who identified hexosephosphoric esters in yeast fermentation juice. The history of the so-called Emden ester from muscle, which we now know is a mixture of two-thirds glucose-6-phosphate (Robison ester) and one-third fructose-6-phosphate (Neuberg ester), goes back to 1914. According to Embden and his coworkers, a crude mixture of yeast sugar phosphates incubated with muscle extract should yield lactic acid as well as phosphoric acid. They were successful in demonstrating such a correlation. The naturally occurring muscle phosphoric ester which they identified a few years later as hexosemonosphate was called *lactacidogen,* lactic-acid generator. This name was changed to *Embden ester* in honor of Embden's work.

In the early twenties, Meyerhof observed that no detectable hexosephosphate in muscle extract is formed from free glucose although Harden's phosphorylation coenzymes as well as inorganic phosphate are present, and although the muscle preparation can form the ester from glycogen. A number of free monohexoses, such as glucose mannose or fructose, added to "muskel brei" (minced muscle) or to muscle extracts did form lactic acid but at a much slower rate than from glycogen or from phosphorylated hexoses. Unlike yeast, adding Harden's cofactors or small amounts of phosphorylated hexoses failed to bring about any stimulation of glucose metabolism. However, if muscle extract was fortified with a heat-labile protein fraction from yeast, an immense stimulation of lactic acid formation from free glucose ensued. This yeast protein fraction Meyerhof called *hexokinase* because it initiated metabolism of free hexose [*q.v.*]. These investigations were carried out only a few years after Banting and Best's discovery of insulin. Hence, Meyerhof quite naturally wondered about the possible relation or even identity of hexokinase and insulin. This line of approach was not successful at that time. Meyerhof was able to provide, however, some evidence that hexokinase catalyzed the phosphorylation of glucose. This was substantiated years later by von Euler and Adler as well as by the Parnas school.

In the intact skeletal muscle, Embden ester occurs at a fairly constant level. In 1930 Lundsgaard showed that these levels are greatly increased during the onset of iodoacetate rigor(3). Later in the thirties, Carl and Gerti Cori added a noteworthy chapter to the story of hexosemonophosphate. Their early studies(4) on the increase of the steady-state level of Embden ester by epinephrine became the starting point for their studies on the phosphorolytic fission of glycogen and the discovery of the enzyme phosphorylase [*q.v.*]. Around 1936, the Coris as well as Parnas suspected the existence of a direct phosphorolytic splitting of glycogen in muscle. The same year the Coris succeeded in isolating α-glucose-1-phosphate from muscle extract(5). This work and its consequences(6) are too well known to warrant further introduction. Enzymatic phosphorolysis of nitrogen carbon bonds was later described for nucleosides, deoxynucleosides(6a), and citrulline [*q.v.;* see also(6b)].

It is now well known that the primary biological role of phosphorylase is to mobilize glucose-1-phosphate from glycogen in order to furnish the energy for metabolism. We know, from the discoveries of Leloir and his coworkers(7), that the synthesis of glycogen in the living cell is brought largely about through an enzymatic system involving the nucleotide, uridine diphosphoglucose (UDP-glucose), which functions as a glucosyl donor.

UDP-glucose is a most interesting derivative of α-glucose-1-phosphate.

The phosphate of the 1-ester is tied in pyrophosphate linkage to the phosphate or 5-uridylic acid. Leloir first discovered UDP-glucose as the crucial "cofactor" in galacto-glucose inversion[7]. Leloir *et al.*[7a] recognized its role as a powerful glucosyl donor in sucrose and glycogen synthesis. UDP-glucose is the physiological biosynthetic donor of terminal glucosyl units in glycogen, rather than glucose-1-phosphate (see section on Precursors of Macromolecules).

The biosynthesis of UDP-glucose as well as glucose-1-phosphate originates with glucose-6-phosphate. The conversion of the 6-ester to the 1-ester is catalyzed by an enzyme, phosphoglucomutase, first described in the Cori laboratory[7b]. It was Leloir [*q.v.*], however, who furthered our understanding of the mechanism of this conversion through the discovery of α-glucose-1, 6-diphosphate and the introduction of some novel ideas about transphosphorylation. These ideas were later modified by Najjar[7] who showed that the enzyme protein itself becomes phosphorylated. It is also known that the enzyme catalyzing the formation of new maltosidic linkages in glycogen from UDP-glucose is more or less dependent on glucose-6-phosphate[7a]. This seems to be an important factor in regulating glycogen synthesis in muscle (see Macromolecules and Regulation).

Adenosinetriphosphate and Guanidinephosphates

It seems appropriate here to return to the middle twenties. In 1927 Embden and Schmidt described the presence of adenylic acid in muscle. The biochemical understanding of phosphate transfer was made possible by the work of Fiske and Subbarow who, in just a couple of years, discovered, isolated, and characterized two new phosphorus compounds of primary biological importance, phosphocreatine and adenosinetriphosphate (ATP). The apparent starting point for the discovery of ATP was the study of nitrogenous compounds in muscle which had been developed into a major field in Folin's laboratory. Fiske and Subbarow were puzzled by the presence of purine nitrogen in the water-insoluble calcium precipitate. They focused all their detective skill on the insoluble fraction and succeeded in crystallizing a silver salt which they characterized in a most elegant manner as adenosinetriphosphate [*q.v.*].*

At the same time, in Meyerhof's Institute, Karl Lohmann, who had described the occurrence of inorganic pyrophosphate in skeletal muscle, was studying the possible relation of this constituent to muscle adenylic acid. Lohmann's starting point may well have been influenced by a 1929 publication of Sacks and Davenport[8] who had made a critical evaluation of Lohmann's work on the occurrence of inorganic pyrophosphate in skeletal muscle[9]. Sacks and Davenport claimed to have shown that the pyrophosphate in muscle is not free but occurs in a bound form. That same year, Lohmann himself, using electrotitration methods, undertook a more physicochemically reoriented search, and this was to become an independent description of "adenylpyrophosphate"[10], identical with Fiske and Subbarow's "adenosinetriphosphate."

The identification of muscle adenylic acid as an entity distinctly different

*Progress was greatly stimulated by Fiske's introduction about 1925[11] of quantitative colorimetric methods for determining orthophosphate and Lohmann's systematic characterization of phosphoric esters by means of their rate of phosphate release in mineral acids (hydrolysis curves at 100°). Another method, determination of enzymatic release of ammonia, also came to play a great role in the development of biochemistry[12].

from yeast adenylic acid by means of a specific enzymatic analysis was accomplished as early as 1928 by Gerhard Schmidt (at that time an important member of Embden's "brain trust"). This work contributed significantly to the development of our concept of ATP and its role in catabolic and anabolic metabolism. Schmidt's work marks one of the earliest instances in which a highly specific enzyme, in this case, 5-adenylic acid deaminase (also first described by Schmidt), was used to identify a compound which is built of three different types of units, a nitrogen base, a pentose, and phosphate. I emphasize this because van Slyke had previously introduced urease for determination of urea in blood plasma and urine. But Schmidt's deaminase [*q.v.*] can discriminate clearly not only between free adenine, adenosine, and adenylic acid but also between two phosphorylated adenosine compounds. In other words, although the enzyme catalyzes a deamination of the 6-amino group of the adenine ring, the presence of ribose and the location of the phosphate ester linkage on the ribose turn out to be equally important in the specificity of the enzyme. The mechanism of reamination of inosinic acid to 5-adenylic acid was not resolved until G. E. Carter discovered 5-adenylosuccinate and the enzyme which catalyzes the interreaction between inosinic acid and aspartic acid. The physiological role of deamination of muscle adenylic acid remained completely obscure for a long time and is still not too well understood.

Subsequent to the recognition of muscle adenylic acid, the biochemistry school of Parnas and Ostern[(14)] in Lvov, Poland, introduced the borate titration method, and by its use Parnas *et al.* were able to furnish strong indication that the phosphate group of muscle adenylic acid is esterified to the terminal 5-hydroxyl of ribose. Parnas' and Ostern's work was done long before the introduction of metaperiodate or lead tetraacetate for dentification of vicinal cis-hydroxyl groups. Thanks to their research we no longer talk about muscle adenylic acid but about 5-adenylic acid.*

*It is ironic to think now that the characterization of muscle adenylic acid as a different entity from yeast adenylic acid contributed indirectly, at least at a certain stage of development, to create an artificial barrier between our concept of the role of the adenine nucleotide of the tissues (5′-AMP, ADP, ATP) and the building blocks of ribonucleic acids. Levene in his monograph *Nucleic Acids*[(17)] gave much space to the description of yeast adenylic acid (it was, of course, not known then that this was a mixture of 2′- and 3′-adenylic acids), but spent rather less space on 5′-adenylic; ADP and ATP are barely mentioned.

Yeast adenylic acid was first obtained by autoclaving yeast ribonucleic acids in strong alkali. This was, of course, not a very gentle way of breaking down nucleic acids. Yet the hydrolytic breakdown of RNA by RNAse likewise yielded yeast adenylic acid (mixture of 2′- and 3′-adenylic acid according to Cohn and Carter), not 5′-adenylic acid. The proper roles of the so-called nucleic acid adenylic acid and 5′-adenylic acid compounds were perhaps first clarified by the observation of Gulland[(18)] around 1939 that certain nucleases will liberate what looked like 5′-nucleotides from RNA. The decisive observation we owe, however, to Cohn, Volkin, and Carter in 1950[(19, 20)], who established the presence of 5′-nucleotides in RNA and DNA. The adenylic and deoxyadenylic acids residues were deaminated by Schmidt deaminase. These observations greatly encouraged belief in 5′-nucleotides or their polyphosphates as precursors of nucleic acids. The work by Buchanan, Greenberg, and Carter pinpointed the biosynthetic pathway through inosinic acid to adenylic acid. Kornberg's discovery of adenyl transferase[(21)] and later observations of uridyl transferases[(22)] gave the clue to the mechanism and further prepared the ground. A few years later, Grunberg-Manago and Ochoa discovered the polynucleotide phosphorylase, and Kornberg, the DNA-synthesizing enzyme (pyrophosphorylase), both of which operate largely according to the same principle. In other words, phosphorylation or polyphosphorylation of 5′-nucleotides provides 5′-nucleotidyl groups of crucial importance in nucleic acid biosynthesis (see Section on Biosynthesis of Macromolecules).

Parnas and Ostern found later that a further phosphorylation of 5-adenylic acid to adenosine diphosphate or triphosphate (ADP, ATP) protected the 6-amino group from catalytical splitting by Schmidt's deaminase. Since, as mentioned, free adenine, adenosine, and 3-adenylic acid are not deaminated, the enzyme specializes only in 5-adenylic acid. A specificity of that order is rarely encountered. Perhaps one might cite Cori's phosphorylase b for which 5-adenylic acid serves as a specific cofactor[15]. Parnas and Ostern exploited the high specificity of Schmidt's deaminase in a study of transphosphorylations. Their studies were forerunners of the discovery of phospho enolpyruvate kinase. Reiss, working in the Parnas institute, later found a specific phosphatase for 5-nucleotides which was called 5-nucleotidase[16].

It is necessary at this point to recall the important observations in which Lohmann [*q.v.*] for the first time described the dephosphorylation and rephosphorylation of ATP. He found that dephosphorylation of phosphocreatine requires a coenzyme, and he identified this coenzyme with adenosine diphosphate (ADP). The latter is phosphorylated to ATP by phosphocreatine. This process can be reversed, especially at alkaline reaction. This one-step phosphorylation and dephosphorylation of the nucleotides dominates many other phosphorylations, as first found in hexokinase[23a]. Phosphorylation of 5-adenylic acid is therefore more involved; it is phosphorylated by ATP to ADP which in turn can be phosphorylated to ATP either by phosphocreatine or by another molecule of ATP, the latter reaction being catalyzed by a specific enzyme called *myokinase*[23b] or more properly *adenylate kinase*.

A guanidinephosphate which occurs in various invertebrates, especially the crustacea, was identified as arginine phosphate[24]. Study and description of phosphocreatine and arginine phosphate among various species make an interesting contribution to comparative biochemistry.

Triosephosphate and Phosphorylations

Let us resume the discussion of carbohydrate metabolism in muscle and follow some early observations by European workers. . . . Incubation of musclebrei (minced muscle) in the presence of sodium fluoride gave rise to an uptake of inorganic phosphate which is partly incorporated into a highly acid-resistant phosphoric ester compound ["Lohmann-Lipmann ester"[25]]. Embden [*q.v.*] and his associates broke the ice and identified this acid-resistant ester as a mixture of alpha glycero-phosphate and phosphoglycerate. This prompted them to suspect that the formation of glycero-phosphate and phosphoglycerate stemmed from triosephosphate.

Since Embden and his coworkers have produced very extensive and detailed papers, I prefer to reprint his first announcement and condensed description which appeared in 1933 in *Klinische Wochenschrift*. At that time, the sequence of glycolysis and fermentation was not completely understood. As we now know, the large accumulation of glycerophosphate is an artifact due to the presence of fluoride which prevents the formation of pyruvic acid from phosphoglyceric acid. Following Embden's discovery of phosphoglycerate came the striking observations by Parnas, Ostern, and Mann [*q.v.*], that ammonia formation in minced muscle is greatly suppressed by the addition of phosphoglycerate. This finding was made intelligible through the subsequent work of Meyerhof and Lohmann[26] and Needham and van Heyningen[27], who showed that phosphoglycerate is converted to phosphopyruvate and that the latter is a phosphoryl donor which can yield 1 molecule of ATP from 1 molecule of 5-adenylic acid. It is now easy to see that extensive phosphorylation of ADP to ATP will indirectly stall the formation of 5-adenylic acid and hence ammonia formation. The enzyme catalyzing transphosphorylation from phosphopyruvate to ADP has since been studied further. The enzyme is strikingly stimulated by potassium ions[28].

Ammonia formation in a whole muscle can also be observed. For instance, in Lundsgaard's 1930 studies on muscles poisoned with mono-iodoacetate, he found that ammonia is formed if the muscle is stimulated to exhaustion and rigor ensues[3]. At that stage, ATP has been converted to inosinic acid. Lundsgaard also found that hexose monophosphate (Embden ester) accumulates if the muscle is structurally intact, but that fructose-1,6-diphosphate accumulates in muscle extracts. The significance of Lundsgaard's studies on muscle metabolism in the iodoacetate-poisoned muscle is closely associated with his discovery of alactic acid muscle contraction and is discussed under Chemical Energetics and Muscular Contraction.

The remainder of this section deals with studies on coupling between fermentation or glycolysis and phosphate uptake. In 1936 Schäffner and Berl[29] found that in yeast extract hexose monophosphate can be formed from glucose and free phosphate, provided that the yeast extract is partly purified and contains both hexose diphosphate and a trace of acetaldehyde. The phosphate esterified is equivalent to the hexose diphosphate available. In crude yeast maceration juice, the same formation of hexose monophosphate from free phosphate goes on, but here a trace of hexose diphosphate can act catalytically. This can be explained by assuming that adenylic acid is present and is phosphorylated by the phosphopyruvic acid formed, and that the ATP formed transfers its phosphate to glucose. A third mechanism was indicated because the catalytic action of hexose diphosphate is not abolished by fluoride. This seems to mean that there is some way of regenerating adenyl pyrophosphate other than the phosphopyruvic mechanism. This third way was interpreted as a direct endothermic reaction between inorganic phosphate and adenylic acid, coupled with the exothermic oxido-reduction between acetaldehyde and triosephosphate.

It must be mentioned here that Lundsgaard's observations of iodoacetate inhibition of lactic acid formation in muscle had bearings on several aspects of carbohydrate metabolism and phosphorylation. Uptake of inorganic phosphate in muscle extract is still possible in the presence of iodoacetate although Adler and von Euler later showed that the oxidation-reduction step was eliminated by this agent. The uptake of phosphate must therefore stem from another source, and this turned out to be muscle glycogen, as suggested by Parnas and by the Coris. It has been mentioned that the Coris succeeded in demonstrating an enzymic phosphorolytic fission by the isolation of α-glucose-1-phosphate (Cori ester). The ester which accumulates in intact muscle is identical with the hexose monophosphate described by Embden (i.e., an equilibrium mixture of glucose-6-phosphate and fructose-6-phosphate). Glucose-6-phosphate can also be formed by the action of hexokinase on glucose if both are incubated with fresh muscle extract. This formation, as explained by von Euler and his coworkers and by Parnas and his group, was due to a direct phosphorylation of the 6-hydroxyl of the hexose (fructose of glucose) by ATP[30].

Oxidation Reductions and Phosphorylations

The availability of pure sugar phosphoric esters, such as glucose-6-phosphate and glyceraldehyde-3-phosphate, enabled Warburg and Christian and Theorell to study the oxidation-reduction steps of respiration and fermentation. In the first instance, Warburg's yellow oxidation enzyme [*q.v*] was shown by Theorell [*q.v.*] to contain flavin in the phosphorylated form. Here, in 1935, we have the first description of a phosphorylated respiratory electron transfer system.

The fermentation coenzyme cozymase was purified by von Euler and Myrbäck; in 1932, they were able to report that the Schmidt-Embden muscle adenylic acid was a constituent of cozymase[32]. During an intensified study of oxidation and fermentation in the middle thirties, Warburg and his coworkers isolated and identified the respiratory coenzyme as well as cozymase as the pyridine (nicotineamide) adenine nucleotides[32a]. Warburg summarized this discovery in a 1936 article [*q.v.*] in which he describes the properties of diphosphopyridine nucleotide (DPN) and triphosphopyridine nucleotide (TPN) and the enzymic oxidation-reductions of these nucleotides. This work is a cornerstone for studies on enzymic oxidation-reduction. It is also the first description of differential enzymatic ultraviolet spectrophotometry as applied to enzyme kinetics, based on the absorption maximum for reduced DPN at 340 mμ.

The key to the biosynthesis of the DPN type of nucleotides was found by Arthur Kornberg [*q.v.*] in 1948. His fundamental work on pyrophosphorolysis also furnished the first illustration of nucleotidyl transfer. In the biosynthesis of DPN, ATP functions as an adenyl donor. The function as adenyl donor has turned out to be the most important attribute of ATP in biosynthetic reactions. For most reactions involved in muscle contraction and recovery, however, ATP acts as a phosphoryl donor.

In 1937, Needham and Pillai[33] described a clear-cut case of coupling between phosphorylation of adenylic acid and oxidation reduction. Shortly afterward, Green, Needham, and Dewan demonstrated the reverse coupling i.e., the reduction of phosphoglycerate to phosphotriose accompanied by a release of phosphate from ATP [*q.v.*]. Meyerhof and his group described related coupled reactions in yeast fermentation. The mechanism of this coupling remained unknown, however, until a purification of the systems involved was accomplished.

Even though the chemical formulation of coupling by oxidation-reduction had yet to be furnished, the role of phosphoglycerate and phosphopyruvate in energy metabolism could nevertheless be formulated in chemical terms. It has already mentioned earlier that Embden pinpointed triose phosphate as the primary hydrogen donor in glycolysis. One should also note that in 1932 H. O. L. Fischer and E. Baer had synthesized glyceraldehyde-3-phosphate chemically. In 1939, Warburg and his group provided the first rigorous chemical formulation of the coupling between dehydrogenation and the uptake of inorganic phosphate[34]. Glyceraldehyde phosphate dehydrogenase was purified and crystallized. In the reaction with DPN a new form of phosphoglyceric acid was isolated; it was called R-*phosphoglyceric acid* [*q.v.*]. As the paper shows, it was inferred that an acid anhydride had been discovered. Ultraviolet spectra from known acid anhydrides, such as acetic acid anhydride, were recorded and similarities were found. Interestingly, the first formulation of this coupling, although subject to later amendments, focused attention on the essential feature: in the dehydrogenation of aldehyde, phosphate and not water is involved. This brings about formation of an un-ionized, substituted carboxylic group, namely, acylphosphate. This fundamental discovery was soon to be followed by Lipmann's [*q.v.*] far-reaching discovery of enzymatic acetylphosphate synthesis in *Lactobacillus Delbrückii.* The latter acid anhydride possesses, as emphasized by Lipmann, an additional important aspect, completely novel at that time, being a double feature, so to speak, donor of phosphoryl ("phosphoryl acetate"), as well as of acetyl ("acetylphosphate"). Lipmann also coined the term *phosphoglycerylphosphate* for the new R-phosphoglyceric acid of Warburg and Negelein.

A few years later Lynen found that the cofactor of acetate activation is a sulphydryl compound[34]. In his monumental Harvey Lecture[34a] Lipmann has also stressed the early discovery by Nachmansohn of a transacetylating cofactor in the central nervous system (see also Nachmansohn[34b]). The presence of a sulphydryl group in coenzyme A raised the question about the importance of this group in the coupled oxidation-reduction of glycolysis. The Coris had meanwhile crystallized muscle phosphoglyceraldehyde dehydrogenase and shown that it incorporated DPN [diphosphopridine nucleotide[35]]. J. Harting and Velick[36] brought into focus the relevance of the sulphydryl group as the primary reactant in this system. They found that adding acetaldehyde to phosphoglyceraldehyde dehydrogenase gave rise to the formation of acetyl phosphate. This work was a forerunner of Racker's well-known studies[37] on the same enzyme in which he suggested the presence of bound glutathione in it. Racker's further formulation was based on Lynen's discovery of sulphydryl as the functional group of coenzyme A[37b] and Stadtman's description of the phosphotransacetylase reaction[37c].* The uptake of phosphate could be considered as a phosphorolytic fission of an acylmercapto linkage. In the glyceraldehydephosphate dehydrogenase reaction which is coupled with phosphorylation, an SH group in the enzyme is supposed to form a mercaptal with the aldehyde group. Dehydrogenation to an acylmercapto linkage ensues. Phosphate splits this linkage to acylphosphate and free SH. The formulation of a C-S phosphorolysis† was ingenious and stimulating.

Bücher[38] continued unraveling biochemical pathways of coupled phos-

*The enzyme, phosphotransacetylase catalyzes the phosphorolysis of acetyl CoA, giving acetylphosphate and CoA. As Stadtman himself emphasizes, this reaction is essentially a phosphorolysis of an acylmercapto linkage in the sense which Lynen had tried to formulate, i.e., the nature of "active" acetate. Hence, Stadtman's discovery paved the way for the new formulation of the Warburg-Negelein reaction.

†We now have examples of specific phosphorolytic fissions of C-O and C-S bond. Enzymatic phosphorytic fissions are, however, not limited to this kind of bonds. Nucleoside phosphorylase[6a] and citrulline phosphorylase [*q.v.*] are examples of C-N phosphorolytic fissions.

phorylations, initiated with the isolation of 3-phosphoglycerylphosphate. He described an enzyme which catalyzed the reversible transfer of phosphoryl from 3-phosphoglycerylphosphate to ADP, giving 3-phosphoglycerate and ATP. It is important to realize that carboxylate groups in this fashion could be phosphorylated to acylphosphates at the expense of ATP. The acylphosphate could then be reduced one step to aldehyde. Simon Black in his biochemical characterization of the aspartate-homoserine pathway[39,40] found an example of a two-step reduction, e.g., from β-aspartylphosphate to homoserine (cf. also[41]).

The Warburg-Negelein-Bücher system accounts from both biochemical and thermodynamic points of view for a crucial step in the resynthesis of carbohydrate from pyruvate (or lactate) at the expense of ATP and oxygen.

The molecular architecture of phosphate and phosphoric esters has been studied by Cruickshank and others, and some of this work has been summarized in an essay by George Wald written in the dedicatory volume for Albert Szent-Györgyi. This essay, which is reprinted here [*q.v.*] sketches some ideas on the origin of life and early evolution. The molecular basis for the biological "fitness" of the phosphorus molecule as coinage for exchange reactions has been attributed to the presence of unoccupied three-dimensional orbitals in that element, permitting an expansion of valences beyond four, combined with a capacity to form multiple bonds. These problems are so well expressed in the Wald essay that we prefer not to try to paraphrase beyond these few remarks, but rather to refer the reader to the reprinted essay.

References

1. Kluyver, A. J., and Donker, H. J. L.: *Chem. Zelle Gewebe,* **13** (1926), 134.
2. Embden, G: *Handbuch Normalen Patholog. Physiol.*, **8** (1925), 1.
3. Lundsgaard, E.: *Biochem. Z.,* **217** (1930), 162.
4. Cori, C. F., and Cori, G. T.: *J. Biol. Chem.,* **79** (1928), 309.

4a. Parnas J. K.: *Ergeb. Enzym. Forsch.*, **6** (1937), 57.

5. Cori, C. F. and Cori, G. T.: *J. Biol. Chem.,* **131** (1939), 397.
6. ———, and ———: *Polysaccharide Phosphorylase* (Les Prix Nobel en 1947). Stockholm: Kungl. Boktryck., P. A. Norstedt and Soner, 1949.

6a. Kalckar, H. M.: *The Harvey Lectures,* **45** (1952), 11.

6b. Slade, H. D., and Slamp, W. C.: *J. Bactievol* **64** (1952), 455.

7. Leloir, L. F.: in *Phosphorus Metabolism*, eds. W. D. McElroy and B. Glass, **1** (1951) 67; Najjar, V. A.: in *The Enzymes,* eds. P. D. Boyer, H. Lardy, and K. Myrbäck, **6** (1962), 161. Springfield, Ill.: Charles C. Thomas, Publisher.

7a. ———, and Cardini, C. E.: in *The Enzymes,* eds. P. D. Boyer, H. Lardy and K. Myrbäck, **6** (1962) 317. New York: Academic Press, Inc.

7b. Cori, G. T., Colowick, S. P., and Cori, C. F.: *J. Biol. Chem.,* **124** (1938), 543.

8. Sacks, J., and Davenport: *J. Biol. Chem.,* **81** (1929), 469.
9. Lohmann, K.: *Biochem. Z.,* **202** (1928), 466.
10. ———: *Biochem. Z.,* **254** (1932), 381.
11. Fiske, C. H., and Subbarow, Y.: *J. Biol. Chem.,* **66** (1925), 375.
12. Parnas, J. K., and Heller, J.: *Biochem. Z.,* **152** (1924), 1.
13. Carter, C. E., and Cohen, L. H.: *J. Amer. Chem. Soc.,* **77** (1955), 499.

14. Klimek, R., and Parnas, J. K.: *Biochem. Z.,* **252** (1932), 392.
14a. ———, and ———: *Z. physiol. Chem.,* **217** (1933), 75.
15. Cori, G. T., Colowick, S. P., and Cori, C. F.: *J. Biol. Chem.,* **123** (1938), 375.
16. Reiss, J. L.: *Bull. Soc. Chim. Biol.,* **16** (1934), 385.
16a. ———: *Bull. Soc. Chim. Biol.,* **22** (1940); 36, see also Heppel, L. T.: *The Enzymes,* eds. Boyer, Lardy, and Myrbäck, **5** (1961), 49. New York: Academic Press, Inc.
17. Levene, P. A., and Bass, L. W.: *Nucleic Acids.* New York: Chemical Catalog Co., 1931.
18. Gulland, J. M., and Jackson, E. M.: *J. Amer. Chem. Soc.* (1938), 1492.
18a. ———, and Walsh, E. O.: *J. Amer. Chem. Soc.* (1945), 172.
19. Cohn, W. E., and Volkin, E.: *J. Biol. Chem.,* **203** (1953), 319.
20. Carter, C. E., and Cohn, W. E.: *J. Amer. Chem. Soc.,* **72** (1950), 2604.
21. Kornberg, A.: *J. Biol. Chem.,* **182** (1950), 779.
22. Munch-Petersen, A., and Kalckar, H. M., Cutolo, E., and Smith, E. E. B.: *Nature,* **172** (1953), 1036.
23. Colowick, S. P., and Kalckar, H. M.: *J. Biol. Chem.,* **148** (1943), 117.
24. Kalckar, H. M.: *J. Biol. Chem.,* **148** (1943), 127.
25. Lipmann, F., and Lohmann, K.: *Biochem. Z.,* **222** (1930), 389.
26. Lohmann, K., and Meyerhof, O.: *Biochem. Z.,* **273** (1934), 60.
27. Needham, D. M., and van Heyningen, E.: *Biochem. J.,* **29** (1935), 2040.
28. Boyer, P. D., Lardy, H., and Phillips, P. H.: *J. Biol. Chem.,* **149** (194), 529.
29. Schäffner, A., and Berl, M.: *Hoppe-Seyler's Z. f. physiol. Chem.,* **235** (1935), 122.
30. von Euler, H., and Adler, Z.: *Hoppe-Seyler's Z. f. physiol. Chem.,* **235** (1935), 122.
31a. Parnas, J. K., Lutwak-Mann, C., and Mann, J.: *Biochem. Z.,* **281** (1935), 108.
31b. Lutwak-Mann, C., and Mann, J.: *Biochem. Z.,* **281** (1935), 140.
32. Myrbäck, von Euler, H., and Hellström, H.: *Z. physiol. Chem.,* **212** (1932), 7.
32a. Warburg, O., Christian, W., and Griese, A.: *Biochem. Z.,* **282** (1935), 157.
33. ———, and Christian, W.: *Biochem. Z.,* **303** (1939), 40.
34. Lynen, F., and Reichert, E.: *Angew. Chemie,* **63** (1951), 47, 490.
34a. Lipmann, F.: *The Harvey Lectures* **44** (1951), 99. Springfield, Ill.: Charles C Thomas Publisher.
34b. Nachmahnson, D.: *J. Neurophysiol.* **6** (1943), 397.
35. Slein, M. W., Cori, G. T., and Cori, C. F.: *J. Biol. Chem.,* **159** (1945), 565.
36. Harting, J., and Velick, S. F.: *J. Biol. Chem.,* **207** (1954), 857, 867.
37. Racker, E., and Krimsky, I.: *J. Biol. Chem.,* **198** (1952), 731.
37b. Lynen, F., and Reichert, E.: *Angew. Chemie,* 1951.
37c. Stadtman, E. A.: *J. Biol. Chem.,* **196** (1952), 535.
38. Bücher, T.: *Naturwiss.,* **30** (1942), 756-57.
38a. ———: *Biochim. Biophys. Acta,* **1** (1947), 292.
39. Black, S., and Wright, N. G.: *J. Biol. Chem.,* **213** (1955), 37.
40. ———, and ———: in *McCollum-Pratt Sympos. Amino Acid Metabolism,* eds. W. D. McElroy and B. Glass. Baltimore, Md.: Johns Hopkins Univ. Press, 1955, 591.
41. Cohen, G. N., and Hirsch, M. L.: *Compt. Rend. Acad. Sci.,* **236** (1953), 1302; ———, and ———: *J. Bactierol.,* **67** (1954), 182.

THE ALCOHOLIC FERMENT OF YEAST-JUICE*

Arthur Harden, D. Sc., Ph. D. and William John Young, M. Sc.

Chemical Laboratory, Lister Institute, London, England
(Communicated by Dr. C. J. Martin, F. R. S.
Received December 8, 1905—Read February 1, 1906)

1. Effect of the Addition of Boiled and Filtered Yeast-Juice on the Fermentation of Glucose Produced by Yeast-Juice

In the course of some experiments on the action of various proteids [in this context, the term essentially means proteins] on the fermentative activity of yeast-juice, it was observed that the alcoholic fermentation of glucose by yeast-juice is greatly increased by the addition of yeast-juice which has been boiled and filtered, either when fresh or after having undergone autolysis, although this boiled liquid is itself incapable of setting up fermentation. Thus, the total fermentation produced by yeast-juice acting on excess of glucose is, as a rule, doubled by the addition of an equal volume of the boiled juice, and a further increase is produced when a greater volume is added, the sugar concentration being kept constant†.

A similar observation was previously made by Buchner and Rapp‡ in a single experiment (No. 265).

The following table embodies a few of the results obtained, the yeast-juice being prepared and the amount of carbon dioxide evolved being estimated by

*A. Harden and W. J. Young, "The Alcoholic Ferment of Yeast-Juice," *Proc. Roy. Soc.* (London), B 77 (1906). Reprinted with the permission of the Royal Society and Cambridge University Press.

†Harden and Young, Preliminary Note, 'Proc. Physiol. Soc.,' 1904, vol. 32, November 12.

‡'Ber.,' 1899, vol. 32, p. 2093.

Table I

EFFECT OF THE ADDITION OF BOILED YEAST-JUICE ON THE TOTAL FERMENTATION OF GLUCOSE BY YEAST-JUICE

No	Juice	Water	Boiled Juice	Glucose	Time	Carbon dioxide
	c.c.	c.c.	c.c.	grammes	hours	gramme
1	25	25	0	5	72	0.137
	25	0	25	5	72	0.378
2	20	20	0	4	44	0.115
	20	0	20	4	44	0.363
3	25	0	0	2.5	40	0.370
	25	0	25	5	40	0.620
4	20	40	0	6	42	0.458
	20	0	40	6	42	0.858
5	25	25	0	5	44	0.346
	25	0	25	5	44	0.709
6	25	25	0	5	48	0.110
	25	0	25	5	48	0.216
7	25	25	0	5	60	0.273
	25	0	25	5	60	0.466
8	25	25	0	5	120	0.424
	25	0	25	5	120	0.959
9	25	25	0	5	72	0.414
	25	20	5	5	72	0.546
	25	10	15	5	72	0.735
	25	5	20	5	72	0.810
	25	0	25	5	72	0.924
10	25	25	0	5	70	0.246
	25	0	25	5	70	0.356
	25	50	0	7.5	70	0.180
	25	0	50	7.5	70	0.431
	25	75	0	10	70	0.141
	25	0	75	10	70	0.515

the method previously employed by the authors.§ In every case the concentration of sugar was kept constant, and both in these and all the fermentation experiments described in this paper, toluene was added as an antiseptic.

In Experiments 1 to 5 the juice added had been autolysed before being boiled; in Nos. 6 to 8 the added juice was boiled as soon as it had been prepared. Experiments 9 and 10 show that each successive addition of boiled juice, from 0.2 to 3 volumes, produces a further increase in the amount of the fermentation.

A similar effect is produced, (1) By the precipitate produced in boiled yeast-juice by the addition of 3 volumes of alcohol (Experiment 1, Table II); (2) By the liquid formed by the autoplasmolysis of yeast, when it is allowed to stand at the air temperature for some time (Experiments 2 and 3, Table II); (3) By the liquid obtained by boiling Buchner's "Aceton-Dauerhefe" with water (Experiment 4, Table II). Further, yeast killed by acetone and ether (Aceton-Dauerhefe) reacts with boiled juice in the same way as does yeast-juice (Experiment 5, Table II).

§Harden and Young, 'Ber.,' 1904, vol. 37, p. 1052.

Table II

EFFECT OF VARIOUS SUBSTANCES IN INCREASING ALCOHOLIC FERMENTATION

No.	Yeast-juice	Addition	Glucose	Time	Carbon dioxide
	c.c.			hours	gramme
1	25	25 c.c. water	5	48	0.110
	25	25 c.c. water + precipitate by 75 per cent. alcohol from 25 c.c. boiled fresh juice	5	48	0.268
	25	Filtrate from 25 c.c. boiled fresh juice + 3 volumes alcohol, made to 25 c.c.	5	48	0.141
	25	25 c.c. water + precipitate from 25 c.c. boiled old juice by 75 per cent. alcohol	5	48	0.286
2	25	25 c.c. water	5	72	0.070
	25	25 c.c. autoplasmolysed yeast-juice, made neutral	5	72	0.189
3	25	25 c.c. water	5	72	0.084
	25	25 c.c. autoplasmolysed yeast-juice, made neutral	5	72	0.172
	25	25 c.c. water	5	72	0.475
4	25	25 c.c. aqueous infusion of 2 grammes Aceton-Dauerhefe	5	72	0.625
5	2 grammes Aceton-Dauerhefe	40 c.c. water	4	48	0.062
	"	20 c.c. water + 20 c.c. boiled juice	4	48	0.136

Table III

DIALYSIS OF BOILED YEAST-JUICE:
25 cc. YEAST-JUICE + 5 GRAMMES GLUCOSE + TOLUENE

No.	Water	Boiled juice	Residue	Dialysate	Time	Carbon dioxide
	c.c.	c.c.	c. c.	c.c.	hours	gramme
1	25	0	0	0	48	0.253
	0	25	0	0	48	0.561
	0	0	25	0	48	0.264
2	25	0	0	0	48	0.268
	0	25	0	0	48	0.497
	0	0	25	0	48	0.276
3	25	0	0	0	72	0.113
	0	25	0	0	72	0.334
	0	0	25	0	72	0.189
	0	0	0	25	72	0.334
4	25	0	0	0	48	0.154
	0	0	0	25	48	0.251

2. Dialysis of the Boiled Juice

The constituent of the boiled and filtered juice to which this effect is due is removed when the liquid is dialysed in a parchment tube, leaving an inactive residue. In the experiments detailed in the following table (Table III, Experiments 1, 2 and 3) the effect of the addition of boiled juice is compared with that produced by the residue and dialysate respectively.

In Experiment 4, the unboiled juice was dialysed, and the fact that the dialysate had a similar effect to a boiled juice shows that the active constituent exists in the original yeast-juice and is not formed during the boiling.

3. Dialysis of Yeast-Juice

The facts above detailed suggested the possibility of dividing yeast-juice into two fractions by dialysis; an inactive residue and a dialysate which, although itself inert, would be capable of rendering this residue active.

This was experimentally realised by filtering the juice through a Martin gelatin filter.*

This method of rapid dialysis was chosen because the yeast-juices at our disposal lost their activity too rapidly to permit of the ordinary process of dialysis through parchment being carried out. Either a 10- or a 7.5-per-cent. solution of gelatin was used to impregnate the Chamberland filter and the filtration was carried out under a pressure of 50 atmospheres.

Only a portion of the juice placed in the filter was actually filtered, the remainder being simply poured out of the case as soon as a sufficient quantity of filtrate had passed through. The residue adhering to the candle, which consisted of a brown viscid mass, was dissolved in water and made up to the volume of the juice filtered. Glucose was then added and one portion incubated at 25° with an equal volume of sugar solution and a second portion with an equal volume of the filtrate or of a boiled juice, containing an equal amount of glucose. Before incubation the carbon dioxide was pumped out of all the solutions. The filtrate was invariably found to be quite devoid of fermenting power, none of the enzyme having passed through the gelatin. The results (Table IV) show that in this way an almost inactive residue can be obtained which is rendered active by the addition of the filtrate (Experiments 1, 2, 3) or a boiled juice (Experiment 4).

The total fermentations observed even in the presence of the filtrate are very low, this being, at all events in part, due to the fact that in this series of experiments the original juices themselves happened to be of low fermenting power.

In a second set of experiments (Table V) a smaller quantity of juice was placed in the filter and the filtration was continued until no more liquid would pass through. The residue was then washed several times by adding water and forcing it through the filter. The time occupied in this process varied greatly with different juices, the limits for the filtration and washing of 50 c.c. of juice, using two filters simultaneously, were about 6 to 12 hours. The carbon

*'Journ. Physiol.,' 1896, vol. 20, p. 364.

Table IV

FILTRATION OF YEAST-JUICE THROUGH
THE MARTIN GELATIN FILTER:
15 c.c. RESIDUE + 3 GRAMMES GLUCOSE + TOLUENE

No.	Water	Filtrate	Boiled juice	Time	Carbon dioxide
	c.c.	c.c.	c.c.	hours	gramme
1	15	0	0	48	0.000
	0	15	0	48	0.035
2	15	0	0	60	0.001
	0	15	0	60	0.051
3	15	0	0	60	0.008
	0	15	0	60	0.064
4	15	0	0	60	0.024
	0	0	15	60	0.282

Table V

FILTRATION OF YEAST-JUICE THROUGH
THE MARTIN GELATIN FILTER

No.	Vol. of juice filtered	Wash water	Residue	Filtrate	Boiled juice	Glucose	Carbon dioxide
	c.c.	c.c.	c.c.	c.c.	c.c.	grammes	c.c.
1	75	200	25	0	0	2.5	10.4
			25	0	25	5	396.3
2	80	260	20	0	0	2	8.3
			20	20	0	4	90.2
3	100	250	25	0	0	2.5	0.4
			25	0	25	5	268
4	50	200	25	0	0	2.5	0.9
			25	0	25	5	192

dioxide was not estimated by absorption in potash as in the previous cases, but was collected and measured over mercury, by means of the apparatus described later on, the object of this procedure being to ascertain not only the total amount of carbon dioxide produced, but the rate and duration of the evolution. The residue was dissolved in water and made to the same volume as the original juice, and the filtrate was evaporated down to the same volume. All the solutions were saturated with carbon dioxide at the temperature of the bath (25°) before the measurements were commenced, and the observations were continued until all fermentation had ceased.

The boiled juice added in Experiments 1, 3 and 4 (Table V) was obtained, by boiling a portion of the same preparation as was used for the filtration. The carbon dioxide is expressed in cubic centimetres under atmospheric conditions.

The process of filtration does not always produce an inactive residue, as on several occasions the residue after very thorough washing has been found to retain a considerable amount of activity. No reason has yet been found for this and it has not yet been ascertained whether it is due to some peculiarity in the particular specimen of juice or in the special filter employed.

It is of interest to note that in Experiment 2 (Table V) the residue alone gave 8.3 c.c. of carbon dioxide in 3 hours, the amount evolved in the last hour being only 0.1 c.c. At the close of this period the liquid still contained the alcoholic enzyme, since on the addition of 20 c.c. of the filtrate, fermentation recommenced and continued for many hours.

These two sets of experiments (Tables IV and V) show that the fermentation of glucose by yeast-juice is dependent upon the presence of a dialysable substance which is not destroyed by heat.

4. Analysis of the Effect of the Addition of Boiled Juice upon the Fermentation of Glucose by Yeast-Juice

In order to compare the course of the fermentation in the presence and in the absence of boiled yeast-juice, experiments were carried out in which the rate of evolution of carbon dioxide was observed in each case throughout the whole period of activity of the juice, which, as a rule, in presence of an excess of sugar, lasts for about 48 to 60 hours.

For this purpose the fermentation was allowed to proceed in a 100 c.c. flask, kept at the constant temperature of 25° by immersion in a thermostat, and connected with an azotometer, in which the gas was collected over mercury. The gas in the fermentation flask was maintained at a constant pressure, as nearly as possible that of the atmosphere, by keeping the mercury in the reservoir at a fixed level, by means of a syphon dipping into a small beaker. The volume of the gas was read on the azotometer without disturbing the mercury reservoir and was reduced to atmospheric pressure by means of a calibration curve. Since yeast-juice readily becomes supersaturated with carbon dioxide, the contents of the flask were vigorously shaken before each reading of the volume of gas. Before the observations were commenced the liquids were brought to the temperature of the thermostat, and were saturated with carbon dioxide. In all comparative experiments the concentration of glucose was the same.

When the rates of evolution of carbon dioxide from (1) a solution of glucose in yeast-juice, and (2) a similar solution to which boiled and filtered yeast-juice has been added are compared, it is found that two phenomena are concerned in the production of the increased fermentation in the presence of boiled yeast-juice.

a. An initial rapid evolution of carbon dioxide is produced, which soon diminishes until a rate is attained which remains nearly constant for several hours and is usually, but not invariably, approximately equal to that given by an equal volume of the same yeast-juice and glucose to which no addition has been made.

b. The fermentation rate diminishes more slowly, so that the fermentation continues for a longer period. The greater proportion of the total increase is usually due to this second phenomenon.

The results obtained in a typical experiment of this kind are shown in Fig. 1. The initial period of the evolution is plotted separately (Curves A′ and B′) on a larger scale.

Curves A and A′ in which the evolution of carbon dioxide is plotted against time represent the course of a fermentation with 25 c.c. yeast-juice +25 c.c. water +5 grammes glucose +toluene. The rate to begin with is 48 c.c. per hour, but rapidly decreases until it becomes equal to 24 c.c. per hour, at which it remains almost constant for about 5 hours, gradually decreasing until, after the expiration of about 40 to 45 hours, fermentation ceases. The total evolution amounted to 369 c.c. under atmospheric conditions.

Curves B and B′ refer to 25 c.c. of the same yeast-juice +25 c.c. of a boiled yeast-juice +5 grammes glucose + toluene. The initial rate is much higher,

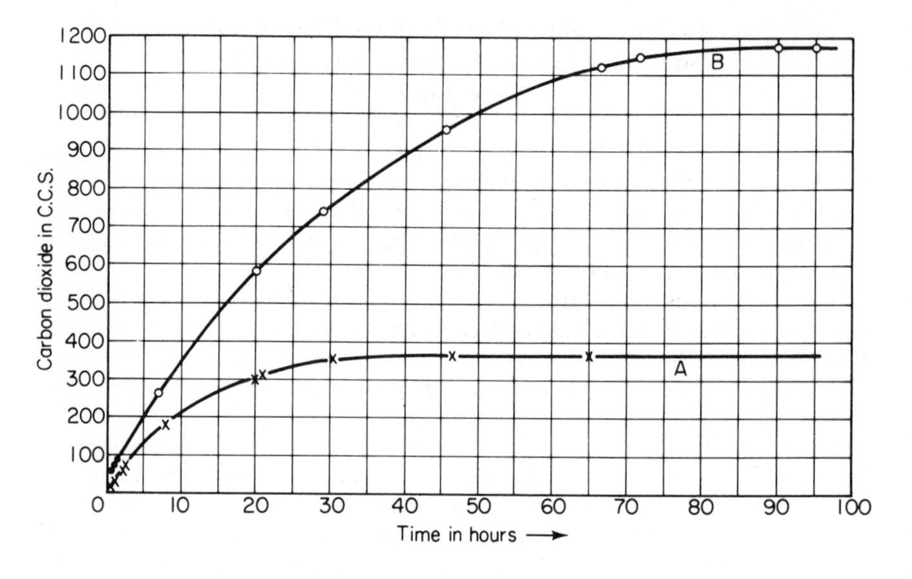

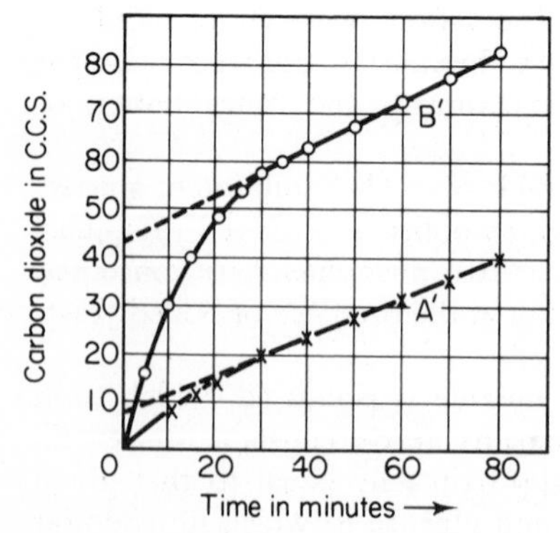

Fig. 1.

168 c.c. per hour, but this falls gradually in the course of 40 minutes to 30 c.c. per hour. This rate of 30 c.c. per hour falls off much less rapidly than that in Experiment A, the fermentation continuing for about 80 to 85 hours and yielding in all 1174 c.c. of carbon dioxide. It is important to bear in mind that these curves represent the gradual disappearance of the fermenting power of the liquid, and not the diminution of the amount of fermentation with diminishing concentration of sugar, an excess of this substance being present throughout.

A comparison of the two curves shows very clearly the two factors involved in the great increase in Experiment B: (1) The initial rapid evolution, and (2) the prolongation of the fermentation.

5. The Initial Period of Rapid Evolution of Carbon Dioxide

This is a very striking phenomenon, and a typical example is illustrated in Fig. 2 in which the curves show the course of the evolution of carbon dioxide

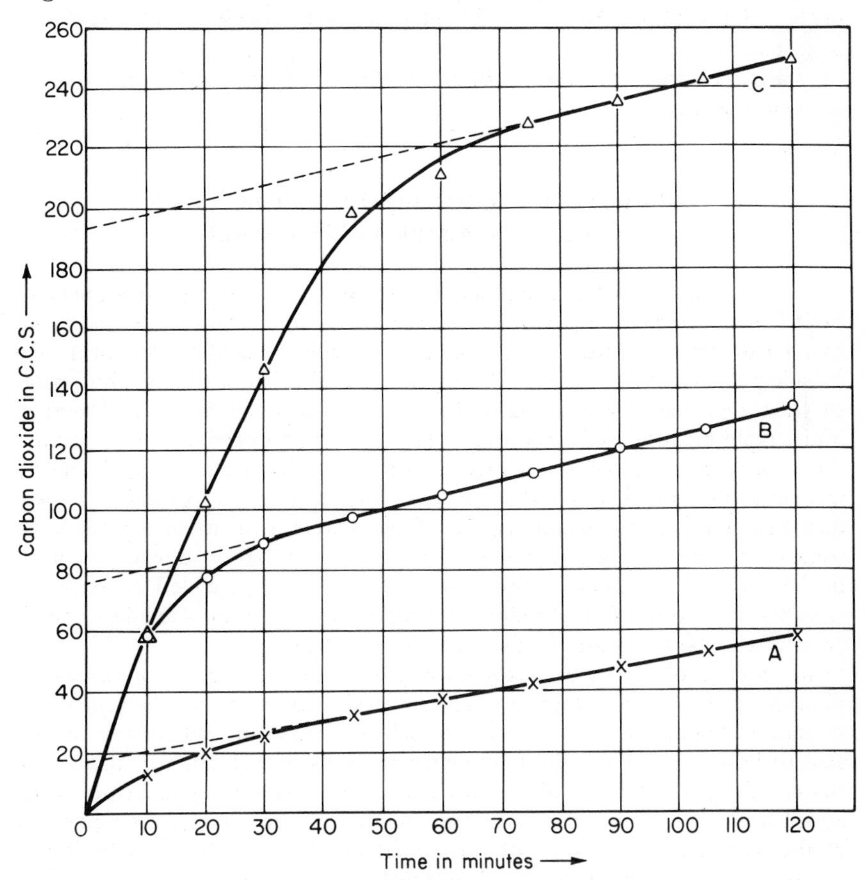

Fig. 2.

(total volume evolved plotted against time) during two hours in the case of:

A. 25 c.c. yeast-juice +75 c.c. water +10 grammes glucose + toluene.

B. 25 c.c. yeast-juice +50 c.c. water +25 c.c. boiled autolysed yeast-juice +10 grammes glucose + toluene.

C. 25 c.c. yeast-juice +75 c.c. boiled autolysed yeast-juice +10 grammes glucose + toluene.

In B and C the initial rates are almost equal (58 c.c. in 10 minutes) and much greater than in A (14 c.c. in 10 minutes). In B the rate rapidly falls off whilst in C it diminishes much more slowly. A similar initial period is also observable in A, but is not nearly so marked.

The extra quantity of carbon dioxide evolved in this initial period may be calculated by subtracting the amount corresponding with the constant rate which is finally attained from the total amount observed. This is done graphically in Fig. 2 by continuing the straight line representing the constant rate back to the axis of ordinates. The following numbers are thus obtained: for A, 16.6; for B, 75.4; for C, 192.9.

The amounts due to the addition of boiled juice are therefore: for 1 volume in B, $75.4 - 16.6 = 58.8$; for 3 volumes in C, $192.9 - 16.6 = 176.3 = 3 \times 58.8$.

The extra amount of carbon dioxide is, therefore, directly proportional to the volume of boiled juice added.

6. Production of the Initial Rapid Evolution of Carbon Dioxide by the Addition of Phosphates

As the result of a large number of attempts to isolate the constituent of boiled juice which brings about the increase in fermentation, it was found that whenever an increase was produced, phosphoric acid in the form of a soluble phosphate was present. The effect of the addition of soluble phosphates to yeast-juice was, therefore, examined and it was found that a well-marked initial rapid evolution of carbon dioxide was thus produced. Since, moreover, the boiled juices employed invariably contained phosphates, precipitable by magnesia mixture, there can be no doubt that it is to the presence of these that this initial phenomenon is due. Quantitative estimations revealed the somewhat surprising fact that the extra quantity of carbon dioxide, evolved in the initial period when a phosphate or a boiled juice is added, corresponds with the evolution of one molecular proportion of carbon dioxide for each atom of phosphorus added in the form of phosphate.

In order to obtain accurate results with solutions of sodium or potassium phosphate, the fact that these absorb carbon dioxide must be taken into consideration. Solutions of the dihydrogen salts of potassium and sodium are too acid to be employed and the monohydrogen salts or a mixture of these with the dihydrogen salts were always used. In every case the liquid before being added to the yeast-juice was saturated with carbon dioxide at the temperature of the bath, and the volume of carbon dioxide liberated by the addition of excess of hydrochloric acid was ascertained in an aliquot portion.

At the close of the fermentation the fermented liquid was acidified and the residual combined carbon dioxide measured, the difference between this and the original amount being subtracted from the amount evolved during the fermentation.

The results are more precise when the yeast-juice employed is an active one, since when the fermenting power of the juice is low the initial period becomes unduly prolonged and the calculation of the extra amount of carbon dioxide is rendered uncertain. The equivalence of the carbon dioxide and phosphate is established by the results contained in the following Table VI. Column 1 gives the observed amount of extra carbon dioxide calculated as described above and reduced to grammes, and Column 2 the equivalent of the phosphate added, this being estimated by precipitation with magnesium citrate mixture in the boiled juice or phosphate solution.

Table VI

EQUIVALENCE OF EXTRA CARBON DIOXIDE EVOLVED DURING THE INITIAL PERIOD, AND PHOSPHATE ADDED

Experiments	Grammes of carbon dioxide		Experiments	Grammes of carbon dioxide	
	Column I—Observed	Column II—Calculated from phosphate		Column I—Observed	Column II—Calculated from phosphate
1	0.090	0.086	8	0.196	0.197
2	0.054	0.055	9	0.066	0.065
3	0.058	0.051	10	0.057	0.061
4	0.060	0.049	11	0.056	0.061
5	0.106	0.112	12	0.059	0.061
6	0.103	0.101	13	0.068	0.070
7	0.113	0.112	14	0.071	0.070

In Experiments 1 to 7 boiled juice was added; in 8 to 14 a solution of sodium or potassium phosphate.

The maximum rate attained during the initial period is from five to eight times as high as the constant rate attained after the evolution of the carbon dioxide equivalent to the phosphate present.

At the commencement of the period when sodium or potassium phosphate solution has been added, the rate only gradually acquires its maximum value and sometimes it only attains this maximum after a considerable interval.

This phenomenon is occasionally observed in the fermentation produced by yeast-juice without the addition of phosphate, and also sometimes occurs, but to a much smaller extent, when boiled juice is added. It is well shown in Curve B, Fig. 3, which represents the fermentation produced by 25 c.c. yeast-juice +25 c.c. of a 0.06 molar solution of sodium phosphate +5 grammes glucose +toluene. The cause of this period of induction has not yet been ascertained.

7. Limit of the Action of Phosphate

If the fermentation in presence of phosphate be allowed to continue until the steady rate is attained and a second quantity of phosphate be then added, a second period of rapid evolution of carbon dioxide sets in and proceeds in a similar manner to the first. This is shown in Curves B and C, Fig. 3, which represent the effect of the successive addition of two quantities of 5 c.c. of 0.3 molar sodium phosphate to 25 c.c. yeast-juice +20 c.c. water, in presence of 10 per cent. glucose. Curve A represents the fermentation in absence of added phosphate. The phosphate solution employed was a mixture of five molecules of NaH_2PO_4 with one molecule of Na_2HPO_4 and no correction for combined carbon dioxide was required. The extra amount of carbon dioxide evolved after each addition is the same, and is equivalent, as already stated, to the phosphate added. The equality is shown graphically in the curve and the equivalence in Experiments 13 and 14, Table VI.

This process cannot, however, be repeated indefinitely, as after a certain limit is reached the reaction no longer occurs and with a large excess the fermentation is stopped. The exact limit appears to vary both with the nature of the phosphate added and with the particular specimen of yeast-juice employed. The greatest amount of carbon dioxide hitherto obtained in this way from 25 c.c. of yeast-juice is about 0.45 gramme (230 c.c.), which was observed

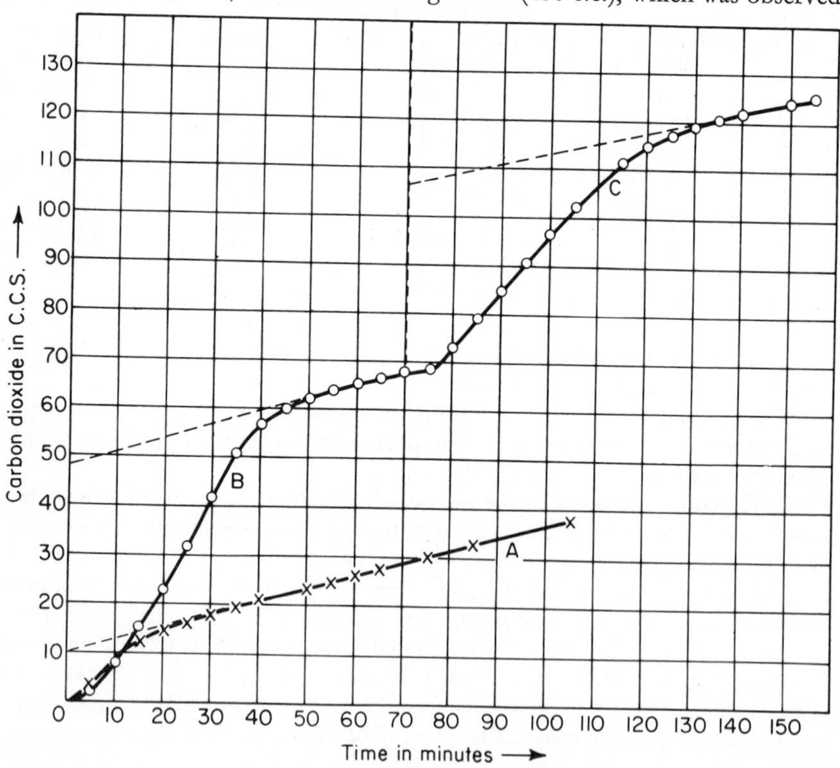

Fig. 3.

on two occasions, once after the addition of four volumes of boiled juice, and again after the addition of 50 c.c. of a solution of a mixed magnesium potassium phosphate yielding with magnesia mixture 1.187 grammes of magnesium pyrophosphate.

When a specimen of yeast-juice has been incubated until it will no longer ferment sugar, it is not affected by the addition of phosphate.

The fact that the extra carbon dioxide calculated in this way is equivalent to the phosphate present, suggests the superposition of two actions. Whether this is to be explained by the presence of two distinct enzymes or simply by the increased activity of a single enzyme remains to be decided.

8. Products of Fermentation in the Presence of Phosphate

The carbon dioxide evolved during the initial period after the addition of a phosphate is the product of a true alcoholic fermentation of the glucose, in which alcohol and carbon dioxide are produced in equivalent amounts. This was proved in the following way. Twenty-five cubic centimetres of a solution containing 2.5 grammes of glucose and 5 c.c. of a 0.3 molar solution of potassium phosphate were added to 25 c.c. of yeast-juice; the mixture was incubated and the carbon dioxide collected and measured.

As soon as the rate of evolution had become constant, a further addition of 10 c.c. of 0.3 molar phosphate solution was made and the fermentation again continued until the rate had become constant. The gas evolved was tested and found to be carbon dioxide. The total amount evolved during the experiment, which lasted for 2 hours 10 minutes, was 163.4 c.c. at 19°.6 and 758.6 mm. or 0.291 gramme, the equivalent of the phosphate added being 0.196 gramme. The liquid was then distilled with steam and the alcohol estimated in the distillate, 1.312 grammes being found to be present. Twenty-five cubic centimetres of the original juice were found to contain 0.983 gramme of alcohol and therefore 1.312 — 0.983 = 0.329 gramme were formed by the fermentation of the sugar. The ratio of alcohol to carbon dioxide produced is therefore 0.329/0.291 = 1.13, which agrees well with the ratio previously found by similar methods for the fermentation of glucose by yeast-juice.* The theoretical ratio is 1.04.

Lactic acid and acetic acid were also estimated in the original juice and after fermentation in presence of phosphate, but only a very small variation was observed. Twenty-five cubic centimetres of juice gave before fermentation 0.122 gramme of zinc lactate and 0.083 gramme of acetic acid, and after fermentation 0.102 gramme of zinc lactate and 0.072 gramme of acetic acid.

9. Fate of the Phosphoric Acid

When the fermented liquid is boiled and filtered almost the whole of the phosphorus present is found in the filtrate, but it is nearly all in a form which is not precipitated by ammoniacal magnesium citrate mixture.

*Harden and Young, 'Ber.,' 1904, vol. 37, p. 1052.

In the following experiment three quantities of 25 c.c. of yeast-juice were taken:—

A. Hot water was added, the solution heated in a boiling water-bath and the coagulate filtered off and well washed.

B. Ten cubic centimetres of a 30 per cent. glucose solution and 10 c.c. of 0.3 molar potassium phosphate solution were added and the liquid at once heated to the boiling point, filtered, and the coagulate washed.

C. The same additions were made as to B and the liquid then fermented until the close of the initial period, after which it was heated and filtered like the others.

The total phosphorus was then estimated in each of the coagulates and in each of the filtrates, and the phosphorus precipitated by magnesium citrate in each of the three filtrates. The estimations of total phosphorus were made by heating with sulphuric and nitric acids until colourless, diluting and precipitating with magnesium citrate mixture in presence of excess of ammonia.

The following were the results obtained, the numbers representing the grammes of magnesium pyrophosphate per 25 c.c. of juice.

Table VII

	A Original juice	B Juice + phosphate not fermented	C Juice + phosphate fermented
Coagulate	0.053	0.057	0.072
Filtrate—			
(*a*) Precipitated by Mg citrate	0.126	0.480	0.070
(*b*) Not precipitated by Mg citrate	0.271	0.282	0.679
Total...........	0.450	0.819	0.821

The amount of phosphate added was equivalent to 0.372 gramme of magnesium pyrophosphate.

A number of other results are given to show the extent to which phosphate is converted into the non-precipitable form by this reaction. All the estimations were made by boiling and filtering the fermented liquid immediately upon the close of the initial period. As before the numbers represent grammes of magnesium pyrophosphate obtained from 25 c.c. of juice.

The form in which this non-precipitable phosphorus is actually present in the fermented liquid, and in the liquid which has been boiled and filtered, has not yet been ascertained with certainty. Experiments which are still in progress, however, appear to indicate that it exists in combination with glucose, probably in the form of a phosphoric ester.

The question as to whether the entire phenomenon of the fermentation of glucose by yeast-juice depends on the presence of phosphates has not yet been definitely decided. The addition of phosphate undoubtedly produces a larger increase in the total fermentation than is simply due to the equivalent amount

Table VIII

CONVERSION OF PHOSPHATE INTO THE NON-PRECIPITABLE FORM BY YEAST-JUICE AND GLUCOSE

	Phosphate added	Precipitable phosphate in filtrate	Non-precipitable phosphate in filtrate
1	0.553	0.066	1.032
2	0.490	0.090	0.832
3	0.250	0.054	0.685
4	0.488	0.091	1.040
5	0.495	0.088	0.881

of carbon dioxide evolved in the initial period. The extent of this increase appears to vary very considerably with different specimens of yeast-juice, but the prolongation of the fermentation is not so great as is caused by boiled fresh juice. This question can only be satisfactorily settled by ascertaining whether the addition of a phosphate to the perfectly inactive residue obtained from a juice by filtration through a gelatin filter is sufficient to restore its fermenting power in the same way as the filtrate or a boiled juice. Experiments on this point are in progress, but no decisive result has as yet been obtained, and all discussion of this point will best be deferred until these are completed.

Various other points of interest raised in the course of the investigation, and the study of the relation of these phenomena to the fermentation of glucose by living yeast, are also occupying our attention.

A short outline of the main conclusions arrived at in the foregoing paper, has been previously published in the form of two preliminary communications, without any experimental details.* After the appearance of these notes, Buchner and Antoni† repeated and confirmed a number of the experiments dealing with the effect of boiled juice and of phosphates on the total fermentation, and with the separation of the juice by dialysis into an inactive residue and a dialysate capable of rendering it active. Buchner and Antoni were able, with the more stable juice at their disposal, to carry out the dialysis in the ordinary way for 24 hours and in this manner to confirm the results obtained by the use of the gelatin filter. Owing to the lack of experimental detail, Buchner and Antoni imagined that in our comparative experiments the concentration of glucose and of enzyme had not been kept constant, and ascribed part of the increase produced by boiled juice to the favourable effect of a diminution in the concentration of the sugar and of the alcohol, which is always present, by dilution with the added boiled juice. The details given above show that neither of these influences had any share in the effects observed by us.

*'Journ. Physiol.,' 1904, vol. 32; 'Proc.,' of November 12; 'Proc. Chem. Soc.,' 1905, vol. 21, p. 189, June 6.

†'Zeit. Physiol. Chem.,' 1905, vol. 46, p. 136.

THE NATURE OF THE "INORGANIC PHOSPHATE" IN VOLUNTARY MUSCLE*

Cyrus H. Fiske and Y. Subbarow

Biochemical Laboratory, Harvard Medical School,
Boston, Massachusetts

Some months ago we† described a colorimetric phosphate method, the special feature of which is the use of a very active agent (aminonaphtholsulfonic acid) for converting the phosphomolybdic acid to its blue reduction product. When we first made use of this method for the determination of inorganic phosphate in protein-free muscle filtrates, shortly after the details had been worked out, we found a marked delay in color production. The time required to reach a constant reading was about thirty minutes, whereas ordinarily the full color (relative to the standard) has developed within four minutes or less. This peculiar behavior appeared to indicate that muscle contains either some substance capable of retarding the color reaction or else a very unstable (presumably organic) compound which liberates o-phosphoric acid while the color is developing. While the course of the color development in muscle filtrates turned out to be quite different from anything which we had seen in testing out the method in the presence of known interfering substances, it is impossible to rely on this point as a means of distinguishing between the two alternatives. Ferric salts, for example, also behave in a way that is unique. A mixture of inorganic phosphate and ferric chloride certainly does not contain a highly unstable organic phosphorus compound, and the delayed color reaction found with muscle filtrates therefore does not constitute conclusive proof of the existence of such a substance in the muscles.

*C. H. Fiske and Y. Subbarow, "The Nature of the 'Inorganic Phosphate' in Voluntary Muscle," *Science,* **65**, 401–403 (April, 1927). Reprinted with the permission of the authors and of the publisher.

†C. H. Fiske and Y. Subbarow, *J. Biol. Chem.* Vol. 66 (375)—1925.

Further study of the course of color development nevertheless did bring out some interesting and suggestive points, notably the fact that the delay is hardly any more pronounced with 10 cc of muscle filtrate, for example, than with 5 cc or less. Every interfering substance which we investigated in the course of our work on the phosphate method, on the other hand, shows a much more marked effect when the phosphorus content of the sample is increased. Although these facts have been in our possession now for more than a year, we have until this time refrained from placing them on record, inasmuch as the phenomena observed could not with any certainty be ascribed to the presence of an organic compound of phosphoric acid until the compound had been isolated, or at least until the organic radicle had been identified. Both these things have now been done, although the isolated substance has not yet been obtained in the pure state, and the outcome appears likely to throw light on a field of biochemistry never before suspected of being in any way related to phosphoric acid.

Muscle filtrates from which all the inorganic phosphate has been removed (by precipitation with barium, silver, etc.), as well as material which has been still further purified, show the same delay in the production of the color. These facts, together with the knowledge that the delayed reaction really is associated with the hydrolysis of an organic compound of phosphoric acid, give real significance to the quantitative data which we have meanwhile been accumulating. Some of these data will now be presented before we proceed to a discussion of the nature of the substance.

The method which we have used for the determination of this unstable form of phosphorus (which we shall for the present designate as "labile phosphorus") differs in no respect from our regular phosphate method, except that readings are taken at brief intervals (every minute, beginning with the third, until the unstable substance is about half hydrolyzed, and thereafter every few minutes until no further change occurs). The growth in color intensity for the first few minutes is practically linear, and the concentration of inorganic phosphate is found by extrapolating back to zero time. The sum of the inorganic and the "labile" phosphorus is calculated from the final reading, and the amount of "labile phosphorus" found by difference. From control analyses of solutions of the purified organic compound, with and without the addition of known amounts of inorganic phosphate, we believe that the results are accurate within 1 or 2 mg of phosphorus per 100 gm of muscle.

The principal results which we have obtained by this method of analysis are as follows: (1) The normal resting voluntary muscle of the cat shows generally about 60 to 75 mg of "labile phosphorus" per 100 gm if removed with the greatest possible care and at once cooled to 0°C. or below. (The further preparation of the material for analysis consists simply in precipitating the protein with ice-cold trichloroacetic acid, filtering and immediately neutralizing the filtrate with sodium hydroxide.) The true inorganic phosphorus under these conditions is only 20 to 25 mg, instead of the 80 to 100 mg, shown by other methods of analysis. (2) After prolonged electrical stimulation, the "labile phosphorus" usually falls to about 20 mg (in one instance to as little as 9 mg), while the inorganic phosphorus is correspondingly increased. (3) Stimulation with the blood supply shut off, in which case complete fatigue ensues, causes complete disappearance of the unstable compound, within the error of analysis. Merely shutting off the circulation for an equal length of time has little or no effect. (4) A period of rest after the muscle has been stimulated is accompanied by resynthesis of the organic compound. The maximum yield of "labile phosphorus" so far observed under these conditions is 46 mg per 100 gm of muscle. The inorganic phosphate, however, falls to the normal level, so in all probability either some of the phosphate has been discharged into the circulation, or else the water content of the muscle has increased.

The compound under consideration is a derivative of creatine. On a small scale we have succeeded in separating it from all other phosphoric acid compounds. On a sufficiently large scale to yield material enough for a complete analysis, the separation is not so readily accomplished by our present methods, and the best products that we have so far procured contain several per cent. of one or more other phosphorus compounds, since the full blue color can not be obtained unless the material is ashed. In spite of this contamination, the phosphate, creatine and base in two salts which have been prepared add up to about 85 per cent. of the weight of the entire substance, indicating the probable absence of a third component. The elementary analysis of amorphous products, unless they are of definitely established purity, proves very little. Our main evidence for the existence of "phosphocreatine" in muscle is of a quite different nature. In the first place, we have found that the "labile phosphorus" can be more or less completely separated by precipitation with a number of different reagents (six in all have been used to date), and that in each case there is precipitated with it one equivalent of creatine. One of these reagents (copper in slightly alkaline solution) has been applied to muscle filtrates prepared under various conditions, and the proportionality between creatine and "labile phosphorus" has never failed. A few typical illustrations of our experience with

	Mg per 100 gm muscle		
	"Labile phosphorus"	Creatine in washed copper precipitate	Molecular ratio of creatine to phosphoric acid
Normal resting muscle	70	280	0.95
do	77	323	0.99
Normal resting muscle:			
(1) Fresh	65	269	0.98
(2) Stood few minutes after removal from body	55	222	0.96
(3) Stood for longer time	32	128	0.95
(4) Analysis of trichloroacetic acid filtrate (from fresh muscle) which had stood for several hours	0	2	
Muscle stimulated for 10 minutes with circulation intact	20	82	0.97
Muscle stimulated to fatigue with blood supply shut off, followed by 1 hour rest period	44	192	1.03

copper precipitation are given in the accompanying table. They serve to show that creatine and the labile form of phosphate are precipitated together, in equivalent proportions, whether the muscle was fresh and in the resting state or whether it had been subjected to manipulations which alter the concentration of the "labile phosphorus." Thus, when the muscle is allowed to stand outside the body, the "labile phosphorus" content progressively diminishes at a fairly rapid rate (it is entirely gone in less than twenty minutes), and at the same rate creatine is set free, as indicated by its failure to be thrown down by copper. If the trichloroacetic acid filtrate prepared from a sample of fresh muscle is kept (unneutralized) for several hours, the "labile phosphorus" has completely disappeared, and the copper precipitate is then virtually free from creatine. The same parallelism holds for muscle which has been stimulated, showing that the liberation of inorganic phosphate during stimulation is associated with the conversion of the creatine to a form in which copper will not precipitate it. Finally, in case the muscle has been stimulated with the blood supply cut off—a procedure which, as stated, leads to the loss of all the "labile phosphorus"—and then permitted to recover, the reappearance of "labile phosphorus" is accompanied by the return of an equivalent quantity of creatine to the condition in which precipitation by copper does take place. Direct evidence for the synthesis of a creatine-phosphoric acid compound during recovery is thereby attained.

Quite aside from its obvious bearing on the mechanism of muscular contraction, the demonstration of "phosphocreatine" in muscle should go far towards providing an explanation for a number of matters which in the past have been obscure. Among these may be mentioned the passage of administered creatine into muscle in spite of the large quantity already there, and the striking difference between resting (living) muscle on one hand and fatigued or dead muscle on the other in their capacity for retaining both creatine and phosphate, as shown by perfusion and dialysis experiments.

CONVERSION OF FERMENTABLE HEXOSES WITH A YEAST CATALYST (HEXOKINASE)*

O. Meyerhof

Kaiser Wilhelm Institut für Biologie, Berlin-Dahlem, Germany

The anhydric polysaccharides, built up of identical building blocks such as one finds in glycogen compounds (classified according to Pringsheim as "hexosanes") are converted equally well to lactic acid in the presence of muscle enzyme. In contrast to this, the free reducing sugars react quite differently in that they are only slightly split. The soluble glycolytic muscle "ferment" resembles, in this respect, the muscle dispersion which has been subjected to freezing with liquid air. Frog muscle extract prepared with an isotonic KCl solution is almost devoid of any activity towards hexoses. An extract prepared with distilled water shows some activity presumably because it is not free of preformed carbohydrate. Figure 1 depicts an experiment using the latter type of extract. Rabbit muscle extract, which in a fresh state often splits glucose at a rate almost comparable to that of glycogen, proves more effective. During storage, however, its ability to split free sugars decreases more rapidly than the ability to split glycogen. The activities of fructose, glucose, and mannose are of the same order of magnitude; maltose and amylobiose show 1/5 of the activity of the former, whereas saccharose and galactose are inactive. An acetone powder made from rabbit muscle extract retains its ability to split glucose as well as glycogen, even after week-long storage of the powder in a vacuum desiccator (compare Fig. 2). Thus, the soluble enzyme system does not lack the capacity for glycolysis of the free hexoses. However, due to the lability of the soluble enzyme system, the potency diminished as compared with the capacity

*Excerpt from O. Meyerhof, "*Die Chemischen Vorgänge im Muskel,*" J. Springer, Berlin (1930), pp. 149–155. [Cf. also, O. Meyerhof, *Biochem. Z.*, **183** (1926).] Translated from the German and reprinted with the permission of Springer-Verlag, Heidelberg, Germany.

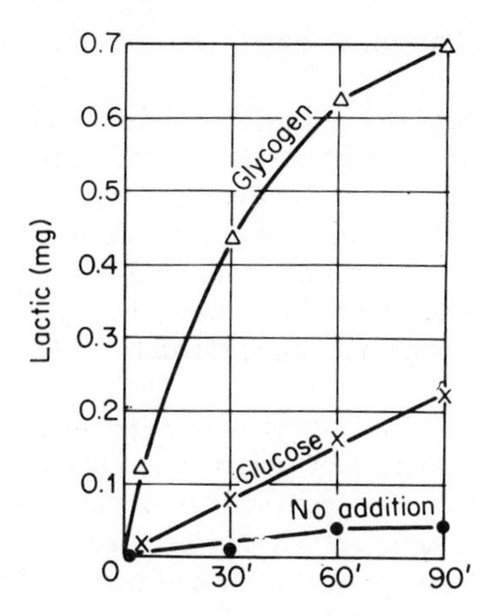

Fig. 1. Splitting of glycogen and glucose by aqueous extract from frog muscle without activator. ●——●, without addition; ×——×, addition of glucose; △——△, addition of glycogen.

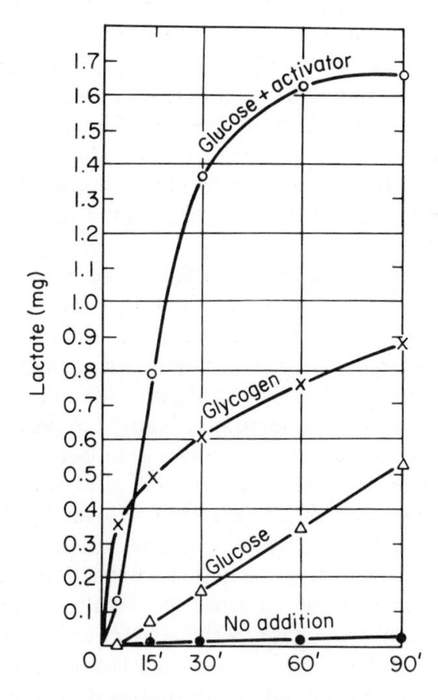

Fig. 2. Splitting of glucose and glycogen by acetone powder aged for three weeks. ●——●, no addition; △——△, with glucose; ○——○, with glucose and activator (hexokinase); ×——×, with glycogen.

to split polysaccharides. Phosphate esterification in the presence of free sugars is almost nil.

Addition of an activator obtained from yeast brings about a complete change of the situation. Under these conditions, lactic acid formation from free sugars is extraordinarily rapid, exceeding by far the lactic acid formation from glycogen, as can be seen from Fig. 2. As an example, 27 mg of acetone powder derived from 1 ml of fresh extract but several weeks old generates within 30 minutes 1.4 mg lactic acid from free sugar. The subsequent decrease in the rate of formation of lactic acid corresponds approximately to the disappearance of the added glucose.

The activator is obtained by autolysis of yeast in the presence of toluene. Subsequently the clear supernatant solution is cooled rapidly and an equal volume of alcohol is added. The precipitate so obtained is dispersed with water and centrifuged. The aqueous extract can be used as it is, or reprecipitated with alcohol. The catalyst is water-soluble and precipitable with 50% alcohol. It is thermo-labile since heating to 50° for only one minute brings about a considerable degree of inactivation. It is sensitive to alkali and acid, but can be preserved for weeks in aqueous solution on ice. The activator has the properties of an enzyme. I propose, therefore, to designate it "hexokinase," since it acts specifically on the fermentable hexoses. It is necessary along with the coenzyme for the described hexose transformation. It is not related to insulin or to Collip's glucokinin; insulin has no effect on any phase of the glucose transformation in solution. A twice-precipitated hexokinase is free of inorganic phosphate; 0.4 to 1 mg of such a preparation is effective in 1 cc of solution. The albuminous precipitate containing the activator undoubtedly consists for the most part of impurities; no special purification has been performed as yet. The activator still converts glucose after further dilution, but at a slower rate. Within a certain range, the initial rate of lactic acid formation is proportional to the amount of hexokinase. Yeast is the only material so far from which it has been possible to isolate the activator.

It is a crucial question whether hexokinase can act only in the presence of the whole enzyme system or whether it is able by itself to transform glucose into a chemically more reactive molecular species. An answer to this question was sought from further experiments in which the enzyme system and the isolated activator were separated by a collodion film and sugar subsequently added. Since neither the enzyme nor the activator can pass through the collodion film, it must be concluded that any increase in lactic acid formation under these conditions must be ascribed to a diffusion of activated glucose from the hexokinase solution into the enzyme solution. However, although lactic acid formation is increased, it is modest as compared to the stimulation observed when activator is added directly. Lactic acid formation also occurs, though more irregularly, when "glucose" is separated by ultrafiltration from the activator solution, and added drop-wise to the enzyme solution. Interpretation of these results is somewhat hampered by the fact that permeable collodion membranes do not always remain impenetrable by hexokinase. Yet it seems likely that a very short-lived, yet reactive glucose is formed.

One will have to renounce the idea that glucose activation only occurs in the structural elements of the cell, because of the fact that hexokinase solution does not contain structural elements. Since considerable glycolytic activity can be observed in cell-free rabbit muscle extracts, the idea cannot be applied on animal hexokinase either. Neither can it be applied to the observations described by Case. Case was able to obtain lactic acid formation from glucose in rabbit muscle extracts (prepared by a method similar to the one described here although using more dilute extracts), not only by adding the yeast activator, but also to a lesser degree by adding tissue suspensions. Tissue suspension of brain, kidney, muscle, blood, liver, and lung proved to be active (activities decreased in the above-mentioned order), while cell-free extracts of these tissues were inactive. These experiments tend to support the view that a hexokinase present in the tissues is involved here as well. Yet its separation from solid tissue components is undoubtedly dependent upon appropriate technical conditions, which are harder to achieve in mammalian tissue than in yeast.

The search for "hexokinase" in yeast autolysates was based on the idea that yeast extract brings about a rapid esterification of the fermentable hexoses with subsequent fermentation. Muscle extract, in contrast, does not show any accumulation of esters even in cases in which amounts of glucose are split. This difference might be due to the fact that a factor essential for glucose esterification is more concentrated in yeast. The validity of this assumption is illustrated by simultaneous observations of lactic acid formation and hexose esterification in activated muscle extract (Fig. 3). The initial rate of glycolysis

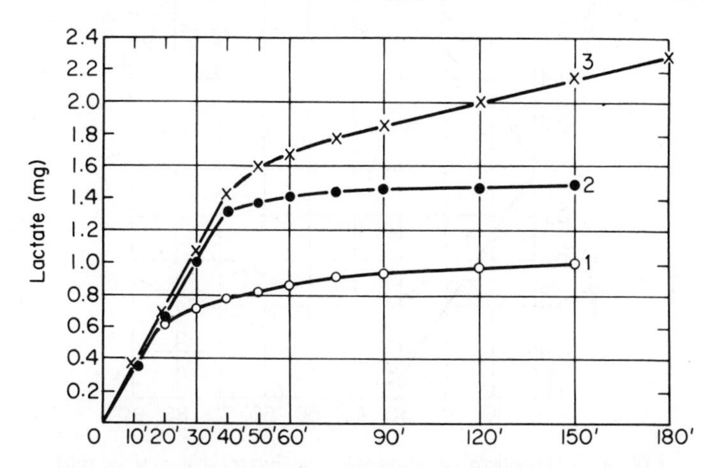

Fig. 3. Lactic acid formation from 3 mg glucose in 1 ml solution with varying amounts of phosphate.

Graph 1: 0.43 mg P_2O_5 corresponding to 0.55 mg lactic acid
Graph 2: 0.80 mg P_2O_5 corresponding to 1.03 mg lactic acid
Graph 3: 1.18 mg P_2O_5 corresponding to 1.51 mg lactic acid

The high initial rates exhibited in all three cases decrease to a constant level as soon as an amount of lactic acid approximately equimolar to that of the phosphate present has been formed. Yet the terminal rate in Experiment 3 is higher than in Experiments 1 and 2 because, in the former case, consumption of sugar by splitting and esterification gives rise to an increase in the phosphate content.

in the presence of glucose is greater than in the presence of glycogen, but decreases rapidly depending on the amount of phosphate present. The first drop in the rate of glycolysis coincides with the complete disappearance of phosphate. Thereafter, and for as long a period as an excess of sugar is present, the inorganic phosphate content is almost nil, but increases again slowly once all the sugar is utilized because the splitting of the stable hexose-diphosphoric acid now becomes dominant. During the period of rapid breakdown about 1 mole of hexose phosphoric acid is formed (mostly diester, but also variable amounts of hexose-mono-phosphoric acid) for each mole of glucose turned into lactic acid. This is illustrated in Fig. 4. Within 30 minutes, all the inorganic phosphate is esterified and all the glucose utilized, partly by being transformed into lactic

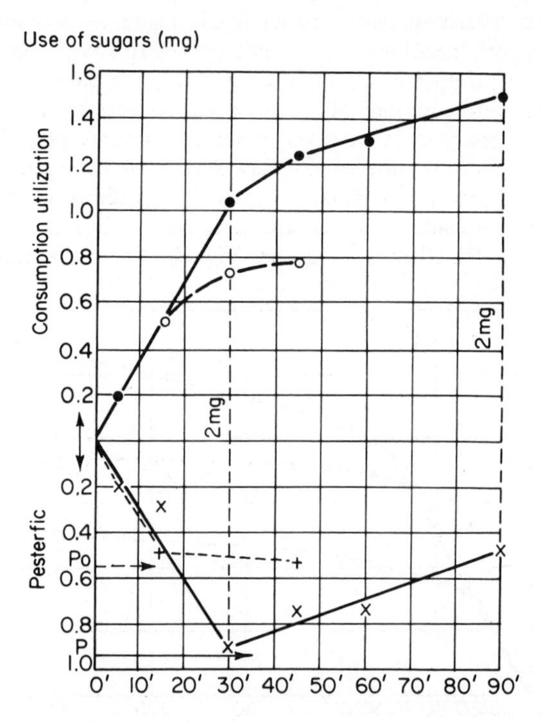

Fig. 4. Metabolism of 2 mg glucose by muscle extract with hexokinase.

●——●, lactic acid formation } with additional phosphate
×——×, phosphate esterification in extract }

(Content: 0.763 mg P_2O_5 equivalent to 0.942 mg sugar. The vertical dotted lines correspond to the consumption of the entire amount of the 2 mg glucose present.)

○——○, lactic acid formation } without additional phosphate
×——×, phosphate esterification }

(Preformed content: P_2O_5, 0.365 mg corresponding to 0.546 mg sugar.) P with solid line ⟶, phosphate content of the first experiment; P_0 dotted line ----→, phosphate content of the second experiment.

acid and partly by being esterified. From this moment on, the inorganic phosphate content increases again slowly, in proportion to the continuing lactic acid formation of this second period. This happens only as a consequence of the breakdown of hexose-di-phosphate into equimolecular amounts of phosphate and lactic acid (Fig. 4). The dotted lines in Fig. 3 represent a second experiment in which a smaller amount of phosphate was used. Here the initial rate is the same as in the previous experiment, but decreases sooner, namely at the moment when the inorganic phosphate has been utilized. The level of phosphate remains nearly nil in the presence of an excess of sugar because of the instantaneous re-esterification of the regenerated phosphate. Since during this rapid period of decay about 1 mole of hexose-diphosphoric acid is formed for each mole of sugar glycolyzed, the amount of lactic acid formed in the presence of an excess of sugar is equivalent to the amount of phosphoric acid present. At the time the rate has decreased, it is therefore possible, through a new addition of phosphate, to bring about an increase in the rate whereby the amount of lactic acid generated again becomes equivalent to the amount of phosphate added. This is demonstrated in Fig. 5. The second increase (curve IIIb) is less steep, apparently due to rapid inactivation of the extract. Fig. 6 shows

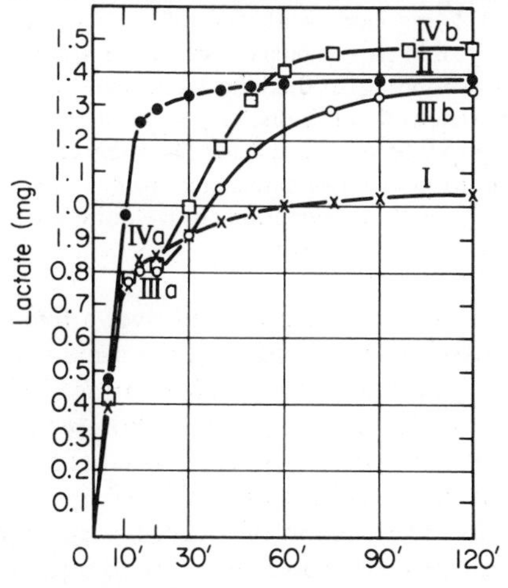

Fig. 5. Kinetics of lactic acid formation from glucose upon a second addition of phosphate (Experiment 9, vii, 1926). Sugar content is 2.8 mg in 0.85 ml.

I, kinetics of glycolysis with 0.688 mg P_2O_5 corresponding to 0.88 mg lactic acid. II, kinetics with 0.99 mg P_2O_5 corresponding to 1.27 mg lactic acid. IIIa and IVa, kinetics during 15 minutes with 0.688 mg P_2O_5 as in I. Subsequently (in IIIa) addition of phosphate (0.35 mg P_2O_5 corresponding to 0.45 mg lactic acid) as illustrated in curve IIIb. Subsequent to IV an addition of 0.43 mg P_2O_5 (corresponding to 0.55 mg lactic acid) as illustrated in curve IVb.

In IIIb and IVb a new increase in the rate ensues until an amount of lactic acid is produced which is equivalent to the second amount of phosphate added. The kinetics for the first 15 to 20 minutes was not recorded in III and IV.

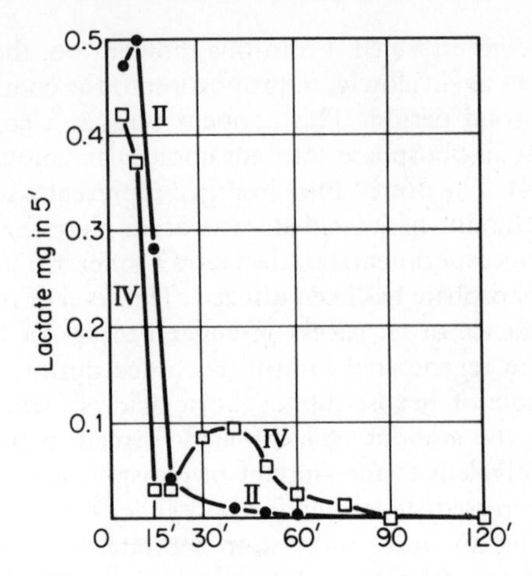

Fig. 6. Rate of lactic acid formation (mg per 5 minutes) derived from curves II and IV in Fig. 5.

●———●, rate in the presence of 0.99 mg P_2O_5 corresponding to 1.27 mg lactic acid.

□———□, first part of curve: rate in the presence of 0.688 mg P_2O_5 corresponding to 0.88 mg lactic acid. Second part of curve: after renewed addition of activator and phosphate corresponding to 0.55 mg lactic acid. The slighter increase upon the second addition of phosphate stems from a rapidly developing inactivation of the enzymatic activity quite apart from the increased dilution of the extracts.

the rate as expressed in mg lactic acid formed over a five-minute period. The other fermentable hexoses are metabolized in the same manner although fructose is metabolized nearly twice as fast as glucose during the "phosphate period." The non-fermentable hexoses such as galactose, however, are not metabolized even in the presence of hexokinase. Glucose metabolism cannot be analyzed by means of fluoride in the same way as it is possible in the case of glycogen. Instead, fluoride inhibits esterification of glucose as strongly as it inhibits the splitting, with the result that instead of an increased formation of hexose-phosphoric acid, a decrease in the accumulation of this ester is observed (as compared with the controls).

ENZYMATIC DEAMINATION IN MUSCLE*

Gerhard Schmidt

Institut für vegetative Physiologie, University of Frankfurt, Germany
(Submitted to the Editors on August 27, 1928)

Recently adenosine phosphoric acid was found as a constituent of muscle and recognized as the biological parent compound of inosinic acid, the substance which Liebig discovered in meat extract. It was shown that minced muscle brings about a vigorous deamination of added adenylic acid. P. Györgyi and H. Rothen(1) found that adenosine can also be deaminated by muscle. Autodigestion of muscle gives rise to a considerable evolution of ammonia, the exclusive sources of which may be considered to be adenosine phosphoric acid and perhaps adenosine. The close connection of this deamination process with muscle function is evident from the fact that the isolated frog muscle regularly gives rise to liberation of ammonia during work. Under proper conditions this formation of ammonia is reversible even in the isolated muscle. J. K. Parnas and W. Mozolowski described ammonia formation prompted by stimulation of isolated frog muscle shortly after the observations by Embden and coworkers; however, they considered any reversibility of this process out of the question.(2) Yet, according to a previously published investigation, the reversibility of the reaction even in isolated muscle is beyond doubt.

On the basis of the results of the above-mentioned studies, it has become possible to correctly interpret important observations by earlier authors concerning purine metabolism of muscle. Richard Burian(3) established through

*Excerpt from G. Schmidt, "Über fermentative Desaminierung im Muskel," *Hoppe-Seyler's Z. f. physiol. Chem.* **179**, 243–249 and 264–269 (1928). Translated from the German and reprinted with the permission of the author by license agreement with Walter de Gruyter & Co., Berlin.

This work was made possible by support from the Rockefeller Foundation and Notgemeinschaft der Deutschen Wissenschaften.

his fundamental studies that a major part of uric acid formation in mammals originates from muscle, especially the working muscle. He found in experiments carried out on himself that after strenuous work the purine bases and later the uric acid content of the urine regularly increased in the first period after the work. Moreover, he was able to show that even a resting perfused dog muscle formed hypoxanthine from an unknown precursor and that by stimulating the muscle this process took place to an increased extent.

At the time Burian performed his experiments, the mechanism of this hypoxanthine formation appeared much more puzzling inasmuch as muscle was considered an organ with only a meager content of nucleic acid. Moreover, W. Jones[(4)] had found that practically all mammalian species are devoid of adenase. Hence the only enzymic process known as a hypoxanthine-generating reaction, the enzymic deamination of adenine, was ruled out with regard to striated muscle. Other pieces of work from that time clearly indicate the necessity of considering alternate sources for the formation of hypoxanthine and adenine in the mammalian organism.

Metabolic studies of the intact animal had led Minkowski[(5)], as early as 1898, to the conclusion that "the process which preferentially determines the fate of purines in the animal organism is the coupling of the purine nucleus with other atom complexes." In feeding experiments this physiologist had found that in the dog uric acid or allantoine originates exclusively from adenine incorporated in the nucleic acid molecule and not from free adenine. By administration of adenine, which has a toxic effect, Minkowski obtained a very peculiar intermediary product of adenine degradation in the form of crystalline deposits which caused kidney infarcts. He erroneously considered the deposits to be uric acid. However, Nikolaier[(6)], in joint work with E. Fischer, later identified the crystals as 6-amino-2, 8-dioxypurine. This pathophysiologically accessory pathway of the oxidation of the adenine ring illustrates in a particularly beautiful way the inertness of the amino group of free adenine in the organism.

According to Minkowski, orally administered hypoxanthine, in contrast to adenine, is converted quantitatively in the dog to the above-mentioned end products of purine metabolism. This illustrates with full clarity that in the biological degradation of adenine the deamination is a process which is decisively dependent on the way in which adenine is chemically bound. Later authors such as Krüger, Schittenhelm and their coworkers[(7)], conducting metabolic experiments on man, failed to observe the diverse behavior of free and bound adenine established in Minkowski's studies on dogs. And yet, the knowledge derived from Minkowski's studies has turned out to be more and more applicable for the study of intermediary metabolism of single organs.

The concept of bound purine received a more exact formulation through the studies by Kossel, Levene, Jones, Steudel, Feulgen, Thannhauser, and others who found that an essential chemical structure of the simple nucleic acids is a glucoside-like link between the purine or pyrimidine bases and the carbohydrate-phosphoric acid ester.

The mentioning of nucleotides in the context to follow will always refer to those containing pentose which are the better known ones, obtainable as pure, beautifully crystallizing substances. According to Thannhauser, the nucleoside derivatives of these substances can be degraded to uric acid in the human organism.

W. Jones[8] and, later, A. Schittenhelm have gathered evidence for the existence of enzymes which exclusively deaminate aminopurines linked to carbohydrate, i.e., exclusive deamination of nucleosides with the free purine bases remaining intact. These observations establish the existence of a far-reaching specificity among nucleine deaminases since it is known from other studies by W. Jones[9] that one has to assume the existence of two separate enzymes for the liberation of ammonia from each of the two aminopurines, adenine and guanine: the latter deaminase is specific for the free purines and the former for the nucleosides. Furthermore, the four enzymes exhibit sharp differences in distribution in various animal species and in different organs. Therefore, one can say that the study of the nucleine deaminases has served as an illustration of the over-all significance of the above-mentioned concept. In view of the numerous previous observations of the degradation of purine, the theoretical objection that the chemical functions carried out by the living organ have been lost by mincing of the tissue should be largely disregarded.

The concept of the extreme specificity of nucleine deaminases, a concept for which W. Jones is the main spokesman, has not remained unchallenged. Even very recently H. von Euler[10] has considered it possible that the different activities shown by various organs does not necessarily indicate the presence of two specific enzymes. It is proposed that the inevitable variety of accompanying impurities might merely simulate the existence of two specific enzymes, whereas, in reality, only one enzyme may be involved. This objection can scarcely be refuted on the basis of investigations of nucleine deaminases performed thus far. This may be due partly to the fact that previous authors have not concentrated on the characterization of the nucleine deaminases by means of enzymic chemical methods. On the contrary, their studies have dealt essentially with differences which can be observed in crude aqueous extracts of various organs during autolysis or during incubation in the presence of exogenous purine derivatives.

A detailed enzymic-chemical investigation of the nucleine deaminases cannot be accomplished by hitherto employed means which involve preparative isolation of the reaction products, i.e., the purine derivatives. In view of this, it is surprising that ammonia assays, which lend themselves so readily to exact micro-determinations, have been employed so rarely for the detection and study of deaminases.

As will be apparent from the present article, ammonia estimations constitute an exact and convenient method for the detection of nucleine deaminases in muscle tissue or in enzyme solutions derived from this tissue. One can indeed say that many investigations have been made possible for the first time by the use of ammonia determination since it is the only method which can provide a quantitative kinetic study of deamination.

By using this method on muscle, results have been obtained which compel one to assume the existence of an even more sharply pronounced specificity within the group of nucleine deaminases than was postulated by Jones himself. In the previous discussion of the deamination of bound purines we have only taken the carbohydrate derivatives, i.e., the nucleosides, into consideration.

It had apparently always been tacitly assumed that phosphoric acid was split off the nucleotide molecule prior to deamination. And yet, one fact has been known for a long time, a fact which points to the existence of a biological deamination of the nucleic acids themselves, namely, the occurence of hypoxanthine phosphoric acid, also called inosinic acid, in meat extracts.

Strangely enough, the recently accomplished clarification of the constitution of this substance, which was discovered by Liebig[11] as early as 1847, did not stimulate any biological thinking along these lines, in spite of the fact that it appeared to be practically impossible that its formation could be a result of preparative processing.

It was largely because of the above-mentioned findings, namely, that adenine nucleotide is an ubiquitous constituent of striated muscle, that Embden was prompted to look into the possible close relation of this substance to inosinic acid and to investigate it with respect to its biological significance. It was successfully demonstrated that adenylic acid is the mother compound of inosinic acid. This constitutes a basis for more precise chemical terms to account for the bound purine in muscle.

At the suggestion of Professor Embden, I have investigated the enzymatic deamination of the free and bound amino purines in muscle. The results of these experiments will be described in the succeeding text.

I. Methods

Preparation of the substrates

The preparation of large quantities of the adenine derivatives from yeast nucleic acid is quite time-consuming and according to my experience is achieved best with the use of yeast nucleic acid from the firm, E. Merck, Darmstadt; the guanine derivatives were obtained in good yield from the preparation from the firm, C. F. Böhringer[12]. Adenylic and guanylic acids were obtained most conveniently according to the modification by W. Jones and M. E. Perkins[13] of the procedure of H. Steudel and E. Peiser[14]. The method described by P.A. Levene and W.A. Jacobs[15] was used for the preparation of the nucleosides. The purines were best isolated, also according to the directions of P.A. Levene, by hydrolysis of yeast nucleic acid in methyl alcohol. Regarding these preparations, it should be pointed out that it is essential for the preparation of nucleotides to decompose the lead precipitates in the cold; otherwise colored impurities adsorbed by the precipitating lead sulfide during its decomposition go into solution again and the crystallization, particularly that of adenylic acid, becomes very difficult.

Muscle adenosine phosphoric acid was obtained according to the directions of G. Embden and M. Zimmerman from rabbit muscle or beef heart[16].

Ammonia produced during deamination was determined according to a previously published article[17] by distillation *in vacuo*. In my experiments the relatively large quantities of ammonia formed made it unnecessary to use a ground glass joint. A distillation period of 20 minutes at 38°C was sufficient for the complete distillation of ammonia from solutions. For the determination of ammonia in minced tissue, the major amount was carried over by distillation in 10 minutes; the remainder was expelled with steam *in vacuo* for a period of another 10 minutes in accordance with the procedure of Parnas. If care is taken that the temperature of the water, acidified with sulfuric acid, does not exceed 70° in the steam generator, no interfering heating in the distillation flask occurs. If solutions of a normality of N/200 were used, the margin of error was ±0.004 mg.

II.

* * *

C. The specificity of adenylic acid deaminase.

The bicarbonate B-extracts deaminate only adenylic acid. The effect of such enzyme solution on a series of other substances was tested. No traces of ammonia production could be observed with the substances studied. The following compounds were tested: glycine, D,L-alanine, L-cystine, D,L-serine, D,L-phenylalanine, L-tyrosine, L-tryptophane, L-histidine, D,L-leucine, D,L-isoleucine, D,L-norleucine, D-glutamic acid, glutamine, D,L-aspartic acid, D,L-asparagine, D,L-alanylglycine, D,L-leucylglycine, taurine, creatine, creatinine, urea, guanidine, adenine, guanylic acid, guanosine, and guanine.

Adenosine is also decomposed by the extracts, but only to an extremely small degree and only during long periods of incubation and by the use of relatively large quantities of enzyme.

In the experiments listed in Table I, six different bicarbonate extracts were studied with regard to their effect on adenylic acid and adenosine; in these

Table I

1	2	3	4	5
Experiment No. and Date	cc Enzyme solution	mg Ammonia formed from		Comments
		Adenylic acid	Adenosine	
		% of theoretical amount		
1 4/2/28	5	0.455 = 92.9%	0.0	3 day extract, incubation: 90 min. at 37°
2 4/2/28	5	0.438 = 89.4%	0.0	24 hr extract incubation: 90 min. at 37°
3 22/2/28	5	0.474 = 96.7%	0.0	Fresh extract, incubation: 90 min. at 37°
4 24/2/28	5	0.449 = 91.6%	0.0	24 hr extract, incubation: 90 min. at 37°
5 6/2/28	10	0.409 = 83.5%	0.042 = 8.6%	Fresh extract, incubation: 240 min. at 38°
6 10/2/28	10	0.450 = 91.8%	0.034 = 7.0%	2 day extract, incubation: 240 min. at 38°

experiments 10 ml of a solution containing 2% $NaHCO_3$ was added to the amount of extract as listed in Column 2. In the first four experiments, during which the incubation period was 90 minutes and the amount of enzyme solution 5 ml, adenylic acid was almost completely deaminated while there was no effect on adenosine; in the last two experiments, during a four hour incubation and with twice the amount of enzyme, a very small degree of deamination occurred. It is unlikely, but not excluded, that more favorable buffering conditions for the deamination of adenosine existed in the last two experiments.

Experiments described below indicate that the slight activity of the bicarbonate B-extracts on adenosine is due to the contamination of an adenosine-splitting enzyme. Since the activity of this contaminant can definitely be suppresed under appropriate conditions, the enzyme extract prepared in the manner described above can be used as a sensitive, highly specific reagent for adenosine phosphoric acid. In fact, this has been done in a previous investigation by H. Rosch and W. te Camp[18].

The far-reaching specificity of the adenylic acid deaminating enzyme has eventually resulted in the discovery of a hitherto unsuspected isomerism. In a paper mentioned above, by G. Embden and M. Zimmerman[16], the adenylic acid isolated from muscle was considered identical, on the basis of melting point, optical rotation, and elementary analysis, with the yeast adenylic acid originally prepared from yeast nucleic acid by Jones and Kennedy[19] and more precisely described by P.A. Levene[20].

The demonstration that enzyme solutions which cause marked deamination of muscle adenylic acid cannot deaminate yeast adenylic acid has provided the first direct clue that one was in fact dealing with two different substances. The results of three such experiments are recorded in Table II. In these

Table II

All experiments performed with acetate buffer (N/15) at *p*H 5.9

1	2	3	4	5
Experiment No.	mg Ammonia formed from			Comments
	10 mg muscle adenylic acid	10 mg yeast adenylic acid	10 mg yeast adenylic acid + 10 mg muscle adenylic acid	
	% of theoretical amount			
1 27/4/28	0.408 = 83.3%	0.0	0.374 = 76.3%	In the experiment with the mixture (Column 4), the neutralization was omitted by error.
2 18/5/28	0.451 = 92.0%	0.0	0.408 = 83.3%	See above comments
3 17/6/28	0.437 = 89.2%	0.0	0.421 = 85.9%	For yeast adenylic acid, one of Prof. P. A. Levene's preparations was used. The mixture was neutralized.

experiments the muscle adenylic acid, the yeast adenylic acid, and an equimolar mixture of both substances were prepared under strictly comparable conditions at *p*H 5.9 (N/15 sodium acetate buffer); in Experiments 1 and 2, the mixture of the two substances was prepared under somewhat different *p*H conditions. In all experiments, incubation was carried out at 37° for 60 minutes.

The experiments clearly demonstrate that the amino group of yeast adenylic acid, in contrast to that of muscle adenylic acid, is completely resistant towards the action of active enzyme solutions. The argument that preparations of yeast adenylic acid contain inhibitory impurities is not valid because, on addition of a mixture of muscle and yeast adenylic acid, almost as much ammonia is given off as in the presence of pure muscle adenylic acid alone. The slight reduction in ammonia formation which occurs on addition of the mixture as compared with that of pure muscle adenylic acid cannot affect this result in principle. Perhaps this reduction is due to the binding of a certain amount of enzyme by yeast adenylic acid; in Experiments 1 and 2 it may be due to the somewhat different *p*H.

There is another objection, namely, that in the preparation of yeast adenylic acid the necessary pre-treatment of the nucleic acid with alkali (sodium hydroxide or ammonia), which is omitted in the isolation of muscle adenylic acid (since it occurs as free mononucleotide), may alter the constitution of yeast adenylic acid thereby making it resistant to the muscle enzyme. In order to test this possibility, I preincubated muscle adenylic acid for 24 hours with 1% sodium hydroxide. As much ammonia was split off by the enzyme from this preparation as from the preparation without alkaline pre-treatment. Therefore, there can be no doubt about the chemical differences of the two substances (muscle and yeast adenylic acid). This will be demonstrated, also by chemical and physical methods, in a subsequent report by G. Embden and G. Schmidt. Through the kindness of Professor P.A. Levene(21), it was possible to confirm the results obtained with a preparation of yeast adenylic acid prepared personally by Professor Levene.

The proof of the different nature of yeast and muscle adenylic acid does not affect the findings regarding the deamination of adenosine in muscle because, according to investigations of P.A. Levene and his associates, it is possible to obtain hypoxanthosine from yeast nucleic acid by chemical degradation and deamination which is completely identical with the inosine of muscle. Therefore, the adenosine component of both adenylic acids must be identical and there remains no other possibility for explaining the chemical differences of the two adenylic acids than assuming that the position of the phosphoric acid bond is different in the two substances. The finding of György(22) that yeast nucleic acid is degraded by minced muscle certainly cannot be due to the action of nucleotide deaminase since the yeast nucleic acid preparations of Merck and Böhringer are not degraded by active B-extracts.

D. Some kinetic experiments regarding the enzymatic deamination of adenylic acid

As yet, only a few experiments have been carried out on the reaction kinetics of the enzymatic deamination of adenylic acid and, indeed, these were performed at *p*H 6.9 which is not the optimum *p*H. In the first experiments

I did not obtain constant values by using the monomolecular reaction equation; however, this was the case when I markedly increased the ratio of substrate to enzyme.

The result of such an experiment, in which 0.5 cc of enzyme solution was added to 10 mg of adenylic acid, is shown in Table III. The upper figures in each column record the amount of ammonia formed from 10 mg of adenylic acid at various times; the lower figures represent the calculated *K*-values. As can be seen, the constancy of these values is satisfactory.

Table III

Temperature 37° (±0.02), phosphate buffer

1		2	3	4	5	6
Date		Time from beginning of experiment				Comments
		10 min.	20 min.	30 min.	40 min.	
1927 30/10	mg Ammonia	0.098	0.192	0.242	0.292	0.5 cc enzyme solution was used for each 10 mg of adenylic acid.
	K (monomolecular)	0.00035	0.00037	0.00034	0.00034	

References

1. *Biochim. Z.,* **187** (1927), 194.
2. *Biochim. Z.,* **184** (1927), 399.
3. *Hoppe-Seyler's Z. f. physiol. Chem.,* **43** (1904/05), 532.
4. cf., C. Vögtlin and W. Jones, *Hoppe-Seyler's Z. f. physiol. Chem.,* **66** (1910), 250.
5. *Arch. f. exper. Pathol.,* **41** (1898), 375.
6. *Z. f. klin. Medizin,* **45** (1902), 359.
7. cf., the condensed article by A. Schittenhelm and K. Harpuder in the *Handbuch d. Biochemie,* ed. by C. Oppenheimer, Vol. 8, Georg Thieme Verlag, Leipzig.
8. See B. S. Amberg and W. Jones, *Hoppe-Seyler's Z. f. physiol. Chem.,* **73** (1911), 407.
9. cf. W. Jones and C. R. Austrian, *Hoppe-Seyler's Z. f. physiol Chem.,* **48** (1906), 110.
10. *Chemie der Enzyme,* Part II, Sec. 2, p. 385. J. Springer, Berlin (1927).
11. *Ann. d. Chem.,* **62** (1847), 317.
12. I am obliged to the firm, C. F. Böhringer, Mannheim-Waldhof for the kind gift of a large amount of yeast nucleic acid.
13. *J. Biol. Chem.,* **62** (1924/25), 557.
14. *Hoppe-Seyler's Z. f. physiol. Chem.,* **127** (1923), 262.
15. *Chem. Ber.,* **43** (1910), 3150.
16. K. Pohle was the first to obtain adenosine phosphoric acid from the heart (unpublished studies).
17. G. Embden, M. Cartensen and H. Schumacher, *Hoppe-Seyler's Z. f. physiol. Chem.,* **179** (1928), 188ff.
18. *Hoppe-Seyler's Z. f. physiol. Chem.,* **175** (1928), 158.
19. *J. Pharm. and Exp. Ther.,* **12** (1918), 253.
20. *J. Biol. Chem.* **41** (1920), 483.

21. Mr. P. A. Levene kindly informs me that for some time he has entertained the view, based on the different stability of the phosphoric acid radical upon acid hydrolysis, that a structural difference might well exist between the inosinic acid and the purine mononucleotides obtained from yeast nucleic acid. He considers it probable that the location of the phosphoric acid bond in yeast purine nucleotide is different from that in inosinic acid. According to his studies, phosphoric acid is esterified with the terminal alcohol group of the pentose of inosinic acid. The findings on which this view is based are contained in a study of M. Yamagawa, *J. Biol. Chem.,* **43** (1920), 339.
22. See reference 12.

PHOSPHORUS COMPOUNDS OF MUSCLE AND LIVER*

Cyrus H. Fiske and Y. Subbarow

Biochemical Laboratory, Harvard Medical School,
Boston, Massachusetts

I. Muscle

Embden and Schmidt[1] have recently made the highly interesting discovery that the adenosine phosphoric acid isolated about two years ago from voluntary muscle[2] is not identical with that obtained from yeast nucleic acid. Among the chemical properties by which the two may be distinguished the difference in resistance to hydrolysis by acid is particularly striking. The muscle nucleotide ("myoadenylic acid") is hydrolyzed (by 0.1 N sulphuric acid at 100°) only about one fifth as rapidly, as measured by the rate at which o-phosphoric acid is split off.[1]

That adenine (in nucleotide combination) is the source of the ammonia formed in muscle during contraction has been amply demonstrated by Embden and his collaborators. The physiological significance of this important work is presumably quite unaffected by the fact, which we now have to report, that myoadenylic acid is not a normal constituent of muscle (except perhaps in traces), but a decomposition product. Our first intimation that this might be the case developed from the observation that when a protein free muscle filtrate is treated with an alkaline solution of calcium chloride[3] a large part of

*C. H. Fiske and Y. Subbarow, "Phosphorus Compounds of Muscle and Liver," *Science,* **70**, 381–382 (October, 1929). Reprinted with the permission of the author and of the publisher.

[1] G. Embden and G. Schmidt, *Z. physiol. Chem.,* 181: 130, 1929.

[2] G. Embden and M. Zimmerman, *Z. physiol. Chem.,* 167: 137, 1927.

[3] This is the first step in the isolation of phosphocreatine (C. H. Fiske and Y. Subbarow, *J. Biol. Chem.,* 81: 629, 1929) where it serves the purpose of removing inorganic phosphate and other products.

the purine nitrogen comes down in the precipitate. The calcium salt of myoadenylic acid is soluble in water, and should consequently—if present—remain dissolved under these conditions.

The purine derivative precipitated by calcium has been isolated by (1) precipitation with mercuric acetate in the presence of 2 per cent. acetic acid, followed (after removal of the mercury) by (2) precipitation with calcium chloride and alcohol from hydrochloric acid solution, which yields an acid calcium salt. By repeating the entire process the acid calcium salt is finally obtained as a microcrystalline precipitate. The yield is not far from quantitative, and accounts not only for most of the purine nitrogen of muscle, but also for most of the acid soluble phosphorus not present as o-phosphoric acid, phosphocreatine or hexose monophosphate. In the case of cat muscle the yield of purified material may be the equivalent of nearly 50 mg of phosphorus per 100 gm of muscle.

The acid calcium salt is not well suited for analytical purposes, owing to the difficulty of removing all the water. It may, however, be converted to a silver salt—by precipitation with silver nitrate from nitric acid solution—and the composition of this product has been found to be $C_{10}H_{13}O_{13}N_5P_3Ag_3$. It contains, in addition to adenine and carbohydrate, three molecules of phosphoric acid, or two more than in adenylic acid. Two of the three molecules of phosphoric acid are readily removed by hydrolysis with acid, and this fact is doubtless sufficient to explain why Embden and Zimmerman[2] obtained a nucleotide (myoadenylic acid) which still retains the one resistant phosphoric acid group.

The new substance includes also the phosphorus which Lohmann[4] believes to be present in the muscle in the form of pyrophosphate, but whether or not it is an ester of pyrophosphoric acid remains to be determined.

II. Liver

The greater part of the organic acid soluble phosphorus of liver may be precipitated from the protein free filtrate, after removing the inorganic phosphate with alkaline calcium chloride solution, by the addition of alcohol. Purification by dissolving in water and reprecipitating with alcohol finally yields a calcium salt, crystallizing in spherulites or aggregates of short needles, and having the composition $C_3H_7O_6PCa.1\frac{1}{2}H_2O$. It is the calcium salt of glycerophosphoric acid, which in spite of text-book statements has not—as far as we have been able to determine—been isolated from animal material before, at least under conditions which preclude its formation from lecithin and related substances.

Experiments are now under way which it is hoped will lead to some procedure for the quantitative estimate of free glycerophosphate in tissue filtrates. From present indications this substance appears to account for at least one third of the total acid soluble phosphorus of the liver.

The properties of the above-mentioned calcium salt, together with the fact that it gives an intense greenish blue color on applying the Denigès codeine

[4] K. Lohmann, *Biochim. Z.*, 202: 466, 1928; 203: 164, 172, 1928.

test after oxidation with bromine,[5] and forms no insoluble double salt with barium nitrate,[6] identify it as α-glycerophosphate. This is of particular interest since all preparations of lecithin so far examined by means of these tests have been found to contain mainly the β form of glycerophosphoric acid.[7]

[5] O. Bailly, *Ann. chim.,* 6: 96, 1916.
[6] P. Karrer and H. Salomon, *Helv. chim. acta,* 9: 1, 1926.
[7] O. Bailly, *Ann. chim.,* 6: 215, 1916; P. Karrer and P. Benz, *Helv. chim. acta,* 10: 87, 1927.

THE CHEMISTRY OF MUSCLE CONTRACTION*

K. Lohmann

Institut für Physiologie,
Kaiser Wilhelm Institut für Medizimische Forschung,
Heidelberg, Germany

The mechanical and thermal energy released during muscle contraction, is, in the last analysis, derived from oxidation, and from carbohydrate oxidation in particular. Nevertheless, an isolated muscle is capable of performing a considerable amount of work even in the absence of oxygen. Thus the fundamental chemical process involved in muscle contraction is not an oxidative, but rather an anaerobic, exothermic one. The exothermic transformation of glycogen into lactic acid was the first to be recognized as a process of this type. This anaerobically formed lactic acid disappears aerobically in the *Pasteur-Meyerhof* reaction, whereby most of the lactic acid is converted into glycogen while a small part of it is burned. (Hill and Meyerhof[1]). Here we have, then, a carbohydrate cycle, although one which is incomplete since the fraction converted into CO_2 and H_2O by combustion is irreversibly removed from the cycle.

Muscle contraction itself, however, is not necessarily bound to the formation of lactic acid (a view already expressed by Embden[2] several years ago). As discovered by Lundsgaard[3], a muscle poisoned with mono-iodoacetate is unable to form any more lactic acid, yet it can still perform a number of twitchings. The chemical process involved here is (according to Lundsgaard) the decomposition of the creatine phosphoric acid found by Eggleton[4] and Fiske[5], into creatine and phosphoric acid. The amount of energy released during such alactacidic contractions, was found to agree, for the most part, with the heat of hydrolysis of creatine phosphoric acid as measured by Meyerhof and

*K. Lohmann, "Über den Chemismus der Muskel Kontraktion," *Naturwiss.* **22**, 409 (1934). Translated from the German and reprinted by permission of the author and of Springer-Verlag, Heidelberg, Germany.

Lohmann[6]. In the normal, unpoisoned muscle, the decomposed creatine phosphoric acid is completely resynthesized in the presence of O_2 (Eggleton[4]), but part of it can also be reconstituted anaerobically (Meyerhof and Lohmann[6]), at the expense of the energy of lactic acid formation ("Creatine Phosphoric Acid Cycle").

What, then, is the relationship between the decomposition of creatine-phosphoric acid and the fundamental process of muscle contraction? The latter must consist of at least two chemical reactions: first, a process by which the contraction is initiated, and then a further step reversing the first, to make relaxation possible. (For this concept it is essentially immaterial whether the contraction initiator or the relaxation process is the energy-yielding reaction—in the first instance, the uncontracted muscle would be in every respect a "quiescent" muscle; in the second, it would be comparable to a tensed spring, which is relaxed during contraction and retensed during relaxation). It seems out of the question that the disintegration and resynthesis of creatine phosphoric acid cannot represent the fundamental muscle processes. For example, in the iodoacetate-poisoned muscle, phosphocreatine is undergoing a hydrolysis, whereas in the normal muscle, phosphocreatine resynthesis proceeds so slowly that it is terminated only after several seconds or even minutes. Yet, both reactions in the fundamental muscle necessarily proceed quite rapidly. Bethe[7] has particularly emphasized this feature in connection with certain insect muscles, in which the entire contraction-relaxation process lasts only 1/500 of a second, giving a margin of only 1/1000 of a second for contraction and relaxation. Although this interval is considerably longer in frog muscle, it still amounts to only a fraction of a second. The fact that we do not have a procedure which permits instantaneous fixing of a muscle constitutes an almost insurmountable obstacle for any direct investigation of the chemistry of a single muscle contraction.

We must, then, find other approaches. The study of the enzymatic creatine phosphoric acid fission[8] has proved to offer such an alternative approach. It was found, for instance, that the creatine phosphoric acid in an aqueous muscle extract is not simply split by a "creatine phosphoric acid-splitting enzyme" into creatine and phosphoric acid, but rather that this cleavage takes place only in the presence of a co-enzyme, specifically adenylpyrophosphoric acid (Lohmann[9]). A true catalysis is involved here, since one part of adenylpyrophosphoric acid brings about the splitting of up to 1000 parts of creatine phosphoric acid in the presence of the specific enzyme. A detailed kinetic investigation of the creatine phosphoric acid cleavage revealed that it proceeds in the following manner.

(1) adenylpyrophosphate = adenylic acid + 2 phosphate + 25,000 cal
(2) adenylic acid + 2 phosphocreatine = adenylpyrophosphate + 2 creatine

---

(3) 2 phosphocreatine = 2 creatine + 2 phosphate + 24,000 cal

The summation of Equations (1) and (2) gives as a balance the decomposition of creatine phosphoric acid into creatine and phosphoric acid (Equation (3)).

This concept is based in particular upon the following findings: both adenylic acid and adenylpyrophosphoric acid are active, the former becoming esterified to the latter. It is possible, furthermore, by prolonged dialysis of a muscle extract, to destroy the adenylpyrophosphoric acid-splitting enzyme

(Reaction (1)), while preserving the esterifying enzyme (Reaction (2)). In such an extract, creatine phosphoric acid is not split in the presence of adenylpyrophosphoric acid, but is split only in the presence of adenylic acid. However, the amount of disintegrated creatine phosphoric acid corresponds strictly to the amount of esterified adenylic acid and the reaction does not proceed beyond that. Conversely, it is possible to inhibit Reaction (2) exclusively by addition of alkali to the extract to a *p*H of 9, leaving Reaction (1) undisturbed. Whereas adenylpyrophosphoric acid is split in such an extract, creatine phosphoric acid is not, neither in the presence of adenylpyrophosphoric acid, nor with adenylic acid.

According to Meyerhof and Lohmann[6], the decomposition of creatine phosphoric acid liberates 11,000 to 12,000 calories per mole; the decomposition of adenylpyrophosphoric acid to 1 mole of adenylic acid and 2 moles of phosphoric acid liberates 25,000 calories. Hence, 12,500 calories are liberated per mole of phosphate. These values inserted into Equations (1) and (3) show that within the accuracy of measurement, Reaction (2) is thermally neutral. In the muscle extract, Reaction (1) is always the slower, hence the rate-determining one, whereas Reaction (2) proceeds considerably faster. This circumstance explains why in the extraction experiment a splitting of only creatine phosphoric acid is observed at first, followed later by the cleavage of adenylpyrophosphoric acid. Similarly with muscle: in the normal muscle, a partial splitting of adenylpyrophosphoric acid is found only after extensive fatigue (Lohmann[11]). After mono-iodoacetate poisoning, cleavage is also observed only after the creatine phosphoric acid has been largely decomposed (usually rather abruptly) (Lundsgaard). This phenomenon can also be observed during the cutting of a fresh muscle (Mozolowski[13]).

All this suggests the hypothesis that creatine phosphoric acid is split in the same fashion in the intact muscle and in muscle extract. Since no free adenylic acid is detectable in fresh muscle, the reactions in the intact muscle must also proceed in the manner described. During a muscle twitch, first adenylpyrophosphoric acid is split. This is then resynthesized at the expense of the creatine phosphoric acid, which in turn is split. Thus, during a muscle twitch, the exothermal cleavage of adenylpyrophosphoric acid takes place before the splitting of the creatine phosphoric acid, which itself proceeds without heat release because of the coupling with an endothermic reaction. Thus the transformation of adenylpyrophosphoric acid constitutes, among the processes demonstrated so far, the first step of a reaction sequence. We are dealing here with a cyclical process which proceeds completely anaerobically in both directions.

Lehnartz[14] has already perceived the possibility, purely on theoretical grounds, that the energy from the fission of creatine phosphoric acid might be used in order to resynthesize the adenylpyrophosphoric acid.

The chemical processes demonstrated so far which are involved during a muscle twitch are summarized in a (schematic) table. The dotted lines indicate the energy coupling of exothermic (left-right) and endothermic (right-left) reactions:

$$\text{I. Adenylpyrophosphate} \underset{\text{anaerobic}}{\overset{\text{anaerobic}}{\rightleftarrows}} \text{Adenylic Acid} + 2\ \text{Phosphate}\ (0.09\ \text{cal})$$

$$\rightleftarrows$$

II. Phosphocreatine	anaerobic $\rightleftarrows$ anaerobic and aerobic	Creatine + Phosphate (0.23 cal)
III. Glycogen (1/n)	anaerobic $\rightleftarrows$ aerobic	2 Lactic Acid (1.2 cal)
IV. Lactic Acid (respectively carbohydrate)	$+3O_2$ $\longrightarrow$	$3\ CO_2 + 3H_2O$ (30 — 60 cal)

The numbers in parentheses indicate the amount of heat released, expressed in calories per gram of fresh frog muscle, when the existing P-bonds are completely disintegrated (I and II); when 0.4% lactic acid is formed (lactic acid maximum) (III); and when 0.5 — 1% carbohydrate is oxidized (IV). Thus, the available "energy-reserve" of the muscle decreases gradually as the corresponding reaction is "located" ever closer in relation to the fundamental process of the muscle twitch. We have, however, no experimental proof that the presence of adenylic acid (though a pharmacologically effective substance), can elicit a muscle contraction, and that relaxation ensues upon its disappearance. We do not intend at this time to make such a specific assumption. These enzymatic experiments give, however, an incentive to a more intensive search for existence of other extractable substances of muscle, by means of more subtle methods than those used heretofore.

It is surely no accident that adenylpyrophosphoric acid acts not only as co-enzyme to the "creatine phosphoric acid-splitting enzyme," but also (along with magnesium and inorganic phosphate), as co-enzyme to the lactic acid-forming enzyme (Lohmann[9]), for which it is equally indispensable. Viewed teleologically, this dual function seems to be a very ingenious arrangement for insuring the orderly sequence of the chemical processes involved in the muscle twitch: in the inert muscle, enzymes, co-enzymes and substrates are separated from each other, but during contraction this spatial separation (which might be called "inactive preliminary stages") is abolished. The contraction brings about a fission of adenylpyrophosphoric acid which in turn imposes a cleavage of creatine phosphoric acid, thereby simultaneously reconstituting adenylpyrophosphoric acid; the latter can now interact as co-enzyme by mobilizing glycogen for lactic acid formation. Lactic acid is, also, the very same substance which intact muscle as well as minced muscle is capable of oxidizing most readily by molecular oxygen.

References

1. Meyerhof, O.: *Die Chemische Vorgänge im Muskel.* Berlin: Springer-Verlag, 1930.
2. Embden, G.: *Hoppe-Seyler's Z. f. physiol. Chem.*, **151** (1926), 209.
3. Lundsgaard, E.: *Biochim. Z.*, **217** (1930), 162.
4. Eggleton, P.: *Biochem. J.*, **21** (1926), 190.
5. Fiske, C. H.: *J. Biol. Chem.*, **81** (1929), 629.
6. Meyerhof, O. and Lohmann, K.: *Biochim. Z.*, **196** (1928), 22.

7. Bethe, A.: *Naturwiss.*, **18** (1930), 678.
8. Lohmann, K.: *Biochim. Z.* (1934), (in press).
9. Lohmann, K.: *Naturwiss.*, **17** (1929), 624.
10. Meyerhof, O. and Lohmann, K.: *Biochem. Z.*, **253** (1932), 431.
11. Lohmann, K.: *Biochim. Z.*, **203** (1928), 172.
12. Lundsgaard, E.: *Arch. di Sci. biol.*, **18** (1933), 232; *Biochim. Z.*, **269** (1934), 308.
13. Mozolowski, W. and Sobczuk, B.: *Biochim. Z.*, **265** (1933), 41.
14. Lehnartz, E.: *Ergeb. Physiol. und exper. Pharmakol.*, **35** (1933), 966.

AN ACTIVATOR OF THE HEXOKINASE SYSTEM*

Sidney P. Colowick and Herman M. Kalckar

Department of Pharmacology,
Washington University School of Medicine,
St. Louis, Missouri

(Received for publication January 13, 1941)

Hexokinase, the yeast factor which activates the fermentation of glucose or fructose in muscle extracts,[1] is an enzyme which is supposed to catalyze the following reaction:[2] adenyl pyrophosphate +2 hexose → adenylic acid + 2 hexose-6-phosphate.

Hexokinase prepared from bakers' yeast by the method of Meyerhof,[1] was precipitated with saturated ammonium sulfate and the precipitate dried. The resulting powder when kept at low temperature retains its activity for at least 2 months. A purification of the hexokinase can be obtained by precipitation with ammonium sulfate between 50 and 65 per cent saturation.

In the presence of (1) hexokinase (crude as well as fractionated), (2) adenosine triphosphate, (3,a) glucose or (3,b) fructose, and (4) magnesium ions the following reaction takes place: adenosine triphosphate + hexose → adenosine diphosphate + hexose-6-phosphate (I).

Both reaction products were identified; glucose-1-phosphate is not the primary reaction product, since the enzyme necessary for its conversion to 6-phosphate is not present.

The reaction is completed within 5 minutes when 0.25 mg. of crude hexokinase protein per cc. is incubated at 30° with 2.5 mg. of glucose and adenosine

*S. P. Colowick and H. M. Kalckar, "An Activator of the Hexokinase System," *J. Biol. Chem.*, **137**, 789–90 (1941). Reprinted with the permission of the authors and of the American Society of Biological Chemists, Inc.

[1] Meyerhof, O., *Biochem. Z.*, **183**, 176 (1927).

[2] von Euler, H., and Adler, E., *Z. physiol. Chem.*, **235** (1935). Meyerhof, O., *Naturwissenschaften*, **23**, 850 (1935).

triphosphate containing 60 γ of labile P, 30 γ of which are transferred to glucose. The progress of this reaction can be followed not only by the disappearance of acid-labile P but also manometrically, since one acid equivalent is liberated when one of the pyrophosphate linkages is broken with the formation of hexose-6-phosphate.

In equation (I) only one of the two labile P groups in the nucleotide is transferred to sugar, a fact which was already indicated in some experiments of Meyerhof and Kiessling.[3] The second labile phosphate group can be transferred to hexose, provided that a heat-stable protein, present in muscle tissue, is added to the hexokinase.

The heat-stable protein in the absence of hexokinase has no activity. When adenosine diphosphate is added instead of triphosphate in the presence of hexokinase, no reaction with glucose or fructose occurs until the heat-stable protein is added; the reaction then proceeds according to the equation: adenosine diphosphate + hexose ⟶ adenylic acid + hexose-6-phosphate (II). The phosphorylated sugar, which has been isolated, contains no hexose diphosphate.

The factor necessary for reaction (II) is a protein, since it is precipitated by trichloroacetic acid, ammonium sulfate, or sodium sulfate and is destroyed by pepsin. The trichloroacetic acid precipitate of the protein when dissolved in dilute sodium hydroxide exhibits full activity, when tested immediately. The protein is inactivated in alkaline solutions but is reactivated by reduced glutathione or cysteine; the latter two substances are inactive when added without the protein. The protein retains most of its activity even after 20 minutes of boiling in 0.1 N hydrochloric acid.

The heat-stable protein is active in very low concentration; 1 γ of purified protein per cc. is able to catalyze the transfer of at least 14 γ of labile nucleotide P in 5 minutes at 30°. The protein concentration was determined by the biuret reaction. The heatstable protein is not identical with insulin.

[3] Meyerhof, O., and Kiessling, W., *Biochem. Z.*, **283**, 83, table VI A (1935).

THE ENZYMATIC ACTION OF MYOKINASE*

Herman M. Kalckar

Department of Pharmacology,
Washington University School of Medicine
St. Louis, Mo.

(Received for publication February 16, 1942)

Myokinase is an acid-stable protein[1] occurring in skeletal muscle; when acting with yeast hexokinase, it brings about the transphosphorylation: adenosine diphosphate + hexose $\longrightarrow$ adenylic acid + hexose monophosphate.[1] It has been found recently[2] that myokinase is also a necessary component in the dephosphorylation of adenosine diphosphate in muscle. Myosin, purified by repeated precipitations, dephosphorylates adenosine *tri*phosphate but not adenosine *di*phosphate.[3] Addition of a few micrograms of myokinase to the system brings about dephosphorylation of adenosine diphosphate.

The mechanism of the myokinase action has so far not been understood. Incubation of adenosine diphosphate with myokinase does not give rise to any change in the labile phosphate. It was found, however, that incubation of adenosine diphosphate with myokinase and adenylic acid deaminase[4] yields ammonia, whereas incubation of the nucleotide with deaminase or myokinase separately does not yield any ammonia. Myokinase apparently forms adenylic acid or some other nucleotide[5] which can be deaminated by Schmidt's deaminase.

Dr. M. Johnson, University of Wisconsin, suggested in a personal communication about 6 months ago that myokinase might catalyze a reversible

*H. M. Kalckar, "The Enzymatic Action of Myokinase," *J. Biol. Chem.*, **143**, 299 (1942). Reprinted with the permission of the American Society of Biological Chemists, Inc.

[1] Colowick, S. P., and Kalckar, H. M., *J. Biol. Chem.*, **137**, 789 (1941).

[2] Kalckar, *Proc. Am. Soc. Biol. Chem.*, in press (1942).

[3] Ljubimova and Pevsner, *Biokhimiya*, **6**, 178 (1941).

[4] Schmidt, *Z. physiol. Chem.*, **179**, 243 (1928).

[5] Kiessling and Meyerhof, *Biochem Z.*, **296**, 410 (1938).

transfer of phosphate from 1 mole of adenosine diphosphate to another, yielding adenosine triphosphate and adenylic acid. This suggestion has now been proved to be very nearly true. Incubation of adenosine diphosphate with myokinase gives rise to formation of small amounts of adenylic acid and adenosine triphosphate. The adenylic acid was crystallized and identified. The presence of adenosine triphosphate was demonstrated by means of hexokinase or reprecipitated myosin. Incubation of adenosine diphosphate with myokinase and deaminase yields adenosine triphosphate and inosinic acid. The table shows that adenosine triphosphate and adenylic acid (or inosinic acid) are formed in approximately equimolar amounts. The conversion of adenosine diphosphate to the two nucleotides does not go very far (about 25 per cent is converted in the presence of myokinase, and about 50 per cent in the presence of both myokinase and deaminase).

Conversion of Adenosine Diphosphate to Adenosine Triphosphate and Adenylic Acid

Adenosine diphosphate (ADP) was incubated with myokinase or deaminase or both. Ammonia was determined on an aliquot of the trichloroacetic acid filtrate, and to the remainder were added barium acetate and sodium hydroxide to alkalize the reaction. The barium precipitate which contains ADP and adenosine triphosphate (ATP) was analyzed for the latter and for pentose by the hexokinase test. The barium supernatant which contains adenylic (or inosinic) acid was analyzed for organic phosphorus.

ADP corresponding to 300 γ labile P; volume of reaction mixture, 0.7 ml. (incubation, 30 min. at 30°). Myokinase, 10 γ protein per ml.; deaminase, 500 γ protein per ml.	Ammonia	Barium precipitate		Barium supernatant	
		P difference calculated from pentose	ATP, disappearance of labile P in hexokinase test	P difference found	P difference calculated from ammonia
	γN	γ	γ	γ	γ
ADP + deaminase	0		15*		
〃 + myokinase	0	−41	53.5	+51	
〃 + 〃 + deaminase..	36	−80	53.0	+83	+80

*Probably there were minute traces of myokinase in the deaminase.

When adenosine triphosphate and adenylic acid are incubated with myokinase, about 20 per cent of the two nucleotides is converted into adenosine diphosphate and then the reaction stops. The phosphate dismutation, 2 adenosine diphosphate $\rightleftharpoons$ 1 adenosine triphosphate + 1 adenylic acid, is apparently not a simple equilibrium. Some other compound must be formed as an intermediate product.

Gustav Embden, about 1933. While studying the effect of inorganic ions (Hofmeister series) on muscle proteins, the "crystal ball," phosphoglyceric acid, revealed the Embden-Parnas-Meyerhof pathway, like snow in a Christmas paperweight.

[The late Gustav Embden was Professor and Director of the Institut für Vegetative Physiologie at the University of Frankfurt from approximately 1920 until 1933. Dr. Gerhard Schmidt was his last student.]

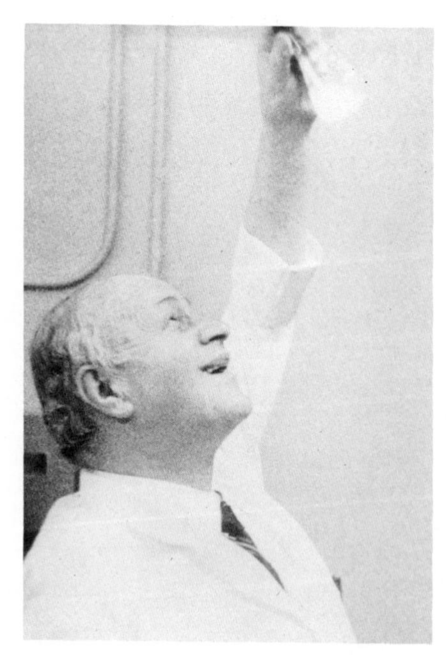

Gerhard Schmidt, Boston, 1965. The nucleohistone gel has become dehydrated to a precipitate.

ON THE INTERMEDIARY REACTIONS OF MUSCLE GLYCOLYSIS*

G. Embden, H. J. Deuticke and G. Kraft

Institut für vegetative Physiologie,
University of Frankfurt, Germany

Some time ago a series of investigations carried out in the Institute demonstrated that in the presence of certain ions, muscle juice or muskel brei [thick muscle dispersions] can bring about a disappearance of inorganic phosphate. On addition of carbohydrate, in the form of glycogen or starch, this disappearance of inorganic phosphate was enhanced and in experiments with muscle juice it was possible to isolate a phosphate ester the properties of which agreed with those of the Harden-Young ester.

Reinvestigation of this problem by Lohmann revealed that this matter is indeed more complex. Lohmann observed that the mixture of phosphorylation products formed, when heated in acid, does not split off phosphoric acid at the same rate as does the Harden-Young ester, but is much more resistant to acid hydrolysis. Therefore, the major portion of the phosphate ester produced in his experiments must consist of a compound different from hexosediphosphate. Isolation experiments led Lohmann to believe that he was dealing with a hexosediphosphate which differed from the Harden-Young ester by its resistance to acid hydrolysis.

Recently, in the course of experiments designed for an entirely different purpose, i.e., the effects of ions on hexosephosphate synthesis, we were fortunate to observe an abundant formation of a beautifully crystallizing barium salt which could be identified as the secondary barium salt of a monophosphoric acid ester of 1-glyceric acid.

It appears to us that further exploration of this finding might shed considerable light on the mechanisms involved in the glycolytic formation of lactic

*G. Embden, H. J. Deuticke, and G. Kraft, "Über Die Intermediären Vorgänge Bei der Glykolyse in der Muskulatur," *Klinische Wochenschrift,* **12,** 213–215 (1933). Translated from the German and reprinted by permission of Springer-Verlag, Heidelberg, Germany.

acid, at least with regard to glycolysis associated with intermediary phosphorylation.

First of all, it was found that under anaerobic conditions, muskel brei produces phosphoglyceric acid to a much greater extent from added hexosediphosphate than from starch. In spite of the inhibition of the dephosphorylation of the hexose esters by certain ions, the splitting of the 6-carbon chain into two 3-carbon fragments apparently is not affected. We are inclined to believe that in the experiments of Lohmann a good portion of the acid-resistant ester also consisted of phosphoglyceric acid.

So far, we have observed the formation of this substance in the presence of sodium lactate, sodium fluoride, and sodium oxalate.

Our assumption that the recently discovered substance is a normally occurring intermediate of muscle glycolysis could be supported by further experimentation. In the course of these studies it was found that fresh muskel brei readily splits added monophosphoglyceric acid into phosphoric acid and pyruvic acid. The latter compound could be isolated without the use of a trapping agent and with a method involving some losses, in a yield of up to 80% of the theoretically possible amount. Under similar conditions, unphosphorylated glyceric acid did not give rise to any detectable amounts of pyruvic acid.

Glyceric acid and pyruvic acid, the latter produced in muscle from the monophosphate ester of the former, are compounds of a higher oxidation level than are the 6-carbon sugars and lactic acid produced during glycolysis. Large amounts of these oxidation products are produced under anaerobic conditions; obviously these oxidations can take place only at the expense of concurrent reductions.

On the basis of the experimental results reported above, we believe that lactic acid production is preceded by the splitting of the 6-carbon chain of one molecule of hexosediphosphate into two molecules of triosephosphate. For fructosediphosphoric acid this reaction should be formulated as follows:*

Equation 1

$$
\begin{array}{l}
CH_2{-}O{-}P(=O)(OH)_2\\
|\\
C{=}O\\
|\\
CHOH\\
|\\
CHOH\\
|\\
CHOH\\
|\\
CH_2{-}O{-}P(=O)(OH)_2
\end{array}
\quad = \quad
\begin{array}{l}
CH_2{-}O{-}P(=O)(OH)_2\\
|\\
C{=}O\\
|\\
CH_2OH\\
\quad +\\
C(=O){-}H\\
|\\
CHOH\\
|\\
CH_2{-}O{-}P(=O)(OH)_2
\end{array}
\quad
\begin{array}{l}
\textbf{Dihydroxyacetone-}\\
\textbf{phosphoric acid}
\end{array}
$$

Fructose diphosphoric acid **Glyceraldehyde-phosphoric acid**

*In agreement with this view, Embden and Jost have shown recently that under the conditions described in this paper, synthetically prepared D-L-glyceraldehyde phosphoric acid produces large quantities of lactic acid.

By means of a dismutation analogous to the Cannizzaro reaction, these two molecules of triosephosphoric acid could be converted into one molecule of glycerophosphoric acid and one molecule of phosphoglyceric acid according to the following equation:

EQUATION 2

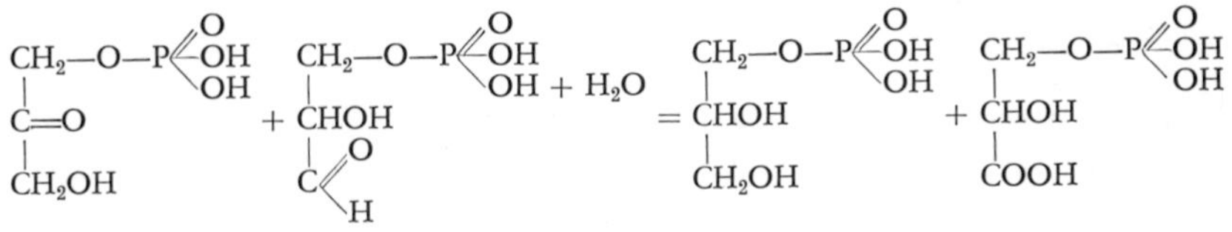

$$\begin{array}{llllll}
CH_2{-}O{-}P(=O)(OH)_2 & & CH_2{-}O{-}P(=O)(OH)_2 & & CH_2{-}O{-}P(=O)(OH)_2 & & CH_2{-}O{-}P(=O)(OH)_2 \\
\vert & & \vert & & \vert & & \vert \\
C{=}O & + & CHOH & + H_2O = & CHOH & + & CHOH \\
\vert & & \vert & & \vert & & \vert \\
CH_2OH & & C(=O)H & & CH_2OH & & COOH
\end{array}$$

Dihydroxyacetone-phosphoric acid **Glyceraldehyde-phosphoric acid** **Glycero-phosphoric acid** **Phosphoglyceric acid**

Of these two hypothetical dismutation products glycerophosphoric acid—whose abundant occurrence in the liver recently has been demonstrated by Fiske—could not be isolated so far.

According to the observations reported above, phosphoglyceric acid is converted into pyruvic acid and inorganic phosphate.

We then proceeded to determine whether pyruvic acid can be reduced to lactic acid with a concurrent oxidation of glycerophosphoric acid to triose-phosphoric acid. Indeed it can be shown that the low production of lactic acid by muskel brei from phosphoglyceric acid and from glycerophosphoric acid can be increased considerably on simultaneous addition of both of these substances. The participation of phosphoglyceric acid in this formation of lactic acid is indicated by the fact that in the experiments where both phosphate esters were added, the concentration of pyruvic acid is much smaller than in those where phosphoglyceric acid was added alone; the participation of glycerophosphoric acid is also supported by the observation that in the experiments with both substances added, the increase in lactic acid is much greater than can be accounted for by the decrease in pyruvic acid.

An example may illustrate this: 50 gm of muskel brei, obtained from a fasted rabbit pretreated with epinephrine, contained 173 mg of lactic acid immediately following its preparation (Experiment A). Two hours after incubation at 40°C in a solution of sodium bicarbonate, the lactic acid content increased to 213 mg (Experiment B_1). Addition of the sodium salt of phosphoglyceric acid resulted only in a slight increase in lactic acid as compared with the control, Experiment B_1 (244 mg, Experiment B_2). Addition of an equimolar amount of sodium glycerophosphate produced only a minimal increase in lactic acid above the control experiment (218 mg, Experiment B_3). When equal amounts of *both* these phosphate esters were added *together* (Experiment 4) lactic acid increased to 402 mg.

While the increased formation of lactic acid in the presence of both substances as compared to that in the presence of phosphoglyceric acid alone amounted to 158 mg, pyruvic acid was reduced under these conditions by only 105 mg, or by an equivalent of 107 mg of lactic acid. The remaining 51 mg of lactic acid must have been formed from glycerophosphoric acid; the ability of this substance to give rise to pyruvic acid in muskel brei has been noted already by Amandus Hahn. This author could not observe the formation of pyruvic acid from glycerol. In our own experiments there was no lactic acid production from added glycerol nor was there any increase in lactic acid when phosphoglyceric acid and glycerol were added together.

Muskel brei quantitatively gives rise to the formation of inorganic phosphate from phosphoglyceric acid, as shown in experiments in which this substance was added alone. By comparison, formation of inorganic phosphate from glycerophosphate, added *alone*, is only minimal; however, this is increased considerably after simultaneous addition of phosphoglyceric acid because, as might be postulated, the reduction of pyruvic acid (produced from phosphoglyceric acid) is brought about at the expense of the simultaneous dehydrogenation of glycerophosphoric acid up to the phosphoglyceric acid stage. A portion of the phosphoglyceric acid formed in this manner is then further converted into pyruvic acid and lactic acid. In any case, the fact that the enzymatic dephosphorylation of phosphoglyceric acid proceeds at a strikingly rapid rate when compared to that of glycerophosphoric acid, it appears to us, is of considerable significance.

Our experimental results obtained so far lead us to the following picture of glycolytic lactic acid production, insofar as glycolysis is associated with phosphorylations (see also Bumm and Fehrenbach).

First phase

Synthesis of hexosediphosphate from one molecule of hexose and two molecules of inorganic phosphate or from one molecule of hexosemonophosphate* and one molecule of inorganic phosphate.

Second phase

Splitting of one molecule of hexosediphosphate into two molecules of triosephosphate (see Equation 1).

Third phase

Dismutation of two molecules of triosephosphate into one molecule of glycerophosphate and one molecule of phosphoglycerate whereby either α- or β-glycerophosphate could be produced depending on whether the hexose diphosphate is a ketose or an aldose (see Equation 2).

Fourth phase

Splitting of phosphoglycerate into inorganic phosphate and pyruvate according to the equation:

*In fresh muscle of those species investigated so far, Embden and Zimmermann have found hexose*mono*phosphate exclusively. These authors used the term "lactacidogen" to designate this precursor of lactic acid. The conversion of hexosemonophosphate into hexosediphosphate proceeds not only on addition of certain salts but also, as shown recently, on grinding the muskel brei with large amounts of Kieselgur (Embden, Jost, and Lehnartz). The absence of hexosediphosphate in fresh muscle may be due to the fact that it is metabolized much more rapidly than is hexosemonophosphate. Meyerhof, *et al.*, have already expressed the view that the first step in the metabolism of hexosemonophosphate is its conversion into hexosediphosphate.

EQUATION 3

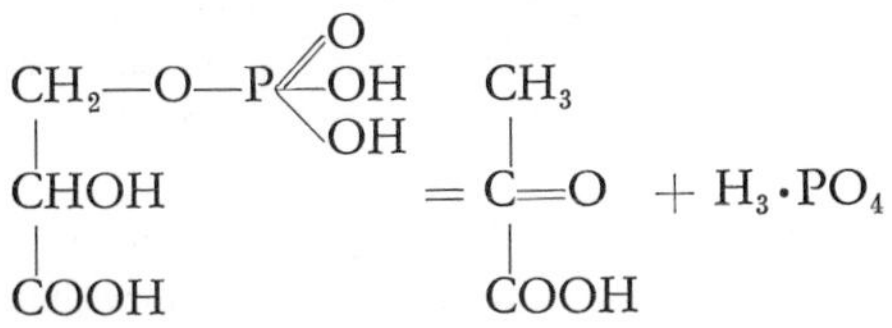

$$
\begin{array}{l}
CH_2\text{—}O\text{—}P(=O)(OH)_2 \\
| \\
CHOH \\
| \\
COOH
\end{array}
=
\begin{array}{l}
CH_3 \\
| \\
C{=}O \\
| \\
COOH
\end{array}
+ H_3 \cdot PO_4
$$

Phosphoglyceric acid — **Pyruvic acid**

According to this equation this split is *not* associated with the intermediate formation of glyceric acid; therefore this process takes place without uptake of water from the medium.

Fifth phase

Reduction of pyruvic acid to lactic acid at the expense of the oxidative formation of triosephosphoric acid from glycerophosphoric acid:

EQUATION 4

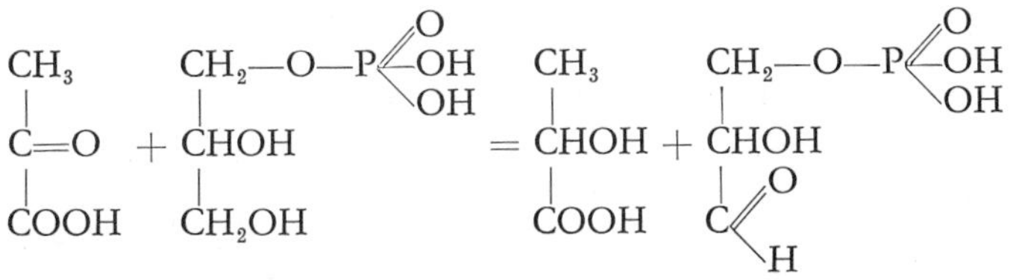

$$
\begin{array}{l}
CH_3 \\
| \\
C{=}O \\
| \\
COOH
\end{array}
+
\begin{array}{l}
CH_2\text{—}O\text{—}P(=O)(OH)_2 \\
| \\
CHOH \\
| \\
CH_2OH
\end{array}
=
\begin{array}{l}
CH_3 \\
| \\
CHOH \\
| \\
COOH
\end{array}
+
\begin{array}{l}
CH_2\text{—}O\text{—}P(=O)(OH)_2 \\
| \\
CHOH \\
| \\
C(=O)H
\end{array}
$$

Pyruvic acid — **Glycerophosphoric acid** — **Lactic acid** — **Triosephosphoric acid**

Triosephosphoric acid then undergoes the conversions described in phases 3 to 5 described above.

As can be seen in this picture of glycolysis, methylglyoxal is missing. This compound is considered generally, and especially by Neuberg, as an intermediate of yeast fermentation and of lactic acid production in animals.

In a paper from our Institute, several years ago, the possible role of methyl glyoxal as a precursor of the *dextrorotatory* lactic acid, formed exclusively in animals, was considered unlikely because methylglyoxal is not converted by animal tissues into pure D-lactic acid* but into a racemic mixture of the optical isomers among which L-lactic acid predominates. Furthermore, Lohmann recently made the important observation that the activity of methylglyoxalase—the enzyme catalyzing the direct conversion of methylglyoxal into lactic acid—requires the presence of glutathione while the enzymatic production of lactic acid from glycogen in muscle can proceed without glutathione.

It is impossible to exclude the possibility that the form of methylglyoxal occurring in the intermediary metabolism differs from the synthetic one and that this allelomorphic methylglyoxal can be metabolized to dextrorotatory

*The dextrorotatory lactic acid belongs sterically to the L-series.

lactic acid in the absence of glutathione. This is the point of view expressed by Neuberg some time ago. On the basis of our findings, however, it appears necessary to consider the scheme formulated above as the main pathway of anaerobic carbohydrate degradation associated with intermediary phosphorylations.

Almost 20 years ago, Embden and Oppenheimer demonstrated that during perfusion of pyruvate through the liver, this compound is converted exclusively into D-lactate; the same should be true for muscle (such studies are now in progress).

In the course of the normal metabolism of various organisms, various asymmetric substances are produced only in one possible stereoisomeric form. Lactic acid might conceivably be an exception to this rule if the formation of the two optical isomers consistently occurred by means of two different pathways. Methylglyoxalase appears to catalyze the formation of L-lactic acid; the recent finding of Embden and Metz that red cell hemolysates convert methylglyoxal exclusively to L-lactic acid as well as the earlier report of Weinmann, from the Neuberg laboratory, of a similar reaction in bacteria, probably can be ascribed in both cases to the action of pure methylglyoxalase, and thus to a *direct* conversion of methylglyoxal into lactic acid.

Since, so far, only dextrorotatory lactic acid has been detected as a product of glycolysis in animal tissues, it appears that methylglyoxal plays no, or only an insignificant, role in the metabolism of animals.

While our scheme of glycolytic lactic acid formation contrasts rather sharply with concepts developed up to the present, it should be pointed out that many observations of other investigators are quite compatible with our viewpoint. We are referring first of all to several findings of Neuberg during his classical studies of yeast fermentation. This applies, for example, to the recent observations of Neuberg and Kobel that under certain abnormal conditions during yeast fermentation equimolar amounts of pyruvate and glycerol are formed and to the much older finding of Neuberg and Kerb that addition of glycerol during the decarboxylation of pyruvate by yeast carboxylase results in the production of ethylalcohol instead of acetaldehyde.

We do not intend to discuss here the question in what manner the pathway postulated for muscle glycolysis applies also to alcoholic yeast fermentation. However, it should be pointed out that on addition of fluoride to yeast, Nilsson obtained an amorphous neutral barium salt and a crystalline strychnine salt which he considered to be phosphoglycerate, although definite identification was not carried out.*

Previously, Greenwald has isolated from red blood cells diphosphoglyceric acid and elucidated its structure; Jost in an investigation published from this Institute has demonstrated that this substance is formed during anaerobic glycolysis in red cells. On the basis of this finding, a pathway for the formation of diphosphoglyceric acid which was similar to the one postulated in this paper was already assumed, although the glycolytic formation of lactic acid *via* phosphoglyceric and pyruvic acids had not been considered at that time.

*Several years ago Neuberg, Weinmann and Vogt, as well as Vogt, prepared monophosphoglyceric acid synthetically and its sparingly soluble crystalline barium salt which has great similarities to the optically active form that we obtained.

J. K. Parnas and his son, Jan, in a Polish countryside setting.

[The late Professor Parnas was Professor and Head of the Institute of Medical Chemistry, Lwow, Poland from 1914 until 1939. Jan is now an eminent surgeon in Poland.]

Dr. Pawel Ostern. The first available 5′ AMP came from the University of Lwow, Poland in 1938 and was rushed to the Coris in St. Louis.

[The late Dr. Ostern was Research Professor with J. K. Parnas at the Institute of Medical Chemistry, Lwow, Poland until 1940.]

Carl and Gerty Cori, Washington University School of Medicine, St. Louis, 1940. The priming reaction is probably the essence of polysaccharide biosynthesis.

[The late Gerty T. Cori and her husband were awarded the Nobel Prize in Medicine in 1947.]

COUPLING OF CHEMICAL PROCESSES IN MUSCLE*

J. K. Parnas, P. Ostern and T. Mann
Institute of Medical Chemistry, University of Lwow, Poland

I.

In the course of investigation of the various factors which contribute to the inhibition of the formation of ammonia in minced muscle, the following observations were made:

In "muskel brei," [thick muscle dispersion] obtained by mixing freshly ground muscle with an equal volume of water at 12 to 14°, there is no formation of ammonia initially. However, when larger amounts of water are used, ammonia formation sets in at once (see Parnas and Mozolowski[1]) and ceases within 2 to 5 minutes. It appears thus that the "traumatic ammonia formation" depends on the dilution, presumably due to the dilution of a factor which inhibits this reaction in the tissue. This factor is inorganic phosphate; it was established that where muskel brei is made in 0.1 M potassium phosphate (pH = 7) the formation of ammonia is suppressed for a period of 30 to 60 minutes. Potassium chloride or borate buffer have no such effect.

The peculiarity of this suppression is that it lasts for only a limited length of time and that subsequently the formation of ammonia is very rapid. It seems, therefore, that this is not an enzyme inhibition but probably an effect on the substrate of deamination. It is noteworthy that adenylic acid added to a thick

*J. K. Parnas, P. Ostern, and T. Mann, "Über die Verkettung der chemischen Reaktion der Muskel," *Biochim. Z.*, **272**, 64–70 (1934). Translated from the German and reprinted with the permission of one of the authors (T. Mann) and of Springer-Verlag, Heidelberg, Germany.

The author is greatly indebted to Professor T. Mann and Dr. C. Lutwak-Mann, University of Cambridge, England, for their valuable suggestions concerning the translation of this paper.

muskel brei is deaminated very rapidly. This is in agreement with the results of studies by Mozolowski and Sobczuk,[2] and is also in accord with the observation made by T. Mann, i.e., that adenylic acid, rather than adenosine triphosphoric acid, is subject to deamination in muscle[3].

A method for the determination of adenosine triphosphoric acid (ATP) has hitherto been lacking. The routine determination based on the liberation of phosphate by acid, usually 1 M HCl ("$(P_7 - P_0) - (P_{30} - P_7)$"), gives only the total content of ribose polyphosphoric acids. We have worked out a method in which the barium salt of the adenosine triphosphoric acid is first precipitated from a trichloracetic acid extract and next dispersed in sulfuric acid and subjected to enzymic deamination by incubation with a muscle suspension prepared according to G. Schmidt.[5] The enzymatically released ammonia is subsequently determined. By means of that method it was found that in muskel brei in which the formation of ammonia had been suppressed, the adenosine triphosphoric acid is stable, and that with progressing ammonia formation the content of ATP decreases. The rate of the deamination is predetermined by the release of phosphate.

The inhibition of ammonia formation by 0.1 M phosphate is completely abolished by the addition of 0.001 N iodoacetic acid or 0.05 N sodium fluoride. Hence, the process involved in the phosphate inhibition of ammonia liberation seems to be lactic acid formation. As is well known, lactic acid formation in aqueous muskel brei proceeds, at first, at a high rate which subsequently declines. In phosphate brei,* the reaction rate is only slightly lower. *By the addition of phosphate to muskel brei it was possible for the first time to observe a formation of lactic acid without simultaneous ammonia formation.* The reverse situation, i.e., formation of ammonia without formation of lactic acid, can, as is well known, be observed under conditions of fluoride[6] or iodoacetic acid[7],[8] poisoning.

In experiments with dispersions containing phosphate and iodoacetic acid, it was possible to observe exactly the same relationship between ammonia formation and the disappearance of adenosine triphosphoric acid as in the earlier experiments carried out in aqueous extract in the absence of iodoacetate.

These observations must be interpreted by assuming that only adenylic acid is deaminated in muscle. In concentrated muskel brei as in the phosphate brei, adenosine triphosphoric acid is continuously regenerated by a mechanism which is ultimately linked with the process of lactic acid formation, and of which the nearest and most important stage is the reaction between adenylic acid and phosphocreatine, discovered by Lohmann[9]. The older experiments by Mozolowski, Mann, and Lutwak (Table IV, Fig. 1[7]) illustrate well the mechanism whereby the ammonia formation begins in iodoacetic acid-poisoned intact muscle upon the breakdown of creatine phosphoric acid. The findings of Lohmann as well as the facts here reported have clarified the mechanism of these processes. The "big" ammonia formation is turned on at the moment when phosphocreatine has broken down and when adenosine triphosphoric acid can no longer be resynthesized.

*[The use of "brei" in phrases other than "muskel brei," such as "phosphate brei" and "fluoride brei," refers to muscle dispersions in dilute aqueous solutions with added salts, such as sodium fluoride or sodium phosphate. Correspondingly, "water brei" refers to muscle dispersions in distilled water.]

We cannot say exactly in which way the rise in phosphate concentration impinges upon these reactions. It is apparently not by an acceleration of lactic acid formation, but, perhaps, rather by promoting the resynthesis of phosphocreatine and adenosine triphosphoric acid. This is supported by direct observations on phosphate enriched muskel breis. In dispersions made from symmetrical muscles one finds, for example, 128 mg% phosphocreatine-P in the phosphate brei, but only 37 mg% phosphocreatine-P in the water brei (3 minutes after grinding).

The inhibition of deamination by phosphate, likewise observed in concentrated muskel brei, wears off after a while. The augmented release of ammonia from adenylic acid which Mann observed in aging muscle extracts is probably a related phenomenon[3]. The missing link in the mechanism at present remains unknown. In the interaction of phosphate liberation and resynthesis of adenosine triphosphoric acid in water brei, about 25% of the originally present adenosine triphosphoric acid is still preserved after one hour, as found by means of the new methods of determination. Ultimately, the remainder is used up by deamination at the intermediary level of adenylic acid, whereby the process of lactic acid formation is brought to a standstill. However, in the presence of iodoacetic acid, the picture changes markedly both in the phosphate brei as well as in the water brei. ATP almost completely disappears after a few minutes. This disappearance is accompanied by a maximal formation of ammonia, as well as a complete splitting of the barium-precipitable acid-labile phosphate residues [i.e., fraction $(P_7 - P_0) - (P_{30} - P_7)$].

The formation of ammonia takes place at the adenylic acid level; it does not form a link in the energy supplying cycle of this substance. That it takes place under physiological conditions seems intelligible only within the context of a previously formulated idea[10], namely, an uncoupling of definite links in the muscle machinery which are rendered idle until the onset of oxidative recovery. The oxidative recovery enables a restitution of the specific constituent of the coenzyme as well as the substrate for glycolysis.

II.

Further experiments have revealed that phosphate muskel brei (an experimental preparation which chemically resembles intact muscle) shows an unequivocal relationship between preservation and restitution of ATP on the one hand and the formation of ammonia on the other. When adenylic acid is added to such a muscle preparation one finds, after 10 minutes incubation (at 12°), that two-thirds can be accounted for as ATP and one-third as ammonia. After 1 hour, ammonia constitutes two-thirds, and ATP one-third of the total. *Adenylic acid in muscle is not stable; it is either resynthesized to ATP or undergoes deamination.* On that basis the role of ammonia formation in muscle chemistry can be envisaged as follows: the liberation of adenylic acid by release of two equivalents of phosphate in the first step of the anaerobic recovery is followed by the Lohmann reaction,[1] in which the adenylic acid is reconverted to ATP

[1] On a different basis this idea was also proposed by Barrenscheen and Filz,[4] who used Schmidt extracts which deaminate adenylic acid but not ATP. The latter investigators have, however, formulated concepts concerning the constitution of ATP with which both Lohmann and ourselves cannot agree.

by interaction with phosphocreatine. The prerequisite for this is that the anaerobic recovery, including carbohydrate breakdown, should be able to proceed in both directions. Should this reaction fail (for instance as in fatigue) then the unreconstituted adenylic acid will be deaminated. *Owing to the high activity of deaminating enzymes in muscle, this deamination is very rapid, and proceeds at the same high rate as that observed in muscle poisoned by iodoacetic acid (cf. III), or fluoride, or simply in water brei. Within a few minutes all adenylic acid, endogenous as well as exogenous, is quantitatively converted to inosinic acid. By suppressing glycolysis, the last link in the sequence of anaerobic recovery processes, fluoride or iodoacetic acid inhibits the synthesis of ATP and adenosine monophosphoric acid thereby falls prey to deamination.*

Under physiological conditions one can imagine the events to proceed in the following way: In the cycle of the anaerobic recovery processes, the first link, the structure-bound ATP is split to adenylic acid. From the breakdown products and in the later steps of the reaction sequence ATP is again rebuilt. These later steps consist, as we may now assume, of reactions in which phosphoric acid residues are transferred to organic linkages of the substances which are to be resynthesized, as, for instance, phosphate transfer from phosphocreatine to adenylic acid and *from an intermediary product of sugar metabolism to creatine.* Ultimately the sequence of the anaerobic recovery processes is coupled to preceding oxidative recovery. But the condition for sugar metabolism is the availability of its coenzyme, ATP. When in a functional, submicroscopic, or visible element of the muscle machinery the anaerobic recovery process begins to fail, deamination of adenylic acid will ensue, thus rendering unavailable the direct precursor of the coenzyme. This, in turn, prevents the anaerobic reaction sequence until the oxibiotic recovery processes can regenerate both the glycolytic substrate as well as the coenzyme precursor.

III.

The relationships between the maintenance of adenosine triphosphoric acid (ATP) on the one hand, and sugar metabolism on the other hand, point toward events which involve ATP biosynthesis. These conclusions can be gained from experimental studies of the ammonia formation in muscle.

The ammonia formation is stopped in phosphate-enriched muskel brei. Addition of fluoride or iodoacetic acid, which divert the sugar metabolism into another pathway, releases the formation of ammonia. One might conceive of a device for testing intermediary products of the sugar metabolism using muskel brei as an experimental tool. Metabolites capable of suppressing ammonia formation could be considered as possible intermediary products of glycolysis. Moreover, further conclusions might be drawn about the role of these intermediary products on the basis of their effects on ammonia formation. For example, if one could show that pyruvic acid or phosphoglyceric acid are capable of suppressing the fluoride-induced ammonia formation, this would have a definite bearing upon the interpretation of their role. We have investigated this question experimentally and obtained quite clear results. Spurious conclusions have been avoided by performing two separate types of measurements on the muskel brei, namely, ammonia determinations as well as separate determinations of the content of adenosine triphosphoric acid.

I. In a phosphate muskel brei, containing fluoride (N/400 *to* N/60 NaF) *the ammonia formation is prevented by addition of either pyruvic acid or phosphoglyceric acid, but not by lactic acid, glycerol phosphate or Harden-Young diester.*

II. In phosphate brei with iodoacetic acid, phosphoglyceric acid is capable of abolishing ammonia formation. However, pyruvic acid or the other above-mentioned substances do not exert such an effect.

As an example of the range of effects, the following experiment might be quoted. A fluoride brei, incubated about 10 minutes at 13°, contained 7.48 mg% NH_3-N and 1.38 mg% hydrolyzable ATP-N in the absence of pyruvic acid. In the same brei, after 10 minutes incubation with added sodium pyruvate (M/50), the corresponding values were 1.60 mg% NH_3-N and 7.02 ATP-N, respectively. Clearly resynthesis of phosphocreatine and adenosine triphosphoric acid is coupled not to glycolysis as a whole, but to certain distinct steps of this process. This in turn leads to the conclusion that in the resynthesis we are confronted not so much by a part of what may be called energy coupling, but rather by a step in the transfer of phosphate residues from molecule to molecule, a reaction similar to that which Lohmann[(9)] discovered in the two-step phosphorylation of adenylic acid by phosphocreatine.

The demonstration that phosphoglyceric acid specifically suppresses the ammonia formation in muskel brei poisoned with iodoacetic acid indicates a significant role of this phosphate ester in the resynthesis processes.

As to the type of underlying metabolic reactions involving phosphoglyceric acid and pyruvic acid in these experiments, we can only say that in fluoride muskel brei both phosphoglyceric acid as well as pyruvic acid are active, the latter being more so. According to the concept of the intermediary course of glycosis in muscle, elaborated by Embden, Deuticke and Kraft[(11)] and Meyerhof and Kiessling,[(12)] the breakdown of phosphoglyceric acid to pyruvic acid is interrupted by poisoning with fluoride. Thus the suppression of ammonia formation by pyruvic acid is presumably due to a synthesis of ATP brought about by the final stage of the formation of lactic acid from pyruvic acid. Yet, the suppression of ammonia formation brought about by phosphoglyceric acid cannot operate via pyruvic acid. This we have clearly established by showing that while there is a complete suppression of ammonia formation in phosphate-fluoride muskel brei by phosphoglyceric acid, there is no accompanying formation of either methylglyoxal or lactic acid.

Ammonia formation in iodoacetate-phosphate-muskel brei is suppressed only by phosphoglyceric and not by pyruvic acid. The iodoacetate inhibition of carbohydrate breakdown is considered to be due to abolishing the dismutation between pyruvic acid and glycerophosphoric esters. We conclude, therefore, that the linking between the synthesis of ATP (via phosphocreatine) takes place at a stage of glycolysis which can be influenced by phosphoglyceric acid as well as pyruvic acid in fluoride muskel brei but only by the former in the iodoacetate brei. We have convinced ourselves that lactic acid formation and methyl glyoxal formation from phosphoglyceric acid in iodoacetate-phosphate muscle brei do not take place even though ammonia formation is suppressed.

The compounds to which the phosphorus transfer via creatin-phosphocreatine to adenylic acid is coupled must be searched for in the sequence between phosphoglyceric acid and pyruvic acid. The existence of further intermediaries at this level follows from the occurrence of diphosphoglyceric acid, and the breakdown of this substance to pyruvic acid, as demonstrated by Neuberg[13]. In any case, the early stages in the breakdown of sugar are not coupled with the resynthesis of the important nitrogenous compounds. This is demonstrated by the observation that neither the Harden-Young ester nor its metabolite dihydroxyacetone phosphate[14] (formed in fluoride-poisoned or iodoacetate-poisoned muskel brei) can affect such a resynthesis. It would be easy to formulate equations illustrating the phosphate transfer from phosphoglyceric acid to creatine. However, such formulations are best postponed as long as the material as well as the energetic basis are still lacking.

References

1. Parnas, J. K. and Mozolowski, W.: *Biochim. Z.*, **184** (1927), 399.
2. Mozolowski, W. and Sobczuk, B.: *Biochim. Z.,* **265** (1933), 41.
3. Mann, T.: *Biochim. Z.*, **266** (1933), 162.
4. Barrenscheen, H. K. and Filz, W.: *Biochim. Z.,* **253** (1932), 422; **250** (1932), 281.
5. Schmidt, G.: *Z. physiol. Chem.,* **179** (1928), 243.
6. St. Chrzaszczewski and Mozolowski, W.: *Biochim. Z.,* **194** (1928), 233.
7. Mozolowski, W., Mann, T., and Lutwak, C.: *Biochim. Z.*, **231** (1931), 290.
8. Embden, G.: *Klin. Wochenschrift,* **9** (1930), 1337.
9. Lohmann, C.: *Naturwiss.,* **22** (1934), 409.
10. Parnas, J. K.: *Annual Review of Biochemistry,* **1** (1932), 443.
11. Embden, G., Deuticke, H. J., and Krafft, G.: *Klin. Wochenschrift,* **12** (1933), 213.
12. Meyerhof, O.: *Nature,* **132** (1934), 337, 373; Meyerhof, O. and Kiessling W.: *Biochim. Z.,* **267** (1934), 313.
13. Neuberg, C., Schuchardt, W., and Vercellone, A.: *Biochim. Z.,* **271** (1934), 221.
14. Meyerhof, O. and Lohmann, K.: *Biochim. Z.,* **271** (1934), 89.

MECHANISM OF FORMATION OF HEXOSEMONOPHOSPHATE IN MUSCLE AND ISOLATION OF A NEW PHOSPHATE ESTER*

C. F. Cori and G. T. Cori

Department of Pharmacology,
Washington University School of Medicine
St. Louis, Mo.

Experiments performed on intact frog muscle indicated that hexosemonophosphate, in contrast to hexosediphosphate, is formed by esterification with inorganic phosphate.[1] A further study of this problem was carried out on minced frog muscle which was almost completely inactivated (in regard to lactic acid formation) by 3 to 4 extractions with distilled water. When such muscle, which contains only 2 to 4 mg. % of organic, acid-soluble P, is incubated anaerobically for 3 hours in isotonic phosphate buffer, the organic P content rises to 8 to 13 mg. % due to the formation of hexosemonophosphate. Addition of small amounts of adenylpyrophosphoric or of adenylic acid greatly enhances the formation of hexosemonophosphate, as shown in Table I. The experiments indicate that hexosemonophosphate is formed from inorganic phosphate and that adenylic acid serves as the mediator of this reaction.

Observations after short periods of incubation showed that the first phosphorylation product is not hexose-6-phosphoric acid (Embden ester), but a new ester which is slowly converted to the 6-phosphoric acid under the conditions of these experiments (Table II). After one and 2 hours of incubation much more organic P is present than can be accounted for on the basis of reducing power, assuming the latter to be due to hexose-6-monophosphate. Hydrolysis in NH_2SO_4 at 100° for 10 minutes revealed the presence of a compound which

*C. F. Cori and G. T. Cori, "Mechanism of Formation of Hexosemonophosphate in Muscle and Isolation of a New Phosphate Ester," *Proc. Soc. Exp. Biol. and Med.*, **34**, 702–705 (1936). Reprinted with the permission of the author and of the publisher.

[1] Cori, G. T., and Cori, C. F., Summaries of Communications, XVth International Physiological Congress, p. 66, 1935.

(unlike the 6-monophosphoric acid) is easily hydrolyzed in acid and which yields equivalent amounts of fermentable sugar and inorganic phosphate.

Table I

EFFECT OF COENZYME ON HEXOSEMONOPHOSPHATE FORMATION IN INACTIVATED MUSCLE

Minced muscle was extracted 3 times with 20 times its weight of distilled water and was incubated anaerobically in isotonic phosphate buffer (pH 7.2) at 20° for 3 hours.

Values in mg. per 100 gm. muscle.

Hexosemonophosphate				
hexose	P found	P calc.	Lactic acid	Additions (per 100 gm. muscle)
15		3	6	Incubated in isotonic KCl
68	10	12	8	
162	32	28	30	6 mg. adenylpyrophosphate P
75	11	13	10	
141	36	24	25	6 mg. adenylpyrophosphate P
68	13	12	10	
264	59	46	26	12 mg. adenylic acid P
48	8	8	10	
119	28	21	12	4 " " " "

Table II

EFFECT OF LENGTH OF TIME OF INCUBATION ON PHOSPHORYLATION

Values in mg. per 100 gm. muscle.

	Hexosemonophosphate fraction			
Time hr.	hexose	P found	P calc.	Additions (per 100 gm. muscle)
0	5		0.9	
1	79	62	14	13 mg. adenylic acid P
2	131	69	23	" " " " "
3	252	67	43	" " " " "

After 10 min. hydrolysis in NH_2SO_4.

		Fermentable Sugar	
	Inorg. P	found	calc.
1	39	219	226
2	45	256	261
3	20	122	116

The new compound was isolated from 4 different preparations. About 100 gm. of minced and washed muscle, to which 120 mg. of adenylic acid had been added, was incubated for one hour in phosphate buffer. After deproteinization with $HgCl_2$, the water-soluble barium salts were precipitated with alcohol. The soluble barium salts consisted of about 65% of the new compound, the remainder being the adenylic acid added and the small amount of hexose-6-phosphoric acid formed. Fractionation of the barium salts was not successful and they were therefore converted to the brucin salts. After removal of most of the adenylic acid by concentrating the aqueous solution of the brucin salts to a small volume, the remainder of the brucin salts was evaporated to dryness. The semi-crystalline residue was extracted with hot methyl alcohol containing 1% ethyl alcohol. On cooling the ester crystallized out in the form of fine needles which soon formed large aggregates. After recrystallization from methyl alcohol, the brucin salt was reconverted to the barium salt and purified further by repeated solutions in water and precipitations with alcohol. The final product contained 1.5% of an ester difficult to hydrolyze in acid as judged by the ratio of easily hydrolyzable to total organic P.

Elementary analyses for C, H, P, and Ba and determination of the sugar formed after hydrolysis (by means of hypoiodite) agreed with a compound of the composition of $C_6H_{11}O_5.PO_4Ba.3H_2O$. $[\alpha]_D^{25} = 75.5°$ (calc. for the anhydrous barium salt in water; conc. 1.26%). During hydrolysis in 0.01 N HCl at 70°, sugar and inorganic phosphate were liberated at the same rate, the percentages being 21.1, 38.3, 62.0 and 76.3 for the 30, 60, 120, and 180 minute periods, respectively. The percentages calculated for a velocity constant of 3.47×10^{-3} were 21.4, 38.1, 61.9, and 76.2, respectively.

The ester (when unhydrolyzed) does not reduce alkaline copper solution and does not react with hypoiodite, and it can be heated to 130° without discoloration, all factors which point to the absence of a free reducing group. We believe that the new ester is an aldose-1-phosphoric acid, the sugar being presumably glucose.

When the new ester is added to frog muscle extract, it is converted in a few minutes to an ester difficult to hydrolyze in acid and possessing reducing power, presumably hexose-6-phosphoric acid. The same change occurs in muscle extract inactivated by dialysis, so that the presence of the coenzyme system does not appear to be necessary for the wandering of the phosphate group.

Summary

In minced and washed frog muscle incubated in phosphate buffer, added adenylic acid transfers inorganic phosphate to carbohydrate resulting in the formation of hexosemonophosphate. The first phosphorylation product proved to be a new ester which was isolated as the crystalline brucin salt and had the properties of glucose-1-phosphoric acid; when added to frog muscle extract it was converted in a few minutes to the Embden ester.

ON THE ACTION GROUP OF THE YELLOW ENZYME*

Hugo Theorell

Kaiser Wilhelm Institut für Zellphysiologie,
Berlin-Dahlem, Germany

(Submitted November 20, 1934)

The author has previously reported in this Journal[1] on the purification and the reversible splitting of the yellow enzyme into a protein component and a yellow action group.

If lactoflavin (obtained through the courtesy of P. Karrer) is used in the reversion experiment as a substitute for the yellow enzyme action group, *no* yellow enzyme is formed. Consequently lactoflavin as described by Kuhn is not identical with the action group of the enzyme. (Yet the action group and lactoflavin show, as we know, the same spectrum and both give upon irradiation in alkaline solution the same pigment, $C_{13}H_{12}N_4O_2$).

If one mixes the pure yellow enzyme with 3 volumes of methanol, a precipitation, largely of the protein component (denatured), takes place, whereas the yellow pigment goes into solution.

At the suggestion of Otto Warburg, I have investigated the possible occurrence of phosphorus in the yellow pigment obtained as described. The amount of pigment was determined by photoelectric methods and the phosphorus of the ash of the solution was determined according to Briggs. As will appear, one atom of phosphorus was found per mole of pigment.

Preparation 1: 196_2 pigment $= 17_\gamma$P (i.e., 1.06 Mol P/Mol pigment)
Preparation 2: 120_2 pigment $= 11_\gamma$P (i.e., 1.1 Mol P/Mol pigment)

In contrast the protein component is free of phosphorus.

Moreover, kataphoretic experiments at *p*H 7.2 showed that the liberated

*H. Theorell, "Über die Wirkungsgruppe des Gelben Ferments," *Biochim. Z.*, **275**, 37–38 (1934). Translated from the German and reprinted by permission of the author and of Springer-Verlag, Heidelberg, Germany.

action group of the enzyme migrates strongly in the direction of the anode and with a velocity to be expected of a monophosphoric ester of the corresponding molecular size ($v = 16 \times 10^5$ cm sec volt). Lactoflavin at the same *p*H did not show any detectable migration.

Hence the action group of the yellow enzyme seems most likely to be a phosphoric ester, a "nucleotide" in which the *pur*ine base is substituted with Karrer's[2] dimethylalloxazin.

References

1. Theorell, H.: *Biochim. Z.*, **272** (1934), 155.
2. Karrer, P., Salomon, H., Schöpp, K., Schlitter, E., and Fritzche, H.: *Helv. Chim. Acta*, **17** (1934), 1010.

Otto Warburg (center) may well be summing up his approach to experimental science: "Do not forget Maxwell's plea to Ampere—'If you have built up a perfect demonstration, *do not remove all traces of the scaffolding* by which you have raised it.'" Listeners are Feodor Lynen (left) and Karl Lohmann (right). Berlin, 1950.

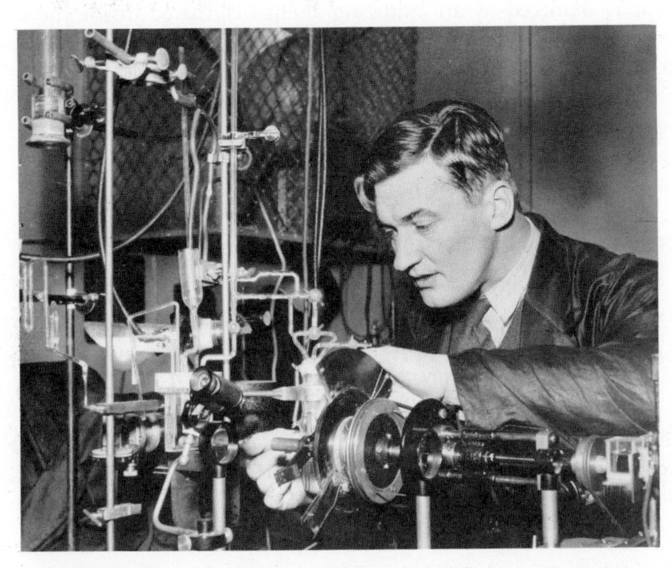

Hugo Theorell, Stockholm, 1936. A demonstration of the Warburg-Keilin concept. Cytochrome *c* in a Warburgian spectrophotometer of 1935.

PYRIDINE, THE HYDROGEN TRANSFERRING CONSTITUENT OF FERMENTATION ENZYMES*

Otto Warburg and Walter Christian

Kaiser Wilhelm Institut für Zellphysiologie,
Berlin-Dahlem, Germany

(Lecture delivered at The International Chemistry Symposium, Zürich, August 1936)

Fermentation is, as Pasteur expressed it, "internal" respiration. In fermentation, sugar is oxidized by an oxidizing agent originated inside the fermenting cell, generated through the fermentation process, in contrast to respiration in which sugar is oxidized by an exogenous oxidizing agent, molecular oxygen.

The oxidizing agent of alcohol fermentation is, as first discovered by Carl Neuberg, acetaldehyde, which oxidizes the sugar to pyruvic acid while itself being reduced to ethyl alcohol, i.e.:

$$1 \text{ hexose} + 2 \text{ acetaldehyde} = 2 \text{ pyruvic acid} + 2 \text{ alcohol.} \quad (1)$$

Since, in the presence of carboxylase

$$2 \text{ pyruvic acid} = 2 \text{ acetaldehyde} + 2 \text{ carbon dioxide,} \quad (2)$$

pyruvic acid and acetaldehyde do not enter into the balance; the addition of Equations (1) and (2) furnishes the equation of the alcohol fermentation:

$$1 \text{ hexose} = 2 \text{ alcohol} + 2 \text{ carbon dioxide.} \quad (3)$$

In the same way—according to the work of R. Nilsson, G. Embden and O. Meyerhof—glycolysis can be characterized as a process by which pyruvic acid oxidizes hexose to pyruvic acid, thereby itself being reduced to lactic acid:

$$1 \text{ hexose} + 2 \text{ pyruvic acid} = 2 \text{ pyruvic acid} + 2 \text{ lactic acid.} \quad (4)$$

*Excerpt from O. Warburg and W. Christian, "Pyridin, der wasserstoffübertragende Bestandteil von Gärungsfermenten," *Helv. Chim. Acta*, **19**, E79 (1936). Translated from the German and reprinted by permission of the editors of *Helvetica Chimica Acta*.

Since

$$2 \text{ pyruvic acid} = 2 \text{ pyruvic acid} \quad (5)$$

pyruvic acid is eliminated from the balance, and by adding Equations (4) and (5) one obtains the equation for lactic acid fermentation:

$$1 \text{ hexose} = 2 \text{ lactic acid} \quad (6)$$

By formulating it in this way we have left out the intermediary phosphorylations as they do not essentially alter the oxidation states of the reacting molecules. Nor have we brought up the question concerning the origin of the aldehyde or the pyruvic acid, both of which are necessary to *initiate* the fermentation. Once these oxidative agents are there, they will constantly be generated by the same reactions which make them disappear, a profoundly ingenious way of nature to maintain life under conditions where molecular oxygen is absent.

If we focus specifically on Equation (1), then we can formulate alcohol fermentation as a process by which the hydrogen of hexose is transferred to aldehyde. In the same way, we can interpret—according to Equation (4)—lactic acid fermentation as being a process in which the hydrogen of hexose is transferred to pyruvic acid.

In this lecture, we are going to demonstrate that fermentation (Equation (1)) can be resolved through the following two reactions:

$$1 \text{ hexose} + 2 \text{ pyridine} = 2 \text{ pyruvic acid} + 2 \text{ dihydropyridine}$$
$$2 \text{ dihydropyridine} + 2 \text{ acetaldehyde} = 2 \text{ pyridine} + 2 \text{ alcohol}$$

Correspondingly, Equation (4) can be resolved into the equations:

$$1 \text{ hexose} + 2 \text{ pyridine} = 2 \text{ pyruvic acid} + 2 \text{ dihydropyridine}$$
$$2 \text{ dihydropyridine} + 2 \text{ pyruvic acid} = 2 \text{ pyridine} + 2 \text{ lactic acid}$$

Hence, fermentations are "pyridine-catalyses" by which, through the conversion of the pyridine ring,

$$\text{pyridine} \longrightarrow \text{dihydropyridine}$$

a transfer of hydrogen ensues.

Pyridine acting in this way is not free pyridine but a pyridine derivative, an amide of nicotinic acid, bound to the fermentation enzyme.

CH
CH C·$CONH_2$
CH CH
N

It is released by acid splitting of the fermentation enzymes.

Reversible Reduction of Simple Pyridine Compounds

Nicotinamide released from the fermentation enzyme cannot be reduced to the reversible dihydro compound and is therefore incapable of acting as a hydrogen transfer agent. Platinated hydrogen reduces nicotinamide irreversibly to the hexahydro compound. In contrast, hyposulfite $Na_2S_2O_4$ does not reduce nicotinamide.

The first simple pyridine compound which we were able to reduce partially and reversibly was a compound related to nicotinamide in which the pyridine nitrogen was pentavalent, namely the methylbetain of nicotinic acid which occurs in nature and was called trigonelline by its discoverer, E. Jahns.

CH
CH C·COO—
CH CH
N
CH_3

Trigonelline is reduced to a dihydro compound by hyposulfite in weak alkaline solution, and can again be dehydrogenated back to the pyridine compound by an enzyme found in nature, the yellow oxidation enzyme.

Another simple pyridine compound which we have been able to reduce partially and reversibly is even more closely related to nicotinamide than the trigonelline, namely the methyliodine derivative of nicotinamide.

CH
CH C·$CONH_2$
CH CH
N
CH_3 I

P. Karrer supplied us with this compound and suggested we test it.

The methyliodine of nicotinamide is reduced by hyposulfite in weak alkaline solution to the dihydro derivative, which, like the partially hydrogenated trigonelline is dehydrogenated back again to the pyridine compound by the yellow enzyme.

According to P. Karrer, the equation of the hydrogenation is as follows:

CH | | | CH |
CH C·$CONH_2$ | | | CH C·$CONH_2$ |
CH CH | + 2H | = | CH CH_2 | + HI
N | | | N |
CH_3 I | | | CH_3 |

Thus the dihydro derivative formed is N-methyl-ortho-dihydropyridine.

When the pyridine compound is hydrogenated to the reversible dihydro derivatives, a characteristic absorption band emerges in the long-wavelength ultraviolet region which we call the "dihydroband." The dihydroband of trigonelline is located at 350 $m\mu$; the dihydroband of Karrer's compound is located at 360 $m\mu$.

The white fluorescence, which is elicited by the reversible dihydro-pyridine compounds when irradiated by ultraviolet light, stems from the dihydroband. Those pyridine compounds which have not become hydrogenated do not fluoresce. Hence, the hydrogenation and dehydrogenation of the pyridine compound is directly *visible* in the light of the analytical quartzlamp.

Experiment 1

Methyl iodine of nicotinamide is dissolved in water, and bicarbonate is added. The solution does not fluoresce. When hyposulfite is added, the white fluorescence, characteristic of the dihydro compound, appears. In order to decompose the excess of hyposulfite, the solution is exposed to oxygen. Yet, the fluorescence remains, i.e., the dihydro compound is not auto-oxidizable. The white fluorescence disappears upon addition of the yellow enzyme; the alloxazine ring of this enzyme removes the 2 hydrogen atoms from the dihydropyridine and thus dehydrogenates it to the pyridine compound. Acidification of a solution of the dihydro compound brings about the disappearance of the white fluorescence. This is not, however, a dehydrogenation process but an irreversible decomposition.

Pyridine Nucleotide

Nicotinamide occurs in nature in nucleotides which we call "Pyridine Nucleotides" in accordance with their chemically active group. One of the pyridine nucleotides containing 3 molecules of phosphoric acid per molecule of nicotinamide we have called "triphosphopyridine nucleotide." We have called another pyridine nucleotide containing 2 molecules of phosphoric acid per molecule nicotinamide "diphosphopyridine nucleotide."

Triphosphopyridine nucleotide was discovered in 1934 in Dahlem. It is a di-nucleotide of molecular weight of about 800 and contains two bases, a purine and a pyridine base. The purine base is adenine, the pyridine base is the amide of nicotinic acid. Adenine constitutes about 17% and nicotinamide 15 to 16% of the compound. Triphosphopyridine nucleotide is widely distributed in the living matter, and it always accompanies another biologically important nucleotide, the cozymase of von Euler.

Cozymase is a component of the coferment of fermentation discovered by A. Harden and W.J. Young in 1904. Based upon the work of H. von Euler and K. Myrback, it was considered an adenine mononucleotide until 1933. Analytical findings after hydrolysis gave the following estimates: 28% adenine and a molecular weight of 350, which is identical to that of adenylic acid.

After the discovery of triphosphopyridine nucleotide, a revision of von Euler's cozymase studies seemed desirable, and was undertaken at Euler's Institute and simultaneously at Dahlem. In the von Euler preparations, a separation of adenylic acid, which had been the main component of his earlier preparations, was accomplished. In Dahlem, cozymase was isolated from red blood cells by a new and simpler procedure.

The results of these supplementary analyses made it clear that cozymase is not an adenine mononucleotide. It is a dinucleotide with molecular weight of about 700 and differs from triphosphopyridine only by containing one less molecule of phosphoric acid. *Hence, cozymase is diphosphopyridine nucleotide.* The content of adenine is not 28% but only 19%, while nicotinamide constitutes 17% of its weight. The correct adenine percentage was confirmed in Stockholm in 1935. The isolation of nicotinamide from cozymase was attempted in Stockholm in 1935, but it was accomplished in 1936 in Dahlem. Concerning the nicotinic acid amide of cozymase, we should like to mention that von Euler is raising claims which we do not acknowledge because of his failure to submit an analysis.

Reversible Hydrogenation of Pyridine Nucleotide

The pyridine ring of pyridine nucleotides is reduced by hyposulfite to a dihydropyridine ring. Like the dihydropyridine derivatives of our model compounds, the dihydro-pyridine-nucleotide can be rehydrogenated by the yellow enzyme. Reduction of pyridine nucleotide, just like reduction of the model substance, brings about an absorption band in the long wavelength ultraviolet region. The reduction band of both pyridine nucleotides is located at 340 $m\mu$ while the reduction band of trigonelline is at 350 $m\mu$ and that of Karrer's derivative is located at 360 $m\mu$.

The dihydropyridine nucleotides, like those of the model substances, show a white fluorescence when exposed to ultraviolet light. The non-reduced pyridine nucleotides do not fluoresce. It is, therefore, possible to observe the hydrogenation and dehydrogenation directly in the light of the analytical quartz lamp. Historically speaking, it may be pointed out that we first observed the partial reversible hydrogenation of the pyridine nucleotides and only subsequently the partial reversible reduction of simple pyridine compounds. The sequence of events thus was the reverse of the chronology in this report.

Experiment 2

Hyposulfite is added to a solution of triphosphopyridine nucleotide, which is non-fluorescent, and the white fluorescence of the dihydro compound appears. In order to decompose the excess of hyposulfite, oxygen is introduced into the solution. Yet the white fluorescence remains, i.e., the dihydro compound is not auto-oxidizable. However, the white fluorescence disappears by addition of the yellow enzyme; the alloxazine ring dehydrogenates dihydropyridine to pyridine. Acidification of a solution of dihydropyridine nucleotide causes an irreversible disappearance of the white fluorescence. This is not a dehydrogenation process but a decomposition of the pyridine.

Experiment 3

Performed like Experiment 2 but using diphosphopyridine nucleotide (cozymase). The phenomena observed are identical to those in the hydrogenation and dehydrogenation of triphosphopyridine nucleotide.

EXPERIMENT 4

When our preparation procedure is used, pyridine nucleotides obtained from red blood cells stubbornly contain adhering impurities which fluoresce beautifully. This fluorescence is blue in acid solution and greenish in alkaline solution. Much work has been done in order to remove this compound, which perhaps is hydrogenated in the position para to the nitrogen. In all likelihood it has no biological significance.

The Attachment of Nicotinamide in Pyridine Nucleotides

Hydrogenation of pyridine nucleotides to the reversible dihydro compound liberates 1 molecule of acid. Since the sole acid component of pyridine nucleotides is phosphoric acid esterified to sugar, the hydrogen ion liberated by the reduction can only be the H-ion of a phosphoric acid esterified to sugar by another linkage.

This suggests that nicotinamide in pyridine nucleotide is linked in an analogous way to that of the Karrer methyl iodide derivatives. Accordingly, the equation of the hydrogenation of nicotinamide in the nucleotide is entirely parallel to the Karrer equation:

```
       CH                            CH
 CH ⁄ ⸌ C·CONH₂               CH ⁄ ⸌ C·CONH₂
   ‖    |                       ‖    |                      OR
 CH ⸌  ⁄ CH                   CH      CH₂                  /
      N      OR + 2H =              N           + HO—P = O
     / \    /                       |                \
   R'   OP = O                      R'                 OH
          \
           OH
```

which makes it immediately understandable why the nicotinamide on the nucleotides reacts with hyposulfite, while nicotinamide liberated from the nucleotides is unreactive towards hyposulfite.

R is a sugar derivative in our formula. R′, as already suspected by Karrer, is likewise a sugar derivative, perhaps even identical with R.

Reversible Hydrogenation of the Pyridine Nucleotide—Protein Compounds (Enzymes)

Although solutions of pyridine nucleotides are reduced by hyposulfite, they are not reduced by those substances which act as biological reductants. Particularly, carbohydrates or their phosphoric acid esters are incapable of reducing pyridine nucleotides in aqueous solutions.

However, if specific proteins present in yeast or other cells are added to the solutions of pyridine nucleotides, the nucleotides and the protein will combine into a compound in which the pyridine ring can be reduced by carbohydrate compounds. These pyridine nucleotide-protein compounds are the hydrogen transmitting enzymes, the active group of which is the pyridine ring.

Hence, in addition to the enzymes whose active group is hemin bound iron, and in addition to the yellow enzyme whose active group is alloxazine, there now emerges a third class of enzymes, the action of which can be accounted for by a simple chemical reaction.

Governed by the nature of the pyridine nucleotide and the specific carrier protein, there exist different hydrogen transferring enzymes. Triphosphopyridine nucleotide combined with a specific protein constitute an enzyme which can dehydrogenate the terminal carbon of hexose monophosphate, forming hexonic acid. Diphosphopyridine nucleotide linked to another specific protein creates the hydrogen-transferring enzyme of alcoholic fermentation. Diphosphopyridine nucleotide and still other specific proteins probably create the hydrogen-transferring enzymes of the other fermentations.

Since the carrier protein does not fluoresce and since the fluorescence of free dihydropyridine nucleotide does not interfere, it is possible to observe the hydrogenation of the enzyme directly under the analytical quartz lamp.

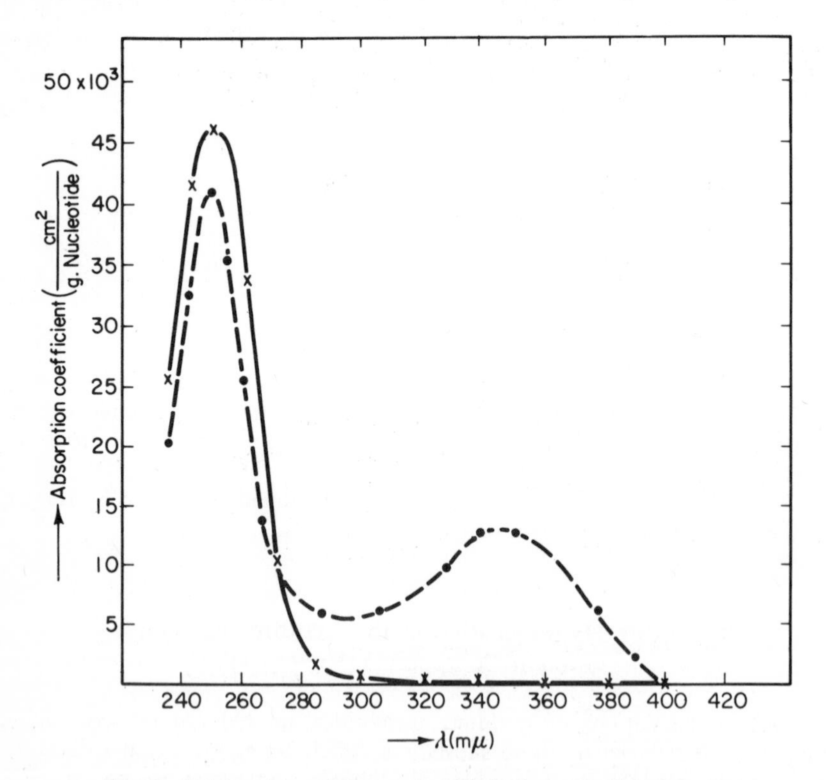

Fig. 1. Triphosphopyridine nucleotide hydrogenated by hyposulfite. ×——×, non-hydrogenated; ●- - - - -● hydrogenated.

Experiment 5

Potassium hexose-monophosphoric acid is added to a solution of triphosphopyridine nucleotide. This does not give rise to fluorescence since free nucleotide is not reduced by hexose-monophosphoric acid. If carrier-protein is added to a solution of triphosphopyridine nucleotide and subsequently potassium hexose-monophosphoric acid, then the white fluorescence of the dihydropyridinenucleotide appears corresponding to the reduction of protein by hexose-monophosphoric acid. A quantity of phosphohexonic acid equivalent to the pyridine of the nucleotide can be found in the solution. If the dihydro compound of the nucleotide is reoxidized to the pyridine compound with oxygen and the yellow enzyme, the initial state of the catalytic system is restored. Thus, with any small amounts of pyridine, any large amounts of hexose-monophosphoric acid can be oxidized to phosphohexonic acid.

Experiment 6

If diphosphopyridine nucleotide (cozymase) is substituted for the triphosphopyridine nucleotide used in Experiment 5 and another specific protein is

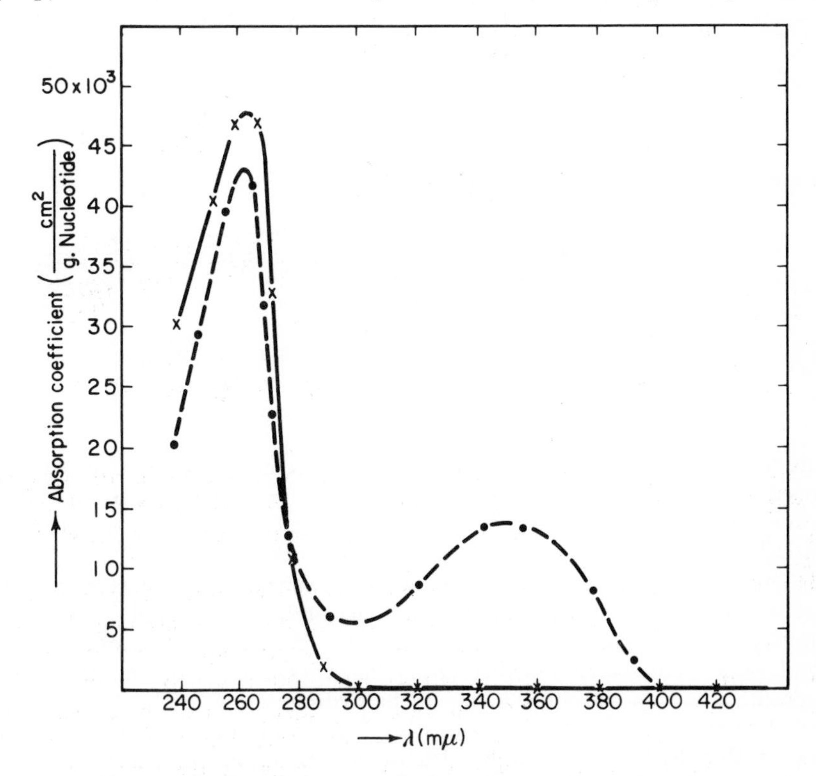

Fig. 1a. Triphosphopyridine nucleotide in the form of enzyme hydrogenated by hexose-monophosphoric acid. ×——×, non-hydrogenated; ●------●, hydrogenated.

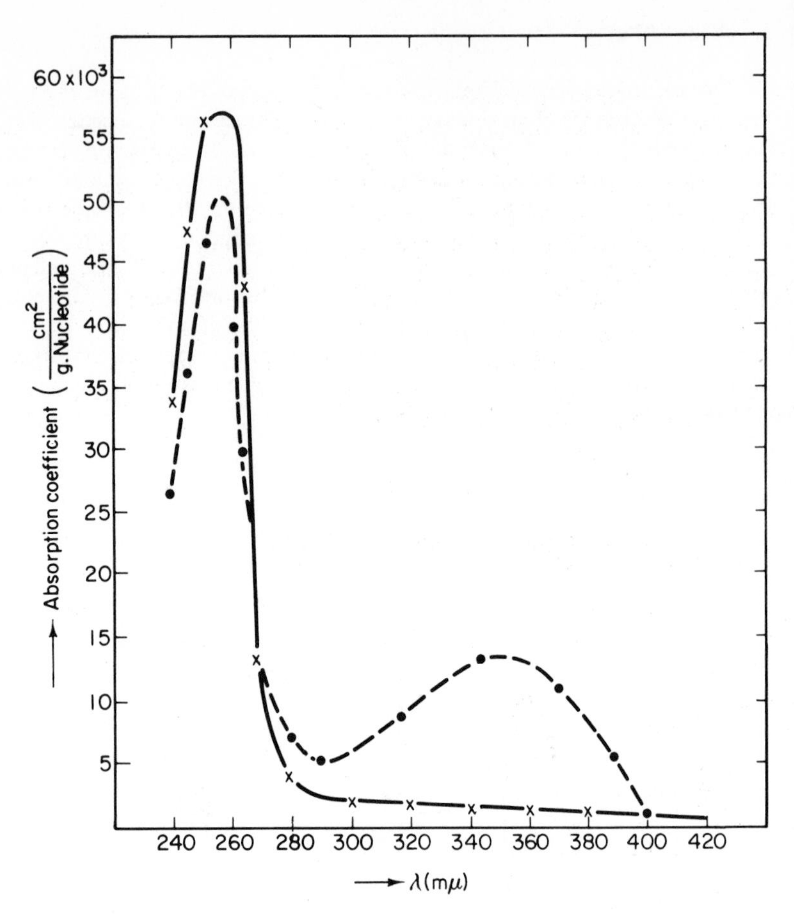

Fig. 2. Diphosphopyridine nucleotide hydrogenated by hyposulfite. × ——— ×, non-hydrogenated; ● - - - - - ●, hydrogenated.

substituted for the carrier protein, the white fluorescence of the dihydropyridine appears upon addition of the hexose-monophosphoric acid. However, phosphohexonic acid is no longer found in the solution but there is now pyruvic acid instead.

That the white fluorescence occurring in the enzyme experiments is actually due to the hydrogenated pyridine of the nucleotide and to nothing else, was proved by the measurement of the ultraviolet spectra appearing on the one hand in the hydrogenation of the pyridine nucleotide in aqueous solution with hyposulfite and on the other hand, in the hydrogenation of the enzymes by carbohydrate. It can be seen by comparing Fig. 1 with Fig. 1a and Fig. 2 with Fig. 2a, that the long wave ultraviolet bands occuring in the reduction of the enzymes conform in position and height with the dihydropyridine bands occurring in the reduction by hyposulfite.

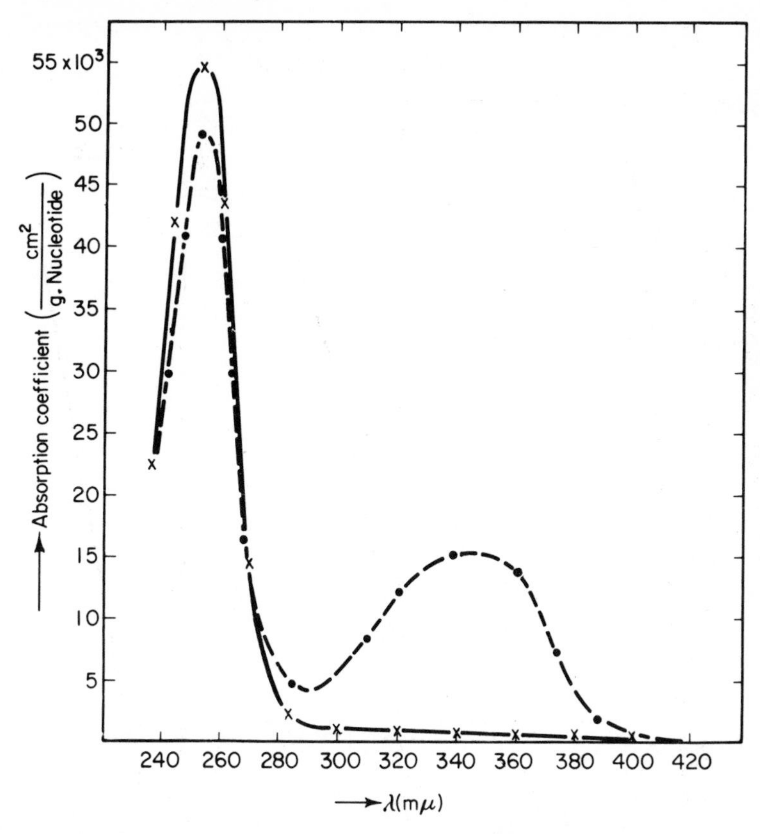

Fig. 2a. Diphosphopyridine nucleotide in the form of enzyme hydrogenated by hexose-monophosphoric acid. ×———×, non-hydrogenated; ●- - - - -●, hydrogenated.

Dehydrogenation by Acetaldehyde

EXPERIMENT 7

Acetaldehyde is added to the white fluorescing solution of the hydrogenated diphosphopyridine nucleotide, formed by reduction of the enzyme with hexose-monophosphoric acid. The fluorescence disappears immediately and the solution contains alcohol: the acetaldehyde has dehydrogenated the dihydropyridine to pyridine and has itself been hydrogenated into alcohol. Consequently, this is the chemical reaction by which alcohol is formed in the alcoholic fermentation.

In order to do a quantitative analysis of this important reaction the enzyme solution in which the nucleotide is hydrogenated by carbohydrate and dehydrogenated by acetaldehyde, is put before the photoelectric cell. The appearance of light absorption upon dehydrogenation is measured in the spectral region

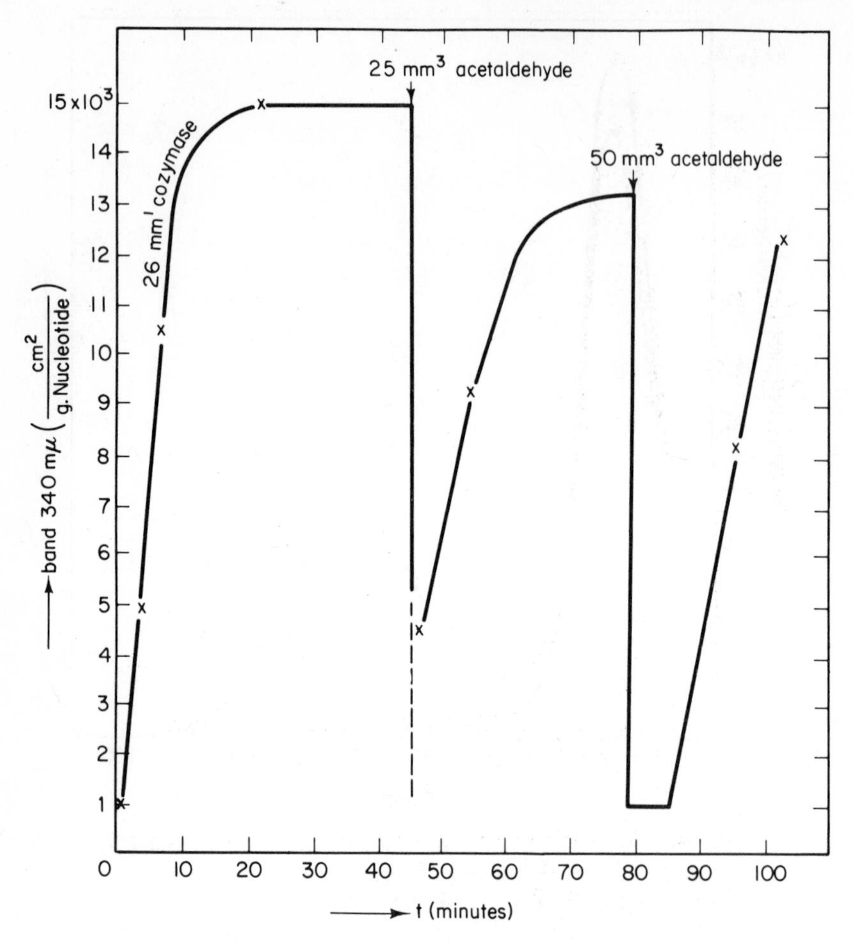

Fig. 3. Hydrogenation and dehydrogenation of the pyridine of the fermentation enzyme. Ordinate: Height of the dihydroband of diphosphopyridine nucleotide (cozymase).

of the dihydrobands at 340 mμ. In Fig. 3, such an experiment is presented graphically.

At time $t = 0$ the hexose monophosphate is added. The light absorption then increases until all the nucleotide present in the enzyme solution has been hydrogenated to the dihydropyridine; acetaldehyde is now added, and the light absorption disappears. This verifies the equation:

$$1 \text{ dihydropyridine} + 1 \text{ acetaldehyde} = 1 \text{ pyridine} + 1 \text{ alcohol}$$

If the enzyme solution contains excess hexose-monophosphoric acid, the light absorption increases again (Fig. 3) after the aldehyde is used up and can be made to disappear again by addition of aldehyde, and so on until all the hexose-monophosphoric acid is fermented.

That the entire fermentation actually proceeds according to the pyridine reactions shown here, is further supported by the Harden-Young experiment in 1904 which shows that if the cozymase is removed from a fermenting liquid, the fermentation stops.

Other Pyridine Catalyses

In conclusion, it should be noted that, according to the works of von Euler, von Wagner-Jauregg and others, most enzymic dehydrogenations proceed only if one of the two coenzymes now recognized as pyridine nucleotides is added to the protein and substrate solution. Although not proved as yet, it seems that all these dehydrogenations can hardly be anything but pyridine catalyses.

SYNTHESIS OF TRIOSEPHOSPHATE*

David E. Green, Dorothy M. Needham
and John G. Dewan

Sir William Dunn School of Biochemistry,
University of Cambridge, England

Triosephosphate and pyruvate in presence of the appropriate mutase and coenzyme I react until all the triosephosphate is oxidized. That is to say equilibrium is reached only when the reaction has proceeded practically to completion. In order to reverse the forward reaction, some means must be found of shifting the equilibrium point. Since cyanide combines with both triosephosphate and pyruvate, the use of this reagent should enable the reaction to proceed in the reverse direction. The reduction of phosphoglycerate to triosephosphate involves the disappearance of an acid. Hence, if the experiment is conducted manometrically, the course of the reaction should be indicated by an absorption of CO_2 from the gas space into the bicarbonate solution containing the reactants. Table I is a summary of a typical experiment. The absorption of CO_2 depends upon the presence of the coenzyme, lactate, phosphoglycerate and the mutase. There is an appreciable blank in absence of coenzyme but this is also true of the forward reaction.

There are a few experimental details that are important in connexion with the synthesis of triosephosphate. The cyanide solution (2 *M*) should be neutralized immediately before the experiment. The neutralized solution should be just colourless to phenolphthalein and distinctly blue to bromothymol blue. The neutralized solution should not be kept more than 2 hr.

*Excerpt from D. E. Green, D. M. Needham, and J. G. Dewan, "Dismutations and Oxidoreductions," *Biochem. J.*, **31**, 2347–2349, 2352 and corresponding figure references and acknowledgements (1937). Reprinted with the permission of the authors and of the editors of *Biochemical Journal.*

One of us (D.E.G.) is grateful to the Ella Sachs Plotz Foundation for a grant. We wish to express our thanks to Mr. S. Williamson for his assistance with some of the chemical preparations.

Table I

REACTION OF LACTATE WITH PHOSPHOGLYCERATE

The following quantities were used; 1 ml. mutase, 1.0 ml. 0.2% coenzyme I, 0.4 ml. *M*/5 phosphoglycerate (natural) 0.2 ml. 2*M* HCN and 0.1 ml. *M dl*-lactate.

	μl. CO_2 absorbed in 10 min.
Mutase + coenzyme + lactate + cyanide + phosphoglycerate	394
Complete system without lactate	60
Complete system without cyanide	0
Complete system without mutase	0
Complete system without phosphoglycerate	0
Complete system without coenzyme	80

Table II

OXIDATION OF MALATE AND α-HYDROXYBUTYRATE BY PHOSPHOGLYCERATE

Details as for Table XXIV.

	μl. CO_2 absorbed in 10 min.
Mutase + coenzyme + cyanide + malate + phosphoglycerate	352
Complete system without phosphoglycerate	0
Complete system without malate	60
Mutase + coenzyme + cyanide + α-hydroxybutyrate + phosphoglycerate	103
As above without phosphoglycerate	0

Not only will lactate react with phosphoglycerate but malate and α-hydroxybutyrate do so as well (cf. Table II). The three reactions may be summarized by equation (19):

$$\left\{\begin{array}{l}\text{Malate}\\ \text{Lactate}\\ \alpha\text{-Hydroxybutyrate}\end{array}\right. + \text{phosphoglycerate} \xrightarrow{\text{HCN}} \left\{\begin{array}{l}\text{oxaloacetate}\\ \text{pyruvate}\\ \alpha\text{-ketobutyrate}\end{array}\right. + \text{triosephosphate.} \qquad (19)$$

Only *l*(+)-lactate and *l*(—)-malate are active as reductants for (+)-phosphoglycerate:

	μl. CO_2 absorbed in 10 min.
l(+)-Lactate	398
d(—)-Lactate	38
l(—)-Malate	236
d(+)-Malate	19

The production of triosephosphate and pyruvate was also demonstrated chemically. The former was estimated as alkali-labile phosphate by the method of Meyerhof & Lohmann [1934]; the latter was isolated as the 2:4-dinitrophenylhydrazone and identified as such by a mixed M.P. with a known sample. Table III compares the absorption of CO_2 with the production of alkali-labile phosphate. The parallelism, although not exact, indicates that the reaction is proceeding according to equation (19).

Table III

CO_2 Absorption and Formation of Alkali-Labile Phosphate

	μl. CO_2 absorbed	mg. alkali labile phosphate produced
Complete system	210	0.70
Without coenzyme	100	0.34
Without cyanide	0	0
Without lactate	0	0
Without phosphoglycerate	0	0

In addition to the production of alkali-labile P, a considerable liberation of inorganic phosphate is observed coincidently with the oxidoreduction. In absence or in presence of iodoacetic acid, the production of inorganic phosphate as well as of alkali-labile phosphate is suppressed. The mechanism whereby the inorganic phosphate is derived from phosphoglycerate is unknown. But it appears likely that the production of inorganic phosphate is dependent upon the oxidoreduction. This observation recalls the results of Meyerhof *et al.* [1937] who found with yeast preparations that dephosphorylation of phosphopyruvate was under certain conditions greatly accelerated by the presence of a trace of hexosediphosphate provided that coenzyme I and an active oxidoreduction system were also present. The production of inorganic phosphate has no direct bearing on the synthesis of triosephosphate. It is not due for example to breakdown of the triosephosphatecyanohydrin. Hexosediphosphate in presence of the mutase solution and cyanide gave rise quantitatively to triosephosphatecyanohydrin. No inorganic phosphate was found.

Thus far we have been unable to produce the following reversals: glycerate + lactate, α-glycerophosphate + phosphoglycerate and α-glycerophosphate + glycerate. There are three possibilities: (1) these reactions are irreversible, (2) cyanide is inhibiting the triose and α-glycerophosphate mutases and (3) cyanide does not fix glyceraldehyde efficiently.

A direct and simple method demonstrating the reduction of phosphoglycerate to triosephosphate lies in the use of oxidoreduction indicators. Glyceraldehyde in presence of the mutase, coenzyme I and the factor at *p*H 7.2 reduced benzylviologen practically to completion (intense blue colour). When the reduction had proceeded to completion, phosphoglycerate was introduced. The blue colour of the reduced form gradually faded and in a few minutes the solution became colourless. Glyceraldehyde is merely a reducing agent for the coenzyme, reduced coenzyme in turn reducing benzylviologen. By addition of phosphoglycerate, the equilibrium is shifted in favour of oxidized coenzyme and hence in favour of the oxidized or colourless form of benzylviologen (for a complete discussion of the theory of the method, cf. Green *et al.* [1937]).

The formation of lactate and inorganic phosphate from hexosediphosphate in muscle is usually regarded as taking place in six stages: (1) formation of triosephosphate, (2) oxidoreduction between triosephosphate and a trace of pyruvate, (3) transformation of 3-phosphoglycerate into 2-phosphoglycerate, (4) transformation of 2-phosphoglycerate into phosphopyruvate, (5) dephosphorylation of phosphopyruvate by reaction with adenylic acid, the pyruvate thus formed taking part in the second reaction and (6) formation of free phosphate and adenylic acid from adenosinetriphosphate. Of these stages, (1), (3) and (4) have been shown by Meyerhof and his collaborators to be reversible. Needham & Pillai [1937] have shown that stage (6) can be reversed by coupling with an energy-yielding reaction. Thus stage (5) is the only one which has not so far been reversed *in vitro*. An investigation of the conditions under which adenylpyrophosphate might phosphorylate pyruvate, or under which pyruvate might react with inorganic phosphate is being undertaken.

These facts lead on to the consideration of a possible path of carbohydrate synthesis from lactate. Given a trace of phosphoglycerate, reaction with lactate would produce triosephosphate and pyruvate. This reaction would proceed to an appreciable degree if the triosephosphate formed were continually and completely removed—as would happen if, for example, the hexosediphosphate formed were being converted into glycogen. As for the pyruvate formed in the oxidoreduction, once it had been phosphorylated (a step at present obscure), its conversion into a further supply of phosphoglycerate to react with the lactate would readily follow. Thus given an energy source for the phosphorylation of pyruvate and for other energy-requiring reactions, it should be possible to convert lactate quantitatively into glycogen.

Summary

1. The properties and reactions of the triose, triosephosphate and α-glycerophosphate mutases have been studied. The activities of all these three mutases depend upon coenzyme I.

2. The mutases all catalyse the oxidation of their substrates by α-ketonic acids. Some, like the triosephosphate mutase, also catalyse the dismutation of the substrate.

3. Mutase action is inhibited almost completely by low concentrations of iodoacetic acid.

4. The coenzyme functions in these oxidoreductions by undergoing a cycle of oxidation and reduction.

5. Mutase systems cannot react with oxygen carriers such as flavin, methylene blue etc. in absence of a thermolabile factor found in skeletal and cardiac muscle extracts. The factor is non-dialysable and sedimented by high speed centrifuging. It is not identical with any of the known carriers.

6. The reaction between lactate and phosphoglycerate yielding triosephosphate and pyruvate has been shown to take place in presence of the triosephosphate mutase system and strong cyanide.

7. The theory of mutase action is discussed.

References

1. Meyerhof & Lohmann (1934). *Biochem. Z.* **271**, 89.
2. Meyerhof, Kiessling & Schulz (1937). *Biochem. Z.* **292**, 25.
3. Green, Dewan & Leloir (1937). *Biochem. J.* **31**, 934.
4. Needham & Pillai (1937). *Biochem. J.* **31**, 1837.

R-DIPHOSPHOGLYCERIC ACID, ITS ISOLATIONS AND PROPERTIES*

Erwin Negelein and Heinz Brömel
Kaiser Wilhelm Institut für Zellphysiologie,
Berlin-Dahlem, Germany
(Submitted August 30, 1939)

If one permits the "oxidizing enzyme" of fermentation to react on triosephosphate [dehydrogenation] an oxidation product is formed[1] which can be converted back [to triosephosphate]. The oxidation product is not identical with 3-phosphoglyceric acid nor with 2, 3-diphosphoglyceric acid.

We have tried, at the suggestion of Otto Warburg, to isolate this oxidation product and have found[2] that it had the composition of a diphosphoglyceric acid. In order to distinguish this substance from the 2, 3-diphosphoglyceric acid,[3] we shall name this new acid R-diphosphoglyceric acid ("R-acid").

A prerequisite for the isolation of the R-diphosphoglyceric acid was the isolation of the oxidative enzyme of fermentation. The reason why the oxidative enzyme of fermentation must be pure is explained elsewhere.[1]

Isolation of the R-diphosphoglyceric acid

A small portion of pyridine nucleotide and the crystallized proteins of the oxidative[4] and reductive[5] enzymes of fermentation were added to a mixture of 3-phosphoglyceraldehyde (Fischer-ester), phosphate and acetaldehyde;

*E. Negelein and H. Brömel, "R-Diphosphoglycerin saüre, ihre Isolierung und Eigenschaften," *Biochim. Z.*, **303**, 132 (1939). Translated from the German and reprinted by permission of Springer-Verlag, Heidelberg, Germany.

[1] O. Warburg and W. Christian, *Biochim. Z.*, **303**, 40 (1939).
[2] E. Negelein and H. Brömel, *Biochim. Z.*, **301**, 135 (1939).
[3] J. Greenwald, *J. Biol. Chem.*, **63**, 339 (1925).
[4] O. Warburg and W. Christian, *Biochim. Z.*, **301**, 221 (1939) and **303**, 40 (1939).
[5] E. Negelein and H. J. Wulff, *Biochim. Z.*, **289**, 436 (1937) and **293**, 351 (1937).

this gives rise to the following reaction brought about by the oxidizing enzyme:

$$\text{3-phosphoglyceraldehyde} + \text{phosphate} + \text{pyridine nucleotide} \quad (1)$$
$$\rightleftarrows \text{R-diphosphoglyceric acid} + \text{dihydropyridine nucleotide}$$

The corresponding reducing enzyme brings about the following reaction:

$$\text{dihydro-pyridine nucleotide} + \text{acetaldehyde} \rightleftarrows$$
$$\text{pyridine nucleotide} + \text{alcohol} \quad (2)$$

By adding (1) and (2), one can eliminate the catalyst, thus obtaining the dismutation equation for the fermentation:

$$\text{3-phosphoglyceraldehyde} + \text{phosphate} + \text{acetaldehyde}$$
$$\rightleftarrows \text{R-diphosphoglyceric acid} + \text{alcohol} \quad (3)$$

Under the conditions of our assay, Reaction (3) goes almost completely from the left to the right.

Crystallization and composition of R-diphosphoglyceric acid

Subsequent to the reaction of *p*H 7.6, according to Equation (3), we acidify with sulfuric acid to *p*H 2.1 and then we precipitate with ten volumes of acetone, whereupon the bulk of the acidic phosphate remains in solution, whereas the acidic salt of R-acid precipitates. By dissolving the precipitate in water and by adding strychnine chloride to the neutralized solution, the sparingly soluble tetra-strychnine salt of the R-acid precipitates in a crystalline form.

For analysis the recrystallized strychnine salt was dried at 60° in high vacuum. W. Lüttgens found:

4.225 mg: 10.047 mg CO_2, 2.315 mg H_2O; 64.85% C, 6.13% H.
4.235 mg: 10.045 mg CO_2, 2.317 mg H_2O; 64.69% C, 6.12% H.
6.760 mg: 0.421 cc N (23.6°, 751 mm); 7.08% N.
6.060 mg: 0.369 cc N (21.0°, 751 mm); 6.99% N.
0.922 mg: 35.9 μg P; 3.89% P (colorimetrically).
0.922 mg: 34.9 μg P; 3.79% P (colorimetrically).

Calculated for $C_{87}H_{96}N_8P_2O_{18}$	65.13% C / 6.04% H / 6.99% N / 3.87% P
Found	64.77% C / 6.12% H / 7.04% N / 3.84% P

Detection and determination of R-diphosphoglyceric acid

In order to identify and to determine R-acid, we use the backward reaction of the reversible Reaction (1), which proceeds sufficiently toward completion if one starts with solutions free of inorganic orthophosphate.

Thus, by adding the specific protein catalyzing Reaction (1) to a solution of R-acid (free of orthophosphate) containing (in excess) reduced pyridine nucleotide, an amount of reduced pyridine nucleotide equivalent to the R-acid is oxidized, causing a decrease in the light absorbency in the longwave UV (absorption band of dihydropyridine). We are thus able to titrate the R-acid by photoelectric measurement of the change in the light absorbency.

Photoelectric measurement

We modified the method of the photoelectric measurement of pyridine reactions (discovered in 1935[1]) by replacing the photoelectric cell of Elster und Geitel [manufacturers] by a corrected photocell, and the fiber electrometer by a multiflex galvanometer constructed by Dr. Bruno Lange, Berlin-Zehlendorf. The [natural] logarithm i_0/i, proportional to the concentration of dihydropyridine, is read off a milky glass pane directly.

The modified method has the advantage of being simpler and the apparatus is less expensive.

Decomposition of R-diphosphoglyceric acid

R-acid decomposes spontaneously in aqueous solution (without participation of a protein) according to:

$$\text{R-diphosphoglyceric acid} + H_2O \rightleftarrows \text{3-phosphoglyceric acid} + \text{phosphoric acid} \qquad (4)$$

The velocity of decomposition depends on the temperature and the *p*H of the solution. We measured it manometrically (expulsion of carbon dioxide from a bicarbonate solution) and by photoelectric titration of the R-acid.

If the concentration of R-acid at the time t is c, and the velocity constant of the reaction is k, then we have:

$$-\frac{dc}{dt} = kc$$

during the whole course of the decomposition reaction.

At 38° and neutral *p*H, we found $k = 0.026 \text{ min}^{-1}$, that is: 2.6% of R-acid is decomposed per minute (halftime of decomposition is 27 minutes).

Asymmetric C atom

Although the optical rotation of R-acid is too small for exact measurement, it is, nevertheless, possible to show that R-acid does contain an asymmetric C atom because phosphoglyceric acid, formed in the course of the spontaneous decomposition of R-acid (without participation of a protein), is optically active. According to Meyerhof and Schulz,[2] we found for phosphoglyceric acid formed from R-acid with ammonium molybdate:

$$[\alpha]_D^{20°} = -675°$$

The constitution of R-acid, therefore, is *not* the following, as one could

[1] O. Warburg, W. Christian and A. Griese, *Biochim. Z.* **282**, 157 (1935). E. Negelein and E. Haas, *Biochem. Z.* **282**, 206 (1935).

[2] O. Meyerhof and W. Schulz, *Biochim. Z.* **297**, 60 (1938).

have imagined:

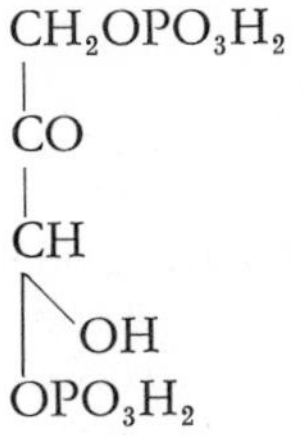

$$
\begin{array}{l}
CH_2OPO_3H_2 \\
| \\
CO \\
| \\
CH \\
| \diagdown OH \\
OPO_3H_2
\end{array}
$$

As possible constitution formulas remain therefore:

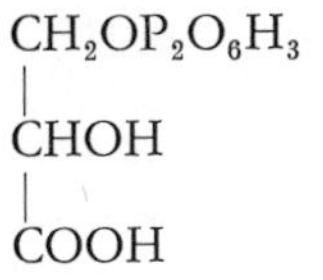

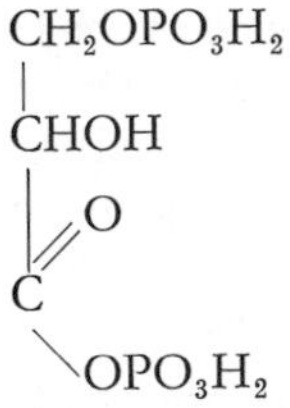

$$
\begin{array}{l}
CH_2OP_2O_6H_3 \\
| \\
CHOH \\
| \\
COOH
\end{array}
\qquad\qquad
\begin{array}{l}
CH_2OPO_3H_2 \\
| \\
CHOH \\
| \diagup\!\!\diagup O \\
C \\
\diagdown OPO_3H_2
\end{array}
$$

(I) **3-pyrophosphoglyceric acid**

(II) **1,3-diphosphoglyceric acid**

Against (I) speaks the fast decomposition in aqueous solution and neutral conditions, at room temperature. Under those conditions neither pyrophosphate nor adenosine triphosphate are decomposed.

In addition, the ultraviolet spectrum of R-acid speaks against (I), for we found an absorption band at λ 215 mμ, which disappears when R-acid is decomposed according to Equation (4). We have found a band of similar height and position in the ultraviolet spectrum of acetic anhydride. This absorption band also disappears when acetic anhydride is hydrolyzed.

We believe, therefore, that R-acid should be assigned the constitutional formula (II)[1], but we cannot decide whether it is monomeric or dimeric since we have not determined the molecular weight. The presence of an asymmetric C atom speaks against the dimeric form.

Experimental Part

3-phosphoglyceraldehyde

3-phosphoglyceraldehyde was synthesized according to H. O. L. Fischer and E. Baer. The aqueous solution, obtained by dissolving the dry residue after reductive cleavage of benzyl-cycloacetalglyceraldehyde-phosphoric acid was used.[2] The clear solution was adjusted to approximately *p*H 5 by adding sodium hydroxide and then stored in a cool place. The Fischer-ester contains both the dextrorotatory and the levorotatory compounds, according to C. V.

[1] A glucose phosphorylated in 1-position was recently discovered by C. F. Cori and G. T. Cori, *Proc. Soc. Exp. Biol. and Med.* **34**, 702 (1936); C. F. Cori, S. P. Colowick and G. T. Cori, *J. Biol. Chem.* **121**, 465 (1937). This acid does not react with the pyridine proteides [e.g., proteins with a prosthetic group] known till now.

[2] H. O. L. Fischer and E. Baer, *Ber. d. Deutsch. chem. Ges.* **65**, 337 (1932).

Smythe and W. Gerischer,[1] but only the dextrorotatory component is reactive.

PREPARATION OF R-DIPHOSPHOGLYCERIC ACID

12.6 cc of a 0.070 M 3-phosphoglyceraldehyde solution were neutralized to a *p*H of about 7.5 by adding normal sodium hydroxide solution. To this, 5.5 cc 0.5 M phosphate solution (9.5 parts Na_2HPO_4 and 0.5 parts KH_2PO_4 and 3.4 cc of a 0.004 M diphosphopyridine nucleotide solution were added. The solution was adjusted to a volume of 36 cc by adding water. Temperature: 18°. Then, 0.34 cc of a molar acetic aldehyde solution, 0.60 cc of a 0.75% solution of the enzyme of the fermentation and 0.60 cc of a 0.60% solution of the reductive enzyme of fermentation were added. Owing to the large protein concentration, we obtained a large turnover in a short time. After 2 and 4 minutes, we added again 0.34 cc of molar acetic aldehyde solution in each case. The total volume of the reaction mixture was finally 38.2 cc; the *p*H was 7.6.

Amounts of reaction partners:

$0.44 \cdot 10^{-3}$ moles D-glyceraldehyde-phosphoric acid.
$2.60 \cdot 10^{-3}$ moles inorganic phosphoric acid.
$0.0136 \cdot 10^{-3}$ moles pyridine nucleotide.
$1.02 \cdot 10^{-3}$ moles acetic aldehyde.

The solution remained at 18° for 25 minutes. Then the solution was acidified by adding about 6 cc sulfuric acid, so that the *p*H was 2.10. (The *p*H should not be below 2.0 and not above 2.2). The acidified solution was *immediately* poured into 445 cc cold acetone (10 volumes). The precipitated substance was centrifuged in the cold, washed once with cold acetone and dried in a desiccator. Since R-acid is unstable in the dried form, it was dissolved soon afterwards in 28 cc of cold water. The insoluble, denatured protein was filtered off. The clear, acid solution was 29 cc; 1 cc of the solution contained $1.24 \cdot 10^{-5}$ moles R-acid. The average yield of acid formed in different experiments was 80-85%, calculated from the initial amount of glyceraldehyde-phosphoric acid.

To 29 cc of the neutral solution, containing $3.6 \cdot 10^{-4}$ moles R-acid, were added 29 cc of a saturated (about 0.1 M) strychnine chloride solution. After a short time, crystallization of the strychnine salt of R-acid started. The solution was kept for 15 hours at 0°, then the crystals were filtered off at 0° and washed with a small amount of cold water. For recrystallization we dissolved the substance in 25 cc of cold water by dropwise addition of the exact amount of N hydrochloric acid. Afterwards, we added 0.2 N sodium hydroxide to the solution, drop by drop, until the first crystals of the strychnine salt appeared. The solution remained for a short while at 0°, then more sodium hydroxide was added, until the solution showed a neutral reaction with litmus paper. After a few hours, the recrystallized strychnine salt was filtered off and washed as previously described.

For the preparation of the strychnine-free R-acid, the recrystallized substance was washed in a separatory funnel with about 10 cc of cold water. The

[1] C. V. Smythe and W. Gerischer, *Biochim. Z.* **260**, 414 (1933).

strychnine was then removed by repeated extractions with chloroform, where, in every case, enough 0.2 N NaOH was added, to keep the *p*H of the aqueous solution slightly alkaline. (*p*H about 7.6). After the separation of chloroform the clear aqueous solution was placed in a vacuum desiccator, in which the residual chloroform was completely removed and the solution was simultaneously brought to a smaller volume. The volume was now 10.3 cc; 1 cc of the solution contained $2.45 \cdot 10^{-5}$ moles R-acid. The yield was $2.52 \cdot 10^{-4}$ moles R-acid, that is 57% of the initial D-glyceraldehyde-phosphoric acid.

In 1 cc of the R-acid, solution $4.75 \cdot 10^{-5}$ moles of total phosphorus and $2.39 \cdot 10^{-5}$ moles labile phosphorus (easily cleavable according to Equation (4)) were found.

$$\frac{\text{total phosphorus}}{\text{labile phosphorus}} = \frac{4.75}{2.39} = 1.99; \qquad \frac{\text{total phosphorus}}{\text{R-acid}} = \frac{4.75}{2.45} = 1.94$$

To store the neutral solution, we kept it frozen in a freezing mixture.

Phosphorus Determination

The phosphorus was determined colorimetrically, according to Briggs, in the way described by M. Martland and R. Robison.[1] The phosphoric acid, cleavable according to Equation (4), is so loosely bound that it is split off by the method used for phosphorus determination, and it is therefore determined directly as inorganic phosphorus. There is no possibility therefore to determine labile phosphorus along with inorganic phosphorus.

Determination of the R-diphosphoglyceric acid

To measure the pyridine reaction photoelectrically, we used the mercury line at λ 334 mμ. We determined the difference between the absorption coefficients of the reduced and the non-reduced pyridine nucleotide in the presence of arsenate[2] (instead of phosphate) with a known amount of 3-phosphoglyceraldehyde. We found: $\Delta\beta$ 334 mμ = $12.6 \cdot 10^6$ (cm²/mol).

For the determination of the R-acid, a plane-parallel quartz cell is used, which has a volume of 3 cc and a lightpath of 0.61 cm. The cell is filled with the reaction mixture, lacking the enzyme protein. The reaction mixture is composed of 1.0 cc of a 0.1 M pyrophosphate solution of *p*H 7.9 (0.6 moles hydrochloric acid to 1 mole sodium pyrophosphate), 0.2 cc of a $2.2 \cdot 10^{-3}$ M reduced pyridine nucleotide (= $0.44 \cdot 10^{-6}$ moles), 0.2 cc of the R-acid solution, which is to be examined (less than $0.40 \cdot 10^{-6}$ moles), and 1.57 cc water. The total volume of the reaction mixture without protein is 2.97 cc.

Logarithm ($\ln i_0/i$) is measured. Then one adds 0.03 cc of a 0.1% protein solution of the oxidizing enzyme of fermentation[3] to the reaction mixture, mixes

[1] M. Martland and R. Robison, *Biochim. J.* **20**, 847 (1926).
[2] D. M. Needham and R. K. Pillai, *Biochim. J.* **31**, 1837 (1937); O. Warburg and W. Christian, *Biochim. Z.* **303**, 40 (1939).
[3] Otto Warburg and W. Christian, *Biochim. Z.* **301**, 221 (1939); **303**, 40 (1939).

and observes the course of the reaction on the galvanometer. The amount of protein is so large that the reaction comes to an end in a short time. Again the logarithm is measured and from its decrease ($\Delta \ln i_0/i$) one can calculate the decrease of reduced pyridine nucleotide by taking the dilution (1%) into account.

In order to examine how far the reaction between R-acid and reduced pyridine nucleotide has proceeded two determinations with different amounts of R-acid solution were carried out. We found for $\Delta \ln i_0/i$ and by using 0.1 cc of the R-acid solution a value of 0.51, and with 0.2cc the value was 1.02. The double amount of reduced pyridine nucleotide is therefore oxidized by doubling the amount of R-acid.

The amount of R-acid (m) present in 0.2 cc of the R-acid solution can be calculated from the indicated values:

$$m = \frac{1}{12.6 \cdot 10^{-6}} \cdot \frac{1}{0.61} \cdot 1.02 \cdot 3 = 0.398 \cdot 10^{-6} \text{ moles}$$

Since the reaction mixture contained only $0.44 \cdot 10^{-6}$ moles of reduced pyridine nucleotide, 90% of it is oxidized. Hence, it follows that under the specified conditions, even in the case of 90% conversion of reduced pyridine nucleotide, the reaction between R-acid and reduced pyridine nucleotide has proceeded sufficiently towards completion.

It should be noted that the presence of strychnine does not interfere with the photoelectric titration of R-acid. It is, therefore, also possible to determine the concentration of a solution of the strychnine salt.

Decomposition of R-diphosphoglyceric acid

The decomposition of R-acid to 3-phosphoglyceric acid and inorganic phosphoric acid occurs with great facility. We do not know any conditions in which R-acid remains unchanged for a longer period of time. In aqueous solution at slightly alkaline reactions *p*H between 7 and 9, the rate of decomposition is the smallest; we found under these conditions a decomposition of 6% in 24 hours at 0°. The slightly alkaline solution is more stable if kept frozen in a freezing mixture. Under these conditions, the loss occurring in 24 hours is only 3%. On the other hand, we lost 77% of a neutralized R-acid preparation that was kept in a desiccator, whereas a dry preparation obtained by acetone precipitation from an acid solution (*p*H 2) only decreased 19% during storage in a desiccator for 24 hours. Crystallized strychnine salt and an amorphous calcium salt obtained by precipitation with alcohol were likewise unstable in the dried form.

We measured the speed of decomposition of R-acid in neutral bicarbonate simultaneously by manometric and photoelectric methods at 38°. The reaction mixture was a $3.75 \cdot 10^{-2}$ molar sodium bicarbonate solution, containing $1.75 \cdot 10^{-6}$ moles of R-acid/cm^3. Cone-shaped pressure gauge vessels were used for the manometric assays; after the acid formation was ended, the retention was measured by addition of tartaric acid. For the photoelectric titration of R-acid, the reaction mixture remained at the same temperature, in a test tube.

By taking the retention (1%) into consideration, the acid formation was

Temperature: 38°C; *p*H 7.2

	Formation of 3-phosphoglyceric acid, manometrically determined		Decrease of R-acid photometrically determined
Time after placing in the thermostat	Flask 1 In main volume: 7.10^{-6} moles R-acid in 4.0 cc of a $3.75 \cdot 10^{-2}$ M $NaHCO_3$-solution. In appendix: 0.2 cc of a 0.01 M tartaric acid (desiccated at 100°), i.e., 2.10^{-6} moles of tartaric acid. In gas volume: 10 vol. % carbon dioxide in argon. $v_F = 4.0$ cc, $v_G = 13.76$ cc	Flask 2 In main volume: 4.0 cc of a $3.75 \cdot 10^{-2}$ M $NaHCO_3$-solution In appendix: $2 \cdot 10^{-6}$ moles tartaric acid, as in flask 1 In gas volume: 10 vol. % carbon dioxide in argon. $v_F = 4.0$ cc, $v_G = 13.91$ cc	Test tube with 8 cc of the same reaction mixture as in flask 1. A gas stream of 10 vol. % of carbon dioxide in argon is passed slowly through the solution. For the photoelectric titration of the R-acid, 0.17 cc samples were withdrawn
	produced carbon dioxide:	produced carbon dioxide:	R-acid found in 0.17 cc of the solution:
0 min.	—	—	$0.298 \cdot 10^{-6}$ moles.
10 min.	27.2 mm^3 (extrap.)	—	0.224
20 min.	46.1 "	—	—
30 min.	60.8 "	—	0.143
50 min.	80.6 "	—	0.090
80 min.	97.0 "	—	0.045
120 min.	106.5 "	—	0.017
150 min.	109.4 "	—	—
190 min.	110.4 "	—	0.003
240 min.	110.8 "	—	—
	Tartaric acid poured in: 86.5 mm^3 CO^2, that is $\frac{86.5}{87.5} = 98.4\%$ of added tartaric acid	Tartaric acid poured in: 87.9 mm^3 CO_2	

found to be

$$\frac{110.8}{0.984} = 112.6 \text{ mm}^3 \text{ or } \frac{112.6}{22.4} = 5.03 \cdot 10^6 \text{ moles},$$

that is, $5.03/7.0 \times 100 = 72\%$ of the initial R-acid. The difference may be explained, at least partially, by the decrease of the dissociation of the secondary dissociation of phosphoric acid.

Table I contains the combined values (in percentages), obtained by the photoelectric measurement of the decrease of R-acid, and those obtained manometrically by measuring the increase of formed 3-phosphoglyceric acid. In the fourth column are the measured values for the velocity constant k of the decomposition reaction:

$$-\frac{dc}{dt} = kc$$

Table I

Time min.	Decrease of R-acid photometrically determined %	Increase of evolving 3-phosphoglyceric acid, manometrically determined %	$\frac{1}{t} \ln \frac{c_0}{c}$ completed from the manometric assay (min^{-1})
10	24.8	(24.5)	(0.028)
20	—	41.6	0.027
30	52.0	54.9	0.027
50	69.8	72.8	0.026
80	85.0	87.6	0.026
120	94.3	96.1	0.027
150	—	98.7	—
180	99.0	99.6	—

Rotation of Polarized Light

We tried to measure the rotation of a neutral $2.45 \cdot 10^{-2}$ M solution of R-acid in a 10 cm tube and in yellow light. The values obtained were so low that it was not possible to read off the values with certainty.

By mixing R-acid with half the volume of a 25% ammonium molybdate solution, it was shown that the speed of cleavage of the R-acid becomes so large—because of the presence of the ammonium molybdate—that within a few minutes all the R-acid is decomposed to 3-phosphoglyceric acid and phosphoric acid. When the rotation is measured soon after mixing with the ammonium molybdate, one can observe an increase of rotation with time which comes to an end point after a few minutes.

In this way we found a rotation of —2.05° for a mixture of 2 cc of a neutral, $2.45 \cdot 10^{2}$ molar solution of R-acid and 1 cc of a 25% ammonium molybdate solution in a 10 cm tube from which the specific rotation of phosphoglyceric acid can be calculated:

$$[\alpha]_D^{20^\circ} = -2.05 \cdot \frac{1}{2.45 \cdot 10^{-2}} \cdot \frac{3}{2} \cdot \frac{1000}{186} = -675^\circ$$

The specific rotation of 3-phosphoglyceric acid was measured for comparison under the same conditions, that is also in the presence of an equivalent amount of inorganic phosphoric acid. After mixing 1 cc of a neutral, $4.90 \cdot 10^{-2}$ molar solution of 3-phosphoglyceric acid +1 cc neutral phosphate solution of the same molarity +1 cc of a 25% ammonium molybdate solution, we observed a rotation of —2.07° in the 10 cm tube, which means a specific rotation of:

$$[\alpha]_D^{20^\circ} = -2.07 \cdot \frac{1}{4.9 \cdot 10^{-2}} \cdot 3 \cdot \frac{1000}{186} = -682^\circ$$

Ultraviolet spectra

We measured the absorption, in the region of 200–370 mμ, of a neutral, aqueous solution of R-acid[1] and the absorption of the same solution after decomposition of the R-acid. For decomposition of the R-acid, the solution was

[1]For the preparation of the R-acid, the strychnine salt was repeatedly recrystallized.

placed for 15 minutes in a water bath at 60°. The completeness of decomposition was measured by photoelectric titration.

The absorption of a neutral, aqueous solution, containing equivalent amounts of 3-phosphoglyceric acid and inorganic phosphoric acid was measured for comparison. The obtained values from the three solutions are shown in Table II, and the absorption spectra are presented in Fig. 1.

The absorption spectrum of R-acid shows an absorption band at 215 mμ and the ascending part of a second one. The absorption bands are so close together that no absorption minimum is detectable because of overlapping of the two bands.

The absorption band at 215 mμ disappears during the cleavage of R-acid while the ascending part of the second band is shifted by about 5 mμ towards the longer wave lengths. The solution of the cleavage products and the solution of an equivalent mixture of 3-phosphoglyceric acid and phosphoric acid show the same absorption spectrum.

The absorption coefficients of acetic anhydride and acetic acid are registered in Table III; the absorption spectra are presented in Fig. 2. Acetic acid was dissolved in water. Since acetic anhydride is hydrolyzed too rapidly in water, we used a solution of acetic anhydride in hexane for the measurement of the absorption. The transparency of the solution was compared with the transparency of hexane.

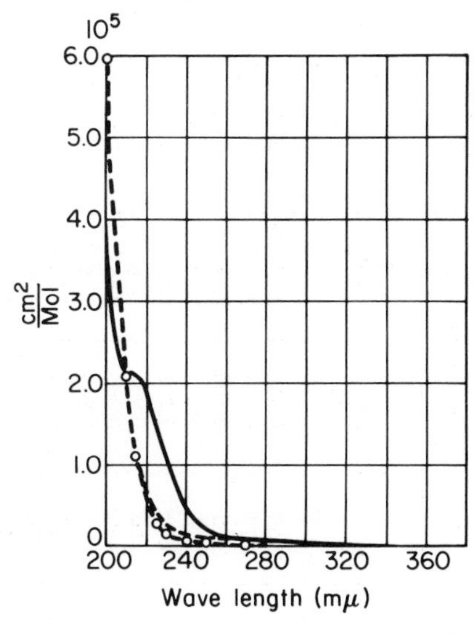

Fig. 1. Absorption spectra of neutral solutions of:
——— R-diphosphoglyceric acid (ordinates: absorption coefficients)
– – – – mixture of the cleavage products (ordinates: sum of absorption coefficients of both cleavage products).
–o– –o– equivalent mixture of 3-phosphoglyceric acid and phosphoric acid (ordinates: sum of absorption coefficients of 3-phosphoglyceric acid and phosphoric acid).

Table II

Wave-length ($m\mu$)	Light pathway (cm)	Solution: $1.38 \cdot 10^{-5}$ moles/cm³ of R-diphosphoglyceric acid (neutralized)						Light pathway (cm)	Solution: $1.4 \cdot 10^{-5}$ moles/cm³ 3-phosphoglyceric acid and $1.4 \cdot 10^{-5}$ moles/cm³ phosphoric acid (neutralized)		
		before cleavage			after cleavage						
		$\frac{i_o}{i}$	$ln \frac{i_o}{i}$	β $\left(\frac{cm^2}{mol}\right)$	$\frac{i_o}{i}$	$ln \frac{i_o}{i}$	β-mixture $\left(\frac{cm^2}{mol}\right)$		$\frac{i_o}{i}$	$ln \frac{i_o}{i}$	β-mixture $\left(\frac{cm^2}{mol}\right)$
200	0.20	2.62	0.962	**$3.49 \cdot 10^5$**	5.05	1.620	**$5.87 \cdot 10^5$**	0.20	5.32	1.669	**$5.96 \cdot 10^5$**
210	0.20	1.79	0.584	**2.12**	1.83	0.602	**2.18**	0.20	1.80	0.584	**2.09**
215	0.617	6.05	1.800	**2.11**	3.03	1.110	**1.30**	1.11	5.56	1.713	**1.10**
220	0.617	5.13	1.635	**1.92**	1.87	0.626	**0.74**	1.11	2.47	0.903	**0.58**
225	0.617	3.60	1.281	**1.50**	1.43	0.361	**0.42**	1.11	1.60	0.468	**0.29**
230	0.617	2.56	0.941	**1.10**	1.26	0.233	**0.27**	1.11	1.28	0.245	**0.16**
240	0.617	1.49	0.396	**0.46**	1.13	0.123	**0.14**	1.11	1.10	0.098	**0.06**
250	0.617	1.18	0.166	**0.19**	1.11	0.100	**0.12**	1.11	1.06	0.057	**0.04**
260	0.617	1.10	0.092	**0.11**	1.10	0.092	**0.11**	—	—	—	—
270	0.617	1.08	0.075	**0.09**	1.08	0.077	**0.09**	1.11	1.02	0.023	**0.01**
280	0.617	1.07	0.064	**0.08**	1.06	0.056	**0.07**				
290	0.617	1.05	0.048	**0.06**	1.05	0.046	**0.05**				
310	0.617	1.03	0.027	**0.03**	1.02	0.023	**0.03**				
330	0.617	1.01	0.010	**0.01**	1.01	0.010	**0.01**				
350	0.617	1.00	0.000	**0.00**	1.00	0.000	**0.00**				
370	0.617	1.00	0.000	**0.00**	1.00	0.000	**0.00**				

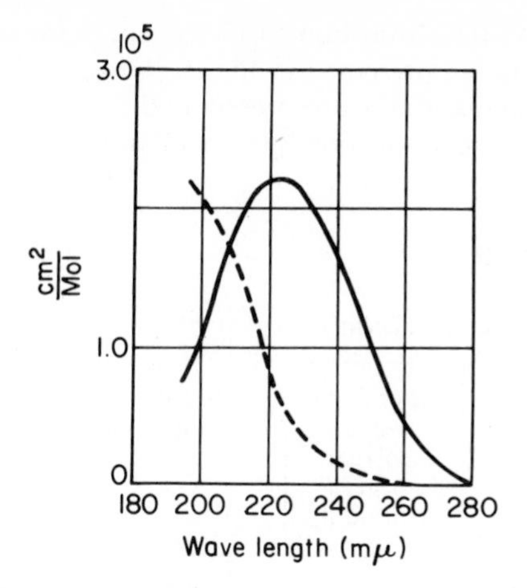

Fig. 2. Absorption spectra of:
——— acetic anhydride, dissolved in hexane (ordinates: absorption coefficients).
– – – – acetic acid, dissolved in water (ordinates: *doubled* values of absorption coefficients).

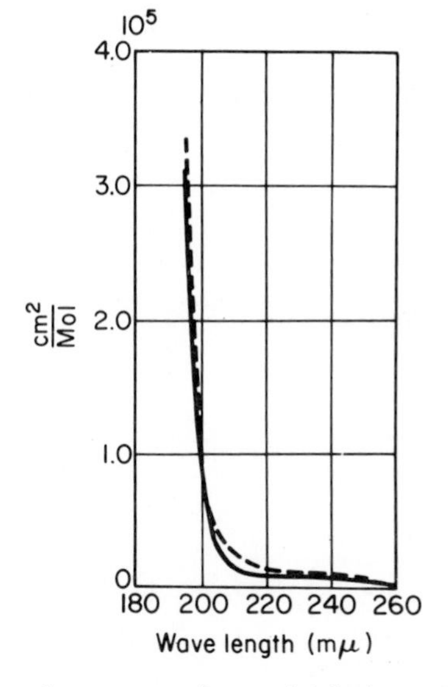

Fig. 3. Absorption spectra of neutral solutions of:
——— pyrophosphate (ordinates: absorption coefficients).
– – – – orthophosphate (ordinates: the *doubled* values of the absorption coefficients).

We measured also the ultraviolet absorption of pyrophosphate and orthophosphate in order to see if the connection of two phosphoric acids, as represented in pyrophosphoric acid, can give rise to a similar absorption band. Table IV shows the absorption coefficients of pyro- and orthophosphate; the absorption spectra are presented in Fig. 3. Pyrophosphate does not possess an absorption band similar to the one present in the spectra of R-acid at 215 mμ. Figure 3 shows that the absorption spectra of pyro- and orthophosphate are almost the same.

Table III

Wave-length (mμ)	Absorption coefficient	
	β of acetic anhydride (solvent: hexane). (cm^2/Mol)	β of acetic acid (solvent: water). (cm^2/Mol)
195	$0.80 \cdot 10^5$	$1.11 \cdot 10^5$
200	1.07	1.04
210	1.86	0.80
220	2.19	0.42
230	2.08	0.16
240	1.67	0.08
250	1.04	0.04
260	0.43	0.00
270	0.12	—
280	0.00	—

Table IV

Wave-length (mμ)	Absorption coefficient	
	β of pyro-phosphate (neutral solution). (cm^2/Mol)	β of ortho-phosphate (neutral solution). (cm^2/Mol)
195	$3.06 \cdot 10^5$	$1.63 \cdot 10^5$
200	1.04	0.74
210	0.15	0.13
220	0.08	0.07
230	0.07	0.06
240	0.09	0.06
250	0.06	0.04
260	0.05	—

A PHOSPHORYLATED OXIDATION PRODUCT OF PYRUVIC ACID*

Fritz Lipmann

Department of Biochemistry,
Cornell University Medical College,
New York, N. Y.

(Received for publication May 16, 1940)

It has been shown that pyruvic acid oxidation is dependent on the presence of inorganic phosphate,[1] but so far it had not been possible to demonstrate the formation of a phosphorylated intermediate. Since pyruvic acid was found to promote adenylic acid phosphorylation,[2] any such intermediate must contain an energy-rich phosphate bond. Phosphopyruvic acid had been excluded by earlier experiments, but recent evidence suggested that acetyl phosphate might be the intermediate.[3] It was found that the phosphate of synthetic acetyl phosphate could be transferred to adenylic acid by an enzyme present in *Bacterium delbrückii*. Tentatively the oxidation was formulated as follows:

$$CH_3 \cdot CO \cdot COOH + H_3PO_4 + O_2 = CH_3 \cdot COOPO_3H_2 + CO_2 + H_2O_2$$

In order to obtain a method for determining acetyl phosphate, a closer study of its stability was undertaken. It developed that at room temperature acetyl phosphate is rapidly broken down above *p*H 8.5 or below *p*H 2, but is fairly stable between *p*H 5 and 7. Hence, the magnesia mixture used for the determination of acid-unstable creatine phosphate is too alkaline for this purpose. It has now been found, however, that inorganic phosphate is completely

*Fritz Lipmann, "A Phosphorylated Oxidation Product of Pyruvic Acid," *J. Biol. Chem.*, **134**, 463 (1940). Reprinted with the permission of the author and of the American Society of Biological Chemists, Inc.

[1] Lipmann, F., *Enzymologia,* **4**, 65 (1937). Banga, I., Ochoa, S., and Peters, R. A., *Biochem. J.*, **33**, 1980 (1939).

[2] Lipmann, F., *Nature,* **143**, 281 (1939). Colowick, S. P., Welch, M. S., and Cori, C. F., *J. Biol. Chem.*, **133**, 641 (1940).

[3] Lipmann, F., *Nature,* **144**, 381 (1939); in Cold Spring Harbor symposia on quantitative biology, Cold Spring Harbor, **7**, 248 (1939).

precipitated at *neutral* reaction with calcium chloride in 30 per cent ethyl alcohol, whereas the calcium salt of acetyl phosphate is soluble under these conditions. Moreover, cold trichloroacetic acid can be used for deproteinization, causing practically no decomposition. Thus the calcium precipitation method can be used to determine acetyl phosphate by difference.

Pyruvate	Fluoride	O_2 consumed	Ca ppt., inorganic P (I)	Direct estimation, inorganic + acid-unstable P (II)	Mg ppt., inorganic + alkali-unstable P (III)	Acetyl phosphate P (II − I)		$\frac{\text{Acetyl P}}{\text{Excess } O_2}$
		c.mm.	*mg.*	*mg.*	*mg.*	*mg.*	*c.mm.*	
−	−	116	1.30	1.32				
+	−	796	0.57	1.30		0.72	486	0.7
−	+	114	1.29	1.31	1.32			
+	+	734	0.52	1.32	1.31	0.77	556	0.9

Experiments were carried out with enzyme solutions obtained from *Bacterium delbrückii*. After pyruvic acid oxidation for about 1 hour, the cooled solution was deproteinized with 2 per cent trichloroacetic acid, the filtrate rapidly neutralized, and the true inorganic phosphate precipitated with $CaCl_2$. As shown by the data presented in the table, large amounts of inorganic phosphate disappear either in the absence or presence of fluoride, and are nearly equivalent to the extra oxygen consumed, in accordance with the equation, due allowance being made for the imperfect stability of the product. The organic phosphate so formed behaves like acetyl phosphate; *i.e.*, it is split readily and completely by the acid required to determine phosphate directly (II in the table) and (in contrast to creatine phosphate) by alkaline magnesia mixture (III). Colorimetric determination of phosphate was carried out according to Lohmann and Jendrassik.[4]

The demonstration here of a phosphate compound of such limited stability, which is formed metabolically, might well necessitate some revision as to what has usually been regarded as inorganic phosphate in cells and tissues.

[4] Lohmann, K., and Jendrassik, L., *Biochem. Z.*, **178**, 419 (1926).

ROLE OF PHOSPHATE IN PYRUVIC ACID DISSIMILATION BY CELL-FREE EXTRACTS OF CLOSTRIDIUM BUTYLICUM*

H. J. Koepsell, Marvin J. Johnson and J. S. Meek

Department of Biochemistry, College of Agriculture,
University of Wisconsin, Madison

(Received for publication April 8, 1944)

The preparation of a vacuum-dried cell-free extract of frozen cells of *Clostridium butylicum* has been described[1]. This extract catalyzes the fermentation of pyruvic acid to acetic acid, carbon dioxide, and molecular hydrogen. The rate of hydrogen evolution during the reaction is proportional to the concentration of added inorganic phosphate. No stable phosphorylation product has been detected.

In this paper the results of a further study of the rôle of phosphate in the reaction are given.

Experimental

Large scale preparation of cell-free extract

To avoid differences in the behavior of small batches of enzyme preparation individually prepared as needed, a large quantity of dry extract was prepared. Cells of *Clostridium butylicum* were grown as follows: 600 liters of medium consisting of 1 per cent commercial glucose, clear aqueous extract of 0.5 per cent malt sprouts and 0.25 per cent fresh pork liver, 0.25 per cent Cuban blackstrap molasses, 0.25 per cent ammonium sulfate, 0.1 per cent phosphoric acid

*H. J. Koepsell, Marvin J. Johnson, and J. S. Meek, "Role of Phosphate in Pyruvic Acid Dissimilation by Cell-Free Extracts of Clostridium Butylicum," *J. Biol. Chem.*, **154**, 535 (1944).

Published also with the approval of the Director of the Wisconsin Agricultural Experiment Station. Supported in part by a grant from the Wisconsin Alumni Research Foundation.

neutralized to pH 6.8 with ammonium hydroxide, and salts,[1] was sterilized, cooled, and inoculated with 30 liters of an 18 hour culture of *Clostridium butylicum* grown in the same medium. After incubation at 37–40° for 18 hours, the cells were harvested by centrifuging. 900 gm. of wet cells were obtained and were immediately frozen.

After 14 days, 100 gm. portions of frozen cells were ground in 200 ml. of ice-cold water and adjusted to pH 6.5 with sodium hydroxide solution. The suspension was allowed to stand in an ice bath for 10 to 15 minutes and then centrifuged several times for 10 to 15 minute periods until a clear amber supernatant liquid could be drained from the cell débris. This liquid was evaporated under a high vacuum while frozen. From 700 gm. of frozen cells thus treated, 28 gm. of dry extract were obtained. Assay indicated recovery of 36 per cent of the total activity of the frozen cells used.

Determination of inorganic phosphate and phosphate esters

For convenience, the nomenclature given in Table I will be used in the discussion of phosphate compounds. These compounds were determined as follows: *True inorganic and labile phosphates* were determined concomitantly by a procedure modified from that of Lipmann[(2)]. This procedure depends on the solubility of the calcium salts of labile phosphates and the insolubility of the calcium salts of true inorganic phosphate in alcoholic calcium chloride solution under controlled conditions, and on the hydrolysis of labile phosphate, if present, to inorganic phosphate during the determination of inorganic phosphate by a modification of the method of Fiske and Subbarow[(3)].

In the modified Fiske and Subbarow method referred to, 1 ml. of acid-molybdate reagent (5 gm. of ammonium molybdate plus 100 ml. of 10 N sulfuric acid diluted to 200 ml. with water) and 1 ml. of reducing reagent (14.25 gm.

Table I

Nomenclature of Phosphate Compounds

Compound	Definition
Inorganic phosphate . .	Phosphate whose calcium salt is insoluble in water-alcohol mixtures under controlled conditions (*e.g.*, NaH_2PO_4)
Labile phosphate.	Phosphate whose calcium salt is soluble in water-alcohol mixtures under controlled conditions, and which is hydrolyzed, if present, during determination of inorganic phosphate by modified Fiske and Subbarow method (*e.g.*, acetyl phosphate)
7 minute phosphate. . . .	Phosphate not hydrolyzed during Fiske and Subbarow determination of inorganic phosphate, but hydrolyzed by exposure to 1.0 N HCl at 100° for 7 minutes (*e.g.*, adenosine triphosphate)
Stable ester phosphate. .	Phosphate not appreciably hydrolyzed by exposure to 1.0 N HCl at 100° for 7 minutes (*e.g.*, hexose diphosphate)

[1] A solution having the following content (in gm. per liter) was prepared: $MgSO_4 \cdot 7H_2O$ 10, $MnSO_4 \cdot H_2O$ 10, $CaCl_2 \cdot 2H_2O$ 10, $FeSO_4$ 0.2, $CuSO_4 \cdot 5H_2O$ 0.1, $Co(NO_3)_2 \cdot 6H_2O$ 0.01, $Na_2B_4O_7 \cdot 10H_2O$ 0.01, $ZnSO_4$ 0.1. 10 ml. of this solution were added per liter of medium prepared.

of sodium bisulfite, 0.25 gm. of 1-amino-2-naphthol-4-sulfonic acid, and 10 ml. of 5 per cent sodium sulfite solution made up in 80 ml. of water, dissolved, and diluted to 250 ml.) are added to 10 ml. of sample solution containing between 0.1 and 1.0 micromole of inorganic phosphate. After 10 minutes at room temperature, the blue color developed is compared in the Evelyn colorimeter at a wave-length of 660 mμ with the color produced by standard samples of inorganic phosphate. Labile phosphate, if present in the sample, is completely hydrolyzed and appears as inorganic phosphate.

The separation of true inorganic and labile phosphates, and their concomitant determination, were effected by the following procedure: The sample solution was adjusted to contain 5 to 10 micromoles of total true inorganic and labile phosphates in 1 ml. of solution at pH 6.5. 1 ml. of this solution was placed in a 15 ml. conical centrifuge tube. To it were added 1 ml. of ammonium acetate buffer reagent (a solution 0.5 M with respect to ammonium hydroxide and 0.05 M with respect to $NaHCO_3$, adjusted to pH 8.5 with acetic acid) and 4 ml. of 1.0 M calcium chloride in 95 per cent ethyl alcohol. The resulting precipitate of calcium carbonate and inorganic phosphate was allowed to flocculate for 5 to 7 minutes. The tube was then centrifuged for about 5 minutes. The supernatant liquid was poured off, and the precipitate was washed on the centrifuge with 2 ml. of alcoholic calcium chloride solution. All of the above operations were carried out in the cold (0–5°) to minimize hydrolysis of acetyl phosphate. The precipitate was dissolved with a drop of concentrated hydrochloric acid and diluted to a suitable volume. In this solution inorganic phosphate was determined by the Fiske and Subbarow method as described; the value thus obtained was assumed to represent the content of true inorganic phosphate in the sample. At the same time, the Fiske and Subbarow method was applied directly to a suitable aliquot of the original sample solution; the value thus obtained was assumed to represent the sum of the true inorganic plus labile phosphate content in the sample solution. Labile phosphate content was then calculated by difference.

Several experiments were conducted to check the accuracy of this separation procedure. The observation by Lipmann[2] that the coprecipitation of calcium carbonate is necessary for complete precipitation and flocculation of calcium phosphate was verified. It was found that inorganic phosphate was precipitated 97 to 100 per cent under the conditions given. In order to determine the action of labile phosphate in this course of procedure, a sample of disilver acetyl phosphate[3] was converted to the sodium salt by shaking in the cold with sodium chloride solution. It is probable that some hydrolysis of the acetyl phosphate occurred during the process. The results of recoveries on this solution are given in Table II. When applied to enzyme reaction mixtures, experience indicated that this separation procedure gave slightly low values for labile phosphate if the ratio of inorganic phosphate to labile phosphate was greater than 4.0, and slightly high values if the ratio was less than 0.3. In the latter case and in other cases when advisable, known quantities of inor-

[2] Private communication.

[3] We are indebted to Fritz Lipmann for generous gifts of disilver acetyl phosphate, and for helpful advice in this work.

Table II

RECOVERY OF ACETYL PHOSPHATE IN PRESENCE OF INORGANIC PHOSPHATE

The sample contained the indicated amounts of added inorganic phosphate and acetyl phosphate preparation in 1 ml. at pH 6.5.

Inorganic phosphate added	Total phosphate added as acetyl phosphate preparation	Total phosphate pptd.	Labile phosphate in acetyl phosphate preparation
micromoles	*micromoles*	*micromoles*	*per cent*
8.0	1.76	8.40	77.3
6.0	3.52	6.88	75.0
5.0	4.40	6.20	72.7
4.0	5.28	5.40	73.5
2.0	7.03	4.07	70.5
0.0	8.80	2.62	70.2

ganic phosphate were added to the sample to bring the ratio to 1.0 to 1.3 before the assay was repeated.

The 7 minute and stable ester phosphates were determined as follows: By calculation from the amount of inorganic phosphate originally added to the reaction mixture, and the amount of true inorganic and labile phosphate found in the mixture after reaction, the approximate amount of 7 minute plus stable ester phosphate in the sample was determined. The sample was diluted to contain approximately 10 micromoles of total phosphates in 1 ml. of solution at pH 6.5. To 1 ml. of this solution, 1 ml. of 2.0 N hydrochloric acid was added, and the solution was placed in a steam bath for exactly 7 minutes and cooled. Inorganic phosphate was then determined in this solution by the modified Fiske and Subbarow method described previously. 7 minute phosphate was calculated by subtracting the sum of labile plus inorganic phosphate, previously determined, from the inorganic phosphate found in this determination. Stable phosphate was calculated by subtracting the sum of inorganic, labile, and 7 minute phosphates from the amount of inorganic phosphate originally added to the reaction mixture.

DETERMINATION OF SILVER-PRECIPITABLE VOLATILE ACIDS

This procedure was applied to enzyme reaction mixtures after protein precipitation with trichloroacetic acid and neutralization or to purified labile phosphate solutions. The sample was adjusted to contain 16 to 20 micromoles of labile phosphate in 2.0 ml. of solution at pH 6.5. To 2.0 ml. of this solution, 13 ml. of 95 per cent ethyl alcohol and 200 micromoles of silver nitrate in 0.2 ml. of solution were added. The resulting precipitate was centrifuged out, washed twice with 4 ml. portions of 95 per cent ethyl alcohol, and taken up in 5.0 ml. of 0.1 N sulfuric acid. This suspension was steamed for 15 minutes in a stoppered tube, cooled, diluted to 30 ml., and subjected to Duclaux distillation.

For this distillation the entire hydrolysate was placed in an all-glass distilling apparatus. The first 5.00 ml. of distillate, containing carbon dioxide, was discarded. The two succeeding 10.00 ml. portions were heated just to a boil, cooled rapidly, and titrated to a phenol red end-point with standard 0.01 N barium hydroxide solution. The content of acetic and butyric acids in the sample was then calculated by the usual method[4]. The rate of distillation was adjusted so that, after rejection of the first 5 ml. of distillate, the following 10 ml. portion was delivered in 7.5 minutes. Before analyses of samples, the apparatus was calibrated with standard solutions of acetic and butyric acids under the same conditions, and consistently gave recoveries of 96 to 100 per cent on mixtures of these two acids.

At all times solutions believed to contain acetyl phosphate were kept cold, in an ice bath or preferably frozen, when possible, to minimize hydrolysis.

Phosphate balances in presence and absence of glucose

In earlier work, attempts to detect the formation of a stable phosphorylation product in pyruvic acid fermentations by cell-free extracts of *Clostridium butylicum* had failed. Apparently any labile phosphates produced had been hydrolyzed during the determination of inorganic phosphate, and had escaped detection. However, application to fermented liquors of the above method for concomitant determination of labile and true inorganic phosphate indicated that labile phosphate was indeed produced, at the expense of inorganic phosphate. Lipmann reported that enzyme preparations used in this work would also catalyze the phosphorylation of added glucose during pyruvic acid fermentation, with the accumulation of stable ester phosphate. The magnitude of these reactions was studied in the following phosphate balances of pyruvic acid fermentation in the presence and absence of glucose.

The reaction was carried out in the usual manner[1]. After reaction, the contents of the flasks were washed into 15 ml. conical centrifuge tubes and diluted to 5 ml. with water. Proteins were precipitated by the addition of 1 ml. of 1.0 M trichloroacetic acid to each tube. After immediate centrifugation, the supernatant liquids were rapidly adjusted to pH 6.5 with 1 N sodium hydroxide solution and diluted to 10 ml. On these solutions the indicated determinations were made. The results are given in Table III.

In the absence of glucose, inorganic phosphate was taken up and appeared as labile phosphate esters. In the presence of glucose, the reaction rate increased 133 per cent; inorganic phosphate was again taken up, but appeared as stable ester phosphate plus 7 minute phosphate instead of as labile phosphates. Because the esters in this case were probably a mixture of hexose phosphate esters whose character is not of great importance to the present study, they were not characterized further.

Indications of character of labile phosphate

As mentioned previously[1], the fermentation of pyruvic acid by cell-free extracts of *Clostridium butylicum* appears to be somewhat similar to the oxidation of pyruvic acid catalyzed by enzyme systems prepared by Lipmann[5] from

Table III

Phosphate Balance of Pyruvic Acid Fermentation in Presence and Absence of Glucose

The Warburg flasks contained 97 micromoles of inorganic phosphate and 30 mg. of dry enzyme powder; pH 6.5. Total volume, 2.8 ml. KOH in the center cup. Reaction time, 60 minutes. All values are expressed in micromoles.

Flask No.	Pyruvate added	Glucose added	Hydrogen evolved	Inorganic phosphate	Labile phosphate	7 minute phosphate	Stable phosphate
1	0	0	−0.4	94	2	3	−2
2	0	0	−0.2	94	3	4	−4
3	50	0	14.4	81	14	4	−2
4	50	0	14.2	81	13	7	−4
5	0	150	−0.3	92	2	2	1
6	0	150	−0.3	91	3	1	4
7	50	150	34.9	53	3	11	30
8	50	150	31.4	54	3	11	29

cells of *Lactobacillus delbrueckii*. The discovery that labile phosphate is formed in the former reaction leads naturally to the supposition that acetyl phosphate, isolated by Lipmann[6] in the oxidation reaction, is also an intermediate labile phosphate ester in the fermentation reaction.

The isolation of labile phosphate from fermentation mixtures for rigid characterization was attempted and was unsuccessful. Pyruvic acid dissimilation by cell-free extracts under the conditions described previously[1] is slow, and the amounts of labile phosphate obtained are too small and solutions are too dilute for such isolation; furthermore, labile phosphate is relatively unstable at incubation temperature. In order to force reasonably rapid reaction, it was necessary to use great excesses of inorganic phosphate in the reaction mixtures. The resulting problem of separating the excess inorganic phosphate from labile phosphate proved difficult. Precipitation of inorganic phosphate with calcium nitrate in the concentrated reaction mixtures which were obtained resulted in coprecipitation and loss of much labile phosphate, and purified solutions of labile phosphate hydrolyzed with such ease that satisfactory separation from inorganic phosphate could not be effected.

However, indications of the character of the labile phosphate were secured by precipitating purified labile phosphate solutions with silver, regenerating the sodium salts, and hydrolyzing these with acid. The volatile acid content of the hydrolysates was then correlated with the labile phosphate content of the sodium salt solution. It was shown that added volatile acids were not precipitated with silver under these conditions, and that no volatile acids could be detected in analogous silver precipitates from blank reaction mixtures in which no fermentation had taken place.

To obtain labile phosphate for this purpose, pyruvic acid fermentation was carried out in the usual manner, but in a Warburg flask of about 115 ml. capacity. The manometer was filled with mercury. The reaction mixture consisted of 4000 micromoles of inorganic phosphate, 800 micromoles of sodium pyruvate, and 500 mg. of dry enzyme powder, at pH 6.5 in a volume of 19 ml.

KOH was placed in the center cup. After 140 minutes incubation the contents of the flask were removed and diluted with washing to 23 ml. To precipitate proteins, 3 ml. of 1.0 M trichloroacetic acid were added, and the suspension was immediately centrifuged. The supernatant liquid was adjusted to pH 6.7 with 1 M sodium hydroxide solution. The total volume of 32 ml. contained 2820 micromoles of inorganic phosphate and 280 micromoles of labile phosphate.

To remove inorganic phosphate, 3000 micromoles of calcium nitrate in 3 ml. of solution were added, and the pH was readjusted to 6.5 with 1 N sodium hydroxide solution. The resulting calcium phosphate precipitate was centrifuged out. Much labile phosphate had coprecipitated. The volume was adjusted to 19 ml. This solution contained 123 micromoles of labile phosphate and 63 micromoles of inorganic phosphate. To 15 ml. of this solution 60 ml. of 95 per cent ethyl alcohol and 550 micromoles of silver nitrate in 1.1 ml. of solution were added. The resulting silver precipitate was washed once with 10 ml. and once with 6 ml. of alcohol. The precipitate was suspended in 10 ml. of 0.044 N sodium chloride solution (containing one-thirtieth of the chloride ion as hydrochloric acid in order to attain pH 6.5 to 7.0 after the cation exchange) and shaken well. The suspension was centrifuged, and the supernatant liquid containing sodium salts of labile phosphate at pH 7.0 was diluted to 12 ml. Assay indicated that this solution contained 59 micromoles of labile phosphate and 22 micromoles of inorganic phosphate. To 6.7 ml. of this solution (33 micromoles of labile phosphate) 205 micromoles of silver ion in 5.7 ml. of silver sulfate solution and 5.0 ml. of 0.1 N sulfuric acid solution were added. The resulting suspension was steamed in a stoppered tube for 10 minutes to hydrolyze the labile phosphates and cooled. The precipitate of silver chloride was centrifuged out; the supernatant liquid gave no further precipitate when tested with a drop of silver sulfate solution. The supernatant liquid was diluted to 40 ml. and subjected to Duclaux distillation.

The sample was adjusted to contain 30 to 40 micromoles of volatile acids, 36 micromoles of silver ion as silver sulfate, and 5.0 ml. of 0.1 N sulfuric acid, in 40 ml. of solution. After 5 ml. of distillate were discarded, three (instead of two) succeeding 10 ml. portions were titrated as before. The apparatus had been previously calibrated with standard solutions of acetic, butyric, valeric, and isovaleric acids under identical conditions.

The titration values obtained in the distillation of the labile phosphate hydrolysate did not conform to the volatility of acetic acid alone; approximately one-third of the volatile acid present was more volatile than acetic acid. Since butyric acid is a normal product of glucose dissimilation by *Clostridium butylicum,* it was believed that butyric acid might also be present. Therefore, the distillation results were recalculated on the assumption that the volatile acid distilled was a binary mixture of acetic and butyric acids. Results of separate calculations involving all possible combinations of two titration values[4] agreed well. Similar calculations based on the assumption that a binary mixture of acetic and of several other more volatile acids was present disagreed widely. The results of these calculations are given in Table IV.

It was, therefore, assumed that the mixture of volatile acids in the labile phosphate hydrolysate consisted chiefly of acetic and butyric acids. Thus the

Table IV

IDENTIFICATION OF SILVER-PRECIPITABLE VOLATILE ACIDS BY DUCLAUX DISTILLATION

Results of Duclaux distillation calculated from titratable acidities of Fractions I, II, and III for the following combinations of volatile aliphatic acids.

Acids assumed present		Micromoles of acid in Ag ppt. calculated from		
		Fractions I and II	Fractions I and III	Fractions II and III
Acetic and butyric	Acetic acid	23.6	24.7	24.3
	Butyric 〃	12.4	12.1	12.0
	Total	36.0	36.8	36.3
Acetic and valeric	Acetic acid	31.7	28.6	27.9
	Valeric 〃	7.7	9.2	12.0
	Total	39.4	37.8	39.9
Acetic and isovaleric	Acetic acid	34.1	29.3	28.9
	Isovaleric acid	6.4	8.5	13.5
	Total	40.5	37.8	42.4

aliquot containing 33 micromoles of labile phosphate also contained, after hydrolysis, 24.2 micromoles of acetic acid and 12.2 micromoles of butyric acid.

To show that the presence of silver-precipitable acetic and butyric acids in the purified labile phosphate preparations was associated with the fermentation of pyruvic acid and that such acids were not present before fermentation, a control experiment was performed. A blank enzyme reaction mixture was prepared containing 2000 micromoles of inorganic phosphate, 400 micromoles of acetic acid, and 250 mg. of dry enzyme powder, at pH 6.5 in 9.5 ml. of solution. This reaction mixture was intended to simulate the fermented enzyme reaction mixture from which the labile phosphate solutions in previous experiments had been prepared. Since no fermentation had occurred in this control mixture, there should have been no accumulation of labile phosphate, and no acetic or butyric acids should be found in analogous silver precipitates. On this mixture an exactly analogous procedure for preparing labile phosphate solution was carried out. Both the labile phosphate and silver-precipitable volatile acid contents of this solution were determined. None of either could be detected.

Silver acetate and butyrate are sparingly soluble in water; thus it might be possible that these salts were merely precipitated as such from the purified labile phosphate preparations, since no separation of labile phosphate from acetic and butyric acids originally occurring in the fermented reaction mixture had been made. That any acetic acid present was not precipitated under the conditions existing during the silver precipitation has been shown by the control experiment just described. That any butyric acid present similarly was not precipitated was shown by a second experiment. A labile phosphate solution was prepared as has been described. A 2 ml. aliquot of this solution had been found to contain 18.5 micromoles of labile phosphate, 11.23 micromoles of silver-precipitable acetic acid, and 4.90 micromoles of silver-precipitable butyric acid. To a second 2 ml. aliquot, 10 micromoles of sodium butyrate in 0.2 ml. of solution at pH 6.5, and to a third 2 ml. aliquot 0.2 ml. of water were added. The solutions were assayed for silver-precipitable volatile acids. It was found that in the second aliquot 8.90 micromoles of acetic acid and 5.03 micromoles of butyric acid, and in the third aliquot 9.36 micromoles of acetic and 4.98 micromoles of butyric acids, were precipitated with silver. The difference between these values is within experimental error, and it was therefore assumed that none of the added butyric acid had been precipitated.

These experiments indicate that the acetic and butyric acids precipitated with silver from purified labile phosphate solutions were not present as such, that their presence is related to the production of labile phosphate in the pyruvic acid fermentation reaction, and that they may be present as acetyl and butyryl phosphates.

Source of silver-precipitable butyric acid

After the indication that acetyl and butyryl phosphates may be formed in the fermentation reaction was established, the chemical mechanism by which silver-precipitable butyric acid arises was studied. The enzyme preparation had a pronounced butyric acid odor; a transphosphorylation reaction between butyric acid contained in the enzyme preparation and acetyl phosphate formed in the reaction, to yield butyryl phosphate and acetic acid, was postulated. This reaction would yield acetic acid and butyryl phosphate.

This possibility was investigated in the experiment outlined in Table V. In the presence of enzyme extract, acetyl phosphate was incubated both with (Flask 1) and without (Flask 2) added butyric acid. In the control (Flask 3), acetyl phosphate was not added to the reaction mixture until after the enzyme preparation had been inactivated; this procedure was intended to prevent the endogenous transphosphorylation evident in Flask 2. The initial reaction mixtures given in Part I of the table were incubated in Warburg flasks as described. After incubation the contents of each flask were removed and diluted to 5.0 ml. by washing. Proteins were precipitated and enzymes were inactivated by adding 1 ml. of 1.0 M trichloroacetic acid. At this time, before the resulting precipitate was centrifuged out, the additions indicated in Part II were made. These additions were intended to produce the same concentrations of butyric acid and labile phosphate in all of the reaction mixtures, so that subsequent silver precipitations would occur under comparable conditions in active and control samples; however, because the acetyl phosphate in the enzyme-containing reaction mixtures of Flasks 1 and 2 was subjected to incubation temperature and the action of phosphatases during incubation, while the acetyl phosphate now added to the reaction mixture of Flask 3 had been kept cold and separate, the labile phosphate level in Flask 3 was higher during the silver precipitation than the level in Flasks 1 and 2. The protein precipitates were immediately removed by centrifuging, and the supernatant liquids were adjusted to pH 6.5 with 1 N sodium hydroxide solution. The solutions were diluted to 8.0 ml. and assayed for labile phosphate and silver-precipitable volatile acids.

The results of this experiment are given in Part III of Table V. No butyric

Table V

Source of Silver-Precipitable Butyric Acid

All solutions adjusted to pH 6.5 before addition. Gas phase, hydrogen. Temperature, 37°. Reaction begun by tipping enzyme suspension from side arm into main compartment. Incubation time, 30 minutes. No KOH in the center cup.

	Flask 1 Active		Flask 2 Endogenous		Flask 3 Control	
Part I. Initial additions						
Volume of reaction mixture	3.3 ml.					
Sodium acetyl phosphate	150 μM		150 μM		0 μM	
″ butyrate	150 ″		0 ″		150 ″	
Enzyme preparation	100 mg.		100 mg.		100 ″	
Part II. Additions after protein pptn. with trichloroacetic acid						
Sodium acetyl phosphate	0 μM		0 μM		150 μM	
″ butyrate	0 ″		150 ″		0 ″	
Water	Sufficient to make up to 7.0 ml.					
Part III. Results						
	micromoles		*micromoles*		*micromoles*	
Gas evolved (as CO_2)	+2.07		+0.44		+5.31	
Phosphate distribution*						
Inorganic phosphate	59.9	60.0	57.5	61.8	32.2	32.3
Labile phosphate	82.0	81.0	78.5	74.5	104.2	104.1
Volatile acids in silver ppt.*						
Acetic acid	34.6	37.8	45.4	45.7	89.0	91.3
Butyric ″	25.1	23.8	13.3	13.1	−1.2	−1.4
Total volatile acids	59.7	61.6	58.7	58.8	87.8	89.9
Micromoles volatile acid per micromoles labile P	0.733	0.756	0.765	0.767	0.848	0.862

*Duplicate determinations are given.

acid was found in the silver precipitate from the reaction mixture in which the enzyme preparation was inactivated before the addition of acetyl phosphate, and the addition of butyric acid to the incubated reaction mixture resulted in a large increase in the amount of silver-precipitable butyric acid. These results indicate that an exchange of the phosphate group of acetyl phosphate and butyric acid had taken place.

It might be assumed that the butyryl phosphate was produced by condensation and reduction of 2 acetyl phosphate molecules. In the above experiment in which acetyl phosphate was incubated with enzyme preparation under molecular hydrogen, this would have been indicated by gas uptake during incubation. Instead there was slight gas evolution. Therefore, unless the necessary reductive hydrogen was provided by a suitable hydrogen donor in the enzyme preparation, such condensation and reduction did not occur.

Discussion

Intermediate phosphorylation apparently occurs in the fermentation of pyruvic acid by cell-free preparations of *Clostridium butylicum*. Inorganic phosphate is taken up in the reaction, and labile phosphate appears. In the presence of glucose, stable ester phosphate is produced in place of labile phosphate. The demonstration of this series of intermediate phosphorylations suggests that fermentative energy utilization in this butyric acid anaerobe is in accord with the current concept[7,8] that the energy of carbohydrate dissimilation by living cells is utilized by means of the generation and hydrolysis of high energy phosphate bonds.

The data of Table III and of the previous paper[1] are consistent with the assumption that the reaction taking place is

$$\text{Pyruvate}^- + H_2PO_4^- \rightleftarrows CO_2 + H_2 + \text{acetyl phosphate}^=$$

In the presence of compounds (such as butyric acid or glucose) capable of functioning as phosphate acceptors, a transphosphorylation occurs. From the estimations made by Kalckar[7] it is apparent that the decomposition of pyruvate into acetate, carbon dioxide, and hydrogen involves, at pH 7 and when reactants and products are present at equal concentrations, a free energy change of approximately 10,000 calories. Although the free energy of formation of acetyl phosphate is not accurately known, generally accepted values for the energy of a carboxyl phosphate bond are also in the neighborhood of 10,000 calories. The experimental data also indicate that we are dealing with an over-all reaction involving very little energy change. A relatively high concentration of inorganic phosphate must be present in order to obtain a satisfactory reaction rate. The reaction rate is also much increased when glucose is the ultimate phosphate acceptor. (The phosphate bond energy in hexose phosphates is much less than that in carboxyl phosphates.)

Both of these observations suggest that the equilibrium position of the reaction is such that maximum reaction velocity in a positive direction is obtained only when reactant concentration is much higher than product concentration.

Rigid characterization of the labile phosphate produced in the pyruvic acid fermentation reaction has not been achieved because of inability to isolate the labile phosphorus compound. However, indications of the presence of acetyl and butyryl phosphates have been obtained. These are as follows: Purified labile phosphate solutions from pyruvic acid fermentation reaction mixtures contain amounts of silver-precipitable acetic and butyric acids which correlate reasonably with the labile phosphate content of the solutions. Similar solutions obtained from blank fermentation reaction mixtures in which no fermentation has taken place do not contain silver-precipitable volatile acids. Added acetic and butyric acids are not precipitated by the silver treatment. Finally, incubation of acetyl phosphate with butyric acid in the presence of enzyme extract results in the formation of silver-precipitable butyric acid in the reaction mixture.

The apparent ability of the enzyme preparation to catalyze transphosphorylation between acetyl phosphate and butyric acid to produce butyryl phosphate suggests that butyryl phosphate may play an intermediary rôle in the production of butyric acid and butyl alcohol by *Clostridium butylicum*.

The production of labile phosphate, apparently acetyl phosphate, is a point of similarity between pyruvic acid fermentation by preparation of *Clostridium butylicum* to acetic acid, carbon dioxide, and hydrogen, and by preparations of *Escherichia coli* to acetic and formic acids[9] and pyruvic acid oxidation by preparations of *Lactobacillus delbrueckii*[5].

Summary

1. The preparation from *Clostridium butylicum* of a large supply of cellfree dried water extract of frozen cells which catalyzes pyruvic acid fermentation is described. 36 per cent of the activity of the frozen cells was recovered in the dried extract.

2. Phosphate balances of the fermentation of pyruvic acid by enzyme extract in the presence and absence of glucose are given. In the absence of glucose, inorganic phosphate is taken up and appears as labile phosphate, but no stable phosyphorylation product accumulates. In the presence of glucose, no labile phosphate accumulates, but inorganic phosphate is taken up and appears as stable ester phosphate.

3. Attempts to isolate labile phosphate from the fermentation mixture were unsuccessful. Silver-precipitable acetic and butyric acids were present in purified labile phosphate preparations. Apparently the labile phosphate is a mixture of acetyl and butyryl phosphates.

4. Silver-precipitable butyric acid, apparently butyryl phosphate, arises by incubation of acetyl phosphate with butyric acid in the presence of enzyme extract. An exchange of the phosphate group of acetyl phosphate apparently occurs between acetyl phosphate and butyric acid to yield acetic acid and butyryl phosphate.

Bibliography

1. Koepsell, H. J., and Johnson, M. J., *J. Biol. Chem.*, **145**, 379 (1942).
2. Lipmann, F., *J. Biol. Chem.*, **134**, 463 (1940).
3. Fiske, C. H., and Subbarow, Y., *J. Biol. Chem.*, **66**, 375 (1925).
4. Virtanen, A. I., and Pulkki, L., *J. Am. Chem. Soc.*, **50**, 3138 (1928).
5. Lipmann, F., in Cold Spring Harbor symposia on quantitative biology, Cold Spring Harbor, **7**, 248 (1939).
6. Lipmann, F., *Federation Proc.*, **1**, pt. 2, 122 (1942).
7. Kalckar, H. M., *Chem. Rev.*, **28**, 71 (1941).
8. Lipmann, F., in Nord, F. F., and Werkman, C. H., Advances in enzymology and related subjects, New York, **1**, 99 (1941).
9. Utter, M. F., and Werkman, C. H., *Arch. Biochem.*, **2**, 491 (1943).

ACETYL COENZYME A AND THE "FATTY ACID CYCLE"*

Feodor Lynen

Institut für Biochemie, University of Munich, Germany

(Lecture delivered May 21, 1953)

The organic matter of the living cell consists mainly of compounds containing the elements carbon, hydrogen, oxygen and nitrogen. In addition, two other important elements are present in somewhat smaller amounts, namely, phosphorus and sulfur. The knowledge of the metabolic functions of organic phosphate compounds and their participation in the energy transformations occuring in the cell is today one of the basic principles of cell physiology[1,2,3].

The involvement of sulfur in the transmethylation reaction has long been known.[4] In recent years it has been found that the same element has great metabolic significance in yet another respect. It is the purpose of this lecture to discuss this latter aspect of the biology of sulfur.

In this lecture I have the privilege of acting as spokesman for a group of students and coworkers at Munich who have shared in this work. Especially I should like to mention Ernestine Reichert, Gabriele Vogelmann, Luise Wessely, Luistraud Rueff, Mathias Bühler, Otto Wieland, Werner Seubert, Helmut Hilz and Karl Decker.

*Excerpt from F. Lynen, "Acetyl Coenzyme A and the 'Fatty Acid Cycle'," *The Harvey Lectures,* **48**, 210–222, 241–242, 244 (1952–53). Reprinted with the permission of the author, of the Harvey Society, and of Academic Press Inc.

The following abbreviations are used: Coenzyme A (reduced), CoA, CoA—SH or $\bar{\text{CoA}}$—SH; *S*-acyl coenzyme A derivatives, *S*-acyl CoA, acyl-S—CoA, acyl-S—$\bar{\text{CoA}}$, or acyl CoA; adenosine triphosphate, ATP; adenosine-5′-phosphate, AMP; pyrophosphate, PP; oxidized and reduced diphosphopyridine nucleotide, DPN^+ or DPN and DPNH; reduced triphosphopyridine nucleotide, TPNH; oxidized and reduced flavin adenine dinucleotide, FAD and $FADH_2$.

"Active Acetate"

The starting point of our experiments was an observation made during the study of the oxidation of acetate by living yeast, a problem which was initiated twenty years ago by the work of Heinrich Wieland and Robert Sonderhoff in Munich[5]. With the use of normal and isotopic acetate, the Munich laboratory had, quite early, accumulated rather strong evidence that acetate oxidation occurs through the citric acid cycle[6]. It was already recognized then that acetate first has to be "activated" in order to enter this cycle[7]. The reason for postulating the existence of "active acetate" was the observation that starved yeast cells obviously have difficulty in using acetate as a substrate for oxidation (Fig. 1). Only after a more or less prolonged lag phase, the oxidation of acetate begins. This can be shortened by addition of small amounts of an easily oxidizable substrate such as alcohol. This indicated to us that the oxidation of acetate required an initial input of energy, this energy being furnished by the simultaneous oxidation of another substrate.

In retrospect it appears that we had drifted into a field where Fritz Lipmann had been active since 1937[8]. His approach to this problem was quite different from our own. He had found that during the oxidation of pyruvate in certain microbial systems a compound was elaborated which had the constitution of acetyl phosphate, an anhydride of acetic and phosphoric acid[9]. Lipmann had the impression that this compound might represent the "active acetate"[10].

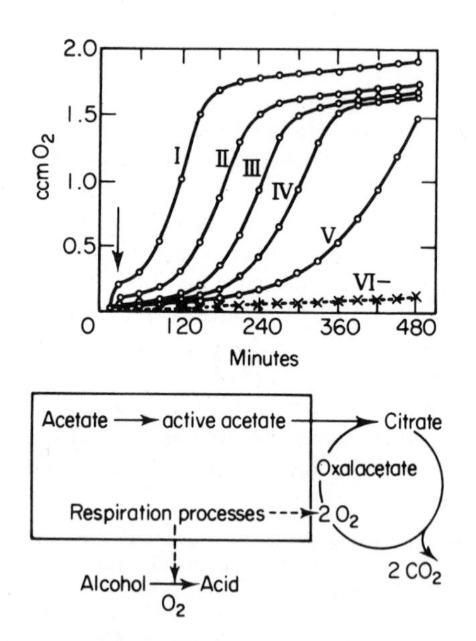

Fig. 1. Effect of small amounts of alcohol on the lag phase during oxidation of acetate by starved yeast[7]. Each vessel contained 100 mg. starved yeast in 3.0 ml. $M/30$ KH_2PO_4; I–V: 50 μM acetate and 368 μg. (I), 184 μg. (II), 92 μg (III) or 37 μg. (IV) ethanol, added at arrow.

But trial experiments, at least in animal tissues and in yeast, gave uniformly negative results. We tried to combine chemically synthesized acetyl phosphate[11] with oxalacetate using yeast preparations. No trace of citrate, however, was formed[12]. Equally negative results were experienced by Lipmann[13]. He employed acetyl phosphate as an agent for the acetylation of aromatic amines. For this process pigeon liver preparations had been found to be especially useful.

The interest of biochemists in "active acetate" increased enormously with the realization that the 2-carbon fragment was involved not only in citric acid synthesis or acetylation of sulfonamide and of choline but also in the synthesis of acetoacetate, fatty acids and steroids. This insight was due in large part to the work of Schoenheimer and his group[14]. The story of "active acetate" became a dramatic chapter of biochemistry. An essential part was presented in 1948 to the Harvey Society by Fritz Lipmann[13]. The first development in this puzzling field came from an experiment of Nachmansohn and Machado[15]. In experiments on nerve activity they were studying the acetylation of choline. They added acetate and ATP together to brain extracts and found that with this combination vigorous acetylation occurred. When Lipmann[16] tried the effect of acetate and ATP in the acetylation system of aromatic amines, an excellent acetylation was obtained.

During these studies on sulfonamide acetylation in pigeon liver, Lipmann[16] observed that a coenzyme was involved. Concurrently Nachmansohn and Berman[17], Feldberg and Mann[18], as well as Lipton and Barron[19], observed the need for an activator in choline acetylation. On purification, which was done in Lipmann's laboratory, this new coenzyme A turned out to be a pantothenic acid derivative[20]. The discovery of this coenzyme opened the way for the solution of the problem of acetate activation. It soon appeared that coenzyme A functions not only in the acetylation process but also much more generally in the synthesis of acetoacetate[21] and citrate[22,23].

To explain the action of this coenzyme, Fritz Lipmann, whose brilliant experiments contributed most to the clarification of this field, suggested that coenzyme A may act as an acetyl carrier[13]. This proposition was solidified through the work of Stadtman[24] and of Ochoa and his group[25]. But the final proof was still lacking in 1950. At that time work on the activation problem was resumed in my laboratory in Munich after a long intermission due to unfavorable conditions.

Isolation of "Active Acetate" from Yeast and Its Identification as S-Acetyl Coenzyme A

Experiments on baker's yeast led us also to the conclusion that the "active acetate" had to be an acetylated coenzyme A. We had observed that during the lag phase (see Fig. 1) the CoA content increases and eventually reaches a plateau[26]. If therefore the idea was right that "active acetate" is a derivative of coenzyme A, yeast cells oxidizing acetate should be a rich source of it. To prove this, we embarked upon the isolation of the compound from such yeast cells[26,27].

Our test for the detection of "active acetate" was based on the enzymatic reaction with sulfanilamide. As shown in Figure 2, in the presence of the enzyme obtained from pigeon liver, the acetylation of sulfanilamide increases with increasing addition of "active acetate." Through use of this procedure for the quantitative assay of "active acetate" the purification of this substance from yeast Kochsaft became possible.

During this work on the isolation, we were led on by a suspicion based on two different observations. We were impressed by the fact that all enzymatic reactions in which coenzyme A was involved required the addition of large amounts of cysteine or glutathione(21, 22, 28, 29), presumably as reducing agents. On the other hand, in the paper by Lipmann(20) on the chemistry of crude CoA preparations, it was noted that such preparations contain considerable amounts of disulfide. It was not quite certain that the sulfur belonged to coenzyme A.

These two observations could be unified by postulating that the sulfur was a part of the coenzyme and that the acetyl carrier function required that the sulfur be present in the reduced form. It was therefore by no means surprising to us that our purified substance turned out to be a thio ester or, as I sometimes prefer to call it, an "acyl mercaptan"(27).

In testing several reagents we found that mercury salts and alkali destroyed the activity of the substance. It is known that both agents hydrolyze thio esters. But more conclusive was an observation which is concerned with the kinetics of the color development in the nitroprusside reaction.

Figure 3 shows that the color development reaches a maximum in 2 to 3 minutes and then rapidly fades. This delayed nitroprusside reaction is typical of acyl mercaptans(30). Calculations of the amount of acyl mercaptan, taken

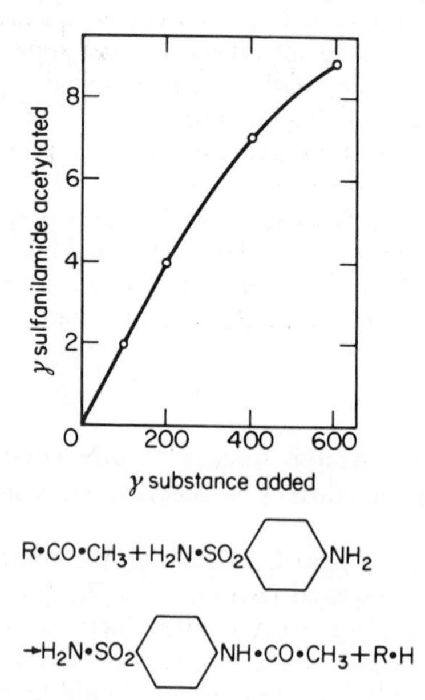

Fig. 2. Enzymatic assay for "active acetate"(26). The substance added was a yeast preparation, containing 9.8% of the pure substance.

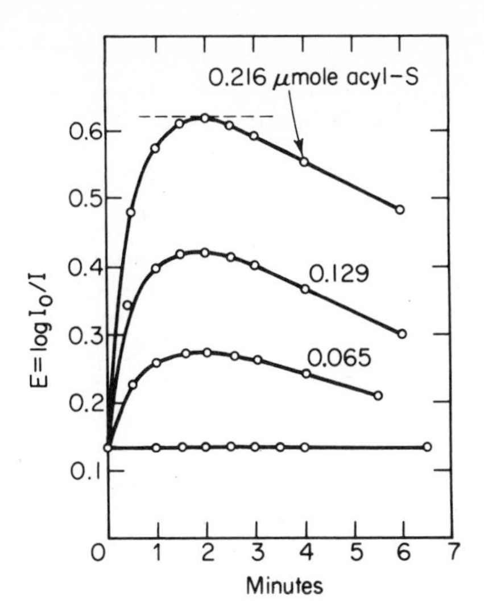

Fig. 3. Kinetics of the color development in the nitroprusside reaction(30). Each sample had a volume of 4.0 ml., containing 0.5% sodium nitroprusside and 27% ammonium sulfate; at zero time 0.3 ml. ammonium hydroxide (density 0.91) was added. $\lambda = 546$ mμ; light path = 1 cm.

from the point of maximum color development, are in excellent agreement with values obtained by sulfanilamide acetylation. To document this a comparison of different preparations of "active acetate," assayed by the two methods, is presented in Figure 4. Furthermore, during incubation of our active material with sulfanilamide and liver enzyme, the acyl mercaptan content decreases in proportion to the amount of sulfanilamide acetylated (Table 1).

At the time of these experiments the idea that the sulfhydryl compound was identical with coenzyme A derived new support from the work of Snell and his group[32, 33] on a growth factor for *Lactobacillus bulgaricus,* originally named "LBF" and now known as "pantethine." This factor they found to be

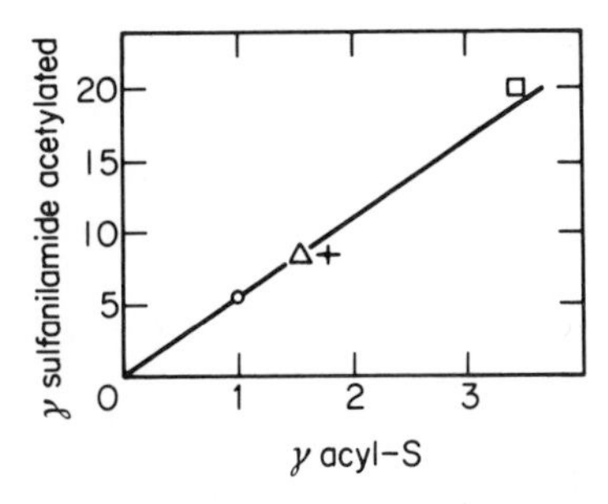

Fig. 4. Identity of acetyl donor with acyl mercaptan(26). Preparations of varying purity were obtained from yeast Kochsaft. ○ = decomposed barium salt; + = charcoal eluate; △, □ = acetone precipitates.

Table 1

ACETYLATION OF SULFANILAMIDE AND CONSUMPTION OF ACYL MERCAPTAN[31]

Each sample contained in a total volume of 0.75 ml., 0.15 ml, $M/5$ pyrophosphate buffer, pH 7.4, 0.1 ml. liver enzyme[26] (= 7.2 mg. protein) and 5 mg. "active acetate" preparation, containing 0.57 μM acetyl CoA; incubated 10 minutes at 35°. Sulfanilamide was determined by the procedure of Bratton and Marshall, acyl-S by the nitroprusside reaction[30], after previous oxidation of the free sulfhydryl groups by iodine.

	Sample 1	Sample 2	
μM sulfanilamide at start	0	1.0	
at end	0	0.62	
μM sulfanilamide acetylated			0.38
μM acyl-S at start	0.57	0.57	
at end	0.46	0.12	
Δ acyl-S at end between samples 1 and 2			0.34

a component of coenzyme A and, as proved by synthesis, a dipeptide-like compound with a peptidic link between the carboxyl group of pantothenic acid and the amino group of thioethanolamine. Therefore it was not difficult to reach the conclusion that the active component in our preparations was the thio ester of acetic acid and coenzyme A.

The formula for "active acetate," based on the recently[34] completed structure analysis of coenzyme A by Lipmann, Baddiley, Novelli and Kaplan, is shown in Figure 5.

The identity of the sulfhydryl component of our active preparations with coenzyme A was finally proved in various ways[26, 35]. The most convincing proof

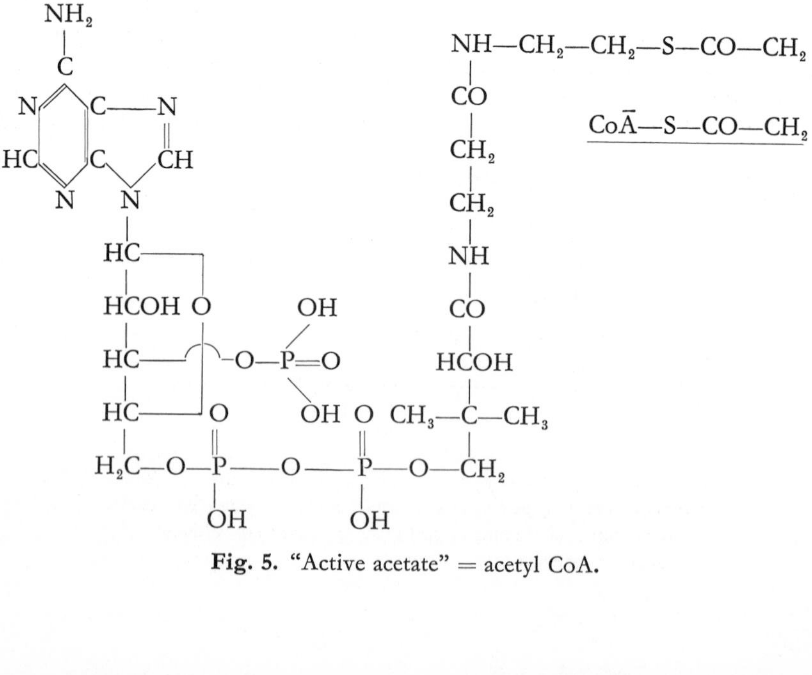

```
   NH2
    |                          NH—CH2—CH2—S—CO—CH3
    C                          |
 N     C——N                    CO
 |     |  ||                   |                    CoA—S—CO—CH3
HC     C  CH                   CH2
  N    N                       |
       |                       CH2
      HC——                     |
       |   |                   NH
      HCOH O      OH           |
       |   |     /             CO
      HC———    —O—P=O          |
       |   |     \            HCOH
      HC———  O    OH O  CH3—C—CH3
       |     ||       ||       |
      H2C—O—P———O———P—O—CH2
             |        |
             OH       OH
```

Fig. 5. "Active acetate" = acetyl CoA.

Table 2

ANALYSIS OF "ACTIVE ACETATE" FROM YEAST, CONTAINING 29% ACETYL CoA[35]

	Calculated	Found
Active acetyl[26]	1.54%	1.54%
S in acetyl-S[30]	1.16%	1.15%
S in disulfide-S or sulfhydryl-S	0	0
CoA units[29]	116 units/mg.	108 units/mg.
Adenine*	4.85%	14.8%
PO_4*	10.3%	18.7%

*The high adenine and phosphate values show that the main impurities of the acetyl CoA preparation were adenine nucleotides.

was that the sulfur content of our best preparations coincided with the amount of CoA and at the same time was present exclusively in the form of a thio ester (Table 2).

The presence of the thio ester grouping in addition to adenine was furthermore recognized by the ultraviolet absorption spectrum (Fig. 6). Highly purified preparations show the strong adenine band at 260 mμ. Inactivation by alkali does not affect the height of the adenine band. This treatment however leads to a change in the short ultraviolet. The decrease in absorption caused by alkali is plotted below as the difference spectrum; the maximum at 233 mμ is due to the acetyl mercaptan group. Absorption by other thio esters in the region of this wave length had been previously observed by Sjoeberg[36]. Stadtman[37] has also made this observation independently and applied it in a beautiful manner to the estimation of acetyl CoA.

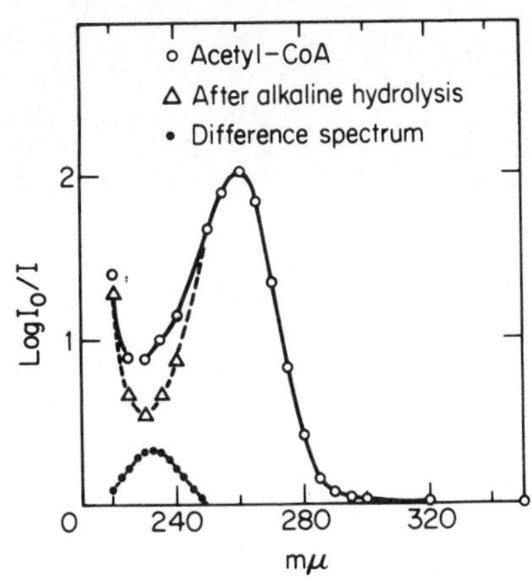

Fig. 6. Absorption spectrum of acetyl CoA before and after alkaline hydrolysis: 1.56 mg. acetyl CoA (40% pure) in 10 ml. of *M*/20 phosphate buffer; pH 7.4; light path = 1 cm. For hydrolysis the acetyl CoA was incubated with *M*/4 KOH for 10 minutes at 20°. Before absorption measurement this solution was neutralized with HCl.

These results also confirmed our ideas that acetyl CoA is actually an acyl mercaptan and that the extraordinary susceptibility to alkali is due to the splitting of the thio ester bond. The formula of acetyl CoA was further substantiated by its successful preparation through acetylation of coenzyme A by Stadtman[38] with the enzymatically catalyzed transacetylase reaction.

Acetyl CoA, the General Metabolic Acetyl Donor

The next step after the isolation of the active compound from yeast was to prove that this substance, acetyl CoA, which acted so well as acetyl donor in the sulfanilamide system, was really the general metabolic acetyl donor.

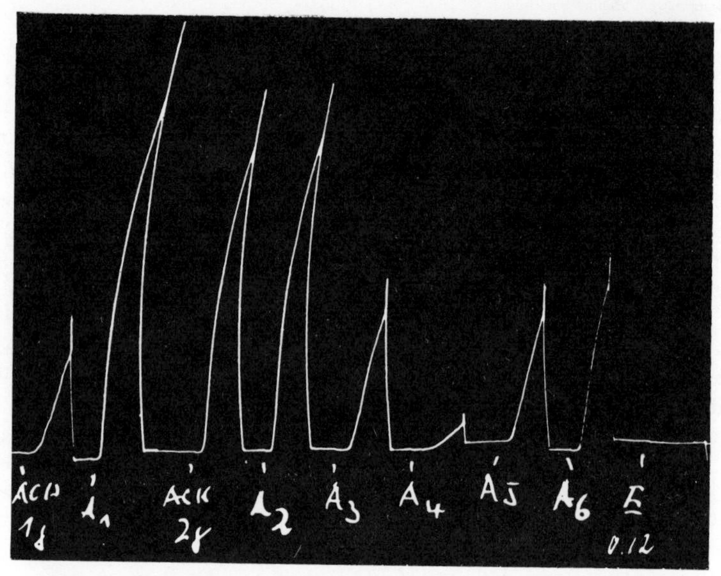

Fig. 7. Synthesis of acetylcholine from acetyl CoA and choline[39]. The figure shows the kymographic record of acetylcholine bioassay, according to Chang and Gaddum[41]. ACH: 1 μg. or 2 μg. acetylcholine added; A_1–A_6: 0.38 ml. reaction mixture added, containing 10 μM pyrophosphate buffer, pH 7.2, 18 μM KCl, 3 μM cysteine, 2 μM choline chloride, 100 μg. eserine sulfate and 0.4 mg. "active acetate" isolated from yeast Kochsaft, containing 0.023 μM acetyl CoA. In addition to these the reaction mixture contained 0.35 μM $MgCl_2$ and 0.12 (A_1), 0.06 (A_2), 0.03 (A_3) or 0.015 (A_4) ml. brain enzyme; A_5 and A_6 contained 0.03 ml. brain enzyme with 3.5 μM (A_6) or without (A_5) $MgCl_2$; E contained 0.12 ml. brain enzyme and 0.35 μM $MgCl_2$ but no acetyl CoA. Prior to the bioassay, the reaction mixtures had been incubated for 40 minutes at 37°. The brain enzyme was the 40–60% acetone fraction of a rabbit brain extract; 0.15 ml. of the enzyme solution corresponded to 100 mg. of brain powder extracted in this way.

Using this preparation, we showed that on incubation with choline and a rabbit brain fraction acetyl choline was formed, as determined by the frog rectus method[39]. Figure 7 shows results of this crucial experiment. The reaction can now be formulated[26, 39, 40] as follows:

$$(1)\quad CoA{-}S{-}CO{-}CH_3 + (CH_3)_3\overset{+}{N}{-}CH_2{-}CH_2{-}OH \longrightarrow CoA{-}SH + (CH_3)_3\overset{+}{N}{-}CH_2{-}CH_2{-}O{-}CO{-}CH_3$$

In a similar manner, citric acid synthesis was demonstrated[42]. As shown in Table 3 the acetyl CoA preparation reacts readily with oxalacetate, in the presence of Ochoa's crystalline "condensing enzyme," to yield equimolecular amounts of citrate and sulfhydryl groups according to the equation:

$$(2)\quad CH_3{-}CO{-}S{-}CoA + O = \underset{}{\overset{CH_2-COOH}{\overset{|}{C}}}{-}COOH + H_2O \rightleftharpoons HO{-}\overset{CH_2-COOH}{\underset{CH_2-COOH}{\overset{|}{\underset{|}{C}}}}{-}COOH + HS{-}CoA$$

The citric acid experiments, carried out in collaboration with Stern and Ochoa, were rather instructive in another respect. By measuring the equilibrium constant for this reaction, we were able to calculate the energy available in the acyl mercaptan bond. In excellent agreement with Stadtman's measurements on the transacetylase equilibrium[38], this energy turned out to be approximately 12,400 calories per mole. Consequently, in the acyl mercaptan bond of acetyl CoA a new type of metabolically active energy-rich bond had been discovered.

Therefore, the assumption that available metabolic energy is exclusively converted into energy-rich phosphate bond had to be considerably modified. It is the acyl mercaptan bond which indeed primarily forms in many cases; the phosphate bond is then secondarily formed through exchange reactions.

Table 3

CITRATE SYNTHESIS FROM ACETYL CoA AND OXALACETATE*

The system contained 50 μM of potassium phosphate buffer, pH 7.4, 4 μM of $MgCl_2$, 5 μM of potassium oxalacetate, 0.10 mg. of crystalline "condensing enzyme," and acetyl CoA from yeast as indicated. Final volume, 1.0 ml.; temperature, 23°; incubation time, 5 minutes.

Experiment No.	Acetyl CoA		Sulfhydryl	Citrate
	mg.	μM†	μM	μM
1	7.8	0.56	0.48	0.46
2	1.0	0.36	0.38	0.34

*From Stern, Ochoa, and Lynen[42].
†Determined by sulfanilamide acetylation.

The new insight into the intimate mechanism of the dehydrogenation of phosphoglyceraldehyde[43, 44, 45], pyruvate[46] and α-ketoglutarate[47] is particulary instructive in this connection because it shows that this principle applies to a whole class of metabolically active SH-compounds including proteins. Since now we have encountered a second type of energy-rich bond in addition to the phosphate bond, we may expect the occurrence of still other such bonds in nature[48, 49].

In the case of acetyl CoA, nature again uses the principle of acid anhydride. Two components of acid properties, the carboxyl and the sulfhydryl groups, condense with the liberation of water. It is in these terms that the reactivity of acetyl CoA as an agent for the acetylation of amino groups or hydroxyl groups in compounds such as sulfanilamide or choline is to be understood. It should be remembered that the preparative chemist carries out acetylations of such groups with acid anhydrides like acetyl chloride or acetic anhydride.

The peculiarity of the acyl mercaptan bond is the fact that in spite of the high energy of hydrolysis it is completely stable in aqueous solution at neutral *p*H and body temperature, a property which permits these compounds to function in cellular metabolism. Under physiological conditions, the energy of the bond is realized only in the presence of specific proteins. Furthermore, the attachment of the acetyl group at the sulfur of the coenzyme creates a second site of reactivity through the labilization of the methyl hydrogen[50]. This becomes particularly evident in the process of citrate and acetoacetate synthesis.

The Acetoacetate Condensation

The acetoacetate condensation deserves special attention because it represents a rather fundamental metabolic reaction. Analogous condensations occur as primary steps in many cases where acetate serves as a building block for larger molecules.

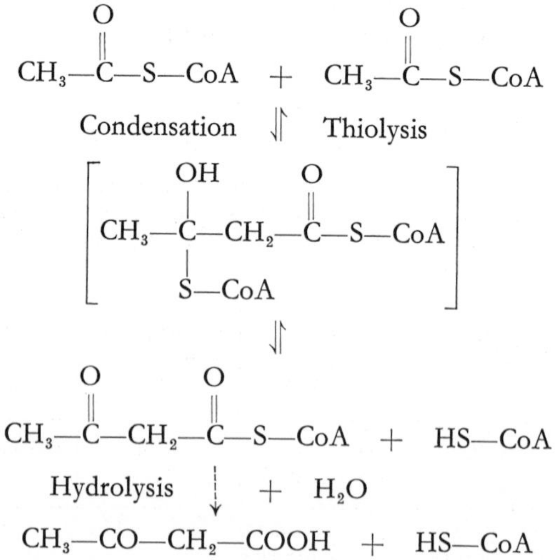

$$CH_3\text{—}\overset{\overset{\displaystyle O}{\|}}{C}\text{—}S\text{—}CoA \;+\; CH_3\text{—}\overset{\overset{\displaystyle O}{\|}}{C}\text{—}S\text{—}CoA$$

$$\text{Condensation} \;\rightleftharpoons\; \text{Thiolysis}$$

$$\left[\; CH_3\text{—}\underset{\underset{\displaystyle S\text{—}CoA}{|}}{\overset{\overset{\displaystyle OH}{|}}{C}}\text{—}CH_2\text{—}\overset{\overset{\displaystyle O}{\|}}{C}\text{—}S\text{—}CoA \;\right]$$

$$\rightleftharpoons$$

$$CH_3\text{—}\overset{\overset{\displaystyle O}{\|}}{C}\text{—}CH_2\text{—}\overset{\overset{\displaystyle O}{\|}}{C}\text{—}S\text{—}CoA \;+\; HS\text{—}CoA$$

$$\text{Hydrolysis} \;\downarrow\; + H_2O$$

$$CH_3\text{—}CO\text{—}CH_2\text{—}COOH \;+\; HS\text{—}CoA$$

Fig. 8.

The most significant consequence of having recognized the thio ester bond in acetyl CoA was that we now could formulate the acetoacetate condensation in chemical terms[26,27]. This was rather important because at the same time insight into the puzzling problem of fatty acid degradation and synthesis was gained[26].

Conclusion

The role of coenzyme A in fatty acid metabolism may be compared to the role of inorganic phosphate in carbohydrate metabolism. The sugar molecule is introduced into the chain of degradations and transformations by conversion into a phosphorylated derivative. Quite analogous to this, fatty acids are linked in the first step to the sulfur of coenzyme A, and only then can the enzymatic oxidation within the cell proceed. It may be no accident that in both cases weak acids like secondary phosphate or the hydrogen sulfide derivative, coenzyme A, are involved.

For medicine, the recent developments in the field of fatty acid metabolism are significant because now the chemical transformations are known and the individual enzymatic steps can be tested separately. Biochemistry thus supplies medicine with facts and with methods for recognizing metabolic failures. Perhaps it is not too fantastic to hope that future research along these lines will generate new weapons against disease.

References

1. Lipmann, F. 1941. *Advances in Enzymology* **1**, 99.
2. Kalckar, H. 1941. *Chem. Revs.* **28**, 71.
3. Lynen, F. 1942. *Naturwissenschaften* **30**, 398.
4. Du Vigneaud, V. 1942/43. *Harvey Lectures, Ser.* **38**, 29.
5. Wieland, H., and Sonderhoff, R. 1932. *Ann. Chem.* **499**, 213.
6. Martius, C., and Lynen, F. 1950. *Advances in Enzymology* **10**, 167.
7. Lynen, F. 1942. *Ann. Chem.* **552**, 270.
8. Lipmann, F. 1937. *Enzymologia* **4**, 65.
9. Lipmann, F. 1940. *J. Biol. Chem.* **134**, 463.
10. Lipmann, F. 1946. *Advances in Enzymology* **6**, 231.
11. Lynen, F. 1940. *Ber.* **73**, 367.
12. Neciullah, N. 1941. Thesis, Munich.
13. Lipmann, F. 1948/49. *Harvey Lectures, Ser.* **44**, 99.
14. Schoenheimer, R. 1942. *The Dynamic State of Body Constituents,* Harvard Univ. Press, Cambridge.
15. Nachmansohn, D., and Machado, A. L. 1943. *J. Neurophysiol.* **6**, 397.
16. Lipmann, F. 1945. *J. Biol. Chem.* **160**, 173.
17. Nachmansohn, D., and Berman, M. 1946. *J. Biol. Chem.* **165**, 551.
18. Feldberg, W., and Mann, T. 1946. *J. Physiol.* (*London*) **104**, 411.
19. Lipton, M. A., and Barron, E. S. G. 1946. *J. Biol. Chem.* **166**, 367.
20. Lipmann, F., Kaplan, N. O., Novelli, G. D., Tuttle, L. C., and Guirard, B. M. 1950. *J. Biol. Chem.* **186**, 235.

21. Soodak, M., and Lipmann, F. 1948. *J. Biol. Chem.* **175**, 999.
22. Stern, J. R., and Ochoa, S. 1949. *J. Biol. Chem.* **179**, 491.
23. Novelli, G. D., and Lipmann, F. 1950. *J. Biol. Chem.* **182**, 213.
24. Stadtman, E. R. 1950. *Federation Proc.* **9**, 233; Stadtman, E. R., Novelli, G. D., and Lipmann, F. 1951. *J. Biol. Chem.* **191**, 365.
25. Ochoa, S. 1951. *Physiol. Revs.* **31**, 56.
26. Lynen, F., Reichert, E., and Rueff, L. 1951. *Ann. Chem.* **574**, 1.
27. Lynen, F., and Reichert, E. 1951. *Angew. Chem.* **63**, 47.
28. Nachmansohn, D., and John, H. M., 1945. *J. Biol. Chem.* **158**, 157.
29. Kaplan, N. O., and Lipmann, F. 1948. *J. Biol. Chem.* **174**, 37.
30. Lynen, F. 1951. *Ann. Chem.* **574**, 33.
31. Lynen, F., and Bühler, M. Unpublished data.
32. Brown, G. M., Craig, J. A., and Snell, E. E. 1950. *Arch. Biochem.* **27**, 473.
33. Snell, E. E., Brown, G. M., Peters, V. J., Craig, J. A., Wittle, E. L., Moore, J. A., McGlohon, V. M., and Bird, C. D. 1950. *J. Am. Chem. Soc.* **72**, 5349.
34. Baddiley, J., Thain, E. M., Novelli, G. D., and Lipmann, F. 1953. *Nature* **171**, 76; Wang, T. P., Shuster, L., and Kaplan, N. O. 1952. *J. Am. Chem. Soc.* **74**, 3204.
35. Lynen, F. Unpublished data.
36. Sjoeberg, B. 1942. *Z. physik. Chem.* **B,52**, 209.
37. Stadtman, E. R. 1953. *J. Cell. and Comp. Physiol.* **41**, Supplement 1, 89; *J. Biol. Chem.* **203**, 501.
38. Stadtman, E. R. 1952. *J. Biol. Chem.* **196**, 535.
39. Wieland, O., and Lynen, F., 1953. *Z. physiol. Chem.*, In press.
40. Korkes, S., Del Campillo, A., Korey, S. R., Stern, J. R., Nachmansohn, D., and Ochoa, S. 1952. *J. Biol. Chem.* **198**, 215.
41. Chang, H. C., and Gaddum, J. H. 1933. *J. Physiol. (London)* **79**, 225.
42. Stern, J. R., Ochoa, S., and Lynen, F. 1952. *J. Biol. Chem.* **198**, 313.
43. Harting, J. 1951. *Federation Proc.* **10**, 195.
44. Racker, E., and Krimsky, I. 1952. *Nature* **169**, 1043.
45. Holzer, H., and Holzer, E. 1952. *Z. physiol. Chem.* **291**, 67.
46. Ochoa, S. 1950/51. *Harvey Lectures, Ser.* **46**, 153.
47. Kaufman, S. 1953. *Federation Proc.* **12**, 704.
48. Stadtman, E. R., and White, F. H., Jr. 1953. *J. Am. Chem. Soc.* **75**, 2022.
49. Wieland, T., and Schneider, G. 1953. *Ann. Chem.* **580**, 159.
50. Lynen, F. 1953. *Federation Proc.* **12**, 683.

Addendum

The mode of action of β-keto thiolase whereby the sulfhydryl group of the enzyme itself accomplishes the thiolysis is further supported by the new experiments of H. Beinert and P. G. Stansly (*J. Biol. Chem.* **204,** 67 (1953)). These authors found the exchange reaction between acetoacetate and labeled acetyl CoA in the presence of soluble enzymes of pig heart to lead to asymmetrically labeled acetoacetate.

DISCUSSION*

Jane Harting[1] *and Britton Chance*
New York University College of Medicine, New York
University of Pennsylvania, Philadelphia

Glyceraldehyde-3-phosphate dehydrogenase has been shown to have a broad specificity and to catalyze the oxidation of a number of aldehydes[1,2]. In connection with this symposium the oxidation of acetaldehyde to acetyl phosphate (*reaction 1*) is of rather special interest.

(1) acetaldehyde + phosphate + DPN $\rightleftharpoons$ acetyl phosphate + $DPNH_2$

The oxidation product, acetyl phosphate, can act as substrate for a second type of reaction, namely, a transacetylation which is also catalyzed by the triose phosphate dehydrogenase[3]. In the second reaction (*reaction 2*) the acetyl group of acetyl phosphate is transferred to CoA and yields acetyl CoA.

(2) acetyl phosphate + CoA $\rightleftharpoons$ acetyl CoA + phosphate

These two reactions, the production of acetyl phosphate and the transacetylation to CoA, can be catalyzed by the crystalline dehydrogenase isolated from either rabbit muscle or yeast. Similar reactions have been observed in bacteria[4,5], but neither of these activities had previously been found for enzymes from mammalian tissues or from yeast. Although the biological significance of these catalyses in mammalian metabolism is still problematical, it is nevertheless interesting that a single enzyme can catalyze not only the formation of acetyl phosphate but also the transfer of this acetyl group to CoA as acceptor.

In transacetylase reactions involving acetyl phosphate, it is generally as-

*J. Harting and B. Chance, "Discussion," *Fed. Proc.,* **12**, No. 3, 714 (1953). Reprinted with the permission of the authors and of the editor of *Federation Proceedings*.

[1] Fellow of the National Foundation for Infantile Paralysis.

sumed that acetyl phosphate first forms an acetyl enzyme complex, and the acetyl group is subsequently transferred to another acceptor, such as CoA[5]. The exchange of acceptors may be schematically represented by the following equation:

$$\text{(3)} \qquad \text{acetyl phosphate} + \text{enzyme} \xrightleftharpoons{\pm\ \text{phosphate}} \text{acetyl}$$

$$\text{acetyl enzyme} \xrightleftharpoons{\pm\ \text{CoA}} \text{acetyl CoA} + \text{enzyme}$$

The three criteria for testing such a hypothesis, as established by Hassid, Doudoroff, and Barker[6], are: *1*) exchange of the phosphate of acetyl phosphate with radioactive inorganic phosphate, *2*) arsenolysis of acetyl phosphate, *3*) transfer of the acetyl group to still other acceptors. These three criteria have been fulfilled in the case of the triose phosphate dehydrogenase[3,7]; therefore, the existence of an acetyl enzyme complex was suggested in the transfer reactions.

It was then decided to attempt a direct spectroscopic identification of the acetyl enzyme intermediate. Since Dr. Racker has focused our attention on the role of the thiol group in this reaction[7], we have investigated the spectra in the ultraviolet region where the known thiol esters have a characteristic absorption peak. No previous spectroscopic identification of compounds of a dehydrogenase and its substrate has been described because the absorption of the protein itself in the ultraviolet is so large that a special spectrophotometric system is required[8].

Upon the addition of acetyl phosphate to triose phosphate dehydrogenase, the formation of an acetyl enzyme complex can be observed. When the spectrum of the acetyl enzyme compound (fig. 1) is compared to that of a typical thiol ester, acetyl glutathione, it may be seen that the curves are not the same. This fact suggests that if the acetyl group is linked to a thiol group on the enzyme it does not have the same properties as a simple thiol ester, or that the linkage to the protein is of some other type. Whatever the nature of the bond, it is clearly of an energy rich type in view of the readily reversible transfer of the acetyl group between compounds such as acetyl phosphate and acetyl CoA.

The acetyl enzyme compound is decomposed by arsenate. This would be

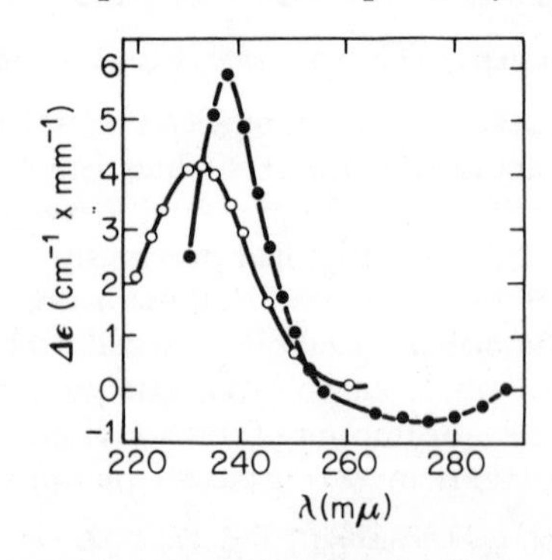

Fig. 1. Curves represent the spectra of the acetyl enzyme complex —●—●— and acetyl glutathione —○—○—.

expected since the enzyme has been shown to catalyze the arsenolysis of acetyl phosphate. In accordance with the previous scheme, this reaction may be written.

$$\text{(4)}\quad \text{acetyl phosphate} + \text{enzyme} \underset{}{\overset{\pm\ \text{phosphate}}{\rightleftarrows}} \text{acetyl enzyme} \overset{\pm\ \text{arsenate}}{\rightleftarrows}$$

$$\text{enzyme} + (\text{acetyl arsenate}) \longrightarrow \text{acetate} + \text{arsenate}$$

The arsenolysis of the enzyme substrate complex supports the view that this complex may be the active intermediate in the transfer reactions.

In summary, an acetyl enzyme compound of triose phosphate dehydrogenase has been identified spectroscopically. With the further identification of enzyme substrate complexes with the dehydrogenases, perhaps a more detailed study of the mechanism of oxidation will be possible.

References

1. Harting, J. *Federation Proc.* **10**: 195, 1951.
2. Nygaard, A. P. and J. B. Sumner. *Arch. Biochem* **39**: 119, 1952.
3. Harting, J. and S. F. Velick. *Federation Proc.* **11**: 226, 1952.
4. Stadtman, E. R. and H. A. Barker. *J. Biol. Chem.* **180**: 1095, 1949.
5. Stadtman, E. R. and H. A. Barker. *J. Biol. Chem.* **184**: 769, 1950.
6. Doudoroff, M., H. A. Barker and W. Z. Hassid. *J. Biol. Chem.* **170**: 147, 1947.
7. Racker, E. and I. Krimsky. *J. Biol. Chem.* **198**: 731, 1952.
8. Chance, B. *Rev. Sci. Instr.* **22**: 619, 1951.

FORMATION OF ACYL AND CARBONYL COMPLEXES ASSOCIATED WITH ELECTRON-TRANSPORT AND GROUP-TRANSFER REACTIONS*

E. Racker

Department of Biochemistry, Yale University,
New Haven, Connecticut

There appears to be a trend in modern biochemistry to abolish the boundaries that have been set up between the different dominions of cellular metabolism. The use of isotopes in the investigation of the metabolism of intact animals has particularly served to emphasize the incredibly rapid flow of intermediates in all directions. Although we may now begin to perceive some of the predominating currents, we have not yet been able to evaluate quantitatively the metabolic reactions within the cells. Isolated attempts have been made to solve this problem. The role of the neoformation of methyl groups as compared to methyl-group transfer, has been investigated with the aid of doubly labeled compounds[1]. An approximate estimate of the relative roles of the shunt mechanism and the Meyerhof-Embden pathway in the utilization of glucose in *E. coli* cells has been made[2]. Investigations into the fate of CO_2 labeled with C^{14} and rate measurements have been performed in the field of photosynthesis[3]. But our tools and approaches in such studies are still few and limited, and the interpretation of data is subject to a good deal of disagreement.

What can we learn about the operation of an enzyme within the cell by studying its action in an isolated state? The most accurate determinations of specific activity, turnover number, and the amounts of enzyme present in cell extracts will at best give us a qualitative answer to its operation within the cell. On the other hand, it may be impossible to analyse the function of an enzyme

*Excerpt from E. Racker, "Formation of Acyl and Carbonyl Complexes Associated with Electron-Transport and Group-Transfer Reactions," in *Mechanism of Enzyme Action,* eds. W. D. McElroy and B. Glass. Baltimore: The Johns Hopkins Press, 1954, pp. 464–470, Refs. 1–22. Reprinted with the permission of the author and of the publisher.

unless it is isolated from the other enzymes; an example of this will be cited later.

In the approach to isolated enzymes a trend to a broader concept has also become apparent. Enzymes which for many years had been classified as hydrolytic have been discovered to catalyze transfer reactions. Probably these transfer reactions participate in both the degradation and the synthesis of cell constituents such as proteins, nucleic acids, and fats.

In energy metabolism, too, the fields seem to have converged. The process of phosphorylation has been found to be "coupled" to the electron-transport mechanism. A recent trend, away from the omnipotent ATP, has emphasized the importance of group-transfer reactions for synthetic processes. Thiolesters, acylimidazole, and sulfonium compounds have received special attention at the higher energy levels, while peptides, amides, esters, glycosides, and nucleosides have been shown to participate in transfer reactions at the lower energy level. Renewed attention has been given in recent years to a reaction that is concerned with the transfer of an aldehyde group to another aldehyde or to an alcohol. The enzymes that catalyze these reactions have been referred to as transketolase and transaldolase. These enzymes too, appear to participate both in synthetic and degradation reactions and are found to be widely distributed in cells of all forms of life.

Since many of these transfer reactions have been reviewed at this and at previous sessions of the Baltimore symposium, the present discussion will be restricted to the formation and utilization of acyl and carbonyl complexes.

The presentation will be divided into three parts: (I) the formation of carbonyl and acyl complexes in oxidative processes; (II) the degradation of acyl complexes; and (III) transfer reactions of acyl and carbonyl complexes.

I. Formation of Carbonyl and Acyl Complexes in Oxidative Processes

1. Glyceraldehyde-3-phosphate dehydrogenase

According to the hypothesis of Warburg and Christian[4], glyceraldehyde-3-phosphate and inorganic phosphate react chemically to yield 1, 3-diphosphoglyceraldehyde. The first indication that this hypothesis was incorrect came from the work of Meyerhof and his associates[5]. From kinetic and equilibria data it was concluded that non-enzymatic formation of the hypothetical 1, 3-diphosphoglyceraldehyde does not take place, but the possibility that such a compound is formed by the enzyme was left open.

A study of the mechanism of the conversion of ketoaldehydes to the corresponding hydroxyacids by glyoxalase[6] has shown that the reaction proceeds in two major steps. The first step was found to involve an interaction between the aldehyde and the sulfhydryl group of glutathione, and results in the formation of a thiolester (III). In the second step the thiolester is hydrolysed to the acid, and the free SH compound is liberated.

```
CH3          CH3          CH3          CH3               CH3
|            |            |            |                 |
C=O          C-OH         C-OH         HCOH              HCOH
|            ||           ||           |                 |
|    O       C=O          C-OH         C=O    OH         COOH
|   //                    |            |
C            +            |       ---------------        +
 \                        |            |
  H          GSH          GS           GS     H          GSH
             (I)          (II)         (III)             (IV)
```

These findings suggested the possibility that a similar mechanism might operate in triose phosphate oxidation and the first variant of a scheme was suggested and is shown below:

```
R               R                    R                R
|  O            |                    |                |  O
| //            |                    |                | //
C               HCOH  DPN            C=O   P1         C
 \              |      ⇌             |     ⇌           \
  H   ⇌         |     DPNH           |                  OPO3H2
+               |                    |                +
SH              S                    S                SH
|               |                    |                |
Enzyme          Enzyme               Enzyme           Enzyme
(I)             (II)                 (III)            (IV)
```

By means of this scheme it was possible to explain the high sensitivity of triose phosphate dehydrogenase to iodoacetate[7] and to visualize a mechanism for the formation of the carboxyl phosphate by phosphorolysis of a thiolester. The identification of acetyl-CoA as a thiolester of coenzyme A[8] and its participation in the oxidation of pyruvate[9, 10] and aldehydes[11, 12] provided a complete analogy to the proposed mechanism. Moreover, the observation[13] that acetaldehyde and acetylphosphate are substrates for glyceraldehyde-3-phosphate dehydrogenase permitted analysis of the reaction with these readily available substrates. It was then demonstrated that several times recrystallized triose phosphate dehydrogenase catalyses a two-step reaction consisting of an oxidative and a phosphorolytic step. The evidence for this mechanism, obtained by the independent work in two laboratories[14, 15], can be summarized as follows:

1. In the absence of a system to reduce DPN, arsenolysis of the acylphosphate was shown to occur, and P^{32} exchanged rapidly with the phosphate of the acylphosphate.

2. When reduced glyceraldehyde-3-phosphate dehydrogenase (obtained by isolation of the enzyme in the presence of Versene, and fully active without addition of reducing agent) was blocked with an excess of IAA, the oxidation-reduction step, which was tested in both directions in the presence of glutathione, was obliterated. However, arsenolysis still took place if glutathione was added to the IAA-treated enzyme.

3. The enzyme catalyzes a transfer from acetylphosphate to either glutathione or CoA. Accumulation of acetylglutathione was demonstrated when acetylphosphate and glutathione were incubated in the presence of enzyme.

4. Thiolesters were slowly hydrolyzed, or in the presence of DPNH were reduced by the enzyme.

5. It was shown that glutathione was a firmly bound component of the enzyme, and the formation of an acyl-enzyme complex in the presence of the substrate was demonstrated. The latter point will be discussed in greater detail.

The concept of an "acyl-enzyme complex" is not very new. It has been customary for enzymologists to retreat to the hilly surfaces of their enzymes when evidence for the existence of a postulated intermediate was not forthcoming. The availability of glyceraldehyde-3-phosphate dehydrogenase in gram quantities[16, 17] made it possible to approach this problem more directly. With acetylphosphate as substrate the formation of an acyl enzyme could be demonstrated by the following procedure[21]:

The enzyme was incubated with an excess of acetylphosphate for a few minutes and then boiled for five minutes at pH 4.5. This procedure precipitated the enzyme and destroyed practically all the acetylphosphate present. The enzyme was centrifuged off and washed three times with water to remove all traces of residual substrate. When such a preparation was treated with hydroxylamine, a soluble hydroxamic acid was found in the supernatant solution and measured colorimetrically[18]. Up to 1.8 equivalents of acyl groups per mole of enzyme were found to be formed by this procedure. When the acyl-enzyme complex was treated with proteolytic enzymes (trypsin was found to be most suitable for this purpose) the hydroxylamine-reactive component was solubilized. Fractionation of the mixture of amino acids and peptides thus formed proved to be difficult because of heavy loss of the thiolester. Small amounts of the thiolester (as determined by the alkaline nitroprusside test) were isolated on several occasions on paper chromatograms. However, final identification of this ester has not yet been achieved.

In the course of studies on the effect of SH-inhibitors on glyceraldehyde-3-phosphate dehydrogenase, it was observed that the enzyme exhibited an absorption at 340 mμ, which was decreased on addition of iodoacetate or *p*-chloromercuribenzoate[14]. After treatment of the enzyme with charcoal, which removes DPN[19], the absorption at 340 mμ dropped and was restored to the original value by the addition of DPN. This reconstructed absorption band disappeared again on addition of IAA, as shown in Fig. 1. When the differences between the spectra with and without IAA were plotted, a rather broad absorption band with a peak at 360 mμ was obtained.

Because of the effect of compounds that react with SH groups, the data were interpreted in terms of a complex formed between DPN and the SH groups of the enzyme. To evaluate the participation of this complex in the reaction catalyzed by the enzyme, the effect of acylphosphate on the absorption was measured. It was found that either 1, 3-diphosphoglyceric acid or acetylphosphate induced a drop in absorption at 360 mμ which was not further affected by addition of IAA. From these data a second variant of the mechanism of aldehyde oxidation was proposed, as shown in Fig. 2.

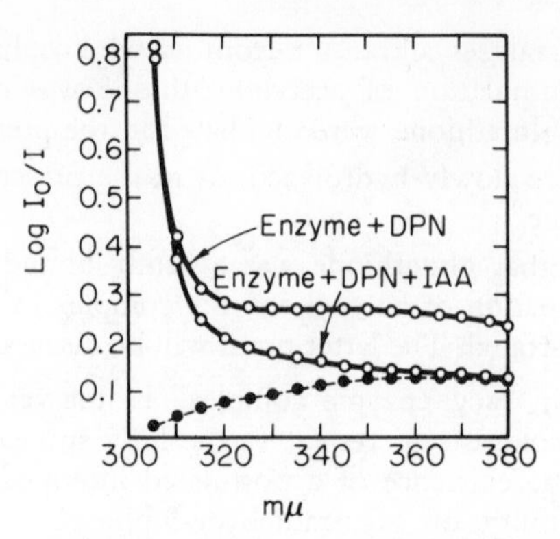

Fig. 1. Absorption spectra of enzyme-DPN complex.

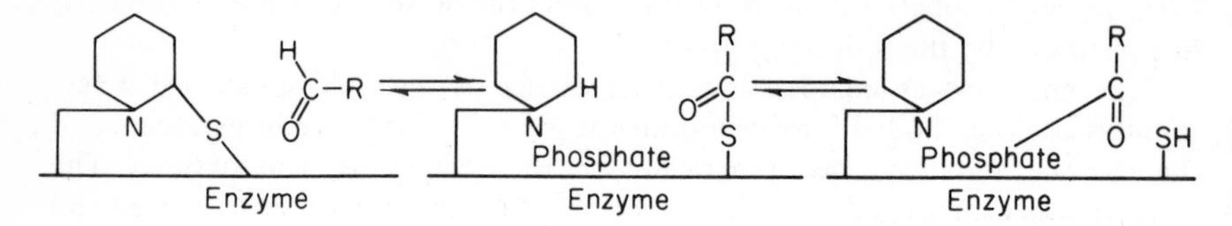

Fig. 2. Aldehydolysis of enzyme-DPN complex.

The two essential points of this theory are (a) a participation of the DPN-S-complex in the mechanism of electron transfer, and (b) a process of aldehydolysis by which that complex is cleaved.

In the above formulation the actual mechanism of electron transport was omitted from the scheme. Two alternatives could be suggested. According to the first, the addition of the SH group to DPN leads to a change of the quaternary to a tertiary nitrogen with the release of a proton, while in the aldehydolysis step the hydrogen of the aldehyde is donated to DPN. According to the second alternative, the hydrogen transfer occurs in the first step and is derived from the SH group, while aldehydolysis leads to the liberation of a proton. Recent data[20, 21] on the effect of pH on the equilibria of the two reactions (the formation of the DPN-enzyme complex and the aldehydolysis) favor the second alternative.

It should also be pointed out that in view of recent experiments on the entrance of hydrogen in the formation of reduced DPN[22], the sulfhydryl should add to carbon 4 of the pyridine nucleus.

References

1. Keller, E. B., J. R. Rachele, and V. du Vigneaud, *J. Biol. Chem.,* **177**, 733 (1949).
2. Cohen, S. S., in *Phosphorus Metabolism,* Vol. 1 (W. D. McElroy and B. Glass, eds.), p. 148, Johns Hopkins Press, Baltimore (1951).
3. Buchanan, J. G., J. A. Bassham, A. A. Benson, D. F. Bradley, M. Calvin, L. L. Daus, M. Goodman, P. M. Hayes, V. H. Lynch, L. T. Norris, and A. T. Wilson, in *Phosphorus Metabolism,* Vol. 2 (W. D. McElroy and B. Glass, eds.), p. 440, Johns Hopkins Press, Baltimore (1952).

4. Warburg, O., and W. Christian, *Biochem. Z.,* **303**, 40 (1939).
5. Meyerhof, O., and P. Oesper, *J. Biol. Chem.,* **170**, 1 (1947).
6. Racker, E., *J. Biol. Chem.,* **190**, 685 (1951).
7. Rapkine, L., *Biochem. J.,* **32**, 1729 (1938).
8. Lynen, F., E. Reichert, and L. Rueff, *Liebig. Ann. Chem.,* **574**, 1 (1951).
9. Chantrenne, H., and F. Lipmann, *J. Biol. Chem.,* **187**, 757 (1950).
10. Korkes, S., A. del Campillo, I. C. Gunsalus, and S. Ochoa, *J. Biol. Chem.,* **193**, 721 (1951).
11. Burton, R. M., *Federation Proc.,* **11**, 193 (1952).
12. Pinchot, G. B., and E. Racker, in *Phosphorus Metabolism,* Vol. 1 (W. D. McElroy and B. Glass, eds.), p. 366, Johns Hopkins Press, Baltimore (1951).
13. Harting, J., *Federation Proc.,* **10**, 195 (1951).
14. Racker, E., and I. Krimsky, *J. Biol. Chem.,* **198**, 731 (1952).
15. Harting, J., and S. Velick, *Federation Proc.,* **11**, 226 (1952).
16. Cori, G. T., M. W. Slein, and C. F. Cori, *J. Biol. Chem.,* **173**, 605 (1948).
17. Krebs, E. G., G. W. Rafter, and J. McBroom Junge, *J. Biol. Chem.,* **200**, 479 (1953).
18. Lipmann, F., and L. C. Tuttle, *J. Biol. Chem.,* **159**, 21 (1945).
19. Taylor, J. F., S. F. Velick, G. T. Cori, C. F. Cori, and M. W. Slein, *J. Biol. Chem.,* **173**, 619 (1948).
20. Velick, S. F., in *Mechanism of Enzyme Action* (W. D. McElroy & B. Glass, eds.), Johns Hopkins Press, Baltimore (1954).
21. Krimsky, I. and E. Racker, unpub. exper.
22. Pullman, M. E., *Federation Proc.,* **12**, 255 (1953).

CARBAMYL PHOSPHATE, THE CARBAMYL DONOR IN ENZYMATIC CITRULLINE SYNTHESIS*

M. E. Jones†, *L. Spector and F. Lipmann*

Biochemical Research Laboratory and
Huntington Memorial Laboratory,
Massachusetts General Hospital and
the Department of Biological Chemistry,
Harvard Medical School
Boston, Massachusetts

The recent work on a phosphorolysis of citrulline in microbial extracts by Knivett,[1] by Slade[2,3] and by Korzenovsky and Werkman[4,5] and more recently by Stulberg and Boyer[6] has greatly advanced the understanding of the mechanism of this reaction. It seemed of considerable promise now to attempt the identification of the probable phosphorylated intermediary in this system which appeared to have a certain similarity to the so-called phosphoroclastic reaction of pyruvate in microbial extracts. In attempts to identify a phosphorylated intermediary, using extracts of *Streptococcus faecalis* R, no reaction between ATP[7] and ornithine could be observed. As shown in Table I, however, on incubation of an equilibrium mixture of ammonium carbonate-carbamate with

This investigation was supported by research grants from the Cancer Institute of the National Institutes of Health, Public Health Service and the Life Insurance Medical Research Fund.

†Research Fellow of the American Cancer Society.

[1] V. A. Knivett, *Biochem. J.,* **50**, XXX (1952); **58**, 480 (1954).

[2] H. D. Slade and W. C. Slamp, *J. Bact.*, **64**, 455 (1952).

[3] H. D. Slade, *Arch. Biochem. Biophys.,* **42**, 204 (1953).

[4] M. Korzenovsky and C. H. Werkman, *ibid.*, **41**, 233 (1952).

[5] M. Korzenovsky and C. H. Werkman, *Biochem. J.,* **57**, 343 (1954).

[6] M. P. Stulberg and P. D. Boyer, THIS JOURNAL, **76**, 5569 (1954).

[7] The following abbreviations are used: ATP for adenosine triphosphate; ADP, adenosine diphosphate; CAP, carbamyl phosphate; P_i, orthophosphate; P_u, unstable phosphate; P_{10} and, phosphate hydrolyzed in 10 minutes with N HCl at 100°.

Table I

FORMATION OF CARBAMYL PHOSPHATE FROM ATP

The complete incubation mixture for the formation of carbamyl phosphate consisted of: 200 μM. tris-(hydroxymethyl)-aminomethane buffer, *p*H 8.5; 5 μM. $MgCl_2$; 25 μM. KF; 0.6 μM. of ADP, *p*H 7.0; 100 μM. ammonium carbonate; 5.1 μM. phosphoenol pyruvate; 10 μM. L-ornithine; 0.01mg. crystalline pyruvate kinase; and 0.5mg./ml. of *Streptococcus faecalis* extract in 1ml. final volume. Vessels were incubated at 30° for 30 minutes. Carbamyl phosphate is that phosphorus which is hydrolyzed by 0.01 *N* HCl in 1 minute at 100°. Citrulline was determined according to Archibald.[8]

		P_i, μM./ml.	P_u, μM./ml.	Citrulline, μM./ml.
1	No enzyme	0.20	0.41[a]	0
2	No phosphoenol pyruvate	0.15	0.05	0
3	No Mg or ornithine	0.32	0.46[a]	0
4	No ornithine	0.45	1.30	0
5	Complete	5.50	0.10	5.1

[a] This blank shows that our hydrolysis procedure decomposes a small fraction of the phosphoenol pyruvate.

ATP, or better, phosphopyruvate + ADP, a relatively stable phosphorylated compound was formed. The compound decomposes only slowly in the Fiske and Subbarow phosphate reagent, but hydrolyzed completely on one minute heating with 0.01 normal hydrochloric acid to 100° and is determined in this manner.

This precursor of the carbamyl group in citrulline has been identified by synthesis as carbamyl phosphate. Carbamyl phosphate is surprisingly easily prepared by mixing dihydrogen phosphate with cyanate in the following manner: 0.1 mole of potassium dihydrogen phosphate and 0.1 mole of potassium cyanate were dissolved in 100 milliliters of water, the solution warmed to 30° for 30 minutes, and then cooled in ice. To the cool solution, an ice-cold solution of 0.3 mole of lithium hydroxide and 0.2 mole of perchloric acid in 83 milliliters of water were added slowly, final *p*H 8.3. A precipitate forms which consists of potassium perchlorate and lithium phosphate. This is removed by filtration. The supernate contains the lithium carbamyl phosphate. This is precipitated by slow addition of an approximately equal volume of ethanol. On reprecipitation with ethanol, dilithium carbamyl phosphate of a purity of 90 to 95 per cent. was obtained which was used for enzymatic tests.

The synthetic compound behaved analogously to the enzymatically formed compound with regard to acid hydrolysis and relative stability in the Fiske—Subbarow molybdate mixture. Citrulline formation from synthetic carbamyl phosphate and ornithine with the microbial enzyme are shown in Table II. It may be noted that a small part of the compound decomposed spontaneously in the absence of enzyme or in its presence if ornithine is omitted. From observations of the previous workers on the phosphorolytic split of citrulline in the presence of ADP with the formation of ATP, the intermediary was expected to react easily with ADP. This is confirmed in the experiment shown in Table III, which shows a rapid reaction. We therefore formulate citrulline synthesis as Magnesium ion is required in reaction (1) but not in (2) (*cf.* Tables III and II).

Experiments with mitochondria have shown that CAP in the animal system

[8] R. M. Archibald, *J. Biol. Chem.*, **156**, 121 (1954).

$$H_2N\cdot\overset{\overset{\displaystyle O}{\cdot\cdot}}{C}\cdot OH + ADP\cdot O{\sim}PO_3^- \longleftrightarrow H_2N\cdot\overset{\overset{\displaystyle O}{\cdot\cdot}}{C}\cdot O{\sim}PO_3^- + ADP \quad (1)$$

$$H_2N\cdot\overset{\overset{\displaystyle O}{\cdot\cdot}}{C}{\sim}O\cdot PO_3^- + NH_2\cdot(CH_2)_3\cdot CHNH_2\cdot COOH \longleftrightarrow H_2N\cdot\overset{\overset{\displaystyle O}{\cdot\cdot}}{C}{\sim}NH\cdot(CH_2)_3\cdot CHNH_2\cdot COOH + HO\cdot PO_3^- \quad (2)$$

Table II

SYNTHESIS OF CITRULLINE FROM CARBAMYL PHOSPHATE

The complete incubation mixture for the carbamyl phosphate system contained: 200 μM. tris-(hydroxymethyl) aminomethane buffer, *p*H 8.5; 10 μM. ornithine, *p*H 8.0; 5 μM. $MgCl_2$; 6.2 μM. CAP (containing 0.8 μM. orthophosphate); and 0.1 mg. protein as the *Streptococcus faecalis* extract in 1 ml. final volume. The vessels were incubated at 30° for 30 minutes.

	P_i, μM./ml.	CAP, μM./ml.	Citrulline, μM./ml.
Complete, zero time	0.8	6.20	0
Complete, incubated	6.9	0	6.3
No Mg	6.8	0	6.3
No ornithine	2.4	4.7	0
No enzyme	2.3	4.5	0

Table III

FORMATION OF ATP FROM CARBAMYL PHOSPHATE

The complete incubation mixture contained: 200 μM. tris-(hydroxymethyl)-aminomethane buffer, *p*H 8.5; 10 μM. $MgCl_2$; 7.20 μM. of ADP, *p*H 7.0; 8.58 μM. of CAP (containing 1.3 μM. inorganic phosphate and 1.5 μM. ammonia[a]); 0.5 mg. protein as the *Streptococcus faecalis* extract in 1 ml. final volume. Vessels were incubated at 30° for 20 minutes. The difference figures represent the difference between the complete system deproteinized at zero time and the reaction vessels.

		ATP,	CAP	
	P_i, μM./ml.	P_{10}, μM./ml.	P_u, μM./ml.	NH_3,[a] μM./ml.
Complete	+0.30	+7.4	−7.35	+7.5
No enzyme	+1.20	0	−1.64	0
No ADP	+1.30	0	−1.32	...
No Mg	+0.86	+3.4	−4.54	+3.2

[a] Ammonia was determined by a Conway distillation procedure[9] using saturated K_2CO_3 to liberate the ammonia from solution. The data reflect the stability of the fixed carbamate to alkali (*cf.* ref. 10).

likewise donates carbamyl to ornithine. The carbamyl-ornithine kinase appears more stable and far more active than the over-all reaction starting with ATP. In the experiment of Table IV, a rapid transfer from CAP is shown to occur while in the same extract, not shown in the table, ATP was inactive. With intact mitochondria, ATP showed in 60 minutes only about one-fifth of the activity shown by CAP in 10 minutes. With similar preparations, Grisiolia

[9] R. B. Johnston, M. J. Mycek and J. S. Fruton, *J. Biol. Chem.*, **185**, 629 (1950).
[10] A. Jensen and C. Faurholt, *Acta Chem. Scand.*, **6**, 385 (1952).

Table IV

Sonorated extract of rat liver mitochondria, 0.1 ml. (0.9 mg. protein), in 1 ml. total volume, *p*H 7.5, incubated 30 minutes at 37°. Otherwise conditions were similar to those in Table II.

	P_i, μM./ml.	CAP, μM./ml.	Citrulline, μM./ml.
Complete, 0 minutes	1.0	4.2	0
Complete, incubated	5.2	0	4.1
No CAP, incubated	0.1	0	0

and Cohen[11] have previously reported on an unstable precursor of the carbamyl group in citrulline.

If aspartic acid is substituted for ornithine in the microbial extracts as carbamyl acceptor, a somewhat slower reaction with carbamyl phosphate is observed, indicating an analogous mechanism for the synthesis of carbamyl aspartate. Formation of this compound from aspartate with ATP and ammonium carbonate or an unstable carbamyl precursor has recently been described in mammalian liver extracts by Lowenstein and Cohen[12] and by P. Reichard.[13]

[11] S. Grisiolia and P. P. Cohen, *J. Biol. Chem.*, **204**, 763 (1953).
[12] J. M. Lowenstein and P. P. Cohen, This Journal, **76**, 5571 (1954).
[13] P. Reichard, *Acta Chem. Scand.*, **8**, 795 (1954).

THE PARTICIPATION OF INORGANIC PYROPHOSPHATE IN THE REVERSIBLE ENZYMATIC SYNTHESIS OF DIPHOSPHOPYRIDINE NUCLEOTIDE*

Arthur Kornberg†

National Institutes of Health, Bethesda, Maryland

(Received for publication November 10, 1948)

A purified enzyme preparation has been obtained from an autolysate of dried brewers' yeast by ammonium sulfate fractionation and isoelectric precipitation which catalyzes the reaction: nicotinamide mononucleotide (NMN) + adenosine triphosphate (ATP) $\rightleftharpoons$ DPN + inorganic pyrophosphate (P-P). In the table are summarized two experiments in which equilibrium was attained starting from the left (DPN synthesis) and the right (ATP synthesis). The equilibrium constant, $K =$ ((DPN)(P-P))/-((NMN)(ATP)), calculated from the data of the two experiments is 0.3 in one case and 0.5 in the other. The concentrations of acid-labile phosphate were unchanged and no orthophosphate was produced. Nicotinamide nucleoside, adenosine diphosphate, and adenylic acid were inactive in DPN synthesis. The reduced form of DPN was split by the purified enzyme preparation in the presence of inorganic pyrophosphate, but triphosphopyridine nucleotide and flavin-adenine dinucleotide were not. The possibility that the latter two nucleotides may participate in analogous reactions with crude enzyme preparations requires further study.

These findings indicate a mechanism for the synthesis of DPN and for the origin and function of inorganic pyrophosphate. Ochoa, Cori, and Cori[1] isolated inorganic pyrophosphate from dialyzed rat liver dispersions in which glutamate, pyruvate, or succinate was being oxidized. It was later identified in washed

*A. Kornberg, "The Participation of Inorganic Pyrophosphate in the Reversible Enzymatic Synthesis of Diphosphopyridine Nucleotide," *J. Biol. Chem.*, **176**, 1475–76 (1948). Reprinted with the permission of the author and of the American Society of Biological Chemists, Inc.

†The valuable technical assistance of Mr. W. E. Pricer, Jr., is gratefully acknowledged.

[1] Cori, C. F., in A symposium on respiratory enzymes, Madison (1942).

Substance estimated†	DPN synthesis,* micromoles per ml.			ATP synthesis,* micromoles per ml.		
	0 min.	60 min.	Δ	0 min.	60 min.	Δ
NMN‡	2.50			0.0		
ATP	2.20	1.36	−0.84	0.0	1.02	+1.02
DPN	0.0	0.74	+0.74	1.50	0.61	−0.89
P-P	0.0	0.83	+0.83	1.80	0.79	−1.01
Orthophosphate	0.11	0.13		0.16	0.09	
Phosphate, acid-labile	4.44	4.50		3.60	3.66	

*For DPN synthesis, 1.0 ml. of reaction mixture contained 50 γ of the enzyme preparation, 0.3 micromole of $MgCl_2$, and 50 micromoles of glycylglycine buffer (*p*H 7.4) in addition to ATP and NMN; for ATP synthesis 1.0 ml. contained 50 γ of the enzyme preparation, 0.75 micromole of $MgCl_2$, and 50 micromoles of glycylglycine buffer (*p*H 7.4) in addition to DPN and P-P. Constant values were reached after 30 to 40 minutes at 38°.

†ATP was estimated spectrophotometrically by triphosphopyridine nucleotide reduction in the presence of glucose, hexokinase, and *Zwischenferment,* and DPN by reduction with the triose phosphate dehydrogenase system. Inorganic pyrophosphate was estimated as orthophosphate after acid hydrolysis of the precipitated and washed manganous salt and acid-labile phosphate as the orthophosphate released after 10 minute hydrolysis in 1 N H_2SO_4 at 100°.

‡NMN was prepared by hydrolysis of DPN with nucleotide pyrophosphatase.[2, 3] After purification the ratio, nicotinamideribose moiety to organic phosphate, was 1.0.

rabbit kidney particles oxidizing glutamate,[3,4] in molds,[5] and in yeast.[6] The present findings suggest that the accumulation of inorganic pyrophosphate in fungi and in tissues may be explained by a sequence of three reactions: (1) the irreversible hydrolysis of DPN by nucleotide pyrophosphatase[2,3] to yield NMN and adenylic acid, (2) the phosphorylation of adenylic acid to ATP in respiration or fermentation, and (3) the combination of NMN with ATP to produce inorganic pyrophosphate and regenerate DPN.

[2] Kornberg, A., *J. Biol. Chem.,* **174**, 1051 (1948).
[3] Kornberg, A., and Lindberg, O., *J. Biol. Chem.* **176**, 665 (1948).
[4] Green, D. E. *et al.,* Abstracts, American Chemical Society, Atlantic City, 26B, April (1947).
[5] Mann, T., *Biochem. J.,* **38**, 345 (1944).
[6] Lindahl, P. E., and Lindberg, O., *Nature,* **157**, 335 (1946).

LIFE IN THE SECOND AND THIRD PERIODS; OR WHY PHOSPHORUS AND SULFUR FOR HIGH-ENERGY BONDS?*

George Wald

The Biological Laboratories, Harvard University,
Cambridge, Massachusetts

It is one thing to ask questions and another to answer them; but of the two, the former is by far the rarer and more difficult—that is, if the questions are of the right sort, capable of being answered, and leading to the next questions. Szent-Györgyi has been having such a dialogue with Nature all his life; from inside, of course, not from outside, since he is part of Nature, like all of us. It is hard to know which is Socrates and which the Athenians, which is gadfly and which stung; but if one stays within earshot, one is almost sure to be stung too.

Some time ago, talking with Szent-Györgyi, I was a little startled to hear him ask, why phosphorus plays the peculiar role it does in organisms. It was a question I had asked myself, as part of a larger question: why any of the atoms that compose organisms occupy the position they do.

This is clearly a proper question, for there is nothing random about the choices. There are the atoms that form the bulk of the organic molecules: C, H, N, and O, and in special instances S and P; there are the major monatomic ions Na^+, K^+, Mg^{++}, Ca^{++}, and Cl^-; there are the trace elements, mostly transition elements, and hence adapted to fill the roles in which we mainly find them, as nuclei and ligands in metallo-organic complexes and oxidoreduction en-

*G. Wald, "Life in the Second and Third Periods; or Why Phosphorus and Sulfur for High-Energy Bonds?" *Horizons in Biochemistry* (Albert Szent-Györgyi Dedicatory Volume), eds., M. Kasha and B. Pullman. New York: Academic Press, Inc., 1962, pp. 127–142. Reprinted with the permission of the author, of the editors and of Academic Press, Inc.

The general lines of this discussion were included in a paper called "Toward a Universal Biochemistry," presented to the American Philosophical Society in Philadelphia at its Fall Meeting in November, 1960.

zymes: Fe, Mn, Co, Cu, Zn. All these elements are relatively light: the 16 elements mentioned fall within the first 30 in the Periodic System. Of the remaining 62 natural elements, only two—I and Mo—have restricted roles as trace elements in certain organisms. By and large, the lightest elements tend to be the most available; but except for the dominant position in the sea of the monatomic ions mentioned above, I think it is clear that something other than simple availability has governed the choice of the elements of which organisms are principally composed.

That "something other" is *fitness,* the peculiar combination of properties that makes these elements most suitable to play the parts that organisms demand of them. As one pursues this argument, it is to realize, I think, that organisms have had little choice among the elements. For the most part, they had to end up as they did. For that reason I think that not only the Periodic System, but the constitution of organisms, is probably about the same everywhere in the universe.

The question we ask is therefore meaningful; and it is within this larger context that I came to ask the question Szent-Györgyi asked: why phosphorus? —except that, as we shall see, I would prefer to ask at once, why phosphorus and sulfur?—for sulfur is part of the same question. And why *not* silicon?— that's part of the question too. It's all a matter finally of being in the Second and not in the Third Period of the Periodic System.[1]

About 99% of the living parts of living organisms are made of the four elements, H, O, N, and C. Some time ago I asked myself, why these?—and after a time thought I knew the answer. *These are the smallest elements in the Periodic System that achieve stable electronic configurations by adding respectively 1, 2, 3, and 4 electrons.* C, N, and O follow one another directly in the Second Period. Next comes fluorine, F; but fluorine, though it too gains a stable electronic configuration by adding one electron, has no part in making organisms, because in any of its properties that might matter for this reason it is outdone by the smaller element, hydrogen. Adding electrons is part of the story, but smallness is the rest. The only thing better than being in the Second Period is to be in the First.[2]

What it means to add electrons is clear enough. Adding electrons by sharing them with other atoms is the mechanism for forming chemical bonds, and hence molecules. But what has smallness to do with it? Two things: (1) the smallest atoms ordinarily form the tightest, most stable bonds; and (2) they alone regularly form stable multiple bonds. Many years ago G. N. Lewis (1923, p. 94) commented that "the ability to form multiple bonds is almost entirely, if not entirely, confined to elements of the first period of eight, and especially to carbon, nitrogen and oxygen." Indeed he suggested that were it not for these elements, the concept of the multiple bond might never have been invented.

[1] There is some equivocation in numbering the periods of the Periodic System. I shall call the First Period H and He; the Second Period running from Li to Ne, and the Third Period from Na to A. I shall speak therefore of C, N, and O as in the Second Period, and Si, P, and S as in the Third.

[2] It should be added that the tendency to add electrons, carried to an extreme, ends in forming anions rather than molecules. Their strong tendency to form anions is one reason why fluorine—the most electronegative of the elements—plays no role in biochemistry, and why chlorine plays its principal role as an ion. This is also why I have paid so little attention to chlorine in the discussion of multiple bonding.

Incidentally the rare occasions when one does encounter multiple bonds outside this trio involve most frequently sulfur and phosphorus.

All of this comes to a head in the comparison of carbon with silicon. In those parts of the earth even remotely available to living organisms, silicon is about 135 times as plentiful as carbon. In the surface layers of the earth, including the atmosphere and hydrosphere, silicon constitutes 16.08% of all atoms, carbon only 0.119%. Coming directly under carbon in the Periodic System, it shares with carbon the property of tending to gain four electrons, and so to form four covalent bonds by sharing electrons with other atoms. Why then is life on earth based upon the relatively rare element carbon rather than on the prevalent silicon?

First, the strength and hence stability of bonding. These are shown in the following table, from which it is clear that the interatomic distance (bond length) is much smaller in a C–C than in an Si–Si bond, and the bond energy of the former is almost twice that of the latter:

	Interatomic distance (Å)	Bond energy (kcal per mole)
C–C	1.54	80–83
Si–Si	2.34	42.2

It is clear that to the degree that the business of making organisms includes the capacity to form tight, stable bonds, and eventually long, stable chains of atoms, carbon has a large intrinsic advantage over silicon.

The importance for organisms of the capacity to form multiple bonds can best be illustrated by comparing CO_2 with SiO_2. In CO_2, carbon is bonded to each of the oxygen atoms by double bonds, each involving the sharing of two pairs of electrons. By this means, each of the atoms of CO_2 achieves the complete octet of outer shell electrons found in the neighboring inert gas, neon. All the combining tendencies are satisfied, and the molecule, free and independent, goes off in the atmosphere as a gas, and readily dissolves in and combines with water, the forms in which living organisms obtain and use it.

In SiO_2, on the contrary, Si is joined to oxygen by single bonds, leaving two unpaired electrons on the silicon and one on each of the oxygen atoms. Unable to pair by forming multiple bonds, these pair instead with the unpaired electrons on neighboring molecules of silicon dioxide. This process, repeated endlessly, ends in a huge polymer, in a sense a huge supermolecule of silicon dioxide. This is the essential structure of quartz—an extraordinarily dense, hard, inert material, which can be broken only by breaking covalent bonds.

Silicon has a third fundamental disability relative to carbon. Like carbon, silicon has a strong tendency to combine with itself to form chains. Potentially it should be possible for silicon compounds to exist in a variety and complexity rivaling the carbon compounds. From the point of view of life, however, the former have a fatal disability: Si–Si bonds are unstable in the presence of water, ammonia, or oxygen.

The reason for this is fundamental to our discussion. In such a Second-Period element as carbon, the outer shell contains one 2*s* and three 2*p* orbitals, each potentially capable of holding a pair of electrons of opposed spin, so completing an octet. (In forming four tetrahedral covalent bonds, the four second-shell orbitals hydridize to form four identical hybrid *sp* orbitals.) The formation of four covalent bonds—in whatever mixture of single and multiple bonds—in a sense completes the octet, filling all available second-shell orbitals, and that is the end of it.

In silicon, however, the outermost electron shell is the third. When the one 3*s* and three 3*p* orbitals have been filled, as the result of chemical combination, the atom has achieved a measure of stability, its electron configuration—though distorted in that the atomic orbitals have been replaced by molecular orbitals—approaching that of the inert gas argon. The completion of an octet of electrons, however, does not saturate the third shell. It possesses in addition five 3*d* orbitals, holding potentially 5 further pairs of electrons. That is, though when the third shell is outermost its stable number of electrons is 8, as in argon, the formation of a fourth shell, as in the Fourth Period, ends in bringing the third shell number from 8 to 18, as in the elements from zinc to krypton. After silicon has in effect completed an octet by forming four covalent bonds, its outermost shell can still potentially accept further electrons in the empty 3*d* orbitals.

The molecules to which silicon chains are most susceptible have the particular characteristics of small size and the possession of lone pairs of electrons. The relatively large interatomic spacing of silicon chains (see above) seems to allow such molecules to come in close enough for their lone electron pairs to occupy empty 3*d* orbitals, and to disrupt the chain. Indeed much the same attack is made on linkages between silicon and other atoms than itself. For example, whereas methane (CH_4) is stable to water and sodium hydroxide, silane (SiH_4) is attacked by these substances to form sodium silicate and gaseous hydrogen ($SiH_4 + 2\ NaOH + H_2O \longrightarrow Na_2SiO_2 + 4\ H_2$).[3]

We can state therefore three powerful reasons, one relative and two absolute, why silicon is unsuitable as a basis for forming organisms: (1) It forms much weaker bonds than carbon, both with itself and with other atoms. (2) Its reluctance to form multiple bonds results in the formation of huge, inert, covalently bonded polymers, so removing all but traces of silicon from circulation. (3) The instability of silicon chains and compounds in the presence of oxygen, ammonia, and water should suffice to disqualify them. I cannot imagine life existing at all apart from water, or going very far without oxygen; and both conditions a priori rule out basing the constitution of living organisms upon silicon.

It is precisely these disabilities of silicon that help to create a special opportunity for phosphorus and sulfur. I am thinking here of the opportunity for these elements to play their most distinctive role in organisms, as agents of group and energy transfer. They have of course other functions, but no others that make such special demands, or that single these atoms out so particularly.

Two properties should most facilitate this type of function: on the one hand the capacity to form a wide variety of linkages of small and large energy

[3] This is of course an example of what is called a nucleophilic attack, and the hydroxyl ion is here a typical nucleophilic reagent. I avoid this special terminology in the present paper in order to concentrate upon the phenomena themselves.

potential, i.e., an adequate coinage; and added to this, an intrinsic instability of linkage, so facilitating exchange. The first property lies in the province of thermodynamics; as for the second, thermodynamics sets the stage, but what actually happens lies in another province, that of reaction mechanisms and kinetics. We will be mainly concerned with this latter type of consideration in discussing the special virtues of phosphorus and sulfur.

The thermodynamic aspects of energy and group transfer—the business of "low-energy bonds" and "high-energy bonds," as biochemists commonly employ these terms—have been carefully explored in the twenty years since Lipmann (1941) first formulated this type of concept (cf. Kalckar, 1946, 1947; Oesper, 1950; Hill and Morales, 1951; George and Rutman, 1960; Huennekens and Whiteley, 1960; Rutman and George, 1961). It is by now well recognized that these are poor terms; that one is dealing here, not with the energy localized in a bond—a bond energy in the strict sense—but with the change in free energy that accompanies a transfer reaction, i.e., with the differences in free energy between the reactants and their products. This free energy change in any given instance is compounded from changes in resonance stabilization, ionization, electrostatic forces—indeed all the consequences of reaction under the particular circumstances in which it occurs. It is in this sense that high- and low-energy compounds are at present defined on the basis of the standard free energy change that accompanies the hydrolysis of a particular bond; a $\Delta F°$ of -1 to -3 kcal per mole characterizes a low- and -5 to -10 kcal per mole a high-energy bond.

By far the widest variety of energy and group transfer reactions in biological systems is carried out by organic phosphates. Sulfur forms three types of "high-energy" complex, one consisting of acyl esters of thiols (e.g., acetyl CoA); another of mixed anhydrides of phosphoric and sulfuric acids [e.g., adenosine-phosphoryl-sulfate for sulfonations (cf. Lipmann, 1958)]; and a third of sulfonium compounds [e.g., S-adenosyl methionine for methylations (cf. Cantoni, 1953, 1960)]. This short list almost exhausts the known categories of "high-energy" compounds. One must add an—as yet—very limited class of acyl imidazoles, in which the acyl group is attached to a nitrogen atom (Stadtman and White, 1953; Stadtman, 1954); and the highly important class of activated amino acids, in which the amino acid carboxyl group is joined in ester linkage with the 2′- or 3′-OH of ribose in the terminal adenylic acid of transfer-RNA (Zachau *et al.*, 1958; Hecht *et al.*, 1959; Preiss *et al.*, 1959). These latter instances are as yet the only ones I know in which group transfers are negotiated by something other than sulfur or phosphorus compounds.

What lends sulfur and phosphorus this special position? Or to put this question a little more in the context of our general argument, what properties do sulfur and phosphorus have that oxygen and nitrogen—their congeners in the Periodic System—lack, that fit them better to perform this type of function?

I think the answer to this question is to be sought in three directions, all multiply interconnected: (1) S and P form more open and in general weaker bonds than O and N. (2) S and P can expand their covalent linkages beyond 4 on the basis of their 3*d* orbitals. (3) To a unique degree among Third-Period elements, S and P retain the capacity to form multiple bonds.

1. The interatomic distances (bond lengths) are larger in covalent linkages of S and P than in those of O and N. In making such comparisons one must consider comparable molecules, since the length of any particular bond varies somewhat with the remaining structure of the molecule. A first approach to this situation is gained in such tabulations of covalent bond radii (Table I) as given by Pauling (1960, pp. 224, 228):

Table I

Covalent Bond Radii (Å)

	C	N	O
Single bonds	0.772	0.74	0.74
Double bonds	0.667	0.62	0.62
	Si	**P**	**S**
Single bonds	1.17	1.10	1.04
Double bonds	1.07	1.00	0.94

The interatomic distances of bonded atoms, obtained by adding together the appropriate bond radii, are reasonably reliable for molecules in which the atoms in question possess the numbers of covalent bonds ordinarily associated with their position in the Periodic System (i.e., C, 4; O, 2; P, 3; etc.), and in which the bonds do not possess too much ionic character or in which the bond order does not depart too greatly from a whole number.

In general the larger bond radii encountered in the Third Period go also with the formation of weaker bonds—bonds of lower energy—than those formed by their congeners in the Second Period. This point was made earlier in comparing Si–Si with C–C. So, for example, the bonds formed by these atoms with hydrogen have the energies shown in Table II (Pauling, 1960, p. 85):

Table II

Bond Energies (kcal per mole)

C–H	98.8	Si–H	70.4
N–H	93.4	P–H	76.4
O–H	110.6	S–H	81.1

The first point in our argument therefore is that in going from the Second to the Third Period, from N and O to P and S, one goes in general to more widely open and weaker bonds, hence bonds intrinsically more susceptible to attack, and more ready to undergo cleavage and exchange reactions.

2. We have already pointed out with reference to silicon that elements of the Third Period possess *d* in addition to *s* and *p* orbitals, and so have place to hold electrons beyond the normal outer-shell octet. It is this property that most distinguishes S and P from their congeners in the Second Period. The five

3d orbitals of S and P could hold potentially 5 pairs of electrons. It seems however that the possibilities of forming stable linkages are more restricted. Ordinarily phosphorus does not go beyond 5 covalent bonds, as in PCl_5, and sulfur does not go beyond 6, as in SF_6 Some formalisms prefer to represent such molecules otherwise than possessing 5 or 6 equivalent covalent bonds; but it is significant that, however represented, they call upon properties not shared by N and O, which do not form molecules of comparable valence.

We can expect therefore that as with compounds of silicon, the presence of unoccupied *3d* orbitals in S and P invites attack by molecules possessing lone pairs of electrons, the lone pairs occupying in an intermediate stage the *3d* orbitals, with an exchange reaction as the eventual result.

3. Sulfur and phosphorus, to a unique degree among Third Period elements, retain the tendency to form multiple bonds, otherwise so characteristic of carbon, nitrogen, and oxygen. Having stressed earlier in this paper that silicon does not markedly display this property, it is interesting to ask why sulfur and phosphorus possess it.

The tendency of elements to form multiple bonds seems to me to be associated somehow with small size as such. I have not been able to find a clear formulation of this view in the literature, nor can I formulate it myself. Coulson (1953, p. 178) ascribes "the experimental fact that multiple bonds are practically confined to the first two rows of the Periodic Table" to increasingly strong repulsive forces between nonbonding electrons as atoms grow larger. If one thinks of the bond as a pair of electrons shared between two atoms, this demands that the bonded atoms approach each other closely. Under such circumstances the electrons not engaged in bonding repel one another strongly, and the larger the number of such electrons the stronger the mutual repulsion. This is thought to be the reason for the relative weakness of bonds between the heavier elements.[4] A multiple bond, since it involves still closer approach between atoms than a single bond, evokes still stronger repulsions. For this reason only atoms of small kernel, containing relatively few electrons, might permit the formation of stable multiple bonds.

This cannot be the whole story, however, for it would seem to imply that in the series Si, P, S, both bond strengths and the tendency to form multiple bonds should decline continuously, since the number of nonbonding electrons and hence the mutual repulsions are rising; whereas just the opposite is the case. The energy values of single bonds are: Si–Si, 42.2; P–P, 51.3; S–S, 50.9; and Cl–Cl, 58.0 kcal per mole; and S and P have more, not less tendency than Si to form multiple bonds.

It may be therefore that a small atomic radius as such promotes the tendency to form multiple bonds. If so, that raises an interesting consideration. The

[4] The large repulsion between nonbonding electrons has also been suggested to account for the abnormally low bond energies of N–N (38.4 kcal per mole), O–O (33.2), and F–F (36.6). N, O, and F form strong bonds with other elements in part because they are so highly electronegative; their bonds tend to have considerable ionic character. The bond energies of P–P (51.3 kcal per mole), S–S (50.9), and Cl–Cl (58.0) are relatively strong, because though the presence of more nonbonding electrons could lead to even stronger interatomic repulsions, these repulsions are relieved by the availability in these elements of *3d* orbitals (cf. Pauling, 1960, p. 144).

atomic radii of the elements do not rise continuously with atomic number; on the contrary, they decline within each period of the Periodic System, jumping to a higher level at the opening of each new period. The reason for this is that the atomic radius is set principally by the number of electron shells, which of course does not change within each period. As one ascends each period, however, the increasing positive charge on the nucleus draws the electrons in closer toward it. For this reason within each period the atomic radii grow smaller as the atomic number rises.

It is not as strange as it first seems therefore that sulfur and phosphorus may make double bonds more readily than silicon, for they are *smaller*. The order of covalent bond radii in Second and Third Period elements is as given in Table III (Pauling, 1960, pp. 224, 228):

Table III

Covalent Bond Radii (Å) vs. Atomic Number

	F	O	N	C	Cl	S	P	Si
Single bond	0.64	0.74	0.74	0.772	0.99	1.04	1.10	1.17
Double bond	0.60	0.62	0.62	0.667	0.89	0.94	1.00	1.07
Atomic number	9	8	7	6	17	16	15	14

Selenium has the same single and double bond radii as silicon, bromine has somewhat smaller radii. All the other elements have larger bond radii. One might suppose from such a series that phosphorus is the largest (not heaviest!) element that readily forms multiple bonds; and that from silicon on only vestiges of this tendency remain.

Occasionally one encounters the notion that it is more rigorous to write the P–O linkages in H_2PO_4 as three single covalent bonds and one dative bond, rather than three single bonds and one double bond[5]. This is however an almost meaningless distinction, and has little basis in fact or theory. I think it arises in part through a supposed analogy with nitrogen, which after forming three covalent bonds, as in trimethylamine, $(CH_3)_3N$, forms a fourth dative bond as in trimethylamine oxide, $(CH_3)_3N \longrightarrow O$. The analogy is not apt, however, since unlike nitrogen, phosphorus can form a fifth covalent bond by accepting a pair of electrons in its 3*d* orbitals. A double bond in addition to three single bonds is clearly possible in this instance. Indeed the measurements of bond length make this the best *single* formula that can be written for phosphoric acid. Similarly sulfuric acid is best written with two single and two double S–O bonds, making six covalences in all (cf. Pauling, 1959, p. 240).

Actually of course no *single* formula properly represents these molecules.

[5] See for example Huennekens and Whiteley (1960; footnote on p. 114): "The P=O double bond (i.e., of phosphoric acid) is more correctly written as a 'semi-polar' double bond: $P \rightarrow O$." Compare this with Pauling's discussion (1960, p. 320), concluding "that the available evidence indicates that the older valence-bond formulas . . . with the double bonds resonating among the oxygen atoms, making them equivalent, and with the bonds considered to have partial ionic character, represent the ions (of the oxygen acids) somewhat more satisfactorily. . . ."

Table IV

BOND LENGTHS IN TETRAHEDRAL OXYANIONS (Å)[a]

Ions:	SiO_4^{4-}	PO_4^{3-}	SO_4^{2-}	ClO_4^-
Single bonds				
Calculated	1.76	1.71	1.69	1.68
Observed	1.63	1.54	1.49	1.46
Contraction	0.13	0.17	0.20	0.22

[a] Cruickshank (1961).

They are resonance hybrids in which all the P–O and S–O bonds are considerably contracted compared with single covalent bonds. Cruickshank (1961) has estimated the amount of such contraction in the tetrahedral oxyanions of Third-Period elements (Table IV).[6]

The single bond lengths were calculated here on the basis of the Schomaker-Stevenson formulation (1941), in which a correction term is introduced for the partial ionic character of the bond, owing to differences in electronegativity of the participant atoms. The further contraction of the bonds observed here is taken to be evidence of distributed double-bond character, and involves in each instance the 3*d* orbitals of the central atom.

Cruickshank (1961) has also estimated the rearrangement of bond lengths that accompanies the attachment of another atom or group to one or more of the oxygen atoms in such ions. He states it to be "a simple empirical rule" that the *average* X–O distances in XO_4 tetrahedral ions remain equal to the "observed" distances in Table IV. The attachment of another atom or group to O may lengthen that X–O bond by amounts up to 0.15 Å, but the other X–O bonds simultaneously contract so as to preserve the average.

So for example in the tetrahedral ion PO_4^{3-}, each P–O bond is 1.54 Å long. In H_3PO_4 however, the P–OH bonds are about 1.57 Å long, the P–O bond 1.52 Å. If instead of H a large organic radical is attached to O, this produces a much larger asymmetry. In serine phosphate, for example, the P–O bond lengths are as in Scheme I.

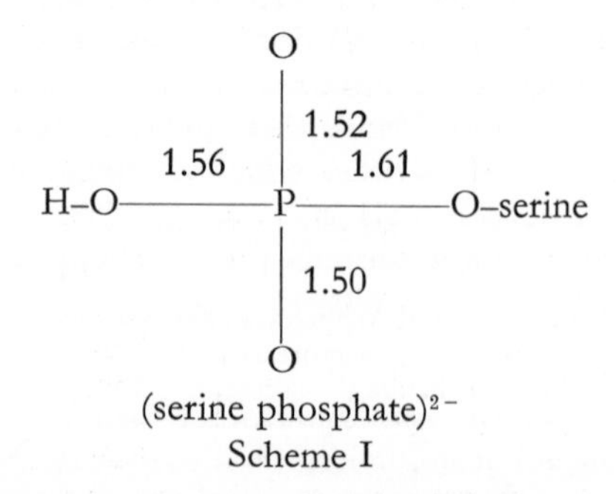

(serine phosphate)$^{2-}$
Scheme I

[6] I am much indebted to Professor William Lipscomb of Harvard University for directing my attention to this paper.

Sulfate exhibits similar relationships. In the ion, SO_4^{2-}, the S–O bonds are 1.47 Å long. An S–OH bond is longer, 1.56 Å for example in $KHSO_4$. In the ethyl sulfate ion, however, (C_2H_5)–O–SO_3–, the S–O(C_2H_5) bond is 1.603 Å long, the three S–O bonds 1.464 Å long.

Molecules in which phosphate P is bonded directly to N, as in phosphocreatine and phosphoarginine, exhibit similar relationships. So for example in the phosphoammonium ion the bond lengths are distributed as shown in Scheme II.

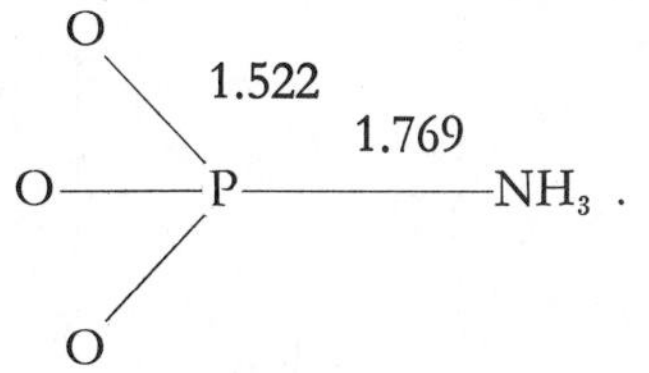

```
    O
     \  1.522
      \         1.769
O------P----------------NH3 .
      /
     /
    O
```

(phosphoammonium)$^-$
Scheme II

In the substituted phosphates and sulfates, therefore, the lengthened bond where an organic or other radical is attached to O makes a preferred opening; the unoccupied *3d* orbitals on P or S provide a berth. Any molecule that can donate a lone pair of electrons should find easy access, and a ready means of attachment to the S or P nucleus. This is a condition that invites attack by water, ROH, RNH_2, and similar molecules; and to some degree this vulnerability must be shared by all organic compounds of sulfur and phosphorus.

This discussion implies that in general the exchange reactions in which organic S and P compounds participate begin—whatever rearrangements may follow—with the addition of a lone electron pair to the S or P nucleus, occupying an empty *3d* orbital. The hydrolysis of an S or P bond for example would begin with the attachment of a lone electron pair of oxygen to S or P. Barring further rearrangements one would expect that in the products of such a hydrolysis, a hydroxyl group derived from water would remain attached to S or P.

Apparently all known enzymatic hydrolyses of organic phosphates do end in this way, as has been shown by carrying out such reactions with H_2O^{18}. All the known kinase reactions, in which molecules with alcohol–OH groups replace water as reagents, also yield this result. In all such reactions the existing P–O or P–N bond is broken, and the –OH or –OR of the attacking molecule remains attached to P (cf. Cohn, 1959).

The reactions of ATP seem to involve a similar mechanism. Here enzymatic hydrolysis (ATPase reaction) and exchange (kinase) reactions with ROH or RNH_2 cleave the terminal phosphate bond, the HO–, RO–, or RN– remaining with the amputated phosphoric acid. In another important class of reactions involving the activation of fatty and amino acids, ATP is cleaved so as to separate inorganic pyrophosphate, yielding the fatty acid or amino acid adenylate as the other product of the exchange (Scheme III). In all these cases the reaction is probably initiated by a lone pair of O or N attaching to P. Why this attack involves the first phosphoric acid in some cases and the terminal phosphoric acid in others is not understood. (No case of pyrophosphoryl transfer has yet been observed.) It should not be forgotten that these are enzymatic reactions,

```
                          OH        OH        OH
                          |         |         |
Adenine-ribose-O——P—O——P——O——P——OH
                          ||        ||        ||
                          O         O         O

                    O                          ..
                    ||   ..               H—OH        ATPase
                   RC—O—H
Fatty and amino      ..                        ..
acid-activating   H  O                    H—OR        Kinases
enzymes           |  ||  ..                    ..
                 RC—C—O—H                 H—NHR       Creatine and
                  |                                   arginine phos-
                  NH2                                 phokinases
```

Scheme III

and the enzyme presumably has much to do with where the attack occurs.

To leave the subject at this point, as I am about to do, is hardly to begin it. However such exchange reactions may in general be initiated, they do not always end as simply as just described. So, for a prominent example, acyl–S–CoA transfers the acyl group so as to leave, not a donator of lone pairs, but a hydrogen atom on S: the product of course is HS–CoA. None of this is especially mysterious, though it does present further problems. The mechanisms of numbers of enzymatic reactions involving the cleavage of organic phosphates have been carefully studied by Cohn, Koshland, and others (cf. Cohn, 1959); and nonenzymatic transfer reactions involving acyl phosphates have recently been explored by Di Sabato and Jencks (1961). Clearly also one should have to invoke quite different mechanisms from those I have discussed to deal with the transfer reactions of acyl imidazoles and the amino acid-ribosyl linkages on transfer-RNA.

To summarize the whole argument: I have tried to find a basis for the biological selection of S and P for group and energy transfer reactions in (1) the fact that they form more open and usually weaker bonds than their congeners in the Second Period, O and N; (2) their possession of 3*d* orbitals, permitting the expansion of their valences beyond four; and (3) their retention of the capacity to form multiple bonds, a property otherwise characteristic of C, N, and O.

The capacity to form multiple bonds contributes principally to the thermodynamics of energy transfer. Particularly when combined, as in P and S, with the possibility of forming 5 and 6 covalent bonds, this introduces a wide range of resonance possibilities among the precursors and products of exchange reactions that greatly increases the variety and extent of the energy changes that can occur. To use an earlier phrase, these properties ensure an adequate coinage.

The relatively wide spacing and weakness of S and P bonds, together with their tendency to add lone electron pairs in their unoccupied 3*d* orbitals, induces an intrinsic instability and vulnerability to attack by other molecules that promote exchange reactions.

I think that as with all the other elements of which organisms are principally composed, sulfur and phosphorus were selected on the basis of fitness: among all the elements of the Periodic System they apparently possess to a unique degree properties that lend themselves to group and energy transfer. It was to find those properties that organisms had to go from the Second into the Third Period.

References

1. Cantoni, G. L. (1953). *J. Biol. Chem.* **204**, 403.
2. Cantoni, G. L. (1960). *In* "Comparative Biochemistry" (M. Florkin and H. S. Mason, eds.), Vol. I, p. 181. Academic Press, New York.
3. Cohn, M. (1959). *In* Symposium on Enzyme Reaction Mechanisms. *J. Cellular Comp. Physiol.* **54**, Suppl. 1, 17.
4. Coulson, C. A. (1953). "Valence." Oxford Univ. Press, London and New York.
5. Cruickshank, D. W. J. (1961). *J. Chem. Soc.* p. 5486.
6. Di Sabato, G., and Jencks, W. P. (1961). *J. Am. Chem. Soc.* **83**, 4393.
7. George, P., and Rutman, R. J. (1960). *Progr. in Biophys. and Biophys. Chem.* **10**, 2.
8. Hecht, L. I., Stephenson, M. L., and Zamecnik, P. C. (1959). *Proc. Natl. Acad. Sci. U.S.* **45**, 505.
9. Hill, T. L., and Morales, M. F. (1951). *J. Am. Chem. Soc.* **73**, 1656.
10. Huennekens, F. M., and Whiteley, H. R. (1960). *In* "Comparative Biochemistry" (M. Florkin and H. S. Mason, eds.), Vol. I, p. 107. Academic Press, New York.
11. Kalckar, H. M. (1946). *In* "Currents in Biochemical Research" (D. E. Green, ed.), p. 229. Interscience, New York.
12. Kalckar, H. M. (1947). *Nature* **160**, 143.
13. Lewis, G. N. (1923). "Valence." Chemical Catalog Co., New York.
14. Lipmann, F. (1941). *Advances in Enzymol.* **1**, 99.
15. Lipmann, F. (1958). *Science* **128**, 575.
16. Oesper, P. (1950). *Arch. Biochem.* **27**, 255.
17. Oesper, P. (1951). *In* "Phosphorus Metabolism" (W. D. McElroy and B. Glass, eds.), Vol. I, p. 523. Johns Hopkins Press, Baltimore, Maryland.
18. Pauling, L. (1959). "General Chemistry," 2nd ed. W. H. Freeman, San Francisco.
19. Pauling, L. (1960). "The Nature of the Chemical Bond," 3rd ed. Cornell Univ. Press, Ithaca, New York.
20. Preiss, J., Berg, P., Ofengand, E. J., Bergmann, F. H., and Diekmann, M. (1959): *Proc. Natl. Acad. Sci. U.S.* **45**, 319.
21. Rutman, R. J., and George, P. (1961). *Proc. Natl. Acad. Sci. U.S.* **47**, 1094.
22. Schomaker, V., and Stevenson, D. P. (1941). *J. Am. Chem. Soc.* **63**, 37.
23. Stadtman, E. R. (1954). *In* "Mechanisms of Enzyme Action" (W. D. McElroy and B. Glass, eds.), p. 581. Johns Hopkins Press, Baltimore, Maryland.
24. Stadtman, E. R., and White, F. H. (1953). *J. Am. Chem. Soc.* **75**, 2022.
25. Zachau, H. G., Acs, G., and Lipmann, F. (1958). *Proc. Natl. Acad. Sci. U.S.* **44**, 885.

PART II

OXIDATIVE PHOSPHORYLATION

THE STIMULATION of phosphorylation by aerobic oxidation of hexose-phosphates was first dealt with in 1930, when V. A. Engelhardt described how esterification of inorganic phosphate in lysed erythrocytes was markedly increased by the addition of methylene blue.[1]

About the same time Lundsgaard found that the number of alactacid twitches which a iodoacetate poisoned muscle is able to perform is greatly increased if the muscle is permitted to respire. In order to insure aerobic conditions, thin flat muscles (like the well-known sartorius muscle) that permit rapid diffusion of oxygen had to be used. Under these conditions, respiration was found to bring about a marked rephosphorylation of creatine.[2] It therefore seemed quite clear that the large increase of working performance brought about by respiration operated through a coupling with phosphorylation of creatine. The energetics of the anaerobic as well as the aerobic coupling will be discussed in the section on energetics. The problem which might be stressed here is the nature of the aerobic metabolism which is able to harness energy for phosphorylation. In the erythrocytes the pathway of oxidative metabolism might well be that of the Embden-Meyerhof pathway with the exception that the pyruvate formed is not converted to lactic acid but instead is further oxidized. In a muscle unable to form lactic acid it seemed more likely that aerobic metabolism proceeded along different pathways. The idea that oxidation of fatty acids may play a role in the iodoacetate poisoned muscle was actually discussed at that time.[3]

Knowledge about the nature of the respiratory chain was developing rapidly subsequent to these developments. As early as the middle and late twenties Otto Warburg,[4] with great precision, described the nature of the hemin catalyst, which he found responsible for the activation of oxygen ("Atmungs-ferment"). About the same time Heinrich Wieland and Torsten Thunberg advanced the ideas of hydrogen activation.[5] An important link between the two systems, cytochrome *c*, was discovered and described by David Keilin in 1925.

Keilin's description of cytochrome offers a fascinating example of the individualistic approach of a physiologist and general biologist who also became a biochemist. With the aid of a simple hand spectroscope, Keilin observed in the thin muscles of the vibrating wings of insects the appearance of a set of characteristic absorption lines corresponding to those of a hemin compound. In his first description of this hemin compound, Keilin paid a special tribute to the zoologist McMunn, who fifty years earlier had observed cellular heme pigments and considered them genuine constituents of the cell rather than degradation products of hemoglobin as had been implied by the biochemist Hoppe-Seyler. These events are described so much better by Keilin himself in his first cytochrome *c* article that we found it particularly pertinent for reprinting [*q.v.*].

In the years between 1934 and 1938 it became evident that certain metabolites could also serve as respiratory "catalysts," i.e., as electron transfer systems—an idea initiated by Albert Szent-Györgyi. Thunberg, Quastel, and others had been describing the succinic acid dehydrogenase. This enzyme catalyzes the dehydrogenation of succinate to fumarate, which can be converted to malate when catalyzed by a specific enzyme, fumarase. Malate dehydrogenase in turn catalyzes the dehydrogenation of malate to oxaloacetate. The latter reaction is reminiscent of the lactate dehydrogenase system. Szent-Györgyi showed, however, that the dicarboxylic acid system (as contrasted to the monocarboxylic system lactate $\rightleftharpoons$ pyruvate) is capable of acting in "catalytic" amounts in hydrogen transfer systems. [*q.v.* his summary of these problems and[6]].

The story of tricarboxylic acids and their function in tissue started much like that of the dicarboxylic acid. In 1936 Knoop and Martius [*q.v.*] synthesized citric acid from oxaloacetate and pyruvate and suggested that this type of condensation could be a prototype for citrate formation in tissues.

Soon afterward, Martius was able to provide evidence that such a reaction took place in various tissues.[7] In one master stroke Hans Krebs [*q.v.*[8]] in 1937 created "a new order," the Krebs cycle. The idea of cyclic group transport (the Krebs urea cycle) had already been introduced by Krebs in 1932. The new Krebs cycle was largely devoted to dehydrogenation, i.e., to hydrogen transfer, yet an acetyl group was actually "carried" and simultaneously was undergoing dehydrogenation to carbon dioxide.

These important events undoubtedly sidetracked scores of biochemists and physiologists. In 1937, when I was a slowly developing physiologist, I became sidetracked to a "new" type of aerobic phosphorylation, "oxidative phosphorylation," when I observed a very peculiar type of phosphate uptake. This type of phosphorylation became manifest if dicarboxylic or tricarboxylic acid were incubated with dispersions of cortex of mammalian kidney to which glucose was added as a phosphate acceptor. Why was it that instead of pursuing the main line, i.e., experiments on muscle or yeast, I found a study of kidney cortex so promising? To begin with, Einar Lundsgaard, with whom I was then studying, was primarily interested in the active transport of carbohydrates. It had been implied from several quarters[9] that active sugar absorption might be driven by a phosphorylation-dephosphorylation cycle in the membrane. Intestine and kidney, both organs of active sugar absorption, were known to be rich in phosphatases. This coincidence had been used as a starting point for discussions (cf.[9]) concerning the importance of phosphatases for active sugar transport.* I was not satisfied with the current concept that phosphatases might catalyze phosphorylations as well as dephosphorylations and set out to scout for a really vigorous phosphorylation system that could "drive" the active transport of carbohydrates.

It was known that the reabsorption of sugars took place in the cortex of the kidney which also was found to have a very high respiration. In any event, my more biochemically oriented studies soon revealed the existence of a type of biological phosphorylation that was strikingly different from that previously described in skeletal muscle and in erythrocytes [*q.v.*[10]]. At this point, let us call this "new" type of phosphorylation aerobic phosphorylation and summarize a few of its important characteristics.

1. The "aerobic phosphorylation" was found to have an absolute requirement for respiration, i.e., it could not operate in the absence of oxygen.

*These views were further encouraged by observations that phlorizine, known to inhibit reabsorption of glucose in the kidney, likewise inhibited absorption of glucose in the intestine and phosphorylations-dephosphorylations in tissue dispersions (cf.[9]).

Glycolysis, therefore, was unable to sustain any trace of this type of phosphorylation.

2. Addition of dicarboxylic or tricarboxylic acid to the kidney cortex dispersion strongly stimulated respiration as well as aerobic phosphorylation.

3. Addition of glucose, glycerol, or 5′-adenylic acid to cortex dispersions (in the presence of phosphate and fluoride) gave rise to a marked accumulation of hexosephosphate, α-glycerophosphate, or adenylpyrophosphate (ATP and ADP), respectively, provided dicarboxylic or tricarboxylic acids were undergoing aerobic oxidation (cf.[10]).

4. Addition of fumarate or malate to cortex dispersion (likewise in the presence of phosphate and fluoride, but in the absence of other phosphate acceptors such as glucose or fructose) brought about a peculiar type of coupled phosphorylation by which fumarate or malate accumulated in the form of phosphoenolpyruvate ("PeP").[10]

Regarding the latter reaction (cf.[10] and reprinted article), it seems appropriate to mention that this type of PeP formation represents (as observed fifteen years later in a series of studies from the laboratory of H. G. Wood) one of the key reactions in CO_2 assimilation (i.e., non-photosynthetic carbon dioxide assimilation). In particular, we refer to the studies by Utter and Kurahashi[11] on the carboxylation of PeP to oxaloacetate coupled with a phosphorylation of guanosine diphosphate to guanosine triphosphate.

Oxidative phosphorylation, or "aerobic phosphorylation" as I called it 30 years ago, shows one more peculiar feature. This was discovered in observing that the extent of phosphorylation of glucose in cortex dispersions which uses succinate as "fuel" is significantly higher than that found in the corresponding system, using fumarate (or malate) as "fuel" (see[10]). The step from succinate to fumarate, therefore, also seemed to be coupled to phosphorylation. About two years later, Colowick, Welch, and Cori[12] subjected the system to a careful analysis and established beyond doubt that the oxidation of succinate to fumarate as an isolated step is indeed coupled to phosphorylation of glucose or ADP.

The next important events in this field were set in motion by Belitser and Tsibakowa and by Severo Ochoa (cf.[13]). Both of these studies revealed coupling between phosphorylation (P) and oxygen (O) of a magnitude which prompted a further remodeling of our thinking. They pointed out that P: O ratios of two or three which they were able to observe could be accounted for only if one cession the energy from the transfer of hydrogen and electrons between carriers could be harnessed for phosphorylation. These first descriptions and discussions of that high P: O ratio in oxidative phosphorylation (which we owe to Belitser and to Ochoa) are both reprinted here [*q.v.*].

The next step of major importance came ten years later when Kennedy and Lehninger [*q.v.*] first established mitochondria as the seat of oxidative phosphorylation as well as of oxidative metabolism of fatty acid of the Krebs cycle. Lehninger's further studies [*q.v.*] demonstrated clearly the segregation between pyridine nucleotides involved in the Krebs cycle oxidations which are clearly intramitochondrial and the pyridine nucleotides operating in other types of catabolism, such as glycolysis, which are extramitochondrial. We are reprinting one of his articles [*q.v.*], but the reader is also referred to Lehninger's monograph, *The Mitochondrion,*[14] for discussions along broad lines on these topics.

It seems that respiration is tightly coupled to phosphorylation. Dinitrophenol is known to inhibit a number of assimilatory reactions and at the same time stimulate oxygen consumption. In 1945 Lardy and Elvehjem suggested that dinitrophenol may act as an uncoupler of oxidative phosphorylation (see excerpt). This was experimentally proven by Loomis and Lipmann [*q.v.*] who demonstrated that dinitrophenol is a strong inhibitor of oxidative phosphorylation in rat liver homogenates.

Subsequently, various approaches toward the study of oxidative phosphorylation were pursued. One approach was directed toward the development of submitochondrial or "soluble" systems able to catalyze oxidative phosphorylation. This approach was pursued by Kielley and by Pinchot and Racker. The other approach was more along the lines of cell physiology, and its development was particularly stimulated by the studies of the Swiss physiologist Raaflaub who described the phenomenon "non-osmotic swelling" of mitochondria in 1953. Raaflaub studied volume changes in mitochondria. Although mitochondria in many ways respond like osmometes, Raaflaub soon realized that many of the volume changes observed in mitochondria must be brought about by forces quite different from the usual passive water-salt transport.[15] He found, for instance, that mitochondria suspended in isotonic mannitol solutions swell if phosphate or succinate is added to the suspensions. Since this type of swelling could not be ascribed to osmotic changes, Raaflaub called it non-osmotic swelling. Non-osmotic swelling could be prevented by addition of ATP. The interpretation of these phenomena varied a good deal. Lehninger[14], for instance, suggested that non-osmotic swelling was related to phenomena observed in myosin. Raaflaub's own interpretations might be considered a little more conservative but of no less interest. They also have brought to focus problems which were to become of great interest for our understanding of the contraction-relaxation processes in muscle [see, for instance,[16] and the section on *Relaxation* later in this book]. In the 1954 article by Raaflaub reprinted here [*q.v.*], the role of calcium ions in the phenomenon of non-osmotic swelling is discussed from the point of the interaction by ATP. Raaflaub demonstrates that ATP can act like calcium complexing agents, especially "glycol complexon" which is a particularly strong complexing agent of calcium ions. On this basis, he poses the question, "Could this 'complexor' phenomenon account for the interplay of calcium ions and ATP on mitochondria and mitochondrial metabolism?" All of Raaflaub's important studies have been published in the German language, and this may have contributed to the fact that they have largely escaped attention. The only reference that I found to the interesting article that is included in this volume was in Swiss biochemical literature, which prompted me to translate it and reprint it here.

The Chemiosmotic Theory of Oxidative Phosphorylation

The discovery of non-osmotic swelling and contraction in mitochondria and the studies on active ion transport through the mitochondrial membrane, which seems otherwise to constitute a formidable barrier between the cytoplasm and the inside of the mitochondrion, were bound to influence our ideas about oxidative phosphorylation. They posed problems of a very different nature from those encountered in studies on fermentation. On the other hand, successes in preserving oxidative phosphorylation in submitochondrial particles encouraged belief in chemical models of oxidative phosphorylation not unrelated to those established for fermentative phosphorylations.

Mitchell's interesting ideas about chemiosmosis and oxidative phosphorylation [*q.v.*] place[17] the mitochondrial membrane with its built-in respiratory electron-hydrogen chain as unseparable from oxidative phosphorylation; oxidative phosphorylation in fragmented subunits of mitochondria constitutes no exception to this thesis. The chemiosmotic theory is of interest not only for our understanding of oxidative phosphorylation in mitochondria, but also for our comprehension of photophosphorylation in chloroplasts, both phenomena occurring in heterogeneous systems.

The chemiosmotic thesis focuses on the existence of ionic gradients generated by oxidations in membranes with a proper "polarity." It is assumed that the catalytic sites of the respiratory enzymes are oriented in such a way that the OH^- ions generated by the oxygen cytochrome system are deposited on one side of the membrane, presumably the inside, whereas the H^+ ions generated by the succinate-flavine system and other dehydrogenases are released on the opposite side of the membrane, presumably the outside[17], [18].

The OH^-, $(H_3O_2)^-$, or larger ions are supposed to be unable to pass through the mitochondrial membrane—or the flux, if any, would be very slow. The same applies to the H^+ ion which may be surrounded by organized water. *Hence respiration per se in such an organized membrane is able to generate a very strong gradient of protons* (H^+) *without the intervention of ATP or other energy-rich phosphate bonds*. The gradient, however, can in turn be used to drive phosphorylation of ADP to ATP or other energy-rich phosphate bonds.

The Mitchell theory envisages generation of "active" orthophosphate (a positive "phosphorylium ion") as driven by the respiratory proton gradient. The observations by Cohn [*q.v.*][19] that ^{18}O-labeled phosphate loses its label to water in the presence of intact mitochondria generating oxidative phosphorylation is of great interest in this context. Hence this "multiple" exchange of water with inorganic orthophosphate is linked to oxidative phosphorylation. The rate of this ^{18}O exchange of phosphate with water can proceed as much as tenfold, or more rapidly than the ^{32}P exchange between phosphate and the terminal phosphoryl group of ATP.[20] Many of these observations could be explained by assuming that orthophosphate donates an OH^- group to the accumulated hydrogen ions; the electrophilic phosphorylium ion then reacts with various nucleophilic constituents, ADP, protein-bound histidine,[21] etc., forming $\sim P$. When the $\sim P$ bonds are subsequently hydrolyzed (for example, by enzymes like ATPase), the released phosphorylium ion will incorporate an OH^- which, if it stems from the outside of the membrane, would contain ^{18}O if the experiment employs ^{18}O-labeled water. A heterogeneous structure like the membrane may even permit a catalysis of the reversible reaction between phosphorylium ions and hydroxyl ions, provided the respiratory chain operates in the way indicated by the chemiosmotic theory (cf.[21a]). Perhaps, such a reversible reaction might best account for the high multiplicity of ^{18}O cycles over those of phosphate as determined by ^{32}P exchange into ATP.

The assumption that each phosphorylation step through the mediation of a phosphate acceptor might "siphon off" one phosphorylium ion and hence also one hydrogen ion might also account for some other observations made on mitochondria. Chance and his coworkers[22], for instance, made the important observation that the availability of phosphate acceptors in respiring mitochondria increased the steady state levels of hydrogen transfer systems towards the more oxidized state. This feature, which is strikingly different from the effects observed in anaerobic pathways, might be partly explained by the chemiosmotic model. Availability of phosphate acceptors is supposed to help "short circuit" the hydrogen ion gradient, and this again affects the oxidation-reduction potentials of flavine and pyridine nucleotide systems relative to the cytochrome systems. The prosthetic groups of the dehydrogenases would tend to become more oxidized. The fact that the cytochrome system follows the same trend, however, is difficult to account for solely on this basis.

One completely aberrant reaction observed in mitochondrial metabolism should be mentioned. Hunter[23] found a type of anaerobic phosphorylation in mitochondria which is driven by an anaerobic dismutation of α-ketoglutarate to oxaloacetate, giving CO_2 and succinate from the former and malate from the latter. Hunter found this dismutation coupled to phosphorylation and the latter insensitive to dinitrophenol. This type of phosphorylation is clearly different in nature from oxidative phosphorylation.

Returning to the discussion of chemiosmosis, an additional potentially highly important application of this type of process ought to be mentioned. The recent observations by Jagendorf [*q.v.*] that acid-base transitions in chloroplasts can elicit ATP synthesis in the dark are most relevant to the problem of photophosphorylations coupled with the various types of "Hill reactions" observed in chloroplasts[24], [25], [26].

If phosphorylation is generated through the development of a gradient, one could well argue that oxidative phosphorylation as well as photophosphorylation are more or less "optional" features of cell function[27]. This might have important bearings on our concept of photosynthesis. Warburg has advanced the idea of a photosynthetic "CO_2-photolyte"[28], [29]. If this form of "active CO_2" plays a role in photosynthesis, one might well consider a direct activation of CO_2, or of carbonic acid, through the development of an acid-base gradient. It should also be borne in mind that Warburg's studies on photosynthesis in chlorella are carried out at a *p*H of 4.

It has recently been shown that a "pulse of H^+" to chloroplasts can bring about luminescence of chlorophyll[30]. The latter observation places phosphorylation at a relatively secondary level as compared with the primary generation of energy. On the other hand, Arnon and his group, using quenching of chloroplast fluorescence as a signal for transformation of radiant energy into chemical energy, have gathered indications in favor of a tight coupling of phosphorylation with the primary photosynthetic process. It was found that addition of phosphate acceptors to chloroplast preparations brought about a marked additional quenching of chlorophyll fluorescence. This quenching was interpreted as an establishment of coupled phosphorylation [*q.v.*].

Since carbon dioxide is required for some of the primary processes in photosynthesis, such as the Hill reaction[31], it would be interesting to see if a quenching of chloroplast fluorescence could be brought about by addition of CO_2 (in the absence of phosphate and phosphate acceptors). If such a type of quenching could be demonstrated, and especially if independent of the presence of phosphate or phosphate acceptors, it would indicate the existence of a link between the primary photoreactions and active carbon dioxide which would seem of relevance in a discussion of the nature of the photolyte.

The implications derived from the idea that phosphorylation may after all merely be an optional feature in cell respiration may be far reaching. At the present time, however, nobody probably feels that our knowledge is sufficient to subject this topic to an exhaustive discussion.

From the point of view of evolutionary concepts, however, Wald, McElroy, and Seliger have made interesting comments. Wald's interest in evolution from "the start" (origin of life) induced him to an assessment of phosphate regarding its fitness as a strategic constituent in the origin of life [*q.v.*]. McElroy and Seliger's discussion is centered around another fascinating aspect of life, bioluminescence.

Generation of Photons

Some very special oxidation systems are so engineered that they are able to generate light. McElroy's remarkable work has taught us that a specific protein, luciferase, isolated in the pure form soluble in water can generate one large quantum of light for each mole of activated substrate (adenylo luciferin) being oxidized by peroxide[32]. This unique efficiency in generating a photon does not even require a membrane system. Remember, however, that the interior of proteins is highly organized, can be highly hydrophobic, and is also capable of undergoing great conformational changes. McElroy has discussed the possibility that bioluminescence processes may represent vestiges of early organizations in evolution meant to detoxify peroxides formed in oxidation[33]. Later, when the respiratory chain came into existence, the membrane structures were supposed to take care of the peroxides in a more economical "earthbound" way, perhaps along the lines previously discussed, eventually resulting in oxidative phosphorylation.

References

1. Engelhardt, V. A.: *Biochim. Z.,* **227** (1930), 16; *Biochim. Z.,* **251** (1932), 113, 343.
2. Lundsgaard, E.: *Biochim. Z.,* **217** (1930), 162; *Biochim. Z.,* **227** (1930), 51.
3. Meyerhof, O., and Boyland, E.: *Biochim. Z.,* **237** (1931), 406.
4. Warburg, O.: *Schwermetalle als Wirkungsgruppen von Fermenten.* Berlin: Saenger, 1948.
5. Thunberg, T. L.: "The Hydrogen-Activating Enzymes of the Cells." *Eighth Sedgwick Memorial Lecture* (given in Woods Hole, Mass., 1929). Baltimore, 1930.
6. Szent-Györgyi, A.: *Z. physiol. Chem.,* **244** (1936), 105.
7. Martius, C., and Knoop, F.: *Z. physiol. Chem.,* **246** (1937), 1.
8. Krebs, H. A., and Johnson, W. A.: *Enzymologia,* **4** (1937), 148; "The Tricarboxylic Cycle," in *The Harvey Lectures,* **44** (1948–49), 165. Springfield, Ill.: Charles C Thomas, Publishers, 1950.

9. Lundsgaard, E.: *Biochim. Z.,* **124** (1933), 209, 221; Rosenberg, Th., and Wilbrandt, W.: *Internat. Rev. Cytol.,* **1** (1952), 65.
10. Kalckar, H. M.: *Enzymologia,* **2** (1937), 47; *Fosforyleringer i Dyrisk Vaev* (with an English Review). Copenhagen: Nyt Nordisk Forlag, A. Busck, 1938.
11. Utter, M. F., and Kurahashi, K.: *Am. Chem. Soc.,* **75** (1953), 758; Kurahashi, K., and Utter, M. F.: *Fed. Proc.,* **14** (1955), 240.
12. Colowick, S. P., Welch, M., and Cori, C. F.: *J. Biol. Chem.,* **133** (1940), 641.
13. Ochoa, S.: *J. Biol. Chem.,* **138** (1941), 751.
14. Lehninger, A. L.: *The Mitochondrion.* New York, Amsterdam: Benjamin, Inc., 1964.
15. Raaflaub, J.: *Helv. Physiol. Pharmacol. Acta,* **11** (1963), 142.
16. Weber, A., and Winicur, J.: *J. Biol. Chem.,* **236** (1961), 3198.
17. Mitchell, P.: "Molecule, Group and Electron Translocation through Natural Membranes," in *The Structure of the Membranes and Surfaces of Cells,* eds. J. K. Grant and D. J. Bell, 142. Cambridge: Cambridge University Press, 1963.
18. ———, and Moyle, J.: *Nature,* **208** (1965), 147.
19. Cohn, M.: *J. Biol. Chem.,* **201** (1953), 735.
20. Mitchell, R. A., Hill, R. D., and Boyer, P. D.: *J. Biol. Chem.,* **242** (1967), 1793; Hinkle, P. C., Penefsky, H. S., and Racker, E.: *J. Biol. Chem.,* **242** (1967), 1788.
21. Suelter, C. H., de Luca, M., Peter, J. B., and Boyer, P. D.: *Nature,* **192** (1961), 43.
22. Chance, B., and Williams, G. R.: *Adv. Enzymol.,* **17** (1956), 92.
23. Hunter, F. E.: *J. Biol. Chem.,* **177** (1949), 361.
24. Arnon, D. I.: *Brookhaven Symp. Biol.,* **11** (1959), 181.
25. Frenkel, A. W.: *Brookhaven Symp. Biol.,* **11** (1959), 276.
26. Hill, R.: "The Biochemists' Green Mansions: The Photosynthetic Electron-Transport Chain in Green Plants," in *Essays in Biochemistry,* eds. P. N. Campbell and G. D. Greville, **1**, 121. London and New York: Published for The Biochemical Society by Academic Press, 1965.
27. Kalckar, H. M.: "High Energy Phosphate Bonds: Optional or Obligatory?," in *Phage and the Origins of Molecular Biology,* eds. J. Cairns, G. S. Stent, and J. D. Watson, 43. Cold Spring Harbor Laboratory of Quantitative Biology, L. I., N. Y., 1966.
28. Warburg, O., and Krippahl, G.: *Hoppe-Seyler's Z. F. physiol. Chem.,* **332** (1963), 225.
29. Vennesland, B.: "Some Flavin Interactions with Grana," in *Photosynthetic Mechanisms of Green Plants,* eds. B. Kok and A. T. Jagendorf, 421. Washington, D. C.: National Academy of Sciences, 1963.
30. Mayne, B. D., and Clayton, R. K.: *Proc. Natl. Acad. Sci.,* **55** (1966), 170.
31. Warburg, O., and Krippahl, G.: *Z. Naturforsch.,* **15b** (1960), 367.
32. McElroy, W. D., and Seliger, H. H.: "Mechanisms of Bioluminescent Reactions," in *Symposium on Light and Life,* eds. W. D. McElroy and B. Glass, 219. Baltimore, Md.: The Johns Hopkins University Press, 1961.
33. ———: "Origin and Evolution of Bioluminescence," in *Horizons in Biochemistry,* Albert Szent-Györgyi Dedicatory Volume, eds. M. Kasha and B. Pullman, 91. New York: Academic Press, 1962.

ON CYTOCHROME, A RESPIRATORY PIGMENT, COMMON TO ANIMALS, YEAST, AND HIGHER PLANTS*

D. Keilin

(ScD., Beit Memorial Research Fellow)
Molteno Institute for Research in Parisitology,
University of Cambridge, England
(Communicated by Sir William Hardy, Sec. R. S.
Received (revised) May 12, 1925)

Introduction

Under the names myohaematin and histohaematin MacMunn (1884–1886) described a respiratory pigment, which he found in muscles and other tissues of representatives of almost all the orders of the animal kingdom. He found that this pigment, in the reduced state, gives a characteristic spectrum, with four absorption bands occupying the following positions: 615—593/567·5—561/554·5—546/532—511/. When oxidized, the pigment does not show absorption bands. In 1887 MacMunn described a method by which it can be extracted in a "modified form" from the muscles of birds and mammals. He found the pectoral muscle of a pigeon to be the most suitable material for the extraction of "myohaematin," in the belief that it was the sole colouring matter of the muscles in a pigeon bled to death. From this "modified myohæmatin" he also obtained other derivatives, such as acid hæmatin and hæmatoporphyrin, and he finally arrived at the conclusion that myo- and histohæmatin are respiratory pigments different and independent from hæmoglobin and its derivatives.

In 1889 Levy carefully repeated MacMunn's experiments in extracting myohæmatin from muscles of birds and mammals and obtained the substance

*Excerpt from D. Keilin, "On Cytochrome, a Respiratory Pigment, Common to Animals, Yeast, and Higher Plants," *Proc. Roy. Soc. B. London,* **98**, 312–324, 337–339 (1925). Reprinted with the permission of the publisher.

described by MacMunn as "modified myohaematin." But he regarded this substance as an ordinary haemochromogen, derived from haemoglobin.

Levy's paper was soon followed by a reply from MacMunn (1889), and by a discussion between this author and Hoppe-Seyler (1890), who fully supported Levy and refused to take into consideration the presence of myohaematin in invertebrates devoid of haemoglobin. As to the four-banded absorption spectrum of myohaematin, which, according to MacMunn, can be seen in a fresh muscle of mammal or bird, Hoppe-Seyler explained it as a mere superposition of bands of oxyhaemoglobin on the surface of the muscle with the bands of reduced haemoglobin of the deeper layer, and possibly also with the bands of a small amount of haemochromogen. Hoppe-Seyler finally dealt with the CO compound which he had obtained from MacMunn's "modified myohaematin" present in the extracted fluid. This compound, which is in all respects similar to the compound obtained from ordinary haemochromogen, Hoppe-Seyler brings as conclusive evidence against the existence of myohaematin as a separate respiratory pigment.

In 1890 MacMunn tried once more to defend his position, but his defence was not even replied to by Hoppe-Seyler, who merely appended to MacMunn's paper a short editorial note, stating that he considered all further discussion as superfluous, MacMunn not having brought any fresh evidence in support of his views. Hoppe-Seyler's note ended the discussion and MacMunn's new respiratory pigment was gradually forgotten. The term myohaematin still made occasional appearances in the literature, but authors mentioning it have seldom seen the pigment or even read MacMunn's original papers; those who have seen the pigment have misunderstood its properties, and have not failed to show that they were aware of Hoppe-Seyler's criticisms, with which they were in full agreement.

Methods

For the spectroscopic examination of living organisms, cells, tissues or their extracts, two instruments have been used—the microspectroscopic ocular of Zeiss and the Hartridge reversion spectroscope. Zeiss's microspectroscope was mainly used for the detection of the pigment when its concentration was very low, and for the examination of opaque tissues, portions of organism, suspensions of cells or turbid fluids. It was also used for examination of rapid oxidation and reduction of the pigment in cells or complete living organisms. For the precise determination of the position of absorption bands, the Hartridge reversion spectroscope gave exceptionally good results. It was, however, slightly modified by Dr. Hartridge for use with the microscope. The main modification consisted in inserting in front of the slit a double-image Wollaston prism.* Calibration of this instrument was obtained by a determination of 14 sharp lines of emission spectra, giving a straight line when plotted. A strong source of light such as the Nernst lamp was used with both microspectroscopes.

*A description of a new reversion microspectroscope will shortly be given by Dr. Hartridge in the 'Journal of the Royal Microscopical Society.'

Distribution of Cytochrome

In the course of my study on the respiration of parasitic insects and worms, I have found that the pigment myo- or histohaematin not only exists, but has much wider distribution and importance than was ever anticipated even by MacMunn. Considering that this pigment is not confined to muscles and tissues, but exists also in unicellular organisms, and further, that there is no evidence that it is a simple hæmatin in the proper sense of the term, the names myo- and histohaematin, given to it by MacMunn, are misleading. In fact, as we shall see later, there is ample evidence that this pigment is not a simple compound, but a complex formed of three distinct haemochromogen compounds, the nature of which is not yet completely elucidated. I propose therefore to describe it under the name of *Cytochrome,* signifying merely "cellular pigment," pending the time when its composition shall have been properly determined. This name, which expresses also its intracellular nature, does not, however, relegate the pigment to any definite compound, an important consideration inasmuch as the properties of various compounds cannot hereafter be ascribed to it without good evidence.

I have found cytochrome in the cells and tissues of a great number of individuals of the following groups and species of animals:—Turbellaria: *Dendrocoela lactea*; Oligochaetes: *Allolobophora chlorotica, Helodrilus caliginosus*; Nematodes: *Ascaris megalocephala, Ascaris suis*; Molluscs: *Limnaea peregra, L. Stagnalis, Helix nemoralis, H. aspersa*; Crustacea: *Oniscus sp., Asellus aquatilis, Cancer pagurus*; Myriapods: *Lithobius forficatus, Geophilus sp.*; Arachnids: *Epeira diademata.* Most of the orders of insects, 40 species of which have been examined. Amongst insects the following is a list of common species, which can be easily recognised and examined:—Diptera: *Musca domestica* (ordinary house fly), *Homalomyia canicularis* (lesser house fly), *Calliphora erythrocephala* (blow fly), *Sarcophaga carnaria, Glossina palpalis* (tsetse fly), *Gastrophilus intestinalis, Eristalis tenax, Anopheles maculipennis* and *Culex pipiens*; Hymenoptera: *Vespa crabro* (hornet), *V. vulgaris* (wasp), *Bombus terrestris* (bumble bee), *Apis mellifica* (honey bee); Coleoptera: *Carabus sp., Melolontha vulgaris, Tenebrio molitor, Dytiscus marginalis*; Lepidoptera: *Pieris brassicæ, Bombyx mori* (silk-worm moth), *Galleria mellonella* (wax moth); Hemiptera: *Notonecta glauca, Corixa sp.*; Orthoptera: *Forficula auricularia, Blatta orientalis.*

The study of this pigment in vertebrates required more complicated manipulation, such as perfusion of their circulatory system, and was therefore confined to a few examples: frogs, pigeons, guinea-pigs and rabbits.

The number and wide range of systematic distribution of the species which show this pigment clearly, and which have been enumerated either by MacMunn or myself, is so great that it may safely be concluded that cytochrome is one of the most widely distributed respiratory pigments. Moreover, cytochrome is not confined to animal cells alone. I have found it, and in great concentration, in cells of bacteria, those of ordinary bakers' yeast, and also in some of the cells of higher plants. To avoid all confusion in the terms which will be used in this paper it is important to mention beforehand that cytochrome (= myohaematin = histohaematin) is a pigment distinct both from blood haemoglobin and from muscle-haemoglobin (= myochrome of Mörner = myoglobin of Günther) or their derivatives. In many cells cytochrome may, however, coexist with haemoglobin.

General Characters of Absorption Spectrum of Reduced Cytochrome

Cytochrome in animal tissues

The best material for the study of the absorption spectrum of cytochrome is provided by the thoracic muscles of the honey bee. Specimens of bees frozen at—7°C. are allowed rapidly to thaw. The head and abdomen are cut off, and by compressing the thorax laterally with the fingers the thoracic muscles are expelled in one mass through the anterior opening of the thorax. The muscles of 2 or 3 bees, compressed between a slide and coverslip and examined with the Zeiss microspectroscope, show clearly a very characteristic absorption spectrum (fig. 1) composed of four bands (*a, b, c, d*), the position of which can be deter-

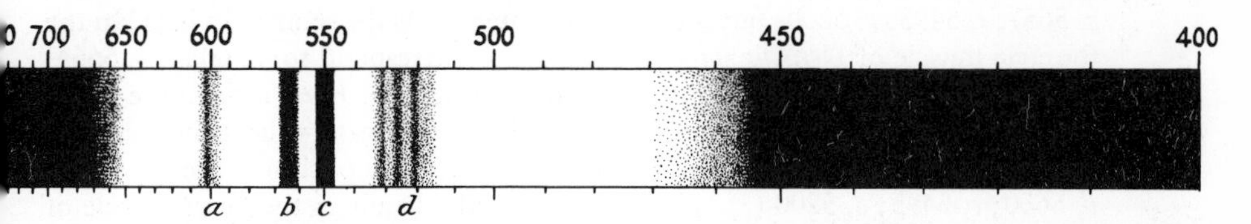

Fig. 1. Absorption spectrum of cytochrome in thoracic muscles of a bee.

mined only with the Hartridge-microspectroscope. For each band I have taken an average of 10 readings and although the pigment was examined *in situ,* the variations between individual readings were only about 7 Ångström units. The position of maximum intensity of the bands in the bees is as follows:—*a*, 6046; *b*, 5665; *c*, 5502; *d*, 5210. The relative width and intensity of the bands, in other words, the general aspect of the spectrum, varies naturally with the thickness of the layer of tissue examined. In moderate concentration the relative width of the band is approximately as follows:—(in $\mu\mu$): *a*, = 614–593; *b* = 567–561; *c* = 554–546; *d* = 531–513. Fig. 1 gives an idea of the absorption spectrum of cytochrome in a layer of muscle 0.65 mm. thick, examined with the Zeiss microspectroscope. It shows that band *a* is very asymmetrical, being darker near the border turned towards the red end of the spectrum; band *b* is symmetrical, but in bees is much lighter than the band *c*, the latter being the strongest band in the spectrum; band *d* which is faint and wide is also asymmetrical, being darker near its short wave end; it also shows a lighter space near the middle, giving to the whole band the appearance of being composed of at least two distinct bands (*x* and *y*) corresponding respectively to 5210 and 5280. In transparent muscles rich in cytochrome, a third very fine band (*z*) could be seen near the green end of the band *d* (532).

The absorption spectrum of cytochrome in other organisms differs very

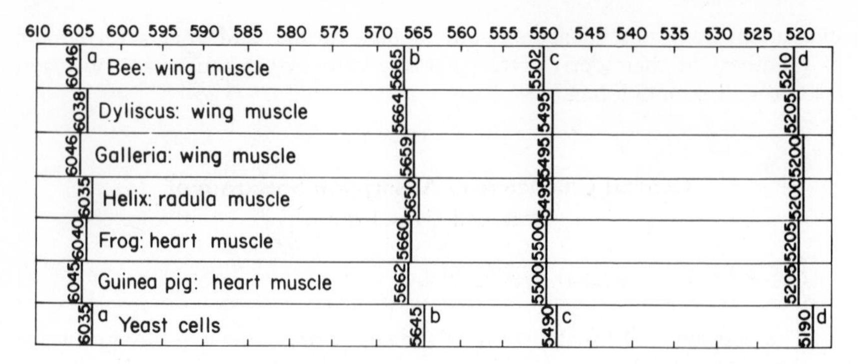

Fig. 2. Positions of the four main absorption bands of cytochrome (*a*, *b*, *c*, and *d*) in various organisms.

little from that of the honey-bee (fig. 2). In the thoracic muscles of a female bumble bee, the position of all the four bands is exactly the same as in a honey bee; band *a* only appearing to be wider and more prominent. In the thoracic muscles of a wax moth (*Galleria mellonella*) the position of the bands is: *a*, 6046: *b*, 5657; *c*, 5495; *d*, 5200; but their relative intensity is the same as in bees. In the thoracic muscle of *Dytiscus marginalis* the bands correspond to: *a*, 6038; *b*, 5664; *c*, 5495; *d*, 5205. In the pharyngeal or radula muscles of *Helix aspersa* the bands are not so easily measured, and band *b* is of approximately the same intensity as band *c*, while band *d* is hardly perceptible. Their positions are: *a*, 6035; *b*, 5650; *c*, 5495; *d*, 5200 (?). The positions of the bands in the heart muscle of a frog are: *a*, 6040; *b*, 5660; *c*, 5500; *d*, 5205. The cytochrome in the heart and other muscles of a guinea-pig shows the bands as follows:—*a*, 6045; *b*, 5662; *c*, 5500; *d*, 5205. In other animals, such as earthworms, crustacea (especially the heart), myriapods (muscles), various insects, different organs of molluscs and birds (pectoral muscles of pigeon), cytochrome was always found with its four characteristic bands, and the variation in the position of these bands was not greater than that which was already shown by the examples mentioned above. It is impossible at this stage to decide whether the small differences observed in the position of the bands of cytochrome in different organisms correspond to a real difference in the composition of this pigment, or are only of the nature of an experimental error, caused by the presence of small amounts of other pigments, such as lipochromes or haemoglobin.

Cytochrome in yeast cells

A slightly wetted fragment of bakers' yeast, compressed between two slides to the thickness of 0.6 mm., and examined with the Zeiss' microspectroscope, shows very clearly the characteristic spectrum of cytochrome, with its four bands: *a*, *b*, *c*, *d*, very similar to those of the thoracic muscles of a bee. Owing to the greater opacity of yeast, the correct position of the four bands is more difficult to determine. This is mainly due to the loss of light which is unavoidable in the Hartridge reversion spectroscope. The insertion of a Ramsden ocular within the tube of the microscope increases the amount of light and makes it possible to obtain a fairly accurate reading. The positions of the four bands in yeast obtained by this method are: *a*, 6035; *b*, 5645; *c*, 5490; *d*, 5190. The positions of these bands in yeast, compared with those of bees or other animals, are slightly nearer the blue end of the spectrum. These differences in the position of the bands are, however, of a much smaller magnitude than those which are known to occur between hæmoglobins of various origins.* We can say, therefore, that cytochrome, in spite of its great range of distribution, shows a very characteristic, easily recognisable and uniform absorption spectrum. This uniformity of the spectrum indicates, moreover, the great similarity in the chemical composition and the properties of this pigment whatever may be its origin.

The four absorption bands of cytochrome are of unequal intensity: thus band *c* is usually the strongest, then come *a* and *b*, and finally *d*, which is faint, and, when the concentration of pigment is low, may easily be overlooked. It is important to note, that this relative intensity of the four bands, although very general, is not so constant as the position of the bands. In some tissues bands *b* or *a* are almost as strong as the band *c*.

When a tissue showing clearly all the four bands is gradually compressed between two slides, the bands become faint and disappear in inverse order to their intensity. During this process the tissue shows a succession of spectra with 3, 2, or 1 absorption bands. This observation shows that the reduction in number of the bands of cytochrome which may be found in tissues of different organisms denotes only a lower concentration of the pigment. Moreover, in the great majority of those cases, the remaining absorption bands can be detected on increasing the depth of tissue examined. This naturally applies to tissues devoid of all pigments other than cytochrome.

The Relative Concentration of Cytochrome in different Tissues and Cells

The relative concentration of the pigment in the tissue is shown by the relative depth of the tissue necessary for the microspectroscope to reveal a clear absorption spectrum with four characteristic bands. A more precise indication is obtained by the thickness of the compressed tissue at which the band *c* is extinguished. The concentration of cytochrome varies not only with the species examined but also, and to a much greater extent, in different tissues of the same animal. In parastic worms, such as *Ascaris lumbricoides,* the highest concentration is found in the spermatozoa and eggs. In a mollusc, *Helix aspersa,* the order of concentration is: (1) pharyngeal or radula muscles; (2) the walls of the stomach; (3) the genital glands and other organs. In crabs the muscles of the heart contain much more cytochrome than the strong muscles of the legs. In insects the thoracic muscles are the richest in this pigment, but in small concentration it is found also in other organs: intestinal wall, brain, genital glands. In frogs the order of concentration is: (1) heart; (2) testes; (3) walls of the stomach; (4) muscles of the body; (5) other organs. In mammals the

*The difference in the position of *a*-band of HbO_2 in *Planorbis* and *Chironomus* is about 31 Å. and that between *Planorbis* and the muscle-hæmoglobin of a mammal may reach as much as 64 Å.

greatest amount of pigment is found in the heart muscle, and then in the muscles concerned with mastication, the diaphragm and the muscles of the legs.

For the study of cytochrome we shall select the muscles of insects and yeast in which it is unmixed with other pigments and in sufficient concentration.

Oxidized and Reduced Cytochrome

The absorption spectrum with four characteristic bands corresponds to the reduced state of cytochrome, while the spectrum of the pigment in its oxidized state, at least in the concentration found in the tissues, shows no distinct absorption bands, but only a very faint shading extending between 520–540 and 550–570 $\mu\mu$. The oxidation and reduction of the pigment can be easily observed in yeast. If a shallow tube (30 mm. high) is half-filled with a suspension of bakers' yeast in water (20 per cent.), and the suspension then examined with the Zeiss microspectroscope, the four absorption bands may be clearly seen; but when the air is rapidly bubbled through the suspension the cytochrome becomes oxidized and the bands disappear. If the current of air is stopped the pigment becomes reduced and the four bands rapidly reappear.

A similar result can be obtained by shaking a 5 c.c. yeast-emulsion in a test-tube and examining it with the microspectroscope. When, instead of air, a current of N_2 is passed through the yeast emulsion, or when the latter is shaken with N_2, the cytochrome remains in a reduced state, showing all the time its characteristic four absorption bands. Similar results are obtained with the thoracic muscles of bees or the striated muscles of a guinea-pig. But in these cases the oxidation and reduction are better seen in the broken-up muscles, which may be spread on a slide. When the muscle is exposed to air the cytochrome is seen in its oxidized form, but when the slide is covered with another slide and the space around the muscle filled with glycerin, the cytochrome becomes reduced and the four bands appear. In these experiments, cytochrome is oxidized by O_2 of the air and reduced by the tissue itself. The oxidation of a reduced cytochrome can easily be obtained without shaking with air, by adding to the tissue a small quantity of potassium ferricyanide or of H_2O_2. On the other hand, when the reducing power of the tissue has been destroyed, cytochrome, which then becomes easily oxidized with air, can be readily reduced by adding to the tissue a small quantity of a reducer such as $Na_2S_2O_4$.

The conditions which determine the oxidation and reduction of cytochrome are not equally influenced by the change of temperature. The oxidation depends upon the rate of diffusion of oxygen into the suspension of yeast on the tissue and is but little affected by temperature; its reduction is due to the chemical activity of the cells which contain the pigment, and therefore has a high temperature coefficient; *e.g.*:—(*a*) At the ordinary room temperature (18°–20°) the oxidized cytochrome of yeast usually becomes reduced by the activity of the cells in six to eight seconds. When the oxidized yeast is kept at 0° to —2°C. complete reduction takes place in 70 seconds to 2 minutes. (*b*) Again, in yeast emulsion kept in a thin layer in a Petri dish at —2° to —4°C. cytochrome usually remains oxidized. Similar results have been obtained with the thoracic muscles of blow-flies and those of bees. These tests show that a low temperature inhibits the reducing power of the tissue more easily than oxidation power of cytochrome.

Action of Narcotics on Cytochrome

When a drop of weak solution of KCN is added to the suspension of yeast, no matter how actively this suspension is shaken with air or pure O_2, the cytochrome remains completely reduced, and continues to show the characteristic bands, which do not differ in the slightest degree from the bands of an ordinary reduced cytochrome. The concentration of KCN which stops oxidation of cytochrome is about *n*/10,000, and a much lower concentration, such as *n*/100,000, inhibits to a great degree the oxidation power of the pigment.

Further and more important, when a drop of KCN is added to the suspension of yeast, kept at a low temperature and previously oxidized by a current of air, the cytochrome becomes immediately reduced, just as if KCN was acting as a powerful reducing agent. In fact, KCN does not act as a reducer, but inhibits the oxidation of cytochrome, while it does not inhibit other oxidation processes which may accompany the reduction of our pigment.

The action of sodium pyrophosphate is similar to that of KCN. It also inhibits the oxidation of cytochrome only, while it does not arrest the reduction of this pigment. Other substances, such as formaldehyde, ethyl alcohol, acetone, and ethyl urethane, act in a very different way. All these substances, even in concentrations which kill the cells of yeast, do not inhibit the oxidation of cytochrome. On the contrary, in such a concentration they completely stop the reduction of cytochrome, which then remains oxidized indefinitely.* In lower concentrations they delay the reduction of oxidized cytochrome, though they do not completely stop it.

If a suspension of yeast in ethyl urethane is shaken with air until the cytochrome becomes completely oxidized, and KCN solution is then added to this suspension, the cytochrome does not become reduced. It remains oxidized because the reducing action of the cells is inhibited, or even destroyed, by urethane, while KCN has no effect on oxidized cytochrome. When the reducing power of the cells is not completely destroyed by the action of urethane, on adding KCN to such suspension a gradual but slow reduction of cytochrome can be observed.

The facts given above show clearly that in relation to the oxidation process in cells in which cytochrome is involved all the inhibitors of oxidation can be separated into two distinct categories, the actions of which are fundamentally different. To one category belong KCN and sodium pyrophosphate; to the other such substances as alcohols, urethane, and aldehydes. The first cate-

*The oxidation of cytochrome in this case differs slightly from ordinary oxidation. In ordinary oxidation all the four bands fade away more or less simultaneously. In presence of urethane or formaldehyde band *c* disappears the first, while the other three bands remain and band *b* seems to be even intensified. On shaking the emulsion for a long time with air, all the bands disappear, band *b* being the last to go.

gory (A) inhibits the oxidation of cytochrome, the second (B) inhibits its reduction.

Diagrammatically this can be represented in the following way:—

$$R \longleftarrow\!\!\!\!\!\!\mid_{B}\!\!- \text{Cytochrome} \longleftarrow\!\!\!\!\!\!\mid_{A}\!\!- O_2,$$

R being the substances which reduce the oxidized cytochrome, B and A indicating the places of rupture in the oxidation system produced respectively by the substances of the corresponding category of inhibitors. This seems to indicate that, at least for oxidation systems similar to that of cytochrome, the problem of the action of narcotics needs further investigation and existing theories require careful revision.

The Natural Behaviour of Cytochrome in Living Organisms

We have previously seen that in a compressed fragment of bakers' yeast or in an emulsion of yeast, cytochrome is usually found in a reduced state. Yet we can hardly speak of natural conditions of life in the case of yeast, which is an organism highly modified by long and constant selection. We have found, however, that while in bakers' yeast the concentration of cytochrome is very high, in brewers' yeast it is very low. This is undoubtedly correlated with the different modes of life of these two categories of yeasts.

In the dissected muscles of a perfused guinea-pig, cytochrome is usually found in a completely, or almost completely oxidized form, and the reduction takes place when the muscles are excluded from the air, *e.g.*, pressed between two slides. The thoracic muscles of insects, on the other hand, even when they are rapidly taken out from the body and immediately examined, almost always show the reduced form of the pigment.

In both these cases we are dealing with portions of tissues excised from an organism and therefore not examined in natural conditions. To understand the function of this pigment it is important to find out in what state this pigment is present in a normal living organism. The main difficulty in answering this question consisted in finding suitable material for such an investigation. One insect, however, the common wax-moth (*Galleria mellonella*) answered the purpose. Several active specimens of this moth were selected out of a large stock bred in the Laboratory, and the thorax of each was carefully cleaned from scales. These specimens were then attached by the ventral surface to a slide (by means of small droplets of gum arabic) and the thorax carefully examined with Zeiss microspectroscope and a strong light.*

The following are the results of these observations:

1. Female of *Galleria* remained very quiet, except for the occasional expulsion of an egg and a somewhat rhythmic movement with ovipositor. The

*In this condition the insects remained alive for long periods, and even after three hours the females, in spite of being attached to the slide, went on ovipositing.

thorax being of a yellow colour showed better the long-wave portion of the spectrum, but no absorption band could be detected.

2. In specimens of males and females which began to struggle constantly, vibrating their wings in efforts to detach themselves from the slides, the bands of cytochrome gradually appeared, band *a* being very clear, and bands *b* and *c* appeared as almost fused into one band.

3. When these specimens ceased to move and stopped the vibration of the wings, the bands became very faint and hardly detectable.

4. In a specimen which showed no absorption bands (the cytochrome being oxidized) a slight pressure exercised upon the thorax made the bands of a reduced cytochrome to appear.

5. Specimens of *Galleria* fixed to the glass bottom of a special air-tight gas-chamber and examined spectroscopically showed no absorption bands. When N_2 or coal gas was passed through the chamber all the four bands of reduced cytochrome appeared very rapidly. When N_2 was cleared with air, the bands rapidly disappeared.

6. When a specimen with oxidized cytochrome was exposed for a few seconds to vapours of KCN all the four bands rapidly appeared and the insect became motionless.

7. The same specimen being brought back into fresh air, the absorption bands of reduced cytochrome gradually faded away and the insect began to show signs of life.

The absorption bands of cytochrome shown by *Galleria* after exertion, still less after intense vibration with the wings, are never so strong as they appear in specimens exposed to pure N_2 or to the vapours of KCN. This fact indicates that in natural conditions cytochrome is in the oxidized form, and that during exertion, however great, cytochrome becomes only partially reduced.

The above experiments with *Galleria* and the previous observations on yeast show that cytochrome acts as a respiratory catalyst, which is functional in oxidized as well as in a partially reduced form. The oxygen is constantly taken up by this pigment and given up to the cells. In the living organism the state of the cytochrome as seen spectroscopically denotes only the difference between the rates of its oxidation and reduction.

These experiments indicate also the rapidity of gas diffusion through the tracheo-spiracular system of an insect. It must be remembered that in winged insects the tracheal system of the thorax is highly developed, the trachea arising from the thoracic spiracles give off numerous branches, which on reaching the thoracic muscles give rise to innumerable small capillary tracheoles. These tracheoles penetrate into the muscular fibres, forming there a rich net of even more minute tracheoles. Ramôn y Cajal (1890), who has carefully investigated the final branches of this system, has described them as being from 0.1 μ to 0.2 μ in diameter. The respiratory movements of insects, when they exist, ventilate only the large trachea, or air sacs connected with it, while the small trachea, to effect the gas-exchange, are dependent on pure diffusion. We can, therefore, easily explain the observations No. 4: the pressure exerted upon the thorax compresses the large trachea, and, therefore, partially cuts off the supply of fresh O_2; this causes partial reduction of the cytochrome and the appearance of the absorption bands.

Concentration of Cytochrome in Relation to Muscular Activity in Insects

We have previously seen, that among all the organisms examined the highest concentration of cytochrome is found in the thoracic muscles of flying insects. This has undoubtedly a connection with the peculiar activity of these muscles. We know in fact from old experiments by Marey (1874) that the wing muscles of insects are capable of producing very rapid contractions. This author found that in the house-fly the wing makes 330 complete vibrations per second. The data given for other insects are: drone fly, 240 vibrations; bee, 190; wasp, 110; hawk moth, 72; dragon-fly, 28; and butterfly, 9. It was also proved that in a fly 330–340 electric shocks per second are required to produce the tetanic contraction of the wing muscle, which shows that the fly's muscle is capable of producing more than 300 separate contractions per second.

This peculiar property of insect-muscle explains the presence in the fibrils of so high a concentration of cytochrome. Moreover, careful examination of wingless insects corroborates this supposition, *e.g.*, the thoracic muscles of the wingless sheep-keds (*Melophagus ovinus*) hardly show the presence of cytochrome. The same applies to the thoracic muscles of the ordinary body louse (*Pediculus humanus*) or bed bug (*Cimex lectularius*). The best example is, however, shown by the winter moth (*Cheimatobia brumata*). The male of this moth, which is provided with well-developed wings and flies well, shows the presence of cytochrome clearly, while the female with reduced non-functional wings scarcely shows the pigment in the muscles. Finally, in cockroaches (*Stylopyga orientalis*) which do not fly, but are good runners, cytochrome is especially localised in the muscles of the legs.

Insects also furnish good material for the study of this pigment during their different stages of development. In the common blowfly cytochrome is already present in the eggs of this fly and in the muscles of the larva. The concentration of the pigment in these muscles is found to be approximately 12 to 15 times lower than that of the thoracic muscles of the adult fly. During the metamorphosis the concentration of cytochrome in the freshly-formed thoracic muscles of the pupa increases with its development. The adult insect, however, does not contain the maximum amount of the pigment immediately after hatching; this is reached during the life of imago, and undoubtedly depends upon the amount and composition of the food supply.*

* * *

Summary

1. Cytochrome is an intracellular respiratory catalyst common to animals, bacteria, yeast and higher plants.

2. In a reduced state it shows a characteristic absorption spectrum with four bands, the position of which in all organisms are approximatively the

*The fat body of these larvae contains on the other hand a haemochromogen-like complex showing three absorption bands, the position of which being approximately 603/558/524.

same: *a*–6046, *b*–5665, *c*–5502, *d*–5210 Å (bees). Band *d* is composed of three secondary bands, *x*, *y* and *z*.

3. In the oxidized form there are no clear absorption bands but only faint shading extending 520–540/550–570.

4. The highest concentration of the pigment is found in the thoracic wing muscles of flying insects, striated muscles of mammals and birds, and bakers' yeast.

5. Cytochrome is easily oxidized with air and reduced by the normal activity of cells or by a chemical reducer.

6. KCN in concentration of *n*/10,000 or sodium pyrophosphate stops the oxidation of cytochrome, but these substances do not prevent the cell reducing the already oxidized form of the pigment.

7. Ethyl urethane, alcohol and formaldehyde do not interfere with the oxidation of cytochrome, but they inhibit the reducing power of the cell, and the cytochrome thus remains oxidized.

8. Under natural conditions in the living animal cytochrome is in the oxidized or in only partially reduced form.

9. The condition of cytochrome as seen spectroscopically in the living organism denotes only the state of equilibrium between the rate of its oxidation and reduction at that particular time.

10. The behaviour of cytochrome in living organisms is dealt with in detail.

11. Evidence is brought forward that cytochrome consists of three haemochromogen compounds (*a*′, *b*′, *c*′) two of which (*b*′ and *c*′) have a haem. nucleus (iron-pyrrol compound) similar to that of haemoglobin.

12. Cytochrome yields as derivatives three haemochromogens which give O_2 and CO compounds.

13. Cytochrome and its derivatives are responsible, at least partly, for the peroxidase reactions in organisms.

14. Cytochrome is distinct from muscle-haemoglobin (= myochrome = myoglobin) and both pigments can be easily seen in the same muscle of a bird or a mammal.

15. Muscle HbO_2 differs from blood HbO_2 only in having the absorption bands slightly shifted towards the red end of the spectrum.

16. The non-coloured portions of plants* show the existence of a haemochromogen-like complex (modified cytochrome) (*a*′, *b*″, *c*′), as well as of cytochrome, both yielding in strong KOH characteristic haemochromogens which give O_2– and CO–compounds.

17. Cytochrome exists in aerobic bacteria and can be oxidized and reduced as in other organisms.

The expenses of this research were defrayed by the Medical Research Council. I wish to thank Mr. J. Barcroft for his constant interest in the progress of this study. My thanks are due to Messrs. M. L. Anson and A. E. Mirsky for the great help they have given me in connection with this investigation. To Dr. I. M. Puri I owe friendly assistance in numerous dissections and the perfusion experiments.

*Cytochrome and its derivatives in plants and bacteria will be dealt with in more detail in a separate paper.

References*

1. Anson, M. L., and Mirsky, A. E., 'Journ. of Physiol,' vol. 60, p. 50 (1925).
2. Cajal, S. R., 'Zeitschr. f. Wiss. Micr.,' vol. 7, p. 332 (1890).
3. Fox, H. M., 'Proc. Chambr. Phil. Soc., Biol. Sc.,' vol. 1, p. 204 (1924).
4. Fisher, H., and Fink, H., 'Hoppe-Seyler's Zeitschr.,' vol. 140, p. 57 (1924).
5. Gola, G., 'Atti R. Accad. dei Lincei Roma,' vol. 24, Part I, p. 1,239 (1915); vol. 24, Part II, p. 289 (1915); vol. 28, Part II, p. 146 (1919). See also 'Chemische Zentralbl.,' Part I, p. 1,249 (1916), and Part I, p. 414 (1922).
6. Hoppe-Seyler, F., 'Zeitschr. f. Physiol. Chem.,' vol. 14, p. 106 (1890).
7. Keilin, D., 'Nature,' vol. 115, No. 2890, p. 446 (1925).
8. Lankester, E. Ray, 'Roy. Soc. Proc.,' London, vol. 21, p. 70 (1872–73).
9. Levy, L., 'Zeitschr. f. Physiol. Chemie,' vol. 13, p. 309 (1889).
10. Lowne, R. T., 'The Blow-Fly,' Part I, p. 206 (1890–92).
11. MacMunn, C. A., 'Phil. Trans.,' vol. 177, p. 267 (1886).
12. MacMunn, C. A., 'Journ. of Physiol.,' vol. 8, p. 57 (1887).
13. MacMunn, C. A., 'Zeitschr. f. Physiol. Chem.,' vol. 13, p. 309 (1889).
14. MacMunn, C. A., *ibid.,* vol. 14, p. 328 (1890).
15. Marey, E. J., "Animal Mechanism," 1874, 'Internat. Sc. Series.' vol. 11, London.
16. Petit, 'Compt. Rend.,' vol. 115, p. 246 (1892).

*Only those references that apply to the excerpted text will be cited.

David Keilin and Thaddaeus Mann at the Molteno Institute. "Keilin continued to find much aesthetic satisfaction in visual spectroscopy and was sorry for those to whom cytochrome is no more than a line drawn by a servo-operated pen." (Quoted from Hartree and Mann).

[The late Professor Keilin was director of the Molteno Institute of Biology and Parasitology, University of Cambridge, 1931–1963 and recipient of the Copley Medal of the Royal Society, 1951.]

OXIDATION AND FERMENTATION*

Albert von Szent-Györgyi, Ph. D., M. D.

Professor of Organic and Biological Chemistry,
University Szeged, Hungary

Whatever a cell does has to be paid for in the currency of energy. If there is no free energy available there is no life. The animal cell does not generate its own energy. It obtains energy ready-made from outside in small parcels which we call foodstuff molecules.† The cell knows two methods of getting out the energy of these molecules; it either fragments them or burns them. The first method we refer to as fermentation, the second, oxidation.

Let us take as an example the lactic acid fermentation in muscle cells. In this process the hexose molecule is fragmented finally into two molecules of lactic acid. The latter contain collectively somewhat less energy than the original hexose molecule. This little difference in energy is the clear gain of the cell. The alternative method is to burn the hexose molecule into CO_2 and H_2O, whereby the bulk of free energy is set free.

Fermentation is much the simpler of the two processes. At the same time it is very uneconomical, since the greatest part of the energy of the hexose molecule remains enclosed in the lactic acid molecules. About thirty times as much energy is liberated by oxidation. Hence fermentation can maintain only the simplest forms of life. There can be little doubt that fermentation is not only the simpler, but also the older process, preceding oxidation in the history of life. The development of more complicated and hence more pretentious

*A. von Szent-Györgyi, "Oxidation and Fermentation," in *Perspectives in Biochemistry,* F. G. Hopkins Dedicatory Volume, eds. J. Needham and D. E. Green, 165–174. London: Cambridge University Press, 1937. Reprinted with the permission of the author, the editors, and of the publisher.

†These parcels are made by the chlorophyll-containing plant cells which have harnessed the radiant energy of the sun.

forms of life became possible only after Nature discovered oxidation by molecular oxygen. This course of events is still reflected in our cells, in which we find oxidation and fermentation intimately mixed and woven into one energy-producing system.

The intimate relations between the two processes has occupied many biochemists since Pasteur discovered their quantitative interdependence, now known as the 'Pasteur Reaction'. Pasteur found that there was some sort of equilibrium between oxidation and fermentation. If oxidation is suppressed by lack of oxygen, fermentation begins. If we promote again oxidation, fermentation is set to rest. The mechanism of this relation has been one of the most attractive puzzles of biochemistry ever since.

We will try to solve this puzzle; but we are still more amibitious, for we will try to answer the deeper problem, how Nature progressed from fermentation to oxidation. Was the process discontinuous in the sense that a new oxidative mechanism was evolved side by side with fermentation, or was the process continuous—the fermentation system being refitted for the new purpose of oxidation?

To approach the problem we must start by summing up the essential facts known about the chemical mechanism of oxidation and fermentation. Following Nature's way we will start with fermentation and will use throughout the example of the muscle cell.*

Every year during the last decade a new and different theory of fermentation claimed our adherence. New ways in which the cell can decompose glucose into lactic acid were constantly being discovered. In the last two years, a new theory was again proposed by Dische, Meyerhof and Parnas, and this time I think it really describes the main road of fermentation.

All theories new and old agree that the first step in the fermentation of sugar must be the splitting of the six-carbon atom chain of hexose into two sugars with three carbon atoms each. The problem is how this triose is converted into lactic acid. The new theory runs as follows: Pyruvic acid, $C_3H_4O_3$, contains the same elements as the triose, $C_3H_6O_3$, only with two less hydrogen atoms, and is thus in a higher state of oxidation. If there is a transfer of two atoms of hydrogen from the triose to pyruvic acid, we say that the triose has been oxidized by the pyruvic acid, and conversely the pyruvic acid has become reduced. Pyruvic acid on reduction yields lactic acid, whereas the triose on oxidation yields pyruvic acid. The process could then start all over again with a new triose molecule, thus:

```
                                   2H |
                                      v
                       Triose ---> Pyruvic ---> Lactic
                       2H |
                          v
             Triose ---> Pyruvic ---> Lactic
             2H |
                v
   Triose ---> Pyruvic ---> Lactic
```

Once begun, this would go on by itself, with the result that the whole of the sugar would be converted into lactic acid.

*No doubt the story is the same in yeast with some variation. Nature knows only a few fundamental principles, which she cleverly adapts to different purposes and circumstances.

Naturally, if triose and pyruvic acid are mixed outside the cell, no reaction occurs. In the cell the reaction is accelerated by specific enzymes called dehydrogenases, which are responsible for activating both triose and pyruvic acid. In the above scheme arrows signify these enzymes.

Of course this story is not complete. We have simplified it to bring out the essence. We have passed over a few things which form perhaps the most exciting chapters of modern biochemistry. That triose does not go over immediately into pyruvic acid, but is first converted into glyceric acid, and only this latter turns into pyruvic, is a minor point. Of primary interest is the role of phosphoric acid, which has no less a function than to transmit the liberated energy from one stage to another. The fact, however, that it is not triose which is fermented, but triosephosphoric acid ester, and that it is not pyruvic acid but pyruvic-phosphoric acid ester which is formed, does not affect our conclusions. Nor does it matter that in the make-up of the dehydrogenases, nucleotides play an important part.

Now let us consider oxidation. Oxidation was thought a few decades ago to be exceedingly simple. There was no problem at all. The organic molecule was burned and energy produced, much as in the steam engine. Since then the problem of oxidation has developed into a wide and fruitful field of scientific enquiry.

Whatever the mechanism of oxidation may be, it must be more involved than that of fermentation, for a new element, oxygen, comes into play. Now the oxygen molecule consists of two oxygen atoms, linked with a double bond. They are stable as they are, and cannot easily be made to do anything at body temperature. Thus for the cell to oxidize a molecule, the first essential is to concentrate the oxygen and at the same time activate it. It was the great achievement of Otto Warburg to discover the catalyst. He called it the 'Atmungsferment', the 'enzyme of respiration'. With the one-sidedness of genius this notable man wished to maintain that the whole of oxidation is affected by this enzyme, reasoning from the fact that if this enzyme is inactivated the whole respiration stops. To-day we know that respiration represents a chain of reactions, and regardless of which link of the chain breaks, the whole chain becomes useless.

The next link was found by D. Keilin, who rediscovered a substance, which he called 'Cytochrome', a substance which, like the 'Atmungsferment', is closely related to haemoglobin. It is this cytochrome which is oxidized directly by the Warburg enzyme. The essential part of this cytochrome molecule is an iron atom which is oxidized from its divalent form into the trivalent ferric form. We will call the system formed by these two catalysts, 'Atmungsferment' and cytochrome, briefly the WK system (Warburg-Keilin).

Cytochrome is not a very powerful oxidizing agent. The labile cell could not work with powerful oxidizing agents except to its detriment. The WK system alone could not oxidize such stable formations as the foodstuff molecules. The great discovery of H. Wieland was to find that in the cell the foodstuff also gets activated for oxidation. We call the catalysts responsible for this activation 'dehydrogenases', for their function is to make the foodstuff molecule give off more easily its hydrogen if there is another substance to take it over. According to current views it is this activated hydrogen of the foodstuff which is oxidized by the WK system, and the oxidation of the foodstuff consists only of a loss of hydrogen. In this way, as I also showed many years ago, the activated hydrogen in the cell is oxidized by the activated oxygen.

It is worth while to reflect for a moment on the amazing implications of this theory of respiration. First, the theory says that the cell knows but one fuel: hydrogen. The foodstuff molecule is but a package of hydrogen and the carbon skeleton of the foodstuff molecule a set of pegs on which the single hydrogen atoms are hung. It also says that the formerly omnipotent oxygen has as such but a very limited role, for it disappears from the scene at the first step and all that is left of it is an electron change on the iron atom of cytochrome.*

From the very beginning of my biochemical studies my mind was bothered by the special position of the four-carbon atom dicarboxylic acids. I was taught that succinic acid was oxidized by most animal tissues at a very rapid rate to fumaric acid. Later I convinced myself that there is in fact no other substance oxidized by tissues as fast as succinate. Ogston & Green showed that the only substance cytochrome could act on was succinate. It was also known that all tissues contained a very powerful enzyme, 'fumarase', which converts fumaric acid to malic, till the relative concentration of both is 1:3. In the same way it converts malic into fumaric. Later on I found that this enzyme is in fact one of the most powerful enzymes known. But what is its function? Nature is not extravagant, and yet neither succinic nor fumaric acid were regarded as among the most important metabolites. Also Thunberg showed that the isomer of fumaric acid, e.g. maleic acid, was a strong and specific poison of respiration. I began to suspect that something must be wrong about the WK-Wieland theory. It might be true, but must be incomplete, and the C_4 dicarboxylic acid must play some very important catalytic role in respiration. So I investigated two things: (1) what happens to the respiration if we cut out the oxidation of succinic acid, and (2) what happens if we increase the minute quantity of fumarate normally present in the tissue. The possibility of inhibiting succinic oxidation in a specific way was opened up by J. H. Quastel, who showed that the oxidation of succinate can be poisoned fairly specifically by the C_3 dicarboxylic acid, malonic acid.

The results were striking. Minute quantities of malonate poisoned respiration almost like cyanide. Fumaric acid strongly increased it. The rapidly declining respiration of tissues *in vitro* could be maintained constant for long periods by fumaric acid. As Baumann & Stare have shown in Keilin's laboratory, even a few γ of fumarate (γ = one millionth part of a gram) were active.

It took several years of hard work to fit the contradictory observations into one theory. The theory is this: the C_4 dicarboxylic acids are a link in the respiratory chain between foodstuff and the WK system. Their function is to transfer the hydrogen of the foodstuff to cytochrome and to reduce by this hydrogen its trivalent iron again to the divalent form. Speaking more precisely, the cytochrome oxidizes off two hydrogen atoms from the succinic acid molecule. By the loss of two hydrogen atoms, the succinic acid is converted to fumaric acid. These two lost H atoms are replaced again by H coming from the foodstuff.

*As a teacher I often find it difficult to explain where the oxygen has gone to. The simplest picture one can give is to say that it has oxidized hydrogen ions present into water. The positive charge of the hydrogen ions serves to charge up the cytochrome from Fe^{++} to Fe^{+++}.

The foodstuff, however, does not give its H immediately to fumaric acid. It gives its two H atoms to oxaloacetic acid, which is also a C_4 dicarboxylic acid. By taking up 2H oxaloacetic turns into malic acid. Malic acid then gives its 2H to fumaric, and thus fumaric is converted to succinic acid. This can again be oxidized by cytochrome, while malic acid, after giving off its 2H becomes oxaloacetic, which can take up H from the foodstuff again, and so the play goes on, H being transmitted all the time from the foodstuff *via* oxaloacetic-malic-fumaric-succinic to the WK system.

The summary of the story is as follows:

$$
\begin{array}{cccccccccccc}
 & & COOH & & COOH & & COOH & & COOH & & & \\
 & & | & & | & & | & & | & & & \\
\text{Food-} & 2H & CH_2 & & CH_2 & 2H & CH & & CH_2 & 2H & & \\
\text{stuff} & \longrightarrow & | & \rightleftarrows & | & \longrightarrow & \| & \rightleftarrows & | & \longrightarrow & \text{Cyto-chrome} — \text{“Atmungs-ferm.”} & —O_2 \\
 & & CO & & HCOH & & CH & & CH_2 & & & \\
 & & | & & | & & | & & | & & & \\
 & & COOH & & COOH & & COOH & & COOH & & & \\
 & & \text{Oxalo-acetate} & & \text{Malate} & & \text{Fuma-rate} & & \text{Succi-nate} & & &
\end{array}
$$

However suggestive the experiments were, it was difficult to believe that such dull substances as C_4 dicarboxylic acids, the formulae of which we find on the first pages of an elementary text-book of organic chemistry, should play such a fundamental role in cell life. On closer examination, however, one becomes more respectful. Biochemistry teaches that carbon atoms which are the neighbours (α) or second neighbours (β position) of a carboxylic group have a special reactivity. Now there is only one group of substances in the whole world which contains two free carbon atoms, which are both at the same time α as well as βC atoms: the group of the C_4 dicarboxylic acids, of which succinic, fumaric, malic and oxaloacetic are the members. Apparently this unique α-β position gives to the two middle C atoms the reactivity which enables them to carry out this catalytic function.

Now if we sum up the whole story of respiration and compare the system of oxidation with that of fermentation, we find no similarity, no fundamental plan in common, and it looks as if they had nothing to do with each other and could be linked only in some secondary manner.

It sometimes happens in research that some apparently insignificant observation unexpectedly changes the whole vista. This happened when we tried in Szeged to answer the uninteresting question, what was the substance, the two hydrogen atoms of which are transferred to oxaloacetic acid? To answer this question we had to develop accurate micro-methods, which would permit us to follow the chemical changes in small quantities of tissues within short periods. Working out such methods means a great amount of very laborious effort. This work, which I would not care to do over again, showed that the substance which was oxidized in the muscle by transferring its hydrogen to oxaloacetic acid, was triose, the same substance which was also the substrate of fermentation.

Now with this new knowledge let us write side by side the simplified schemes of both processes, fermentation and oxidation:

$$\text{Triose} \xrightarrow{2H} \text{Pyruvic acid} \longrightarrow \text{Lactic acid}$$

$$\text{Triose} \xrightarrow{2H} \begin{array}{c}\text{Oxaloacetic}\\ \text{acid}\end{array} \longrightarrow \begin{array}{c}\text{Malic}\\ \text{acid}\end{array} \xrightarrow{2H} \begin{array}{c}\text{Fumar.}\\ \text{acid}\end{array} \longrightarrow \begin{array}{c}\text{Succin.}\\ \text{acid}\end{array} \xrightarrow{2H} \text{WK—}O_2$$

This scheme reveals at once a great similarity of plan in the two processes. In both two hydrogen atoms are taken off from the triose, which is thus oxidized into pyruvic acid. The difference is only that in fermentation these two hydrogen atoms are taken over by pyruvic acid, while in oxidation it is oxaloacetic acid which takes them over. In oxidation there is also an annex to this, the WK system, which has the function of unloading the hydrogen from the dicarboxylic acids.

The fundamental unity of both processes is not revealed until we incorporate the real chemical formulae in the scheme:

$$\text{Triose} \xrightarrow{-2H} \begin{array}{c} CH_3 \\ | \\ CO \\ | \\ COOH \end{array} \xrightarrow{+2H} \begin{array}{c} CH_3 \\ | \\ HCOH \\ | \\ COOH \end{array}$$

$$\text{Triose} \xrightarrow{-2H} \begin{array}{c} COOH \\ | \\ CH_2 \\ | \\ CO \\ | \\ COOH \end{array} \xrightarrow{+2H} \begin{array}{c} COOH \\ | \\ CH_2 \\ | \\ HCOH \\ | \\ COOH \end{array} \xrightarrow{-H_2O} \begin{array}{c}\text{Fumar.}\\ \text{acid}\end{array} \xrightarrow{+2H} \begin{array}{c}\text{Succin.}\\ \text{acid}\end{array} \xrightarrow{-2H} \text{WK—}O_2$$

If we compare the formulae of oxaloacetic acid and pyruvic acid on one side, malic acid and lactic acid on the other side, we realize the remarkable fact that oxaloacetic acid is but a carboxypyruvic acid, and malic acid is but a carboxy-lactic acid, and we also realize what happened millions of years ago, when Nature discovered oxidation: it simply fitted some of the pyruvic acid molecules with carboxyl groups and completed the old system of fermentation with a new one, the WK chain, having no other function than to free the reduced oxaloacetic acid from its two hydrogen atoms, transferring them on to oxygen.

It often happens in research that we cannot predict what we have to expect, how Nature will do, or has done this, that, or the other thing. Once we discover it, we usually find that the way Nature reaches its purpose is the only possible way, and yet, in spite of its simplicity, the most admirably ingenious way. Such is the case with the carboxylation of the pyruvic acid. By this very simple change, the molecule becomes admirably suited for the highly specific new function of a catalyst. Lactic acid (reduced pyruvic acid) becomes resynthesized into carbohydrate, and is thus but one member of a big metabolic cycle. But by carboxylation the molecule becomes unfit for resynthesis. The molecule is taken out from the metabolic cycle, stabilized, and enabled to fulfil the function of a catalyst. At the same time the new carboxyl group introduced lends to the two middle carbon atoms that very high reactivity which is necessary for the catalytic function. The carboxyl group itself is a highly electropolar group with very marked affinities. Thus Nature provides the pyruvic acid molecule by the second carboxyl group with a second handle, by which its activating enzyme can hold it tight. As Quastel found, both carboxyl groups are important for the adsorption of the C_4 dicarboxylic acids on the protoplasmic surface. Only this high adsorbing power enables the traces of C_4 dicarboxylic acids present in the cell to saturate the enzyme and carry the whole respiration, the most powerful function in cell life.

To sum up, the catalytic function of C_4 dicarboxylic acids in cell respiration allows us to understand the relation of oxidation to fermentation; it gives us a glimpse of what happened millions of years ago and what is going on at present in our cells. It gives us an explanation of the quantitative interdependence of both energy-liberating processes. If there is oxygen present, then succinic acid will be oxidized to fumaric, and malic to oxaloacetic, and the two hydrogen atoms of the triose will be transferred on to this latter substance, which again transmits them to the WK system, and so on to oxygen. This process we call respiration. If there is no oxygen present, there will be no oxaloacetic acid formed, the hydrogen of the triose will be transmitted to pyruvic acid, and this process we call fermentation. Its end-product is in the muscle lactic acid, in the yeast alcohol, which is nothing but a lactic acid which has lost its carboxyl group.

At the same time we can see that what happened millions of years ago was first the fitting of a few pyruvic-acid molecules with an extra carboxyl group, secondly the building of a bridge from the two hydrogen atoms of succinic acid to oxygen (the WK system). Thus Nature found its way from Fermentation to Oxidation.

This simple conclusion gave me the greatest mental satisfaction which I have experienced in my life as a scientific worker. That is why I now offer it to Sir F. G. Hopkins. I would like to end by giving expression to my most sincere desire that oxidation and fermentation alike may long continue to interact harmoniously in the cells of our beloved teacher.

References

1. Ogston, F. J. and Green, D. E. *Biochem. J.* 1935, **29**, 1983.
2. Quastel, J. H. *Biochem. J.* 1926, **20**, 166.
3. Stare, F. J. and Baumann, C. A. *Proc. roy. Soc. B.* 1937, **121**, 338.

ON THE FORMATION OF CITRIC ACID*

F. Knoop and C. Martius
University of Tübingen, Germany
(Submitted July 29, 1936)

When considering which products of the intermediary metabolism participate in the formation of citric acid—a constituent frequently encountered in nature—the more obvious choices as participants are perhaps oxalacetic acid and acetic acid. An experiment of that kind has already been performed, although it brought us no result. However, if one replaces the acetic acid with the much more reactive pyruvic acid and then allows it to react with oxaloacetic acid in a soda-alkaline solution at low temperature and subsequently oxidizes the mixture with hydrogen peroxide, according to Hollemann, the ketonic acid formed is oxidized in the α position, to the next lower acid. After 20 to 30 hours it would be possible to isolate citric acid as a Ca salt in about 35% yield. It is identified through the pentabromoacetone derivative.

$$\underset{\substack{|\\ COOH}}{HOOC{-}CH_2{-}CO} + H{-}CH_2{-}CO{-}COOH = HOOC{-}CH_2{-}\underset{\substack{|\\ COOH}}{COH}{-}CH_2{-}CO{-}COOH$$

$$+H_2O_2 = HOOC{-}CH_2{-}\underset{\substack{|\\ COOH}}{COH}{-}CH_2{-}COOH$$

Data in the literature suggests that the physiological synthesis occurs in the same manner. Investigations based on the reaction discovered are currently in progress. It would be interesting to determine if the organism can bring about a fission in a reversed direction. In that case the first phase of the decom-

*F. Knoop and C. Martius, "Über die Bildung von Citronensäure," *Z. physiol. Chem.*, **242**, 204 (1936). Translated from the German and reprinted by license agreement with Walter de Gruyter & Co., Berlin.

position—similar to sugar—would be a non-oxidative breakdown which must give rise to the formation of oxalacetic acid. It is a well known fact that the latter is readily converted into pyruvic acid. So far, the latter and acetic acid have been isolated as decomposition products. It is difficult to find a simpler way to account for the formation of pyruvic acid. We might also put this question to a test.

THE INTERMEDIATE METABOLISM OF CARBOHYDRATES*

H. A. Krebs; M. A., Cambridge; M. D., Hamburg

Lecturer in Pharmacology,
University of Sheffield, England

Considerable advances have recently been made in the analysis of the intermediate metabolism of carbohydrates, and results have been obtained that are of interest not only to the biochemist but also to the clinician. From this work, a new conception of the origin of diabetic acidosis has arisen, which has led Koranyi and Szent-Györgyi (1937) to suggest a new method of treatment, to which reference has already been made (*Lancet,* July 17th, 1937, p. 155, and 24th, p. 200). Since the original investigations are scattered over many journals, it may be of interest to summarise recent developments in this field.

The perfection of new methods of analysis has been the decisive factor in advancing our knowledge, the essential feature of the new technique being the use of isolated surviving tissues. We now know how to keep tissues alive in vitro for considerable periods of time and are thus enabled to study their metabolism under controlled conditions. In addition, modern methods of microchemical analysis reduce the amount of tissue necessary for exact experiments to a few milligrammes and so allow large numbers of experiments to be carried out with relative ease. The manometric methods of Warburg for the microdetermination of gaseous exchange play a prominent part in these methods.

The Catalytic Effect of Succinic Acid and Related Substances

Szent-Györgyi reported experiments in 1935 and 1936 that suggested that succinic acid and its derivatives fumaric acid, malic acid, and oxalo-acetic acid catalytically promote oxidations in muscle tissue. Conclusive proof of this

*H. A. Krebs, "The Intermediate Metabolism of Carbohydrates," *Lancet,* **2**, 736–38 (1937). Reprinted with the permission of the author and of the editor of the *Lancet*.

catalytic effect was presented by Stare and Baumann in December, 1936. These workers showed that minute quantities of these substances suffice to bring about an increase in respiration and that the increase is a multiple of the amount of oxygen necessary for the oxidation of the added substance. Moreover, the added substance is not used up but can be subsequently detected in the medium. There thus remains no doubt that succinic acid and related substances can act as catalysts in respiration.

The respiratory quotient of the additional respiration induced by the substances was found to be near 1.0, which indicates that they promote the oxidation of carbohydrates.

The close relationship between the four substances mentioned is made clear as follows:

REACTION 1

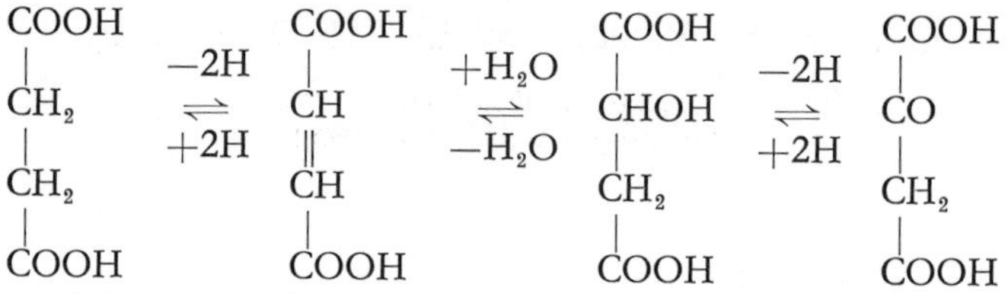

$$\begin{array}{ccccccc}
\mathrm{COOH} & & \mathrm{COOH} & & \mathrm{COOH} & & \mathrm{COOH} \\
| & -2\mathrm{H} & | & +\mathrm{H_2O} & | & -2\mathrm{H} & | \\
\mathrm{CH_2} & \rightleftharpoons & \mathrm{CH} & \rightleftharpoons & \mathrm{CHOH} & \rightleftharpoons & \mathrm{CO} \\
| & +2\mathrm{H} & \| & -\mathrm{H_2O} & | & +2\mathrm{H} & | \\
\mathrm{CH_2} & & \mathrm{CH} & & \mathrm{CH_2} & & \mathrm{CH_2} \\
| & & | & & | & & | \\
\mathrm{COOH} & & \mathrm{COOH} & & \mathrm{COOH} & & \mathrm{COOH} \\
\textbf{succinic acid} & & \textbf{fumaric acid} & & \textbf{malic acid} & & \textbf{oxalo-acetic acid}
\end{array}$$

Catalytic Effect of Citric Acid

The next step was the discovery that citric acid also acts as a catalyst (Krebs and Johnson 1937a). Added to muscle in minute quantities it accelerates the oxidation of carbohydrates in the same way as succinic acid. The experimental analysis of this effect has revealed not only the mechanism of the catalytic action of citric acid but also that of succinic acid and related compounds. Furthermore it has led to the elucidation of the main steps in the oxidative breakdown of carbohydrate.

Fate of Citric Acid in the Animal Body

Citric acid has long been known to be readily oxidised in living tissues; yet the details of its intermediate metabolism remained obscure until March. 1937, when Martius and Knoop discovered that α-ketoglutaric acid is a product of oxidation of citric acid:

REACTION 2

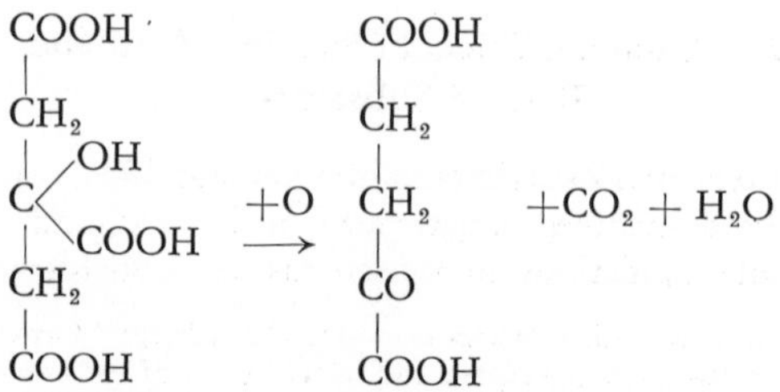

$$\begin{array}{lccc}
\mathrm{COOH} & & \mathrm{COOH} & \\
| & & | & \\
\mathrm{CH_2} & & \mathrm{CH_2} & \\
|\ \diagup\mathrm{OH} & & | & \\
\mathrm{C} & +\mathrm{O} & \mathrm{CH_2} & +\mathrm{CO_2} + \mathrm{H_2O} \\
|\ \diagdown\mathrm{COOH} & \longrightarrow & | & \\
\mathrm{CH_2} & & \mathrm{CO} & \\
| & & | & \\
\mathrm{COOH} & & \mathrm{COOH} & \\
\textbf{citric acid} & & \boldsymbol{\alpha}\textbf{-ketoglutaric acid} &
\end{array}$$

This reaction, which was unexpected on general chemical grounds, involves no doubt a complicated series of intermediate reactions.

We have recently been able (Krebs and Johnson 1937a) to confirm and to supplement the results of Martius and Knoop. The method by which it has proved possible to demonstrate Reaction 2 depends on the use of specific poisons or "inhibitors." In unpoisoned tissue, oxidations generally go to completion, so that only end-products and no intermediates appear. Arsenious oxide is a poison which inactivates specifically the enzymes concerned with the removal of ketonic acids whilst it does not inhibit the oxidation of citric acid. A poison of this kind thus makes it possible to isolate intermediate products which do not normally accumulate.

The fate of α-ketoglutaric acid in the body is well known. This substance has long been of physiological interest since it arises as an intermediate in the breakdown of glutamic acid, of proline and of histidine. It yields, on oxidation, succinic acid and carbon dioxide.

REACTION 3

$$\begin{array}{ccccc} COOH & & COOH & & \\ | & & | & & \\ CH_2 & & CH_2 & & \\ | & +O & | & & \\ CH_2 & \longrightarrow & CH_2 & & +CO_2 \\ | & & | & & \\ CO & & COOH & & \\ | & & & & \\ COOH & & & & \end{array}$$

α-ketoglutaric acid **succinic acid**

Taking Reactions 2 and 3 together we can pass from citrie acid to succinic acid, and this reaction can be directly demonstrated if malonic acid is added. This substance specifically inhibits the oxidation of succinic acid, but does not check the breakdown of citric and α-ketoglutaric acids, and in its presence the conversion of citric acid into succinic acid by respiring muscle is experimentally realised.

Mechanism of the Catalytic Actions

If a substance acts as a specific catalyst, whether in simple chemical reactions or in metabolic processes, we must conclude that it takes an active part in some intermediate stages of the process catalysed. It is characteristic of a catalyst that it is used in a primary stage, but regenerated in a secondary step, and the analysis has to pay attention to the elucidation of the "catalytic cycle"—viz.: (*a*) the reaction in which the catalyst is used up; (*b*) the reaction in which it is regenerated. Szent-Györgyi (1935, 1936) suggested the following scheme to explain the catalytic effects of succinic, fumaric, and oxalo-acetic acids.

REACTION 4

oxalo-acetic acid + "hydrogen of organic molecules" ⟶ succinic acid

REACTION 5

succinic acid + O_2 ⟶ oxalo-acetic acid + H_2O

The balance-sheet of these two reactions is the oxidation of "hydrogen of organic molecules," and the catalytic effect is assumed, in this scheme, to be due to an alternate oxidation and reduction of the catalyst. The nature of the organic radical from which the hydrogen atoms are furnished is not defined in detail by Szent-Györgyi.

There is no doubt that both Reactions 4 and 5 do in fact take place in respiring muscle and that a certain part of the carbohydrate molecule is oxidised by this mechanism. Quantitative experiments, however, indicate that the rate of Reaction 4 is much too slow to account for the whole catalytic effect. The scheme, moreover, does not explain the significance of Reactions 2 and 3.

We must therefore look for a second mechanism in order to explain the catalytic effects fully. The presence in tissues of highly active enzymes for Reactions 2 and 3 suggests that citric acid itself may take part in the catalytic cycle. Reactions 3, 2, and 1 would then be part of the cycle. This idea immediately raises the question as to how the cycle would be completed, that is to say, whether citric acid can be regenerated from one of its breakdown products.

Experiments show that the postulated reaction does in fact occur. If oxalo-acetic acid is added in excess to muscle, citric acid is synthesised in large quantities. This can best be demonstrated in anaerobic experiments, since the reaction does not require molecular oxygen. The removal of citric acid, on the other hand, is an oxidation by molecular oxygen; citric acid thus accumulates anaerobically.

The synthesis of citric acid from oxalo-acetic acid is brought about by a condensation with a second substance, the chemical nature of which is not yet known. It is certain that the second substance is derived from carbohydrate and very probable that it is pyruvic acid. The condensation to citric acid can be formulated in the following way (Reaction 6).

Reaction 6

$$
\begin{array}{l}
\mathrm{COOH}\\
\,|\\
\mathrm{CO}\\
\,|\\
\mathrm{CH_3}\\
+\\
\mathrm{OH}\\
\,|\\
\mathrm{C{-}COOH}\\
\,\|\\
\mathrm{CH}\\
\,|\\
\mathrm{COOH}
\end{array}
\xrightarrow{-\mathrm{H_2O}}
\begin{array}{l}
\mathrm{COOH}\\
\,|\\
\mathrm{CO}\\
\,|\\
\mathrm{CH_2}\\
\,|\\
\mathrm{C{-}COOH}\\
\,\|\\
\mathrm{CH}\\
\,|\\
\mathrm{COOH}
\end{array}
\xrightarrow{+\mathrm{O}}
\begin{array}{l}
\mathrm{CO_2}\\
+\\
\mathrm{COOH}\\
\,|\\
\mathrm{CH_2}\\
\,|\\
\mathrm{C{-}COOH}\\
\,\|\\
\mathrm{CH}\\
\,|\\
\mathrm{COOH}
\end{array}
\xrightarrow{+\mathrm{H_2O}}
\begin{array}{l}
\mathrm{COOH}\\
\,|\\
\mathrm{CH_2}\\
\,|\;\;\diagup\mathrm{OH}\\
\mathrm{C}\\
\,|\;\;\diagdown\mathrm{COOH}\\
\mathrm{CH_2}\\
\,|\\
\mathrm{COOH}
\end{array}
$$

pyruvic acid + enol oxalo-acetic acid	**oxalo-mesaconic acid**	**"cis-" aconitic acid**	**citric acid**

This scheme though supported by experimental evidence is in part hypothetical and one abstains therefore from discussing details; but it should be emphasised that the net effect, the synthesis of citric acid in the presence of oxalo-acetic acid, is an experimental fact.

The results discussed in this section can be summarised by the following (Reaction 7).

This scheme—the "citric acid cycle"—makes clear the manner in which citric acid and the C_4-dicarboxylic acids act as catalysts. The substances undergo alternate formation and decomposition and the net effect of the cycle is the oxidation of carbohydrate.

Reaction 7

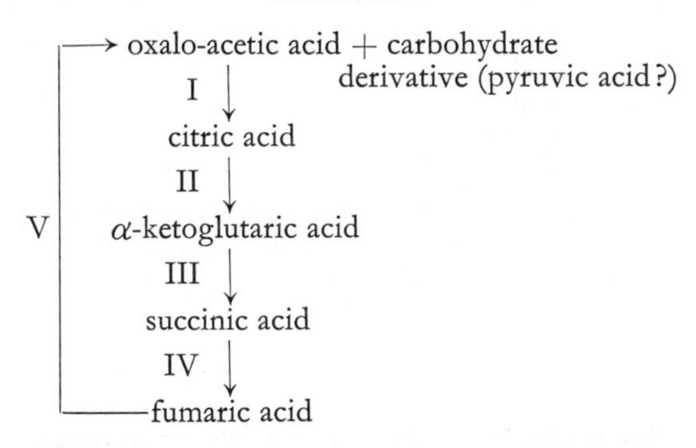

The relationship between the citric acid cycle and Szent-Györgyi's cycle—represented by Reactions 4 and 5—becomes clear when the following considerations are taken into account. Each step of the citric acid cycle is an oxidative process. In certain cases the oxidation is brought about more less directly by molecular oxygen, for instance step IV, but in other cases the oxidation is primarily brought about by oxalo-acetic and fumaric acids, for instance step I. The hydrogen atoms in Reaction 4 are those set free in step I of Reaction 7 and possibly those involved in the conversion of carbohydrate into pyruvic acid—i.e., in reactions prior to step I.

Reaction 7, while necessarily incomplete, outlines the principal pathway of the oxidative breakdown of carbohydrate in muscle tissue.

Alternative Paths of Carbohydrate Breakdown

There is a second pathway which, though quantitatively less significant, has a special interest because it results in the formation of ketone bodies. The intermediate stage from which the two pathways fork is probably pyruvic acid. This substance has long been known to give rise to increased ketone body formation if added to liver in excess (Embden and Oppenheimer 1912). We have recently found (Krebs and Johnson 1937b, 1937c) that this conversion may also take place in other tissues especially in muscle. The mechanism of this reaction is probably as follows.

Reaction 8

$$\underset{\textbf{pyruvic acid}}{CH_3 \cdot CO \cdot COOH} \xrightarrow{+O} \underset{\textbf{acetic acid}}{CH_3 \cdot COOH} + CO_2$$

REACTION 9

$$
\begin{array}{ccccl}
CH_3 & & CH_3 & & \\
| & & | & +O & \\
COOH & & CO & \longrightarrow & CH_3\cdot CO\cdot CH_2\cdot COOH \\
+ & -H_2O & | & -CO_2 & \text{(aceto-acetic acid)} \\
CH_3 & \longrightarrow & CH_2 & & \\
| & & | & +H_2O & \\
CO & & CO & \longrightarrow & CH_3\cdot CH(OH)\cdot CH_2\cdot COOH \\
| & & | & -CO_2 & (\beta\text{-hydroxybutyric acid}) \\
COOH & & COOH & & \\
\text{acetic acid} & & \text{aceto-} & & \\
+ & \longrightarrow & \text{pyruvic} & & \\
\text{pyruvic acid} & & \text{acid} & &
\end{array}
$$

The rate of these reactions is small compared with the rate of the citric acid cycle, but they are of importance for the explanation of diabetic acidosis. It is now evident that carbohydrates as well as fats may be the source of ketone bodies.

Szent-Györgyi's Theory of Diabetic Acidosis

Szent-Györgyi suggests that the supply of oxaloacetic acid is diminished in diabetes; consequently carbohydrate cannot be oxidised to the normal extent since the catalyst required is not available. Pyruvic acid, instead of reacting according to Reaction 7, will therefore follow Reactions 8 and 9, giving rise to an increased ketone body formation. By feeding succinic acid, Korányi and Szent-Györgyi (1937) claim to have diminished the degree of acidosis in diabetes.

There is no direct evidence in favour of this view except its theoretical possibility and the clinical reports in five cases of diabetic acidosis successfully treated by succinic acid. Lawrence and his colleagues (1937), on the other hand, report the complete failure of the succinic acid treatment in the two cases which they have treated, and they warn against undue optimism. More work, both clinical and experimental, must be carried out before a final decision can be made.

According to Reaction 7 citric acid should have the same effects as succinic acid in relieving acidosis, and it would certainly merit clinical trial on the grounds of both greater palatability and cheapness.

References

Embden, G., and Oppenheimer, M. (1912) *Biochem. Z.* **45**, 186.

Korányi, A., and Szent-Györgyi, A. v. (1937) *Dtsch. med. Wschr.* **63**, 1029.

Krebs, H. A., and Johnson, W. A. (1937a) *Enzymologia,* **4**, 148.

——— (1937b) *Biochem. J.* **31**, 645.

——— (1937c) *Ibid,* p. 772.

Lawrence, R. D. (1937) *Lancet,* July 31st, p. 286.

——— McCance, R. A., and Archer, N. (1937) *Brit. med. J.* July 31st, p. 214.

Martius, C., and Knoop, F. (1937) *Hoppe-Seyl. Z.* **246**, i.
Stare, F. J., and Baumann, C. A. (1936) *Proc. Roy. Soc.* B, **121**, 338.
Szent-Györgyi, A. v. (1935) *Hoppe-Seyl. Z.* **236**, 1.
——— (1936) *Ibid,* **244**, 105.

PHOSPHORYLATION IN KIDNEY CORTEX*

Herman Kalckar

Universitetets Medicinsk-Fysiologiske Institut,
Copenhagen, Denmark

(Submitted as doctoral thesis, Copenhagen, 1938)

* * *

In order to observe a phosphorylation in the respiring tissue system it is necessary to inhibit the phosphatases in the tissue; this is possible by means of fluoride in concentrations of about M/20.

In systems poisoned with fluoride the phosphate esters accumulate and a great decrease in the inorganic phosphate is observed.

The sensitivity of tissue phosphatases to fluoride is very much greater than that of the usual phosphatase preparations, which are prepared from autolyzed tissue. A phosphatase preparation prepared after the method of Albers & Albers[1] for instance shows no inhibition at all after the addition of fluoride. The explanation is that most of the phosphatases from autolyzed tissue are "alkaline" phosphatases which are fluoride insensitive, whereas the fresh extract contain about equal amounts of "alkaline" and " acid" (*p*H optimum 5.5) fluoride sensitive phosphatases.

Fluoride also exerts a definite inhibition on tissue respiration. In order to get an optimal accumulation of phosphoric esters it is therefore important not to exceed a fluoride concentration of about M/20. The RQ seems to decrease in the presence of fluoride.

The phosphate, which probably is primarily combined to a codehydrase, is transferred to a phosphate acceptor which may be of a different nature.

*Excerpt from the English review in H. Kalckar, *Fosforyleringer i Dyrisk Vaev,* eds. Nyt Nordisk Forlag and Arnold Busck, 126–128 and pertinent references. Copenhagen, 1938. Reprinted with the permission of the author and of the publisher.

[1] Albers, H., & Albers, E.: Zs. f. physiol. Ch. 282, 189–1935.

Carbohydrates for instance act as phosphate acceptors; glucose and fructose are phosphorylated most intensely, galactose and arabinose at a slower rate.

* * *

The accumulated product of the glucose and fructose phosphorylation has been isolated as barium salt and proved to be identical with fructose-diphosphate (Harden-Young-Ester).

Glycerol is intensely phosphorylated to an acid resistant ester, probably *α-glycerophosphate*.

Adenylic acid is phosphorylated to adenyl pyrophosphate.

It seems however that other substances than those mentioned may act as phosphate acceptors.

As already stated by Krebs,[2] the addition of alanin and glutamic acid greatly increases the respiration of rat kidney slices. This is also the case in extracts of kidney cortex from rabbits and cats. Not only amino acids, but also fumaric acid and malic acid, greatly increase the oxygen consumption. This effect on the respiration most likely may be looked upon as a stabilization of the extract respiration. It may be that the dicarboxylic acids act as hydrogen transfer systems; but a great part is certainly oxidized as a substrate. Citric acid acts in the same manner.

As I have stated in a previous work, phosphorylation (measured as ester accumulation) in kidney cortex is proportional to the oxygen consumption. This proportionality appears not only in experiments on inhibition of respiration (for instance by cyanide), but also in experiments on stimulation of respiration with dicarboxylic acids and alanin. In an experiment with glycerol as phosphate acceptor the sample containing glutamic acid exhibited a double respiration and a quadrupled phosphorylation compared with the control. The relatively larger increase in phosphorylation than in oxidation may be explained by assuming that after addition of glutamic acid a greater part of the respiration is coupled with phosphorylations.

On the basis of our knowledge concerning the coupling between oxidation-reductions and phosphorylations in the oxidation of triosephosphate by cozymase, it is reasonable to suppose that the coupling between oxygen consumption and phosphorylation in kidney also depends on a nucleoproteid system* either the pyridinnucleotides (Warburg[3] 1934–36) or the flavin-adenin-nucleotide (Warburg & Christian,[4] 1938).

Remarkable, however, is the fact that the dehydrogenation of succinic acid to fumaric acid also stimulates phosphorylation. According to Szent Györgyi,[5] this dehydrogenation consists of a transfer of hydrogen from succinic acid to cytochrome *c*. Cytochrome *c* is not a nucleotid like cozymase, however, but a haemin system. It is difficult to explain how a haemin system should be able to transfer phosphate.

In extracts from washed tissue pulp an intense phosphorylation very fre-

*In contrast to muscle extracts the coupling in kidney extracts is not inhibited by arseniate at all.

[2] Krebs, H. A.: Biochem. J1. 29, 1620 & 1951–1935.

[3] Warburg, O. & Christian, W.: Biochem. Zs. 287, 291–1936.

[4] Warburg, O. & Christian, W.: Biochem. Zs. 298, 150–1938.

[5] Szent Györgyi, A.: Zs. f. physiol. Ch. 249, 211–1937.

quently takes place if dicarboxylic acids (for instance fumaric acid) are added to the extract. Fumaric acid undoubtedly increases the phosphorylation not only through a stimulation of the respiration; but also by forming a phosphate acceptor. Even extracts which must be considered quite free from sugars show a marked phosphorylation if fumaric acid or malic acid is added. In these cases fumaric acid or more likely a product formed by oxidation of fumaric acid must be assumed to act as phosphate acceptor. The properties of the ester formed under these conditions agree completely with those of *phosphopyruvic acid* Lohmann & Meyerhof[6]):

EXPERIMENT 11/6 1938

Rabbit kidney cortex is minced and washed in 3 volumes ice cooled water; the washed tissue then is ground with sand and extracted with an equal volume of sec. phosphate. 1.2 ccm extract is used. The mixture contains 0.2% fluoride. Temp. 37°—gas: 100% oxygen.

The extract mixtures are incubated for 30 min.

	cmm. O_2	$P_{dir.}$	$P_{alcal.}$	$P_{iod.}$	$P_{merc.}$	P_{30}
Initial sample		1.08	1.14	1.11	1.10	
Incub. without substrate	230	0.96	1.05	1.00	1.00	1.14
Inc. with 30 mgm. fumarate	432	0.64	0.70	0.96	0.96	1.13
Inc. with 30 mgm. glucose	165	0.58	0.71	0.60	0.60	

$P_{dir.}$: phosphate determined directly.
$P_{alcal.}$: phosphate after 20 min. incubation in 1-N. NaOH.
$P_{iod.}$: phosphate after 20 min. incubation in N/10 alcaline iodine.
$P_{merc.}$: phosphate after 10 min. incubation in neutral $HgCl_2$.
P_{30}: phosphate after 30 min. hydrolysis in 1-N. HCl at 100°.

Whereas the ester formed by glucose exhibits typical properties of fructose diphosphate, for instance the liberation of phosphate in 1-N. NaOH, the ester formed in the sample with fumaric acid exhibits the typical properties of phosphopyruvic acid: liberation of phosphate in N/10 alkaline iodine and in $HgCl_2$ solutions. (Lohmann & Meyerhof.)

Probably the malic acid is oxidised to oxalacetic acid, which, decarboxylated and phosphorylated, yields phosphopyruvic acid.

It is only possible to get an accumulation of phosphopyruvic acid if the extracts are poor in other phosphate acceptors, especially carbohydrates; otherwise the latter will compete with the ketoacids as phosphate acceptors.

[6] Lohmann, K. & Meyerhof, O.: Biochem. Zs. **273**, 60–1934.

THE MECHANISM OF PHOSPHORYLATION ASSOCIATED WITH RESPIRATION*

V. A. Belitser and E. T. Tsybakova

Laboratory of Physiological Chemistry,
University of Moscow, U.S.S.R.

(Submitted June 10, 1939)

Synthesis of adenosinetriphosphate and phosphagen (phosphocreatine) takes place in the muscle at the expense of the energy derived from either glycolysis or cell respiration. The glycolysis mechanism for synthesis of phosphoric esters has been studied in detail during the last years. It has been established that oxidation-reduction between phosphotriose and codehydrogenase I, later identified as DPN, one of the intermediate stages of glycolysis, is coupled with phosphorylation. This oxidation-reduction probably also takes place in the oxidation of hexose. On the basis of these findings one is led to presume that in respiration, as in glycolysis, phosphorylation is connected with a "glycolytic" oxidation-reduction.

From a number of indirect findings, however, it is apparent that some oxidizing processes may be linked with phosphorylation without having any direct connection with glycolysis. Braunshteyn and Severin[1,2] showed that oxidation of pyruvic acid, ketobutyric acid and glutamic acid, as well as of alanine, brings about a stabilization of adenosine triphosphate in nucleated erythrocytes. Grimlund[3] found that oxidation of lactic acid, pyruvic acid, and succinic acid increases the working capacity of a muscle in which glycolysis is obliterated. This was previously found to be true in the case of lactic acid, by Meyerhof and his coworkers,[4] who also stated that lactic acid oxidation brings about stabilization of phosphagen in muscles poisoned by iodoacetate.

*V. A. Belitser and E. T. Tsybakova, "The Mechanism of Phosphorylation Associated with Respiration," *Biokhimiya* (U.S.S.R.), 4, 516–534 (1939). Translated from the Russian by R. Ernsberger with the assistance of S. Black and by N.I.H. and reprinted with the permission of the authors and of the U.S.S.R. Academy of Sciences.

It can be presumed that stabilization of phosphorylated esters is the result of their resynthesis; then, on the basis of the investigations quoted above, a "strictly respiratory" synthesis of these esters might well be inferred. This interpretation has received additional and more direct support. Kalckar[5] has described a "respiratory" synthesis of phosphorylated hexoses in kidney cortex preparations, under conditions in which glycolysis is arrested. About the same time one of us[6] found that minced muscle tissue under aerobic conditions can sustain a synthesis of phosphagen from creatine and inorganic phosphate even after poisoning with monobromo-acetic acid.

Our goal in this investigation was to systematically study "respiratory" synthesis of phosphorylated esters and to determine its mechanism. Some of the results in this area collected up to the summer of 1938 have already been published in a short communication[7]. We record here in more detail our basic, unpublished material.

1. The Object: Determination of the Effect of Various Conditions of the Medium on the Synthesis of Phosphagen Coupled with Respiration

For the purpose of this study it was necessary to find a tissue which would permit the study of oxidative phosphorylation processes using selected exogenous respiratory substrates. In order to observe a clear effect of exogenous substrates, it is necessary that the respiratory phosphorylation due to oxidation of preformed substrates be insignificantly small, under the experimental conditions chosen. We investigated the capability of muscle tissue from various animals to synthesize phosphagen after preliminary washing in water or saline. Various substances, particularly lactic acid, served as respiratory substrates. The majority of species of muscle investigated (rabbit, guinea pig, rat, and frog) were not able to synthesize phosphagen after subsequent mincing and washing. Only pigeon breast muscle and rabbit heart muscle, which evidently possess a more stable enzymic system of respiration, were still capable of respiratory synthesis of phosphagen after certain types of washing.

Muscle tissue freshly cut, chilled and cleared of connective and fat elements was minced on ice with scissors to the consistency of thin pulp and then washed with chilled 0.15 M phosphate solution of *p*H of 7.0–7.2 or in a mixture of 9 parts of 0.9% NaCl and one part of the phosphate solution (40 ml of solution for 5–7 gm. of pigeon muscle, or tissue from one rabbit heart). The tissue was left in the solution at 0°C for 20 minutes and was stirred several times. The liquid was then decanted, and the tissue dried quickly on filter paper and placed in a chilled Petri dish. It was divided into several portions, (usually 200 mg each) one or two of which were immediately fixed with chilled trichloracetic acid (end concentration 4%); the rest were placed in Warburg flasks containing phosphate buffer *p*H 7.5, 0.2 mg cozymase*, 0.14 mg magnesium in the form of $MgCl_2$ and 12 mg creatine; the substrates under investigation were added in amounts to give a concentration of 0.05 M or 0.025 M.

*Cozymase [DPN] was prepared from beer yeast by the Meyerhof and Ohlmeyer method[9] with some modifications by Ochoa. Our preparation, as judged by its activity in sampling with apoenzyme, contained about 60% of pure cozymase.

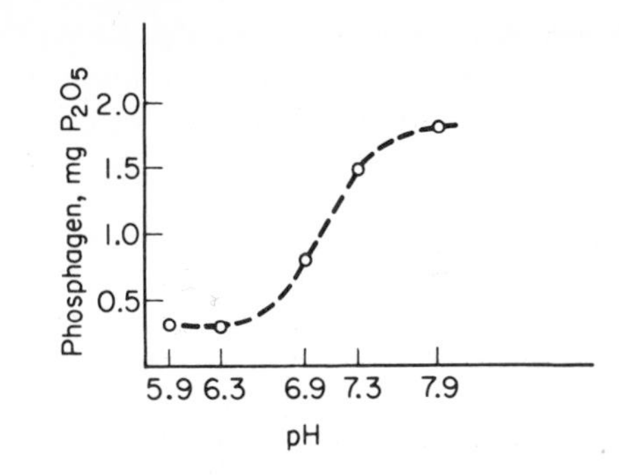

Fig. 1. Phosphagen synthesis in pigeon muscle as a function of the *p*H of the medium.

In order to reach isotonicity KCl or NaCl were added. The total volume was 2 ml. Experiments were conducted at 20°C. Fig. 1 shows the relation of phosphagen synthesis to the *p*H of the medium.

2. Phosphagen Synthesis in the Oxidation of Pyruvic Acid and Other Substrates of Respiration

We have already reported on the synthesis of phosphagen associated with the oxidation of lactic and malic acid in a previous study[7]. One of the other substrates which we studied initially was pyruvic acid. The influence of this acid on the synthesis of phosphagen in aerobic and anaerobic conditions is shown in Fig. 2.

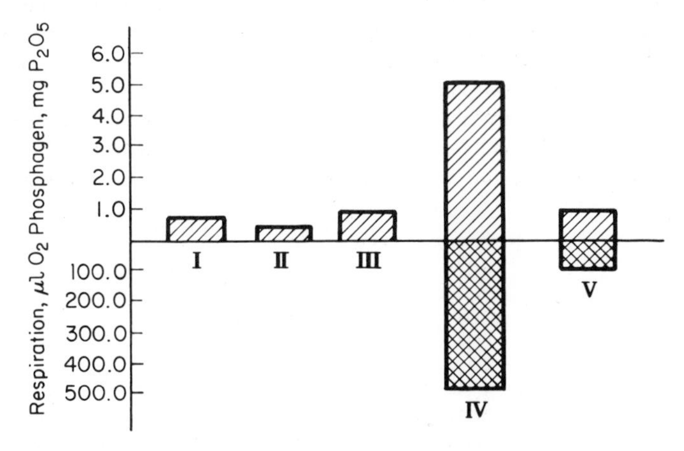

Fig. 2. Synthesis of phosphagen in the presence of pyruvic acid (pigeon muscle). I: before incubation; II: in N_2 without pyruvic acid; III: the same with pyruvic acid; IV: in O_2 with pyruvic acid; V: the same without pyruvic acid. In this and in the following illustrations phosphagen synthesized is given in mg of P_2O_5 per gram of tissue and respiration is given in μl of O_2 per 30 min. per gram of tissue.

Considerable phosphagen synthesis occurred under aerobic conditions but only in the presence of pyruvic acid. Obviously the essential process governing the synthesis was oxidation of pyruvic acid. The insignificant anaerobic synthesis of phosphagen in the presence of pyruvic acid may be ascribed to a dismutation of this acid. Lipmann[11] found some phosphorylation connected with dehydrogenation and dismutation of pyruvic acid, working with an enzymic preparation of *Bact. Delbrückii.*

To clarify whether, under our experimental conditions, complete oxidation of pyruvic acid takes place or whether processes leading to the accumulation of incompletely oxidized products predominate, we undertook the following experiments. After 10 to 12 minutes of preliminary measurement of tissue respiration in the absence of substrates, a small, well-defined amount of neutralized pyruvic acid* was added from a side bulb of the experimental vessel; water was added to the control vessels. The increased respiration produced by the addition of pyruvic acid was determined. Observations were continued until the respiration rate in the samples with added pyruvic acid had decreased to that of the control samples. Knowing the additional amount of oxygen consumed by the samples with pyruvic acid during the whole period of increased respiration, as well as the quantity of added pyruvic acid, it is easy to determine whether or not complete oxidation of pyruvic acid has occurred (complete oxidation of 2 moles of this acid requires 5 moles of oxygen).

In our experiments the quantity of excess oxygen consumed was equal to from 60 to 85% of the amount theoretically required for complete oxidation of the added pyruvic acid. The rate of consumption of excess oxygen is shown in Fig. 3:

Since the pyruvic acid added to the tissue could be oxidized not only at the expense of the additionally consumed oxygen, but also at the expense of the oxygen of "endogenous respiration", the additional respiration gives only the lower limit of that quantity of oxygen which was actually used for oxidizing the acid. Hence at least 60 to 85%, i. e., a major part of the added pyruvic acid, has undergone complete oxidation.

We investigated a number of other respiratory substrates besides pyruvic acid. From all the collected data, it appears that any substrate which is more or less intensively oxidized in muscle tissue gives rise to phosphagen synthesis

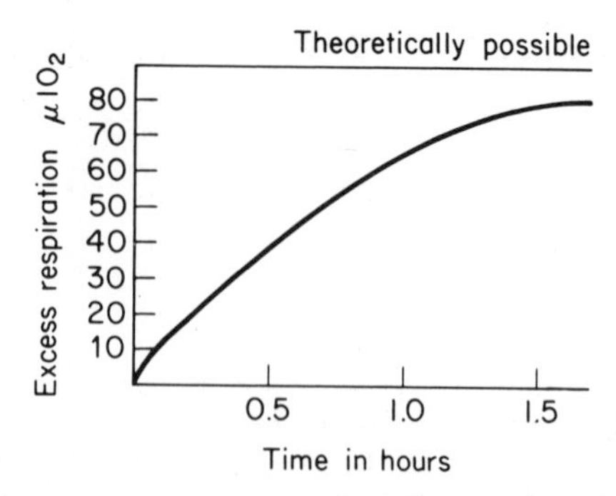

Fig. 3. Excess oxygen consumption during oxidation of pyruvic acid (pigeon muscle).

*Pyruvic acid formed in solution was determined by the Fromageot method[12].

upon oxidation. Among these substrates are the following acids: citric, keto-glutaric [normal keto-pyrotartaric acid], succinic, fumaric, malic, lactic and pyruvic. Acetic acid does not cause any increased respiration or phosphagen synthesis. The activity of the various substrates is illustrated in Table 1.

Table 1

SYNTHESIS OF PHOSPHAGEN LINKED WITH OXIDATION OF VARIOUS SUBSTRATES

Dates of experiment	Tissue	Substrate	Respiration in μl O_2 per 1 gm tissue for 30 minutes		Increase of phosphagen in mg of P_5O_2 per 1 gm tissue	
			without substrate	with substrate	without substrate	with substrate
1939						
April 1	Rabbit heart	Citric acid	263	399	4.00	7.25
April 19	" "	Fumaric acid	95	386	0.20	5.62*
May 5	" "	α-ketoglutaric acid	120	540	0.45	3.40*
April 7	" "	Succinic acid	206	956	2.30	4.26
1938						
June 3	Pigeon muscle	Malic acid	280	420	0	1.84
May 16	"	Lactic acid	252	387	0	1.56
Oct. 26	"	Pyruvic acid	214	420	0	2.34
May 10	"	Acetic acid	170	153	0	0

*In the presence of 0.02 M NaF

Because the data in Table 1 were collected from a number of different experiments, showing a considerable variation in ability of tissue to synthesize phosphagen, it is necessary to refrain from a quantitative comparison of activity of the various substrates.

In experiments with unwashed frog muscle tissue, we repeatedly noted a direct relationship between increased respiration with the help of creatine—"creatine effect"—and the quantity of synthesized phosphagen. In experiments with washed muscle tissue of warm-blooded animals we did not find such regularity. Increased respiration was insignificant (not more than 40%) and frequently completely absent. Apparently in these experiments the oxidatively coupled phosphorylation present was not rate limiting for respiration (see Sections 6 and 7). This circumstance does not, of course, rule out the possibility that "autonomous" regulation of tissue respiration depends on phosphorylation coupled to respiration in comparatively undamaged muscle, and particularly, as can be judged from indirect evidence, in whole muscle.

3. Participation of Various Phosphate Acceptors in Processes of "Respiratory" Phosphorylation

In glycolytic metabolism phosphagen is not the primary but the secondary acceptor of phosphate which is carried over from the adenylic acid system. It would be natural to expect the same in respiration. In the case of "respiratory" phosphorylation, addition of an excess of adenylic acid should result in synthesis of adenosine-polyphosphoric acids. We have conducted such experiments. Phosphorylation of adenylic acid was estimated by the increase of easily hydrolyzed phosphorus (7′-P). As an example we give below the results of one of our experiments in which monoidoacetate was used in order to eliminate glycolysis.

In order to determine inorganic phosphate, as well as phosphate split off during hydrolysis, in this and related experiments, the phosphate concentration of the medium was decreased eight times. The total volume of liquid in the experiment vessels was decreased to 1.1 ml; the portion of washed heart muscle was 200 mg. The medium contained cozymase, magnesium and some samples also contained 0.05 M of lactate. Neutralized adenylic acid (5 mg) was added from a side bulb.

Table 2

ADENYLIC ACID AS A PHOSPHATE ACCEPTOR

Adenylic acid Lactate	− −	− +	+ −	+ +
Dates	Increase in 7′—min. hydrol. P, in mg of P_2O_5 per 1 gm of tissue			
1939				
Jan 21	0	0.28	0.30	0.72
Jan 23	0.85	—	1.10	1.55
Jan 26	0.10	0.35	0.25	0.80

Formation of easily hydrolyzed phosphate in samples with added adenylic acid was actually observed. However, the synthesis was insignificant. We did not investigate the reason for this limitation. It is possible that deamination of adenylic acid plays a part in this, or that there is inhibition of further synthesis by adenosinetriphosphate. Increased respiration was often observed on addition of adenylic acid, though this effect, as in the case of creatine, was not great (respiration increased about 30%). However, in contrast to the prolonged effect of creatine, the effect of adenylic acid continued for only 10 to 20 minutes.

In general, the results of the investigation supported the assumption that adenylic acid is capable of playing the role of acceptor of phosphate in phosphorylations linked with respiration. There is nothing to prevent one from assuming that the adenylic system is a primary acceptor of phosphate in the "respiratory" synthesis of phosphagen.

Inasmuch as tissue glucose can be used in ground muscle as a substrate of glycolysis (Laquer and Meyer[13]), it seemed plausible that glucose instead of creatine could be used as a secondary acceptor of phosphate. We judged the phosphorylation of glucose by the appearance of the sparingly hydrolyzable phosphoric ester. Experiments were conducted on previously washed heart muscle subsequently poisoned with monoiodoacetic acid. The expected increase of the sparingly hydrolyzable phosphate in the presence of glucose actually took place, but the synthesis amounted to less than that in the case of adenylic acid and was scarcely significant. Some esterification of inorganic phosphate may also take place without the addition of phosphate acceptors although the rate of this synthesis is small and varies considerably. The nature of the products which are synthesized should be determined by additional studies.

Table 3

SYNTHESIS OF PHOSPHAGEN AND DECREASE OF INORGANIC PHOSPHATE IN "RESPIRATORY" PHOSPHORYLATION, mg P_2O_5 (WITHOUT ENUMERATION PER gm OF TISSUE)

Experiment number	Prior to incubation			After incubation: With lactate, without creatine			After incubation: With lactate, with creatine		
	Inorganic phosphate	Phosphagen	Sum	Inorganic phosphate	Phosphagen	Sum	Inorganic phosphate	Phosphagen	Sum
1	1.18	0.05	1.28	1.28	0.06	1.34	0.88	1.32	1.32
2	0.68	0.08	0.76	0.75	0.10	0.84	0.57	0.26	0.83
3	0.60	0.05	0.65	0.42	0.03	0.45	0.08	0.36	0.44
4	0.64	0.03	0.67	0.48	0.05	0.53	0.15	0.32	0.47

In comparison with the synthesis of phosphagen, synthesis of other phosphate esters plays only a secondary role. Table 3 records the data of experiments in which, parallel with the calculation of the synthesis of phosphagen, the general decrease of inorganic phosphate was taken into account.

In a number of experiments the quantity of synthesized phosphagen more or less corresponded to the quantity of the vanished inorganic phosphate and, consequently, phosphagen was the sole end product of phosphorylation. In some experiments the quantity of synthesized phosphagen did not correspond to the quantity of inorganic phosphate which disappeared; hence synthesis of other phosphate esters had also occurred. Yet even in these cases synthesis of phosphagen predominated. Data so far collected permit us to view creatine as the most active phosphate acceptor in muscle tissue.

4. Effect of Poisons on Phosphorylation Linked with Respiration

MONOIODOACETIC ACID

From one of our earlier studies, it appeared that "respiratory" synthesis of phosphagen in frog muscle tissue (unwashed) is possible even when glycolysis is completely arrested by monobromo acetic acid. We are now convinced that the same view may be applied to the case of synthesis, associated with oxidation of pyruvic acid in washed heart muscle. In our experiments creatine was added to muscle tissue following a half-hour incubation of the tissue with monoiodoacetic acid.

Monoiodoacetic acid not only eliminates glycolysis but also markedly inhibits respiration even in the presence of added substrate (lactic and pyruvic acid). Hence, it is not surprising that phosphagen synthesis is decreased. Yet, the synthesis still remained quite distinct.

Since the inhibition of glycolysis by monoiodoacetic acid is considered to be due to an elimination of glycolytic oxidation-reduction, these experiments support the existence of a "purely respiratory" phosphorylation.

Sodium fluoride

Inasmuch as NaF has relatively little influence on oxidation-reduction, and does not inhibit glycolytic phosphorylation, it could be expected that synthesis of phosphagen in our experiments would not be inhibited by the fluoride ion. Fig. 5 shows the data relating to respiration and phosphagen synthesis in the presence of NaF.

These findings show that NaF greatly restrains phosphagen synthesis in samples without added substrate; at a concentration of 0.015 M, NaF almost completely stops this synthesis. In samples with added substrate, though there is some delay of respiration, phosphagen synthesis is almost unimpaired.

With decreased respiration, phosphagen synthesis remains high, and so one can speak about the increased effectiveness of respiration in the presence of NaF. This, however, is probably explained not by actual increase of phosphorylation but by inhibition of dephosphorylation, which protects the product of synthesis from reverse dissociation (cf. 5).

The fact that NaF is capable of interrupting respiratory coupling of phos-

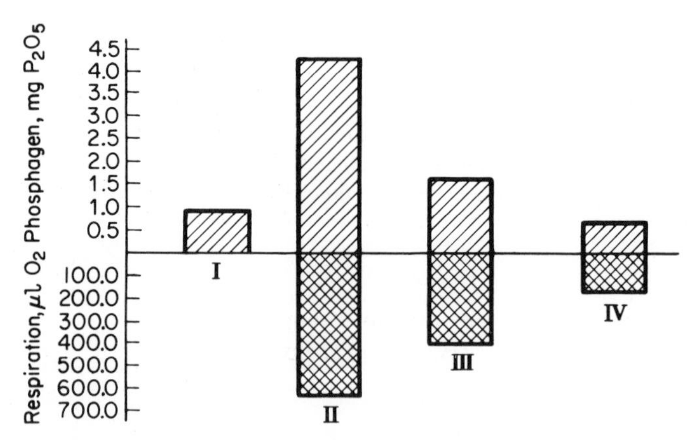

Fig. 4. Effect of monoiodoacetic acid on respiration and on synthesis of phosphagen (pigeon muscle). I: before incubation; II: pyruvic acid without monoiodoacetic acid; III: the same with monoiodoacetic acid; IV: monoiodoacetic acid without pyruvic acid.

phagen synthesis in the samples to which no exogenous substrate is added was subsequently applied in a study of the effect of different substrates. Elimination of "endogenous" synthesis which, despite the uniformity of tissue washing methods, fluctuated considerably from experiment to experiment, created a favorable basis for investigating the role of added substrates.

Arsenic acid

The effect of arsenic acid on respiration and on the synthesis of phosphagen in frog muscle tissue has been reported previously[14]. In experiments with washed rabbit heart muscle, arsenic acid depressed phosphagen synthesis relatively little. Even at a concentration of 0.1 M phosphorylation did not cease completely; 0.005 M had practically no effect; 0.05 M gave 50% inhibition.

It is interesting to note that, in contrast to experiments with frog muscle, no respiratory activation was observed. This result agrees with the fact that "the creatine effect" was weak or absent. An increase in respiration by arsenic acid, seemingly due to a delay in phosphorylation, is manifest only in cases where the rate of respiration is limited by the process of phosphorylation. The weakness or absence of a "creatine" effect indicates that this limitation in the experiment (with washed tissue) was absent or present to an insufficient degree.

Arsenous acid

Arsenous acid strongly inhibits respiration and the phosphorylation linked with it. At a concentration of 0.002 M this poison almost completely stops synthesis of phosphagen in the presence of the substrates: lactic, malic, fumaric, citric and α-ketoglutaric acids; succinic acid alone presents an exception which will be discussed later. In our experiments it was not possible to determine

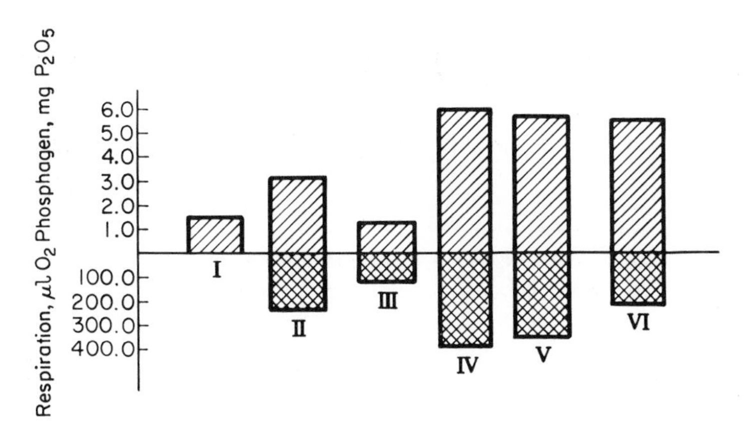

Fig. 5. Effect of NaF on respiration and synthesis of phosphagen (rabbit heart). I: before incubation; II: without a substrate, without NaF; III: the same with 0.015 M NaF; IV: with lactate, without NaF; V: with lactate, 0.015 M NaF; VI: with lactate, 0.05 M NaF.

the exact differences in the action of arsenite in the oxidation of ketoacids and ketoglutaric acid.

Oxalic acid

Oxalic acid is a glycolytic poison which arrests the intermediate conversion at the phosphopyruvic acid stage[15]. Oxalic acid was used in the form of its potassium salt in concentrations of 0.1 and 0.02 M. Both concentrations were sufficient to completely suppress "endogenous" phosphagen synthesis (in samples without added substrates). In the presence of lactate, 0.02 M oxalate solution reduced phosphagen synthesis approximately 50% and 0.1 M reduced it about 90%. Suppression of respiration was also observed, in general, proportionally to decreased synthesis. Hence oxalate like fluoride preferentially influences the "endogenous" synthesis, whereas in the presence of added substrate the suppression is less pronounced. Yet, in contrast to fluoride, oxalate decreases respiration approximately to the same degree as does phosphorylation.

5. Quantitative Relationship between Consumed Oxygen and Phosphagen Synthesis (Efficiency of Respiratory Phosphorylation)

In our previous experiments the ratio of moles of synthesized phosphagen to moles of consumed oxygen amounted to between 5 and 8 by allowing for the excess of consumed oxygen[6]. The applicability of such a calculation is debatable, however, since it was not proved that phosphagen synthesis is tied exclusively to the "surplus" respiration, i.e. respiration resulting from added creatine. On the contrary, it is possible that "basic respiration" (i.e., that portion of respiration that occurs in the absence of creatine) also participates in the synthesis. Our experiments with washed muscle, in which regular synthesis of phosphagen is frequently observed despite the predominant absence of excess respiration, point to this. If phosphagen synthesis in our earlier experiments is related not to "excessive" but to ordinary respiration then the coefficient is 1–3 rather than 5–8.

Determination of the actual quantitative relationship between respiration and phosphagen synthesis is interesting not only for its own value but also in connection with the question of the mechanism of respiratory phosphorylation. Striving to obtain precise figures, we changed the experimental arrangement so as to decrease, as much as possible, basic respiration and to fix more precisely the moment of beginning and ending of synthesis.

Experiments were conducted at 0°C. A Warburg apparatus bath contained a mixture of water and snow, which was replenished at intervals throughout the experiment, with an excess of snow present at all times; 16 mg of dry creatine was placed in one of the side bulbs of the vessel. In the other bulb we put 0.2 ml of 40% trichloracetic acid. The experimental specimen was unwashed minced frog muscle. The medium contained 0.037 M phosphate, *p*H 7.6; 0.15 mg Mg and 0.13 M KCl. The total volume was 2 ml, the tissue weight was 200 mg, and the gas was O_2.

After equilibration of the temperature (which required no less than half an hour) respiration was determined. Subsequently, creatine from the side bulbs was brought into the experimental vessel. Respiration measurement continued for no less than one hour. Then trichloracetic acid from the second bulb was added, thus fixing the tissue in order to stop biological reactions. Phosphagen determination was done in the usual manner.

It should be noted that addition of the dry creatine from the side bulbs caused a small but definitely perceptible increase of gas pressure. This increased pressure, depending most likely on decreased solubility of O_2 under the influence of dissolving creatine, was measured in a separate vessel containing all ingredients except tissue, and was taken into account as a correction.

The coefficient:

$$\frac{\text{(synthesized phosphagen)}}{\text{(consumed oxygen)}},$$

in calculations involving the oxygen consumed in respiration in the period of synthesis, varied in different experiments between 3.8 and 4.3. If calculated on the basis of oxygen consumed in surplus, the coefficient amounted to 5.0–7.3.

It seems most reasonable to assume that the part of respiration directly resulting in synthesis is greater than the "surplus", but smaller than the total respiration. Hence, the figures 5–7.3 are greater than the actual ones and the figures 3.8–4.3 are smaller. Until further refinement is attained, it is, therefore, necessary to consider possible variations of the coefficient in the range of 4 to 7.

Table 4

COEFFICIENT OF SYNTHESIS (EXPERIMENT OF MARCH 23, 1939)

		Without incubation	Incubation	
			with creatine	without creatine
Content of phosphagen	in mg P_2O_5	0.195	0.56	0.21
	in μM "	2.74	7.88	2.95
Consumption of oxygen	in μl O_2		27.5	12.5
	in μM "		1.23	0.55
"Excess" consumption of oxygen	in μl O_2		15.0	
	in μM "		0.67	

Coefficient of synthesis:
- for total respiration $\frac{7.88 - 2.74}{1.23} = 4.2$
- for "excess" respiration $\frac{7.88 - 2.95}{0.67} = 7.3$

6. Phosphagen Synthesis Associated with Oxidation of Succinic Acid to the Stage of Fumaric Acid

In the usual experimental set-up phosphagen synthesis in the presence of succinic acid is no greater than in the presence of other substrates, although oxygen consumption is considerably more intensive. On this basis it could be presumed that the first stage of succinic acid oxidation (on which the intensive consumption of oxygen depends) is not coupled with phosphorylation, but that synthesis depends on the subsequent slower stages of oxidation. Experiments have proven, however, that there is no sufficient basis for concluding that there is a lack of coupling of phosphorylation with oxidation of succinic acid to the stage of fumaric acid.

In comparing phosphagen synthesis during oxidation of succinic and malic acids, we at first used both substrates in equal concentrations (0.025 M). Then, in order to verify that phosphagen synthesis in the presence of succinic acid may be ascribed exclusively to the oxidation of intermediate formation of malic acid, we lowered the concentration of malic acid, using such quantities of it which could, in our estimation, be expected to arise from succinic acid (upon further oxidation) during the course of an experiment. The maximal number of moles of malic acid which could have arisen as an intermediatic product, subject to further oxidation does not exceed the number of moles of oxygen concurrently consumed (in our experiments about 12 μM to 200 mg of tissue).

Figure 6 shows that phosphagen synthesis in the presence of succinic acid is somewhat higher than in the presence of malic acid. The difference, although not great, is nevertheless quite real. The surplus synthesis in the presence of succinic acid should be regarded as relating to the first step of oxidation.

Further substantiation of the existence of phosphorylation linked with the first step of succinic acid oxidation was shown by experiments with tissue poisoned with arsenous acid. These experiments revealed that the only substrate the oxidation of which showed relatively little sensitivity to arsenous acid was succinic acid. On oxidation of this acid, in the presence of arsenous acid, phosphagen synthesis is present to almost the same degree as in nonpoisoned tissue. Fumarate and malic acid under the same circumstances (and the same concentrations), like other substrates, oxidizes slowly, and if synthesis is present it is very weak. When succinic acid is oxidized in tissue poisoned with As_2O_3,

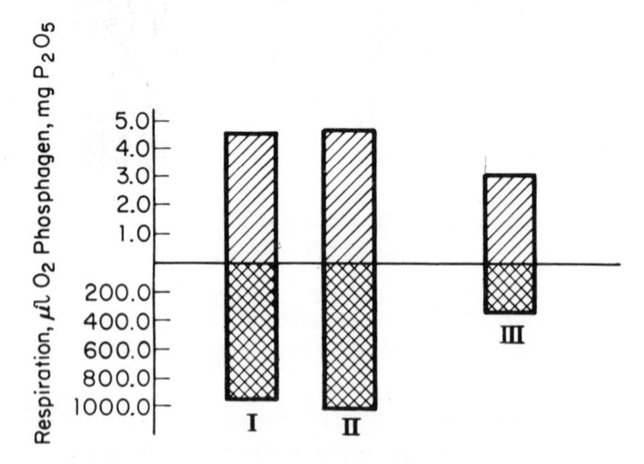

Fig. 6. Respiration and phosphagen synthesis during oxidation of succinic and malic acids. I: succinic + malic acids; II: succinic acid; III: malic acid.

the synthesis of phosphagen is increased further in the presence of NaF. The latter apparently augments the effectiveness of respiratory synthesis.

Since synthesis of phosphagen linked with more extensive oxidation is practically excluded during poisoning of tissue with arsenite, any significant synthesis of phosphagen taking place in the presence of succinic acid can only be attributed to oxidation to the stage of fumaric acid.

In several experiments in which the original concentration of inorganic phosphate was decreased, we detected not only the synthesis of phosphagen but also the decrease of inorganic phosphate. Results of one such experiment are illustrated in Fig. 8.

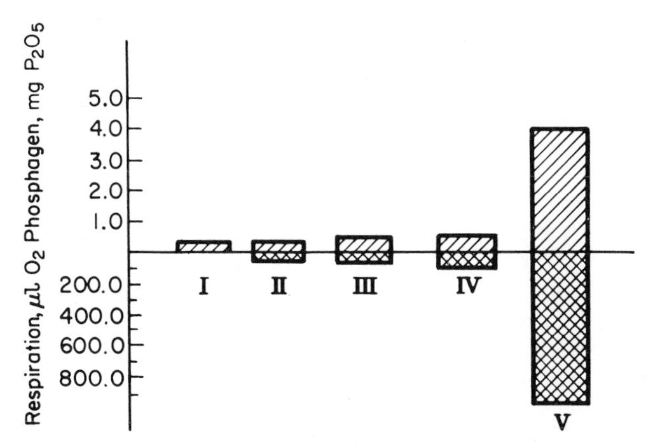

Fig. 7. Phosphagen synthesis during oxidation of malic, fumaric and succinic acids in the presence of arsenous acid and NaF (rabbit heart). I: prior to incubation; II: without substrate; III: with malic acid; IV: with fumaric acid; V: succinic acid.

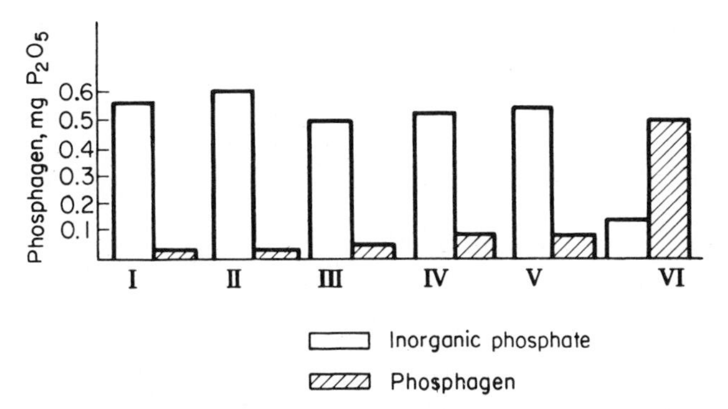

Fig. 8. Uptake of inorganic phosphate and synthesis of phosphagen during oxidation of succinic acid and fumaric acid, in the presence of NaF and arsenous acid. I: before incubation; II: without substrate, with creatine; III: with fumarate, without creatine; IV: with fumarate and with creatine; V: with succinic acid without creatine: VI: the same with creatine.

As in experiments with other substrates, esterification of substrates in the absence of creatine was insignificant. Without substrate, even in the presence of creatine, esterification was not observed. Phosphate esterification was intensive only in samples containing succinic acid and creatine simultaneously. Such conditions frequently brought about almost complete disappearance of inorganic phosphate.

Since only the final links in the chain of intermediate respiratory oxidation-reductions (with participation of cytochrome, Warburg's oxidase and molecular oxygen) operate in the oxidation of succinic acid it might have been expected that part of the intermediate oxidation-reductions, linked with phosphorylation, would be absent here and hence the ratio of phosphagen synthesis to oxygen consumption would be found to be low. In an experiment conducted at 0°C this proportion was 0.6. The maximum value found in one of the experiments (20°C) was 1.2. If the process of succinic oxidation includes one oxidation-reduction coupled with phosphorylation, then the carryover of two atoms of hydrogen could be accompanied by synthesis of one mole of phosphagen. We consider this amount the probable limit of effectiveness of synthesis linked with oxidation of succinic acid to the stage of fumaric acid. To judge whether phosphorylation is coupled with the oxidation-reduction taking place in the "iron system" (system of cytochromes + Warburg's oxidase) we made use of p-phenylenediamine which, as is known, directly without participation of dehydrogenase, reduces cytochrome. In our experiments, in the presence of p-phenylenediamine used as substrate, we observed particularly intensive consumption of oxygen (more intensive than with succinic acid) but phosphagen synthesis was absent. Such a result indicates that oxidation-reductions in the "iron system" are not linked to phosphorylation.

7. Discussion of Results

According to our earlier data, the coefficient of aerobic phosphagen resynthesis (the relation of the quantity of synthesized phosphagen to the quantity of consumed oxygen, in moles) for whole resting muscle in stationary condition varies between 3 and 4[16]. As calculated for excess respiration during aerobic rest, this coefficient averages 5.2[17]. In the present study, values of the same order, i.e., 4 to 7, were found for the synthesis of phosphagen from added creatine.

110,000 calories are released by the consumption of 1 molecule of oxygen. From this quantity of heat, with a coefficient of synthesis of 5 then 57,500 calories (11,550 × 5) are used for phosphagen synthesis. With a maximal coefficient of 7 correspondingly 80,500 calories would be used. This means that more than half (up to 2/3) of the energy of respiration may be utilized in muscle tissue for the synthesis of phosphagen.

It is not difficult to compute how many molecules of phosphagen are synthesized in complete oxidation of one hexose residue of glycogen (or one molecule of hexose). For the oxidation of this, the necessary consumption of oxygen amounts to 6 molecules of O_2($C_6H_{10}O_5 + H_2O + 6O_2 = 6CO_2 + 6H_2O$) and 24 to 42 molecules of phosphagen are synthesized (for each molecule of O_2 there are no less than 4 and no more than 7 molecules of phosphagen). On the average it can be expected that oxidation of a molecule of hexose synthesizes about 33 molecules of phosphagen. This is 10 times greater than the quantity of phosphagen which may be synthesized in glycolysis. (Each hexose residue of glycogen forms 2 molecules of lactic acid, during the course of which 3 molecules of inorganic phosphate are esterified. These latter in the terminal transesterification to creatine can give 3 moles of phosphagen.)

This explains the fact that phosphagen synthesis at the expense of glycolysis only partially compensates "spontaneous" disintegration, but respiratory synthesis compensates it with a surplus.

The problem of the mechanism of the "respiratory" phosphorylation is bound intimately to the problem of the mechanism of cellular respiration.

Without going into a discussion of the various schemes of intermediate substrate transformations, we will merely point out that in complete oxidation a molecule of hexose undergoes 12 different steps of dehydrogenation (these dehydrogenations may be labelled primary). Subsequently every one of the 12 pairs of hydrogen atoms goes, in turn from one "carrier" (intermediate hydrogen acceptor) to another, and in the end, after a number of intermediate oxidation-reductions, is oxidized by molecular oxygen.

Among the number of "primary" dehydrogenations it is known that four may be linked with phosphorylation: this is the dehydrogenation of 2 molecules of phosphotriose and the further dehydrogenation of 2 molecules of pyruvic acid (Lipmann[(11)]). Nothing is known at present about the nature of other primary dehydrogenations, and whether these are coupled with phosphorylation. Even if coupling with other primary dehydrogenations were present, it would be possible to ascribe to this only the esterification of 12 molecules of phosphate (considering that each oxidation-reduction would bind 1 molecule of phosphoric acid). Perhaps the explanation of the high coefficients of "respiratory" phosphorylations must be based on the possibility that not only primary, but also some intermediate oxidation-reductions are coupled with phosphorylation. If, in the route of hydrogen transport, there were one oxidation-reduction linked with phosphorylation, then the upper limit of respiratory coupled synthesis increases to 24 molecules of esterified phosphate. If there were two such reactions it would involve 36 molecules. The latter value approximately corresponds to the average effectiveness of respiratory synthesis found in our experiments. Consequently, it may be assumed that in addition to the primary oxidation-reductions coupled with phosphorylation, two or more oxidation-reductions that participate in hydrogen transport are coupled with phosphorylation. It can now be presumed that one of those intermediate oxidation-reductions linked with phosphorylation appears to be oxidation in the "fumaric system," i.e., to the extent that dehydrogenation of succinic acid was shown to be linked with phosphorylation.*

Other types of other couplings linked with phosphorylation, in the chain

*The intimate mechanism of oxidation of succinic acid linked with phosphorylation is not yet completely clear. Oxidation of succinic acid, as is known, does not demand participation of coenzymes; we have become used to considering coenzymes as indispensable participants in the oxidation-reductions linked with phosphorylation. According to recent investigations by English authors,[(18)] the system for dehydrogenation of succinic acid may turn out to be more complicated than was previously thought.

of reactions of hydrogen transport, besides the one mentioned, remain to be demonstrated.

Apparently the effectiveness of respiratory phosphorylation differs for different cells. In the case of nucleated erythrocytes, for instance, it is low; the synthesis coefficient, according to Engelhardt,[19] is one. The difference in effectiveness may be explained by the fact that the number of respiratory oxidation-reductions linked with phosphorylation varies for different cells.

The coupling between respiration and phosphorylation is apparently accomplished only under definite conditions, and is not necessarily mutually compulsive. Respiratory oxidation-reductions can also proceed as an "idle run," i.e., without esterification. Glycolytic oxidation-reduction, on the contrary, is coupled mutually with phosphorylation. Based on these facts it should be possible to distinguish two types of oxidation-reductions linked with phosphorylation: unconditionally coupled and conditionally coupled.

The previously expressed supposition that the coupling between respiration and phosphorylation depends on particular factors of enzymic nature has not as yet been experimentally proven.

Conclusions

1. Pigeon breast and rabbit heart muscles minced, homogenized, and depleted of endogenous substrates by washing in a phosphate solution, have turned out to be convenient preparations for a study of phosphorylation coupled with respiration.

2. A study was made of the influence of *p*H and other conditions of the medium on "respiratory" phosphorylation.

3. Phosphagen synthesis coupled with respiration was found during oxidation of citric, α-ketoglutaric, succinic, fumaric, malic, lactic and pyruvic acids.

4. In phosphorylation linked with respiration, the following compounds can serve as phosphate acceptors: creatine, adenylic acid and glucose; endogenous acceptors are also present. Of all these substances creatine is phosphorylated to the highest degree. In the presence of a sufficiently large quantity of creatine, most of the decrease in inorganic phosphate, if not all, can be accounted for by the synthesis of phosphagen.

5. The action of the poisons—monoiodoacetic acid, sodium floride, arsenic, arsenous and oxalic acids—on "respiratory" phosphorylation was studied.

6. When the total respiration was considered, the ratio of the number of moles of phosphagen synthesized to the number of moles of oxygen consumed at 0°C ranged between 3.8 and 4.3; when only the "excess" respiration was considered, the ratio ranged between 5.0 and 7.3.

7. Aerobic oxidation of succinic acid to the stage of fumaric acid is coupled with phosphorylation. This can best be proved in experiments with tissue poisoned with arsenous acid, which arrests the further oxidation of fumaric and malic acids, without preventing the oxidation of succinic acid.

8. The high efficiency of respiratory synthesis (in oxidation of a molecule of hexose, no less than 24 molecules of phosphagen are synthesized) shows that phosphorylation is not only coupled with the primary steps of dehydro-

genations of substrate molecules, (like dehydrogenation of phosphotriose or of pyruvic acid; the process of complete oxidation of the hexose molecule has to include 12 primary dehydrogenations), but also with certain oxidation-reductions which participate in further hydrogen transport. In this latter category apparently belongs the oxidation of succinic acid.

References

1. Braunshteyn, A. E.: "Report to the XV International Physiological Congress in Leningrad" (1935).
2. Severin, V. A.: *Biokhimiya*, **2** (1937), 60.
3. Grimlund, K.: *Skand. Arch. Physiol.*, **73** (1936), 109.
4. Meyerhof, O. and Boyland, E.: *Biochim. Z.*, **237** (1931), 406; Meyerhof, O., Gemmill, Ch., and Benatato, G.: *Biochim. Z.*, **258** (1933), 371.
5. Kalckar, H.: *Enzymologia*, **2** (1937), 47.
6. Belitser, V. A.: *Biokhimiya*, **2** (1937), 334; **3** (1938), 80.
7. Belitser, V. A.: *Byulleten' Experimental'noy Biologii i Meditsiny*, **7** (1939), 111.
8. Euler, H., and Adler, E.: *Z. physiol. Chem.*, **246** (1937), 83.
9. Meyerhof, O., and Ohlmeyer, P.: *Biochim. Z.*, **290** (1937), 334.
10. Ochoa, S.: *Biochim. Z.*, **292** (1937), 68.
11. Lipmann, F.: *Nature*, **143** (1939), 281.
12. Fromageot, C., and Desnuelle, P.: *Biochim. Z.*, **297** (1935), 174.
13. Laquer, F., and Meyer, P.: Quoted from Meyerhof, in *Chemische Vorgänge im Muskel.* Berlin: 1930.
14. Belitser, V. A.: *Biokhimiya*, **4** (1939), 498.
15. Lohmann, K., and Meyerhof, O.: *Biochim. Z.*, **273** (1934), 60.
16. Belitser, V. A., Zyukova, M. A., and Fal'k, A. Ya.: *Biokhimiya*, **2** (1937), 28, 38.
17. Meyerhof, O., and Nachmansohn, D.: *Biochim. Z.*, **222** (1930), 1.
18. Hopkins, F. G., Lutwak-Mann, C., and Morgan, E. J.: *Nature*, **143** (1939), 556.
19. Engelhardt, W. A.: *Biochim. Z.*, **251** (1932), 343.

Ochoa, Wood, and Carson: "With heavy carbon . . ." "But with radioactive carbon . . ."
(Courtesy of the University of Wisconsin Press.)

V. A. Belitser and Herman Kalckar reminiscing about 1946, the "banner year" of oxidative phosphorylation. This picture was taken at the Akademy Nayk, Kiev, U.S.S.R., in 1960.

NATURE OF OXIDATIVE PHOSPHORYLATION IN BRAIN TISSUE*

Severo Ochoa

Department of Biochemistry
University of Oxford
England

(Received for publication August 5, 1940)

A connexion between oxidation of pyruvic acid and esterification of inorganic phosphate with hexosemonophosphate to hexosediphosphate in brain preparations has been recently described[1]. I have now found that glucose can be substituted for hexosemonophosphate and the phosphorylation product is still hexosediphosphate. The necessity of 'adenine nucleotide' for the above reaction as well as the dependence of the oxidation on the presence of both 'adenine nucleotide' and inorganic phosphate[2] led to the assumption that the oxidation of pyruvic acid was linked up to a phosphorylation of adenylic acid to adenosinepolyphosphate, which would then transfer its labile phosphate groups to hexosemonophosphate or glucose. This is supported by the fact that either glucose or hexosemonophosphate can be phosphorylated to hexosediphosphate anærobically by transfer of phosphate from phosphopyruvic acid in the presence, but not in the absence, of adenylic acid. Half of the phosphopyruvate phosphate is set free and the other half is esterified with glucose or hexosemonophosphate. This is the same ratio as was found by Ostern, Guthke and Terszakovec[3] for the transfer of phosphate from adenosinetriphosphate to hexosemonophosphate in muscle extract. The phosphorylation of glucose under these conditions had not been so far observed in animal tissues.

The reactions taking place with glucose can be formulated as follows:

*S. Ochoa, "Nature of Oxidative Phosphorylation in Brain Tissue," *Nature*, **146**, 267 (1940). Reprinted with the permission of the author and of the editors of *Nature*.

[1] Ochoa, *Nature*, **145**, 747 (1940).

[2] Banga, Ochoa and Peters, *Biochem. J.*, **33**, 1980 (1939).

[3] *Z. physiol. Chem.*, **243**, 9 (1936).

(1) 4 phosphopyruvic acid + 2 adenylic acid = 2 adenosinetriphosphate + 4 pyruvic acid.

(2) 2 adenosinetriphosphate + 1 glucose + 2 H_2O = 1 hexosediphosphate $+2H_3PO_4$ + 2 adenylic acid.

The sum of equations (1) and (2) is the overall reaction (3):

(3) 4 phosphopyruvic acid + 1 glucose + $2H_2O$ = 1 hexosediphosphate + $2H_3PO_4$ + 4 pyruvic acid.

Recently Colowick, Welch and Cori[4] concluded that the oxidative phosphorylation of glucose in kidney extracts, previously studied by Kalckar[5], is linked up with the oxidation of succinic to fumaric acid. In brain dispersions oxidation of succinate does give rise to phosphorylation of glucose. However, if oxidation of the pyruvate—arising by subsequent oxidation of malate to oxaloacetate and breakdown of the latter—is checked by arsenite, the ratio of atoms phosphorus esterified to molecules oxygen taken up is only about a half of that obtained when pyruvate is oxidized (in the absence of arsenite) as shown in the accompanying table. Arsenite, whilst fully inhibiting both oxidation of pyruvate and the accompanying phosphorylation, is without effect on the phosphorylation of glucose by transfer of phosphate from phosphopyruvic acid.

1.5 ml. dispersion from pigeon brain (dialysed 6.5 hours) to 2.3 ml. with additions including phosphate buffer *p*H 7.3 (0.025*M*), Mg^{++} (0.2 mgm.), adenylic acid (0.0007 *M*), glucose (10 mgm.), and NaF (0.02*M*). 35 min. in air at 38°. (O_2 uptake measured during the last 30 min.).

	No further addition	Na fumarate (0.005*M*) +Na pyruvate (0.013*M*)		Na succinate (0.03*M*)	
		No arsenite	0.008*M* Na arsenite	No arsenite	0.008*M* Na arsenite
μl O_2 uptake	0	320	19	276	178
mgm. P esterified	—	1.20	0.00	0.77	0.23
atoms P/moles O_2	—	2.78	0.00	2.01	0.94

The above suggests that the oxidative phosphorylation may be of a twofold nature; half of it being connected with the dehydrogenation of pyruvic acid as suggested previously[1], the other half with the transfer of hydrogen catalysed by dicarboxylic acids. This view would seem to be supported by the fact that when pyruvate is oxidized by brain dispersions, in the presence of fumarate, the P/O_2 ratio in the first few minutes is 4 atoms phosphorus to 1 molecule oxygen. It would be difficult to understand how the removal of 2H would give rise to an uptake higher than 1 atom phosphorus unless their further catalytic transfer is also linked up with phosphorylation.

I am indebted to Prof. R. A. Peters for his interest and to the Nuffield Trustees and the Rockefeller Foundation for grants in aid of this work.

[4] *J. Biol. Chem.*, **133**, 359 and 641 (1940).
[5] *Biochem. J.*, **33**, 631 (1939).

OXIDATIONS AND REDUCTIONS*

H. A. Lardy and C. A. Elvehjem

University of Wisconsin
Madison, Wisconsin

* * *

Action of agents which decrease efficiency of energy utlization

A number of agents exist which have the property, when added in appropriate concentrations to various tissues, of increasing the rate of exergonic reactions while simultaneously decreasing the energy utilizing functions; the

Table I

AGENTS WHICH AFFECT THE COUPLING OF OXIDATION-REDUCTION REACTIONS WITH ENERGY UTILIZATION

Agent	Process stimulated	Process inhibited	Reference
Dinitrophenol (DNP)	Respiration and glycolysis	Maintenance of phosphocreatine	126
DNP	Respiration and glycolysis	Assimilation	127
DNP	Respiration and glycolysis	Sperm motility	128
Azide	Yeast fermentation	Assimilation	129
Chloral hydrate or chloretone	Respiration	Assimilation and luminescence	130
Gramicidin	Respiration	Assimilation, P uptake	131
Dysentery toxin	Hydrolysis of ATP		132

*Excerpt from H. A. Lardy and C. A. Elvehjem, "Oxidations and Reductions," *Annual Review of Biochemistry,* **14**, 16–17 (1945). Reprinted with the permission of the authors and of the publisher.

result is of course an increased heat production. The action of several of these agents is summarized in Table I. Hotchkiss (131) recently found that dinitrophenol (DNP) prevented phosphate uptake by bakers' yeast respiring in glucose and that the bacteriostatic action of gramicidin may also be related to its effect in preventing phosphate uptake. The concept of high energy phosphate as the intermediate energy carrier between oxido-reductions and energy utilization, and as the mechanism controlling the rate of oxidation and glycolysis (133) invites the hypothesis that those agents which speed up metabolism and at the same time decrease the energy available for work or assimilation act by allowing oxidations to occur without phosphorylation* or actually cause dephosphorylation of high energy phosphate. Early evidence for this was provided by the finding that DNP caused a decrease of phosphocreatine in frog muscle (126). DNP was also found to increase the rate of hydrolysis of ATP added to minced rat muscle (134).

* * *

References

126 Ronzoni E. and Ehrenfest E., *J. Biol. Chem.* **115**, 749–68 (1936).
127 Pickett M. J. and Clifton C. E., *J. Cell. Comp. Physiol.* **22**, 147–65 (1943).
128 Lardy H. A. and Phillips P. H., *J. Biol. Chem.* **149**, 177–82 (1943).
129 Winzler R. J., *Science* **99**, 327–28 (1944).
130 McElroy W. D., *J. Cell. Comp. Physiol.* **23**, 171–92 (1944).
131 Hotchkiss R. D., *Advances in Enzymology* **4**, 153–99 (1943).
132 Braun A. D. and Ratner M. Y., *Biokimiya* **7**, 171–79 (1942). *Chem. Abstracts* **38**, 154 (1944).
133 Johnson M. J., *Science* **94**, 200–02 (1941).
134 Lardy H. A., unpublished data.

*Compare the effect of arsenate on the oxidation of triosephosphate.

REVERSIBLE INHIBITION OF THE COUPLING BETWEEN PHOSPHORYLATION AND OXIDATION*

W. F. Loomis† *and Fritz Lipmann*

Biochemical Research Laboratory
Massachusetts General Hospital and the Department of Biological Chemistry,
Harvard Medical School, Boston, Massachusetts
(Received for publication, February 13, 1948)

Clifton[1] was among the first to show that dinitrophenol (DNP) in low concentrations completely blocked synthetic reactions without interfering with oxidation. Other workers have shown that this drug inhibits nitrogen assimilation,[2] growth and differentiation,[3] the formation of adaptive enzymes,[4] and Hotchkiss[5] has reported preliminary data showing that DNP prevents phosphate uptake by respiring yeast cells. These results would appear to indicate that DNP acts on the basic mechanism in the cell by which phosphate bond generation is coupled to oxidative reactions.

During a study of this coupling mechanism, it was observed that 5×10^{-5} to 2×10^{-4} M DNP prevented phosphorylation without affecting or with slightly stimulating oxidation.

Concentrations of DNP as low as 5×10^{-6} M were found to lower markedly the P:O ratio, an effect that could be reversed by washing out the DNP with fresh buffer. Furthermore, it was found that DNP could "replace" inor-

*W. F. Loomis and Fritz Lipmann, "Reversible Inhibition of the Coupling between Phosphorylation and Oxidation," *J. Biol. Chem.*, **173**, 807–808 (1948). Reprinted with the permission of the authors and of the American Society of Biological Chemists, Inc.

†Fellow in the Medical Sciences, National Research Council.

[1] Clifton, C. E., in Nord, F. F., and Werkman, C. H., Advances in enzymology and related subjects, New York, **6**, 269 (1946).

[2] Winzler, R. J., Burk, D., and du Vigneaud, V., *Arch. Biochem.*, **5**, 25 (1944).

[3] Clowes, C. H. A., and Krahl, M. E., *J. Gen. Physiol.*, **20**, 145 (1936).

[4] Spiegelman, S., *J. Cell. and Comp. Physiol.*, **30**, 315 (1947).

[5] Hotchkiss, R. D., in Nord, F. F., and Werkman, C. H., Advances in enzymology and related subjects, New York, **4**, 153 (1944).

Table I

All samples contained 1.0 cc. of an enzyme preparation similar to that of Green *et al.*,[6] prepared by centrifuging a rabbit kidney homogenate in KCl-$NaHCO_3$ buffer and washing the residue twice with fresh buffer. To this was added 0.1 cc. of yeast hexokinase, 0.0067 M $MgCl_2$, 0.013 M NaF, 0.00067 M adenosine-5-phosphate, 0.02 M phosphate buffer of pH 7.2, 0.0167 M fructose, and 0.01 M Na glutamate as substrate. Identical control cups were prepared, into which acid from a side arm was tipped at the beginning of the experiment to provide the initial level of inorganic phosphate. Temperature, 25°; gas phase, air; time, 6 minutes.

Additions	Oxygen uptake	Phosphate uptake	P: O ratio
	microatoms	*micromoles*	
None	8.0	17.5	2.2
1.8×10^{-4} M DNP	7.9	1.3	0.2

ganic phosphate, which otherwise is a compulsory component of this system. It appears that the phosphate-deficient system is strongly stimulated by DNP, while the complete system responds only with a slight stimulation (see Table II).

Table II

Temperature, 25°; gas phase, air; time, 30 minutes.

Phosphate, *M*	0	0	2×10^{-2}	2×10^{-2}
DNP, *M*	0	8×10^{-5}	0	8×10^{-5}
O_2, *micromoles*	5.1	17.9	18.6	21.0

These results indicate that DNP reversibly uncouples phosphorylation from oxidation, an effect that can also be obtained with atebrin (mepacrine) in 10^{-3} M concentration. Although sodium azide can lower the P:O ratio, it cannot replace phosphate in the system and is, in slightly higher concentration, a powerful inhibitor of respiration as well. DNP does not inhibit respiration except in high concentration.

[6] Green, D. E., Loomis, W. F., and Auerbach, V. H., *J. Biol. Chem.*, **172**, 389 (1948).

OXIDATION OF FATTY ACIDS AND TRICARBOXYLIC ACID CYCLE INTERMEDIATES BY ISOLATED RAT LIVER MITOCHONDRIA*†

Eugene P. Kennedy‡ *and Albert L. Lehninger*

Departments of Biochemistry and Surgery,
University of Chicago, Illinois

(Received for publication March 2, 1949)

Beginning with the fundamental observation of Warburg in 1913[1] it has been a general finding that the more highly organized enzyme systems of animal tissues responsible for oxidation of metabolites by molecular oxygen are associated with the insoluble particulate portion of the cell. Among the several approaches which have been used to study the morphology and composition of such catalytically active particulate material, the most fruitful has been the differential centrifugation technique for separation of nuclei, mitochondria or "large granules," and other substructures developed by Bensley and his school[2] and refined by Claude[3,4] and Hogeboom, Schneider, and Pallade[5,6]. Considerable work on the composition and enzymatic activity of the various particulate fractions has been described by the Rockefeller group[7,8], Schneider[9], and other investigators. For instance, quantitative assays have revealed that most of the succinoxidase and cytochrome oxidase activity is present in the mitochondria or "large granules"[7].

In this laboratory studies have been made on the enzymatic oxidation of fatty acids to acetoacetate and also via the Krebs tricarboxylic acid cycle. These complex and highly organized reactions take place in suspensions of particulate material separated from rat liver homogenates by centrifugation[10,11]. Certain

*Excerpt from E. P. Kennedy and A. L. Lehninger, "Oxidation of Fatty Acids and Tricarboxylic Acid Cycle Intermediates by Isolated Rat Liver Mitochondria," *J. Biol. Chem.*, **179**, 957–60, 964–67, 969–72 (1949). Reprinted with the permission of the authors and of the American Society of Biological Chemists, Inc.

†This investigation was supported by grants from the American Cancer Society (recommended by the Committee on Growth of the National Research Council), Mr. Ben May, Mobile, Alabama, and the Nutrition Foundation, Inc.

‡Nutrition Foundation Fellow in Biochemistry.

observations on the properties of this enzyme system[11] suggested that the activity was to some extent dependent on osmotic factors, and Potter, on the basis of measurements of "cytolysis quotients," suggested that the activity was present only in intact cells[12].

With the publication of what appears to be a definitive method for the isolation of mitochondria or "large granules" by Hogeboom, Schneider, and Pallade[5,6], it was possible to demonstrate that mitochondria isolated by this method bear all the demonstrable fatty acid oxidase activity of whole rat liver. The particulate material isolated by this method is stated to be homogeneous and identical in morphology and vital staining characteristics with the mitochondria of the intact cell[6]; mitochondria isolated by the earlier procedures of Bensley and Hoerr[2] and Claude[3] apparently represent partially damaged forms without these properties.

Since the publication of our preliminary note on the localization of fatty acid oxidase activity in these particles[13], Schneider has published a confirmatory report[14]. This paper is concerned with the experimental details of the basic experiments.

Experimental

Analytical methods

Octanoate was determined by the method of Lehninger and Smith[15], acetoacetate by a modification of the method of Greenberg and Lester[16], and citrate by the method of Speck, Moulder, and Evans[17]. Manometric measurements of oxygen uptake were made at 30° in Warburg vessels of conventional design with air as the gas phase. Flasks were equilibrated for 5 minutes prior to closing of the taps. Determinations of inorganic and total phosphorus were made according to the method of Gomori[18] and partition of the phosphorus of the enzyme preparations was carried out according to methods described by Schneider[19] and Schmidt and Thannhauser[20]. Radioactivity measurements were made on thin layers of aqueous solutions by means of a Geiger-Müller counting tube and recording apparatus of standard commercial type. Separations of esterified phosphate for these measurements were performed as described elsewhere[21].

Preparation of mitochondria from rat liver

The procedure of Hogeboom, Schneider, and Pallade[5,6] was used for the preparation of the particulate fractions of rat liver. Normal adult albino rats of Sprague-Dawley stock were used throughout this study. The animals were killed by decapitation and exsanguinated. The livers were quickly removed and chilled in cracked ice. All operations during the preparation of the fractions were carried out in a room maintained at 2° and all reagents and apparatus were previously chilled. The fresh, chilled rat liver was homogenized in 9 volumes of cold 0.88 M sucrose in a glass homogenizer of the type described by Potter and Elvehjem[22]. Nuclei, whole cells, stroma, and erythrocytes were removed by three successive centrifugations, each of 3 minutes duration, at about 1500 × *g* in the Sorvall model SP centrifuge. The mitochondria were then sedimented from the cleared supernatant by centrifugation in a Sorvall model SS-1 centrifuge at 18,000 × *g* for 20 minutes. The sedimented mitochondria were washed by resuspension in 10 volumes of 0.88 M sucrose, followed by resedimentation for 20 minutes at 18,000 × *g*. The supernatant was carefully decanted and the washed mitochondria were taken up in sufficient ice-cold 0.15 M KCl or water (about 5.0 ml. for each gm. of whole tissue used as starting material) to yield a suspension containing about 1 mg. of total nitrogen per ml. The final concentration of KCl in the fatty acid oxidase test system was about 0.05 M, a value shown to be near the optimum for fatty acid oxidation in a previous study(11). In experiments in which the mitochondria were taken up in distilled water, sufficient KCl was added to the test flasks to provide a final concentration of about 0.05 M.

Throughout these fractionations, it was found essential that low temperatures be maintained in order to preserve enzyme activity. We have found that the Sorvall angle centrifuges are especially well adapted for this purpose, since the temperature rise during centrifugation in the cold room is held to a minimum. The International refrigerated centrifuge has also been used with complete success. Although these fractions can be obtained at higher temperatures, their ability to oxidize fatty acids and Krebs cycle intermediates then becomes greatly attenuated or lost, probably because of enzymatic destruction of as yet unidentified cofactors.

Microscopic examination showed that the mitochondria so prepared were free of whole cells, nuclei, and débris, confirming the work of Hogeboom *et al.* who have stated that this procedure yields morphologically intact mitochondria free of extraneous elements(6). We have found that these preparations are contaminated to a small degree with erythrocytes. These extraneous elements may be removed by taking up the unwashed pellet of mitochondria which had been sedimented once in 10 volumes of 0.88 M sucrose as described above, and subjecting the suspension at this point to two or three preliminary sedimentations at low speed (2000 × *g*), each of 5 minutes duration. The main bulk of the mitochondria, now freed of red blood cells, is then sedimented by means of a 20 minute centrifugation at 18,000 × *g*. This procedure also reduces the desoxypentose nucleic acid phosphorus content of the mitochondria preparations to vanishingly small values. The phosphorus distribution in the mitochondria is discussed more fully in a later section of this paper.

To avoid the necessity of a high speed centrifuge for the preparation of mitochondria, we have also used an abridged procedure which yields preparations of mitochondria which are entirely satisfactory for the study of the enzyme systems involved in this report. The 0.88 M sucrose extract of rat liver, freed of nuclei and whole cells exactly as described above, is sedimented at 2400 × *g* for 30 minutes at 0° in the Sorvall model SP angle centrifuge. The supernatant is decanted and the mitochondria are then washed by resuspension in 10 volumes of 0.15 M KCl and resedimented by centrifugation for 7 minutes at 2400 × *g*. While the yield of mitochondria obtained by this procedure is not so large as in the standard procedure, the material appears to be identical in composition and enzymatic activity. A second abridged procedure has also been used for preparing mitochondria. The 0.88 M sucrose extract of rat liver, after removal of nuclei, etc., by preliminary centrifugations as outlined by Hogeboom *et al.*, is treated with 0.1 volume of 1.5 M KCl and allowed to stand in an ice bath for 5 to 10 minutes. The addition of salt causes agglutination of a large part of the mitochondria and they are now sedimentable in 5 to 10 minutes at 2000 $\times g$. The sedimented material can then be washed with 0.15 M KCl solution to free it of sucrose. Such material appears to be identical in enzymatic behavior with the material obtained by the original method of Hogeboom *et al.* and is obviously more convenient to prepare.

All experiments reported in this paper were done with mitochondria prepared by the original method of Hogeboom *et al.* with or without the additional low speed centrifugations to remove extraneous erythrocytes.

* * *

OXIDATION OF KREBS TRICARBOXYLIC ACID CYCLE INTERMEDIATES IN MITOCHONDRIA

Since the saline-washed particulate material previously studied contains all the enzymes involved in the Krebs cycle and since the mitochondria had previously been shown to contain considerable succinoxidase activity[7], it was of interest to determine whether isolated mitochondria also possessed

Table I

ACTIVITY OF SUBCELLULAR FRACTIONS OF RAT LIVER IN OXIDATION OF INTERMEDIATE COMPOUNDS OF KREBS CYCLE

The flask contents were as follows: glycylglycine buffer, *p*H7.2, 0.033 M; adenosine triphosphate, 0.0005 M; cytochrome *c*, 1×10^{-5} M; $MgSO_4$, 0.005 M; 0.1 ml. of orthophosphate containing P^{32}, 359,000 counts per minute (the phosphate esterification data are in Table III). The final concentration of substrates was 0.01 M in each case, except for oxalacetate and pyruvate which were added together at a concentration of 0.005 M each. Mitochondria and nuclear precipitate fractions were added, so that each flask contained an amount of material derived from 225 mg. of fresh wet liver tissue. The flasks containing supernatant were tested with the material derived from 90 mg. of tissue. The final volume was 3.0 ml.; incubation for 40 minutes with air as gas phase.

Fraction	Substrate	Oxygen uptake
		μM
Mitochondria	Citrate	7.1
	α-Ketoglutarate	6.3
	Pyruvate + oxalacetate	7.1
	None	0.18
"Nuclear ppt."	Citrate	1.9
	α-Ketoglutarate	1.7
	Pyruvate + oxalacetate	0.98
	None	0.0
Supernatant	Citrate	0.54
	α-Ketoglutarate	0.0
	Pyruvate + oxalacetate	1.4
	None	0.31

the enzymatic equipment necessary for the oxidation of pyruvate and other intermediates of the Krebs cycle. The results of a typical experiment in Table I indicate that intermediate compounds of the Krebs cycle are readily oxidized by suspensions of rat liver mitochondria. The nuclear and microsome fractions showed slight activity, which may have been due to contamination of these fractions by mitochondria. The substrates tested in this experiment were pyruvate plus oxalacetate, citrate, and α-ketoglutarate. These oxidations represent key enzymatic steps of the Krebs cycle. In addition, these mitochondrial preparations are capable of catalyzing the condensation of oxalacetate and pyruvate to yield citrate. The experiment summarized in Table II shows aerobic citrate formation from pyruvate when malate served as a source of oxalacetate.

Table II

CITRATE FORMATION FROM MALATE AND PYRUVATE IN MITOCHONDRIA

The flask contents were as follows: 0.05 M KCl, 0.005 M $MgSO_4$, 0.01 M phosphate, *p*H 7.4, 0.0005 M adenosine triphosphate, and 10^{-5} M cytochrome *c*. The final concentration of malate and pyruvate was 0.01 M in each case, and the final volume was 3.0 ml. Each flask contained mitochondria suspended in 0.15 M KCl equivalent to about 1 mg. of enzyme N. The time of incubation was 65 minutes at 30° with air as the gas phase.

Substrate	Oxygen uptake	Pyruvate used	Citrate formed
	μM	μM	μM
Pyruvate only	10.1	18.4	3.9
Malate only	10.7		2.3
Pyruvate + malate	13.5	9.4	10.4
None	0.5		0.0

ESTERIFICATION OF PHOSPHATE COUPLED TO OXIDATION IN MITOCHONDRIAL PREPARATIONS

The oxidation of fatty acids and of the intermediate compounds of the Krebs cycle proceeds with the release of considerable amounts of energy. It is now well known that energy released during oxidations over the Krebs cycle may be recovered in part by coupled esterification of inorganic phosphate. More recently it has been shown that oxidation of octanoate by particulate rat liver preparations also caused coupled esterification of inorganic phosphate[11]. In order to determine whether the enzymatic equipment necessary for esterification of phosphate coupled to Krebs cycle oxidations and to fatty acid oxidation is present in purified mitochondria, such oxidations were carried out in the presence of inorganic phosphate labeled with P^{32}. At the completion of the incubation, the carrier-diluted inorganic phosphate was removed from the neutralized trichloroacetic acid filtrates by repeated magnesia precipitation[21] and the radioactivity of the esterified phosphorus fractions determined. The data are presented in Table III. It can be seen from these data that both Krebs cycle oxidations and octanoate oxidation in the suspensions of mitochondria cause extensive incorporation of the P^{32} into the esterified fraction.

Table III

ESTERIFICATION OF PHOSPHATE COUPLED TO OXIDATIONS IN MITOCHONDRIA

The conditions of Experiment 1 (Krebs cycle oxidations) were exactly as described for experiments summarized in Table I. In Experiment 2 (fatty acid oxidation) vessels contained 0.005 M $MgCl_2$, 0.01 M glycylglycine buffer, *p*H 7.4, 1×10^{-5} M cytochrome *c*, 0.001 M adenosine triphosphate, 0.0001 M malate, 0.05 M KCl, 0.001 M octanoate, and inorganic orthophosphate labeled with 157,000 counts per minute of P^{32}. Octanoate was omitted in the control vessel. The time of incubation was 20 minutes at 30°.

Experiment No.	Substrate	O_2 uptake	Esterified P	P_{32} esterified
		μM	γ	*per cent*
1	None	0.18	24.2	0.67
	Citrate	7.1	106	31.3
	α-Ketoglutarate	6.3	113	39.3
	Pyruvate + oxalacetate	7.1	113	32.6
2	None (0.0001 M malate present)	0.5	37	3.2
	Octanoate	4.5	121	27.8

PHOSPHORUS DISTRIBUTION IN PURIFIED MITOCHONDRIA

Previous workers in describing the chemical constitution of the mitochondria[4, 23] have emphasized the high content of phospholipide in these structures and the fact that they contain nucleic acid of the pentose nucleic acid type. The distribution of phosphorus in a typical preparation of mitochondria made by the standard method of Hogeboom *et al.*[5] is presented in Column 1 of Table IV. Characteristically high values of phospholipide phosphorus, and the predominance of pentose nucleic acid, with only small amounts of desoxypentose nucleic acid, are to be noted. These figures are in fair agreement with those published by Hogeboom *et al.*[6] and Schneider[9]. Their data did not include direct measurement of both nucleic acids on the mitochondrial fraction but did show that almost all of the desoxypentose nucleic acid was in the first nuclear precipitate. Our direct analysis of the mitochondria shows the presence of appreciable amounts of DNA,[1] which may be due to contamination by other morphological elements. In Column 2 is given the phosphorus distribution in another preparation of mitochondria which had been subjected to more extensive removal of extraneous elements by repeated centrifugation at low speed prior to the resedimentation of the washed particles at high speed as already described. It can be seen that this procedure has reduced the amount of DNA and raised the ratio of PNA:DNA from 11.8 to 41.6. This ratio represents the analytical limit of the methods of Schneider and Schmidt and Thannhauser for measuring pentose nucleic acid and DNA in our hands. Preparations with very low DNA values thus obtained were found to be active in the oxidation of fatty acids. It is difficult to determine whether the last trace of DNA phosphorus in the preparations studied is analytically significant. It is of course

[1]DNA, desoxypentose nucleic acid; PNA, pentose nucleic acid.

Table IV

Distribution of Phosphorus in Mitochondria Derived from 0.88 M Sucrose Homogenates of Rat Livers

Total nucleic acid phosphorus was determined by the method of Schneider (19). Pentose nucleic acid was differentiated from desoxypentose nucleic acid by the method of Schmidt and Thannbauser (20). In Column 1 are listed values obtained for mitochondria isolated by the original method of Hogeboom *et al.* (6). In Column 2, values are given for such mitochondria which had been freed of erythrocytes and other extraneous elements by the modification described in the test.

Fraction	Per cent of total P	
	Method of Hogeboom *et al.* (1)	Modified method (2)
Acid-soluble P	21.3	15
Lipide P	56	56
Total nucleic acid P	19.3	21.3
"Protein" P	6.0	7.2
Desoxypentose nucleic acid P	1.5	0.5
Pentose nucleic acid P	17.8	20.8
PNA-P: DNA-P	11.8	41.6

conceivable that trace amounts of DNA are present normally in the mitochondria.

* * *

Intracellular distribution of some glycolytic enzymes

The finding that highly organized respiratory systems are localized in the mitochondria raises the question of the intracellular location of other organized enzyme systems. In contrast to the respiratory systems, which are associated with particulate matter, the glycolytic enzymes all appear to be readily soluble and several have been crystallized. It was therefore of interest to examine the different fractions of rat liver for their ability to catalyze the oxidation-reduction reactions of glycolysis. The system studied involved fructose-1, 6-diphosphate as substrate, aldolase, triose phosphate dehydrogenase, diphosphopyridine nucleotide, and arsenate to "decouple" the phosphorylation step, and lactic dehydrogenase and pyruvate as hydrogen acceptors. Fluoride was added to inhibit enolase and the end-point measured manometrically indicated the formation of 3-phosphoglyceric acid, which causes CO_2 liberation from a bicarbonate buffer. The conditions used were found to give approximately linear results with varying concentrations of an extract of rabbit muscle. When the different fractions were assayed for the presence of the three enzymes involved in the reaction, the mitochondria and the nuclear precipitate contained only a very small fraction of the activity, whereas the supernatant, containing all the soluble material of rat liver as well as difficultly sedimentable particles ("microsomes," etc.), contained 82 per cent of the activity shown by the orig-

Table V

ACTIVITY OF SOME GLYCOLYTIC ENZYMES IN LIVER FRACTIONS

In the oxidation-reduction system assay the Warburg vessels contained 0.03 M fructose-1, 6-diphosphate, 0.002 M arsenate, 0.02 M sodium fluoride, 0.02 M sodium pyruvate, 0.048 M $NaHCO_3$, 0.02 M nicotinamide, and 0.001 M diphosphopyridine nucleotide. The liver fractions in amounts specified were tipped in from the side arm after temperature equilibration. Total volume, 2.0 ml.; gas phase, 95 per cent N_2-5 per cent CO_2; temperature, 30°; time of incubation, 1 hour.

Fraction	Weight of whole liver from which fraction was derived	CO_2 liberated	Oxidation-reduction activity*	Aldolase activity†
	mg.	*c.mm.*	*per cent of total*	*per cent of total*
Nuclear..................	60	18	5.4	3
Mitochondria	60	10	3.0	1
Supernatant..............	20	90	81.7	96
Whole homogenate	20	110	(100)	(100)

*The $Q_{CO_2}^{N_2}$ (30°) of the whole liver homogenate under those conditions was 27.5.
†Q_{HDP} (24) at 38° of whole liver homogenate was 56.

inal unfractionated homogenate (see Table V). Also shown in Table V is the distribution of aldolase in the different fractions, measured by the method of Sibley and Lehninger[(24)]. It is seen that the particulate fractions contain only a very small fraction of the total aldolase of rat liver, 96 per cent being present in the soluble fraction.

Although assay of all the individual enzymes of glycolysis in these fractions may eventually show that one or more of these enzymes are present in the mitochondria, it appears certain that the mitochondria do not possess the complete enzymatic machinery for the conversion of glycogen or glucose to lactic acid at a rate comparable to the rate of respiration of these bodies. LePage and Schneider have recently shown that particulate fractions of rat tumors or rabbit liver have little or no ability to glycolyze glucose, most of the activity being in the soluble fraction[(25)].

In addition to these data, it has been found by Friedkin in this laboratory (*cf.*[(26)]) that isolated mitochondria are capable of incorporating inorganic phosphate labeled with P^{32} into pentose nucleic acid, phospholipide, and an unidentified acid-insoluble "phosphoprotein" fraction coupled to the oxidation of substrates of the Krebs tricarboxylic acid cycle. It would therefore appear that these bodies are also capable of at least one type of reaction leading to synthesis of these intracellular materials.

Discussion

The data reported in this paper show that the complex enzyme systems responsible for the oxidation of fatty acids and Krebs cycle intermediates and esterification of phosphate coupled to these oxidations are localized in that fraction of rat liver which consists of morphologically intact mitochondria or "large granules," almost completely free of other formed elements. As has been pointed out, it is not possible with our present knowledge of these complex systems to assay individual enzymes of these systems quantitatively and it is therefore conceivable that other elements such as the nuclei, "microsomes," or soluble material may be capable of many of the enzymatic transformations involved in the over-all reactions studied. The striking fact is that all the individual enzymes concerned in these complex systems should be found in one species of morphological element. These findings in some measure justify the early views of Altmann[27] that these bodies are fundamental biological units and possess a certain degree of autonomy and certainly, together with the considerable work already done on their enzymatic and chemical composition by the Rockefeller school, Schneider, and others, provide considerable basis for the apt designation "intracellular power plants" conferred on the mitochondria by Claude.

Although the mitochondria appear to be the major site of these activities, it would appear from our examination *in vitro* that these bodies are not completely autonomous with respect to their respiratory behavior, since they must be supplemented with certain cofactors such as adenosine triphosphate and Mg^{++}. It appears likely that in the cell there is a rapid interchange of these factors, substrates, and inorganic phosphate between the cytoplasm and the mitochondria. It also would appear that these bodies are dependent on the cytoplasm for certain preparatory metabolic activities such as glycolysis, since, as our data show, they are almost completely lacking in glycolytic activity.

Claude has found that isolated mitochondria are quite sensitive to changes in osmotic pressure[4]. Adverse osmotic conditions may therefore be responsible for the inactivity of the mitochondria in catalyzing fatty acid oxidation in hypotonic reaction media; when the concentration of neutral salts or nonelectrolytes approximates isotonicity, the system shows maximum activity. No attempts have been made to determine whether the mitochondria are morphologically intact in all stages of the enzymatic reaction.

It is also of some interest that mitochondria are capable of causing oxidation-coupled incorporation of labeled inorganic phosphate into nucleic acids and phospholipides of these structures (*cf.*[26]). The work of Hill and Scarisbrick[28] and Warburg and Lüttgens[29] on the photochemical activity of isolated chloroplasts or granules derived therefrom provides some indication that highly organized enzyme systems are localized in analogous structures of plant cells.

The localization of organized respiratory activity in mitochondria poses some problems in connection with the separation and purification of the individual enzymes involved. The difficulties in rendering such enzymes as cytochrome oxidase and succinoxidase soluble are well known[30,31]. Recent work in this laboratory indicates that the separation from mitochondrial preparations in soluble form of simple dehydrogenase proteins which might be expected to be readily soluble is also quite difficult and a variety of drastic procedures has failed to release any significant amount of such proteins into soluble form.

Summary

Morphologically homogeneous mitochondria ("large granules") separated from rat liver dispersions by the hypertonic sucrose method of Hogeboom, Schneider, and Pallade contain essentially all the measurable activity of the liver in the oxidation of fatty acids. Likewise, the integrated reactions of the Krebs tricarboxylic acid cycle are found in this fraction. Esterification of inorganic phosphate accompanies these oxidations in purified preparations of mitochondria. These bodies have insignificant glycolytic activity. "Mitochondria" prepared by other methods involving saline or water as the dispersing media are inactive in these reactions, possibly because of osmotic damage.

Bibliography

1. Warburg, O., *Arch. ges. Physiol.,* **154**, 599 (1913).
2. Bensley, R. R., and Hoerr, N. L., *Anat. Rec.,* **60**, 251 (1934).
3. Claude, A., *Proc. Soc. Exp. Biol. and Med.,* **39**, 398 (1938).
4. Claude, A., *J. Exp. Med.,* **84**, 51, 61 (1946).
5. Hogeboom, G. H., Schneider, W. C., and Pallade, G. E., *Proc. Soc. Exp. Biol. and Med.,* **65**, 320 (1947).
6. Hogeboom, G. H., Schneider, W. C., and Pallade, G. E., *J. Biol. Chem.,* **172**, 619 (1948).
7. Hogeboom, G. H., Claude, A., and Hotchkiss, R. D., *J. Biol. Chem.,* **165**, 615 (1946).
8. Schneider, W. C., Claude, A., and Hogeboom, G. H., *J. Biol. Chem.,* **172**, 451 (1948).
9. Schneider, W. C., *J. Biol. Chem.,* **165**, 585 (1946).
10. Lehninger, A. L., *J. Biol. Chem.,* **164**, 291 (1946).
11. Lehninger, A. L., and Kennedy, E. P., *J. Biol. Chem.,* **173**, 753 (1948).
12. Potter, V. R., *J. Biol. Chem.,* **163**, 437 (1946).
13. Kennedy, E. P., and Lehninger, A. L., *J. Biol. Chem.,* **172**, 847 (1948).
14. Schneider, W. C., *J. Biol. Chem.,* **176**, 259 (1948).
15. Lehninger, A. L., and Smith, S. W., *J. Biol. Chem.,* **173**, 773 (1948).
16. Greenberg, L. A., and Lester, D., *J. Biol. Chem.,* **154**, 177 (1944).
17. Speck, J. F., Moulder, J. W., and Evans, E. A., Jr., *J. Biol. Chem.,* **164**, 119 (1946).
18. Gomori, G., *J. Lab. and Clin. Med.,* **27**, 955 (1942).
19. Schneider, W. C., *J. Biol. Chem.,* **161**, 293 (1945).
20. Schmidt, G., and Thannhauser, S. J., *J. Biol. Chem.,* **161**, 83 (1945).
21. Lehninger, A. L., *J. Biol. Chem.,* **178**, 625 (1949).
22. Potter, V. R., and Elvehjem, C. A., *J. Biol. Chem.,* **114**, 495 (1936).
23. Chantrenne, H., *Biochim. et biophys. acta,* **1**, 437 (1947).
24. Sibley, J. A., and Lehninger, A. L., *J. Biol. Chem.,* **177**, 859 (1949).
25. LePage, G. A., and Schneider, W. C., *J. Biol. Chem.,* **176**, 1021 (1949).
26. Friedkin, M., and Lehninger, A. L., *J. Biol. Chem.,* **177**, 775 (1949).
27. Altmann, R., Elementaroranismen, Leipzig (1890).
28. Hill, R., and Scarisbrick, R., *Nature,* **146**, 61 (1940).
29. Warburg, O., and Lüttgens, W., *Naturwissenschaften,* **32**, 161, 301 (1944).
30. Wainio, W. W., Cooperstein, S. J., Kollen, S., and Eichel, B., *J. Biol. Chem.,* **173**, 145 (1948).
31. Hogeboom, G. H., *J. Biol. Chem.,* **162**, 739 (1946).

PHOSPHORYLATION COUPLED TO OXIDATION OF DIHYDRODIPHOSPHOPYRIDINE NUCLEOTIDE*†

Albert L. Lehninger

Departments of Biochemistry and Surgery,
University of Chicago
Chicago, Illinois

(Received for publication November 30, 1950)

It has been shown in this laboratory that aerobic incubation of dihydrodiphosphopyridine nucleotide ($DPNH_2$) with a properly supplemented suspension of particulate elements of rat liver caused extensive incorporation of inorganic phosphate labeled with P^{32} into a form having the properties of the acid-labile group of adenosinetriphosphate (ATP)[1,2]. Such incorporation did not occur under nitrogen, in the absence of Mg^{++} or ATP, or when the oxidized form of the nucleotide (DPN) was substituted for $DPNH_2$. It was concluded that oxidative phosphorylation had occurred coupled to the passage of electrons from $DPNH_2$ to oxygen via the cytochrome system, but no conclusion could be drawn from these data concerning the quantitative relationship between phosphorylation and oxidation. These findings were substantiated by subsequent work in which the DPN-linked β-hydroxybutyric dehydrogenase was used to generate $DPNH_2$ continuously from DPN and the substrate[3,4]. In the latter case it was not only possible to observe P^{32} uptake coupled to this DPN-linked oxidation, but it was also found possible to measure the P:O ratio directly in non-isotopic experiments with adenylic acid as phosphate acceptor[4]. A sufficiently large proportion of the measurements yielded values over 2.0 (uncorrected for dephosphorylation losses of known and relatively large magnitude) to warrant the tentative conclusion that the actual ratio is 3.0.

In this paper experiments are described in which highly purified $DPNH_2$

*A. L. Lehninger, "Phosphorylation Coupled to Oxidation of Dihydrodiphosphopyridine Nucleotide," *J. Biol. Chem,* **190**: 345–359 (1951). Reprinted with the permission of the author and of the American Society of Biological Chemists, Inc.

†This investigation was supported by research grants from the Nutrition Foundation, Inc., and from the National Institutes of Health, United States Public Health Service.

was used as the test substrate to determine directly in non-isotopic experiments the amount of phosphorylation occurring when electrons pass from $DPNH_2$ to oxygen by measurement of phosphate uptake. These experiments show in a direct way that most of the phosphorylation observed during the DPN-linked oxidation of β-hydroxybutyrate to acetoacetate can be accounted for as being coupled to the oxidation of $DPNH_2$. In addition, a number of observations were made which have a bearing on the question of the existence of permeability or "availability" barriers in the organized cellular components (mitochondria or "large granules") in which phosphorylating oxidations are believed to take place.

Experimental

The DPN and $DPNH_2$ preparations employed in this study were highly purified materials. The DPN preparations were upwards of 87 per cent pure. A combination of the counter-current purification of Hogeboom and Barry[5] and the acid methanol-ethyl acetate treatment of Warburg and Christian[6] sufficed to achieve this purification, starting from commercial samples of 45 to 66 per cent purity (Schwarz Laboratories, Inc.,[1] and Sigma Chemical Company products were used). It was found that most commercial preparations caused very troublesome emulsification in the phenol-water system of Hogeboom and Barry, reducing both yield and purity below figures obtained by the latter. In some cases emulsions could not be completely broken after hours of centrifugation. It was possible to obviate this difficulty by first subjecting the commercial DPN to the acid-methanol treatment as described by Warburg and Christian[6], which not only removed emulsifying impurities but also achieved some purification, bringing the purity up to about 70 per cent in several preparations. This material was then subjected to the four transfer phenol-water distribution of Hogeboom and Barry[5]. The DPN recovered from Tubes 0 and 1 of this procedure was combined and reprecipitated with alcohol. The contents of Tube 2 were not used, since this material was of lower purity and represented only a small portion of the total DPN. In our experience the second sixteen tube distribution (water-phenol-chloroform system) described by Hogeboom and Barry did not raise the purity of the contents of Tubes 0 and 1 more than about 3 per cent, and it was therefore not used. This finding may, however, be true only for DPN of the purification history outlined above.

The two techniques combined in the sequence described gave yields of over 70 per cent of the starting DPN. The purity of several preparations, assayed by the reduction method of Gutcho and Stewart[7] (by use, however, of the constant 6.22×10^6 sq. cm. per mole[8], rather than the lower constant 8.5 sq. cm. per mg. (5.64×10^6 sq. cm. per mole) employed by Gutcho and Stewart[7], Hogeboom and Barry[5], and LePage[9]), ranged from 87 to 93 per cent on a dry weight basis. The loss in weight over P_2O_5 at 60° under a high vacuum for 4 hours varied from 5.4 to 7.2 per cent. These values are

[1] We wish to thank Mr. David Schwarz of the Schwarz Laboratories, Inc., for his cooperation in providing samples of DPN subjected to different histories of purification.

approximately equal to the purity of the best preparations described by Hogeboom and Barry[5], if we allow for the fact that the latter authors used a lower extinction coefficient.

Preparations of $DPNH_2$ made by reduction with hydrosulfite and isolated as the disodium salt, as described by Ohlmeyer[10], never approached in our hands the purity obtained by him or by Drabkin[11], even when the DPN used as starting material was over 90 per cent pure. Sodium sulfite and sulfate were the major contaminants. A variety of modifications was examined with the hope of improving the separation of the sodium salt of $DPNH_2$ from the inorganic salts. Also three different hydrosulfite samples were tested without success in improving the purity of the final product. Since both sulfite and sulfate have been found in unpublished experiments to have significantly large uncoupling effects on oxidative phosphorylation in concentrations in which they would be introduced as impurities in the $DPNH_2$, it was necessary to obtain material free of these contaminants. This was achieved by isolating $DPNH_2$ from Ohlmeyer's reduction medium as the barium salt. Since the barium salt of $DPNH_2$ is very soluble in water and in 50 per cent ethanol, while barium sulfate and sulfite are essentially insoluble in 50 per cent ethanol, it was possible to achieve this separation. The details of a typical reduction and isolation follow.

In a flask, 500 mg. of DPN (purity, 86 per cent; total of 662 μM) were dissolved in 41.0 ml. of 1.3 per cent $NaHCO_3$, and 250 mg. of sodium hydrosulfite were quickly added with swirling. The contents were then gassed with 95 per cent N_2-5 per cent CO_2. The vessel was closed and kept at 25° for 2 hours. At the end of this time the contents were vigorously gassed with oxygen for 15 minutes to oxidize excess hydrosulfite. The slightly yellow solution contained 650 μM of $DPNH_2$ as determined by absorption at 340 mμ, with use of the molar extinction coefficient 6.22×10^6 sq. cm. per mole[8]. The solution was then chilled to 0° and to it were added dropwise with stirring 4.0 ml. of 2 M barium thiocyanate. After addition was complete, the *p*H of the suspension was brought to 7.5 by addition of about 0.4 ml. of 2.5 N NaOH. The precipitated barium salts were removed by centrifugation and discarded. (This precipitate contains the bulk of the sulfite and sulfate.) Barium thiocyanate was chosen, since this barium salt is quite soluble in ethanol-water mixtures and the presence of contaminating thiocyanate in the final product can be easily detected colorimetrically as the ferric complex. Barium iodide or bromide is also appropriate. To the clear supernatant fluid was added an equal volume of cold ethanol. After standing 20 minutes, the slight flocculent precipitate was centrifuged and discarded. The clear, slightly yellow supernatant solution contained 579 μM of $DPNH_2$. An additional 5 volumes of cold ethanol were added to precipitate the barium salt of $DPNH_2$, which was recovered by centrifugation and washed and dried with absolute ethanol and finally ether. A total of 522 μM of $DPNH_2$ was recovered in the form of a slightly yellow powder weighing 535 mg., indicating a purity of 85 per cent on the basis of the composition $C_{21}H_{27}O_{14}N_7P_2Ba \cdot 4H_2O$. This material was found to give a faint test for thiocyanate and contained considerable barium carbonate. It was dissolved in 12.0 ml. of ice-cold CO_2-free H_2O, and some insoluble material was removed by centrifugation. Then 13.0 ml. of H_2O were added, followed by 25.0 ml. of ethanol, and the solution was clarified by centrifugation in the cold. The barium salt of $DPNH_2$ was then precipitated by slow addition of 140 ml. of cold ethanol and washed and dried with ethanol and ether. The material was dried over calcium chloride. It weighed 476 mg. and contained 501 μM of Ba $DPNH_2 \cdot 4H_2O$, indicating a purity of 92 per cent and an over-all yield of 75 per cent. Yields in other preparations were generally lower (65 to 69 per cent). The substance was faintly yellow in color and readily soluble in water to give a clear solution. On drying in a high vacuum over P_2O_5 at 60° for 4 hours, it lost 7.45 per cent of its weight, slightly less than calculated for 4 moles of H_2O per mole of $DPNH_2$. Drying at 100° caused a loss of weight corresponding to nearly 7 moles of H_2O. The material contained less than 0.004 μM of—SCN per micromole of $DPNH_2$, estimated colorimetrically as the ferric complex. The material contained carbonate or bicarbonate as measured manometrically in a Warburg vessel by tipping acid into an aqueous solution of the material in CO_2-free water. The evolved gas was completely absorbable by NaOH. Calculated as $BaCO_3$, this impurity amounted to 3.9 per cent. Analysis, found, N 11.4, P 7.01 per cent (on the basis of desiccator-dry weight, as given above); calculated for $C_{21}H_{27}O_{14}P_2Ba \cdot 4H_2O$, N 11.23, P 7.10 per cent. From the analytical and spectrophotometric data presented it appears permissible to conclude that the preparation is about 92 per cent pure and that about half of the 8 per cent impurity is $BaCO_3$. It is possible that the rest may be moisture not removed at 60°, but labile at 100°.

Six preparations have been made by this method and they showed purities, uncorrected for content of $BaCO_3$ and tightly bound water, from 85.7 to 96.8 per cent. The ratio of the molar extinction coefficient at 340 mμ to that of 260 mμ, the peak of the adenine absorption, was found to vary from 0.348 to 0.368 at *p*H 7.4. The latter ratio was that of the best preparation and corresponds to that given by the preparation of Drabkin[(11)]. The molar extinction coefficient at 260 mμ of the best preparation was 16.9×10^6 sq. cm. per mole. The molar extinction coefficient at 400 mμ of the best preparation was less than 0.02×10^6 sq. cm. per mole. There was no significant indication of the presence of the monohydro form as judged by the absorption at 380 mμ and higher wavelengths. The 340 mμ absorption was completely discharged by treatment with lactic dehydrogenase and an excess of sodium pyruvate.

The method described above has also been applied to impure specimens of DPN of 0.45 to 0.66 per cent purity. However, no great purification was achieved by reducing such material and isolating $DPNH_2$ as the barium salt. Since adenylic acid, a frequent contaminant, also forms a water-soluble alcohol-insoluble barium salt, it would be expected to appear in the final product. The purer preparations described above appeared rather stable in the cold over calcium chloride. Preparations held in a desiccator for a year at room temperature showed a decline of the 340:260 ratio to about 0.26 and decline of enzymatic assay values to about 70 per cent purity. Only a small part of this loss was due to autoxidation to DPN.

For use in the experiments to be described the barium salt was dissolved in water (10 to 20 μM per ml.) and the barium was precipitated by a 100 per cent molar excess of 0.1 M phosphate buffer, pH 7.8. After chilling for a half hour, the barium phosphate was centrifuged. There was no measurable adsorption of $DPNH_2$ on the barium phosphate. Such stock solutions, containing

10 μM of $DPNH_2$ and 10 μM of inorganic phosphate at pH 7.4, were found to be stable for some days in the frozen state. Such solutions did not decolorize methylene blue anaerobically. They slowly reduced cytochrome *c* non-enzymatically over a period of 90 minutes at room temperature.

The adenosinediphosphate used in these experiments was obtained from the Sigma Chemical Company, and, although analytical data indicated the presence of adenylic acid, the preparations were of sufficient purity for these experiments.

The preparations of rat liver enzyme were made exactly as described before[12]; they were taken up in either ice-cold 0.15 M KCl or distilled H_2O as indicated and used immediately. In the text the two types of preparations are referred to as "KCl suspension" and "H_2O suspension." These suspensions are known to contain both mitochondria and nuclei.

DPN was determined enzymatically in neutralized trichloroacetic acid filtrates with yeast alcohol dehydrogenase[2][13] and ethanol at pH 8.8[14]. $DPNH_2$ was determined enzymatically with yeast alcohol dehydrogenase and acetaldehyde as acceptor at pH 7.2 in clear filtrates obtained when the reaction medium was heated for 1 minute in a boiling water bath. The loss on heating was found to be about 4 per cent under these conditions and the data given are corrected for such losses.

Factors affecting rate of oxidation of $DPNH_2$

The rate of oxidation of $DPNH_2$ by suspensions of rat liver particles supplemented with the components used in the oxidative phosphorylation experiments (Mg^{++}, $HPO_4^{=}$, KCl, adenosinediphosphate (ADP), NaF, cytochrome *c*, and glycylglycine buffer) depended in an unexpected but reproducible way on the history of the enzyme preparation and the concentration of cytochrome *c*. At a cytochrome *c* concentration of 1.5×10^{-5} M (this is a saturating concentration for maximal activity of the succinoxidase system in rat liver homogenates[15]), $DPNH_2$ was oxidized several fold faster by liver particles which had been suspended in H_2O for a period of 5 minutes at 0° prior to the incubation than by particles which had been held in 0.15 M KCl. In such experiments KCl was added to the flask containing the water-treated particles to yield the same final salt concentration in both vessels; therefore the only difference was in the exposure of the enzyme particles to the H_2O and KCl media for 5 minutes before addition to the vessels. The endogenous oxygen uptakes (absence of added $DPNH_2$) were insignificant. This effect of pretreatment of the enzyme particles on the rate of oxidation of $DPNH_2$ was reproducibly observed, and oxygen uptake curves of a typical experiment are shown in Fig. 1. It will be seen that the half period of complete oxidation with the H_2O suspension was about 2.5 minutes; that with the KCl-suspension was about 11 minutes. The amount of oxygen ultimately taken up in all experiments of this type was within about 5 per cent of that calculated from the light absorption at 340 mμ of the stock solution of $DPNH_2$ employed. The error was always in the direction of a slightly smaller oxygen uptake, and this deficiency may

[2] Kindly donated by Dr. Simon Black.

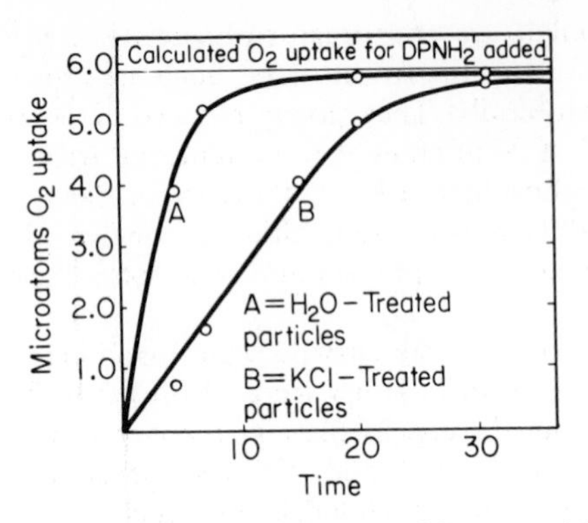

Fig. 1. The effect of water treatment of liver particles on the rate of $DPNH_2$ oxidation. The Warburg vessels contained final concentrations of 0.005 M $MgCl_2$, 0.075 M KCl (the total concentration of KCl in each vessel compensated for that introduced with enzyme suspension), 0.002 M ADP, 0.003 M HPO_4^-, 1.5×10^{-5} M cytochrome *c*, 0.02 M glycylglycine buffer, pH 7.4, 0.03 M NaF, and 5.90 μM of $DPNH_2$ (determined spectrophotometrically) in a total volume, including enzyme, of 2.0 ml. Enzyme suspension (0.30 ml., particles derived from 50 mg. of whole liver) was present in the main chamber with other components for 5 minutes temperature equilibration at 21°; $DPNH_2$ was tipped in from side arm at end of equilibration period. The endogenous oxygen uptake in presence of DPN or in absence of nucleotide was 0.6 microatom of oxygen at 40 minutes. The values in graph have been corrected for the small endogenous changes. Time in minutes.

have been due to non-oxidative destruction of some of the $DPNH_2$ (see below). The cytochrome *c* in the vessels in these experiments was very largely in the reduced form until near the end of the oxidation, as judged by visual observation of the characteristic pink color of ferrocytochrome *c*.

The rate of oxidation of $DPNH_2$ depended also in an unexpected way on the concentration of cytochrome *c*. In Fig. 2 is shown the effect of cytochrome *c* concentration on the rate of oxidation of $DPNH_2$ by the "KCl enzyme." In Fig. 3 is shown the effect of cytochrome *c* concentration on the rate of oxidation of β-hydroxybutyrate in control experiments on the same enzyme preparations. The first point of interest is that in the absence of *added* cytochrome *c* the rate of oxidation of $DPNH_2$ is virtually nil; yet, in the absence of added cytochrome *c* β-hydroxybutyrate is oxidized rapidly, though not maximally, by the same enzyme preparation under very similar conditions. These enzyme preparations will also readily oxidize Krebs cycle intermediates and fatty acids in the absence of added cytochrome *c*. There appears to be sufficient cytochrome *c* within the particles for oxidation of such substrates, but this cytochrome does not appear to be readily available to externally added $DPNH_2$. Oxidation of $DPNH_2$ did take place when cytochrome *c* was added to the vessels, and, as the cytochrome *c* concentration was increased to very high levels (0.00015 M), the $DPNH_2$ was oxidized at progressively higher rates, which, in the instance of the highest level, was so high that it could not be measured accurately. Under

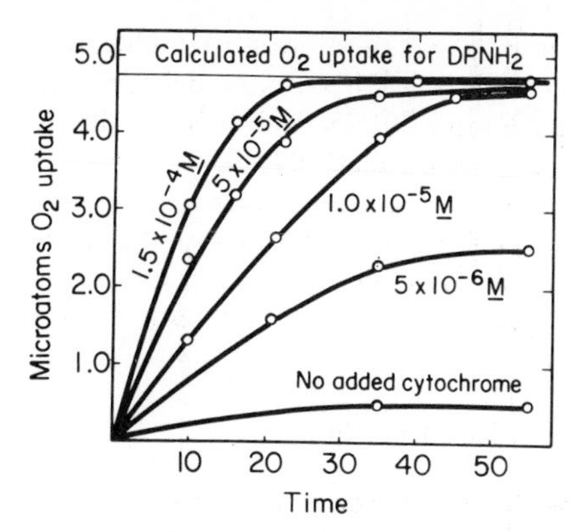

Fig. 2. Effect of cytochrome *c* concentration on rate of $DPNH_2$ oxidation. The experimental conditions were as described in Fig. 1. The KCl suspension of liver particles was used. Temperature, 19°. Time in minutes.

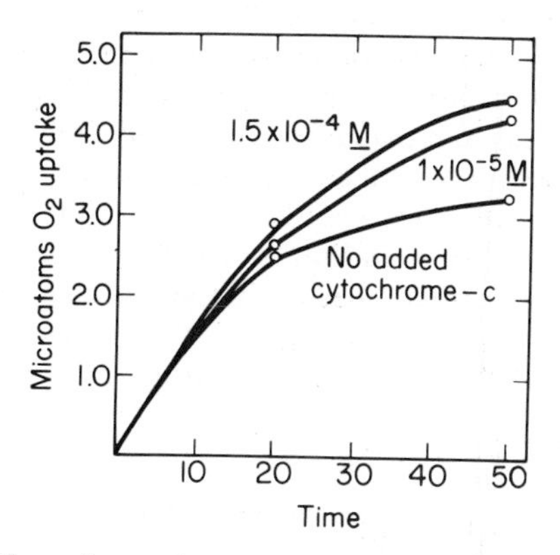

Fig. 3. Effect of cytochrome *c* concentration on the rate of oxidation of β-hydroxybutyrate. The experimental details are as described in Fig. 2, except that $DPNH_2$ was substituted by 0.02 M *dl*-β-hydroxybutyrate plus 0.001 M DPN. Temperature, 19°. Time in minutes.

the same circumstances the rate of oxidation of β-hydroxybutyrate did not increase at all with increase in cytochrome concentration above the level of about 1×10^{-5} M cytochrome *c*, indicating that the system was already saturated with cytochrome *c*. The succinoxidase system under these conditions is likewise saturated at about 1×10^{-5} M cytochrome *c*. At all levels of cytochrome *c* in the $DPNH_2$ experiments the cytochrome was essentially in the reduced state throughout the period of active oxidation, as judged visually. When the H_2O suspension was tested with the higher levels of cytochrome *c* (0.00015 M), the oxidation of $DPNH_2$ was substantially complete in a matter of a minute.

These findings are perhaps difficult to explain in detail in terms of mitochondrial structure on the basis of present knowledge, but they strongly indicate the existence of permeability or other structural barriers to free diffusion of $DPNH_2$ to the active centers of the highly organized mitochondrial body.

In this connection, it may be mentioned that Huennekens and Green[16, 17] have obtained evidence that the DPN-linked dehydrogenases of mitochondrial systems contain tightly bound DPN which behaves differently toward autolytic enzymes than does "externally added" DPN.

Although the experiments outlined above served to clarify a number of apparently erratic and discordant preliminary experiments on $DPNH_2$ oxidation and phosphorylation, they also presented the possibility that demonstration of coupled phosphorylation might likewise be dependent on structural factors.

P:O RATIO OF $DPNH_2$ OXIDATION

The observations already described on the factors affecting the rate of $DPNH_2$ oxidation served as a guide in choosing conditions for studying the P:O ratio of the oxidation, since it was desirable to have a slow enough oxidation for accurate measurements of oxygen uptake and at the same time not so slow that a non-oxidative destruction of $DPNH_2$ or DPN became a serious problem. The correlation of the rate of disappearance of $DPNH_2$ and the rate of appearance of DPN, the oxidation product, with the rate of oxygen uptake was undertaken. To this end balance experiments were performed in which oxygen uptake was measured manometrically, and $DPNH_2$ disappearance and DPN formation were measured enzymatically with alcohol dehydrogenase in heated filtrates from identical vessels. The data obtained from one of four such experiments are presented in Fig. 4. It will be seen that the rate of disappearance of $DPNH_2$, measured spectrophotometrically, agrees almost exactly with the rate of oxygen uptake throughout most of the reaction course. The rate of formation of DPN also coincides fairly well with the rate of oxygen

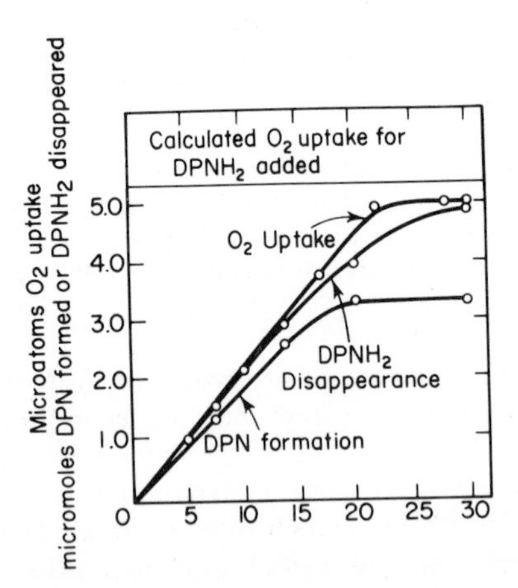

Fig. 4. Correlation of oxygen uptake, $DPNH_2$ disappearance, and DPN appearance during oxidation of $DPNH_2$. O_2 uptake measured manometrically at 22° in system exactly as described for Fig. 1, with KCl suspension and 10^{-5} M cytochrome *c*. $DPNH_2$ and DPN measurements made on identical vessels stopped at the times (in minutes, abscissa) indicated. Analytical details in the text.

uptake until about 50 per cent of the $DPNH_2$ is oxidized, and then falls off. A virtually identical picture was obtained in a similar experiment with H_2O suspension instead of KCl suspension. The falling off of DPN levels is due to selective destruction of the oxidized nucleotide, as shown by a control experiment. In this experiment $DPNH_2$ and DPN (0.003 M) were incubated in separate vessels in the absence of added cytochrome *c* under nitrogen. In 20 minutes 54 per cent of the DPN, measured enzymatically in the neutralized trichloroacetic acid filtrate, disappeared, whereas only 13 per cent of the $DPNH_2$ disappeared, as measured spectrophotometrically in heated filtrates. The mechanism of destruction was not studied in detail. It is known that the DPNase acting on the nicotinamide-ribose linkage is specific for DPN and does not attack $DPNH_2$[18]. Some evidence was obtained that the pyrophosphatase type of cleavage[19] (which is more active on $DPNH_2$ than on DPN) takes place in these preparations, since disappearance of enzymatically active $DPNH_2$ occurs at a greater rate than the disappearance of 340 mμ absorption. Since dihydronicotinamide mononucleotide has the same extinction coefficient as $DPNH_2$ at 340 mμ, but is not oxidized by alcohol dehydrogenase[19], this finding indicates that some cleavage of $DPNH_2$ at the pyrophosphate bond occurred.

The experiment charted in Fig. 4 demonstrated that the spectrophotometric measurement of $DPNH_2$ disappearance was equivalent to manometric measurement of oxygen taken up under the special conditions outlined (at least until some 75 per cent of the $DPNH_2$ had disappeared) and therefore simplified study of the P:O ratio of $DPNH_2$ oxidation, since the relatively cumbersome manometric method could be dispensed with in favor of a simpler, much more sensitive spectrophotometric determination. In measurements of the P:$DPNH_2$ ratio pairs of identical vessels were prepared containing Mg^{++}, KCl, $DPNH_2$ (or DPN, or β-hydroxybutyrate + DPN, or H_2O), $HPO_4^=$, ADP, NaF, cytochrome *c* and glycylglycine buffer. At zero time enzyme suspension was added to each of the vessels. At the specified times one vessel was fixed with trichloroacetic acid and the other with cold 1 per cent Na_2CO_3. Phosphate exchanges and DPN were measured in the trichloroacetic acid filtrate (DPN is stable in acid but not in base) and $DPNH_2$ disappearance was measured spectrophotometrically in the alkali-fixed medium after centrifugation. Oxidation of $DPNH_2$ is effectively stopped by this method.[3] The data of a series of experiments are shown in Table I. It will be seen that in each case inorganic phosphate disappears concurrently with oxidation of $DPNH_2$. No inorganic phosphate, or only negligible amounts, dissappear when DPN is substituted for $DPNH_2$ or when pyridine nucleotides are omitted entirely. The P:$DPNH_2$ ratios observed (assumed equal to the P:O ratios) vary from

[3] Although $DPNH_2$ is stable to concentrations of Na_2CO_3 used, it was found necessary to remove the particulate material by centrifugation in the cold immediately after addition of Na_2CO_3. The supernatants were then brought to pH 7.2 for enzymatic assay of $DPNH_2$. It was found that the liver particles catalyzed the "pyrophosphate" cleavage of $DPNH_2$ in the presence of Na_2CO_3 (pH 10.7), as shown by a decline in $DPNH_2$ content when measured with alcohol dehydrogenase not accompanied by a decline in 340 mμ absorption of the suspension, indicating that the inactivation process took place without affecting the dihydropyridine moiety. This destruction was enzymatic; omission of particles or addition of heated particles caused no significant decline in $DPNH_2$ with time. The effect of the particles appeared to be more pronounced at pH 10.7 than at neutrality.

a low of 0.40 to a high of 1.89 in the series of ten individual measurements. It is also seen that in general the H_2O-treated preparations yield higher P:$DPNH_2$ ratios than those suspended in KCl. The difference is not due to a significant difference in phosphatase activity, since control experiments established that there were no significant differences between the preparations in rate of dephosphorylation of ATP. Insufficient data are at hand to conclude that the H_2O suspensions are actually more effective in phosphorylation, but in any event a significant difference could be expected if the permeability or availability of $DPNH_2$ to active phosphorylating sites may vary with preparative history, as has already been shown for the oxidation.

In Experiments 4, 4A, 5, and 5A a comparison is made of efficiency of phosphorylation during $DPNH_2$ oxidation and during the oxidation of β-hydroxybutyrate to acetoacetate with the same enzyme preparation under as similar conditions as could be arranged. It is seen that the P:$DPNH_2$ ratios are somewhat lower than the P to acetoacetate ratios, but the former are high enough to account for most of the phosphorylation shown by the complete β-hydroxybutyrate-O_2 system. If any allowance can be made for structural barriers to the entry of $DPNH_2$ to phosphorylating sites, and this appears reasonable, it is possible that all the phosphorylation may be accounted for. However, two major factors appeared to rule against completely satisfactory experimental accounting of the phosphorylations. The first of these is the fact that the H_2O-treated particles generally yield lower P:O ratios than untreated particles when β-hydroxybutyrate or Krebs cycle intermediates are employed as substrates, whereas they appear to give, from the limited data available, somewhat higher P:O ratios when $DPNH_2$ is the substrate. Furthermore, the P:O ratios tested with β-hydroxybutyrate as substrate have been found to decline faster with time in H_2O-treated preparations. Secondly, the enzyme particle concentration is quite low in these experiments and under these conditions P:O ratios with β-hydroxybutyrate as substrate have not been as high as can be realized with higher enzyme concentrations. It may also be pointed out here that the use of much higher concentrations of $DPNH_2$ in order to achieve greater oxidative and phosphorylation exchanges leads to lower P:O ratios. High concentrations of $DPNH_2$ or DPN of high purity have been found to uncouple phosphorylation quite significantly. At 0.005 M DPN, P:O ratios of the β-hydroxybutyrate system are depressed some 20 to 50 per cent without significant inhibition of the rate of oxidation.

In two other experiments with $DPNH_2$ as substrate, not listed in Table I, no net phosphorylation was observed. An old preparation of $DPNH_2$, originally 88 per cent pure but assaying 38 per cent (enzymatic assay) at the time of the experiment, was found to be readily and completely oxidized by the liver particles, with P:$DPNH_2$ ratios, in two measurements, of 1.12 and 0.78.

Discussion

The direct determinations of the P:$DPNH_2$ ratios reported in this paper fully bear out the tentative conclusions reached in earlier experiments in which P^{32} was used as a tracer to follow phosphorylative events. Since the tracer technique gave no quantitative information on the P:O ratio, the experiments

Table I

MEASUREMENT OF P: $DPNH_2$ RATIOS

Duplicate vessels (see the text) contained 0.005 M $MgCl_2$, 0.075 M KCl (compensated for KCl introduced with the enzyme), 0.002 to 0.004 M ADP, 0.02 M glycylglycine buffer, pH 7.4, cytochrome *c*, orthophosphate, and $DPNH_2$ in the concentrations indicated, and 0.03 M NaF. Each vessel received 0.30 ml. of the indicated particle suspension (particles derived from 50 mg. of wet weight whole rat liver) to make a total volume of 2.0 ml. Temperatures in the different experiments ranged from 17–24°. Experiments 4A and 5A contain data on control experiments with the enzyme preparations of Experiments 4 and 5. In these, 0.02 M β-hydroxybutyrate and 0.001 M DPN replaced $DPNH_2$. Analytical methods described in the text.

Experiment number	Enzyme type	Cytochrome *c*	Time	Orthophosphate	$DPNH_2$	P: $DPNH_2$
		M	*min.*	*μM*	*μM*	
1	KCl	1×10^{-5}	0	9.96	9.44	
			13	5.98	5.76	1.08
			20	4.14	1.24	0.71
	($DPNH_2$ omitted)		0	9.01		
	(〃 〃)		20	11.42		
2	H_2O	4×10^{-6}	0	4.72	4.94	
			8	2.81	3.92	1.89
			15	1.96	3.13	1.52
	(0.003 M DPN		0	7.43		
	instead of $DPNH_2$)		17	7.27		
3	KCl	1.5×10^{-5}	0	4.83	5.24	
			10	3.99	4.16	0.78
			16	3.50	2.91	0.40
	(0.003 M DPN		0	6.42		
	instead of $DPNH_2$)		14	6.94		
4	H_2O	5×10^{-6}	0	4.83	5.48	
			7	2.45	3.68	1.32
			11	1.55	2.70	1.18
5	〃	3×10^{-6}	0	6.92	4.84	
			10	4.84	3.43	1.48
			14	3.80	2.61	1.40

β-Hydroxybutyrate controls

				P	Acetoacetate	P: AcAc
4A	H_2O	1.5×10^{-5}	0	8.24	0.30	
			30	3.25	3.64	1.50
5A	〃	3×10^{-6}	0	9.01	0.16	
			32	4.15	3.23	1.54

described here rule out the permissible interpretation that the P^{32} incorporation observed earlier may have corresponded to relatively minor phosphate exchanges. These experiments therefore show in a direct way that phosphorylation sites do exist above the primary coenzyme level in the electron transport chain, as had been suggested on thermodynamic grounds by Belitser and Tsibakova[20], Lipmann[21], and others.

It should be pointed out here that the P:DPNH$_2$ and P:AcAc ratios recorded in this paper have not been corrected for dephosphorylation losses. Such losses are relatively high when ADP is used as acceptor. Previous work has indicated the magnitude of such losses under very nearly the same conditions as were used in these experiments[4].

Attempts to obtain a more satisfactory and complete accounting for all the phosphorylations coupled to oxidation of β-hydroxybutyrate and DPNH$_2$ did not appear profitable in view of the permeability factor demonstrated, the lability of phosphorylating activity of the H_2O suspension, the considerable phosphate "leaks," and the difficulty of arranging completely similar experimental conditions for such comparisons. In addition, the large amount of highly purified DPNH$_2$ required has been a deterrent not only to such comparative experiments but also for fuller examination of the factors of permeability and the optimal conditions for DPNH$_2$-linked oxidative phosphorylation.

The possible existence of permeability barriers in the mitochondria to substances like DPNH$_2$ or perhaps to larger molecules like cytochrome *c* may be a serious obstacle to further experimentation designed to identify the phosphorylating loci in the electron transport chain. However, the existence of such barriers poses some new questions of interest regarding the intracellular physiology of metabolic exchanges between the mitochondria and the soluble portion of the cytoplasm.

Summary

In this investigation conditions have been determined for the direct demonstration that orthophosphate uptake occurs coupled to the aerobic oxidation of high concentrations of dihydrodiphosphopyridine nucleotide (DPNH$_2$) by suitably supplemented suspensions of washed particles of rat liver. The preparation of a highly purified barium salt of DPNH$_2$ has been described. The rate of oxidation of DPNH$_2$ by molecular oxygen in suspensions of washed particles of rat liver has been found to depend on the preparative history of the particles and on cytochrome *c* concentration. Particles pretreated under hypotonic conditions oxidized DPNH$_2$ at much higher rates than did untreated particles, indicating the existence of an osmotically sensitive permeability or availability barrier of the particles to DPNH$_2$. With untreated particles little DPNH$_2$ oxidation occurred unless cytochrome *c* was added, although the same particles unsupplemented with cytochrome *c* readily oxidized substrates such as β-hydroxybutyrate or succinate. The "native" cytochrome *c* within the particles appears therefore to be largely unavailable for oxidation of externally added DPNH$_2$. In balance experiments the rate of DPNH$_2$ disappearance paralleled exactly the rate of oxygen uptake over a large part of the reaction course. By measuring DPNH$_2$ disappearance spectrophotometrically and the phosphate exchanges colorimetrically, values of the P:O (equal to P:DPNH$_2$) ratio of DPNH$_2$ oxidation have been determined. Values as high as 1.89, uncorrected for dephosphorylation losses, have been obtained. Comparative experiments showed that electron transport from DPNH$_2$ to oxygen very nearly accounts for all the phosphorylation observed when β-hydroxybutyrate undergoes aerobic oxidation to acetoacetate via the DPN-linked β-hydroxybutyric dehydrogenase. The data presented provide direct support for predictions based on thermodynamic considerations that a large part of phosphate uptake during oxidative phosphorylation takes place during electron transport between primary coenzyme and oxygen.

The author wishes to acknowledge the assistance of Dr. E. P. Kennedy and Dr. S. S. Barkulis and Sylvia Wagner Smith in the purification of various DPN samples.

Bibliography

1. Friedkin, M., and Lehninger, A. L., *J. Biol. Chem.,* **174**, 757 (1948).
2. Friedkin, M., and Lehninger, A. L., *J. Biol. Chem.,* **178**, 611 (1949).
3. Lehninger, A. L., *J. Biol. Chem.,* **178**, 625 (1949).
4. Lehninger, A. L., and Smith, S. W., *J. Biol. Chem.,* **181**, 415 (1949).
5. Hogeboom, G. H., and Barry, G. T., *J. Biol. Chem.,* **176**, 935 (1948).
6. Warburg, O., and Christian, W., *Biochem. Z.,* **287**, 291 (1936).
7. Gutcho, W., and Stewart, E. D., *Anal. Chem.,* **20**, 1185 (1948).
8. Horecker, B. L., and Kornberg, A., *J. Biol. Chem.,* **175**, 385 (1948).
9. LePage, G. A., *J. Biol. Chem.,* **168**, 623 (1947).
10. Ohlmeyer, P., *Biochem. Z.,* **297**, 66 (1938).
11. Drabkin, D. L., *J. Biol. Chem.,* **157**, 563 (1945).
12. Lehninger, A. L., and Kennedy, E. P., *J. Biol. Chem.,* **173**, 753 (1948).
13. Racker, E., *J. Biol. Chem.,* **184**, 313 (1950).
14. Kornberg, A., *J. Biol. Chem.,* **182**, 779 (1950).
15. Schneider, W. C., and Potter, V. R., *J. Biol. Chem.,* **149**, 217 (1943).
16. Huennekens, F. M., and Green, D. E., *Arch. Biochem.,* **27**, 418 (1950).
17. Huennekens, F. M., and Green, D. E., *Arch. Biochem.,* **27**, 428 (1950).
18. McIlwain, H., and Rodnight, R., *Biochem. J.,* **45**, 337 (1949).
19. Kornberg, A., and Pricer, W. E., Jr., *J. Biol. Chem.,* **182**, 763 (1950).
20. Belitser, V. A., and Tsibakova, E. T., *Biokhimiya,* **4**, 518 (1939).
21. Lipmann, F., in Green, D. E., Currents in biochemical research, New York, 137–148 (1946).

ETHYL ALCOHOL OXIDATION AND PHOSPHORYLATION IN EXTRACTS OF *E. COLI**

G. B. Pinchot and E. Racker

Department of Microbiology
New York University College of Medicine, New York

An enzyme system has been found in *E. coli* extracts which is capable of esterification of inorganic phosphate during alcohol oxidation. The crude extract can be stored for several weeks in the deep freeze without marked loss of activity. When centrifuged for 3 hours at 18,000 g, most of the activity remains in the supernatant.

The crude enzyme preparation was made from strain B *E. coli* cells, grown in neopeptone broth for 18 hours. The organisms were washed twice in distilled water and broken with a Raytheon sonic disintegrator, after which the cellular debris was removed by a short period of high-speed centrifugation.

DPNH oxidase activity was determined spectrophotometrically by measuring the disappearance of DPNH absorption at 340 mμ. The specific activity of extracts made from cells grown on neopeptone broth was three to five times that of preparations made from organisms grown on a defined medium with glucose as the carbon source. Oxidation of reduced DPN was not inhibited by 2×10^{-3} M KCN, whereas 2×10^{-4} M caused 80% inhibition in a pigeon heart muscle preparation prepared according to the method of Keilin and Hartree[(2)]. 2×10^{-2} molar BAL did not inhibit the bacterial enzyme, while according to Slater this concentration causes complete inhibition of heart muscle preparations[(4)]. These findings demonstrate some of the differences between the bacterial and animal oxidizing systems.

*G. B. Pinchot and E. Racker, "Ethyl Alcohol Oxidation and Phosphorylation in Extracts of *E. coli*," in *Phosphorus Metabolism*, eds. W. D. McElroy and B. Glass, **1**, 366–369. Baltimore: The Johns Hopkins Press, 1951. Reprinted with the permission of the authors and of the publisher.

Phosphorylation was studied by following oxygen and phosphate uptake in a mixture of glycylglycine and phosphate buffer at pH 7.2. 100 micromoles of alcohol, an excess of crystalline yeast alcohol dehydrogenase, and 0.75 micromoles of DPN were added to the enzyme preparation. The high energy phosphate formed was trapped by adding 100 micromoles of glucose, purified yeast hexokinase, $MgCl_2$ (7×10^{-3} M), and 1.5 micromoles of ATP. 10^{-2} M NaF was added to inhibited ATPase and enolase. At the end of the experiment the reaction was stopped with trichloroacetic acid, and the disappearance of inorganic phosphate was measured. The oxidation and phosphorylation were largely dependent on the addition of DPN to the reaction mixture, as shown in Table 1.

Table 1

THE DEPENDENCE OF PHOSPHORYLATION ON THE ADDITION OF DPN

Enzyme Preparation / Substrate	Crude Extract Alcohol	Crude Extract Alcohol
DPN	—	+
Oxygen consumption in microatoms	2.80	11.25
Phosphate taken up in micromoles	3.01	11.84
P/O ratio	1.07	1.05

The complete system contained in each vessel: 0.5 ml. *E. coli* extract, 0.5 ml. of 0.5 M glycylglycine buffer, pH 7.2; 20 micromoles of inorganic phosphate, 100 micromoles of alcohol, 0.75 micromoles of DPN, excess alcohol dehydrogenase, $MgCl_2$ (7×10^{-3} M), and water to make a final volume of 3.0 ml. In the side arm were placed excess hexokinase, 100 micromoles of glucose, 1.5 micromoles of ATP and NaF (10^{-2} M f. c.). The center well contained 0.2 ml. of 40% KOH.

It was then observed that the oxidation of acetaldehyde contributed considerably to the phosphate uptake, resulting in P/O ratios often as high as 1.5. This was found to be due to an aldehyde dehydrogenase which is present in the *E. coli* extracts, similar to the enzyme in *Clostridium kluyveri* described by Stadtman and Barker[5]. As shown in Table 2, this enzyme in *E. coli* is dependent on the presence of coenzyme A preparations for its activity in respect to hydrogen transfer as well as acetyl phosphate formation. Treatment with Dowex 1 according to the method of Chantrenne and Lipmann[1] resulted in nearly complete loss of dehydrogenase activity, which was fully restored by the addition of coenzyme A preparations (kindly supplied by Dr. S. Korkes). The large amounts of acetyl phosphate formed relative to oxygen uptake was probably due to a dismutation, since these crude *E. coli* preparations contain some alcohol dehydrogenase activity.

When the crude *E. coli* enzyme was treated with Dowex-1, then precipitated at 50% ammonium sulfate saturation, and dialyzed for 2 hours, the preparation was still capable of esterification of inorganic phosphate if an alcohol dehydrogenase preparation and alcohol were added (Table 3). Pretreatment of the *E. coli* extract with iodoacetate (10^{-3} M) effectively depressed endogenous glycolysis of such preparations without impairing significantly the uptake of

Table 2

EFFECT OF COENZYME A PREPARATION ON ACETALDEHYDE DEHYDROGENASE OF *E. coli*

	Micromoles	
	O_2 uptake	Acetylphosphate formed
Dowex-treated enzyme	0	< 0.3
Dowex-treated enzyme + acetaldehyde	0.2	< 0.3
Dowex-treated enzyme + acetaldehyde and Co A	6.15	9.5
Dowex-treated enzyme + Co A	0.15	< 0.3

The system contained in each vessel 0.5 ml. of the *E. coli* enzyme, 0.2 ml. of 1 M K phosphate buffer, pH 7.4 and 0.75 micromoles of DPN; where indicated 8.4 units of coenzyme A and 30 micromoles of acetaldehyde. Final volume was 3 ml. and the center well contained 0.2 ml. of 40% KOH.

Table 3

THE EFFECT OF DOWEX-1 TREATMENT ON PHOSPHORYLATION AND ACETALDEHYDE OXIDATION

Enzyme preparation	0.50% Dowex, 1	Ammonium 2	Sulfate Precipitate 3
Substrate added	None	Alcohol	Acetaldehyde
Oxygen uptake in microatoms	1.43	6.63	1.46
Phosphate uptake in micromoles	0	4.91	0
P/O ratio	0	0.74	0

The vessels contained the complete system as outlined in Table 1. Vessel #3 contained 30 micromoles of acetaldehyde, and no alcohol dehydrogenase was added in vessels #1 and #3.

inorganic phosphate. Although treatment of the extracts with Dowex-1 and iodoacetate would seem to eliminate acetyl phosphate and glycolysis as donors of energy-rich phosphate, a final conclusion cannot yet be drawn from these preliminary observations.

Several findings which still obscure the picture should be mentioned. Dinitrophenol, known to inhibit phosphorylation in animal tissues, has no effect on the bacterial system. Two attempts, in which large amounts of reduced DPN have been used to determine whether phosphate uptake occurs during its oxidation, have been negative. However, in view of Lehninger's findings[3] of a toxic factor in DPNH preparations, these experiments cannot be considered conclusive. Furthermore, with some preparations of alcohol dehydrogenase and hexokinase no phosphate esterification was obtained. This could have been due to an inhibitor present in the preparation. Such an inhibitor of phosphorylation was actually found to be present in fractions of baker's yeast obtained during hexokinase purification. On the other hand, it is equally possible that the positive results may have been due to contamination of the "crystalline enzyme preparations" with small amounts of glycolytic enzymes. Since treatment of the alcohol dehydrogenase with iodoacetate was avoided because of the known sensitivity of this enzyme to the poison, glycolysis still remains a possibility to explain the phosphorylating activity after Dowex-1 treatment.

In conclusion, it may be said that two phosphorylating reactions can be demonstrated in *E. coli* extracts both of which are dependent on the addition of DPN. One was shown to be due to the formation of acetyl phosphate. The other may be due to the oxidation of reduced DPN, but a more rigid exclusion of a glycolytic mechanism or another unknown pathway of phosphorylation must be awaited, in view of the negative results with reduced DPN.

References

1. Chantrenne, H., and F. Lipmann. 1950. *J. biol. Chem.,* **187**: 757.
2. Keilin, D., and E. F. Hartree. 1947. *Biochem. J.,* **41**: 500.
3. Lehninger, A. L. 1951. *J. biol. Chem.,* **190**: 345.
4. Slater, E. C. 1950. *Biochem. J.,* **46**: 484.
5. Stadtman, E. R., and H. A. Barker. 1948. *J. biol. Chem.,* **174**: 1039.

COMPLEXING AGENTS AS CO-FACTORS IN ISOLATED MITOCHONDRIA*

J. Raaflaub

Institute of Physiological Chemistry
University of Zürich, Switzerland
(Submitted November 18, 1954)

Liver mitochondria which are suspended in appropriate media retain their compact structure for long periods of time as well as their ability to catalyze the oxidation of products of intermediary metabolism (such as members of the citric acid cycle, etc.)[1]. A number of observations have shown that this metabolic activity is lost when the mitochondria swell[2,3]. A detailed presentation of the chemical and physico-chemical changes associated with this swelling has been previously published[3,4]. It should be mentioned here, however, that swelling is retarded by ATP† but accelerated by split products of ATP (e.g., phosphate). The mechanism by which ATP retards the mitochondrial swelling has so far remained obscure. In this paper I shall report on experiments which clarify this effect of ATP. Slater and Cleland have made the important discovery that "complexon" (EDTA) stimulates the oxidation of α-ketoglutarate by heart mitochondria and prevents such morphological changes as swelling. According to these authors EDTA acts by chelating those calcium ions which otherwise interfere[5]. Since it is well established that polyphosphates can complex a number of polyvalent cations[6], it seems possible that the similar effects of ATP on liver mitochondria and of EDTA on heart muscle mitochondria depend on the metal-complexing capacities of these substances. The question then arises

*J. Raaflaub, "Komplexbildner als Cofaktoren isolierter Zellgranula," *Helv. Chim. Acta,* **38**, 27–37 (1955). Translated from the German and reprinted with the permission of the author and of the editors of *Helvetica Chimica Acta.*

†The following abbreviations are used: ATP: adenosinetriphosphate, Complexon: ethylenediamine tetra acetate equals "Complexon III" of B. Siegfried and Co., Inc. in English-American literature also called EDTA or versene; Glycolcomplexon: Compound V, Table 1.

whether synthetic complexing agents similar to EDTA can replace ATP as co-factors in mitochondrial respiration. This question could be tested experimentally with a number of synthetic complexing agents made available to us through the kindness of Prof. G. Schwarzenbach.

Results

We first examined the effect of EDTA on the swelling of liver mitochondria. EDTA, as well as the other synthetic chelating agents in Table 1, inhibited non-osmotic swelling (for differentiation of osmotic vs. non-osmotic swelling see Ref. 4). It appears from Fig. 1 that EDTA "complexon" (cf. Table 1) was more effective and its action of longer duration than that of ATP (Fig. 1). Apparently certain polyvalent cations are involved in non-osmotic swelling. Earlier studies had shown that certain divalent cations in concentrations of 10^{-3} to 10^{-4} M inhibited swelling of diluted mitochondrial suspensions. This effect was attributed to a reduction in the permeability of the mitochondrial surface. This finding seems at first inconsistent with the observation that swelling is also inhibited

Table 1

A SUMMARY OF THE COMPLEXING AGENTS INVESTIGATED

```
HOOC—CH2                 CH2—COOH
        \               /
       +HN—CH2—CH2—NH+
        /               \
-OOC—CH2        I        CH2—COO-
```

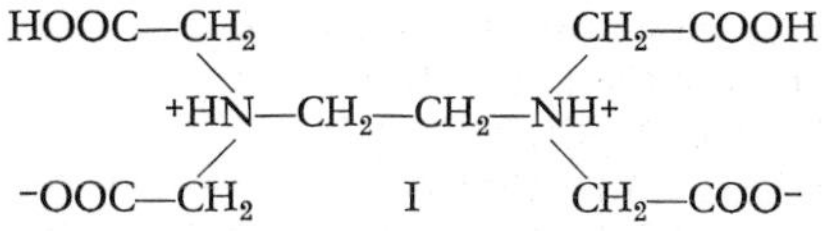

Ethylenediamine N, N′ tetra acetic acid (Complexon III) [or EDTA]

```
              ,CH2—COOH
         —NH+
              `CH2—COO-
              ,CH2—COOH
  II     —NH+
              `CH2—COO
```

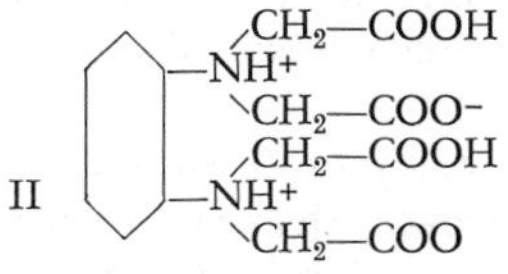

1, 2 Diaminocyclohexan-tetra acetic acid

```
-OOC—CH2            CH3            CH3             CH2—COO-
        \            |              |             /
       +HN—CH2—CH2—NH+—CH2—CH2—NH+—CH2—CH2—NH+
        /                                         \
-OOC—CH2            III                            CH2—COO-
```

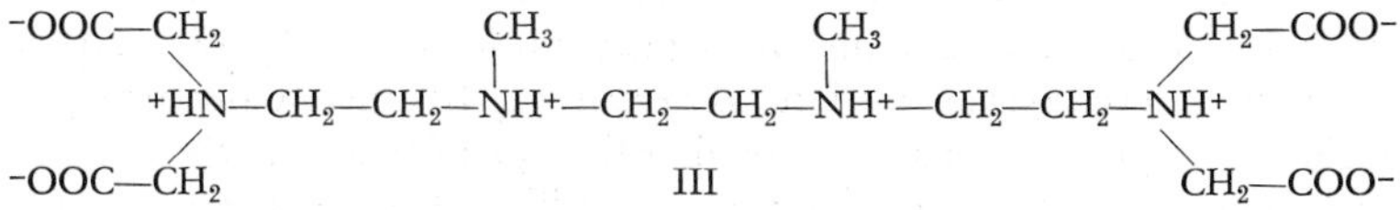

N′, N″ Dimethyl-triethylamine —N, N‴-tetra acetic acid

```
      O               CH2—COOH
      ||             /
  HO—P—O—CH2—NH+
      |              \
      O-   IV         CH2—COOH
```

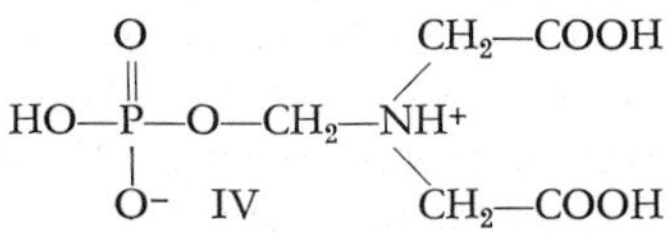

Amino methyl phosphoric acid-N-di acetic acid

```
HOOC—CH2                                              CH2—COOH
        \                                            /
       +HN—CH2—CH2—O—CH2—CH2—O—CH2—CH2—NH+
        /                                            \
-OOC—CH2                    V                         CH2—COO-
```

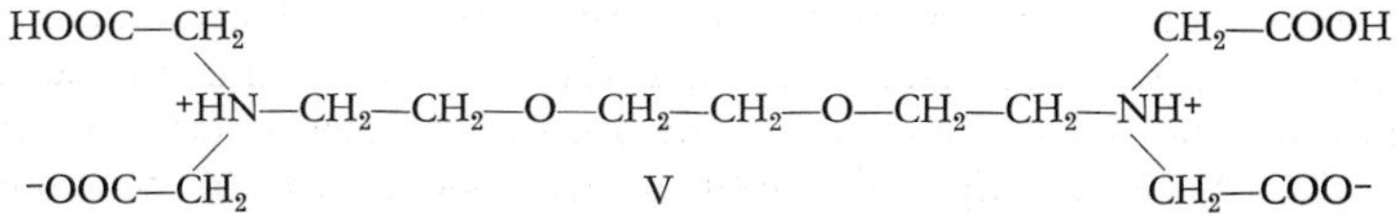

Ethyleneglycol-bis-β-aminoethyl ether. —N, N′-tetra acetic acid (abbrev. glycol complexon)

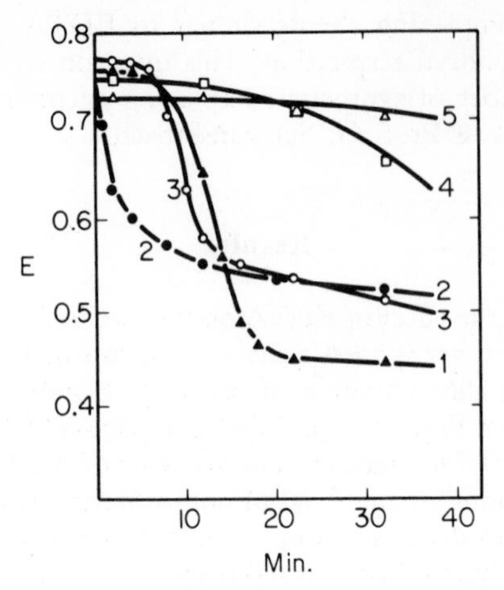

Fig. 1. Non-osmotic swelling of mitochondria in dilute suspension. Influence by addition of divalent cations (Mg^{++}, Ca^{++}) and by complexing agents (EDTA complexon, ATP). Incubate: Mitochondrial stock suspension, "M_1" (see experimental section), diluted 1 to 200 by an isotonic mannitol solution containing sodium phosphate 10^{-3} M, *p*H 6.9. Temperature 20°.

1. Control without further additions.
2. Addition of $CaCl_2$ 2×10^{-4} M.
3. Addition of $MgCl_2$ 2×10^{-4} M.
4. Addition of ATP 10^{-3} M.
5. Addition of EDTA complexon 10^{-3} M.

The extinction E (ordinate) is inversely proportional to the degree of swelling of the mitochondria.

by calcium and magnesium complexers such as EDTA. Examination of the curves in Fig. 1 shows, however, that the divalent cations exert two effects. (1) The degree of swelling (cf. ordinate E) which is the most extensive in the absence of any addition of divalent cations proceeds to a somewhat lesser extent in the presence of calcium or magnesium ions (Fig. 1). (2) The rate of swelling (cf. abscissae) is affected in a different way by the two divalent cations. Addition of magnesium ions does not alter the rate of swelling whereas calcium ions bring about a marked acceleration of swelling (Fig. 2). The more concentrated the mitochondrial suspension employed the more prominent is the kinetic effect. If the stock suspension M_1 is diluted only 1: 3, the addition of calcium ions alone is sufficient to produce rapid and extensive swelling, whereas magnesium ions merely exert an inhibitory effect. Moreover, at the mitochondrial concentrations used in metabolic studies (stock M, diluted 1: 10) this calcium effect is particularly noteworthy in view of the fact that calcium ions bring about a premature cessation of respiration (Fig. 3). A certain amount of calcium is apparently associated with isolated mitochondria; this may be already bound intracellularly or fixed during the process of isolation (as suggested by the work of Slater and Cleland). Optimal metabolic activity of long duration depends therefore on the complexing of these calcium ions. In both heart muscle

Table 2

CHEMICAL DATA OF THE COMPLEXING AGENTS INVESTIGATED

Compounds (see Table 1)	1 pK_1 $\left(K_1 = \frac{(H)(H_3Y)}{(H_4Y)}\right)$	2 pK_2 $\left(K_2 = \frac{(H)(H_2Y)}{(H_3Y)}\right)$	3 pK_3 $\left(K_3 = \frac{(H)(HY)}{(H_2Y)}\right)$	4 pK_4 $\left(K_4 = \frac{(H)(Y)}{(HY)}\right)$	5 $\log K_{CaY}$ $\left(K_{CaY} = \frac{(CaY)}{(Ca)(Y)}\right)$	6 $\log K_{MgY}$ $\left(K_{MgY} = \frac{(MgY)}{(Mg)(Y)}\right)$	7 $\frac{(CaY)}{(Ca)}$
I	2.0	2.7	6.16	10.26	10.59	8.69	$4.0 \cdot 10^1$
II	2.4	3.5	6.12	11.70	12.5	10.3	$7.9 \cdot 10^1$
III	3.05	5.15	8.99	10.54	9.45	4.31	$0.8 \cdot 10^1$
IV	2.0	2.25	5.57	10.76	7.18	6.28	$0.14 \cdot 10^1$
V	2.1	2.65	8.85	9.46	11.0	5.21	$4.8 \cdot 10^3$

Column 1–4: acid dissociation constant
Column 5–6: constants for complexing agents with respect to Ca and Mg
Column 7: The binding of calcium under the conditions of milieu corresponding to the respiration experiment in the Warburg manometer, i.e. (CaY)/(Ca) is calculated for the case that $pH = 7.0$, $\Sigma Y = 10^{-3}$ M, $\Sigma Mg = 3 \times 10^{-3}$ M and $\Sigma Ca < \Sigma Y$.

The constants presented in the table are, strictly speaking, valid for only 20°C and an ionic strength $\mu = 0.10$ although the milieu in Warburg manometer provides for 38°C and an ionic strength of $\mu =$ app. 0.07. Yet this should scarcely affect the calculated coefficients (CaY)/(Ca). Further data on the complexing agents of Table 2 can be found in the articles by G. Schwarzenbach and coworkers for I[1], for II[2], for IV[3], for III and for V.[4]

[1] Schwarzenbach, G., and Ackermann, H.: *Helv. Chim. Acta,* **30** (1947), 1798.
[2] Schwarzenbach, G., and Ackermann, H.: *Helv. Chim. Acta,* **32** (1949), 1682.
[3] Schwarzenbach, G., Ackermann, H. and Ruckstuhl, P.: *Helv. Chim. Acta,* **32** (1949), 1175.
[4] Senn, H., Diss. Universität Zürich 1954.

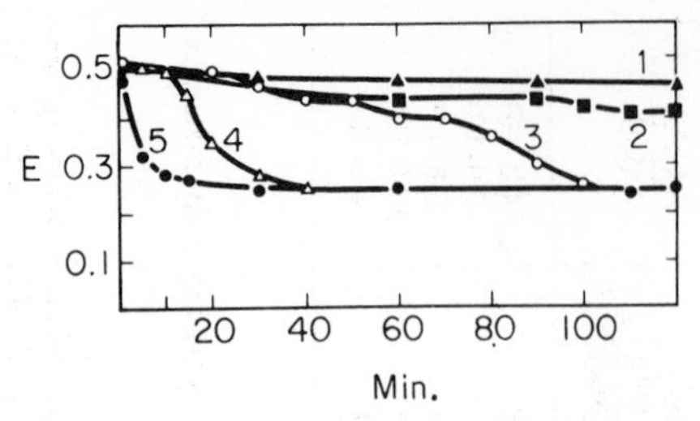

Fig. 2. The influence by divalent cations (Mg, Ca) and EDTA complexon on the non-osmotic swelling of mitochondria in concentrated suspension. Incubate: Stock mitochondrial suspension M_1 diluted 1 to 3 in isotonic mannitol solution, *p*H 6.7 (sufficiently buffered by outward-diffusing metabolite). Temperature 20°.

Additions:

1. EDTA complexon 10^{-3} M.
2. $MgSO_4$ 4×10^{-4} M.
3. Control without additions.
4. $CaCl_2$ 10^{-4} M.
5. $CaCl_2$ 4×10^{-4} M.

The proper dilution of mitochondria for photometry is prepared by mixing 9.9 ml isotonic mannitol solution for each 0.1 ml of 1 to 3 diluted mitochondrial suspension prior to the determination of each single point on the graph. The extinction E is inversely proportional to the degree of swelling of the mitochondria.

Table 3

The Effect of Different Complexing Agents on the Oxygen Consumption of Isolated Liver Mitochondria

Substrate	with ATP	with glycol complexon	with EDTA complexon	without complexing agents
malate or fumarate	471	325	76	54
pyruvate	556	258	106	101
citrate	686	296	292	178
α-ketoglutarate	514	197	62	31
glutamate	532	232	89	58

The figures represent oxygen consumption in mm^3 in 120 minutes in the presence of different complexing agents and substrates. In addition, the incubates routinely contain ATP, glycol "complexon" in a concentration of 10^{-3} M, and the sodium salts of the substrate in a concentration of 10^{-2} M (see experimental section). The experiments with different substrates have been conducted with mitochondria from different animals. Thus the comparable figures of the table are to be found, strictly speaking, in only the horizontal direction of the table.

mitochondria and liver mitochondria (cf.[5]) the oxidation of α-ketoglutarate is prolonged by the addition of EDTA to the usual incubation medium. These

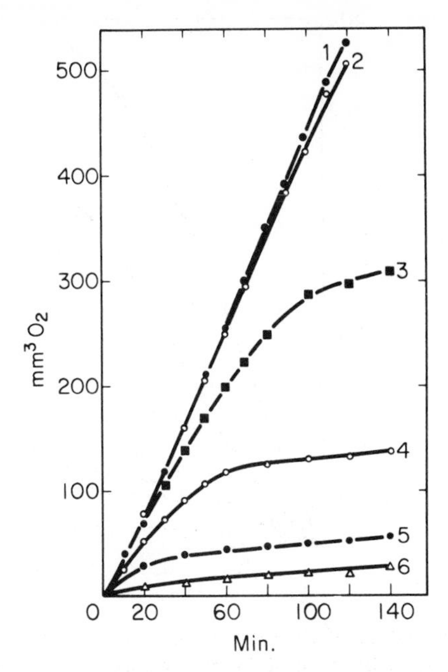

Fig. 3. The effect of calcium and EDTA complexon on the oxygen consumption of liver mitochondria. Substrate: α-ketoglutarate 10^{-2} M. In addition, the incubate routinely contains the following ingredients (see also the experimental section):

1. ATP 10^{-3} M + EDTA complexon 10^{-3} M.
2. ATP 10^{-3} M + EDTA complexon 10^{-3} M + $CaCl_2$ 3×10^{-5} M.
3. ATP 10^{-3} M.
4. ATP 10^{-3} M + $CaCl_2$ 3×10^{-5} M.
5. ATP 10^{-3} M + $CaCl_2$ 10^{-4} M.
6. Blank without substrates.

results support the work of Slater and Cleland and show the similar behavior of mitochondria from various tissues with regard to morphologic and functional changes and insofar as these alterations are affected by calcium and EDTA.

If it is true that the action of ATP is due to its capacity to bind calcium, as suggested above, it should be possible to substitute for it synthetic chelating agents. Table 1 lists the most important tested complexing agents. The first four acids could not replace ATP; conversely, in the presence of glycol "complexons" (Formula V, Table 1) the oxidation of malate and fumarate proceeds almost as well as with ATP (apart from an initial phase where ATP is clearly superior)—(Fig. 4). The oxidation of pyruvate, citrate, ketoglutarate and glutamate is diminished when ATP is replaced by glycol complexons but it is greater than that obtained with other complexing agents or in the absence of all complexing agents (Table 3).

Table 3 gives O_2 utilized in mm^3 per 120 min., in the presence of various complexing agents and substrates. The usual medium has in addition 10^{-3} M ATP, or complexing agent, and 10^{-2} M sodium salts of substrates. Experi-

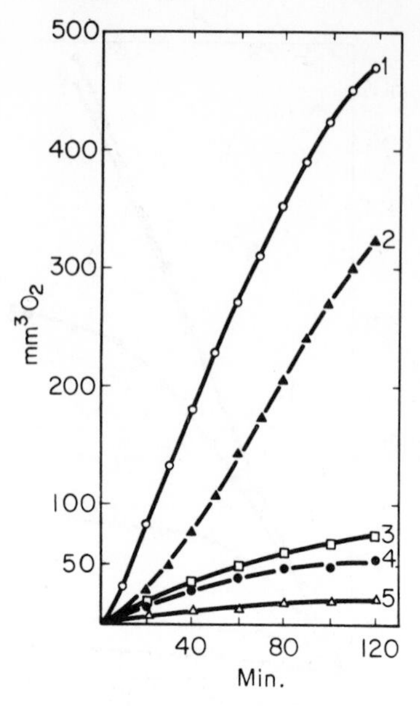

Fig. 4. The effect of different complexing agents on oxygen consumption of liver mitochondria. Substrate: Malate 10^{-3} M. In addition, the incubates contain the following ingredients:

1. ATP 10^{-3} M.
2. Glycol complexon 10^{-3} M.
3. EDTA complexon 10^{-3} M.
4. No additions.
5. Glycol complexon 10^{-3} M without substrate (glycol).

ments with various substrates were performed on mitochondria of different animals.

Referring to the oxidation of citrate, it is important to keep in mind the metal-complexing activity of other components of the incubation medium, particularly the substrates. The oxidation of citrate, in a medium containing only Mg^{2+} and phosphate buffer, is substantially greater than that of other members of the citrate cycle (malate, ketoglutarate, etc.) under identical conditions (Fig. 5). This is probably due to the appreciable binding of calcium by citrate (affinity constant $K = 10^{3.2}$).

Discussion

Since the fundamental work of F. Lipmann[7], the biochemical effects of ATP have been attributed to its "high-energy phosphate bonds". On the other hand, the metal-complexing capacities of ATP have been examined by only a few investigators[8, 9, 10] and with only rather specific objectives. It is well recognized that isolated mitochondria require ATP (or AMP, which is rapidly phosphorylated to ATP) as cofactors, but there has been no explanation for this requirement. Considerable progress in this direction came from the realiza-

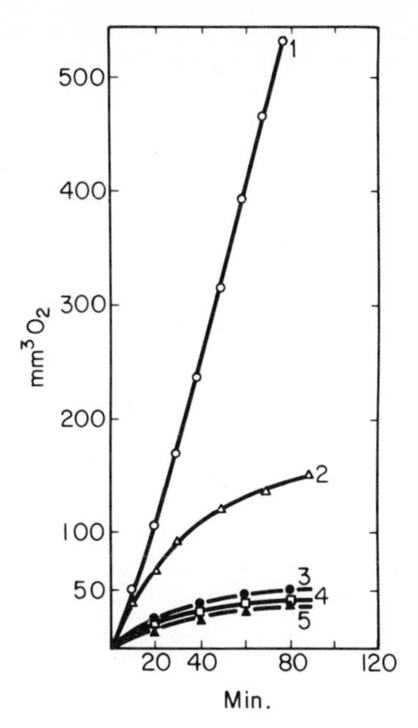

Fig. 5. The oxygen consumption of liver mitochondria with different substrates without specific complexing agents. To the usual reaction mixtures (see experimental section) were added:

1. Citrate 10^{-2} M + ATP 10^{-2} M.
2. Citrate 10^{-2} M.
3. Malate 10^{-2} M.
4. Pyruvate 10^{-2} M.
5. α-ketoglutarate 10^{-2} M.

tion that ATP delays the structural changes in mitochondria (such as swelling) and also that metabolic activity is correspondingly diminished[3, 4, 11]. Once Slater and Cleland had shown that the addition of EDTA to the ATP-containing incubation medium stabilized the structure and metabolism of the mitochondria from heart muscle, it seemed reasonable to look for a common, complexometric basis for the action of ATP and EDTA. However, Slater and Cleland[11] considered the ATP effect due to the "high energy" phosphate: "Both the spontaneous transformation (non-osmotic swelling) and the enzymic inactivation are greatly slowed down by versene, ATP, or ADP. It seems likely that a common factor, possibly the level of some energy-rich phosphorus compound in the sarcosome, determines the onset of both morphological and biochemical changes."

One of the synthetic chelating agents of the EDTA type, glycol complexons (Table 1), can replace ATP as a cofactor for mitochondrial respiration. It is almost as effective as ATP in the oxidation of malate and fumarate but only partly so in the oxidation of pyruvate, α-ketoglutarate, citrate, and glutamate. This points toward the complex forming action of ATP since glycol complexon, acting in the same fashion, is not an energy source.

Isolated liver mitochondria are unstable because of bound calcium ions.

They therefore require complexing agents as cofactors which bind calcium but which do not interfere with the metabolism of the synthetic complexing agents. The glycol complexon is closest to ATP. This is believed to be due to the special complexometric properties of glycol complexon. As shown in Table 2, this reagent has a very high affinity for calcium but much less for magnesium. At *p*H 7 there is almost no magnesium binding.

It can be calculated that in the magnesium-rich media in which mitochondrial respiration is measured, calcium is strongly bound by glycol complexon (cf. last column (CaY)/(Ca) Table 2). It will require the discovery of other complexing agents which are as effective or more effective than glycol-complexon to find out whether factors other than calcium-binding are also of importance. Certainly our data do not imply that mitochondrial respiration is independent of ATP. After all functionally capable mitochondria contain 0.1 μ mole adenosine phosphate per mg nitrogen.[4] In the presence of glycol complexon any added ATP should then be quite ineffective.

Our experiments illustrate that there are ATP effects where ATP acts not as a source of "high-energy phosphate" but as a complexing agent. This is not an exceptional case, as will be demonstrated later. We are still far from a complete understanding of the manifold roles of ATP. However, until now, the energetic aspects of ATP have been emphasized excessively. This is already apparent from the fact that the affinity constants of the complexes of ATP with alkaline earth cations and other ions are not fully known. Distefano and Neuman give a dissocation constant of 8.7×10^{-5} [9] for the ATP-calcium complex. Spicer implies considerable affinity for Ca^{2+} and Mg^{2+} on the basis of the displacement of electrometric titration curves of ATP in the presence of these ions, although no quantitative data are provided[12]. Recent measurements by Melchior show that K^+ and Na are bound only weakly by ATP[13]. A rigorous study on the complexing action of biochemically important polyphosphates is still lacking.

Experimental Section

Chemicals

The synthetic complexing agents III to V of Table 1 were made available to us through Professor G. Schwarzenbach, Chem. Inst., Univ. of Zürich.

Experimental animals

White rats, fasted 16–24 hours before onset of experiments.

Preparation of mitochondria

The liver mitochondria were isolated according to a previously published method[3]. The concentrated stock solution obtained in this way in an isotonic mannitol solution (0.3 M) was called "M_1".

Determination of the swelling of the mitochondria

The photometric method employed has been described previously[3]. The recording takes place in suspensions diluted 200 to 300 fold. Decrease of extinction (respectively decreased scattering) means swelling of mitochondria.

Determination of oxygen consumption

This takes place in the usual fashion, the manometric method of Warburg, at 38° with air in gas phase. The medium contains (final concentrations in 3 cm^3 incubate) K-phosphate 0.008 M; $MgSO_4$ 0.003 M; mannitol 0.18–0.24 M; mitochondria "M_1", diluted 1 to 10; the *p*H of the mixture is 7.1. The concentrations of the additional ingredients (substrate, complexing agents) are stated in the legends to the illustrations. The complexing acids are neutralized to a *p*H of 7.0 before they are added to the medium. When the complexing takes place with the magnesium ions of the medium at *p*H 7.0 an acidification occurs. The *p*H change is of no concern if the concentration of complexing agents is essentially lower than that of the phosphate buffer since small *p*H variations have no influence on the oxygen consumption by the mitochondria.

This work was made possible by support of *Fritz Hoffmann-La Roche Stiftung zur Forderung wissenschaftlicher Arbeitsgemeinschaften in der Schweiz*.* For this support we want to express our best thanks. I should like to add my best thanks to Professor F. Leuthardt and Professor G. Schwarzenbach for the support which they have rendered me and for their valuable stimulation and criticism. I am indebted to Miss T. Leupin for the careful performance of a major number of experiments.

Summary

Ethylenediaminetetra-acetic acid and other chelating agents of the same type inhibit the swelling of liver mitochondria and enhance their oxygen consumption. This is probably due to complex formation of the calcium ions.

In the presence of one of the five chelating agents tested (glycol complexon, see Table 1) the respiration of mitochondria proceeds fairly well without the addition of ATP. This suggests that ATP, as a cofactor of isolated mitochondria, acts as a complexing agent.

References

1. Leuthardt, F., and Mauron, J.: *Helv. Physiol. Pharmacol. Acta,* **8** (1950), 386. Green, D. E., Sympos. sur le cycle tricarboxylique IIe Congr. Internat. de Biochimie Paris, 1952.
2. Harman, J. W.: *Exp. Cell Res.,* **1** (1950), 382.
3. Raaflaub, J.: *Helv. Physiol. Pharmacol. Acta,* **11** (1953), 142, 157.

*FH-LaR Fund for Development of Scientific Research Communities in Switzerland.

4. Brenner-Holzach, O., and Raaflaub, J.: *Helv. Physiol. Pharmacol. Acta,* **12** (1954), 243.
5. Slater, E. C., and Cleland, W. K.: *Biochem. J.,* **55** (1953), 566.
6. Van Wazer, J. R., and Campanella, D. A.: *J. Amer. Chem. Soc.,* **72** (1950), 655.
7. Lipmann, F.: *Adv. Enzymol.,* **1** (1941), 99.
8. Mandl, J., Grauer, A., and Neuberg, C.: *Biochim. Biophys. Acta,* **8** (1952), 654.
9. Distefano, V., and Neuman, W. F.: *J. Biol. Chem.,* **200** (1953), 759.
10. Hers, H. G.: *Biochim. Biophys. Acta,* **8** (1952), 424.
11. Cleland, K. W., and Slater, E. C.: *Quart. J. of Microscop. Sci.,* **94** (1953), 329.
12. Spicer, S. S.: *J. Biol. Chem.,* **199** (1952), 301.
13. Melchior, N. C.: *J. Biol. Chem.,* **208** (1954), 615.

A STUDY OF OXIDATIVE PHOSPHORYLATION WITH INORGANIC PHOSPHATE LABELED WITH OXYGEN 18*

Mildred Cohn

Department of Biological Chemistry
Harvard Medical School
Boston, Massachusetts

Since most reactions of phosphate compounds of biochemical interest involve the rupture of a carbon-oxygen or a phosphorus-oxygen bond, the detailed mechanism of such reactions can be elucidated by investigations with O^{18}. The first such study undertaken included a well defined group of reactions of glucose-1-phosphate(1). The results of that study which are relevant to the current study of oxidative phosphorylation are (a) that in hydrolysis reactions catalyzed by phosphatases, cleavage occurred between phosphorus and oxygen; and (b) that in phosphorolysis reactions catalyzed by muscle and sucrose phosphorylases which involve the reversible formation of an organic phosphate compound from inorganic phosphate, the cleavage occurred between carbon and oxygen.

It seemed possible that the extension of this approach to the reactions involved in oxidative phosphorylation might lead to some useful information. If inorganic phosphate labeled with O^{18} is taken up in organic linkage by the formation of a carbon-oxygen linkage, as in the phosphorylase reactions, the oxygen bridging the carbon and phosphorus is thus labeled with O^{18}. Should the organic phosphate now be cleaved by the rupture of the phosphorus-oxygen bond, the organic moiety remaining would contain O^{18}. Furthermore, should inorganic phosphate be formed in such a reaction, it would be diluted with normal oxygen. This leads (a) to the possibility of identifying phosphorylated intermediates which are too unstable or too low in concentration to be isolated

*M. Cohn, "A Study of Oxidative Phosphorylation with Inorganic Phosphate Labeled with Oxygen 18," in *Phosphorus Metabolism,* eds. W. D. McElroy and B. Glass, **1**, 374–376. Baltimore: The Johns Hopkins Press, 1951. Reprinted with the permission of the author and of the publisher.

as such, by establishing the presence of O^{18} in the dephosphorylated product; and (b) to the possibility of detecting reactions otherwise unobservable by following the O^{18} content of the inorganic phosphate. The advantage of labeling phosphate with oxygen rather than phosphorus is apparent from the foregoing discussion.

Some exploratory experiments using oxygen-labeled phosphate have been done on the phosphorylation coupled with the oxidative decarboxylation of α-ketoglutarate. The first experiment, which was done in collaboration with F. E. Hunter with a preparation of washed rat liver particles, showed that during the course of oxidation of α-ketoglutarate, the inorganic phosphate initially labeled with O^{18} did indeed lose most of its O^{18}. This necessarily implies that in some reaction a phosphorus-oxygen cleavage had occurred. In subsequent studies designed to uncover the nature of this reaction, the system contained rat liver mitochondria prepared in isotonic sucrose by the method of Kielley and Kielley[(2)]. This preparation has a relatively low ATP-ase activity, and adenylic acid may be used as the acceptor of phosphate. It was found that in the malonate-inhibited oxidation of α-ketoglutarate, about 90% of the O^{18} initially present in the inorganic phosphate had been replaced by normal oxygen. When no substrate was added to the system either aerobically or anaerobically, there was a loss of about 25% of the O^{18} from the inorganic phosphate. An anaerobic experiment with α-ketoglutarate yielded the same results as the control without substrate, a result indicating that the observed reaction of inorganic phosphate was not due to a reversible addition on the carbonyl group of α-ketoglutarate. It could also be shown, by a study of the dismutation reactions of α-ketoglutarate with oxalacetate or ammonia, that the reaction causing the loss of O^{18} from inorganic phosphate was not associated with the phosphorylation at the substrate level. Since one must conclude that this reaction of inorganic phosphate is involved only in the phosphorylation accompanying electron transport and since 2,4-dinitrophenol is known to uncouple oxidation and phosphorylation, its effect was tried on this system. 2,4-dinitrophenol was found to inhibit the reaction completely.

An analysis of the quantitative aspects of the results leads to the conclusion that the inorganic phosphate has cycled through a phosphorus-oxygen cleavage reaction 7 to 9 times. This follows from the fact that only 1/4 of the O^{18} is replaced in each cycle, and 90% of the initial O^{18} would be replaced only after 8 cycles. Under the experimental conditions of these experiments, this would correspond to the formation and breakdown of 300 to 500 μmoles of organic phosphate linkages when a maximum of 30 μmoles of α-ketoglutarate are simultaneously oxidized to succinate, carbon dioxide, and water. From considerations of energetics, it is hardly likely that this could be due to hydrolysis of high-energy phosphate bonds.

There is one other aspect of the α-ketoglutarate reaction which has been investigated, namely, the question of the existence of a phosphorylated intermediate of succinate. Although there is no evidence of a reversible reaction of inorganic phosphate involving a phosphorus-oxygen cleavage in phosphorylation at the substrate level, this by no means excludes the possibility of the formation of a phosphorylated intermediate of succinate, with a subsequent cleavage of the phosphorus-oxygen bond in a transfer reaction to some acceptor.

If this were the case, one might expect to find O^{18} in the succinic acid formed in the reaction. The succinic acid isolated in the malonate-inhibited oxidation of α-ketoglutarate was found to contain O^{18}. In fact, 80% of the theoretically expected amount of O^{18} was found, based on the assumption that one of the four oxygen atoms in succinic acid would contain the amount of O^{18} corresponding to the average value of the O^{18} in the inorganic phosphate during the course of reaction. In the same reaction with added 2,4-dinitrophenol, however, the O^{18} content was only 40% of the calculated value.

In conclusion, it has been demonstrated that with the use of O^{18}-labeled phosphate, reactions hitherto unobservable can be detected. These preliminary results indicate that this approach, at worst, may lead to some interesting speculation and, at best, may succeed in removing some of the reactions involved in oxidative phosphorylation from the realm of speculation into the realm of experimentally demonstrable mechanisms.

References

1. Cohn, Mildred. 1949. *J. biol. Chem.*, **180**: 771–781.
2. Kielley, Wayne W., and Ruth K. Kielley. 1951. *Fed. Proc.*, **10**: 207.

COUPLING OF PHOSPHORYLATION TO ELECTRON AND HYDROGEN TRANSFER BY A CHEMI-OSMOTIC TYPE OF MECHANISM*

Peter Mitchell

Chemical Biology Unit, Zoology Department
University of Edinburgh, Scotland

At present, the orthodox view of the coupling of phosphorylation to electron and hydrogen transfer in oxidative and photosynthetic phosphorylation stems from knowledge of substrate-level phosphorylation[1,2]. It is based, consequently, on the idea that water is expelled spontaneously between two chemical groups, *A* and *B*, by the formation of a strong bond (of low hydrolysis energy), and that subsequent or simultaneous oxidation or reduction of *A–B* to *A–B** can result in a weakening of the bond, popularly written $A \sim B^*$, so that adenosine triphosphate (ATP) can be synthesized by coupling the opening of $A \sim B^*$ to the closing of the 'high-energy bond' between adenosine diphosphate (ADP) and phosphorus through group transfer systems of appropriate substrate and oxido-reduction-carrier specificities[3-5]. There are a number of facts about the systems catalysing oxidative and photosynthetic phosphorylation that are generally acknowledged to be difficult to reconcile with this orthodox (chemical) view of the mechanism of coupling. For example: (*a*) The hypothetical 'high-energy' intermediates (for example, reduced diphosphopyridine nucleotide (DPNH) ~ ?, reduced flavin adenine dinucleotide (FADH) ~ ?, cytochrome ~ ?) are elusive to identification[4,6]. (*b*) It is not clear why phosphorylation should be so closely associated with membranous structures[7-9]. It has sometimes been assumed that digitonin-treated mitochondrial 'particles' can couple oxidation to phosphorylation without membranes;[10] but it is doubtful whether this assumption can be justified, for such 'particles'

*P. Mitchell, "Coupling of Phosphorylation to Electron and Hydrogen Transfer by a Chemi-Osmotic Type of Mechanism," *Nature,* **191**, 144–148 (1961). Reprinted with the permission of the author and of the editors of *Nature*.

give only poor respiratory control and contain much lipid[11], which may well exist as a leaky membrane[12]. (*c*) Coupling may vary with the stress, causing variation of respiratory control without corresponding variation in phosphorus/oxygen quotient[4,13]—a phenomenon difficult to explain in terms of molecular stoichiometry. (*d*) Hydrolysis of external ATP by mitochondria causes reduction of internal DPN, accentuated by the oxidation of succinate[14]. There is disagreement as to what complex assumptions offer the better explanation[15]. (*e*) Uncoupling can be caused at all three hypothetical oxido-reduction sites in mitochondria by agents that do not share an identifiable specific chemical characteristic[4] (for example, dinitrophenol, dicoumarol, salicylate, azide). (*f*) Unexplained swelling and shrinkage phenomena accompany the activity of the phosphorylation systems[7,9a,11,16].

Structural features have been invoked as the causes of the departure of the phosphorylation systems from 'ideal' behaviour[4,14,17]; but, as Green has pointed out in a far-sighted paper[18], although the structural features have been recognized as playing an important part in the catalytic activity of multi-enzyme systems, they have so far been treated rather conservatively. The structural (or supramolecular) features have, in fact, generally been regarded only as modifiers of the basic chemical type of coupling process outlined above. The general conception of enzyme-catalysed group translocation that we have been developing in my laboratory for some years[19] offers a more radical approach to the problem; but this has not, so far, been made use of by those working in the field of oxidative and photosynthetic phosphorylation.

The purpose of this article is to suggest that in view of the difficulties confronting the orthodox chemical conception of coupling in oxidative and photosynthetic phosphorylation, one might now profitably consider the basic requirements and potentialities of a type of mechanism that is based directly on the group translocation conception. This type of mechanism differs fundamentally from the orthodox one in that it depends absolutely on a supramolecular organization of the enzyme systems concerned. Such supramolecularly organized systems can exhibit what I have called chemi-osmotic coupling[19c,20], because the driving force on a given chemical reaction can be due to the spatially directed channelling of the diffusion of a chemical component or group along a pathway specified in space by the physical organization of the system[20]. We shall consider chemi-osmotic coupling between the so-called ATPases on one hand and the electron and hydrogen transfer chain on the other: mediated by the translocation of electrons and the elements of water across the membrane of mitochondria, chloroplast grana, and chromatophores.

The first basic feature of the chemiosmotic coupling conception is a membrane-located reversible 'ATPase' system[20b] shown very diagrammatically in Fig. 1. This system, which may include lipids and other components as well as proteins, is assumed to be anisotropic so that the active centre region (indicated by the dotted circle) is accessible to OH' ions but not H^+ ions from the left (inside organelle), and to H^+ ions but not OH' ions from the right (cytoplasm side of organelle). The active centre, like that of phosphokinases in general, is assumed to be relatively inaccessible to water as water. To illustrate the hydrolytic activity of the system, the OH' ion is depicted as diffusing from left to right and finally combining at the active centre region with the terminal

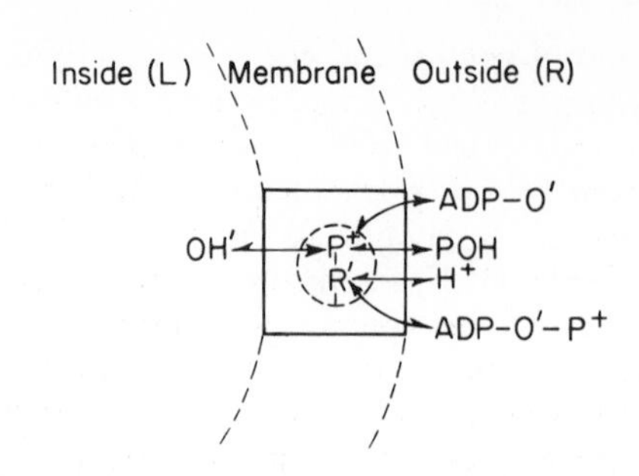

Fig. 1. Anisotropic reversible 'ATPase' system located in an ion-impermeable membrane between aqueous phases L and R.

phosphorylium (P^+) derived from the ATP (ADP—O′—P^+) giving inorganic phosphate (POH) and ADP having a terminal ionized oxygen (ADP—O′), assuming the right-hand phase to be at about *p*H 7 and at electrical neutrality. The elegant conception of the transfer of phosphoryl as phosphorylium ion is due to Lipmann[21]. The system, of course, catalyses hydrolysis equilibrium, and the reverse or phosphokinetic activity of the 'ATPase' is indicated by the single barbs on the arrows of Fig. 1. In the case of the phosphokinetic activity, the OH′ ion is depicted as passing down a free-energy gradient towards the left from an inorganic phosphate group (POH) that passes to the active centre region from the right. The phosphorylium ion (P^+), created by the withdrawal of the OH′, is attacked by the negative atom (—R') in the active centre region, and the phosphorylium ion is then donated to the terminal oxygen of ADP—O′ to give ATP. The chemi-osmotic coupling hypothesis depends thermodynamically on the fact that in such an anisotropic ATPase system, the electrochemical activity of the water at the active centre ($[H_2O]_c$), which determines the poise of the hydrolysis equilibrium in the ATP/ADP system, would be given, not by the product of $[H^+] \times [OH']$ in the aqueous phases L or R, but by the product $[H^+]_R \times [OH']_L$ (where [ ] stands for electrochemical activity, and R and L for right- and left-hand phases respectively). The ratio of the electrochemical activity [ATP]/[ADP] (including all ionic forms) can be raised, consequently, and the ATPase activity can be reversed to give an ADP phosphokinase activity proportional to the lowering of $[H_2O]_c$, in accordance with the mass-action law for hydrolysis equilibrium, written to include the elements of water as follows:

$$\frac{[\mathrm{ATP}]}{[\mathrm{ADP}]} = \frac{[\mathrm{P}]}{K_1[\mathrm{H_2O}]_c} \tag{1}$$

The electrochemical activity of a component in a certain place defines absolutely the escaping tendency of the particles of the component due both to the chemical and to the electrical pressure to which the particles are subject in that place at equilibrium. Since

$$K_2 = \frac{[\mathrm{OH}']_L \times [\mathrm{H^+}]_L}{[\mathrm{H_2O}]_L} \tag{2}$$

and K_2 is independent of the medium because we are using electrochemical activities, we can describe the electrochemical activity of the water at the active centre of the 'ATPase' system as follows:

$$[H_2O]_c = [H_2O]_{aq.} \times \frac{[H^+]_R}{[H^+]_L} \tag{3}$$

where $[H_2O]_{aq.}$ stands for the electrochemical activity of water in the aqueous physiological media of phases L or R, and is equivalent to about 55.5 M water.

By the definition of the electrochemical activity:

$$\frac{[H^+]_L}{[H^+]_R} = 10^{pH_{R-L}} \times 10^{(mV._{L-R})\frac{F}{2303RT}} \tag{4}$$

where pH_{R-L} is the pH of phase R minus that of phase L; $mV._{L-R}$ is the membrane potential in millivolts, positive in phase L; R is the gas constant; F is the faraday; and the factor (F/2303 RT) is approximately 1/60. It can be seen from equation (4) that the ratio $[H^+]_L/[H^+]_R$ is multiplied by a factor of 10 for each pH unit more negative on the left, relative to the right, and for each 60 mV. membrane potential, positive on the left. Equations 1 and 3 show that the ratio [ATP]/[ADP] at equilibrium is determined by $[H^+]_L/[H^+]_R$ as follows:

$$\frac{[ATP]}{[ADP]} = \frac{[P]}{K_1[H_2O]_{aq.}} \times \frac{[H^+]_L}{[H^+]_R} \tag{5a}$$

When the right-hand phase (representing the cytoplasm) is the region of the zero or reference potential, the electrochemical activity ratio of total ATP to ADP will be nearly the same as the corresponding concentration ratio, and [P] will correspond approximately to the inorganic phosphate concentration. Thermodynamic data[22] show that at pH 7 and at physiological temperatures the 'hydrolysis constant' as usually defined or the product $K_1[H_2O]_{aq.}$ is approximately 10^5; and when [P] is at the physiological level of 10^{-2} M, equation (5a) can be written:

$$\frac{[ATP]}{[ADP]} \simeq 10^{-7} \frac{[H^+]_L}{[H^+]_R} \tag{5b}$$

Thus, the [ATP]/[ADP] equilibrium can be poised centrally through the anisotropic 'ATPase' by making the ratio $[H^+]_L/[H^+]_R$ about 10^7. Equation 3 shows that this could be done, for example, by poising the left side 2 pH units below and 300 mV. above the right side. Such is the basic thermodynamic conception of the mechanism of reversal of the 'ATPase' activity. In kinetic terms, the reversal of the 'ATPase' activity can be understood by regarding the electrochemical activity gradient of hydrogen and hydroxyl ions across the active centre region of the 'ATPase' as the cause of the donation of the phosphorylium ion to the negative acceptor atom (—R′) by the simultaneous withdrawal of an OH′ ion from the inorganic phosphate down a steep gradient to the left and of an H^+ ion from —RH to the right to ionize —RH or to prevent H^+ from competing with phosphorylium ion for the acceptor —R′. Dehydration is accomplished by using the high $[H^+]$ region as a sink for OH′ and the high [OH′] region as a sink for H^+. It should be understood that the phosphorylation of ADP can be strictly described as dehydration only when the standard state is less than pH 6 (see Lipmann[21]). Since, however, the hydrogen atom of the terminal hydroxyl group of ADP, involved in phosphorylation, is dissociated reversibly as an H^+ ion, when the standard state is taken as pH

7, the phosphorylation process is mainly that of dehydroxylation. From the kinetic point of view, the fundamental processes involved are dehydroxylation + deprotonation, or dehydroxylation. It is relevant to note that phosphorylation is not directly caused by raising $[H^+]$, but is due to depression of $[OH']$. For this reason, the hydrogen ion depicted in phase R of Fig. 1 (and correspondingly in Figs. 2 and 3) is shown as equilibrating with the active centre region of the 'ATPase' system, but not as being withdrawn stoichiometrically in relation to the withdrawal of OH'.

The second basic feature of the chemi-osmotic coupling conception is the electron and hydrogen translocation system, which is assumed to create the gradient of electrochemical activity of H^+ and OH'. Unlike the conception of the anisotropic 'ATPase', the idea of the anisotropic *o/r* system is not new, but stems from the work of Lund[23], and Stiehler and Flexner[24], and was first stated explicitly by Lundegårdh[25] more than 20 years ago. Lundegårdh's idea was more exactly defined in relation to ion transport by Davies and Ogston[26], and was elaborated by Conway[27]. It has been excellently reviewed by Robertson[28].

Figure 2 illustrates how the electron transfer can affect the ratio $[H^+]_L/[H^+]_R$. The electron translocation and 'ATPase' systems are depicted as being placed in opposition in a charge-impermeable membrane. The hydrogen ions generated on the left and the hydroxyl ions generated on the right by the electron translocation system dehydrate ADP and inorganic phosphate (now simply denoted by P) to form ATP by withdrawing hydroxyl ions to the left and hydrogen ions to the right through specific translocation paths in the 'ATPase' system as described here. Conversely, of course, the effect of the back pressure of ATP hydrolysis is to force the *o/r* system towards reduction on the left (inside the organelle). Note that the stoichiometry is 2 ATP per O, as in succinate oxidation by mitochondria. As shown in this simple diagram, the dehydrating force driving the phosphokinetic activity of the 'ATPase' system would be due largely to the chemical potential differential of the H^+ and OH' ions across the membrane, which would have to show as a *p*H difference of some 7 units across the membrane, acid on the left when the [ATP]/[ADP] equilibrium was poised centrally (see equations (5*b*) and (4)). We assume,

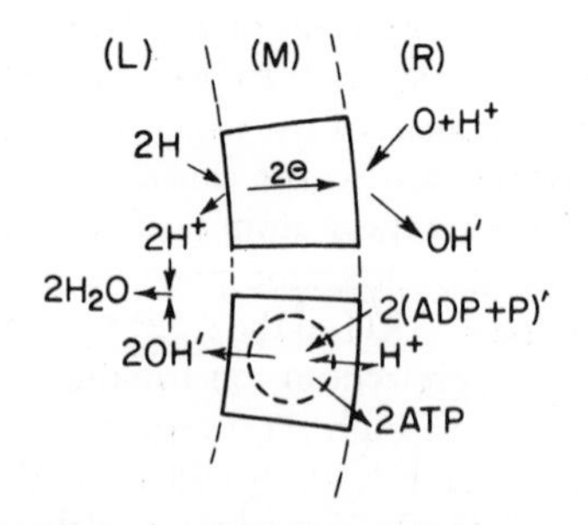

Fig. 2. Electron transport system (above) and reversible 'ATPase' system (below) chemiosmotically coupled in a charge-impermeable membrane (*M*) enclosing aqueous phase *L* in aqueous phase *R*.

however, that exchange diffusion carriers, as defined by Ussing[29], are present in the membrane and that they will allow strictly coupled one-to-one exchange of H^+ against K^+ or of OH′ against Cl′, for example. The *p*H differential would thus tend to be reduced to a relatively small figure and would be equivalently replaced by a membrane potential as described by equation (4).

When the oxido-reduction and phosphorylation systems are in chemi-osmotic equilibrium, one ATP molecule will be produced per electron translocated across the membrane. The relationship between the *o/r* potential span (ΔE) of the electron and hydrogen translocation system and the poise of the [ATP]/[ADP] equilibrium will be given as follows:

$$\frac{[\mathrm{ATP}]}{[\mathrm{ADP}]} = \frac{[\mathrm{P}]}{K_1[\mathrm{H_2O}]_{\mathrm{aq.}}} \times 10^{\frac{\Delta E.F}{2303RT}} \tag{6a}$$

or approximately:

$$\frac{[\mathrm{ATP}]}{[\mathrm{ADP}]} = \frac{[\mathrm{P}]}{10^5} \times 10^{\frac{\Delta E}{60}} \tag{6b}$$

ΔE being in millivolts. It should be understood that ΔE, as defined here, is equivalent to the free-energy change in the electron-translocating system per electron translocated. Assuming that ATP synthesis were occurring when the [ATP]/[ADP] ratio was poised centrally at *p*H 7 and in the presence of 10^{-2} *M* inorganic phosphate, the *o/r* span, ΔE, would have to be about 420 mV. This, of course, being the equilibrium potential, represents the minimum *o/r* span of the electron and hydrogen translocation system required to drive ATP synthesis under the conditions specified above. The span between the succinate —fumarate couple and oxygen at 76 mm. mercury pressure is about 750 mV. —well above the minimum ΔE of 420 mV.

In practice, the mitochondrial or chloroplast membrane across which the chemi-osmotic coupling may be organized would allow a certain amount of ion leakage, and the translocation paths for H^+ and OH′, connecting the internal and external phases (*L* and *R*) with the active centre region of the 'ATPase', would not be expected to have absolute specificity for H^+ and OH′ respectively. Consequently, equations (6*a*) and (6*b*) would represent the practical state of affairs in the most tightly coupled systems. On 'loosening' the membrane system, or if 'uncoupling' were effected by catalysing the equilibration of H^+ and charge across the lipid of the membrane with reagents like dinitrophenol or salicylate, the relationship between the poise of the [ATP]/[ADP] ratio and ΔE in the steady state would be described by the inequality:

$$\frac{[\mathrm{P}]}{K_1[\mathrm{H_2O}]_{\mathrm{aq.}}} \leqslant \frac{[\mathrm{ATP}]}{[\mathrm{ADP}]} \leqslant \frac{[\mathrm{P}]}{K_1[\mathrm{H_2O}]_{\mathrm{aq.}}} \times 10^{\frac{\Delta E.F}{2303RT}} \tag{6c}$$

The outer terms of equation (6*c*) represent the extreme values of the [ATP]/[ADP] ratio from complete uncoupling on the left to complete coupling on the right.

Figure 3 shows, in principle, a rather fuller description of oxidative phosphorylation, in which I have included the *o/r* components *FP* and *Q*, tentatively identified with flavoprotein and quinol-quinone systems respectively. The main aim is to illustrate how a stoichiometry of 3 ATP per O can readily be obtained for substrate (SH_2) oxidation through DPN, by the obligatory trans-

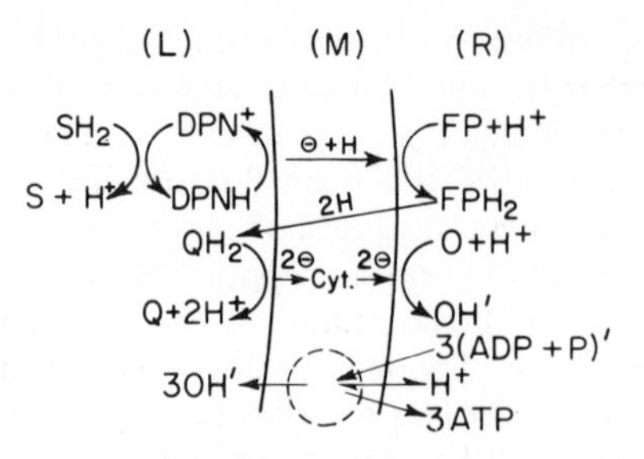

Fig. 3. Diagram of chemi-osmotic system for coupling phosphorylation to the oxidation of substrate (SH_2) through DPN, *FP* (tentatively identified with flavoprotein), *Q* (tentatively identified with a quinone) and the cytochromes (*Cyt.*). The other conventions are as in Figs. 1 and 2.

port of one (net) hydrogen atom inwards per O, owing to the spatial arrangement of the electron and hydrogen transfer chain and the zero, one, or two hydrogen-transfer characteristics of the carriers involved. In this system, the span of both parts of the *o*/*r* chain across the membrane would have to be poised against the same ratio $[H^+]_L/[H^+]_R$ at equilibrium, according to equation (5*a*). Using the conventions of Dixon[2], the $-\Delta F$ value for the DPN/DPNH couple at about *p*H 5 would be some 3,000 cal. and the $-\Delta F$ for *FP* (corresponding to an E_0' (*p*H 7) of —60 mV.) (ref. 30) would be about 17,000 cal., giving a span of 14,000 cal.; equivalent to a ΔE of 600 mV. Assuming the $-\Delta F$ of the *Q* system to be about 24,000 cal., corresponding to an E_0' (*p*H 7) of + 100 mV., as in the ubiquinone system[(31)], the span from the *Q* system to oxygen at 76 mm. mercury pressure would be equivalent to a ΔE of about 750 — 100 = 650 mV. The tendency of the two *o*/*r* values to drift together and the exact magnitude of the composite *o*/*r* potential would, of course, depend on many factors that it would be premature to consider here. It will suffice to point out at present that the value of 600–650 mV. for ΔE is appropriately above the required minimum of 420 mV. and that the proposed system is thus in accord with the thermodynamic facts.

The above basic chemi-osmotic conception can be applied to photosynthetic phosphorylation with the difference that the electron and hydrogen translocation are seen as being driven, not by the affinity of oxygen for the hydrogen atoms and electrons, but by the energy of the absorbed photons, according to the type of mechanism described by Calvin[(9b)]. It can readily be shown that in a chemi-osmotically coupled system for non-cyclic photophosphorylation, the photon-activated movement of 2 electrons and 2 hydrogen atoms outwards through the membrane of the grana would produce one O_2 and 2 ATP molecules. Similarly, in non-cyclic photophosphorylation, the skew of $[H^+]_L/[H^+]_R$, and thus the synthesis of ATP, could be caused by the photon-activated passage of equal numbers of hydrogen atoms and electrons in opposite directions across the membrane of the grana or chromatophores.

The facts that were listed at the beginning of this article as being difficult to reconcile with the orthodox chemical conception of the mechanism of coupling phosphorylation to electron and hydrogen transfer can now be reconsidered in relation to the chemi-osmotic coupling hypothesis: (*a*) The elusive character of the 'energy-rich' intermediates of the orthodox chemical coupling hypothesis would be explained by the fact that these intermediates do not exist. (*b*) According to the chemi-osmotic coupling hypothesis, the differential of the electrochemical activity of the hydrogen and hydroxyl ions across the membrane, generated by electron transport, causes the specific translocation of hydroxyl and hydrogen ions from the active centre of the so-called ATPase system, thus effectively dehydrating ADP + P. The charge-impermeable membrane would therefore be an absolute requirement for tight coupling. (*c*) Coupling would be expected to vary with the extent of leakiness or strain in the membrane, determined, of course, by the osmotic and electrical stress. (*d*) The internal components of mitochondria such as DPN would tend to be reduced by the high electrochemical activity of H^+ caused by hydrolysis of external ATP, which would withdraw OH' ions from the inside in competition with the electron-transport system. This effect would be accentuated by oxidation of succinate (which would raise the internal value of $[H^+]$) but not, of course, by substrates such as acetoacetate that directly oxidize DPNH. (*e*) Uncoupling would be caused by lipid-soluble reagents, such as DNP, salicylate, azide, and ammonia, catalysing equilibration of H^+ or OH' and charge across the membrane. (*f*) According to the chemi-osmotic type of hypothesis, the coupling of phosphorylation to electron and hydrogen translocation would cause considerable electrical and mechanical stress in the membrane across which coupling was effected. Complex swelling and shrinkage effects would therefore be expected to accompany the activity of the system.

It is evident that the basic features of the chemi-osmotic coupling conception described here and elsewhere[20] are in accord with much of the circumstantial evidence at present available from studies of oxidative and photosynthetic phosphorylation. This simple hypothesis also has the merit that it represents the result of carrying to its logical conclusion the present trend towards recognizing the equivalent status of supramolecular and molecular features in the channelling of chemical processes in living organisms[18]. Further experimental support for the chemi-osmotic coupling conception may best be sought by attempting to characterize separately each of the three hypothetical basic elements of which the system is thought to be built: (1) the anisotropic 'ATPase' system which I have defined above; (2) the anisotropic *o*/*r* system of the type originally defined by Lundegårdh; (3) the specific charge-impermeable membrane in which the systems 1 and 2 are supposed to be orientated in opposition. Work along these three lines is proceeding in my laboratory.

In the exact sciences, cause and effect are no more than events linked in sequence. Biochemists now generally accept the idea that metabolism is the cause of membrane transport. The underlying thesis of the hypothesis put forward here is that if the processes that we call metabolism and transport represent events in a sequence, not only can metabolism be the cause of transport, but also transport can be the cause of metabolism. Thus, we might be inclined to recognize that transport and metabolism, as usually understood by biochemists, may be conceived advantageously as different aspects of one and the same process of vectorial metabolism[20a, 32].

I am indebted to the Nuffield Foundation for grants in aid of this work.

It is also a pleasure to thank Dr. G. D. Greville, Prof. D. Keilin, Dr. R. Hill, Dr. Jennifer Moyle, Sir Rudolph Peters, and other colleagues for very helpful discussion and criticism.

References

1. Lipmann, F., *Adv. Enzymol.,* **1**, 99 (1941); *Currents in Biochemical Research,* edit. by Green, D. E., 137 (Intersci. Pub., Inc., New York, 1946).
2. Dixon, M., *Multienzyme Systems* (Cambridge Univ. Press, 1949).
3. Slater, E. C., *Nature,* **172**, 975 (1953). Chance, B., and Williams, G. R., *Adv. Enzymol.,* **17**, 65 (1956). Myers, D. K., and Slater, E. C., *Nature,* **179**, 363 (1957). Dawkins, M. J. R., Judah, J. D., and Rees, K. R., *ibid.,* **182**, 875 (1958).
4. Slater, E. C., and Hülsmann, W. C.; Chance, B.; Lehninger, A. L., Wadkins, C. L., and Remmert, LeM. F., in *Ciba Found. Symp. Regulation Metabolism,* edit. by Wolstenholme, G. E. W., and O'Connor, C. M., 58, 91 and 130, and associated discussion (Churchill, Ltd., London, 1959).
5. Arnon, D. I., *Nature,* **184**, 10 (1959). Hill, R., and Bendall, F., *ibid.,* **186**, 136 (1960).
6. Slater, E. C., in *Biological Structure and Function,* First IUB/IUBS Joint Symp., Stockholm, September 1960, edit. by Goodwin, T. W. (Academic Press, Inc., New York, in the press).
7. Slater, E. C., *Symp. Soc. Exp. Biol.,* **10**, 110 (1957).
8. Zeigler, D. M., Linnane, A. W., and Green, D. E., *Biochim. Biophys. Acta,* **28**, 524 (1958).

9a. Lehninger, A. L., in *Biophysical Science: A Study Program,* edit. by Oncley, J. L., *et al.,* 136 (John Wiley & Sons, Inc., New York, 1959). (b) Calvin, M., *ibid.,* 147. (c) Vatter, A. E., and Wolfe, R. S., *J. Bacteriol.,* **75**, 480 (1958).

10. Cooper, C., and Lehninger, A. L., *J. Biol. Chem.,* **219**, 489 (1956).
11. Lehninger, A. L., *et al., Science,* **128**, 450 (1958).
12. Lehninger, A. L. (personal communication, 1960).
13. Hoch, F. L., and Lipmann, F., *Proc. U.S. Nat. Acad. Sci.,* **40**, 909 (1954). Lipmann, F., in *Enzymes: Units of Biological Structure and Function,* edit. by Gaebler, O. H., 444 (Academic Press, Inc., New York, 1956).
14. Chance, B., *Nature,* **189**, 719 (1961).
15. Krebs, H. A., Hopkins Memorial Lecture, March 1961, *Biochem. J.* (in the press).
16. Ernster, L., and Lindberg, O., *Ann. Rev. Physiol.,* **20**, 13 (1958). Beechey, R. B., and Holton, F. A., *Biochem. J.,* **73**, 29 (1959). Lehninger, A. L., *Ann. N.Y. Acad. Sci.,* **86**, 484 (1960). Emmelot, P., *et al., Nature,* **186**, 556 (1960). Emmelot, P., *ibid.,* **188**, 1197 (1960).
17. Green, D. E., *Adv. Enzymol.,* **21**, 73 (1959). Lehninger, A. L., in *Biological Structure and Function,* First IUB/IUBS Joint Symp., Stockholm, September 1960, edit. by Goodwin, T. W. (Academic Press, Inc., New York, in the press).
18. Green, D. E., *Symp. Soc. Exp. Biol.,* **10**, 30 (1957).

19a. Mitchell, P., *Symp. Soc. Exp. Biol.,* **8**, 254 (1954); (b) *Nature,* **180**, 134 (1957); (c) in *Structure and Function of Subcellular Components,* Sixteenth Symp. Biochem. Soc., February 1957, edit. by Crook, E. M., 73 (Cambridge Univ. Press, 1959); (d) Mitchell, P., and Moyle, J., *Nature,* **182**, 372 (1958); (e) *Proc. Roy. Phys. Soc., Edinburgh,* **27**, 61 (1958).

20a. Mitchell, P., in *Biological Structure and Function,* First IUB/IUBS Joint Symp., Stockholm, September 1960, edit. by Goodwin, T. W. (Academic Press, Inc., New York, in the press); (b) *Biochem. J.,* **79**, 23 P (1961).

21. Lipmann, F., in *Molecular Biology,* edit. by Nachmansohn, D., 37 (Academic Press, Inc., New York, 1960).
22. Atkinson, M. R., Johnson, E., and Morton, R. K., *Nature,* **184**, 1925 (1959).
23. Lund, E. J., *J. Exp. Zool.,* **51**, 327 (1928).

24. Stiehler, R. D., and Flexner, L. B., *J. Biol. Chem.,* **126**, 603 (1938).
25. Lundegårdh, H., *Lantbr. Hogsk. Ann.,* **8**, 233 (1940).
26. Davies, R. E., and Ogston, A. G., *Biochem. J.,* **46**, 324 (1950).
27. Conway, E. J., *Internat. Rev. Cytol.,* **2**, 419 (1953).
28. Robertson, R. N., *Biol. Rev.,* **35**, 231 (1960).
29. Ussing, H. H., *Nature,* **160**, 262 (1947); *Physiol. Rev.,* **29**, 127 (1949).
30. Kuhn, R., and Boulanger, P., *Ber.,* **69**, 1557 (1936).
31. Morton, R. A., *Nature,* **182**, 1764 (1958).
32. Mitchell, P., in *Membrane Transport and Metabolism,* Symposium, Prague, August 1960, edit. by Kleinzeller, A., and Kotyk, A. (Academic Press, Inc., New York, in the press).

ATP FORMATION CAUSED BY ACID-BASE TRANSITION OF SPINACH CHLOROPLASTS*†

André Tridon Jagendorf and Ernest Uribe‡
McCollum-Pratt Institute and Biology Department
The Johns Hopkins University, Baltimore, Maryland
(Communicated by W. D. McElroy, November 3, 1965)

Previous work[1-5] has shown the existence of a high-energy condition of isolated chloroplasts, caused by illumination in the absence of phosphate or ADP. This state is inferred from the ability to form ATP in the postillumination darkness. Accompanying the condition is an apparent uptake of hydrogen ions[6,7] together with an excretion of Mg^{++} or other cations[8]. A feasible interpretation, stemming from the "chemi-osmotic" hypothesis for the mechanism of phosphorylation in double membrane containing organelles[9,10], is that illumination causes the uptake of hydrogen ions into the inner space of the grana disk double membranes. The resulting inequality in hydrogen ion electrochemical activity across the membrane is postulated to be, in itself, the high-energy condition, able to drive the formation of ATP.

If this interpretation is correct, then the same high-energy condition should be formed artificially, entirely in the dark and without electron transport, by loading the inner space of the grana disk membranes with protons. If this could be accomplished by placing chloroplasts in an acid medium, when first returned to pH 8 they might be expected to make some ATP due to the pH gradient across the membranes. The operational formation of ATP in this manner, entirely in the dark, was discovered and noted briefly earlier[4]. The present

*A. T. Jagendorf and E. Uribe, "ATP Formation caused by Acid-Base Transition of Spinach Chloroplasts," *Proc. Natl. Acad. Sci.*, **55**, 170–177 (1966). Reprinted with the permission of the authors and of the National Academy of Sciences.

Publication 455 of the McCollum-Pratt Institute. This work was supported in part by grant GM-03923 from the National Institutes of Health, and by a Kettering Research Award.

†Abbreviations used in this paper include CMU for p-chlorophenyl-1, 1-dimethylurea; DCMU for dichlorophenyl-1, 1-dimethylurea.

‡National Science Foundation postdoctoral fellow.

paper represents an extension and further exploration of this phenomenon of "acid-bath" dark phosphorylation by chloroplasts.

Materials and Methods

Chloroplasts were prepared from market spinach as described previously[2]. After the first centrifugation, they were resuspended in: 1.0 *M* sucrose, 0.01 *M* NaCl, and 5% dimethylsulfoxide, then frozen and stored under liquid nitrogen. (The use of dimethylsulfoxide has been found to aid in the preservation of photophosphorylation activity during freezing and thawing.) The chloroplasts were thawed before use and diluted to a concentration of 0.133 mg of chlorophyll per ml with: 0.80 *M* sucrose, 0.02 Tris pH 8.0 (at 0°C), 0.01 *M* NaCl, then centrifuged. They were then resuspended in 0.01 *M* NaCl alone, to the same concentration as in the previous step, and allowed to stand 20 min at 0°C for breakage. The broken chloroplasts were collected by centrifuging at 10,000 × *g* for 10 min and finally resuspended in 0.01 *M* NaCl at 0.50 mg of chlorophyll per ml. These broken and washed chloroplasts were used routinely, although occasional experiments gave identical results when suspension in sucrosedimethylsulfoxide and freezing were omitted. All preparative steps were carried out at or close to 0°C.

The routine reaction procedure consisted of two successive stages, both carried out at 0°C: (*a*) exposure of chloroplasts to an acid pH, then (*b*) simultaneously raising the pH and adding ADP and phosphate to permit phosphorylation to occur. In the first stage chloroplasts containing 0.25 mg of chlorophyll were injected into 0.40 ml of buffer at pH 4.0 (or other as noted) containing 27 mμmoles of DCMU. The total volume at this point was 0.9 ml. After 60 sec, the acidified chloroplasts were taken up in a syringe and injected into a second test tube containing in 0.9 ml volume: Tris, 100 μmoles; ADP, 0.2 μmole; inorganic phosphate, 2.0 μmoles; $MgCl_2$, 5 μmoles; carrier-free radioactive phosphate with 5×10^5 counts per min, and enough NaOH to neutralize the buffer used in the acid stage. All reaction mixture components (exclusive of, and prior to addition of the NaOH) were at pH 8.0 or some other as indicated. Since the NaOH was neutralized by addition of the acid chloroplast mixture, the final pH was that of the buffer in the second stage of the reaction. This final pH was checked directly in each experiment. The phosphorylation reaction (in final volume of 1.8 ml) was allowed to proceed for 15 sec, then stopped by the addition of 0.2 ml of 20% trichloroacetic acid. After centrifugation of denatured chloroplasts, the esterified radioactive phosphate was determined by the method of Avron[11], or occasionally adsorbed to charcoal, washed, and counted[2]. Chlorophyll was determined by the method of Arnon[12]. In measuring ATP formation by the firefly luciferase assay[13], three-times-recrystallized enzyme was used with reduced firefly luciferin as substrate.

Results

In the first experiments on this phenomenon, great precautions were taken to exclude light; these included a 1–2-hr preincubation of the chloroplasts in complete darkness before acidification. Subsequent work showed, however, that adding the following inhibitors had absolutely no effect on the yields of ATP: 30 μM CMU, 10 or 30 μM DCMU, 50 μM o-phenanthroline, or 10 μM simazine. After this, DCMU was added routinely, at a final concentration of 3×10^{-5} *M*. Under these conditions (strong inhibitor of noncyclic electron transport, and no redox dye present), the chloroplasts were found to be completely inert to either room light or 5000 ft-c of white light; this was true even in the second stage with ADP and P^{32} present. Because of this, it has been possible to perform the experiments routinely in room light in an ice bath.

Formation of ATP due to making chloroplasts first acid, then alkaline, is shown in Table 1. There is substantial agreement between ATP formation as measured by luciferase, and by the incorporation of P^{32} followed by isobutanol extraction of the inorganic phosphate. The 20 per cent discrepancy between the two methods is under further investigation. It can be seen that ATP formation is absolutely dependent on achieving a pH below 7 in the acid stage, and on added phosphate and ADP in the alkaline stage. The 25 per cent yield in the absence of added Mg^{++} is presumably due to internal magnesium ions. Further identification of the product as ATP comes from the fact that all of the incorporated counts can be adsorbed on charcoal, even if the charcoal has been previously coated with stearic acid(14). Once adsorbed, the counts are removed in boiling 1 *N* HCl at a rate identical to the hydrolysis of known ATP adsorbed under similar conditions.

The amount of ATP that is formed is highly dependent on the nature of the acid used to bring the chloroplasts down to pH 4. Figure 1 illustrates the yields obtained when using only HCl to adjust the pH, or when using various concentrations of glutamate, of succinate, or of phthalate buffers at pH 4. HCl in this experiment gave a yield of 17 mμmoles per mg chlorophyll, and in other experiments varied between 8 and 18. The same yield is obtained from 3 mM of glutamate. When checking the effects of other buffering anions, therefore, they were routinely added over a background of either 3 or 5 mM glutamate to control the pH. Phthalate, the first additional buffer tried, increases the yield from 17 up to 60 mμmoles per mg chlorophyll, but an excess is inhibitory. Inhibition depends on the pH; the same concentrations at pH 4.5 do not inhibit, and the concentration curve is less sharp. Succinate provides higher yields of ATP, with no sign of inhibition up to 15 mM. The largest amount of ATP

Table 1

ATP FORMATION DUE TO ACID-BASE TRANSITION

Reaction mixture	Acid pH	Luciferase Assay: Total	Luciferase Assay: Net†	P-Molybdate Extraction: Cpm	P-Molybdate Extraction: Total	P-Molybdate Extraction: Net
Complete*	3.8	141‡	129‡	2200	166‡	163‡
"	7.0	12	—	45	3	—
—PO4	3.8	12	—	—	—	—
—ADP	3.8	4	—	47	3	—
—Mg	3.8	60	48	630	48	45
—Chloroplasts	3.8	7	—	50	3	—

*Complete, in this case, indicates an acid stage at pH 3.8 with 10 mM succinate and 3×10^{-5} *M* DCMU; and phosphorylation stage at pH 8.0 with components as indicated in *Methods* section.

†Net refers to the yield of ATP after subtracting the control of chloroplasts brought only to pH 7 before the phosphorylation stage.

‡mμmoles ATP per mg chlorophyll.

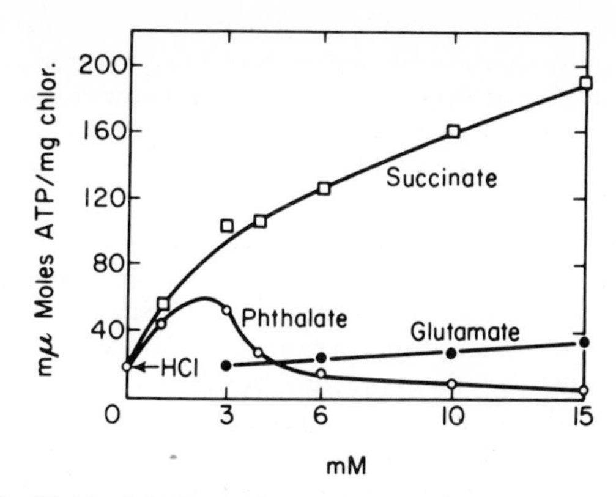

Fig. 1. Yield of ATP as a function of acid concentration. Acid stage pH was 4.0, caused by addition of either 0.1 ml of 5 mM HCl, or by glutamate at the concentrations shown. With succinate or phthalate, 3 mM of glutamate was present in all reactions to achieve primary control of pH, and the other organic acid effects are over and above those due to pH 4 in itself.

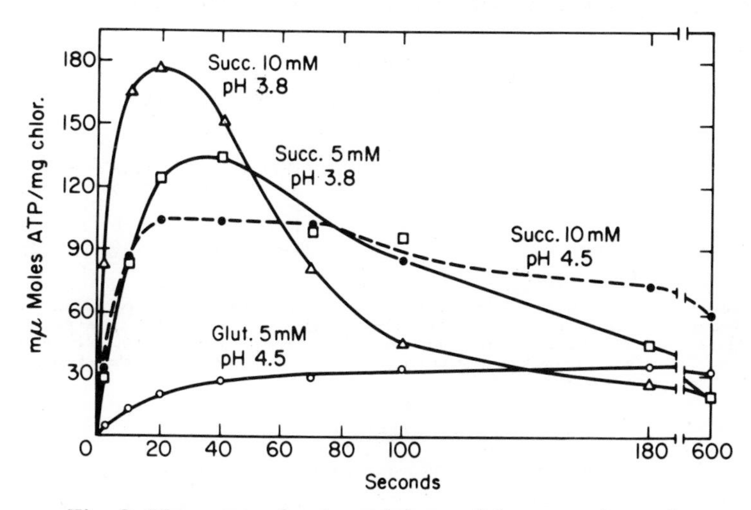

Fig. 2. Time course for rise and decay of the energetic condition at pH 3.8 or 4.5. Glutamic or succinic acids used at the concentrations shown; conditions as in *Methods* section.

formed in the experiment of Figure 1 was 189 mμmoles per mg chlorophyll; in other experiments yields up to 240 (i.e., 43 μmoles per gram of protein) have been seen routinely, and higher yields very occasionally.

Time courses for the rise of the energetic conditions in acid are shown in Figure 2. Glutamate gives a rise to 85 per cent of the maximum in about 40 sec, then a very slow increase to slightly higher yields by 3 min. Succinate addition produces a time course which reaches a maximum by 20 sec. With succinate, the half time for the rise is 8 sec for 3 mM and 2 sec for 10 mM at pH 3.8, and

4 sec for 10 mM at pH 4.5. Thus the effect of higher succinate or the lower pH is not to prolong the time course, but to give a faster initial rate and a higher plateau. The succinate curves in this experiment show unusual instability; in most experiments the high-energy condition was stable for at least 60 sec at pH 3.8. It is clear that any instability of the high-energy potential is greater as the pH drops below 4.5.

Time courses at pH 8 are shown in Table 2. By omitting the ADP and P^{32} for varying lengths of time the rate of decay can be estimated; at pH 8 the decay reaction appears to be complete in about 6 sec and seems to have a half life of between 1 and 2 sec. Phosphorylation of ADP is complete in about 4 sec.

Besides succinate, the following acids have shown good activity at 1–3 mM: o-phthalic, p-phthalic, barbituric, fumaric, dimethylglutaric, lactic, and glycolic. Dicarboxylic acids with hydroxyl groups have been either weakly active or inhibitory, and all monofunctional carboxylic acids tested have been strongly inhibitory. Further studies of the structural requirements for induction of ATP formation, or for inhibition, are currently in progress.

Reversibility of acid induction is shown in Table 3. It is possible for one batch of chloroplasts to become charged at pH 4, discharged at pH 6.5, and then almost fully recharged when returned to pH 4 for the second time. This whole sequence is accomplished before addition of ADP and phosphate.

Table 2

KINETICS OF DECAY OF ACID-INDUCED X_E AT pH 8, AND OF PHOSPHORYLATION IN THE ADP STAGE

Reaction	Yield of ATP at Time (in sec)*					
	0	2	4	6	30	60
pH 8 Decay	41	7	1.7	1.1	0.4	—
ADP + P → ATP	3	27	48	47	42	47

*ATP yield in mμmoles/mg chlorophyll. The acid stage contained 10 mM phthalate at pH 4.6. The alkaline decay was measured by injecting the acid chloroplasts into Tris at pH 8, then waiting the indicated number of seconds before adding ADP, Mg, and P^{32}, followed by a standard 15 sec with these reagents before adding TCA. The phosphorylation time course was determined by injecting acidified chloroplasts into the complete phosphorylation stage reaction mixture, and waiting the indicated number of seconds before stopping the reaction with TCA.

Table 3

REVERSIBILITY OF INDUCTION OF X_E BY ACID

Acid	pH in Successive Minutes			
	1	2	3	ATP
Glutamic (5 mM)	4.0	—	—	15.6*
	4.0	6.5	—	0.0
	4.0	6.5	4.0	13.7
Succinic (5 mM)	4.0	—	—	49.7
	4.0	6.5	—	2.8
	4.0	6.5	4.0	33.6

*mμmoles ATP per mg chlorophyll. The pH, initially 4.0, was adjusted up to 6.5 by adding a measured amount of Tris as the pure base, then readjusted back to 4.0 by adding a measured amount of the free acid. The phosphorylation stage components were added only after the last of the pH transitions noted in the first three columns; thus, the discharge at pH 6.5 represents instability of X_E near neutral pH. ATP formation occurred on addition of the usual phosphorylation stage ingredients at pH 8 for 15 sec, except that in this particular experiment the succinic and glutamic acids were not neutralized by extra NaOH.

Table 4 shows that ATP formation in this system can be inhibited by the known uncouplers of photosynthetic phosphorylation. Ammonia, atebrin, CCP, Triton-X-100, phenylmercuric acetate, and uncoupling by EDTA are all reasonably effective. On the other hand, a better yield was found anaerobically than aerobically (160 vs. 109 mμmoles ATP per mg chlorophyll).

The yield of ATP as a function of pH in the acid stage is shown in Figure 3, for the case of glutamic acid. Very similar curves, although with much higher yields, are obtained when using succinic acid. Note that the yield goes up as the pH drops. No distinct pH optimum has been observed because at pH 3.5 the chloroplasts aggregate severely and it is no longer possible to use the usual syringe procedures. Both the cutoff point on the alkaline side and the total yields are raised when the pH of the phosphorylation stage is higher.

Phosphorylation stage pH curves are shown in Figures 4 and 5. There is

Table 4

INHIBITION OF ATP FORMATION BY UNCOUPLERS OF PHOTOPHOSPHORYLATION

Inhibitor	Yield	% Inhibition
None	239*	—
NH_4Cl, 2 mM	61	75
Atebrin, 0.4 mM	57	76
PMA, 50 μM	59	75
CCP, 10 μM	135	43
Triton, 0.27 mg	110	54
EDTA	18	92

*mμmoles ATP per mg chlorophyll. Acid stage contained 10 mM succinate at pH 4.0, ADP stage pH was 8.3. PMA refers to phenylmercuric acetate, CCP to carbonyl cyanide m-chlorophenylhydrazone, and EDTA to ethylenediamine tetraacetic acid. The concentrations are those in the acid stage; in the phosphorylation stage they would be twice as dilute. Chloroplasts were preincubated for 5 min at 0°C with the CCP and PMA; in the case of EDTA they were resuspended in 1 mM EDTA in place of the 10 mM NaCl during their preparation, and washed free of the EDTA and coupling factor before use.

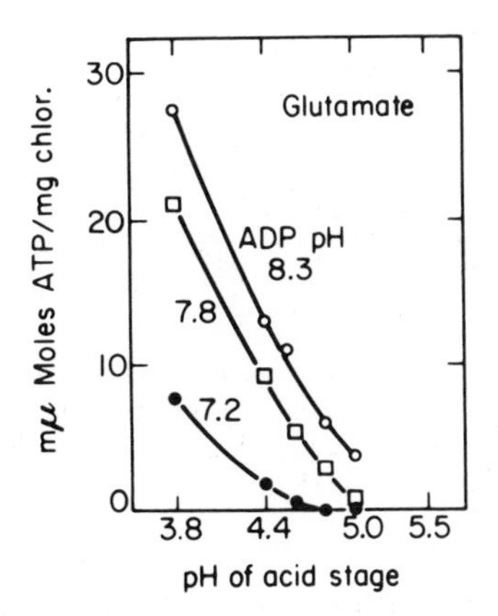

Fig. 3. Yield of ATP as a function of acid stage pH. Five mM glutamate was used in the acid stage. Similar curves, although with higher yields, were found when using succinate, between 1 and 10 mM.

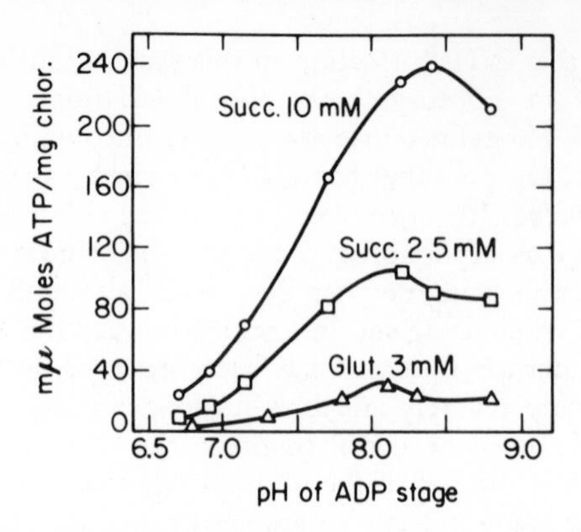

Fig. 4. Phosphorylation stage pH curves, with differing amounts of succinate (pH 4.0 in each case) in the acid stage. Each amount and pH of succinate was neutralized by an appropriate amount of NaOH, and the pH of the phosphorylation mixture was checked at 0°C.

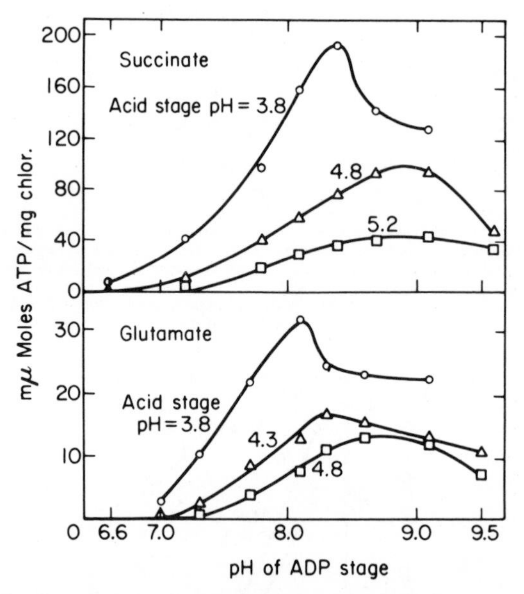

Fig. 5. Phosphorylation stage pH curves, with either succinate (top set of curves) or glutamate (bottom set) in the acid stage, at different acid stage pH's as indicated. The acids were neutralized with NaOH in the ADP mix in all cases and the final pH was checked at 0°C.

a distinct optimum, but its position shifts depending on conditions in the acid stage. With decreasing yields due to lowering the succinate concentration from 10 to 2.5 mM (Fig. 4), the optimum phosphorylation stage pH goes *down*, from 8.4 to 8.1. By contrast, if the yield is reduced by using an insufficiently acid pH in the first stage (Fig. 5), the pH optimum of the phosphorylation stage *rises* from 8.1 to 8.6 (glutamate, *lower curves*) or from 8.4 to 9.0 (succinate, *upper curves*). Not only the optimum, but also the cutoff point on the acid side, is forced up about 0.6 pH units as the acid stage pH is raised from 3.8 to 4.8. In short, formation of ATP seems to depend not only on the pH in each stage, but also to a large extent on the actual size of the *difference* in pH between the two stages. The minimal pH difference for synthesis of measurable ATP ranges from 2.2 to 2.9 units, and the optimum from 4.0 to 4.5 units.

Discussion

The experiments reported here show that ATP is formed by chloroplasts depending only on an artificial transition from acidic to basic medium. This ATP synthesis occurs certainly without the aid of light, and quite possibly without any electron transport. Nevertheless, it seems that a large part of the ordinary photophosphorylation machinery must be in operation, as shown by sensitivity to a representative group of uncouplers (Table 4). Especially significant is severe inhibition by EDTA treatment, known to remove the chloroplast "coupling factor"[15]. The uncoupler effects, the kinetics of the pH 8 decay reaction, and of the phosphorylation proper (Table 2) are virtually identical to those seen previously for the high-energy conditions, X_E, induced by light[2,16]. Light-induced X_E in turn had been closely related to a light-induced rise in pH of the medium, indicating acidification of the interior of the chloroplasts[6,7].

No effort has yet been made to find the absolute maximal yield of ATP. Occasional experiments have shown values higher than 240 mμmoles per mg of chlorophyll and, as noted with respect to the light X_E[5], the potential yield might be twice as much as the observed one. In any event it is clear that the current procedures lead to formation of at least 1 ATP per 4 chlorophylls, or 100 ATP molecules per cytochrome *f*.

A new factor in the present work is the important function of added organic acids, especially dicarboxylic acids. The basal level, caused by pH 4 in itself or by glutamate, is increased about 15-fold by including succinate at 10 mM. The function of these acids is far from clear; however, the lack of specificity (e.g., o- and p-phthalic acids, succinate, barbiturate, glycolate) makes it unlikely that they are serving as substrates for an enzymatic reaction.

The thermodynamic basis for formation of ATP in these experiments is not known. The data we have obtained can be used to assess the likelihood of a number of alternative, speculative explanations:

a. At the lower pH a covalent anhydride bond involving chloroplast membrane components might conceivably form spontaneously and reversibly (as, for instance, in the equilibrium between succinic acid and succinic anhydride). On returning to pH 8.3 this could be at a high energy level, and by a series of group transfers lead to ADP phosphorylation. Decrease of the free energy level of pyrophosphate and thioester bonds with decreasing pH was noted considerably earlier.[17]

b. There might exist two electron carriers whose redox potential relationship reverses itself on going from pH 3.8 to 8.3. ATP formation in this case would be coupled to anaerobic electron transport at the higher pH:

at pH 3.8: $A + BH_2 \longrightarrow B + AH_2$

at pH 8.3: $B + AH_2 + ADP + P \longrightarrow A + BH_2 + ATP$

However, three of our observations tend to rule out both of these possibilities, at least in their simplest form. These are: (1) the very high yields of ATP, higher than that of any known bound electron transport component. Even plastoquinone A is present at the level of only 40 moles/cytochrome *f*; and all of the fatty quinones together come to probably 80 moles/cytochrome *f*. As noted above, ATP yields are certainly 100 moles/cytochrome *f*, and may be much higher. (2) The important effect on yield of a rather nonspecific group of added organic acids is entirely inconsistent with possibility (*b*) above, and consistent with (*a*) only if the organic acid added were to participate in formation of an anhydride bond. (3) The shift in the pH curve corresponding to different acid stage pH's (Fig. 5) is not readily consistent with any model in which a static amount of a chemical or redox component is formed in the acid stage, and conversion to ATP depends only on the pH optima for the chemical reactions in the last stages of phosphorylation. Note especially that it is the most abnormal pH (3.8) which produces a phosphorylation reaction pH optimum (8.1 or 8.3) closest to that seen normally for photophosphorylation.

On the other hand, these facts are, on the whole, consistent with (*c*) the "chemiosmotic" hypothesis for the mechanism of phosphorylation, as elaborated by Peter Mitchell[9,10]. In this model, account is taken of the inner and outer regions of the grana disk membranes. When placed in acid, protons would be expected to penetrate the interior space depending on the external pH. Presumably, cations would be simultaneously displaced to preserve electrical neutrality. Added organic acids could penetrate in the undissociated form (without the need to expel cations) and provide an internal reservoir of dissociable protons, thereby accounting for the higher yields. On raising the external pH to 8 or more, a proton gradient would be present from inside to outside, and this in itself would have thermodynamic potential. The height of the potential would depend on the ratio of pH's inside and outside, and this could account for the requirement of a distinct pH difference between the acid stage and the phosphorylation reaction. Indeed, the alkaline displacement of the phosphorylation reaction pH curve due to raising the acid stage pH (Fig. 5) was one of the predictions of this model, and the experiments were performed in order to test it. The actual yield of ATP at any pH, and therefore the shape of the curve, probably reflects a balance between the pH dependence of the relevant chemical reactions, and the thermodynamic effect which is a matter of the change in pH from the previous acid stage.

Given a pH differential across the grana disk membrane, its potential might be translated into ATP formation via an anisotropic, membrane-bound, reversible ATPase as discussed by Mitchell[9,10]. Alternatively, one might imagine a pair of membrane-bound, phosphorylation-coupled electron carriers with potential sensitive to pH (see above) able to "face" alternatively the acidic inside and alkaline outside regions. By continuous movement from one side to the other, a continuing coupled electron transport might be achieved, with a final yield of ATP higher than the net amount of either carrier. Although we consider this a less likely alternative, none of our present evidence can rule it out.

Summary

Chloroplasts form a limited amount of ATP without illumination or oxygen if made first acid, then basic. The yields are greatly increased by having an appropriate organic acid present in the acid stage. Lack of specificity of the acid suggests it does not serve as a substrate for a specific enzyme. The highest yields obtained regularly have been 1 ATP for every 4 chlorophylls (i.e., 100 ATP per cytochrome *f*, or 40 μmoles ATP per gram of protein). Formation of ATP is sensitive to known uncouplers of photosynthetic phosphorylation, and the kinetics of either decay of the intermediate or of phosphorylation at pH 8 are the same as those for the high-energy condition induced by illumination at pH 6. The yield of ATP depends in part on the actual pH differential between the two experimental stages. The data are suggested to be consistent with a model in which the high-energy condition consists of a pH gradient across the grana disk membranes.

The devoted assistance of Marie Smith is gratefully acknowledged, as well as help from Miss Rosemary Kayser with the luciferase assays.

References

1. Shen, Y. K., and G. M. Shen, *Sci. Sinica (Peking),* **11**, 1097 (1962).
2. Hind, G., and A. T. Jagendorf, these PROCEEDINGS, **49**, 715–722 (1963).
3. Hind, G., and A. T. Jagendorf, *Z. Naturforsch.,* **18b**, 689–694 (1963).
4. Hind, G., and A. T. Jagendorf, *J. Biol. Chem.,* **240**, 3195–3201 (1965).
5. Jagendorf, A. T., and G. Hind, *Biochem. Biophys. Res. Commun.,* **18**, 702–709 (1965).
6. Neumann, J., and A. T. Jagendorf, *Arch. Biochem. Biophys.,* **107**, 109–119 (1964).
7. Jagendorf, A. T., and J. Neumann, *J. Biol. Chem.,* **240**, 3210–3214 (1965).
8. Dilley, R. A., and L. P. Vernon, *Arch. Biochem. Biophys.,* **111**, 365–375 (1965).
9. Mitchell, P., *Nature,* **191**, 144–148 (1964).
10. Mitchell, P., in *Regulation of Metabolic Processes in Mitochondria,* ed. E. Quaglieriello, E. C. Slater, S. Papa, and J. M. Tager (Amsterdam: Elsevier Press, 1966), in press.
11. Avron, M., *Biochim. Biophys. Acta,* **40**, 257–271 (1960).
12. Arnon, D. I., *Plant Physiol.,* **24**, 1–15 (1949).
13. McElroy, W. D., in *Methods in Enzymology,* ed. S. P. Colowick and N. Kaplan (New York: Academic Press, 1955), vol. 2, pp. 851–856.
14. Alm, R. S., *Acta Chem. Scand.,* **6**, 1186–1193 (1952).
15. Avron, M., *Biochim. Biophys. Acta,* **77**, 699–702 (1963).
16. Hind, G., and A. T. Jagendorf, *J. Biol. Chem.,* **240**, 3202–3209 (1965).
17. Burton, K., and H. A. Krebs, *Biochem. J.,* **54**, 94–105 (1953).

QUENCHING OF CHLOROPLAST FLUORESCENCE BY PHOTOSYNTHETIC PHOSPHORYLATION AND ELECTRON TRANSFER*

Daniel I. Arnon, Harry Y. Tsujimoto and Berah D. McSwain

Department of Cell Physiology
University of California, Berkeley
(Read before the Academy April 28, 1965)

Fluorescence in photosynthesis represents that portion of absorbed radiant energy which is not converted into chemical energy (or heat) but is re-emitted as radiation. Many investigators have used fluorescence measurements to probe into the mechanism of photosynthesis, particularly the relation between photon capture and the primary photochemical events. The extensive and complex literature on this subject is reviewed by Rabinowitch(1).

Research centered on fluorescence of whole cells or of chlorophyll in solution has led so far to few direct, experimentally demonstrable relations between fluorescence and the primary energy conversion reactions of photosynthesis. The fraction of absorbed light which is re-emitted during photosynthesis by whole cells as fluorescence is very low, of the order of one per cent or less(1). Chlorophyll solutions in organic solvents give much higher fluorescence yields, of the order of 20–30 per cent, but have no photochemical activity that can be directly related to photosynthesis.

Among recent investigations of fluorescence in photosynthesis are investigations of fluorescence in isolated chloroplasts(2–4), the microscopic organelles which contain all of the photosynthetic pigments in green plants and which retain photosynthetic activity when properly removed from the cell. The aim of this investigation was to use fluorescence measurements on chloroplasts to

*D. I. Arnon, H. Y. Tsujimoto, and B. D. McSwain, "Quenching of Chloroplast Fluorescence by Photosynthetic Phosphorylation and Electron Transfer," *Proc. Natl. Acad. Sci.*, **54**, 927–934 (1965). Reprinted with the permission of the authors and of the National Academy of Sciences.

This work was aided by grants from the National Institutes of Health, Office of Naval Research, and the Charles F. Kettering Foundation.

test the view that photosynthetic phosphorylation, independent of carbon assimilation, constitutes the primary energy conversion process of photosynthesis[5,6]. According to this view, photosynthetic phosphorylation should act as a quencher of chloroplast fluorescence.

This article presents evidence that fluorescence of spinach chloroplasts was quenched under experimental conditions that give rise to cyclic or noncyclic photophosphorylation, the two subdivisions of photosynthetic phosphorylation[6]. Cyclic photophosphorylation is a photochemical reaction which produces only ATP[7], whereas noncyclic photophosphorylation is a photochemical reaction that produces simultaneously and in a stoichiometric relation, oxygen, ATP, and a strong reductant, reduced ferredoxin[8,9]. Of the three products of photosynthetic phosphorylation, oxygen is set free whereas ATP and reduced ferredoxin jointly form the assimilatory power that drives carbon assimilation.

Methods

A schematic diagram of the apparatus used for fluorescence measurements is shown in Figure 1. A 1200-watt projection lamp was used to give a monochromatic beam which served as the exciting light. The beam was focused through a system of lenses and filters on the cuvette containing the sample. An interference filter was used to isolate a band of 650-mμ light (half-width, 10 mμ). The fluorescence emitted by the sample first passed through a blocking filter to remove stray light, and then entered the slit of a Bausch and Lomb 500-mm monochromator. The fluorescence was detected at the exit slit of the monochromator by an EMI 9558 B photomultiplier tube and the signal was displayed on a digital voltmeter (Fig. 1). The spectrum of the fluorescence was measured at intervals of 5 mμ, in the range from 670 to 760 mμ.

A standard lamp, calibrated by the National Bureau of Standards, was used to correct fluorescence readings for the transmission characteristics of the optical

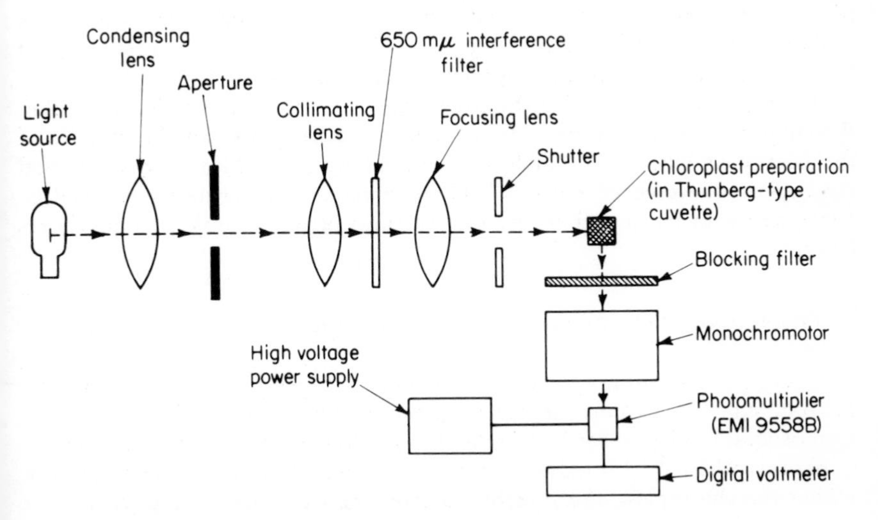

Fig. 1. Diagram of apparatus for fluorescence measurements.

measuring system and for the spectral response of the photomultiplier tube (method described by Stair *et al.*[10]) The fluorescence measurements were made at room temperature with intensities of exciting light (approximately 1.8 μeinstein/min) similar to those under which photophosphorylation was measured in some other experiments. However, the same general effects were also observed at much lower light intensities of approximately 0.005 μeinstein/min.

Thunberg-type cuvettes contained a chloroplast preparation or extracted chlorophyll (in 80% acetone). The concentration of chlorophyll used was selected to minimize errors from reabsorption of fluorescent light. A "control" fluorescence spectrum was recorded first, and then a second fluorescence spectrum was measured after tipping into the cuvette the various components placed in the sidearm. In the experiments with chloroplasts, the cuvettes were rendered anaerobic by 6 cycles of evacuating with a mechanical vacuum pump and gassing with argon. In the experiments with extracted chlorophyll, the cuvettes were made anaerobic by 1 cycle of gassing with argon. The chloroplast preparation consisted of "broken" chloroplasts (C_{1S2}) prepared by our usual method.[11]

Results and Discussion

Fluorescence of chlorophyll solutions and chloroplasts

It was deemed of interest, prior to investigating the effect of photophosphorylation on fluorescence of chloroplasts, to compare their fluorescence with the fluorescence of an equal amount of chlorophyll in an acetone solution. Figure 2 shows that with the same exciting light, the fluorescence of chlorophyll in solution was markedly greater than that of the same amount of chlorophyll remaining within the structure of chloroplasts. The fluorescence of the chlorophyll solution exhibited a broad peak at about 725 mμ and a very sharp peak at about 678 mμ. The corresponding but strikingly lower peaks of chloroplast fluorescence showed maxima around 740, 720, and 685 mμ.

Figure 2 demonstrates that the arrangement of chlorophyll molecules within the chloroplast structure provides in itself a mechanism(s) for the transfer of the energy of chlorophyll molecules excited by light. However, a residual chloroplast fluorescence remains, and in subsequent experiments we shall be concerned with its quenching by the reactions of photosynthetic phosphorylation.

Quenching by cyclic photophosphorylation

Cyclic photophosphorylation in chloroplasts depends on the addition of a catalyst, e.g., menadione or methyl phenazonium methosulfate (phenazine methosulfate) which, according to the electron flow theory of energy conversion in photosynthesis[6,12], catalyzes, in light, an internal or "cyclic" electron flow within the chloroplast and thereby liberates the free energy required for the

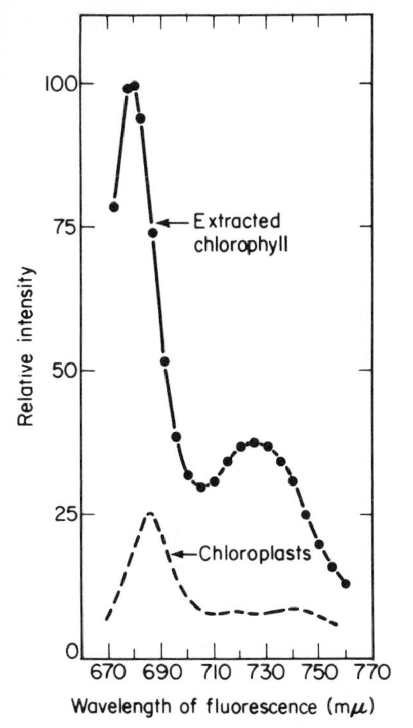

Fig. 2. Fluorescence of chlorophyll in solution and in chloroplasts. The chlorophyll solution (final volume 3.5 ml) contained 140 μg of chlorophyll extracted in 80% acetone. The chloroplast preparation (final volume 3.5 ml) included chloroplasts (equivalent to 140 μg of chlorophyll), and the following in μmoles: Tris buffer, pH 8.3, 100; $MgCl_2$, 5; and Na ascorbate, 20.

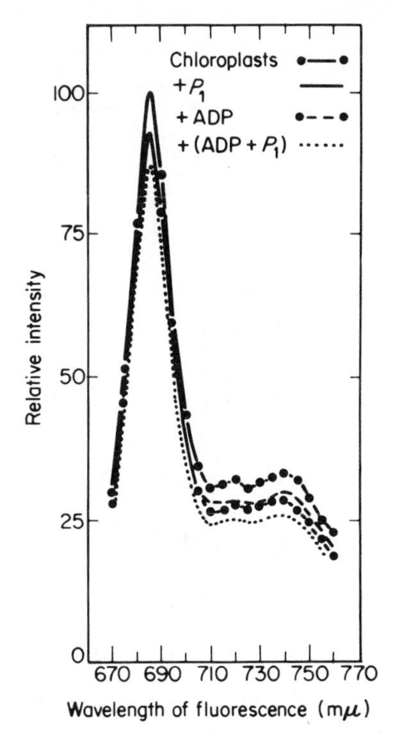

Fig. 3. Quenching of chloroplast fluorescence by ADP and inorganic phosphate. The chloroplast preparation was as given for Fig. 2. Five μmoles of ADP and 5 μmoles of K_2HPO_4 were added as indicated.

synthesis of ATP. The light-induced cyclic electron flow is envisaged as occurring from excited chlorophyll to the catalyst, from the catalyst to a chloroplast-bound electron carrier (e.g., a chloroplast cytochrome), and thence back to chlorophyll[6,12]. Recent findings point to chloroplast ferredoxin, a water-soluble, iron-containing protein, as the endogenous catalyst of cyclic photophosphorylation in chloroplasts[13]. Without the addition of ferredoxin (which is normally lost in the preparation of "broken" chloroplasts) or of a substitute catalyst, there can be no effective cyclic electron flow, and hence little ATP will be formed.

Since the degradation of the energy of a molecule excited by photon capture can occur by electron transfer to an appropriate electron acceptor molecule[14], the catalyst of cyclic photophosphorylation should, by establishing a cyclic electron flow, quench chloroplast fluorescence. As shown in Figure 3, inorganic phosphate or ADP, when added singly, produced a small quenching of chloro-

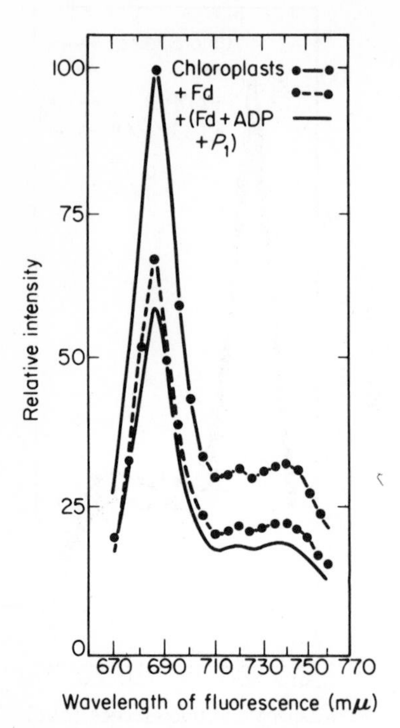

Fig. 4. Quenching of chloroplast fluorescence by ferredoxin, ADP, and inorganic phosphate. The chloroplast preparation was as given for Fig. 2. Three mg of ferredoxin, 5 μmoles of ADP, 5 μmoles of K_2HPO_4 were added as indicated.

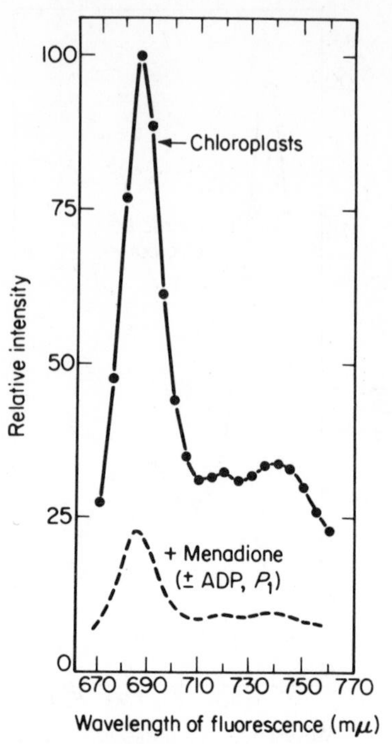

Fig. 5. Quenching of chloroplast fluorescence by menadione, ADP, and inorganic phosphate. The chloroplast preparation was as given for Fig. 2. Three tenths μmole of menadione, 5 μmoles of ADP, and 5 μmoles of K_2HPO_4 were added as indicated.

plast fluorescence. A somewhat greater quenching resulted when ADP and inorganic phosphate were added together. These effects suggest that a low level of cyclic electron flow was enhanced by these additions, probably because a trace of ferredoxin still remained in the chloroplast preparation. This residual cyclic electron flow would be accelerated, and hence give greater quenching, when ADP and inorganic phosphate are added jointly rather than singly. The acceleration of the cyclic electron flow by the addition of ADP and inorganic phosphate would be analogous to a similar effect that was first observed in noncyclic electron flow[8].

A marked quenching of chloroplast fluorescence by the addition of ferredoxin is shown in Figure 4. This effect, which is contrary to the lack of quenching of chloroplast fluorescence by ferredoxin ("PPNR") reported by Kok[3], is in agreement with the role of ferredoxin as a catalyst of cyclic electron flow[13]. A further quenching of chloroplast fluorescence was observed upon adding ADP and inorganic phosphate (Fig. 4), an observation which again supports the idea that the cyclic electron flow catalyzed by ferredoxin is accelerated by concomitant phosphorylation.

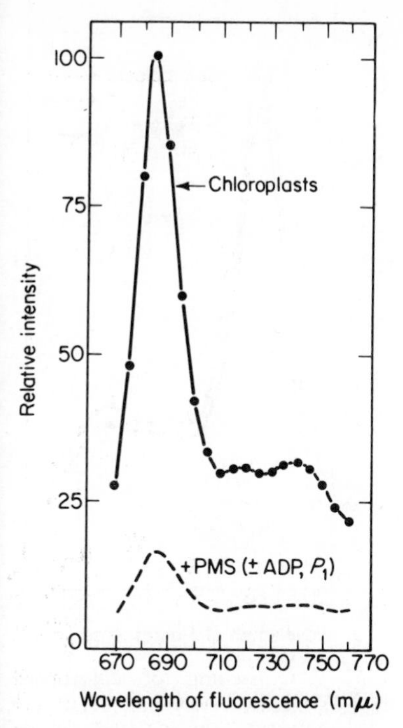

Fig. 6. Quenching of chloroplast fluorescence by phenazine methosulfate (PMS), ADP, and inorganic phosphate. The chloroplast preparation was as given for Fig. 2. One tenth μmole of PMS, 5 μmoles of ADP and 5 μmoles of K_2HPO_4 were added as indicated.

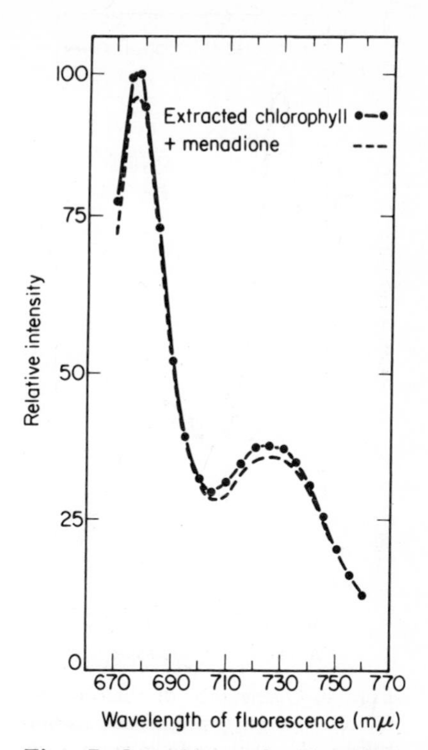

Fig. 7. Quenching of chlorophyll fluorescence by menadione. The chlorophyll preparation was as given for Fig. 2. Three tenths μmole of menadione was added as indicated.

The most striking quenching of chloroplast fluorescence was obtained from the addition of menadione or phenazine methosulfate (Figs. 5 and 6)—two widely used catalysts of cyclic photophosphorylation. In these two cases the marked quenching effect was not further enhanced by the addition of ADP and inorganic phosphate. These results suggest that menadione and phenazine methosulfate catalyze an extremely effective electron flow, more effective in fact, than that catalyzed by ferredoxin, which we now consider to be the physiological catalyst of cyclic photophosphorylation[13]. It is possible that the greater quenching effectiveness of menadione and phenazine methosulfate results from their ability to bypass a slow step in the cyclic electron pathway—a step which is included in the cyclic electron flow catalyzed by ferredoxin and which is sensitive to inhibition by antimycin A and dinitrophenol[13,15].

The marked quenching of chloroplast fluorescence by menadione and phenazine methosulfate led us to examine whether these two compounds would perhaps quench the fluorescence of chlorophyll in solution. However, as Figures

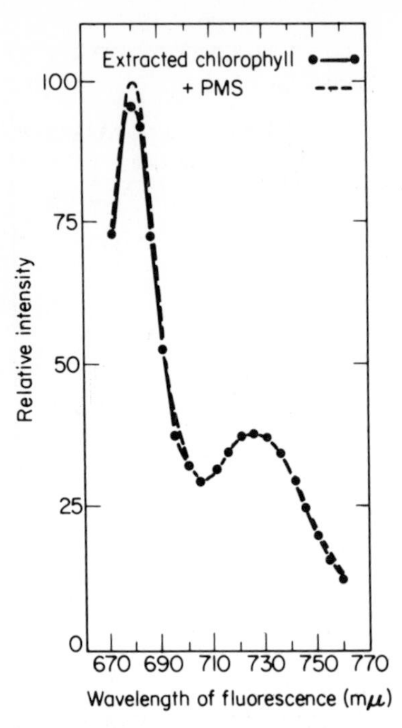

Fig. 8. Quenching of chlorophyll fluorescence by phenazine methosulfate. The chlorophyll preparation was as given for Fig. 2. One tenth μmole of PMS was added as indicated.

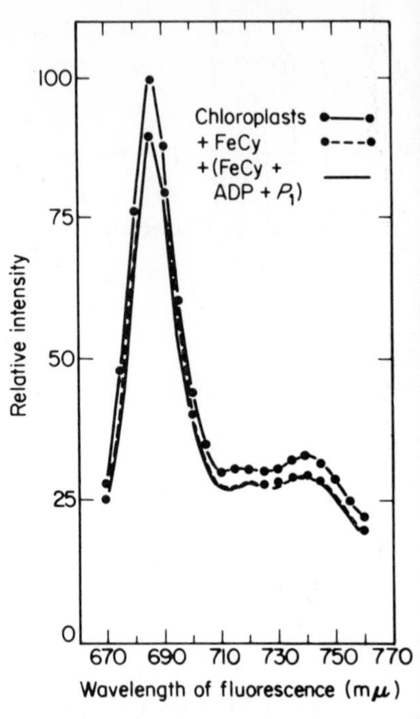

Fig. 9. Quenching of chloroplast fluorescence by ferricyanide, ADP, and inorganic phosphate. The chloroplast preparation was as given for Fig. 2. Ten μmoles of ferricyanide (final concentration 2.9×10^{-3} *M*), 5 μmoles of ADP, and 5 μmoles of K_2HPO_4 were added as indicated.

7 and 8 show, their effect on the fluorescence of chlorophyll in solution was negligible.

Quenching by noncyclic photophosphorylation

In noncyclic photophosphorylation, ATP formation by chloroplasts depends on a transfer of electrons from water to an external acceptor. This type of photophosphorylation provided the first direct experimental evidence[8], as distinguished from conjecture, that photophosphorylation is coupled to a light-driven electron flow in chloroplasts.

The physiological electron acceptor in noncyclic photophosphorylation was thought to be TPN[8] but is now known to be ferredoxin[9], which in turn transfers electrons to TPN via two dark reactions[16]. In the absence of ferredoxin and TPN, illuminated chloroplasts can transfer electrons to such artificial electron acceptors as ferricyanide, and such variants of noncyclic electron flow are also coupled with ATP formation[8]. Thus, noncyclic photophosphorylation affords an opportunity to test the quenching of chloroplast fluorescence by

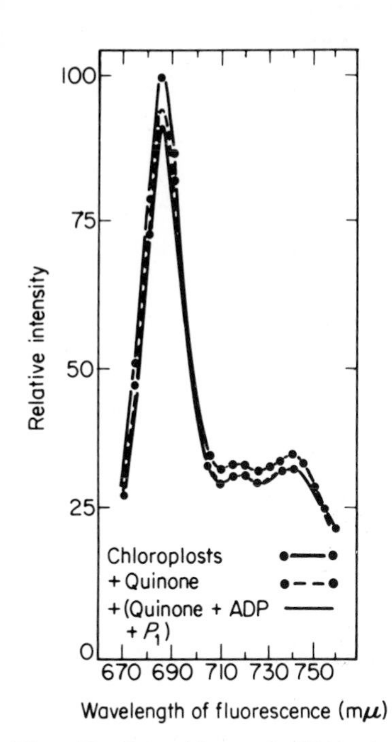

Fig. 10. Quenching of chloroplast fluorescence by benzoquinone, ADP, and inorganic phosphate. The chloroplast preparation was as given for Fig. 2. Five μmoles of benzoquinone (final concentration $1.4 \times 10^{-3}\,M$), 5 μmoles of ADP, and 5 μmoles of K_2HPO_4 were added as indicated.

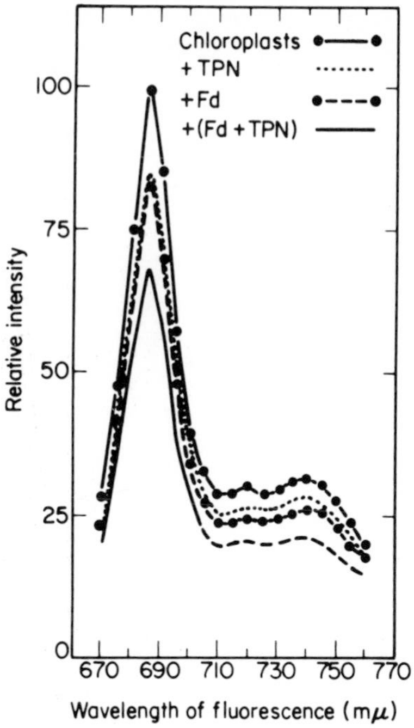

Fig. 11. Quenching of chloroplast fluorescence by TPN and ferredoxin. The chloroplast preparation was as given for Fig. 2. Five μmoles of TPN, and 0.2 μmole of ferredoxin were added as indicated.

electron transfer to ferredoxin or its nonphysiological substitutes in the presence or absence of ADP and inorganic phosphate.

Figure 9 shows that, under the experimental conditions, $2.9 \times 10^{-3}\,M$ ferricyanide gave only a slight quenching of chloroplast fluorescence (compare refs. 2 and 3). No further quenching was observed on adding ADP and inorganic phosphate. Very similar results were obtained with benzoquinone (Fig. 10).

Unlike ferricyanide and benzoquinone, the addition of ferredoxin and TPN gave a more pronounced quenching of chloroplast fluorescence (Fig. 11). The addition of TPN alone produced some quenching, probably because the chloroplast preparation contained residual ferredoxin that permitted some transfer of electrons from excited chlorophyll to TPN. The addition of the small amount of ferredoxin used to catalyze noncyclic electron flow (0.2 mg versus 3.0 mg used for cyclic; see legend for Fig. 4) gave a similar amount of quenching, but much greater quenching was obtained when ferredoxin and TPN were added together.

Figure 12 shows a marked additional quenching of chloroplast fluorescence

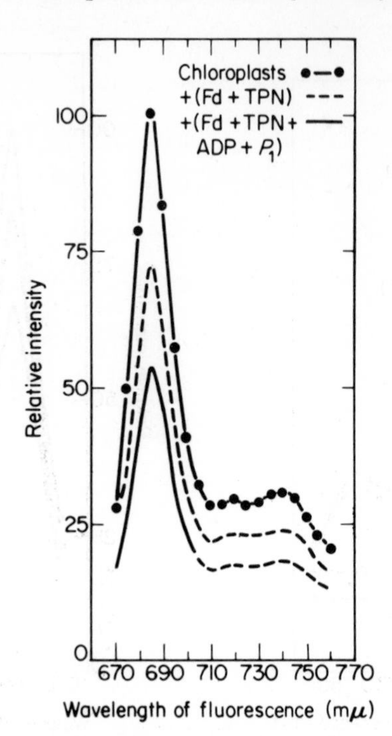

Fig. 12. Quenching of chloroplast fluorescence by TPN, ferredoxin, ADP, and inorganic phosphate. The chloroplast preparation was as given for Fig. 2. Five μmoles of TPN, 0.2 mg of ferredoxin, 5 μmoles of ADP, and 5 μmoles of K_2HPO_4 were added as indicated.

that resulted from the further addition of ADP and inorganic phosphate to the ferredoxin-TPN system. This is consistent with earlier findings that the rate of noncyclic electron flow is markedly increased by concomitant phosphorylation[8,17].

Concluding Remarks

The quenching of chloroplast fluorescence by cyclic and noncyclic photophosphorylation is consistent with their characterization as the primary photochemical reactions in photosynthesis[6,12]. The energy of captured photons may be dissipated as fluorescence or may generate an electron flow which yields the chemical energy stored in the pyrophosphate bonds of ATP and in the reducing potential of ferredoxin.

Of special interest is the striking quenching of chloroplast fluorescence by the cofactors of cyclic photophosphorylation. These results make it seem likely that cyclic electron flow may be an important mechanism for degrading the energy of excited chlorophyll *in vivo* when the normal path of complete photosynthesis is, for one reason or another, blocked. Arnon *et al.*[8] suggested that during the diurnal closure of stomata in leaves, when gas exchange and hence normal photosynthesis is impeded or altogether stopped, cyclic photophosphorylation could occur and form extra ATP for metabolic purposes, for example, for the activation of amino acids for protein synthesis.

The quenching of chloroplast fluorescence provides a direct and sensitive technique for relating, in a photosynthetically active system, the excitation of chlorophyll by light with the primary conversion reactions.

We thank William Ufert for excellent technical assistance.

References

1. Rabinowitch, E. I., *Photosynthesis and Related Processes* (New York: Interscience, 1951, 1956), vol. 2, parts 1 and 2.
2. Lumry, R., B. Mayne, and J. D. Spikes, *Faraday Soc. Discuss.,* **27**, 149 (1959).
3. Kok, B., in *Photosynthetic Mechanisms of Green Plants,* NAS–NRC Pub. no. 1145 (1963), p. 45.
4. Brody, S. S., M. Brody, and J. H. Levine, *Biochim. Biophys. Acta,* **94**, 310 (1965).
5. Arnon, D. I., M. B. Allen, and F. R. Whatley, *Nature,* **174**, 394 (1954); Arnon, D. I., F. R. Whatley, and M. B. Allen, *J. Am. Chem. Soc.,* **76**, 6324 (1954).
6. Arnon, D. I., in *Light and Life,* ed. W. D. McElroy and B. Glass (Baltimore: The Johns Hopkins Press, 1961).
7. The following abbreviations are used: ATP, adenosine triphosphate; ADP, adenosine diphosphate; Pi, orthophosphate; and TPN, triphosphopyridine nucleotide.
8. Arnon, D. I., F. R. Whatley, and M. B. Allen, *Science,* **127**, 1026 (1958).
9. Arnon, D. I., H. Y. Tsujimoto, and B. D. McSwain, these PROCEEDINGS, **51**, 1274 (1964).
10. Stair, R., R. G. Johnston, and E. W. Halbach, *J. Res. Natl. Bur. Stds.,* **64A**, 291 (1960).
11. Whatley, F. R., and D. I. Arnon, in *Methods of Enzymology,* ed. S. P. Colowick and N. O. Kaplan (New York: Academic Press, 1963), vol. 6, p. 308.
12. Arnon, D. I., *Nature,* **184**, 10, 1613 (1959).
13. Tagawa, K., H. Y. Tsujimoto, and D. I. Arnon, these PROCEEDINGS, **49**, 567 (1963).
14. Bowen, E. J., and F. Wokes, *Fluorescence of Solutions* (London: Longmans Green, 1953).
15. Tagawa, K., H. Y. Tsujimoto, and D. I. Arnon, these PROCEEDINGS, **50**, 544 (1963).
16. Shin, M., and D. I. Arnon, *J. Biol. Chem.,* **240**, 1405 (1965).
17. Davenport, H. E., *Biochem. J.,* **77**, 471 (1960).

ENZYMATIC SYNTHESIS OF ADENYL-OXYLUCIFERIN*

W. C. Rhodes and W. D. McElroy

McCollum-Pratt Institute and Biology Department
The Johns Hopkins University, Baltimore, Maryland

In the activation of amino acids, acetic acid, and firefly luciferin by adenosine triphosphate (ATP), pyrophosphate and the corresponding acyladenylates are presumably formed[1]. The synthetic adenylates in all three cases are biologically active. Unfortunately, the enzymatic synthesis of these acyl-adenylates has not been demonstrated. Presumably this is due to the tight binding of the adenylates to the enzyme. Recently we have been able to measure the equilibrium constant for the association of adenyl-oxyluciferin (L-AMP) and firefly luciferase. As expected, it was found that the equilibrium constant of the reaction

$$\text{L-AMP} + \text{E} \rightleftharpoons \text{E-L-AMP}$$

was approximately 2×10^9. Apparently a similar tight binding with enzyme exists with regard to the amino acid and acetic acid adenylates. If this is true, it is not surprising that it has been difficult to demonstrate the enzymatic formation of the adenylates of these compounds. However, the intense fluorescence of oxyluciferin makes it possible to measure the disappearance of quantities which are less than the amount of enzyme added. Because of this sensitivity of the fluorescence technique, we have been able to measure the enzymatic formation of L-AMP in the following reaction:

$$\text{ATP} + \text{L} + \text{E} \rightleftharpoons \text{E-L-AMP} + \text{POP} \qquad (1)$$

*W. C. Rhodes and W. D. McElroy, "Enzymatic Synthesis of Adenyl-oxyluciferin," *Science,* **128**, 253–254 (August, 1958). Reprinted with the permission of the authors and of the American Association for the Advancement of Science.

The measurement of L-AMP synthesis depends on the following differences in properties between chemically synthesized L-AMP and oxyluciferin: (i) L-AMP has a fluorescence intensity which is only 2 percent of the intensity of oxyluciferin. (ii) At a *p*H of 3 oxyluciferin is extractable into ethyl acetate, whereas L-AMP is not. The solubility of the oxyluciferin in ethyl acetate depends upon the dissociation of the carboxyl group. (iii) When paper chromatography and the solvent system described in Table 1 are used, L-AMP has an R_f of 0.55, while oxyluciferin has an R_f of 0.40.

The results of an experiment in which oxyluciferin, luciferase, $MgSO_4$ and ATP were incubated at *p*H 7.1 are shown in Table 1. The reaction was initiated with ATP. Following the addition of ATP there was an immediate decrease in fluorescence corresponding to the disappearance of oxyluciferin approximately equal to the amount of luciferase present. When the reaction mixture was acidified and extracted with ethyl acetate, the excess oxyluciferin was removed from the reaction mixture. The nonextractable, weakly fluorescent material in the aqueous phase had an R_f of 0.55 which is identical to that of synthetic adenyloxyluciferin. Alkaline hydrolysis gives an increase in the fluorescent intensity and a product with an R_f of 0.4. This, and other evidence which has been recently published, indicates clearly that the material in the aqueous phase is adenyl-oxyluciferin(2).

By fluorescence measurements, it has been possible to determine equilibrium concentration of oxyluciferin from which the equilibrium constant for reaction 1 can be calculated. It should be noted that in this reaction the enzyme must be considered in the equilibrium determination. Measurements made on the forward and reverse reactions both give a value of K approximately equal to 2.0×10^5. By using this constant, along with the dissociation constant for luciferase-L-AMP complex, the free energy of hydrolysis of L-AMP can be calculated. If it is assumed that the free energy of hydrolysis of the second

Table 1

Results

The reaction mixture contained the following: oxyluciferin, 3.2 mμmole; $MgSO_4$, 10 μmole; luciferase, 2.0 mμmole; ATP, 4 μmole; brought to final volume of 2.0 ml with 0.1 *M* trismaleate buffer, *p*H 7.1.

Reaction mixture	Fluorescence* intensity	R_f
−ATP	80	0.40
Complete	30	0.55†

*Arbitrary units; excitation, 360 mμ; emission, 540 mμ.
†After ethyl acetate extraction; paper Whatman No. 3-MM; solvent, 30: 70 (by volume) mixture of 1*M* ammonium acetate and 95 per cent ethanol, *p*H 7.5.

phosphate bond in ATP is the same as for the terminal phosphate bond(3), the free energy of hydrolysis is obtained by adding the following reactions:

	$\Delta F^{\circ\prime}$(Kcal)
ATP + L + E $\rightleftarrows$ E-L-AMP + POP	− 7.2
E-L-AMP $\rightleftarrows$ E + L-AMP	+12.6
net ATP + L $\rightleftarrows$ L-AMP + POP	+ 5.4
POP + AMP $\rightleftarrows$ ATP + H_2O	+ 7.7
net L + AMP $\rightleftarrows$ L-AMP + H_2O	+13.1

These calculations were made on the assumption that Mg^{++} has no appreciable effect on the equilibrium of the activation reaction. This point is now being investigated. It is not surprising that the free energy of hydrolysis of L-AMP is considerably higher than that of ATP, since coenzyme A reacts readily with L-AMP to produce AMP and oxyluciferyl-coenzyme A.

The biological significance of the tight binding of acyl adenylates to the enzyme is not clear. However, it should be pointed out that the present evidence indicates that such tight binding restricts the reactivity of the adenylates. For example, L-AMP is capable of reacting with SH compounds such as coenzyme A and cysteine to form the corresponding oxyluciferin derivatives and adenylic acid. However, cysteine is capable of reacting only nonenzymatically with L-AMP, whereas coenzyme A-can react with L-AMP only when it is bound to luciferase[4].

Note added in proof: M. Karasek *et al.* have recently demonstrated the enzymatic synthesis of tryptophan-AMP[5].

References and Notes

1. M. B. Hoagland *et al., J. Biol. Chem.* **218**, 345 (1956); J. A. DeMoss *et al., Proc. Natl. Acad. Sci. U.S.* **42**, 325 (1956); P. Berg, *J. Biol. Chem.* **222**, 1015 (1956); E. W. Davie *et al., Arch. Biochem. Biophys.* **65**, 21 (1956); W. D. McElroy and A. A. Green, *Arch. Biochem. Biophys.* **64**, 257 (1956).
2. R. L. Airth, W. C. Rhodes, W. D. McElroy, *Biochim. et Biophys. Acta,* in press.
3. T. H. Benzinger and R. Hems, *Proc. Natl. Acad. Sci. U.S.* **42**, 896 (1956).

4. This report is contribution No. 232 of the McCollum-Pratt Institute. The work was supported in part by a grant from the National Science Foundation.

5. M. Karasek *et al., J. Am. Chem. Soc.* **80**, 2335 (1958).

PART III

ENERGETICS OF MUSCLE CONTRACTION

THE PRECEDING chapter illustrated some highlights of the chemistry and the metabolism of cells; ATP, a common and essential constituent of all cells, was first discovered in muscle, where it occurs in relatively high concentrations. Likewise, the metabolic pathway of lactic acid formation (glycolysis) was first elucidated in muscle. The contraction process represents a transformation of metabolic (chemical) energy into mechanical energy.

About a hundred years ago, Helmholtz demonstrated, by means of a simple thermopile, a correlation between heat and mechanical response in muscle contraction. Soon after, chemists began to search for substances which might serve as the source of muscular energy. They did not then realize that the muscle is a chemodynamic machine. Hence, the mechanism could be comprehended only through the combined efforts of scientists in thermodynamics and chemistry ("chemical energetics"). Around 1900, biochemists were still inclined merely to analyze dead or at least nonfunctional material. Shortly afterward, in 1907, Fletcher and Hopkins bridged the gap between the old-fashioned chemistry of muscle and the modern biochemical study of muscle function, especially the metabolism of contraction. As mentioned previously, Fletcher and Hopkins were the first to demonstrate the marked increase in lactic acid production during anaerobic contraction and they also succeeded in showing the disappearance of lactic acid during oxidative recovery.

Fletcher and Hopkins found that during electrical stimulation, muscle accumulates lactic acid until a maximum is reached, whereupon, in the presence of oxygen, the lactic acid gradually disappears. They were inclined to explain the limit for lactic acid generation on the basis of the exhaustion of a precursor present in limited amounts. The breakdown of this precursor was believed to cause fatigue, and the restoration of the precursor was supposed to bring about recovery.

An expressive summary and retrospect of the fundamental work of Fletcher and Hopkins can be found in Sir Frederick Gowland Hopkins' 1921 Herter Lecture at Johns Hopkins University. Some passages are quoted here:

> . . . in 1907 it was my privilege to join forces with Fletcher in an attack upon the problem. Although the results of our somewhat laborious research have failed to affect the teaching of some textbooks we have the satisfaction of knowing that they have been the acknowledged point of departure for recent important studies.
>
> We found that the confusion in the literature as to the quantitative relations of lactic acid in muscle were wholly due to faulty technique in dealing with the tissue itself. When the muscle is disintegrated as a preliminary to extraction for analytical purposes, the existing equilibrium is entirely upset. Interacting factors are brought into abnormal relations and the processes of change are greatly accelerated. A fresh muscle had been supposed to contain as much acid as one in rigor simply because it had been produced in the former by the treatment which had preceded estimation. The biochemist had not sufficiently remembered the instability of his material. Fletcher and I, however, found it quite easy, by means of a simple method, not only to avoid starting the changes which led to the formation of lactic acid, but to arrest them at any point during their progress, and thus establish their time relations.
>
> We were able to show that the accumulation of lactic acid in muscle occurs only in the conditions of anaerobiosis. With a proper oxygen supply it fails to accumulate at all. . . .

The next point established by Fletcher and myself is of fundamental importance. If a muscle which, by exposure to anaerobic conditions, has accumulated lactic acid, be placed in oxygen, the acid is removed. The occurrence of this removal under the influence of oxygen is significant in all that follows. It has been fully confirmed by the later work of Parnas and Meyerhof.[1]

Ten to fifteen years later, Otto Meyerhof developed the idea of the lactic acid cycle. This was based on the role of the oxidative recovery succeeding anaerobic muscular work. It was found that one-fifth of the accumulated lactic acid was being oxidized completely and the remaining four-fifths were being reconverted to glycogen[(1)].

Lactic acid formation and oxygen debt were also studied quantitatively by A. V. Hill and his coworkers who trained themselves as subjects for muscle experiments. Krogh's work on the capillaries in muscle had shown that, during and shortly after a work period, the capillary blood flow is greatly increased[(2)], but the increase in capillary flow is not instantaneous (See Part X). It was therefore logical to consider contractions performed during a sudden burst of heavy muscular work as anaerobic work. The relationships were in fact close to those found by Meyerhof in the isolated frog muscle. The Meyerhof cycle therefore was formulated as stated. Earlier the thermal data contributed to support this scheme[(3)], which brings us to a discussion of the origin of those data.

In the early twenties, Meyerhof and A. V. Hill made bold attempts to link data from chemical and biochemical studies with those derived from thermal measurements and from the recording of mechanical work[(3)]. Hill developed techniques which enabled him to obtain precise records of the *rate* of heat generation in a muscle, right from the onset of stimulation, during a single contraction to the succeeding relaxation[(4)]. This novel approach led Meyerhof to initiate a new field, "*Energetik,*" i.e., collection of "biothermal" data with special reference to heat data (ΔH) for glycolysis and for enzymatic phosphorylations and dephosphorylations. Biothermal data were obtained on well-defined biochemical step reactions which are not readily reversed, and for which (before the introduction of radioactive tracers) equilibrium constants could be gathered only through thermal data. As examples, one might mention the determination of the heat of hydrolysis of ATP and phosphocreatine[(5)]. Before entering this field, Meyerhof and his group had conducted a series of quantitative studies on the heat changes brought about by the enzymatic formation of lactic acid from glycogen. These calorimetric data were compared with similar older data. Meyerhof realized that a main contribution to heat formation in glycolysis stems from the heat of neutralization derived from the binding of H^+ with the basic groups of muscle proteins. As described in Meyerhof and Lohmann's 1927 article [q.v.]*, a few months after the discovery of phosphocreatine, the ΔH of lactic acid formation in intact muscle was found to be considerably greater than that measured in muscle extract. This difference could be only partly accounted for in terms of heat of neutralization. The

[1]F.G. Hopkins, "The Chemical Dynamics of Muscle," *Bull. Johns Hopkins Hosp.,* **32**, 321 (1921).

*See material under these authors' names in this section.

other factor considered was the presence of phosphocreatine in intact muscle. In muscle extract, this labile ester was, as described by the Eggletons, largely split before lactic acid formation had reached its peak. Meyerhof was therefore led to study the thermodynamics of phosphoric esters with special reference to the ΔH of hydrolysis of phosphocreatine and other phosphoric esters.

Through their work on the ΔH of the dephosphorylation of phosphocreatine, ATP, and other phosphoric esters, Meyerhof and his coworkers actually initiated a new phase of their important thermochemical studies. These unique studies, which were conducted during the decade 1927–37, form the basis for the concept of "high-energy phosphate bonds," a well-known biochemical phrase coined by Lipmann in 1941. Meyerhof's calorimetric measurements of enzymatic phosphocreatine dephosphorylation showed a ΔH of hydrolysis for this ester strikingly greater than that observed for ordinary phosphate ester bonds. Acid-catalyzed as well as enzyme-catalyzed dephosphorylation of phosphocreatine elicits approximately 12,000 calories per mole of P liberated, whereas enzymatic dephosphorylation of hexose-6-phosphate or glycerophosphate yields fewer than 3,000 calories per mole of P liberated. It was of the greatest interest when Meyerhof and Lohmann found that each pyrophosphate bond broken by enzymatic dephosphorylation of ATP yielded approximately 10,000 calories per mole of P released.

The discovery of phosphoenolpyruvate by Lohmann and Meyerhof brought to light another "high-energy phosphate bond"[6, cf. 6a]. This was the phosphoryl donor present in muscle responsible for suppressing ammonia formation from 5-adenylic acid by addition of phosphoglyceric acid, as described by Parnas, Ostern, and Mann (see the article reprinted in an earlier chapter). Lipmann [*q.v.*], in his 1941 review, subjected the thermodynamics of this system to a very interesting analysis.

The enzymatic deamination of 5-adenylic acid to inosinic acid produces as much as 6,000 calories per mole of ammonia liberated[5]. The ΔH and ΔF of ATP hydrolysis have undergone some quantitative revisions in later years[7], but these revisions have not altered the general perspectives first developed by Meyerhof and his coworkers.

These new parameters played a crucial role in a unique and most fruitful development which we owe to Einar Lundsgaard[8]. Lundsgaard discovered a new type of muscle contraction in which no lactic acid is produced ("alactacid" muscle contractions). This was brought about by iodoacetate poisoning. His discovery prompted a drastic revision of the theories of muscle contraction. Since it was clear that lactic acid formation is several steps removed from the process of muscular contraction, it became evident that although, usually, glycolysis is an energetically important process, it is not a crucial reaction in muscular contraction either *in vivo* or *in vitro*.

Although Lundsgaard's work was essentially unprecedented, in one sense it had had forerunners. As early as 1922, Embden had described *phosphate* liberation during muscular contraction; some years later, he described the so-called delayed lactic acid formation, i.e., lactic acid formation after the termination of a tetanic contraction[9]. This observation received little attention at the time, but it proved to be crucial, as Lundsgaard pointed out later.

Pohl, in 1887[10], had described rigor due to halogenated acetates. In 1923 Schwartz and Ochsman reported that this rigor is "alactacid" in some cases, but they emphasized that this is so only when the animals (frogs) are kept absolutely quiet ("un calme parfait") during the interval between injection of the poison and the development of rigor. This statement may have discouraged a wider interest in these brief reports; indeed, they were actually completely forgotten until Lundsgaard's work appeared in 1930 and Bethe resurrected the old observations to be used as ammunition against Meyerhof's lactic acid theory.

Lundsgaard's discovery of the iodoacetate effect was most certainly independent of these observations. Lundsgaard pursued his discovery along fundamental lines influenced by the ideas of the Meyerhof cycle and by the development of energetics. This was probably the reason for his strong attraction to the thermodynamic approach of Meyerhof's school. The accumulation of exact thermochemical data and the existence of quantitative myothermal data had created a favorable climate for any young physiologist with ability to observe, and with vision for fundamental science. Furthermore, Fiske and Subarrow's discovery of phosphocreatine constituted a sound chemical basis for Meyerhof's, as well as for Lundsgaard's studies on "Energetik der Muskel Kontraktion." It was therefore significant that Lundsgaard, although he encountered the effect of monoiodoacetate by a mere accident, recognized its greater theoretical consequences and pursued them along quantitative lines.

Let me try to sketch briefly the circumstances which led Lundsgaard to the new concept that muscle is able to perform contractions without the formation of lactic acid. During his work on a dissertation dealing with the specific dynamic effect of amino acids and other metabolites, he became interested in iodinated derivatives of amino acids. At that time (around 1928–29), thyroxine was believed to be an iodinated derivative of tryptophane. Monoiodoacetate was one of the close analogues of iodinated glycine which lent itself readily to organic synthesis.

In an attempt to study the effect of iodoacetate on the metabolism of rabbits, Lundsgaard found that even small amounts of this compound were highly toxic and brought about a state of general muscle contracture shortly before the death of the animals. He also discovered that this contracture developed without the usual generation of lactic acid; it was an "alactacid" contracture. Perhaps the most decisive experiment in his series of studies on the basis for "alactacid" contraction was the preparation of iodoacetate "conditioned" muscles without any significant alterations in their chemical composition.

Since it is difficult or impossible to poison bigger muscle *in vitro* with iodoacetate, Lundsgaard had to use another device. He performed a preparatory operation on frogs in which the nerves to one or both hind legs were cut before injection of the toxic iodoacetate into the dorsal lymph sac. The paralyzed hind legs represented a source of resting muscle, which did not develop any contracture as long as it remained unstimulated. As soon as the nonparalyzed forelegs developed contracture, the resting muscles of the hind legs (usually gastrocnemius) were isolated and mounted into a myograph. This technique made it possible to study the energetics of "alactacid" contractions. The outcome of these studies also makes it clear why one can circumvent the development of abnormal events in resting muscles. The observations *in vitro* on properly pretreated frog gastrocnemius were as follows:

A gastrocnemius (or a thin muscle like sartorius) was dissected from an iodoacetate-injected frog, and the muscle was strapped in an isometric myograph and stimulated with single electric shocks. This muscle was able to perform a number of single contractions. Yet its capacity was limited to 50 to 75 contractions, whereas a normal gastrocnemius can perform at least a couple of hundred single contractions. Moreover, the gastrocnemius from the iodoacetate-injected frog terminated its contractions in a state of striking rigor. Note that an isolated gastrocnemius can be considered to be under anaerobic conditions even in air because conditions for diffusion of oxygen into the tissues are poor. The same experiments could be performed in nitrogen with the same results. In this case, too, there was no trace of lactic acid formation, even after rigor had developed. Lundsgaard now pursued the matter into novel territory: the energetics of muscular contraction.

A pair of gastrocnemii from a frog paralyzed in both legs by nerve section was poisoned with iodoacetate. Both muscles were mounted on an isometric myograph. One muscle was left at rest; the other was stimulated moderately, i.e., not to the stage of rigor, and the mechanical work performed was recorded. Both phosphocreatine and lactic acid were then determined quantitatively. It turned out that phosphocreatine was split during the performance of work, and to a much greater extent than Meyerhof and his associates had found.

Lundsgaard succeeded in finding proportionality between the quantity of mechanical work performed and the amount of phosphocreatine split. In order to understand the magnitude of this second discovery, it is necessary to say a few words about the preceding work by Meyerhof. Meyerhof's finding that dephosphorylation of phosphocreatine is a strongly exothermic reaction alerted him to the possibility that this type of phosphate compound might be important to the energetics of muscular contraction. It was not possible, however, to detect any proportionality between phosphocreatine splitting and mechanical work performed. At the beginning of a series of single contractions, phosphocreatine breakdown predominated over lactic acid formation; later in the series of contractions, this was reversed. This type of change in the ratio between phosphate liberated and lactic acid formed had already been noted by the Eggletons[5] and was confirmed by Meyerhof and Lohmann in 1927. By 1930 Meyerhof and Nachmansohn[cf. 5] were inclined to renounce the concept that phosphocreatine is involved in muscle energetics. Lundsgaard's successful demonstration of the proportionality between phosphocreatine splitting and mechanical work in the iodoacetate muscle saved Meyerhof's 1927 concept. Meyerhof wrote a *Nachtrag* (addendum) to his almost completed monograph, *Die Chemische Vorgänge im Muskel*[cf. 5] acclaiming Lundsgaard's work.

Lundsgaard's next move was to try to correlate chemical events in the unpoisoned isolated muscle with the performance of mechanical work. This was first done under anaerobic conditions, using single stimuli as well as tetanic stimuli. New meaning was put into old observations by expressing chemical as well as mechanical events in the same "dimension," *calories,* using data from Meyerhof's thermochemical studies on muscle extracts. The outcome of this refined work is illustrated in Lundsgaard's article [*q.v.*]. (The survey which appeared in Danish had been given in English, in 1931, as a lecture to students at the University of Oxford.)

The "complementation" of phosphate bond energy by glycolysis constitutes the first demonstration of a coupling between the formation of high-energy phosphate bonds and glycolysis. As a part of his remarkable work. Lundsgaard also described a coupling between formation of high-energy phosphate and respiration. A thin sartorius from a frog injected with iodo-acetate, suspended in an oxygen-saturated Ringer's solution, can perform much more mechanical work than in nitrogen under the same conditions. Apparently iodoacetate is a specific poison for anaerobic energy metabolism. These findings also applied to alcohol fermentation [cf. von Euler *et al.*[10a]].

The iodoacetate experiments furnished the first demonstration of the biological role of high-energy phosphates. This role became apparent ten years later from Engelhardt's and Szent-Györgyi's experiments on myosin and ATP. Yet Lundsgaard consistently emphasized that his experiments showed merely that the breakdown of high-energy phosphate is closer to the contraction process than is glycolysis or respiration, and that it would be premature to state whether the breakdown represents processes intimately connected with the contraction process itself.

The phosphoric esters with a high ΔF of hydrolysis include phosphoenol-pyruvate, the two amidinephosphates (phosphocreatine and phosphoarginine), the polyphosphorylated nucleosides (especially ATP and ADP), and the acyl-phosphates. They are all donors of "phosphoryl." In some cases, the OH-ions of water act as the nucleophilic acceptor; this is the case in myosin ATP^{ase} [[11] see later]. Acylphosphates, as well as amidinephosphates, can phosphory-late ADP to ATP with great efficiency. The early studies by Lipmann, Nach-mansohn, Ochoa, and Stadtman revealed that "phosphorylacetate," alias acetyl-phosphate, can serve equally well as a phosphoryl donor and as an acetyl donor. Carbamylphosphate, alias "phosphoryl carbamate," provides another example of that kind[12]. The function of ATP, UTP and other nucleotide triphosphates in the biosynthesis of macromolecules depends on their ability to donate a "nucleotidyl" moiety rather than a "phosphorylium" group (see section on macromolecules).

In this section, excerpts from Lipmann's masterful article formulating the concept of "high energy phosphate bond" are presented. As pointed out in the anniversary volume, which celebrates the 25th anniversary of this article[13], the terminology introduced by Lipmann (including the squiggle sign "$\sim$") may well be considered slang. Some physical chemists may have considered it a disturbing sort of biology jargon when it appeared. Over the 25 years I think most of us felt that the "$\sim$" term, "bond energy," was a fruitful catalyst indeed. Essentially what was meant by the expression "phosphate bond energy" (which relates to an aqueous milieu) referred to the amount of free energy released if a hydroxyl ion of water, acting as a nucleophilic agent, replaces a phosphoric ester group in a linkage. The physico-chemical basis for the great differences in the ΔF of hydrolysis of phosphoric ester linkages has been subjected to discus-sions from various points of view. Entropy problems were brought to bear in explaining the "$\sim$P" character of the phosphoenolpyruvate (cf. Lipmann [*q.v.*]). Hindrance of resonance "opposing resonance" has been invoked[14] as a factor contributing to the "$\sim$ bond" character of acid anhydrides (acyl-phosphates, acyl-adenylates, acyl-mercapto compounds) or phosphoryl amidines (presumably including phosphoryl carbamate). Charge effects have been cited as predominant factors in the ∼P character of pyrophosphates; however, also in this case, opposing resonance was deemed of potential importance (see[15]). Some of these considerations were discussed in the first chapter, others will be dealt with in the section on myosin, ATP, and contraction mechanisms.

References

1. Meyerhof, O.: *Pflügers Arch gesamt. Physiol.,* **175** (1919), 20; **182** (1920), 284.
2. Krogh, A.: *The Anatomy and Physiology of Capillaries.* New Haven, Conn.: Yale Univ. Press, 1922.
3. Hill, A. V., and Meyerhof, O.: *Ergeb. Physiol.,* **20** (II) (1923), 1.
4. ———: *J. Physiol.,* **43** (1911), 1; *Proc. Roy. Soc. London B.,* **126** (1938), 136; *Proc. Roy. Soc. London B.,* **136** (1949), 211.
5. Meyerhof, O.: *Die Chemische Vorgänge in Muskel.* Berlin: Jul. Springer, 1930.
6. Lohmann, K., and Meyerhof, O.: *Biochem. Z.,* **272** (1934), 60.

6a. Needham, D. M., and van Heyningen, W. E.: *Biochem. J.* (1935), 2040.

7. Burton, K., and Krebs, H. A.: *Biochem J.,.* **54** (1953), 94; Podolsky, R. J., and Kitzinger, C.: *Fed. Proc.,* **14** (1955), 115; Benzinger, T. H., and Kitzinger, C.: *Fed. Proc.,* **13** (1954), 11; Alberty, R.A., *J. Biol. Chem.,* **243** (1967), 1337.
8. Lundsgaard, E.: *Biochem. Z.,* **217** (1930), 162; *Biochem. Z.,* **227** (1930), 51; *Biochem. Z.,* **233** (1931), 322.
9. Embden, G.: *Klin. Wochenschrift,* **3** (1924,) 1393; ———, and Schwartz, E.: *L. physiol. Chem.,* **176** (1928), 231.
10. Pohl, J.: *Arch. exp. Pathol. Pharm.,* **24** (1887), 142–50; Schwartz, A., and Ochsmann, C. R.: *Soc. Biol.,* Paris, **91** (1925), 275; and **92** (1925), 169; Bethe, A.: *Naturwiss.,* **18** (1930), 678.

10a. Euler, H. von, Nelsson, R., and Zeile, K.: *Z. physiol. Chem.,* **194** (1931), 53.

11. Clarke, E., and Koshland, D. E.: *Nature,* **171** (1953), 1023; Koshland, D. E., and Clarke, E.: *J. Biol. Chem.,* **205** (1953), 917.
12. Jones, M. E., Spector, L., and Lipmann, F.: *J. Amer. Chem. Soc.* (1955), 77, 819.
13. Kalckar, H. M.: "Lipmann and the Squiggle" in *Current Aspects of Biochemical Energetics,* eds. E. P. Kennedy and N. O. Kaplan, 1. New York: Academic Press Inc., 1966.
14. Kalckar, H. M.: *Chem. Rev.,* **28**, 71 (1941), 71–178.
15. P. Oesper: "The Chemistry and Thermodynamics of Phosphate Bonds," in *Phosphorus Metabolism,* eds. W. D. McElroy and B. Glass, **1**, 523–536. Baltimore: The Johns Hopkins Press, 1951.

F. G. Hopkins ventured to answer the question, which was posed to him by the philosophy of Alfred North Whitehead, "Has the modern biochemist, in analyzing the organism into parts, so departed from reality that his studies no longer have biological meaning?" In a dialogue with himself Hopkins wrote: "So long as his analysis involves the isolation of events, and not merely of substances, he is not in danger of such departure."

[Sir Frederick Gowland Hopkins (1861–1947) was Professor of Biochemistry at the University of Cambridge. One could almost call him the dean of English biochemists in the 1930's. He was a Nobel Laureate and President of the Royal Society. The above quotations stem from his lecture at the Harvard Tercentenary Conference on Arts and Sciences (*Science*, **84** (1936), 258)].

LACTIC ACID IN AMPHIBIAN MUSCLE*

W. M. Fletcher, Fellow and Tutor of Trinity College and
F. Gowland Hopkins F.R.S., Fellow and Tutor of
Emmanuel College, University Reader
in Chemical Physiology

The Physiological Laboratory, University of Cambridge,
England

For a generation it has been recognised that there are means available within the body by which the acid products of muscular activity may be disposed of, and there is already a large body of well-known evidence which indicates that this disposal of acid products—whatever the site of it may be—is most efficient when the conditions for oxidative processes are most favourable, and that it is incomplete when these conditions are unfavourable.

The observations to be described in this paper were undertaken in the hope of determining whether within a muscle itself means exist for an oxidative control of its own acid formation, or for the alteration or destruction of acid which has been formed, either there or by muscular activity elsewhere in the body. With this in view we examined in the first place the effect of an abundant supply of oxygen upon the development of lactic acid in a surviving excised muscle, and upon the stability of the acid within the muscle after its formation. For it has long been known that a surrounding atmosphere of oxygen has the effect of preserving the irritability of a resting excised muscle[1] and of delaying the course of fatigue when contractions have been performed[2], and it has been shown that the oxygen atmosphere may indefinitely delay the onset of rigor mortis in a resting or fatigued muscle[3]. Further we know that muscle entering

*Excerpt from W. M. Fletcher and F. G. Hopkins, "Lactic Acid in Amphibian Muscle," *J. Physiol.,* **35**, 247–250, 296–302 (1907). Reprinted with the permission of the publisher.

An account of the chief experimental results described in this paper was given to the Physiological Society, May 12, 1906.

[1] Liebig. *Arch. f. Anat. Phys. u. Wiss. Med.* p. 393. 1850. (Humboldt's experiments in 1797 are given here.)

[2] Ludwig and Schmidt. *Ludwig's Arbeiten.* Leipzig, 1869.

[3] Fletcher. This *Journal,* XXVIII. p. 474. 1902.

the state of rigor as a result of fatigue, may actually be recalled to a flaccid resting condition by immersion in oxygen gas[4].

Since an acid reaction of the muscle is, as most agree, a constant mark of the fatigued condition and a constant condition of the state of rigor, it is at once suggested that oxygen, when easily available, either restrains by some guidance of chemical events the yield of acid within the muscle, or is able to remove it after its production. And although the direct removal of lactic acid in the presence of oxygen by combustion or otherwise, does not take place under simple chemical conditions out of the body, yet an enquiry into the possibility of its occurrence within a muscle appeared very advisable in view of the facts that the resting survival yield of CO_2 by an excised muscle is at its minimum under anærobic conditions and is greatly increased in the presence of oxygen, and that the special yield of CO_2 due to contractions of the muscle is similarly increased in oxygen and may indeed be absent altogether in an oxygen-free atmosphere[5].

Our earliest experiments gave decisive evidence that lactic acid within an excised surviving muscle is actually diminished in amount, or even wholly eliminated, after exposure of the muscle to abundant free oxygen.

A study of this removal of acid can only be based upon a knowledge of the rate of survival acid-production both in resting and in contracting muscle, under anærobic and aerobic conditions; and before dealing with the question of oxidative removal it will be necessary in the first place to give an account of the main facts of this acid production. For it is notorious that, quite apart from the question of oxidative removal of lactic acid—which has not previously, we think, been examined—there is hardly any important fact concerning the lactic acid formation in muscle which, advanced by one observer, has not been contradicted by some other. Abundant lactic acid formation is said to accompany the process of natural rigor in a surviving muscle (du Bois Reymond, Ranke, Boehm, Osborne), but this is denied (Blome, Heffter): it is said to accompany contraction, and to mark the advance of fatigue (Heidenhain, Ranke, Werther, Marcuse), but this is also denied (Astaschewsky, Warren, Monari, Heffter). Indeed, it may be said that since Ranke wrote in 1865, no description of the elementary facts of lactic formation in muscle, despite the fundamental importance of the subject, has been generally accepted.

We believe that the present confusion is not in chief, if at all, a result of the technical difficulties of lactic acid estimation, but that it is due to the difficulties inherent in the extractive treatment of an irritable muscle. For it is clear that in such a case no treatment for the extraction from muscle can be accepted which, acting itself as a stimulus, has among its effects an increase of the acid to be estimated.

Detailed criticisms of many previous observations will be given later in their place, but it will be convenient to refer briefly here to the main fallacies which underlie the methods of extraction hitherto described.

As solvents of lactic acid, water and alcohol have been used for extraction, and in both cases the muscle must be reduced to small pieces, by cutting or grinding, in order to ensure complete extraction. Now it will be shown that chopping an irritable muscle is, as might be expected, an acid-producing

[4] *Ibid.* p. 480.

[5] Fletcher. This *Journal,* XXVIII. pp. 354 and 488. 1902.

stimulus; and it is obviously fallacious to consider that the extract of a muscle after such a treatment represents its previous condition. This applies in all cases where water is used for extraction; for water, whose virtue in this connection is that it is not itself a stimulant, by tolerating the maintenance of muscle irritability, allows time to pass after the chopping, during which, as we shall show, there occurs great augmentation of the acid yield.

For the purpose of avoiding rapid survival changes, some observers have used boiling water for the quick destruction of irritability before mincing has begun; but in these cases no analysis has been made of the effects of heating as such, and it has been left open to doubt whether sudden heating accounts for a lactic acid yield greater or less than that due to cutting injuries.

Others, again, have employed alcohol—an excellent extractive of lactic acid—and they have hoped that the alcohol in rapidly killing the muscle might not only maintain the *status in quo ante,* but allow subsequently a non-stimulating, cutting up or mincing of the muscle for the completion of extraction,—and this mincing, incidentally, is aided by the coagulative effect of alcohol. But a special dilemma is presented in the use of alcohol: the muscle must be chopped either before or after the immersion. If chopped before, in order to bring the alcohol at once to every small part, then the fallacy due to the chopping stimulation is introduced and no advantage gained by the use of alcohol. If, on the other hand, the intact muscle is bodily immersed in alcohol, without injury, then it does not escape a special stimulant action of alcohol which has not previously been recognised. We shall show that an irritable muscle taken whole, and simply immersed in strong alcohol, may in a short time suffer an eight-fold increase of its acid contents. This special effect is increased with an increase of temperature, and the use of hot alcohol, which only results in the evils of alcohol action being condensed to a shorter time-interval, must be altogether disallowed. It is a surprising fact that the dangers belonging to the use of water for extraction are not half so great as those arising from an uninformed reliance upon the use of alcohol. In view of the fallacies attending the use either of water or of alcohol for the extraction which must precede lactic acid estimation, we think, on a review of the literature, that it is not too much to say that we have as yet no trustworthy comparisons of the lactic acid content of resting and active muscle respectively, and that perhaps in no recorded observations has a genuinely resting muscle been available for examination.

We hope to show that in our experiments we have avoided these dangers of alcohol stimulation,—the magnitude of whose effects we never suspected before trial,—by using ice-cold alcohol, which has no appreciable stimulant action, while it retains its killing and coagulative influence. Immersion in alcohol has been followed by immediate rapid and thorough grinding (with sand) of the muscle ice-cold, in ice-cold alcohol. Of this and of the methods of experiment and estimation a detailed account will now be given.

* * *

Concluding Remarks and Summary

The proof obtained early in the course of this research that the lactic acid content of muscle is profoundly affected by the nature of the treatment received before or during extraction, has enabled us, we believe, to explain some of the contradictions in the statements of others about the fundamental relations of acid production.

Our experiments leave no doubt that in the survival processes which precede the disappearance of irritability there is a steady increase not only of total acidity in the muscle but of lactic acid itself. Equally certain is it that in acid production during fatigue lactic acid takes a large and probably predominant share. The necessity for reinvestigating such fundamental questions as these, before proceeding to the closer study of the phenomena which was our intended task, has given this paper essentially the nature of a preliminary communication. We propose to postpone therefore a full discussion of the bearing of our present results.

It may not be out of place however to consider briefly whether our data are to be reconciled with some current views as to lactic acid production in muscle.

We have given proof that the survival processes in excised unstimulated muscle lead from the moment of excision onwards to a steady accumulation of lactic acid, which, under most conditions, ceases entirely with loss of irritability. The increase of acid in the intact muscle is most rapid under anærobic conditions, is slower in air, and is not to be observed (at any rate for long periods after excision) in an atmosphere of pure oxygen. In the unstimulated muscle the production is, for the greater part of the survival period, very nearly proportional to the lapse of time; but stimulation produces an acceleration which may convert the curve of production velocity from a linear type into one showing exponential characters. Partial disintegration of the muscle represents a strong stimulus, inducing this acceleration to a marked degree, and a want of recognition of the rapidity of the change so induced has led many observers to ascribe much too high values to the lactic acid content of "resting" muscle. This is an error which has necessarily prevented the ascription of right values to the changes occurring during survival processes or fatigue. Exposure of the intact fibres to poisons, such as chloroform or coal gas, also accelerates the velocity of production, and, as has been shown in Part II of this paper, the action of alcohol in this respect is so marked that its uninformed use as a solvent has introduced large errors into many published estimations of lactic acid.

Our experiments make it clear that the excised but undamaged muscle when exposed to a sufficient tension of oxygen has in itself the power of dealing in some way with the lactic acid which has accumulated during fatigue. While the fibres are recovering from fatigue and regaining irritability in an atmosphere of pure oxygen, their content of lactic acid is greatly reduced. As already stated, exposure to pure oxygen also inhibits the production of the acid in fresh resting muscle. In air a slowing of the rate of production is seen, but exposure to an atmospheric tension of oxygen does not inhibit the process of formation.

There is no reason to suppose at present that an increase of oxygen tension has any influence more special than that of accelerating the penetration of the gas into the muscle mass. If, as can hardly be doubted, there occur, in the surviving tissue, processes (encouraged by anærobic conditions) which lead to acid production, opposed to others (demanding oxygen) which make, possibly, for actual inhibition of production, and, certainly, for removal after production, then it is clear that whether we shall observe an accumulation of the product or a balance, or, as a third alternative, removal of the product when formed, will depend upon the rate at which oxygen is supplied. There is every reason to believe that a sufficient tension of oxygen (not reached in air) partially restores to an excised muscle one normal asset otherwise lost on excision, a supply, namely, of the gas sufficiently rapid to turn the balance, from an accumulation to a removal, of the particular breakdown product under consideration.

Hoppe-Seyler and his co-workers were, as is well known, the first to emphasize the importance of deficient oxidation as a factor in inducing the appearance of lactic acid in the intact animal. The facts in this connection are compatible with the view that the substance is a true intermediate product of metabolism, but one which undergoes further change with such rapidity, when oxygen supply is normal, that deficiency in oxygen supply is a necessary condition for its appearance in appreciable amount. The facts may however be taken to indicate that lactic acid is not a normal metabolite at all, but an alternative, asphyxial, product. To judge from references made to Hoppe-Seyler's view we are not sure that all writers have been quite clear as to exactly what it implied, though the author himself seems to have indicated a belief in the latter of the above alternatives.[6]

The proof that muscle possesses in itself the requisite chemical mechanism for the removal of lactic acid when once formed, and that it is not in this respect wholly dependent upon the circulation, indicates, we think, that the substance is a product in its normal metabolism.

Our experiments show that a disappearance of some thirty per cent. of the lactic acid of fatigue occurs during the first two hours of exposure to pure oxygen; for the removal of fifty per cent. ten hours may be necessary. It must of course be realised that the completely fatigued excised muscle, depending upon the surrounding atmosphere instead of a circulation, is so far from being in a physiological condition that these results give no indication of the rate by which lactic acid might be dealt with under normal conditions in the body. It is clear, from familiar evidence, that even in the body large and abnormal amounts, produced under exceptional circumstances, are removed from the muscles by the circulation; but we think that our experiments establish the existence of a power to deal with the metabolite locally, which cannot but serve the muscle under normal physiological conditions. Considerations of energy supply point strongly, as has been urged in previous paragraphs, to the value of such local utilisation. One very striking circumstance has been brought to light by our experiments. We have shown that the removal of lactic acid under the influence of oxygen only occurs when the fibres are irritable and intact. Partial disintegration of the tissue entirely prevents its occurrence.

Our experiments have not so far been concerned with the nature of the precursors of lactic acid; but the results may be held to have some indirect bearing on this matter. Numerous studies have been made by others upon excised muscle, in the endeavour to relate the disappearance of glycogen to the appearance of lactic acid in fatigue and in rigor. The results have been contradictory, but in the well-known experiments of Boehm there was no indication that, in rigor, the acid arises from glycogen. Most of the existing work upon this point (including Boehm's) was done before the importance of the pancreatic function in carbohydrate metabolism had been recognised. The possibility must now be reckoned with that the excised organ, if it does

[6] Hoppe-Seyler. *Berichte* xxv., Ref. 685.

exhibit carbohydrate metabolism, may be doing so in the absence of a normal factor.

The influence of pancreatic extracts upon irritable surviving muscles might, conceivably, be studied by placing them in Ringer's fluid containing such extracts; and in view of Otto Cohnheim's results with ground up muscles, it is desirable that this possibility should be tested. Immersion of course tends to complicate the phenomena somewhat, because of the diffusion of products into the fluid. But if a proportionately small amount of fluid be used, sufficient just to cover the muscles, equilibrium inside and outside the fibres (in the case of substances for which the muscle is permeable) seems to be rapidly attained. At any rate we have found that on bubbling oxygen through Ringer's fluid containing fatigued muscles, there is (after, maybe, a preliminary increase) a final disappearance of a large proportion of the lactic acid, coincident with return of irritability to the muscles. In such experiments the lactic acid was estimated in the muscles and the surrounding fluid taken together. If the pancreatic co-agent can enter the muscle, it may prove more profitable to study the relation between carbohydrate and lactic acid when extracts of pancreas are added to Ringer's fluid as suggested, than to experiment in this connection with the muscle alone.

Since (to judge from the best available data) the average quantity of glycogen in frogs' muscle is, at most, not much more than commensurate with the amount of lactic acid which can be formed during survival processes, it may perhaps be thought that a velocity of formation which shows no *minus* acceleration is incompatible with an exclusive derivation of the latter from the former substance. The results given in the last section, moreover, may be taken to indicate that an excised muscle is capable of producing, under special conditions, a quantity of lactic acid quite out of proportion to its glycogen content. We do not wish, however, before further data are obtained, to insist too much on considerations of this sort, and we do not claim that our present results bear with any special force upon the question of the nature of the precursors of lactic acid.

We feel however that they stand in the way of any direct application of Stoklasa's conceptions concerning lactic acid production to the case of surviving irritable muscle. As is well known this author ascribes to every living cell an anærobic mechanism, the action of which precedes oxidation processes. It brings about a change of carbohydrate into lactic acid, followed by removal of CO_2 and the formation of alcohol. The process is identical with the fermentation of sugar by the yeast-cell and the two stages of change are induced by two specific enzymes, consecutive in action, corresponding with those described in yeast by Buchner and Meisenheimer. Whenever therefore we estimate the gradual increase of lactic acid in surviving muscle, we are following, on this view, the accumulation of a product intermediate between sugar and alcohol. If our evidence for the linear relation to time exhibited by this accumulation be accepted, mathematical consideration will show that it cannot concern a product intermediate between two reactions of an exponential character. The chemical nature of the supposed reactions makes it unlikely that they would proceed on other than exponential lines; but it is, of course, possible that, in the muscle, each is so conditioned as to proceed without change of velocity. The rate of accumulation of the intermediate product will then be linear, and will depend on their relative velocity. Granting this, however, it is still not easy to see why the disappearance of lactic acid from muscle should be encouraged by a free supply of oxygen, if the enzymic change from lactic acid to alcohol is, as on Stoklasa's assumption, essentially an anærobic process. Finally, if the normal fate of lactic acid is to form alcohol by simple cleavage under the influence of an enzyme, it seems even more difficult to understand why the process which leads to its disappearance from muscle should cease so completely upon partial disintegration of the tissue.

Stoklasa's experimental evidence for the occurrence of alcoholic fermentation in the expressed juice of plant-cells, especially as presented in his most recent paper[7], is convincing. But in the case of surviving muscle fibres, even if we assume the presence of agencies necessary to induce alcoholic fermentation, these would not, we think, suffice for a full explanation of the phenomena.

1. Estimations have been made of the lactic acid yielded by the leg muscles of frogs under various conditions. Special precautions have been taken to avoid errors due to the manipulative treatment of irritable muscle. The estimations have been made by gravimetric determinations of zinc sarcolactate according to the well-known method, used with certain modifications.

2. Freshly excised resting muscle is found to yield very small quantities of lactic acid, and these small amounts are possibly not more than can be accounted for by the unavoidable minimum of manipulation prior to extraction.

3. A large increase of the yield of lactic acid is found as the result of mechanical injury, of heating, and of chemical irritation. A large increase of acid is found, in particular, to accompany the immersion in alcohol of resting muscle in bulk, and this, with the other effects due to destructive treatment, is shown to have important bearings upon the choice of methods for extraction.

4. Lactic acid is spontaneously developed, under anærobic conditions, in excised muscles. The course of this survival development of acid has been followed by series of successive estimations. During the survival periods of subsisting irritability, and not after, equal increments of acid arise in equal times. After complete loss of irritability the lactic acid yield remains stationary.

5. Fatigue due to contractions of excised muscle is accompanied by an increase of lactic acid. The amount of acid attainable by severe direct stimulation is found, with notable constancy, to be not more than about one half of that reached in the production of full heat-rigor, or by the action of other destructive agencies than heat.

6. In an atmosphere of oxygen there is no survival development of lactic acid, for long periods after excision.

From a fatigued muscle, placed in oxygen, there is a disappearance of lactic acid already formed. The course of this disappearance has been followed by successive estimations in similar groups of muscles exposed to oxygen for different time intervals: it proceeds at first rapidly, then more slowly, and in general reaches a level about one half of the original yield of the fatigued muscle.

This disappearance of lactic acid due to oxygen does not occur, or is masked, at supra-physiological temperatures (*e.g.* at 30°C.). It is not found in muscle which has suffered mechanical injury: one essential condition for this effect

[7] *Zeitsch. f. physiol. Chem.* Bd. 1. p. 303. 1907.

of oxygen appears to be the maintenance of the normal architecture of the muscle.

7. The amount of lactic acid produced in full heat-rigor (at 40°–45°C.) is constant for similar muscles. This "acidmaximum" of heat-rigor is not affected by a previous appearance within the excised muscle of lactic acid due to fatigue, nor by a previous disappearance of acid in the presence of oxygen, nor by alternate appearances and disappearances several times repeated.

8. In an appendix a new colour test for lactic acid is described.

THE TRANSFORMATION OF ENERGY IN MUSCLE*

Otto Meyerhof

Kaiser Wilhelm Institut für Biologie
Berlin-Dahlem, Germany

The problem of the activity of muscle has accompanied physiology, so to speak, from its cradle. This is not astonishing, for the era of physiology coincides with that of technical development. Our science is hardly a hundred years old. In the muscle, nature has produced a machine, so startling and at the same time so perfect, that the explanation of its mechanism could give satisfaction not only to the searching mind, but also promise a rich harvest to the technical progress of mankind. And this problem is so clearly a physical one that even the vitalist admits the possibility of its being solved by means of inorganic natural science. For it is nothing else but the question, how chemical energy in the animal body is transferred into mechanical work. As a fact, the Heilbronn physician, Robert Mayer conceived the law of the conservation of energy first from work of the human body[21]. Helmholtz demonstrated with the thermopile, constructed by himself, the relation between heat and mechanical response of the muscle[4]. At the same time, the chemists were searching for substances which might serve as "the source of muscular energy." Here, therefore, at a decisive point of physiology, the exact physical view first asserted itself against the claims of vitalism.

Within the last decades, however, interest for our subject lagged again. The most productive and progressive minds were those who first got tired of the matter. No wonder, for on the one hand by studying the muscle exclu-

*Excerpt from O. Meyerhof, *Chemical Dynamics of Life Phenomena.* Philadelphia: J. B. Lippincott Company (1925), pp. 61–82, 101–108. Reprinted with the permission of the publisher.

sively from its physical qualities, it had been overlooked that we have to do with a chemodynamical machine. The comprehension of the mechanism could therefore only be arrived at by the combined efforts of chemistry, energetics, and physics. On the other hand, the biochemist who analyzed only the dead material was helpless in view of the progress of dynamics. Thus experimental analysis stagnated, going round in a circle, and all the theories of muscular function were condemned to fruitless speculation.

Only fifteen years ago the work of Fletcher and Hopkins from the Cambridge Laboratory,[2] discussed in the last lecture, formed the first bridge between the biochemistry of muscle and its performance of work. They noticed in connection with anaërobic contraction and rigor the appearance of lactic acid, and its disappearance on oxidative recovery. This work was followed by the excellent investigations of A. V. Hill[8, 9, 18] "on heat production in muscle," and a series of further studies from Germany and England. Finally the transformation of energy in the working muscle was explained in its chief features.

Since Hermann, it has been known that the muscle can work even if deprived of oxygen, and on complete inhibition of oxidation by poisoning with cyanide. The latter statement was confirmed in a striking manner by the investigations of Weizsäcker[33, 34] on the heart of the frog, where the rate of the work amounted to 60 per cent. on complete stopping of the oxygen consumption by cyanide. This discovery was in curious contrast to the undeniable fact that in the end the combustion of foodstuffs supplies the energy for the muscular work. The connection between oxidation and work could therefore only be quite indirect. The first light was thrown upon this by the investigations of Hopkins and Fletcher, who found out that the muscle accumulates lactic acid anaërobically on electric stimulation. This goes on until a maximum is reached when it is no longer excitable, whereupon the lactic acid formed can disappear in oxygen and the muscle regains its excitability. This relation immediately gave rise to a series of new questions, starting with that of the maximum of lactic acid. The English authors themselves were inclined to explain this maximum on the following grounds: There exists a precursor of lactic acid of a limited amount; its breakdown causes the fatigue of muscle, and its restoration the recovery. This restoration, they supposed, occurred through the reconversion of lactic acid into its precursor. To be sure, oxidations were to have their part in it, but only for the supply of energy. Lactic acid itself was not to be the fuel, but only the lubricating oil, or another part of the machine, and its presence was considered as being outside the energetic metabolism. This whole supposition was founded chiefly on the observation that the same rigor maximum was obtained, regardless of whether the muscle was put directly into the state of rigor, or after it had been repeatedly fatigued and had again recovered in oxygen. Thus a definite amount of a reconvertible precursor of lactic acid seemed to be present in the muscle. This whole theory, however, was outstripped by later investigations. On more delicate analyses of these relations another picture presented itself. In the first place, the maximum of fatigue did not show itself so constant as Fletcher and Hopkins had assumed. That it fluctuated according to the nutritive conditions of the frogs and the temperature of stimulation, could still be explained by the supposition that under these circumstances the amount of the available precursor changes. But this explanation did not account for the circumstance that on fatigue with single induction shocks a maximum is reached 50 per cent. higher than with tetanic stimulation. Finally I found convincing evidence for the fact that this maximum does not depend on the existence of a limited amount of precursor, but on the accumulation of lactic acid in the interior of the muscle itself. The obtainable maximum of lactic acid is considerably increased if the isolated gastrocnemius of a frog is stimulated to fatigue in a Ringer's solution containing more alkali and carbonate than usual, resulting from the addition of a mixture of sodium carbonate and sodium bicarbonate. This maximum may now amount to as much as 0.5 per cent. of lactic acid, in reference to the weight of muscle, instead of 0.34 per cent. At the same time a large part of lactic acid has passed into the surrounding solution. The amount in the interior of muscle is changed little. That the quantity of precursor is of no importance here is also seen from the fact that the contents of glycogen decreases in proportion to the lactic acid formed, while the amount of lower carbohydrates does not change. Even on complete fatigue, the supply of glycogen is not yet exhausted.

If we increase the fatigue maximum considerably in the alkaline Ringer's solution, thus removing the lactic acid, we increase simultaneously the whole work which a muscle can perform anaërobically. A gastrocnemius of 1 gm. in weight and 30 mms. in length can in this way develop totally as much as 160 kilograms of isometric tension, while under ordinary conditions only 120 kilograms are produced when we sum up all the single isometric twitches to total fatigue. I named this ratio isometric coefficient of lactic acid (gm. tension/mg. lactic acid). For I found as a general rule that a fixed relation exists under normal conditions between the lactic acid formed upon stimulus and the developed isometric tension. This fixed ratio, which is also independent of the temperature, assuredly proves that the production of lactic acid is connected with the mechanical response. But while the unfatigued muscle with a definite amount of lactic acid always develops a proportional amount of tension, the efficiency of lactic acid is lowered with fatigue and also with incomplete narcosis. This seems important for the elucidation of the more delicate events in the contractile mechanism.

A still larger problem is connected with the disappearance of lactic acid in the oxidative recovery. Hill recognized the principle very accurately indeed when he compared the activity of muscle with the function of an accumulator[6, 8]. The electric energy which a charged accumulator delivers on closing the circuit originates in the end in the supply of energy in being charged. According to Hill, this charging is done in muscle during the recovery period, when by the expenditure of oxidations potential energy is accumulated. A certain amount of this is liberated on stimulation during contraction, just as in an accumulator on closing the circuit. Taking another simile, we may compare the working of muscle with a clockwork. In the recovery period the clock is wound up. Each stimulus liberates one stroke, and the single strokes are not distinguished from each other until the clockwork has run down. The cause for its having run down is clear to us already; it is the accumulation of lactic acid. But in what does the winding up consist? To be sure, in the removal of lactic acid.

But as to how this is going on, opinions have differed for some time. Hill has derived his conception of an accumulator from his fundamental discovery that the heat of muscle is not evolved simultaneously with the contraction but in two phases of approximately the same amount—the first, so-called "initial heat," during the twitch; the second, so-called "delayed heat," during the oxidative recovery. The latter, however, appears chiefly in oxygen only, and Hill justly connected it with the oxidative disappearance of lactic acid. He went even further and reasoned from his measurements and those of his co-worker Peters[7], that this disappearance of lactic acid could not be attributed to its complete burning up. For in some not quite conclusive experiments the authors had also measured the heat produced by the formation of lactic acid on rigor and anaërobic stimulation and calculated it at about 450 cals. per 1 gm[5,30]. If in the recovery period about as much heat would be evolved, there would result 900 cals. per 1 gm. of lactic acid. The combustion heat of lactic acid, however, is more than 3600 cals., so here 1 gm. of lactic acid cannot burn completely. Nevertheless, this theory was finally adopted by the English investigators for some time[3], for Hill's and Peters' measurements were not quite exact and on the other hand Parnas[29], also in the Cambridge Laboratory, thought to have proven that consumption of oxygen and disappearance of lactic acid in the recovery period corresponded with each other. Indeed, it was very difficult to make Parnas' results accord with those of Hill and Peters.

In reality, however, the relation is a different one[22,23,25]. It is true that after the anaërobic stimulation the oxygen intake is strongly increased for a definite period. This increase is closely connected with the disappearance of lactic acid, and the intake of oxygen above the resting value, which we may call "excess oxygen," only keeps on as long as lactic acid can be proven in muscle. But the total amount of this excess oxygen is only sufficient to oxidize about one-fourth of the disappeared lactic acid. As a matter of fact, the rest, *i.e.,* three-quarters, are reconverted quantitatively into glycogen. Therefore in this recovery period a process is going on which we know already as partial process of muscle respiration. The other part, however, the reconversion of glycogen into lactic acid, belongs to the working phase.

The significance of this coupling between oxidation and reconversion can be understood completely from the energetic standpoint; 1 gm. of lactic acid is produced from 0.9 gm. of glycogen. The combustion heat of glycogen amounts to 4191 cals. according to Stohmann. Emery and Benedict stated higher values[1]; but an accurate determination of the combustion heat of glycogen, in Mr. Roth's Institute at Brunswick, Germany, made upon my suggestion, completely confirmed the correctness of the old value of Stohmann[31]. It was 4188 cals., *i.e.,* 3770 cals. per 0.9 gm. The combustion heat of lactic acid should be 3661 cals. according to Luginin. I found, however, 3601 cals. in a new determination by means of zinc salt, for diluted acid, with which we are concerned here[25]. The repetition in Mr. Roth's Institute brought forth exactly the same value (average 3603 cals.). After the work was finished, Prof. Hill drew my attention to the older determination by Emery and Benedict, which amounted also exactly to 3601 cals. for diluted acid[28]. Therefore the difference between the combustion heat of glycogen and diluted lactic acid amounts to almost

170 cals.* If, now, nothing else would take place but the evolution of lactic acid from glycogen during the working phase and its oxidation during the recovery period, then only 5 per cent. of the energy would be liberated during the working period and 95 per cent. during recovery. This would make an impossible machine. But the process is entirely different. In the first place, I found in a large number of determinations, that in muscle, on the formation of 1 gm. of lactic acid, not 170 cals., but 380 are liberated. This number is not perfectly accurate on account of the great technical difficulties of measurement. But since it represents the mean of numerous determinations without systematic errors, it may be wrong by only a few per cent. As far as accuracy goes, the heat formation is the same under different temperatures. Before inquiring into the source of the excess of 200 cals. above the difference of the combustion heat, the energy balance of the recovery period will be calculated first.

As is shown in Table II of the last lecture (p. 55),** and has been proven by my experiments, about 4 molecules of lactic acid disappear when one is burned; then 3772/4 = 943 cals. must result for 1 gm. of reacting sugar (= 0.9 gm. of glycogen). Since we have determined about 380 during the working period, the rest, *i.e.,* 560 must be looked for during the recovery. Accordingly, 40 per cent. of heat would appear during the working phase and 60 per cent. during recovery. I could establish this very well, at least in the order of magnitude, with a technique first used by Parnas. By direct measurement of the total heat of recovery in a fatigued muscle and comparing this heat with the consumption of oxygen I found: 1, that the heat of recovery formed above the resting value was about the same as the heat of anaërobic fatigue; 2, that this heat, in relation to the intake of oxygen, was less than corresponds to the combustion of carbohydrate. For 5 cals. per c.c. O_2 must be formed on the combustion of carbohydrates. But on the average of the recovery period there were only 3.5 cals. per c.c. O_2, and in all exactly as much heat was lacking here as had appeared during anaërobic fatigue.

We will now compare this result with that which Hill[6], and later on Hill and Hartree[13], have obtained by direct myothermic measurements. As mentioned before, in Hill's previous works it had been stated that the proportion of initial to delayed heat is about unity. Therefore this agrees with our results. But a still more accurate comparison is now of interest. For, as can easily be seen, the proportion of restitution heat to fatigue heat must be given by the quotient of the disappearing molecules of lactic acid to the oxidized ones. In an ordinary coupled reaction, taking place in a homogeneous system, this proportion should be constant. But muscle consists of a heterogeneous system, and for this very reason we cannot expect such a fixed stoichiometrical relation, which does not exist either. Indeed, I found, for instance, in minced muscle

*This calculation would be changed considerably, however, if the latest statements of Slater[32] concerning the combustion heat of glycogen from sea-mussels would be confirmed on muscle glycogen from vertebrata (frog or rabbit). According to Slater there exists a definite glycogen hydrate with the empirical formula $(C_6H_{12}O_6)_n$, which gives a combustion heat of 3883 cals., 110 cals. more than the value of Stohmann and Roth-Ginsberg, calculated for the same formula. In this case some of the following arguments had to be modified, as no "gap" of unexplained calories would remain.

**Table referred to is in portion of the book that has not been reprinted.

Table I

BALANCE OF THE RECOVERY

l. a. disappeared / l. a. oxidized	4/1	5/1	6/1
Total heat per 1 gm.	943	754	630
Recovery heat per 1 gm.	560	374	250
Oxidative heat / Anaërobic heat	1.5: 1	1: 1	0.63: 1

that evidently only one of five molecules of lactic acid is oxidized. Thus 3770/5 = 754 cals. would belong to the exchange of 1 gm. of sugar. Upon the restitution phase devolve therefore 754 — 380 = 374 cals.; the proportion of restitution heat to fatigue heat would be 1. In the same way it is found that, if only one of six molecules of lactic acid is burned, this proportion would be 0.63:1. The more molecules of lactic acid can be reconverted by the oxidation of one single molecule, the better the machine is working. Therefore this figure gives the scale for the efficiency of recovery, it expresses how much of the oxidation energy is used for endothermic reactions, for the charging of the accumulator. From our measurements there followed an efficiency of 40 per cent., while it would be 60 per cent. in the case discussed last. There is no doubt now, after comparing Hill's figures with mine, that the efficiency is better the less tired the muscle is, and the more quickly it can recover. In this point the myothermic technique has an exceedingly great advantage over the calorimetrical and chemical ones. Hill and Hartree obtain an average efficiency of 52 per cent. for the recovery, to which corresponds a proportion of 5.2 molecules of disappearing lactic acid to 1 molecule oxidized. We may therefore correctly assume that in the optimum even 6 molecules of lactic acid may disappear by the combustion of 1, which would mean an efficiency of recovery of 60 per cent.

Hill's and Hartree's results could not be gained except on the basis of a more delicate analysis of the time course of anaërobic heat[9, 10, 13]. With brilliant skill the two English investigators succeeded in analyzing accurately the time course of heat in isometric contraction per 0.1 second, and could even distinguish four different anaërobical phases. Three of these belong to the contraction itself, and may be designated together as initial heat. A certain part of this heat is liberated during development of tension, another part during the state of tension, and the third part during relaxation. (Fig. 1.) This relaxation heat, which may form as much as one-third of the entire heat, is very interesting. Its physicochemical significance will be touched upon later. In the present connection the fourth phase is of special interest to us. This phase does not belong to the initial heat, but is a delayed heat. It is related to the partial recovery which the muscle shows after contraction even in absence of oxygen. We may consider it as an anaërobic restitution heat, according to Hill and Hartree. It corresponds to 30 per cent. of the initial heat. Therefore under anaërobic conditions the proportion of the initial heat to restitution heat is found to be about 1:0.3. In presence of oxygen, however, it was found to be

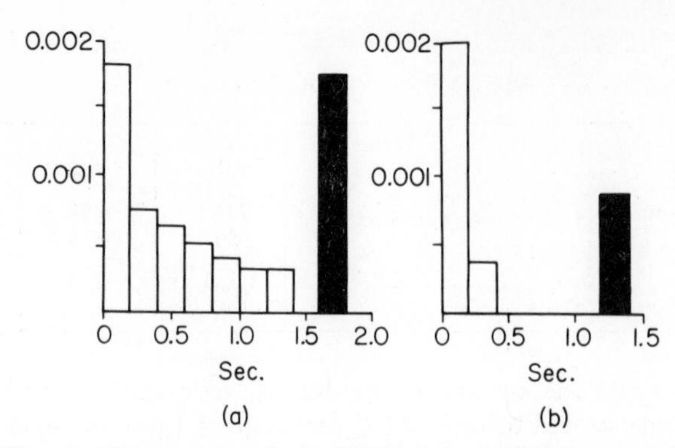

Fig. 1. Time curve of the initial heat analyzed per 0.1 sec. according to Hill and Hartree. (a) Long stimulus, (b) short stimulus. The black area represents the heat of relaxation.

Table II

HEAT-PRODUCTION IN THE LIBERATION AND REMOVAL OF 1 GM. OF LACTIC ACID IN A MUSCLE IN OXYGEN.

Phase.	Relative value.	Absolute value calories.
Initial anaërobic	1.0	295
Delayed anaërobic	0.3	85
Total anaërobic	1.3	380
Total delayed heat	1.5	440
Delayed anaërobic	0.3	85
Difference, delayed-oxidative	1.2	355

1:1.5. Since the anaërobic restitution heat, whatever its share may be, must also be contained in the oxidative restitution heat, there follows from it. If we fix the initial heat at 1, the heat caused by oxidation amounts to $1.5 - 0.3 = 1.2$, and the total anaërobic heat $1.0 + 0.3 = 1.3$. Based on this calculation, we obtain, as do Hartree and Hill, the following relations, taking as standard the value of 380 cals. for formation of 1 gm. of lactic acid. (Table II.)

We now turn to the question, What chemical and physicochemical processes are happening during the contraction act? We must first of all consider the fact that under all circumstances, even in oxygen, lactic acid is formed and only disappears after the contraction has ceased. This fact, important for any theory, is proven by the exactly equal rate of initial heat in the absence and presence of oxygen, and by the slow rise of the oxidative recovery heat, several seconds after the relaxation is over. This is easily understood because the chemical process during activity corresponds here with that during rest. The slow rise towards a maximum is probably caused in this way: The lactic acid, formed during contraction, only gradually increases the resting respiration to the height of recovery oxidation.

The question, What is happening during the contraction act? must be connected with the analysis of the contraction heat which we have measured at 380 cals. per 1 gm. of lactic acid. The difference between the combustion heats of glycogen and lactic acid amounts, as stated before, to only 170 cals. per 1 gm. How does this discrepancy come about? I found out the following facts[24]: In comparing the formation of heat and lactic acid with each other, not in the working muscle but in the minced muscle tissue suspended in a phosphate solution, we obtain pretty exactly 200 cals. per 1 gm. instead of 380. At the same time the lactic acid passes into the phosphate solution. Now, the neutralization heat of lactic acid with biphosphate amounts to 19 cals. per 1 gm. To this is added the heat of cleavage of glycogen into lactic acid, equal to 170 cals. This makes in all 190 cals., which agrees within the limit of experimental error with the value measured, *i.e.,* 200 cals.

Now it can be shown that also in the intact muscle the heat is diminished in the same degree when the lactic acid can escape from it. We first compare the formation of heat and lactic acid during the resting anaërobiosis in unskinned frog's limbs at short intervals. We find the heat here the same as on activity, *i.e.,* 380 cals. per 1 gm. of lactic acid. Indeed, the initial and final conditions of the muscular system are here exactly the same as in the working muscle. The heat therefore agrees in both, although all intermediary links connected with the contraction are lacking. But the heat is different in skinned frog's limbs kept for some length of time in an alkaline carbonated Ringer solution. During the time of the experiment about 50 per cent. of the lactic acid passed into the surrounding fluid. In this case, only 280 cals. per 1 gm. of lactic acid were formed, the heat production decreased from 380 calories at the

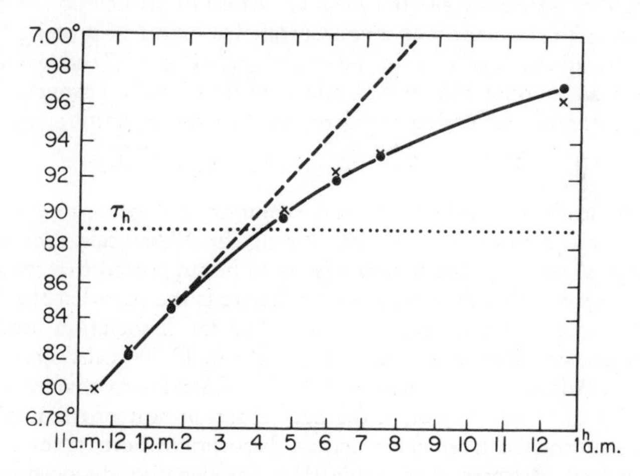

Fig. 2. Heat production of skinned frog legs in alkaline Ringer solution. x—x measured temperature rise; full line, corrected temperature rise; dotted line, theoretical value of temperature rise with constant heat formation. The slope of the curve corresponds to the diffusion of the lactic acid from the muscle into the solution.

beginning to 220 cals. at the end, corresponding to the gradual passing of the acid into the Ringer solution. (Fig. 2.) The special evolution of heat, however, which the lactic acid produces in the intact muscle, if remaining in the tissue, has been proven to be due to the hydrogen ion. If we allow an acid to penetrate into a living muscle from the outside—I used valerianic acid—we obtain without any formation of lactic acid a heat of the same order of magnitude as is produced by the reaction of lactic acid with the substance of living muscle.

What is now really the cause of this heat? It is due for the largest part to nothing but the peculiar manner of the buffering of lactic acid in the muscular tissue. For, as can be shown by calculation, the amount of phosphate and carbonate in muscle is quite insufficient to neutralize the accumulated lactic acid to the extent to which neutralization actually occurs. The real buffer, however, which neutralizes the hydrogen ions of lactic acid is the tissue protein. In this buffering there occurs an unionization of protein. A heat is produced which may be called "unionization heat"; that means the reverse dissociation heat of protein. This heat, which could be found with the free amino-acids as well as with isolated protein, is of exceedingly large amount. We first prepare a solution of glycocoll with caustic soda, *e.g.*, a glycocoll mixture according to Sörensen's technique. Then a small amount of hydrochloric acid or lactic acid is added and the reaction will take place as given in the formula.

$$Na^{+}G^{-} + H^{+}L^{-} = Na^{+}L^{-} + (GH) + 11{,}000 \text{ cals.*}$$

The hydrion concentration hardly changes at all during this process. From the completely dissociated sodium glycocoll is formed the weak undissociated acid glycocoll, and during this reaction we measure a molar heat of a little over 11,000 cals. This is nothing else but the reverse heat of dissociation of glycocoll. We now buffer in the same way a concentrated solution of protein, free from basic salts, by addition of caustic soda to about *p*H 8.** Then we add lactic acid in such an amount that the H concentration hardly changes. Thus the reaction is evidently occurring as represented in the equation:

$$Na^{+}P^{-} + H^{+}L^{-} = Na^{+}L^{-} + (PH) + 12{,}600 \text{ cals.***}$$

This reaction leads to a still greater heat evolution, with the protein of muscle in presence of ammonium salts, amounting to 12,650 cals. per molecule. By a series of arguments, the hypothesis could be supported that we have also to do in this case with a dissociation heat. There is the rule that the difference between the neutralization heat of an acid and its dissociation heat is equal to the well-known dissociation heat of water, *i.e.,* 13,700 cals. This is exactly the case here. Adding caustic soda instead of hydrochloric or lactic acid to the protein buffer to *p*H 8, we obtain the neutralization heat of the weak protein acid which, subtracted from the negative dissociation heat, gives exactly the dissociation heat of water. (See Table III.) The negative dissociation heat of protein found here represents the largest dissociation heat of any acid known at all. It might be due to the fact that on the unionization of amino-acids and protein there is taking place a formation of an internal ammonium salt, or at

*G: glycocoll anion, L: lactic acid.
**Concerning the formation of protein-salts see Loeb's book[19].
***P: protein anion.

Table III

Protein.	Addition.	Heat per mol.	Therefrom heat of dissociation of water.
Albumin	Lactic acid	−12,370	13,720
	NaOH	+ 1,370	
Muscle protein....................	NaOH	+ 1,080	13,680
	Lactic acid	−12,600	
Muscle protein....................	NaOH	+ 1,025	13,725
	Lactic acid	−12,700	

least a linkage of hydrogen and nitrogen through partial valencies according to the scheme sketched below.*

$$\begin{array}{l} CH_2 - NH_2 \\ | \\ COONa \end{array} + H^+ \longrightarrow \begin{array}{l} CH_2 - NH_3 \\ | \qquad / \\ COO \end{array} + Na^+$$

$$\text{or} \longrightarrow \begin{array}{l} CH_2\text{–}NH_2 \\ | \qquad \vdots \\ COO\text{–}H \end{array} + Na^+$$

This fact can easily be shown. As is known, the amino-acids react with formaldehyde according to the equation:

$$\begin{array}{l} R.CHNH_2 \\ | \\ COOH \end{array} + HCOH = \begin{array}{l} R.CHN.CH_2 \\ | \\ COOH \end{array} + H_2O$$

The methylene compound thus formed has lost the amphoteric qualities of amino-acids, and at the same time the large negative dissociation heat has disappeared.

In protein probably the changes from the enol form to the keto form occur besides, as has been discussed by Dakin and Pauli.

$$\underset{\text{enol}}{\begin{array}{l} —C = N— \\ \;\;| \\ \;ONa \end{array}} + H^+ \longrightarrow \underset{\text{keto}}{\begin{array}{l} —C—N—Na^+ \\ \;\;\| \;\;\; | \\ \;\;O...H \end{array}}$$

The dissociation heat of 12,600 cals. yields 140 cals. per 1 gm. of lactic acid. We learned before that of the 380 calories of anaërobic contraction, 170 are due to the splitting up of glycogen into diluted lactic acid. There remain 210 cals., of which 140 can be explained by the unionization of protein. Only 70 cals.

*In assuming an internal ammonium salt, the reaction between the amino and carboxyl groups should correspond to the neutralization of a rather strong organic acid with ammonia. We find indeed an ionization heat of the same order of magnitude as this neutralization heat. In alcoholic solution the amino-acids show an increase of the acid and also of the basic dissociation. Löffler and Spiro[20] have therefore assumed that on the formation of the molecule no partial affinities would be saturated in alcohol, as happens in water. But I found that the ionization heat of glycocoll is the same in alcohol and in water. Therefore the supposition of the formation of an internal ammonium salt seems to be the simpler thing[27].

remain unexplained. I suggest that this too may finally be traced to the dissociation of protein, and that peculiar conditions, still unknown, are responsible for the increase of dissociation heat of protein in muscle. If this is correct, the entire anaërobic heat would be due to the cleavage of glycogen into lactic acid, to the dilution of lactic acid, and its interaction with alkali protein. We could then boldly assert that the capacity of muscle for the performance of anaërobic work, dealt with in the beginning, depends on the alkali combined with protein, the amount of which would determine the height of the fatigue maximum of lactic acid.

We may now ask whether these results can be used for the explanation of the four phases of anaërobic heat production, as found by Hartree and Hill. Some indirect points could be found, especially from the comparison of the temperature coefficients of the single contraction phases with the chemical processes, indicating that the liberation of lactic acid occurs during the moment of shortening, while the interaction of the lactic acid with protein belongs to the relaxation.

According to Hill and Hartree, the development of tension has a temperature coefficient of 2.5 per 10°, the relaxation, however, one of 3.6[11,12]. Corresponding to it I found a temperature coefficient of 2.4 of the lactic acid formation in minced muscle, where besides the chemical reaction only the ionic interchange occurs. But in the intact muscle, where the formed acid reacts with protein, we have a coefficient of 4[24]. This increase of the coefficient might therefore be brought into relation with the interaction of muscle protein. Hill's relaxation heat mentioned before must indeed be imagined as having been produced by a superposition of different chemical and physical phenomena and is finally due to that part of the total energy of contraction which has been transformed into mechanical or elastic tension. However, the unionization of protein evidently forms its basis. We thus obtain a lucid picture of the relaxation process; it depends upon the depression of acidity in the "shortening places" of muscle. In reversed manner we shall make the hydrions responsible for the onset of shortening.

On the other hand, it is not possible for the present to find a satisfactory explanation for the anaërobic restitution heat, the existence of which can no longer be doubted after Hill's and Hartree's measurements. I have proposed to explain it by the unionization of protein, progressing in stages. I think that the protein immediately adjacent to the shortening places could be again dissociated by accepting alkali from other more distant proteins with stronger dissociation heat. A certain restitution would be made possible by it. We must, however, not overlook the fact that we have to deal here with an hypothesis made *ad hoc*.

However, one general point of view seems to be of interest here. The muscle machine itself consists doubtless of protein, while carbohydrate represents the fuel by the burning of which the machine is actuated. Somehow, the oxidation energy of the fuel must take part in the mechanism of the machine itself. With this unionization of protein we have found a point where this is actually the case. In consequence of the coupling of oxidation with the resynthesis of lactic acid, the endothermic restoration of glycogen is accomplished. Moreover, what is still more necessary for the activity of the machine, the alkali is again set free, which causes the endothermic involuntary dissociation of the proteins. The charging of the accumulator, according to Hill, the storing of potential energy, is therefore to be looked for in the conversion of lactic acid into glycogen and in the restoration of the alkali protein.

Doubtless the great dissociation heat of proteins plays in other ways too an important role in the animal body. Hill and Brown[15] recently discovered that with its help the reaction heat of carbonic acid in the blood can be perfectly explained. The process occurs in this way:

$$H^+HCO_3^- + Na^+(Hb)_n^- \longrightarrow Na^+HCO_3^-(H(Hb)_n) \; (Hb_n = \text{Hemoglobin})$$

The process is accompanied by the evolution of 12,100 cals., of which 11,500 must be counted as belonging to the dissociation heat of the weak protein acid $H^+(Hb)_n^-$. There is no doubt that wherever carbonic acid and other acids appear in the tissue, and where protein is working as a buffer substance, this reaction heat obtains.* In reversed manner conclusions can be drawn from the measurement of this heat on the condition of protein in tissue. But the pursuance in this direction would carry us too far from the subject in hand, namely, the energetics of muscle.

These investigations have not only a theoretical interest, but also give us practical explanations about muscular exercise of the healthy and the sick human body. As an example I mention only the practical application which Hill and Lupton[14] have recently made of our various results in basing upon them an exact physiology of exercise. They studied on themselves the consumption of oxygen and the output of carbon dioxide during steady, very violent exercise, especially during flat running. They found that with increasing speed of running, the consumption of oxygen and the output of carbon dioxide first increased to very high values. The respiratory quotient remains exactly unity, corresponding to the exclusive combustion of carbohydrate. But finally a limit is reached, above which the intake of oxygen cannot possibly rise. This is about 4 L. of oxygen per minute. It is, however, possible to increase the speed beyond this limit for a short time. If they determine the respiratory exchange after such exhausting effort, they find for several minutes the intake of a certain amount of "excess oxygen" beyond the rest values before and after. The total amount of this excess oxygen enables us to calculate the quantity of lactic acid which was accumulated at the moment when strength gave out. We assume with Hill and Lupton for the healthy man the most favorable efficiency of recovery oxidation, where the burning of 1 molecule of lactic acid is combined with the removal of 6 molecules *in toto*. In this way we find, in an extreme case, in an athlete of 72 kgm. of weight, 107 gm. of lactic acid accumulated, which, with an approximate muscular weight of 25 kgm., correspond to an amount of 0.4 per cent. of lactic acid in muscle. This agrees with the highest values I have found as maximum in the isolated gastrocnemius of frog and guinea-pig. The study of the output of carbon dioxide is also interesting. As soon as the excess oxygen exceeds a certain amount, the respiratory quotient rises above unity and can reach values up to 2.6. It is caused by the driving off of carbon dioxide by lactic acid. In this way can be established how large a part of lactic acid decomposes carbonate, and how large a part

*In this context, "obtains" presumably means "is encountered."

combines with protein. I calculate that in an experiment of the author's, on an accumulation of 58 gm. of lactic acid, 8 gm. of carbon dioxide were driven off, while 50 gm. must have interacted with protein. For theoretical reasons it is probable that the efficiency of energy-exchange considerably decreases as soon as the lactic acid drives off CO_2 from the blood, instead of unionizing the proteins. Also the ratio

$$\frac{\text{removed lactic acid}}{\text{oxidized lactic acid}}$$

in the recovery state in man could be determined by indirect methods in Hill's laboratory by Long and Lupton[17, 18]. During and immediately following severe prolonged exercise, lactic acid enters the blood, displacing CO_2 from bicarbonate. During the later stages of recovery from exercise, when bicarbonate is practically the only buffer restored, the CO_2 retained in any interval gives a measure of the lactic acid removed, and the O_2 used in excess of the resting value, a measure of the oxidized lactic acid. The average ratio determined in this manner is about 4:1, the same as in my experiments on the frog. Higher values with an average of 6.6 were found on calculating the removal of lactic acid on analyses of venous blood. But this method seems to be evidently less accurate because the lactic acid is probably not uniformly distributed throughout the water phase in the body, as the authors must assume.

It is doubtless very interesting to continue working in this line, studying more intensely the efficiency of muscular work under the circumstances mentioned, the buffer capacity of man, and other points in connection with it.

Bibliography*

1. Emery and Benedict: The Heat of Combustion of Compounds of Physiological Importance, *Am. Journ. Physiol.,* **28**, 301 (1911).
2. Fletcher, W. M., and Hopkins, F. G.: Lactic Acid in Amphibian Muscle, *Ibid.*, **35**, 247 (1907).
3. Fletcher, W. M., and Hopkins, F. G.: The Respiratory Process in Muscle and the Nature of Muscular Motion, *Proc. Roy. Soc. B.,* **89**, 444 (1917).
4. Helmholtz: Über die Wärmeentwicklung bei der Muskelaktion, I, *Müller's Archiv.* (1847).
5. Hill, A. V.: The Heat Production of Surviving Amphibian Muscles, during Rest, Activity, and Rigor, *Journ. Physiol.,* **44**, 466 (1912).
6. Hill, A. V.: The Energy Degraded in the Recovery Processes of Stimulated Muscles, *Journ. Physiol.,* **46**, 28 (1913).
7. Hill, A. V.: The Oxidative Removal of Lactic Acid, *Proc. Physiol. Soc.* in *Journ. Physiol.,* **48**, X (1914).
8. Hill, A. V.: Die Beziehung zwischen der Wärmebildung und den im Muskel stattfinden chemischen Prozessen, *Ergeb. Physiol.,* **15**, 340 (1916).
9. Hill, A. V.: Mechanism of Muscular Contraction, *Physiol. Reviews,* **2**, 310 (1922).
10. Hill, A. V., and Hartree: The Four Phases of Heat Production of Muscle, *Journ. Physiol.,* **54**, 84 (1920).
11. Hill, A. V., and Hartree: The Regulation of the Supply of Energy in Muscular Contraction, *Journ. Physiol.,* **55**, 133 (1921).

*These references have been renumbered to accommodate the excerpted portion of the article.

12. Hill, A. V., and Hartree: The Nature of the Isometric Twitch, *Journ. Physiol.,* **55**, 389 (1921).
13. Hill, A. V., and Hartree: The Recovery Heat Production in Muscle, *Journ. Physiol.,* **56**, 367 (1922).
14. Hill, A. V., and Lupton: Muscular Exercise, Lactic Acid, and the Supply and Utilization of Energy, *Quart. Journ. Med.,* **16**, 135 (1923).
15. Hill, A. V., and Brown: The Oxygen Dissociation Curve of Blood and Its Thermodynamical Basis, *Proc. Roy. Soc. B.,* **94**, 292 (1923).
16. Hill, A. V., Long, and Lupton: Lactic Acid in Human Muscle, *Journ. Physiol.,* **57**, *Proc.* 44 (1923).
17. Hill, A. V., Long, and Lupton: The Removal of Lactic Acid during Recovery from Muscular Exercise in Man, *Journ, Physiol.,* **57**, *Proc.* 47 (1923).
18. Hill and Meyerhof: Über die Vorgänge bei der Muskelkontraktion, Erster Teil, der Mechanismus der Muskelkontraktion, *Ergeb. Physiol.,* **20** (II), 1 (1923).
19. Loeb, J.: Proteins and the Theory of Colloidal Behavior, New York and London, 1922.
20. Löffler, W., and Spiro, K.: Über Wasserstoff und Hydroxylionen-Gleichgewicht in Lösungen, II; *Helvetica Chemica Acta,* **2**, 533 (1919).
21. Mayer, R.: Die Mechanik der Wärme, Ostwald's Klassiker der exakten Wissenschaften, Nr. 180.
22. Meyerhof, O.: Notiz über Eiweissfällungen durch Narkotika, *Biochem. Zeitschr.,* **86**, 325 (1918).
23. Meyerhof, O.: Die Energieumwandlungen in Muskel. III. Kohlehydrat und Milchsäureumsatz im Froschmuskel, *Pflüger's Arch. ges. Physiol.,* **185**, 11 (1920).
24. Meyerhof, O.: Die Energieumwandlungen im Muskel. VI. Über den Ursprung der Kontraktionswärme, *Pflüger's Arch. ges. Physiol.,* **195**, 22 (1922).
25. Meyerhof, O.: Die Verbrennungswärme der Milchsäure, *Biochem. Zeitschr.,* **129**, 594 (1922).
26. Meyerhof, O.: Über die Rolle der Milchsäure in der Energetik des Muskel, *Naturwissenschaften,* **35**, 691 (1920).
27. Meyerhof, O.: Not yet published.
28. Meyerhof, O., and Weber, H.: Beiträge zu den Oxydationsvorgängen am Kohlemodell, *Biochem. Zeitschr.,* **135**, 558 (1923).
29. Parnas: Über das Wesen der Muskelerholung, *Zentr. f. Physiol.,* **30**, 1 (1915).
30. Peters: The Heat Production of Fatigue and Its Relation to the Production of Lactic Acid in Amphibian Muscle, *Journ. Physiol.,* **47**, 243 (1913).
31. Roth and Ginsberg: Dissertation, Braunschweig (1923).
32. Slater: The Heat of Combustion of Glycogen, with Special Reference to Muscular Contraction, International Congress of Physiology, Edinburgh, 1923, and *Journ. Physiol.,* **57**, *Proc. B.,* **38** (1923).
33. Weizsäcker: Arbeit und Gaswechsel am Froschherzen, *Pflüger's Arch. ges. Physiol.,* **141**, 452 (1911).
34. Weizsäcker: II, Wirkung des Cyanids, *Pflüger's Arch. ges. Physiol.,* **147**, 135 (1912).

THE SIGNIFICANCE OF PHOSPHORUS IN MUSCULAR CONTRACTION*

P. Eggleton and M. G. Eggleton

Department of Physiology and Biochemistry,
University College, London, England
(Submitted January 5, 1927)

An examination of the very extensive literature dealing with the function of phosphorus compounds in the chemical mechanism of muscular contraction reveals so many contradictory statements that it is evident that the technique in use must be subject to some serious fault. Since we have found what is probably the main cause of the discrepant results obtained in this field, it seems desirable to communicate our results without delay.

There appears to be in muscle tissue an organic phosphorus compound which, by reason of its great instability in acid solution, has been confused hitherto with inorganic phosphate, to which it gives rise in the course of the estimation of inorganic phosphates by the methods of Embden or of Briggs, or by any method involving the use of mineral acid. The confusion is increased by the fact that this substance, the organic phosphorus compound which we have designated 'phosphagen,' is intimately connected with the chemical mechanism of contraction; the estimation, therefore, of 'inorganic' phosphate by the above methods is hopelessly misleading, since by them one measures the sum total of two substances which vary independently in amount. It is possible, by avoiding the use of acid solutions, to estimate true inorganic phosphate, since 'phosphagen' appears to be stable in neutral or slightly alkaline solution. The following table, which concerns the gastrocnemius muscle of the frog, illustrates the changes in the amount of phosphate and 'phosphagen' in a muscle subjected to different treatments:

*P. Eggleton and M. G. Eggleton, "The Significance of Phosphorus in Muscular Contraction," *Nature,* **119**, 194–95 (1927). Reprinted with the permission of the authors and of the editors of *Nature.*

	Resting.	Rapidly Fatigued.	Heat Rigor.	Incubation in $NaHCO_3$.	
				Without NaF.	With NaF.
Inorganic phosphate	20	50	90	110	20
'Phosphagen'	65	25	0	0	0
Sum total	85	75	90	110	20

The figures are given as milligrams of phosphorus per 100 gm. of muscle, and are representative of a number of experiments. The third row of figures corresponds to the 'inorganic' phosphate as estimated by the ordinary methods.

These facts suffice to explain many of the anomalies to be found in the literature. We will deal here with only one of these. Embden and his co-workers have noted a disappearance of inorganic phosphate in a suspension of minced muscle in sodium fluoride solution. They have attributed this to a synthetic action of the fluoride ion. Comparing the first and last columns in the above table there is an apparent synthesis of 65 mgm. of 'inorganic' phosphate into something not estimated by the Embden technique. The real state of affairs, however, seems to be that the true inorganic phosphate is not affected at all. Either the fluoride ion catalyses the conversion of 'phosphagen' into some acid-stable compound, or else it poisons a catalyst which normally causes its transformation into inorganic phosphate. Rapidly induced fatigue produces a similar but smaller effect.

One further point of interest worthy of note here is that while unstriated muscle contains about the same amount of genuine inorganic phosphate as does striated (skeletal) muscle, it appears to contain no 'phosphagen.' The heart muscle of the frog also gives about 20 mgm. of inorganic phosphorus per 100 gm. of muscle, together with a slight but definite amount of 'phosphagen.' One is tempted to correlate the 'phosphagen' content of a muscle in its resting condition with its ability to respond to sudden demands for violent activity.

Whilst we have at present no definite knowledge of the nature of this substance, it seems quite possible that it may be the unstable ('active') hexose monophosphate, the existence of which was inferred by Meyerhof in interpreting the phenomena of glycolysis in cell-free muscle extracts, and the fermentation of sugar by yeast.

THE ORIGIN OF HEAT OF CONTRACTION*

O. Meyerhof and K. Lohmann

Kaiser Wilhelm Institut für Biologie,
Berlin-Dahlem, Germany
(Submitted July 13, 1927)

The generation of heat brought about by muscle activity can be separated into two phases. The first phase, that of the contraction, coincides with the formation of lactic acid from glycogen. The second phase, which requires oxidation, brings about a resynthesis of glycogen from the bulk of lactic acid.

During the first anaerobic phase an average of 390 calories are liberated per gram of lactic acid produced (caloric coefficient, "c.Q", of lactic acid). The difference in the combustion heat of the original material and that of the end product, respiratory hydrated glycogen and lactic acid (in solution), amounts to only 185 calories, according to our own determinations. A splitting of glycogen to lactic acid by minced muscle or by soluble muscle enzymes yields exactly this amount of calories. In intact muscle an additional contribution of calories stems from a neutralization of charges of protein base. If the reaction goes to completion this extra heat would amount to 140 calories per gram lactic acid. However, only half of the lactic acid formed is neutralized by basic proteins; the other half is neutralized by phosphate[(2)]. On the basis of these contributions one should expect a generation of only 80 calories; this leaves 120 calories, or about one-third of the c.Q., unaccounted for.

The findings by Fiske (Harvard Medical School) and by P. and G.P. Eggleton in London have shown that the major part of phosphate in muscle exists in a highly labile bond with creatine called "phosphagen" (by the latter two authors). During the development of fatigue this compound undergoes gradual

*O. Meyerhof and K. Lohmann, "Über den Ursprung der Kontraktionswärme," *Naturwiss.*, **15**, 670 (1927). Translated from the German and reprinted with the permission of the authors and of Springer-Verlag, Heidelberg, Germany.

fission, whereas during aerobiosis reconstitution ensues[3]. We have obtained the compound from muscle; a product of about 90% purity was selected for a study of the breakdown process. The most likely possibility as regards a constitution formula is that of a phosphoramide, i.e. $(CH)_2 \cdot O \cdot P \cdot NH \cdot C(:NH) \cdot N(CH)_3 \cdot CH_2 \cdot OOH$ in which each molecule of phosphagen contains one molecule of creatine and one of phosphoric acid. The choice of formula emerged as a result of data of molecular weight and electrotitration curves before and after splitting and from kinetics data of splitting in acid. For each gram of inorganic phosphate liberated in the enzymatically catalyzed fission about 150 calories are liberated. One must realize, however, that during muscle activity the ratio of phosphagen splitting to lactic acid formation does not remain constant but increases with the development of fatigue, an observation first made by the Eggletons and confirmed by us. Right at the onset the ratio gH_3PO_4/g lactic acid amounts to about 1.5. For an intermediate degree of fatigue (corresponding to a formation of lactic acid) of about 0.2% approximately 0.75 mg phosphate are released for each mg lactic acid. This provided for an additional amount of heat, 120 to 130 calories, contributing to "c.Q." for which a complete account can therefore be given. The coefficient, on the other hand, cannot be a constant due to the fact that the contribution of phosphagen splitting decreases. Such a decrease concomitant with lactic acid accumulation has already been observed for advanced degrees of fatigue. A corresponding increase for very brief stimuli remains to be demonstrated.

The energy supply for anaerobic muscle work evidently does not stem exclusively from carbohydrate breakdown, as hitherto assumed, but even to a considerable extent from the breakdown of a creatine compound. This can be applied directly only for the vertebrae muscle since the non-vertebrae lack phosphagen[4]. However, we found in crayfish muscle a substance showing chemical and physiological properties analogous to phosphagen; it seems to replace the latter in muscle activity. The special role of phosphagen and the connection of its breakdown with lactic acid formation is still obscure. Yet, its significance is illustrated by the almost exhaustive splitting, or at least labilization, during the culmination of a maximal prolonged contraction. Moreover, upon relaxation, a prompt more or less incomplete regeneration or "restabilization" ensues. The quantitative values for the breakdown applicable for the caloric computations are of course not to be referred to the moment of maximal contraction but rather to the stable state of a relaxed partially fatigued muscle, subsequent to a series of contractions.

References

1. Review: Hill, A. V., and Meyerhof, O.: *Asher Spiro "Ergebn. d. Physiologie,"* **22** (1923), 300; Meyerhof, O.: *Handbuch. d. Physiologie*, **8** (1924), 476.
2. Meyerhof, O., and Lohmann K.: *Biochem. Z.,* **168** (1926), 128.
3. Fiske, C. H., and Subbarow, Y.: *Science,* **65** (1927), 401; Eggleton, P. and G. P.: *Biochem. J.*, **21** (1927), 190; *J. Chem. and Industr.,* **46** (1927), 485; *J. Physiol.,* **68** (1927), 1955.
4. Eggleton, P.: Private communication.

THE SIGNIFICANCE OF THE PHENOMENON "ALACTACID MUSCLE CONTRACTIONS" FOR AN INTERPRETATION OF THE CHEMISTRY OF MUSCLE CONTRACTION*

Einar Lundsgaard

Universitetets Medicinsk-Fysiologiske Institut,
Copenhagen, Denmark

(Lecture given at the Department of Biochemistry,
University of Oxford, England, June 1931)

Before I proceed to the main topic of this lecture—my personal experimental research in the field—I shall first give you a review of the concept of the biochemistry of muscle contraction which was generally accepted about two years ago. In order to facilitate your understanding of the later development to be presented subsequently, I shall even use a fairly large part of my allotted time to summarize the development of muscle chemistry up to about 1929.

The fact that a muscle is able to perform normal contractions over a considerable period of time when placed in an oxygen-free atmosphere, or if poisoned with cyanide, shows that oxidation processes are not necessary for the normal course of muscle contraction. Hence the oxidation processes can have no direct connection with "the mechanism of contraction." It seems reasonable to assume that this very complicated mechanism must always be one and the same, independent of the presence or absence of oxygen. In general, it can be stated that if a reaction which normally takes place in a muscle can be eliminated without affecting contraction, then this reaction cannot possibly be directly connected with the contraction mechanism, by which term I mean the physicochemical alterations which bring about the shortening or the development of tension.

On the other hand, the fact that a muscle which is supplied with oxygen is able to perform considerably more work than a muscle which is kept under anaerobic conditions shows that the oxidation processes, although they have

*Einar Lundsgaard, "Betydningen af Faenomenet maelkesyrefrie Muskelkontraktioner for Opfattelsen af Muskelkontraktiones Kemi," *Danske Hospitalstidende,* **75** (1932), 84–95. Reprinted with the permission of the author and of the publisher.

no direct connection with the contraction mechanism, nevertheless play an important role as a source of energy for muscular work.

The importance of lactic acid formation for muscles was clearly demonstrated for the first time in Fletcher and Hopkins' works. You undoubtedly know how lactic acid accumulates in a muscle which is kept under anaerobic conditions, and how this formation of lactic acid increases considerably when the muscle is stimulated, showing that lactic acid formation is in some way or other connected with the contraction process. You will remember that Fletcher and Hopkins have already observed that lactic acid accumulated in a muscle during anaerobic work disappears when oxygen is supplied to the muscle. Meyerhof later demonstrated in some elegant experiments that under these conditions the largest part of the lactic acid is regenerated into glycogen. Through these experiments we have gained some insight into how the energy of oxidation can be transferred to the processes which take place in the anaerobically working muscle.

In anaerobic conditions—and in the following presentation, except when otherwise specified, I shall be describing anaerobic muscle contractions—Meyerhof studied the relation between work and lactic acid formation "*M*" [The word for "lactic acid" in German is "Milchsäure."] and found it relatively constant. I use here the expression "work," but I really ought to say "mechanical reaction of the muscle," measured (according to Meyerhof and Hill) by the product of the muscle length (L) and the tension developed (T). This relation TL/M and the corresponding relation for tetanic contractions TLD/M (where D is the duration of the tetanic contraction) were found, in numerous experiments performed under different conditions, to be quite constant. At the same time Meyerhof found a very constant relation between the formation of heat in a muscle and the simultaneously occurring formation of lactic acid: about 380 calories of heat was produced per gram of lactic acid formed. Therefore, the result of these two groups of experiments seems to agree with one of Hill's important findings, namely, that the relation between "work" and heat production is constant.

However, there was a discrepancy between the heat liberation per gram of lactic acid formed in the living muscle and the heat formation which would be expected from calculations using calorimetric data of the conversion of glycogen to lactic acid (plus the neutralization by buffers, chiefly protein, of the acid formed). These processes can be assessed to give altogether only about 280 calories per gram of lactic acid. Therefore, at this point of the development, it was necessary to assume either that lactic acid formation in the living muscle actually yields a greater amount of heat than can be accounted for, or that lactic acid formation is accompanied by an exothermal process running exactly parallel, which could furnish the remaining approximately 100 calories per gram of lactic acid formed.

The constant relation between lactic acid formation and work made it probable that lactic acid formation, besides being a very important source of energy which can be utilized anaerobically, was also directly connected with the contraction mechanism. Numerous attempts have been made to establish a contraction theory based upon the assumption that it is the sudden increase of lactic acid concentration which brings about the physical alteration in the muscle during contraction. It must be mentioned, however, that Embden (*Z. physiol. Chem.* **176**: 231, 1928) always maintained that he could show that lactic acid formation is a relatively slow process which does not coincide with the contraction phase and which therefore could have no connection with the contraction mechanism. However, the experimental material on which Embden based his opinion was not recognized by other workers as being convincing.

A few years ago (1927) Eggleton and Eggleton and also Fiske and Subbarow discovered a hitherto unknown chemical compound in the muscles. Eggleton gave this new substance the name of phosphagen, and it was shown by Fiske and Subbarow that this new compound consisted of one molecule of creatine and one molecule of phosphoric acid (phosphocreatine). Therefore the name phosphocreatine is used as much as the name phosphagen. In the following presentation I shall use the English designation "phosphagen." Eggleton showed that phosphagen is decomposed during anaerobic work and is rebuilt when the muscle is supplied with oxygen. Meyerhof and Lohmann (*Biochem. Z.* **196**: 22, 1928) have shown that the disintegration of phosphagen into its two components is an exothermic process which, when brought about by a strong acid, yields a heat formation of 120 calories per gram of liberated phosphoric acid. In experiments with enzymatic disintegration a somewhat higher heat development was found, i.e., 150–160 calories per gram of phosphoric acid set free. Meyerhof at first was inclined to assume that phosphagen disintegration was the unknown process which ran parallel with lactic acid formation and gave the previously mentioned additional 100 calories per gram of lactic acid formed. However, later investigations seemed to render this assumption untenable. Meyerhof and Nachmansohn (*Naturwiss.* **16**: 726, 1928) found that the rate of phosphagen disappearance does *not* proceed parallel with the rate of lactic acid formation and consequently does not parallel the work performed. At the beginning of a series of twitches, the ratio tension × phosphagen disintegration is much smaller than it is in the later sections of the series. This means that the amount of phosphagen disappearing per unit of work is largest during the first part of a series of twitches and then it decreases, first rapidly and later more and more slowly. If the formation of lactic acid is constant per unit of work, whereas this new exothermal process decreases constantly throughout a series of contractions, it follows that the total heat formation per unit of work must be greater during the beginning of a series of twitches than later on in the series; but from Hill's experiments we know that the ratio tension × length/heat is extremely constant throughout a longer contraction series. Therefore, Meyerhof had to assume that in the living muscle no real exothermal hydrolysis of the phosphagen takes place, and proposed the hypothesis that in the living muscle the phosphagen is only transformed into a labile form which, in our analyses, cannot be distinguished from the real decomposition products. Thus, the phosphagen conversion lost its place among the energy-yielding processes in the muscle. This was further justified experimentally in another way. Nachmansohn showed that a resynthesis of phosphagen can take place even under anaerobic conditions. If two symmetrical muscles, suspended in nitrogen, are stimulated with a 5-second tetanus, and then one of the muscles is frozen in liquid air exactly at the moment when the electric current is interrupted, and the other muscle 1 to 2 minutes later, it is found that the muscle that was frozen immediately contains much less phosphagen than does the other. Since it is necessary to assume that exactly the same phosphagen decomposition took place in the two muscles during the entirely uniform stimulation, this result can be explained only by assuming that part of the phosphagen disappeared during the contraction, and was regenerated in the first minutes after the relaxation. Now Meyerhof (*Naturwiss.* **15**: 670, 1927) had asserted on repeated occasions that during the first minutes after a tetanic contraction no exothermal processes take place, especially no lactic acid formation. Consequently there could be no energy available for a resynthesis of phosphagen, and consequently no real endothermal resynthesis could have occurred, but only a "stabilization" of the presumed labile form.

After this relatively long introduction I shall proceed to the main subject of my lecture, namely, my own experimental results in these fields, which you have kindly asked me to summarize for you.

I had made the observation (*Biochem. Z.* **217**: 162, 1930) that monoiodoacetic acid has a peculiar poisonous effect. The main symptom of the poisoning is a pronounced rigor or contracture of all striated muscles. I decided to examine this condition more closely by using for the experiments frogs injected with the poisonous substance into the dorsal lymph sac.

If the nerves to one hind leg of a frog are cut before the poisoning is performed, rigidity will gradually develop in all muscles except in those paralyzed by section of the nerves. If these paralyzed muscles are stimulated either directly or through the distal end of the cut nerves, development of the characteristic rigor occurs in these muscles also. However, no formation of acid took place in the rigid muscles. They showed about the same *p*H as did the normal resting muscles, and closer examinations revealed that the lactic acid concentration in the rigid muscles did not in the least exceed normal resting values. Even in muscles which are first protected against the development of rigor by section of the nerves and which are subsequently stimulated, the lactic acid content is not increased, in spite of the fact that such muscles are able to perform quite an appreciable amount of work before rigor develops, and that the muscles no longer react to stimulation. Moreover, it was possible to demonstrate that muscles which are poisoned with iodoacetic acid and then electrically stimulated until exhaustion no longer contain phosphagen. Now such muscles can perform only 100–200 single twitches and a normal muscle, after corresponding work, would still contain 70–75% of the phosphagen found at rest. In other words, the consumption of phosphagen has increased in the poisoned muscles at the same time that the formation of lactic acid has stopped completely.

It was, as mentioned, assumed by Meyerhof that in the anaerobically working normal muscle phosphagen is not at all broken down but only transformed into a labile form. It could be clearly demonstrated, however, that a real hydrolysis takes place in the poisoned muscles since the phosphagen-phosphoric acid does not appear as free phosphoric acid. In poisoned muscles the phosphoric acid which has disappeared from phosphagen has entered into a new combination and is found as hexose phosphate. Hence it is absolutely certain that in the poisoned muscles a real breakdown of phosphagen has taken place, and consequently equally certain that a considerable amount of energy must have been liberated by this breakdown. Since the formation of hexose phosphate yields only an insignificant amount of heat, I concluded from my experiments that the energy necessary for the work performed by a poisoned muscle must originate from phosphagen breakdown.

Moreover, the hypothesis (*Biochem. Z.* **217**: 162, 1930) was advanced that in normal muscles also the phosphagen undergoes a real breakdown and that consequently part of the heat liberated during a series of contractions originates from phosphagen breakdown. This hypothesis is based upon experiments in which the chemical changes in poisoned muscles which were stimulated to exhaustion were compared with the chemical changes in normal muscles stimulated to a corresponding extent. In the poisoned muscles I had found a phosphagen breakdown of about 1.70 mg. phosphoric acid per gram of muscle, which, when Meyerhof's figure of 120 calories per gram of phosphoric acid set free is used, should yield a heat production of 204×10^{-3} calories per gram of muscle. In the normal muscles I found a phosphagen breakdown of only 0.50 mg. phosphoric acid, but at the same time a lactic acid formation of approximately 0.55 mg. per gram. This should yield an amount of energy of 154×10^{-3} calories from lactic acid formation (280 calories per gram of lactic acid) and 60×10^{-3} calories from phosphagen breakdown, 214×10^{-3} calories altogether. This agreement seems to show: (1) that the liberated energy in the poisoned muscles originates from the phosphagen breakdown, and (2) that heat formation in normal muscles originates partly from lactic acid formation and partly from a breakdown of phosphagen.

Since the block in lactic acid formation seems to lead to an increased phosphagen breakdown, the additional hypothesis was advanced that the apparently smaller phosphagen breakdown in normal muscles is due to a continuous resynthesis of phosphagen brought about by means of energy stemming from the formation of lactic acid. According to this hypothesis, the energy production necessary for the muscle contraction should always and exclusively originate from phosphagen breakdown. In normal muscles part of the disintegrated phosphagen should be immediately resynthesized using energy originating from lactic acid formation. Hence, the net phosphagen breakdown determined in normal muscles is found to be less than the actually occurring breakdown [the rate of] which should be equal to that found in poisoned muscles. Such an assumption presupposes a very high efficiency of the resynthesis of phosphagen. According to the hypothesis from the above-mentioned figures it will be seen that in normal muscles, an amount of phosphagen corresponding to 1.2 mg. H_3PO_4 and to 144×10^{-3} calories should be resynthesized and the energy available from the lactic acid formation totals 154×10^{-3} calories. Later I shall return to this theory and also to the previously mentioned older experimental results which seem to be at variance with it.

The assumption that the contraction energy in the poisoned muscles originates from phosphagen breakdown is strongly supported by later experiments which I have performed in Meyerhof's Institute (*Biochem. Z.* **227**: 51, 1930). In these experiments the "mechanical reaction" (tension $\times$ length) was measured and compared with the phosphagen breakdown occurring simultaneously. The amount of work performed was made to vary as much as

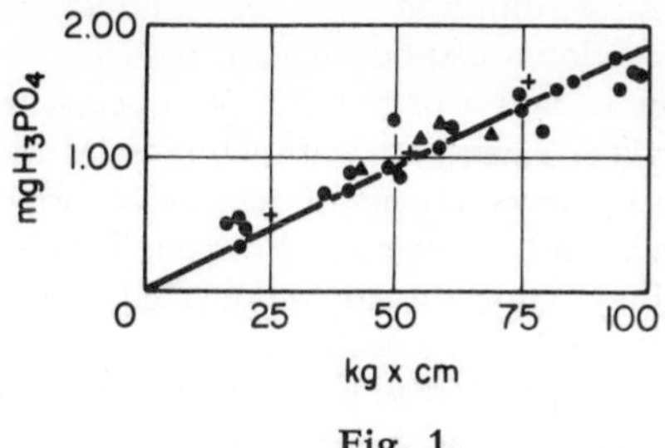

Fig. 1.

possible from experiment to experiment. The result of these experiments is reproduced in the curve of Fig. 1. It will be seen that phosphagen breakdown proceeds almost proportionally with the mechanical reaction as would be expected if the phosphagen breakdown supplied the contraction energy. A corresponding result was obtained in experiments with tetanic stimulation of poisoned muscles.

During my work in Heidelberg I was able to show that a poisoned muscle which is stimulated in oxygen is able to perform a greater amount of work with a smaller phosphagen breakdown than is the symmetrical muscle poisoned in the same manner but stimulated in nitrogen. This observation was later corroborated by experiments recently published from Hill's laboratory. The observation seems to demonstrate that even if the lactic acid formation is blocked, an aerobic resynthesis of the phosphagen can take place, thereby making the oxidation energy available to the contraction mechanism. Therefore, this observation makes possible the assumption that a normal muscle, in which oxidation is sufficiently rapid relative to the rate of work, is able to perform work without any intermediary lactic acid formation.

In accordance with the concept that lactic acid formation in normal muscles supplies energy to an anaerobic resynthesis of phosphagen, it can be shown that the anaerobic resynthesis of phosphagen directly observable after a tetanic contraction does not occur in poisoned muscles.

The formation of heat in muscles, in which formation of lactic acid is stopped, has been measured by various investigators. According to Fischer (*Pflügers Arch. gesant Physiol.* **226**: 500, 1931) and to Hill and Hartree (*Proc. Royal Soc. London B.* **109**: 267, 1931), the initial heat of a single twitch is normal, as is the tension developed. The distribution of the initial heat between contraction and relaxation also is normal according to Hartree. The total quantity of heat which is liberated in a poisoned muscle stimulated to exhaustion is about 150 calories per gram of phosphoric acid set free, and thus comes very close to the figure given by Meyerhof for the heat generation in phosphagen hydrolysis, i.e., 120 calories in acid hydrolysis and 150–160 calories in enzymatic splitting. Simultaneously with the development of rigor additional heat liberation occurs. This "contracture heat" seems, in any case, to be partly due to a breakdown of the adenyl pyrophosphate. The splitting of this muscle component, which I shall not discuss here in detail, seems to take place to an appreciable degree during development of the contracture but not before. Therefore the results of experiments on heat formation in the poisoned muscles also generally support the assumption that the contraction energy in such muscles stems from a breakdown of phosphagen.

Studies concerning the behavior of muscles poisoned with iodoacetic acid have proved that no difference exists between such muscles and normal muscles with regard to: initial heat, development of tension, contraction time, latent period and action potentials, and, I may add, chronaxi. This is a clear indication that lactic acid formation is not directly connected with the contraction mechanism, but must be regarded as a secondary process.

Let me now return to the previously mentioned older experimental results which seem to be at variance with the assumption of a genuine energy-yielding phosphagen breakdown in the normal muscle. The fact that phosphagen breakdown in iodoacetate-poisoned muscles represents the process which furnishes the contraction energy certainly makes this highly probable. Yet it will nevertheless be necessary to examine more closely the difficulties of maintaining this novel assumption.

You will remember that the first objection was that the phosphagen decomposition in normal muscles does not run parallel with the work performed, but is greater per unit of work during the first part of a series of twitches than in the subsequent parts. If lactic acid formation is constant in relation to the work performed, the sum of these two processes—and this means the total initial heat—consequently should be greater per unit of work during the first part of a series of twitches than later in the series. Hill's experiments show, however, that the ratio tension $\times$ length/heat is constant. Now, experiments which I recently finished show that the lactic acid formation per unit of work is not at all constant throughout a longer series of twitches. The lactic acid formation per unit of work is much smaller in the first part of such a series and increases constantly throughout the series, corresponding to the gradually decreasing phosphagen breakdown. Figure 2 gives a schematic picture of the result of these experiments. The uppermost figure represents the phosphagen breakdown per unit of work throughout a series of twitches. The second figure, which is drawn in blocks as the uppermost, represents the lactic acid formation per unit of work. The two curves are drawn in the same scale with respect to the energy liberation in the two processes. Thus, you can see that an increase occurs in the energy originating from the lactic acid formation which corresponds to the decrease in the energy derived from the breakdown of phosphagen. If the two curves are added up, as has been done in the lowest figure, one obtains a quite constant energy production per unit of work, which corresponds to Hill's findings, i.e., that the heat formation throughout a series of twitches is constant in relation to tension $\times$ length. It can be seen clearly from these figures that this constant heat formation is obtained only if the phosphagen decomposition is also taken into consideration as an energy-yielding process.

I think it can be stated now that the previously mentioned objection to the hypothesis that the phosphagen decomposition is a truly exothermal hydrolysis also in the normal muscle, because of the above-mentioned experiments, has rather been turned into a strong support of this hypothesis.

A second objection to the assumption that the phosphagen conversion is an energy-yielding process in the normal muscle was based on the anaerobic

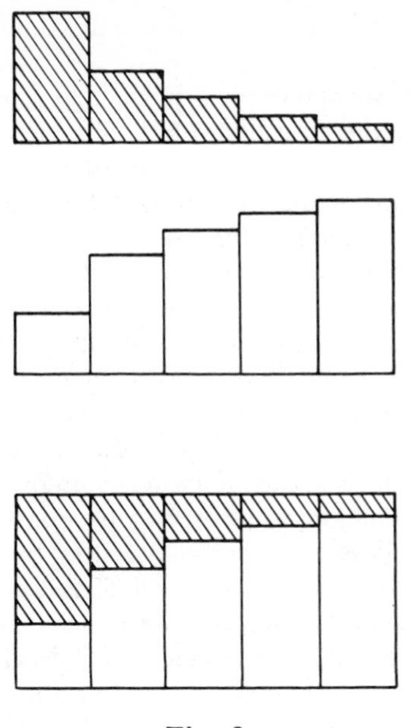

Fig. 2.

resynthesis of phosphagen after a tetanic contraction. According to Meyerhof, no energy should be available for this process which obviously must be endothermal if the decomposition is a true exothermal hydrolysis. My own findings (*Biochem. Z.* **233**: 322, 1931)—which were later confirmed by experiments made in Meyerhof's Institute—showed that during the same interval after a tetanic contraction in which the above-stated resynthesis of phosphagen takes place, a considerable amount of lactic acid is formed. If two symmetrical muscles which are suspended in nitrogen are stimulated with a 5-second tetanus and then one of them is frozen at the moment the stimulation ceases and the other 1–2 minutes later, the following results can be found at the analysis of the two muscles: in the muscle frozen at cessation of the stimulation, a phosphagen breakdown of 1.3 mg. phosphoric acid per gram (156×10^{-3} calories) and a lactic acid formation of 0.20 mg. per gram (56×10^{-3} calories) had occurred; in the symmetrical muscle frozen a minute later, a phosphagen breakdown of 0.50 mg. per gram (60×10^{-3} calories) and a lactic acid formation of 0.55 mg. (154×10^{-3} calories) was found. Thus, in the short interval between the freezing of the two muscles, formation of 0.35 mg. lactate per gram occurs, yielding an energy production of 98×10^{-3} calories; but at the same time resynthesis of 0.80 mg. phosphagen-phosphoric acid per gram has occurred in which 96×10^{-3} calories must be bound. This amount of energy approximates that which is available from the formation of lactic acid. I consider these experiments very convincing. At the same time, they show phosphagen breakdown as a rapid process and lactic acid formation as a relatively slow process, which is secondary in relation to phosphagen breakdown; they also show that resynthesis of phosphagen occurs in connection with lactic acid formation, and they demonstrate the great efficiency of this process, since the amount of energy which is available from lactic acid formation approximates that which is bound in phosphagen resynthesis.

The heat formation which takes place in the first minutes after the cessation of a tetanic contraction, the so-called "delayed anaerobic heat," amounts to only a very small fraction of the initial heat (the heat which is formed during the contraction). Now, if considerable lactic acid formation takes place in this interval, as is the case, then the fact that the delayed heat formation is so small can become comprehensible only if it is assumed that a very large part of the heat generated by lactic acid formation is bound in an endothermal process.

All these experimental results seem to speak in favor of the previously mentioned theory that the *contraction energy is supplied directly and exclusively by the phosphagen breakdown, while the energy from the lactic acid formation is transferred to the phosphagen by bringing about a resynthesis of this compound.* This resynthesis, which at the beginning of a series of twitches amounts to only a relatively small fraction of the disintegrated phosphagen, seems to become more and more complete as the breakdown [series of twitches] proceeds. In other words, the amount of phosphagen, which *remains* hydrolyzed per unit work performed, decreases gradually, whereas the total amount which is broken down obviously increases, at first rapidly but later more and more slowly.

When I say that the contraction energy is supplied directly by the breakdown of phosphagen, it is only to emphasize that this process seems to be more closely connected with the contraction mechanism than is lactic acid formation. How close the connection is, we do not know at the present time.

I believe that at present the energy-yielding processes in the muscle can be illustrated as I have tried to do in the diagram (Fig. 3). We have a small energy reservoir in the form of phosphagen, which contains 0.25–0.3 calories per gram of muscle. Next is a second, larger reservoir—the capacity to form lactic acid—which contains 0.8–0.9 calories per gram, and, finally, there is a very large reservoir (drawn in Fig. 3 to smaller scale than the others), which represents the oxidation processes. In the diagram its content is listed as 35 calories per gram, which corresponds to the heat of combustion of the amount of glycogen normally present in a muscle. If the circulation is intact, this last reservoir is, practically speaking, unlimited, because in such a case new supplies

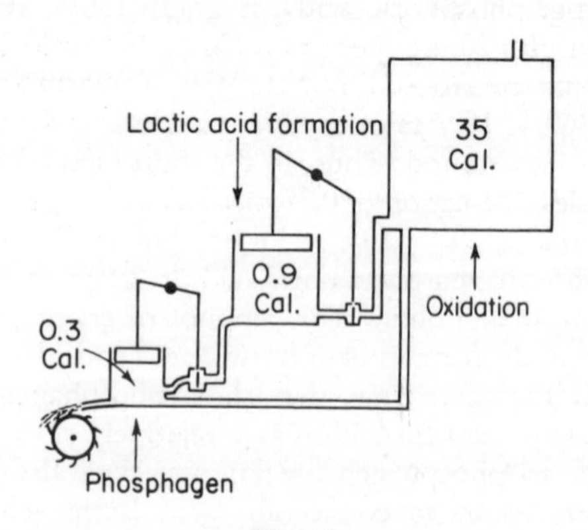

Fig. 3.

of carbohydrates can be brought to the muscle from outside. Besides, it is an open question whether substances other than carbodydrates can be used in the oxidation processes. Personally, I am inclined to believe that this is the case.

The reservoirs are illustrated as being connected with one another by means of "pipe lines" (see Fig. 3). When the muscle begins to work, energy flows out of the phosphagen reservoir and operates the contraction mechanism more or less directly. When the content in the phosphagen reservoir falls to a certain level, the connection with the reservoir of "lactic acid formation" is established and this connection opens up more and more, as is indicated by the lever mechanism. The level in the phosphagen reservoir will therefore fall, at first rapidly and later more and more slowly, as stated previously. If the outflow from the phosphagen reservoir is very rapid, as it is during a tetanic contraction, the content of the reservoir will be substantially reduced because the inflow from the "lactic acid reservoir" cannot keep pace with the outflow, and only after cessation of the tetanic contraction does the phosphagen reservoir fill up (phosphagen resynthesis after a tetanus). When the level of the "lactic acid formation reservoir" drops—i.e. when lactic acid is formed—the connection with the oxidation reservoir opens up, and energy flows in from this and fills up the lactic acid reservoir. This corresponds to the aerobic resynthesis of the glycogen. If the outflow from the lactic acid reservoir is faster than the inflow from the oxidation reservoir, which is limited by the velocity with which air can flow in through the opening at the top of the reservoir which represents the oxygen supply of the tissue, then the level of the lactic acid reservoir falls, that is to say, lactic acid accumulates and the muscle works partly anaerobically.

As I said previously, it is possible that a direct connection exists between the oxidation reservoir and the phosphagen reservoir, so that this can be filled directly from the oxidation energy without any intermediary lactic acid formation. An intermediary lactic acid formation can be thought to be entirely absent when the outflow from the phosphagen reservoir is no faster than the ability of the direct inflow from the oxidation reservoir to keep up the level.

The connections with the oxidation reservoir can be shut off by placing the muscle in nitrogen or by poisoning it with cyanide. Then only those energy quantities which are contained in the phosphagen reservoir and lactic acid reservoir are available. The connection between these two last-mentioned reservoirs can be shut off by poisoning with iodo-acetic acid. In such a case only the limited quantity of energy in the phosphagen reservoir is available.

In closing, I would like to emphasize that these results, about which I have spoken to you, justify maintaining exactly the same view of lactic acid formation which I expressed at the beginning of my lecture about the processes of oxidation. Evidently, since lactic acid formation can be arrested without loss of contraction ability, this process cannot be directly connected with the contraction mechanism. Nevertheless, the fact that a muscle in which lactic acid formation proceeds normally under anaerobic conditions is able to perform a much greater total amount of work than is an alactacid muscle shows that this process is of the greatest importance as a source of energy. Therefore, most of the older experiments which refer to lactic acid formation as an energy-yielding process will naturally not be affected by these newer observations.

Taken at Einar Lundsgaard's home with the Nachmansohns and other colleagues, Copenhagen, 1950.

METABOLIC GENERATION AND UTILIZATION OF PHOSPHATE BOND ENERGY*

Fritz Lipmann

Biochemical Research Laboratory
Massachusetts General Hospital
and the Department of Biological Chemistry
Harvard Medical School,
Boston, Massachusetts

I. Historical Introduction

For a long time after its discovery by Harden and Young, phosphorylation of hexose in alcoholic fermentation was thought to be significant only as a means of modeling the hexose molecule to fit it for fermentative breakdown. However, as the outcome of intensive study of the intermediate reactions in fermentation and the relation between muscular action and metabolism, it later became evident that the primary phosphate ester bond of hexose changes metabolically into a new type of energy-rich phosphate bond[1]. In this bond large amounts of energy made available by the metabolic process accumulate. The recent recognition that in nature there occurs a widespread utilization of such phosphate bonds[2,3] as energy carriers, necessitates a still further revision of the earlier view concerning the biological significance of phosphate turn-over. During various metabolic processes phosphate is introduced into compounds not merely, or at least not solely, to facilitate their breakdown, but as a prospective carrier of energy. To outline the metabolic generation and the circulation of this peculiar type of chemical energy is the primary purpose of this paper.

Through the discovery of creatine phosphate (Fiske and Subbarow[4], Eggleton and Eggleton[5]) a compound of unusual properties was recognized

*Excerpt from F. Lipmann, "Metabolic Generation and Utilization of Phosphate Bond Energy," *Advances in Enzymology*. Vol. 1. New York: Interscience Publishers, Inc., 1941, pp. 99–121, 158–59. Reprinted with the permission of the author and of the publisher.

as component of the chemical make-up of cells. The more or less pronounced breakdown of creatine phosphate during muscular contraction early suggested its connection with energy supply. Interest in the compound became stronger after Meyerhof and Suranyi(6) found that unexpectedly large amounts of heat were released by enzymatic decomposition. The biochemistry of the energy-rich phosphate bond was, in fact, herewith opened. Progress, however, was slowed by the conception then current as to the mechanism of muscle action, connecting contraction rigidly with glycogen breakdown to lactic acid. A profound revision of this conception became unavoidable when Lundsgaard(7) showed that anaerobic contraction proceeded qualitatively, although not quantitatively, undisturbed after complete blocking of glycolysis by iodoacetic acid. He found "*a*-lactacid" contraction accompanied by a quite pronounced breakdown of creatine phosphate, exhaustion of the muscle being coincident with exhaustion of combined creatine. During the contraction period proportionality between creatine phosphate breakdown and action—measured as tension—was found. Using the heat data of Meyerhof, Lundsgaard(8) calculated the tension-heat quotient of Hill (Tl/H)(9). He found practical agreement with the quotient calculated earlier for normal muscle where heat of glycogen breakdown was compared with tension(10). In other words, equal amounts of ultimate heat energy, irrespective of its origin from either creatine phosphate or glycogen, did the same amount of mechanical work. By this finding of Lundsgaard the applicability of phosphate bond energy for the driving of the muscle machine was established.

With normal muscle, the Eggletons(5) and Fiske and Subbarow(4) had already found that creatine phosphate, when largely decomposed during a long series of contractions, was reconstituted quite rapidly during recovery in oxygen. Anaerobically likewise creatine phosphate was resynthesized very effectively at the expense of glycolysis (Nachmansohn(11)). Studying anaerobic resynthesis under most favorable conditions, Lundsgaard(12) found a remarkable efficiency of glycolysis. By breakdown of one-half mole of glucose to lactic acid approximately two moles of creatine phosphate were reformed. Comparing the heats of both reactions each about 24,000 cal but of opposite sign it could be concluded that the total heat energy of glycolysis was utilized for conversion into phosphate bond energy. The free energy of glycolysis might be in fact somewhat greater than 24,000 cal (Burk(13)).

The availability of the energy-rich phosphate bond ($\sim$ph) in absence of glycolytic or combustion energy and the ease and effectiveness with which glycolysis and combustion energy could be converted into $\sim$ph, suggested(12) that the energy utilized in the mechanical set-up of muscle under all circumstances was derived from energy-rich phosphate bonds, supplied constantly by glycolytic or oxidative foodstuff disintegration. The manner in which this supply took place remained, however, entirely obscure.

The study of intermediate reactions in glycolysis and fermentation with tissue and yeast extracts furnished the first explanation of the chemistry of such energy transfers.

The understanding of the transfer mechanism in anaerobic glycolysis still left much unexplained as to how creatine-phosphate could be synthesized in purely aerobic metabolism, especially in the presence of iodoacetic acid(8, 14).

The creatine in muscle must be considered as a natural trap or storehouse for ∼ph. Every metabolic process utilizable for the rebuilding of creatine phosphate must generate energy-rich phosphate bonds. A partial explanation developed when it was found that keto acid oxidation, undoubtedly occurring to some extent in aerobic carbohydrate breakdown, can furnish energy-rich phosphate, which, when brought over to creatine, would reform creatine ∼ph (Lipmann[15]). A more general study of purely oxidative phosphorylation, found to occur abundantly in extracts of kidney and liver, was initiated by the work of Kalckar[16] and is being continued in Cori's laboratory[17]. Here indications are found that present knowledge of the chemistry of generation and transfer of phosphate bonds is far from complete. More and more clearly it appears that in all cells a tendency exists to convert the major part of available oxidation-reduction energy into phosphate bond energy.

The metabolism of muscle is an almost unique case in nature of a straightforward utilization of chemical energy. Here the need of organization into a uniform type is understandable. In all other cells the energy problem is much more complex. If, as in growth, foodstuff is transformed into protoplasm, the comparison of the free energies of starting material and final product frequently does not show appreciable difference, *i.e.,* storage of energy may be insignificant. The extra "energy of synthesis" needed here is used only in such a manner as to force chemical processes to go in desired directions. Ways and means by which phosphate bond energy is utilizable for such general cell purposes are recognizable and partly understood, and shall be discussed in due course.

II. Definition of the Term "Group Potential"

As pointed out, the energy derived from metabolic processes is utilized to force desirable synthetic processes. Groups, such as phospho-, acetyl-, and amino-, are brought by metabolic mechanisms intermediately into positions from where they easily can be carried into desired places. More or less energy is lost or better used up because of special paths adopted in forming these groups. These biologically interesting linkages designed to transfer groups with loss of energy will be called "weak" linkages based on the usual chemical nomenclature with respect to cleavage processes. If, with cleavage, large amounts of energy can be made free (negative change in free energy: $-\Delta F$), the tendency to burst the linkage is relatively great: thus a weak linkage (small affinity). If little energy is freed with cleavage, or energy has even to be furnished, the linkage is called strong (large affinity). Now, very often the biochemist and likewise the synthetic organic chemist is not interested to talk so much about the weakness of the linkage by which a group is bound as about the energy accumulated in the linkage. Instead of emphasizing the negative, the escape of energy through cleavage, he wants to emphasize the positive, the largeness of the energy present in the linkage before cleavage, which determines the *group potential,* the escaping tendency of the group. If in organic chemistry a group is to be brought over into a desired position, compounds with this group in energy-rich linkage (high group potential) are commonly used for the purpose. Acetyl chloride or acetic anhydride is used for acetylation, dimethyl sulfate or diazomethane for methylation, and so forth. Here, it is not the weakness of the linkage but the push, the tension of the group, whereon attention is focussed, although both refer to equivalent attributes. Such clarification seems desirable since useful terms like energy-rich linkage and group potential will be unfamiliar to workers used to the common nomenclature. Attention must be called to the fact that the free energy change does not necessarily measure the group potential but empirically parallelism between both magnitudes may be common.

In the case of the paired hydrogen "group" the familiar term O/R potential designates a group potential. The reaction

$$\text{compound-H}_2 \longrightarrow \text{compound} + (2\epsilon + 2\text{H}^+)$$

is defined by the O/R potential of the compound which expresses the group pressure of the paired hydrogen. Michaelis[18] even showed the possibility of subdividing the over-all potential of the pair into the single potentials of the individual hydrogens.

III. Group Potential of Phospho-Organic Compounds

It appears already from the preceding discussion that phospho-organic compounds with widely different group potentials are found in nature. Evidence shall be given here that two rough groups can be distinguished: one larger group with low potentials and a second group with high potentials.

1. Ester phosphate

To the first group belong all compounds where the phosphate residue is linked to an alcoholic hydroxyl: phospho-hexoses, -pentoses, and -trioses, phospho-glycerols and -3- or -2-glyceric acids, phospho-choline, phospho-serine, etc. In contrast with the energy-rich phosphate bond, designated with ~ph, the energy-poor phosphate ester bond will be designated with -ph in this paper. In many respects phosphate esters behave very much like the esters of alcohols and organic acids, for which, from equilibrium measurements with or without enzyme, the change in free energy with reaction:

$$\text{ester} + \text{H}_2\text{O} \longrightarrow \text{alcohol} + \text{acid}$$

was calculated to be around -1000 cal ($K = 4$–5)[19]. The relation between K, the hydrolysis constant, and $\Delta F°$, the change of standard free energy, is (20):

$$\Delta F° = -RT \ln K = -4.58\, T \log K.$$

As pointed out the same numerical value but with reversed sign gives some measure of the group potential of the linkage.

Only in the case of phospho-glycerol are equilibrium measurements for an actual member of this group available. The equilibrium point for two concentrations of glycerol (50 and 75 vol. %) was determined by Kay[21] with intestinal phosphatase as catalyst. From both sides nearly identical end-points were reached indicating true equilibrium. A value of 40 was found for K, at 38° and pH 8.5:

$$\Delta F^\circ = -4.58 \times 311 \times \log 40 = -2280 \text{ cal.}$$

for:

$$\text{glycerophosphate} + H_2O \longrightarrow \text{glycerol} + \text{phosphate.}$$

For an entirely different type of phosphate ester, the Cori ester phospho-1-glucose, the following considerations lead to an approximate value. Phospho-1-glucose is in equilibrium with glycogen and inorganic phosphate (Cori[(22)]):

$$\text{phospho-1-glucose} \rightleftharpoons \text{glycogen} + \text{phosphate}, (K = 5^{(23)}).$$

Therefore the "glucosidic" ester linkage must be approximately equivalent to the glucosidic glucose linkages in glycogen.

Emil Fischer showed that with high concentrations of glucose and galactose the disaccharide isolactose was synthesized with kefir lactase as catalyst. With appreciable synthesis the hydrolysis constant for the disaccharide cannot be much larger than 100, and ΔF°, therefore ca. —3000 cal. Similar or smaller values were indicated for various glucosides (cf. Veibel[(24)]). That indicates that a value not far different from that for phospho-glycerol should be true for phospho-1-glucose. However, this ester must be on the upper side of the group level at the 3000 cal range since the enzymatic change of the phospho-group from 1- to 6-position was found to be irreversible (Cori[(22)]). This irreversibility may be due, however, to an irreversible change in the rest of the glucose structure following phosphate transfer rather than to the higher potential of the phosphate group in 1-position.

From these considerations the range of 2–4000 cal is to be assumed for the group potential of the phosphate group, when esterified with an alcoholic hydroxyl.

2. Energy-rich phosphate bonds

The result obtained in the preceding paragraph allows us to consider the large and otherwise quite inhomogeneous group of ester-phosphates as uniform, as far as the phosphate bond is concerned. We may now contrast this group with the group of compounds containing ~ph. Here we meet at least four different linkage types, P-O-P, N-P, carboxyl-P, enol-P, the chemistry of each of which will be discussed in the following chapter. The fact, however, that a reversible interchange of ~ph takes place between the members of this group simplifies the discussion of the bond energies, which are on a higher but again uniform level. Only phospho~pyruvate might be treated with a certain reserve since phosphate transfer from this compound to the other members seems to date to be irreversible. This linkage may be at a definitely higher level than the average[(1)]. Uniformity is further indicated by the fairly uniform heat content of the enumerated linkages, 8–12,000 cal (Meyerhof and Schulz[(25)]). However, abundant evidence shows how hazardous it is to take the heat values indiscriminately as a true indication of the free energies which alone determine the direction and extent of a reaction. In the equation relating heat and free energy:

$$\Delta F = \Delta H - T\,\Delta S,$$

the additional entropy factor, $T\,\Delta S$, influences unpredictably the actual value of ΔF, to the extent of up to 10,000 cal. In the present case the following discussion will show that corroborative evidence can be brought forward for an approximate equivalence of heat and free energy:

(*a*) The mechanical efficiency, when measured by the tension-length-heat quotient, is the same in iodoacetic acid muscle, *i.e.,* with N$\sim$P in creatine $\sim$ph as the only source of energy, as it is in normal muscle[8, 10]. The mechanical effect compared with the heat in normal anaerobic muscle is very high, up to 50%, cf. *e.g.*[26]. Since mechanical work measures the free energy, at least 50% of the known heat calories represent convertible energy.

(*b*) In the sequence of intermediate reactions in glycolysis and fermentation the energy-rich phosphate bond in phospho$\sim$pyruvate is formed by dehydration of phospho-2-glycerate:

$$\begin{array}{lcl} CH_2OH & & CH_2 \\ \cdot & -H_2O & \| \\ CHO{-}ph & \rightleftharpoons & C{\cdot}O{\sim}ph \\ \cdot & +H_2O & \cdot \\ COOH & & COOH \end{array}$$

This reaction is freely reversible, *i.e.*, no energy is required to bring about this peculiar transformation. The peculiar difference between the two compounds is that on the glycerate side the phosphate linkage is an ordinary ester linkage and as such is low in energy, but on the pyruvate side becomes an energy-rich enol$\sim$ph linkage. This illustrates a mechanism by which such a transformation takes place: Although the total of energy over the whole compound is equal on both sides of the equilibrium, the intramolecular energy distribution is changed by dehydration in such a way that a much larger part is concentrated now in the organo-phosphate linkage. Besides illuminating the nature of the process these considerations lead to a hitherto unnoticed possibility of calculating the energy present in such a bond. The total change in free energy (ΔF) with the reactions (*a*) and (*b*):

$$\begin{array}{lcll} \text{ph-glycerate} & \longrightarrow & \text{phosphate} + \text{pyruvate} & (a) \\ \downarrow\ \uparrow & & & \\ \text{ph}\sim\text{pyruvate} + H_2O & \longrightarrow & \text{phosphate} + \text{pyruvate} & (b) \end{array}$$

must be the same, since ph-glycerate and ph$\sim$pyruvate are energy-equivalent. As we know from phosphoglycerol, to which phosphoglycerate is entirely comparable with regard to ph-linkage the removal of ester phosphate by hydrolysis alone, *i.e.,* by formation of glycerate, means a loss of ca. 3000 cal ($\Delta F_1 = -3000$ cal) only. But dephosphorylation of ph$\sim$pyruvate presumably occurs with a much larger loss. This difference must be made up by loss of energy with dehydration of glycerate to pyruvate (ΔF_2). Starting with ph-glycerate (reaction (*a*)) the total reaction can be divided into two partial reactions:

$$\begin{array}{ll} \text{I:} & \text{ph-glycerate} + H_2O \longrightarrow \text{glycerate} + \text{phosphate};\ \Delta F_1 \\ \text{II:} & \text{glycerate} \longrightarrow H_2O + \text{pyruvate};\ \Delta F_2 \\ \hline \text{I + II:} & \text{ph-glycerate} \longrightarrow \text{phosphate} + \text{pyruvate};\ \Delta F_1 + \Delta F_2 = \Delta F_{total} \end{array}$$

The dehydration of free glycerate to pyruvate must be a largely exergonic* reaction (occurring with loss of free energy) in contrast to the dehydration of ph-glycerate to ph$\sim$pyruvate, *where through attachment of the phosphate residue this energy, dissipated otherwise, is retained in the enol$\sim$ph bond.* If we add to it the ΔF_1 with hydrolysis of the ester-phosphate linkage (reaction I), the sum, $\Delta F_1 + \Delta F_2$, represents ΔF with hydrolysis of the energy-rich enol$\sim$ph linkage because, as shown by equation (*a*) and (*b*), it must be the same as with cleavage of phosphoglycerate to pyruvate and phosphate. With ΔF_1 now estimated, ΔF_{total} is calculable upon ΔF_2 being determined. Unfortunately this determination can only be carried out in a rough way. But since at present this procedure represents the only means to determine ΔF for the splitting of a phosphate linkage of this type, which is highly desirable, the calculation was carried out, using equation:

$$\Delta F_2 = \Delta H_2 - T\,\Delta S_2. \qquad (1)$$

Here ΔH_2 is given by the difference of the heats of combustion of pyruvic and glyceric acid, and ΔS_2 can be calculated from the entropy data given by Parks and Huffman[27] and by Kelley (in[13]).

The heat of combustion of liquid pyruvic acid, as determined by Blaschko[28], is 279,000 cal. Unfortunately, no experimental data are available for glyceric acid. An apparently very reliable value, however, was obtained by the use of the calculation method of Kharasch[29]. The reliability of this method is shown by the excellent agreement between calculated and determined values for pyruvic acid and for glycerol[30].

Pyruvic acid:	determined	279,000 cal,	calculated	280,000 cal
Glycerol:	"	397,000 cal,	"	397,200 cal

The method of calculation is based on the assumption that for every electron between C and C, and between C and H 26,050 cal are generated with combustion. To the basic electron value "structural correction factors" are added. Both pyruvic and glyceric acids contain 10 of the described electrons. To the basic value of 10 × 26,050 cal are to be added 19,500 cal for C:O next to COOH in *pyruvic* acid. In *glyceric* acid, however, the structural factors are 13,000 cal for CH_2OH and 13,000 cal for CHOH next to COOH, a total of 26,000 cal to be added to 260,500 cal.

Pyruvic acid		Glyceric acid	
10ε	260,500 cal	10ε	260,500 cal
C:O	19,500 cal	CHOH	13,000 cal
		CH_2OH	13,000 cal
	280,000 cal		286,500 cal

The method shows clearly that the replacement of two separate hydroxyls by one carbonyl, the structural difference caused by dehydration, involves an appreciable decrease of the heat of combustion. We think it probable that ΔH_2 (equation[1]) of 6500 cal, as calculated for the structural difference be-

* Coryell (169) introduced recently the terms "exergonic" and "endergonic" specifically for reactions occurring with negative and positive change in free energy, to contrast with exo- and endothermic designating now exclusively heat change.

tween the compounds, represents the most reliable value to be obtained at present. We prefer this to the slightly higher value of 7500 cal, obtained by subtraction of the value for glyceric acid as calculated (286,500 cal) from pyruvic acid determined (279,000 cal).

To obtain the entropies, S, of the compounds the atomic entropies of the atoms in their respective positions were added together. The numerical values were taken from Burk's paper (13), who obtained them from Dr. K. K. Kelley by personal communication: C, -13.4, O in terminal OH of COOH or CH_2OH, 0.9; O in secondary OH, -4.6; O in carbonyl and carboxyl, 24.4, H, 11.3.

$$S(CH_3CO \cdot COOH) = (3 \times -13.4) + (4 \times 11.3) + 0.9 + (2 \times 24.4) = 54.7$$

$$S(CH_2OH \cdot CHOH \cdot COOH) = (3 \times -13.4) + (6 \times 11.3) + (2 \times 0.9) + 24.4 - 4.6 = 48.9$$

$$\Delta S_2 = 54.7 - 48.9 = 5.8$$

Now, ΔF_2, the change of free energy with dehydration of glyceric to pyruvic acid, can be found (for T 298°) from $\Delta F_2 = \Delta H_2 - T\Delta S_2 = -6500 - 298 \times 5.8 = -8250$ cal. This represents the change for the pure liquid acids (undissolved). It is assumed that no appreciable differences in dilution and neutralization occur between the two compounds. Such assumption seems justifiable, because both are infinitely soluble in water and their acidic strength is approximately the same.

The final result, the change of free energy with decomposition of phosphoglycerate ($\rightleftharpoons$ phosphopyruvate), ΔF_{total}, is then obtained by addition of the changes with dephosphorylation of phosphoglycerate, ΔF_1, and with dehydration of glycerate to pyruvate, ΔF_2.

$$\Delta F_{total} = \Delta F_1 + \Delta F_2 = -3000 - 8250 = -11{,}250 \text{ cal}$$

It is to be noticed that an appreciable part of ΔF is made up by the increase in entropy from glyceric to pyruvic acid, due to the formation of a carbonylic group. A partially but not wholly similar situation was some years ago discussed by Burk[13], who pointed out that in compounds of the same molecular formula the formation of the carbonyl group corresponded to an entropy ($T\Delta S$) of some -7500 cal per mol. In the present case the value is, or may be regarded as less (ca. -2000 cal), because elimination of water is also involved.

The heat change determined by Meyerhof and Schulz[25] for the reaction phosphoglycerate = phosphate + pyruvate was ΔH: $-$ 8250 cal and for phosphopyruvate + $H_2O \longrightarrow$ phosphate + pyruvate, ΔH: $-$ 8450 also practically identical. It appears that ΔF is about -3000 cal larger than ΔH, as is to be expected from the positive value of ΔS. This makes us confident that the calculated value is very nearly right. Therefore, it becomes more easily understandable that the heat change with decomposition of enol-phosphate is appreciably lower than with creatine phosphate and adenosine polyphosphate (see Table I), although accounting for irreversibility of phosphate transfer the free energy change with the latter compounds should even be lower than with enol-phosphate.

It is thus concluded that the average energy present in an energy-rich phosphate linkage amounts to 9000–11,000 cal, and of an ester linkage to around 3000 cal. The metabolic relation between the two groups of organo-phosphates

Table I

THERMODYNAMIC DATA FOR HYDROLYSIS OF PHOSPHO-ORGANIC COMPOUNDS

No.	Compound	ΔH cal	Reference	ΔF cal	Reference
1	Phosphoglycerol	....	....	− 2,350	Equilibrium constant, Kay[21]
2	Phosphopyruvate	− 8,450	Meyerhof and Schulz[25]	−11,250 (2–3)	Calculation, this paper
3	Phosphoglycerate (hydrolysis to phosphate any pyruvate)	− 8,250	〃 〃 〃		
4	Phosphocreatine	−10,700	Meyerhof and Schulz[25]	(−10,000) (4–9)	Approximation from phosphopyruvate; mutual equilibrium
5	Phosphoarginine	− 7,700	〃 〃 〃		
6	Adenosine polyphosphate (per easily hydrolyzable P)	−12,000	Meyerhof and Lohmann[32]		
7	Phosphoglyceryl phosphate	....	....		
8	Acetyl phosphate	(− 8,000)	Difference of heat of combustion for acetic acid and acetic anhydride[29]		
9	Amido phosphate, $(OH)_2OP{\sim}NH_2$	−14,000	Meyerhof and Lohmann[31]		

in the cell laboratory is to be compared with the relation of acid anhydride to acid ester in the organic chemical laboratory. Even the numerical difference of the group potential is approximately the same in both cases, *e. g.*, ca. 8000 cal for the acetyl group (see Table I). The differences in energy levels of the energy-rich anhydride and the energy-poor ester drives the group into the ester bond.

IV. Chemistry and Distribution in Nature of Energy-Rich Phosphate Bonds

1. General survey

In three of the four bond types to be discussed, the high group potential is caused by the anhydric nature of the bond. Anhydrization takes place either between two phosphates (P-O$\sim$P) or between a phosphate and a carboxyl (O:C·O$\sim$P) or acidic enol (C:C·O$\sim$P). In guanidine phosphates (N$\sim$P), however, the direct connection between phosphorus and nitrogen is held responsible for the high group potential by Meyerhof and Lohmann[31] (see No. 9 in Table I).

Formulae for Compounds

N$\sim$P

NH
$HN\cdot\ddot{C}\cdot N\cdot CH_2\cdot COOH$
$\wr$ $\dot{C}H_3$
PO_3H_2

creatine phosphate

NH
$HN\cdot\ddot{C}\cdot NH\cdot(CH_2)_3\cdot CHNH_2\cdot COOH$
$\wr$
PO_3H_2

arginine phosphate

$:\dot{C}—O\sim P$

$CH_2:C\cdot COOH$
$\dot{O}$
$\wr$
PO_3H_2

phosphoenol pyruvic acid

$CH_2\cdot CHOH\cdot C:O$
O $\dot{O}$
PO_3H_2 $\wr$
PO_3H_2

phosphoglyceryl phosphate

$CH_3\cdot C:O$
$\dot{O}$
$\wr$
PO_3H_2

acetyl phosphate

P—O$\sim$P

NH_2
$\dot{C}$
N C—N———$C_6H_8O_4\cdot\ddot{P}\cdot O\sim\ddot{P}\cdot O\sim\ddot{P}\cdot OH$ (each P bearing O above and $\dot{O}H$ below)
HC C—N >CH
N

adenosine tri-phosphate

In the formulae represented, the energy-rich linkage expressed by $\sim$ is placed between oxygen and phosphorus. This decision was made because by reversible shifting of $\sim$ph from adenosine polyphosphate to creatine (partial Parnas-reaction[33,34]) the break can only occur between O and the terminal -PO_3H_2, which changes place with an hydrogen in the NH_2 of the guanidine group of creatine to form $HN \sim PO_3H_2$ (compare formulae).

In Table II the commonly used quantitative procedures for determination of the compounds with energy-rich phosphate bond are listed. These methods not only reflect the behavior of the substance but, in most cases, also the manner by which the substance was discovered.

Table II

CUSTOMARY METHODS OF DETERMINATION

No.	Compound	Analytical Denotation	Determined as:
1	Creatine phosphate	Acid-unstable P	Difference between colorimetric P and the P found by alkaline Mg[35] or Ca[4] precipitation (= true inorganic P)*
2	Arginine phosphate	Acid-unstable P	Same procedure adapted to slightly greater acid stability[31]*
3	(ad-ph)$\sim$ph$\sim$ph	Easily hydrolyzable P	Difference between colorimetric P immediately and after 7 min. hydrolysis with *n*-HCl at 100°[36]
4	Phosphopyruvate	......	P, mineralized by hypoiodite treatment[1]
5	Acetyl phosphate	Acid-, alkali-unstable P	Difference between colorimetric P and neutral Ca precipitate[3] (= true inorganic P)*
6	Phosphoglyceryl phosphate	Acid-, alkali-unstable P	See[2]

*The test solution for colorimetric determination contains 0.5 *n*-H_2SO_4 and 2.5% ammonium molybdate (Fiske and Subbarow[37]). The rate of decomposition of Nos. 1, 5, and 6 is greatly accelerated in presence of molybdate. Decomposition in plain acid is much slower. In dilute trichloroacetic acid at 0° practically no decomposition occurs in short periods.

Acid-base changes with phosphorylation

Generally speaking, the attachment of phosphate to an organic molecule is accompanied by increase in acidity since the organo-phosphoric acid is always a stronger acid than ortho phosphoric acid. The complicated change of the acid base equilibrium needs careful consideration because of manifold physiological and methodological implications.

In the physiological range of *p*H ortho phosphate is present in solution in two forms, as primary and as secondary salt:

I			II	
O^-	(1)		O^-	(1)
O:P OH	(2)		O:P O^-	(2)
OH	(3)		OH	(3)
*p*H 5		$\longrightarrow$	*p*H 8	

In both forms, between *p*H 5 and 8, the third hydrogen is entirely undissociated, the first being completely eliminated. It is only the dissociation of the second

H which changes within this range of *p*H. On phosphorylation, always the undissociated third OH group enters the organic linkage. Solely from the replacement of a "homoiopolar" bonded H by an organic radical, no change of the acid-base equilibrium would be expected. However, as pointed out, a more or less pronounced change occurs because in the organo-phosphate the second OH always becomes more acidic. Or, with transition from inorganic to organic phosphate, part of form I is changing to form II. The paradoxical situation results that with the disappearence of a potentially acid group, the reaction becomes more acidic; and in the reverse direction becomes more alkaline. The rise with phosphorylation of the dissociation constant (decrease of *p*K) of the second OH is shown in Table III.

Table III

SECOND ACID DISSOCIATION CONSTANTS OF ORGANO-PHOSPHATES

Compound	pK_2	Reference
Ortho phosphate	6.8	
Hexose-di-phosphate	6.29	Meyerhof and Lohmann[38]
Fructose-6-phosphate (Neuberg ester)	6.11	" " " "
Glucose-6-phosphate (Robison ester)	6.10	" " " "
Creatine phosphate	4.5	Fiske and Subbarow[4]

At *p*H 7 with one mole of phosphate combining with hexose or creatine, respectively, 0.25 or 0.4 equivalents of "acid" appear. With the reverse reactions corresponding amounts of base are liberated. The pronounced alkalization taking place with creatine phosphate breakdown was measured in living muscle by Lipmann and Meyerhof[39] and proved the actual occurrence of this reaction in the living organ.

Only with adenosine polyphosphate the change is not paradoxical. Synthesis is accompanied by disappearance and decomposition by appearance of "acid" in solution. At *p*H 7.1 per mol of "pyro" phosphate formed ca. 0.5 equivalents of acid disappear (Lohmann[40]). Because in adenylic acid the third OH group has already undergone esterification anhydrization occurs between the partly dissociated acidic second OH of adenylic acid and the third OH in ortho-phosphate. A net decrease of "acid" therefore results with the reaction:

$$\text{ad—O—}\underset{\ddot{\text{O}}}{\overset{\text{O}-}{\dot{\text{P}}}}\cdot\text{O}^- + \text{HO}\cdot\underset{\ddot{\text{O}}}{\overset{\text{O}-}{\dot{\text{P}}}}\cdot\text{O}^- \longrightarrow \text{ad—O—}\underset{\ddot{\text{O}}}{\overset{\text{O}-}{\dot{\text{P}}}}\cdot\text{O}\cdot\underset{\ddot{\text{O}}}{\overset{\text{O}-}{\dot{\text{P}}}}\cdot\text{O}^- + \text{OH}^-$$

2. ADENOSINE POLYPHOSPHATE

The P-O$\sim$P linkage, although attached to an organic residue, is in most respects equivalent to the P-O$\sim$P link in inorganic pyrophosphate. The rate of hydrolysis in hot acid and the heat of decomposition are the same for free and bound pyrophosphate. The pyrophosphate first isolated from muscle was the inorganic form[36], then assumed by Lohmann to be a cell constituent. Subsequently, however, Davenport and Sacks[41] showed that in untreated muscle filtrate no inorganic pyrophosphate could be demonstrated with a specific colorimetric method. Reinvestigating the problem, Lohmann[42] eventually isolated adenylpyrophosphate, of which the inorganic pyrophosphate had been a breakdown product. He found that the Ba salt of adenylpyrophosphate splits off the pyrophosphate on standing, especially in alkaline solution. After demonstration of stepwise dephosphorylation[40] the preferable designation adenosine polyphosphate instead of adenylpyrophosphate came into use for the whole group and, specifically, the designations: adenosine-mono-, -di-, and -tri-phosphate. As abbreviations, ad-ph, ad-ph$\sim$ph, ad-ph$\sim$ph$\sim$ph shall be used in this paper. The maximum of two P-O$\sim$P linkages is formed per mole of adenosine. The first phosphate is linked to ribose in an ordinary ester linkage which is, in contrast to the other two, an energy poor linkage. The P-O$\sim$P linkages are readily hydrolyzed with hot acid and frequently are spoken of as easily hydrolyzable phosphate (Table II).

The wide distribution of adenosine polyphosphate in nature is shown in Table IV.

Frequently compounds have been isolated with the properties of a diadeno-

Table IV

ADENOSINE POLYPHOSPHATE IN TISSUES

Tissue	Mg. % P		Reference
	Polyphosphate P	Inorganic P	
Yeast	80	160	Lohmann[43]
Muscle, frog	24	...	" "
Muscle, rabbit, white	32	...	" "
Muscle, rabbit, red	21	...	" "
Muscle, crab	32	...	" "
Heart, rabbit	9	50	" "
Uterus, rabbit	8	..	" "
Liver, rabbit	7	35	" "
Kidney, rabbit	6; 10	45; 18	Lohmann[43]; Eggleton[44]
Spleen, rabbit	17	36	Lohmann[43]
Brain, rabbit	5	7	" "
Nerve, rabbit	6	..	Gerard and Tupikova[45]
Red blood cells, pig	9	9	Lohmann[43]
Embryo, rat	9	17	" "
C 180 tumor, mouse	8	..	Franks[46]
L R 10 tumor, rat	11.5	25	Boyland[47]
Jensen sarcoma	14	29	Lohmann[48]

The listed figures represent easily hydrolyzable phosphate corrected for interfering substances. Actual isolation of the adenosine polyphosphate has been carried out with muscle[42], heart[48], and blood[49,50]. Yeast and muscle contain the largest amounts. The content in spleen is high. Malignant tumors contain fairly large quantities in comparison to other parenchymatous tissues. The compound has been demonstrated in bacteria, plant seeds, and arbacia eggs[43]. It must be considered a universally present cell constituent and thus phosphate transfer should be considered a universal metabolic reaction.

sine polyphosphate (Deuticke[51], Ostern[52], Warburg and Christian[53]). More recently Kiessling and Meyerhof[54] isolated and carefully studied a polymer obtained from yeast to which they assign the composition of a di-adenosine-tetraphosphate:

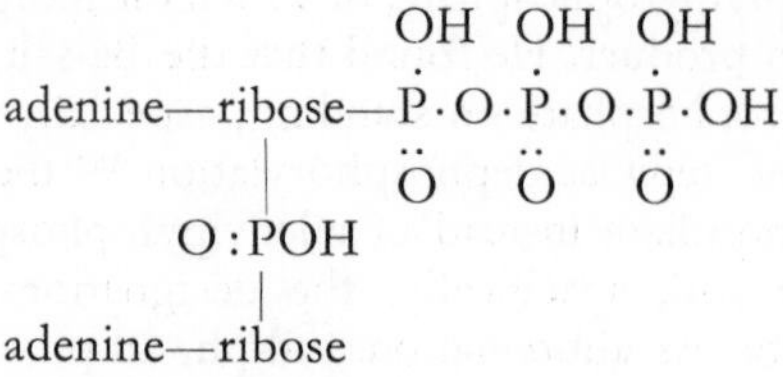

```
                             OH  OH  OH
                              ·   ·   ·
      adenine—ribose—P·O·P·O·P·OH
                                 |   ··   ··   ··
                                 |   O    O    O
                      O:POH
                                 |
      adenine—ribose
```

Only two of the phosphates are easily hydrolizable. Evidence is offered for the existence of a phosphate bridge between the two riboses. The compound is of interest because the linkage between the adenylic acids is very alkali-sensitive, much like the inter-nucleotidic linkages in nucleic acid. No physiological difference could be detected between the mono- and dinucleotide.

The presence of polyphosphate chains in compounds of biological importance should stimulate interest in the numerous and peculiar products formed by anhydric polymerization of orthophosphates partly in combination with ammonia. Stokes[55] made a very interesting study of polymerization products of amido phosphoric acid which he called meta-phosphimino acids and which had the general formula $(HPO_2(:NH))_x$. They correspond to the meta phosphoric acids $(HPO_3)_x$. According to Stokes the phosphimino acids exist in two forms, (*1*) as ring structures of great stability, (*2*) as open chains of small stability. By hydrolysis (*1*) is converted into (*2*). The especially stable ring structure $(HPO_2(:NH))_4$, (I) and the corresponding unstable open chain (II) are shown below.

```
        (I)                                   (II)

         H
         O
        /P\
       /  ··  \
    HN   O   NH
    /          \                  OH         OH         OH         OH
HOP:O        O:POH      HO—P—NH—P—NH—P—NH—P—NH2
    \          /                  ··         ··         ··         ··
    HN   O   NH                  O          O          O          O
         ··
        \P/
         O
         H
```

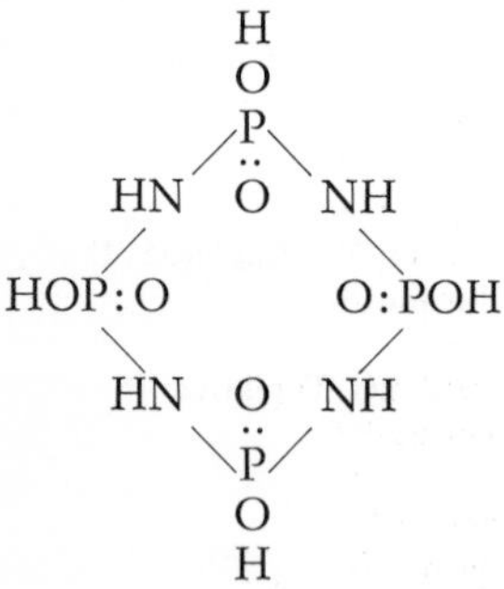

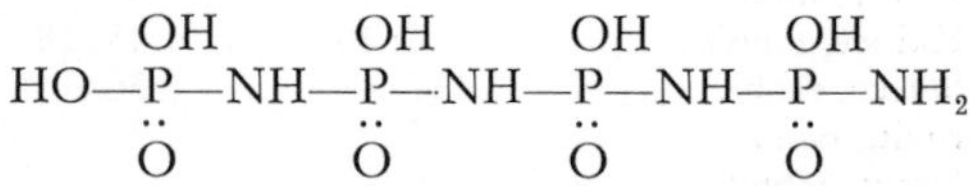

An analogous relation should exist between nitrogen-free compounds like tri-metaphosphoric acid, $(HPO_3)_3$, and tri-phosphoric acid (Neuberg and Fischer[56], Huber[57]), the inorganic homologue of ad-ph~ph~ph.

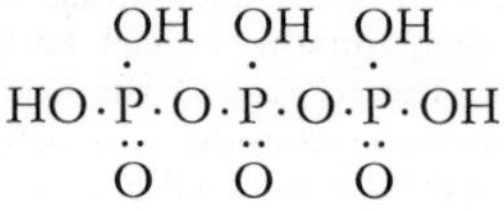

```
      OH  OH  OH
       ·   ·   ·
HO·P·O·P·O·P·OH
      ··   ··   ··
      O    O    O
```

By the increase of stability with ring closure it is suggested that the opening and closing of the ring has a large effect on the intramolecular energy distribution. This might have some bearing on the problem of utilization of P∼N and P-O∼P linkages in muscular contraction. Furthermore, the similarity between the P-NH-P and the P-O-P bond in these structures is remarkable.

3. Phosphoguanidine linkages (phosphagens)

The ontogenetically newer creatine phosphate in vertebrates and the "older" arginine phosphate in invertebrates are not, as is adenosine polyphosphate, substances universally occurring in cells. Their almost exclusive presence in muscular and nervous organs[44,45], including the electric organ of fishes (Kisch[63]), suggests connection with the speed of action required by these organs. Their function as reservoirs of energy easily to be released is readily confirmed by experimental facts.

Creatine phosphate

In Table V the distribution in tissues of creatine P is shown. The substance was isolated only from muscular tissue (Fiske and Subbarow[4]). The figures represented in the table were obtained by determination of "acid unstable" P (see Table II). Therefore it must be emphasized that the significance of all figures below 3 mg. % as found in some parenchymatous organs remains doubtful. Likewise, the demonstration of labile phosphate in tumor tissue, interesting in itself, does not, as pointed out by Franks[46], give conclusive proof for the presence of phospho∼creatine. Boyland[47] confirmed Franks's result and furthermore showed a content of 25 mg. % of creatine in rat tumor. He used, however, the colorimetric method for creatine determination which gives not quite convincing results for such small amounts (see Miller and Dubos[58]).

Table V

Creatine Phosphate in Tissues

Tissue	Mg. % P		Reference
	Creatine P	Inorganic P	
Muscle, frog	54	22	Gerard and Tupikova[45]
Muscle, cat	60	26	Fiske and Subbarow[4]
Muscle, amphioxus	37	72	Meyerhof[61]
Brain, dog	12	. .	Gerard and Tupikova[45]
Nerve, frog	9	7	" " " "
Heart, normal rat	4–7	. .	Bodansky[62]
Heart, hyperthyroid rat	1–3	. .	" "
Stomach, rabbit	2–5	25–32	Eggleton and Eggleton[44]
Testicle	0.6–2.6	8–12	" " " "
Uterus, rabbit	1.4	11.6	" " " "
C 180 tumor, mouse	1.5–2.7	. .	Franks[46]
L R 10 tumor, rat	2.5	25	Boyland[47]
Jensen sarcoma	1.2	22	" "

Table VI

FREE AND PHOSPHORYLATED CREATINE IN TISSUES

Tissue	Total creatine, mg. %	% Phosphorylated
Muscle, frog	495	52
Nerve, frog	146	40
Brain, rat	230	20

In Table VI the values for free and combined creatine are compared in tissues high in phosphocreatine. The table is taken from a paper by Gerard and Tupikova[59].

In muscle over half of the total creatine is combined with P, and about 70% of the available P is found combined with creatine. That makes only a slight surplus of creatine over phosphate. In nervous tissue likewise total creatine is fairly equivalent to available + combined phosphorus. As is noteworthy, testis is an organ known to contain large amounts of creatine [up to 200 mg. % (Hunter[60])] but none or only doubtful amounts of phosphocreatine. It seems to me a question of great interest if the constantly occurring decay of creatine to creatinine might be related to the linking of creatine to phosphate in the body. Comparison between the formulae of creatinine and phospho~creatine makes such a theory quite tempting:

HN:C—N(H)

$H_3C \cdot N \cdot CH_2 \cdot C:O$

HN:C—N(H)—P:O(OH)(OH)

$H_3C \cdot N \cdot CH_2C:O$
OH

In both compounds the free amino group of the guanidino part is combined with an acidic group. With the tension present in the P-N linkage a change to the more stable and similar C-N linkage, offered by the presence of carboxyl in the molecule, seems suggested.

Arginine phosphate

In search for a compound similar to phosphocreatine in invertebrate muscle, known not to contain creatine, Meyerhof and Lohmann[31] found phosphoarginine. In chemical behavior phosphoarginine is very similar to phosphocreatine[31]. Physiologically the compounds are perfectly analogous (Baldwin and Needham[65], Ochoa[64]). Data for phosphoarginine content of invertebrate muscles are represented in Table VII.

Practically no free arginine was demonstrable in the fresh oxygenated muscle. Free and total arginine were estimated with arginase. Phosphoarginine is not attacked by arginase.

4. PHOSPHOENOL PYRUVIC ACID

Embden[66] discovered that phosphoglyceric acid, earlier isolated by Nilsson[67], was fermented to pyruvic acid and phosphate:

$$CH_2OPh \cdot CHOH \cdot COOH \longrightarrow CH_3 \cdot CO \cdot COOH + H_3PO_4.$$

Table VII

ARGININE PHOSPHATE IN INVERTEBRATE MUSCLE

Animal	Mg. % P		Free Arginine	Reference
	Inorganic P	Arginine P		
Crab	35	61	..	Meyerhof and Lohmann(31)
Pecten	42	84	0	Meyerhof(61)
Sipunculus	57	171	Trace	Meyerhof(61)
Holothuria tuberculata	16	46	0	Meyerhof(61)

Subsequently this complex reaction was analyzed by Lohmann, Meyerhof and Kiessling. Embden's reaction was found to occur in three steps:

$$CH_2OPh \cdot CHOH \cdot COOH \rightleftharpoons CH_2OH \cdot CHOPh \cdot COOH$$
$$\rightleftharpoons CH_2 : COPh \cdot COOH + H_2O \longrightarrow CH_3 \cdot CO \cdot COOH + H_3PO_4.$$

First, Parnas, Ostern, and Mann[33] observed that a transfer of the phosphate present in phosphoglyceric acid to adenylic acid occurs in muscle whereby the energy-rich phosphate bond in ad-ph~ph~ph is formed. Lohmann and Meyerhof[1] then found that the removal of phosphate from phosphoglyceric acid occurs only when adenylic acid is present to accept the phosphate. A partial change took place, however, also in absence of adenylic acid, yielding an ester which was less stable against acid hydrolysis. Eventually the transformation product was isolated and identified as phosphopyruvic acid containing phosphate in an energy-rich bond[1]. Later on, with Kiessling[68], the first transformation product, phospho-2-glyceric acid, was isolated. In the absence of adenylic acid the following equilibrium mixture was obtained in muscle extract.

	%	Enzyme
d (−) phospho-3-glyceric acid	58.5	phosphoglycero-mutase[68] (58.5 ↔ 12.5)
d (+) phospho-2-glyceric acid	12.5	enolase[1] (12.5 ↔ 29)
phospho~enolpyruvic acid	29	

SPECIFIC REACTIONS OF PHOSPHOENOL PYRUVIC ACID[1]

Reagent	Products of Reaction
Br_2	$CH_2Br \cdot CO \cdot COOH + HBr + H_3PO_4$
I	$CH_3CI_3 + CO_2 + 3HI + H_3PO_4$
$HgCl_2$	$CH_3CO \cdot COOH + H_3PO_4$ (catalytic)
n-HCl, 100°	Complete hydrolysis after 60 minutes

With the analysis of the stepwise degradation of phosphoglyceric acid Lohmann and Meyerhof were able to define specifically the effect of fluoride long known to be a powerful poison of fermentation and glycolysis. They found that the enolase reaction is very sensitive to fluoride. This explained the accumulation of phosphoglycerate which had been shown to occur in Nilsson's reaction[67].

$$\begin{array}{c}2\text{ acetaldehyde} + \text{glucose} + 2H_3PO_4 \longrightarrow \text{fluo} \mid \text{ride} \\ 2\text{ ethylalcohol} + 2\text{ phosphoglyceric acid} \downarrow \ldots\ldots\end{array}$$

Fluoride inhibition, thus presenting a means to interrupt the chain of transformations of phosphorylated intermediates at a definite stage, but without interfering with the central O/R reaction, became a powerful tool for the analysis of intermediate processes in fermentation. Fluoride inhibition which is shown by various fermentations and by glycolysis in various tissues (Dickens[69]) indicated early the general occurrence of phosphorylation in fermentation processes. More recently Werkman[70] showed in a variety of fermentations with various organisms that there, like in alcoholic fermentation, phosphoglyceric acid accumulation occurs under similar conditions. Yeast (Effront[71]), however, and propionic acid bacteria (Wiggert and Werkman[73]) can by subculturing with rising concentrations of fluoride become adapted to the poison. Wiggert and Werkman[73] showed, furthermore, that fluoride-fast propionic bacteria, in contrast to normal, did not attack phosphoglyceric acid. This was interpreted to show that a non-phosphorylating fermentation mechanism operates in the fluoride-fast organisms. It seems, however, not unlikely that the fluoride-fast organism has become impermeable to fluoride during the period of adaptation, and that a change of permeability had likewise caused the non-fermentability of phosphoglyceric acid. Effront[72] reports for fluoride treated yeast that it had changed greatly in appearance and in composition, *e.g.*, Ca content was raised from 1.65 to 4.1 in arbitrary units. Runnström's[74] experiments are also suggestive in this direction. He showed that the poisoning effect of fluoride in normal yeast depends greatly on the state of the cell. Aerobically and with addition of glucose, inhibition was invariably less or absent. Anaerobically or without substrate, however, the poisoning effect was greatly exaggerated. Runnstöm interprets this effect as a change of permeability for fluoride with active metabolism.

5. Phosphoglyceryl phosphate

A diphosphoglyceric acid was isolated by Negelein and Brömel[2] as the product of enzymatic oxidation of phosphoglyceraldehyde (Fischer ester[75]) + phosphate. This compound was not identical with Greenwald's 2–3-diphosphoglyceric acid[76]. The anhydric nature of one phosphate linkage was assumed because of its instability even at neutral reaction. This assumption was corroborated by the showing of an ultraviolet absorption band at 240 Å, analogous to acetanhydride and disappearing after decomposition of the labile ester linkage. The compound was called by Negelein and Brömel R-diphosphoglyceric acid or 1–3-diphosphoglyceric acid. We have suggested the more descriptive name of phosphoglyceryl phosphate[77].

6. Acetyl phosphate

A phospho organic compound of stability conditions analogous to synthetic acetyl phosphate was shown to be formed by enzymatic oxidation of pyruvate + phosphate (Lipmann[3]). It was synthesized by a modification of the method of Kämmerer and Carius[78] for the preparation of triacetyl phosphate (unpublished). Instead of using Ag_3PO_4, a mixture of Ag_3PO_4 and

$2H_3PO_4$, and acetyl chloride was used. Lynen[79] described recently a more reliable synthesis by passing through the dibenzyl phosphate. As an acid anhydride, the compound is quite stable in water.

Analogously, succinyl phosphate would be formed by oxidation of ketoglutarate + phosphate. We synthesized this compound from succinyl chloride + Ag_3PO_4 and found it very similar to acetyl phosphate (unpublished).

* * *

Bibliography

1. K. Lohmann and O. Meyerhof, *Biochem. Z.,* **273**, 60 (1934).
2. E. Negelein and H. Brömel, *Ibid.,* **303**, 132 (1939).
3. F. Lipmann, *J. Biol. Chem.,* **134**, 463 (1940).
4. C. H. Fiske and Y. Subbarow, *J. Biol. Chem.,* **81**, 629 (1929).
5. P. Eggleton and G. P. Eggleton, *Biochem. J.,* **21**, 190 (1927).
6. O. Meyerhof and J. Suranyi, *Biochem. Z.,* **191**, 106 (1927).
7. E. Lundsgaard, *Ibid.,* **217**, 162 (1930).
8. E. Lundsgaard, *Ibid.,* **227**, 51 (1930).
9. A. V. Hill, *Physiol. Revs.,* **2**, 339 (1922).
10. O. Meyerhof, E. Lundsgaard, and H. Blaschko, *Naturwissenschaften,* **18**, 787 (1930).
11. D. Nachmansohn, *Biochem. Z.,* **196**, 73 (1928).
12. E. Lundsgaard, *Ibid.,* **233**, 322 (1931).
13. D. Burk, *Proc. Roy. Soc.,* **B104**, 153 (1929).
14. A. V. Hill and J. L. Parkinson, *Ibid.,* **B108**, 148 (1931).
    O. Meyerhof and E. Boyland, *Biochem. Z.,* **237**, 406 (1931).
    O. Meyerhof, Ch. L. Gemmill, and G. Benetato, *Ibid.,* **258**, 371 (1933).
15. F. Lipmann, *Nature,* **143**, 281 (1939).
16. H. Kalckar, *Biochem. J.,* **33**, 631 (1939).
17. S. P. Colowick, M. S. Welch, and C. F. Cori, *J. Biol. Chem.,* **133**, 359 (1940).
18. L. Michaelis and M. P. Schubert, *Chem. Revs.,* **22**, 437 (1938).
19. M. Berthelot and St. Gilles, *Ann. chim. phys.,* **66**, 225 (1863).
20. G. N., Lewis and M. Randall, "Thermodynamics and the Free Energy of Chemical Substances," New York, 1923.
21. H. D. Kay, *Biochem. J.,* **22**, 855 (1928).
22. C. F. Cori, *Cold Spring Harbor Symposia,* **7**, 260 (1939).
23. G. T. Cori, C. F. Cori, and G. Schmidt, *J. Biol. Chem.,* **129**, 629 (1939).
24. S. Veibel, *Enzymologia,* **2**, 124 (1936).
25. O. Meyerhof and W. Schulz, *Biochem. Z.,* **281**, 292 (1935).
26. D. Burk, *J. Phys. Chem.,* **35**, 432 (1931).
27. G. S. Parks and H. M. Huffman, "The Free Energies of Some Organic Compounds," New York, 1932.
28. H. Blaschko, *Biochem. Z.,* **158**, 428 (1925).
29. M. S. Kharasch, "Heats of Combustion of Organic Compounds," *Bureau of Standards J. of Research,*, 2 (1929).
30. Stohmann and Langbein, *J. prakt. Chem.,* (2) **45**, 305 (1892).
31. O. Meyerhof and K. Lohmann, *Biochem. Z.,* **196**, 49 (1928).
32. O. Meyerhof and K. Lohmann, *Ibid.,* **273**, 73 (1934).
33. J. K. Parnas, P. Ostern and T. Mann, *Ibid.,* **272**, 64 (1934).
34. H. Lehmann, *Ibid.,* **281**, 271 (1935).

35. K. Lohmann, *Biochem Z.,* **194**, 306 (1928).
36. K. Lohmann, *Ibid.,* **202**, 466 (1928).
37. C. H. Fiske and Y. Subbarow, *J. Biol. Chem.,* **66**, 375 (1925); see also K. Lohmann and L. Jendrassik, *Biochem. Z.,* **178**, 419 (1926).
38. O. Meyerhof and K. Lohmann, *Ibid.,* **185**, 113 (1927).
39. F. Lipmann and O. Meyerhof, *Ibid.,* **227**, 84 (1930).
40. K. Lohmann, *Ibid.,* **282**, 120 (1935).
41. H. A. Davenport and J. Sacks, *J. Biol. Chem.,* **81**, 469 (1929).
42. K. Lohmann, *Biochem. Z.,* **233**, 460 (1931).
43. K. Lohmann, *Ibid.,* **203**, 164 (1928).
44. G. P. Eggleton and P. Eggleton, *J. Physiol.,* **68**, 198 (1929).
45. R. W. Gerard and N. Tupikova, *J. Cellular Comp. Physiol.,* **13**, 1 (1939).
46. W. R. Franks, *J. Physiol.,* **74**, 195 (1932).
47. E. Boyland, *Ibid.,* **75**, 136 (1932).
48. K. Lohmann and Ph. Schuster, *Biochem. Z.,* **282**, 104 (1935).
49. H. K. Barrenscheen and W. Filz, *Ibid.,* **250**, 281 (1931).
50. K. Lohmann and Ph. Schuster, *Ibid.,* **294**, 183 (1937).
51. H. J. Deuticke, *Pflügers Arch. ges. Physiol.,* **230**, 537 (1932).
52. P. Ostern, *Biochem. Z.,* **270**, 1 (1934).
53. O. Warburg and W. Christian, *Ibid.,* **287**, 291 (1936).
54. W. Kiessling and O. Meyerhof, *Ibid.,* **296**, 410 (1938).
55. Stokes, *Z. anorg. Chem.,* **19**, 36 (1899).
56. C. Neuberg and H. A. Fischer, *C. R. Carlsberg,* **22** (Volume jubilaire Sörensen), 366 (1938).
57. H. Huber, *Z. angew. Chem.,* **50**, 323 (1937).
58. B. F. Miller and R. Dubos, *J. Biol. Chem.,* **121**, 447 (1937).
59. R. W. Gerard and N. Tupikova, *J. Cellular Comp. Physiol.,* **12**, 325 (1938).
60. A. Hunter, "Creatine and Creatinine," London, 1928.
61. O. Meyerhof, *Arch. sci. biol.* (Italy), **12** (Botazzi-Festschrift), 526 (1928).
62. M. Bodansky, *J. Biol. Chem.,* **109**, 615 (1935).
63. B. Kisch, *Biochem. Z.,* **225**, 183 (1930).
64. S. Ochoa, *Biochem. J.,* **32**, 327 (1938).
65. E. Baldwin and T. Needham, *Proc. Roy. Soc.,* **B122**, 197 (1937).
66. G. Embden, *Klin. Wochschr.,* **12**, 213 (1933).
67. R. Nilsson, *Arkiv Kemi Mineral. Geol.,* **A10**, No. 7 (1930).
68. O. Meyerhof and W. Kiessling, *Biochem. Z.,* **280**, 99 (1935).
69. F. Dickens and F. Simer, *Biochem. J.,* **24**, 1301 (1930).
70. C. H. Werkman, *Bacteriol. Rev.,* **3**, 187 (1939).
71. T. Effront, *Compt. rend.,* **117**, 559 (1893); **118**, 1420 (1894); **119**, 169 (1894).
72. T. Effront, "Moniteur Scientifique," 1905, p. 19.
73. W. P. Wiggert, and C. H. Werkman, *Biochem. J.,* **33**, 1061 (1939).
74. J. Runnström and E. Sperber, *Biochem. Z.,* **298**, 340 (1938).
75. H. O. L. Fischer and H. Baer, *Ber.,* **65**, 337 (1932).
76. J. Greenwald, *J. Biol. Chem.,* **63**, 339 (1925).
77. F. Lipmann, *Cold Spring Harbor Symposia,* **7**, 248 (1939).
78. Kämmerer and Carius, *Ann.,* **131**, 165 (1864).
79. F. Lynen, *Ber.,* **73**, 367 (1940).

* * *

PART IV

IS THE PHOSPHATE CYCLE PHYSIOLOGICAL?

THE REVISION of the Meyerhof cycle may also have bearings on what might be called the *Lundsgaard cycle*. Lundsgaard's success in advancing the idea of phosphocreatine splitting as a part of energy metabolism in muscle was based partly on ideas stemming from the Meyerhof cycle. Although phosphocreatine breakdown seems to be small during a series of single contractions (especially when compared with the splitting occurring in an iodoacetate-poisoned muscle), it nevertheless takes place and is merely concealed by resynthesis coupled to lactic acid formation.

Yet, for a long time, direct demonstration of a phosphate cycle proved difficult. This raised the question of whether there were any indications of a physiological phosphate cycle. Carl Neuberg[1] and others had been able to demonstrate lactic acid formation from the unphosphorylated compound, methyl glyoxal. Sacks[2] challenged the entire concept of phosphate cycles in intact cells. He even questioned whether phosphate compounds, like phosphocreatine and ATP, had functions other than mere buffering. Although Sacks' experiments and conclusions could well be seriously questioned, they nevertheless stimulated biochemists to settle some of the questions which go back to Embden's discovery of "lactocidogen." The challenge was taken up by Parnas[3] and his coworkers and by Meyerhof[4]. To solve the problem they chose a new and powerful approach introduced by Hevesy—the use of radioactive phosphorus. The paper by Korzybski and Parnas [q.v.] furnished one of the first demonstrations of the existence of a turnover of phosphorus compounds in the intact muscle. Any demonstration of its correlation with muscular contraction was, as I had the opportunity to ascertain for myself later[5], complicated by the fact that the permeation of phosphate into the intact muscle is rate limiting.

In a general way, the early studies by Lynen[6], on the role of phosphorylations and of inorganic phosphate steady-state levels in the Pasteur reaction of intact cells has further illustrated the crucial importance of phosphorylations in cell metabolism.

References

1. Neuberg, C., and Kobel, M., *Biochem. Z.,* **203** (1928), 463.
2. Sacks, J., *Amer. J. Physiol.,* **129** (1940), 227.
3. Korzybski, J. T., and Parnas, J. K., *Z. physiol. Chem.,* **258** (1938), 195.
4. Meyerhof O., Ohlmeyer, P., Gentner, W., and Maier-Leibnitz, W., *Biochem. Z.,* **298** (1938), 396.
5. Kalckar, H. M., Dehlinger, J., and Mehler, A., *J. Biol. Chem.,* **154** (1944), 275; Furchgott, R. F., and Shorr, E., *Fed. Proc.,* **2** (1943), 13.
6. Lynen, F., *Naturwiss.,* **30** (1942), 398.

SOME OBSERVATIONS ON THE TURNOVER OF THE PHOSPHOROUS ATOMS OF ADENOSINE TRIPHOSPHORIC ACID, IN THE LIVING ANIMAL, BY THE USE OF RADIOACTIVE ^{32}P-LABELED PHOSPHORUS*

T. Korzybski and J. K. Parnas

Laboratoire du Chimie, Faculté de Médecine,
Université Lwow, Poland

The authors have attempted to determine the mode of exchange of the phosphorus atoms of adenosine triphosphoric acid, in living rabbit muscle, by injection of radioactive inorganic phosphate ($Na_2H^{32}PO_4$) into the bloodstream. In the living animal the turnover of the phosphorous atom of adenylic acid is very slow, whereas the turnover of the other two atoms of phosphorus is very rapid.

We have attempted to clarify the mode of exchange, in living rabbit muscle, of the atoms of phosphorus contained in adenosine triphosphoric acid (= ATP A). In a previous study† we found that the phosphorus atom of adenylic acid is exchanged with inorganic phosphorus which has been introduced into the bloodstream and marked with radioactive ^{32}P. This exchange, however, is very slow, as compared with the exchange, in living muscle, of inorganic phosphorus, of the phosphorus of phosphocreatine, and of sugar esters, or of the two labile phosphorus atoms of adenosine triphosphoric acid.

The relationship between the atoms of phosphorus contained in a molecule of adenosine triphosphoric acid and those in the products of the splitting of this molecule is indicated in Scheme I.

We have injected rabbits with minute quantities (53 mg) of Na_2HPO_4 containing traces of $Na_2H^{32}PO_4$. After 30, 60, and 120 minutes we sacrificed

*T. Korzybski and J. K. Parnas, "Some Observations on the Turnover of the Phosphorus Atoms of Adenosine Triphosphoric Acid, in the Living Animal, by the Use of Radioactive ^{32}P-Labeled Phosphorus," *Bull. Soc. Chim. Biol.*, **21**, 713–716 (1939). Translated from the French and reprinted with the permission of the publisher.

†T. Korzybski and J. K. Parnas, *Z. physiol. Chem.*, **258** (1938), 195–204.

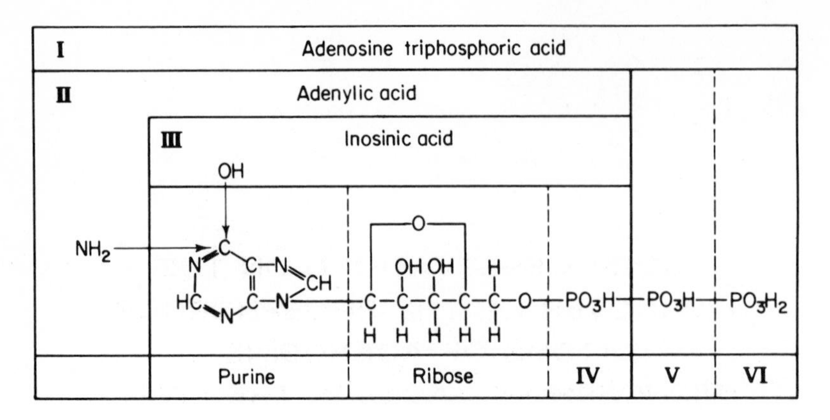

Scheme I

the animals and immediately prepared the phosphate compounds from the muscles, taking great care that the pulp obtained by grinding the muscles at low temperature should be fixed as soon as possible in 10% trichloroacetic acid. The method of preparation of the phosphate fractions is outlined in Scheme II, and the numbers of the fractions correspond to those indicated in Scheme I.

It was important to obtain the following phosphorus fractions: phosphorus from the inorganic phosphates of the tissue (VI); phosphorus from the labile groups of ATP acid, which is obtained as barium pyrophosphate, then transformed into $MgNH_4PO_4$ (V); phosphorus from the adenylic acid (IV), contained in adenosine triphosphoric acid, which we have isolated and purified by means of the barium salt of inosinic acid. The origin of the inosinic acid is certain: it was prepared from the barium precipitate of adenosine triphosphoric acid; this precipitate was subjected to alkaline hydrolysis by $Ba(OH)_2$, and the soluble product was deaminated by muscle adenylic deaminase. The deaminated product was then precipitated by lead acetate, purified as crystalline barium inosinate, and finally converted, after combustion, into $MgNH_4PO_4$ (see Scheme II)*.

Table 1 gives the number of radioactive disintegrations per milligram of $MgNH_4PO_4$, corresponding to the activities of phosphorus atoms IV and V

Table 1

Phosphorus fraction obtained from muscles of a rabbit sacrificed after:	30 min.	60 min.	120 min.
IV. Phosphate of adenylic acid	0	0.03	0.03
V. Phosphate of labile groups of ATP A	0.95	0.62	0.87
VI. Inorganic phosphate	1.2	0.76	1.01

*The barium pyrophosphate contains a considerable amount of adsorbed adenylic acid, resulting in losses in the final yield of inosinic acid. This impurity does not change the final result, since inosinic acid is not hydrolyzed in the course of the ten minutes used for hydrolysis of the pyrophosphate. One must nevertheless remove by recrystallization of the final $MgNH_4PO_4$ the Mg salt of the inosinic acid which is adsorbed onto it. (See: Hevesy, Baranowski, Guthke, Ostern and Parnas, *Acta Biologiae Experimentalis,* **12** (1938), 39.

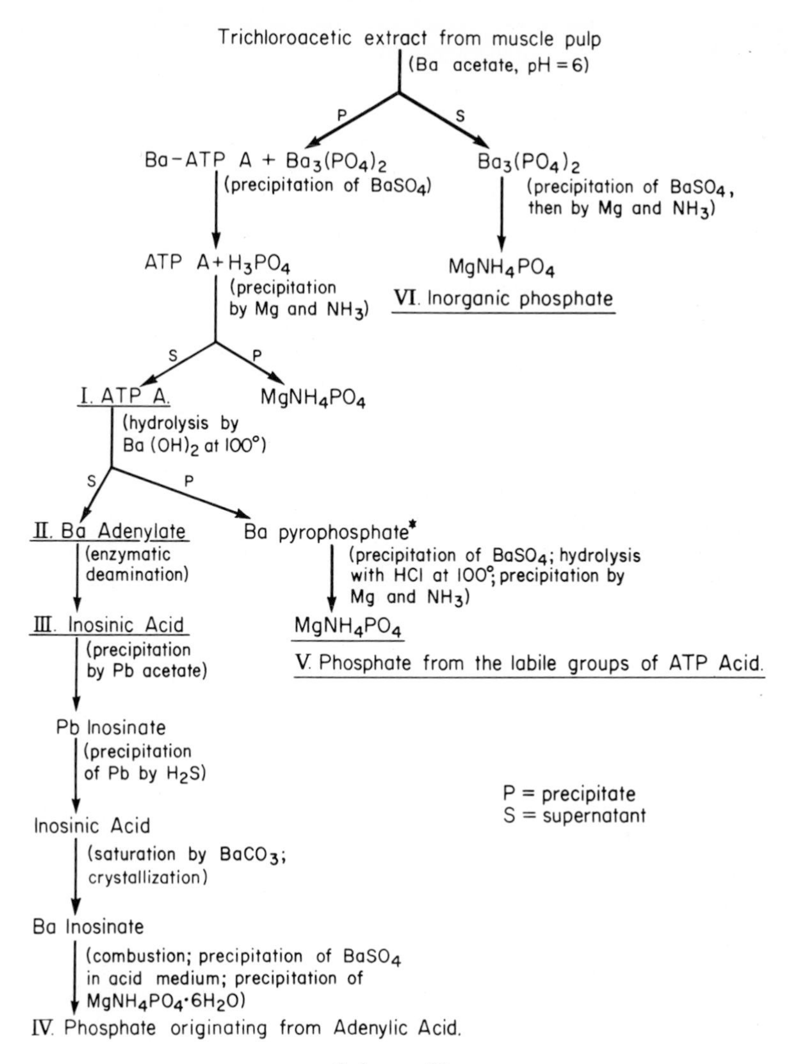

Scheme II

contained in the same molecule of adenosine triphosphoric acid. It also gives the activities of inorganic phosphorus (VI) present at the same time in the same tissue.

It seems that an equilibrium is established within 30 minutes between the phosphorus atom of the inorganic phosphate of muscle and the labile phosphorus atoms of adenosine triphosphoric acid. During this same period activity of the phosphorus of adenylic acid is zero. After one hour and after two hours presence of a measurable activity indicates that a slight turnover of the phosphorus of adenylic acid has occurred, but these figures are not very precise, owing to the difficulty of measuring accurately such low levels of radioactivity. Taken as a whole, these figures nevertheless prove very definitely that adenosine triphosphoric acid has a very rapid turnover of its two phosphorus atoms which are considered as labile, and which can be split off by chemical means as pyrophosphoric acid; the turnover of the phosphorus atom of adenylic acid, however, takes place extremely slowly.

This is exactly what one might expect, on the basis of experiments done with isolated muscles, and with preparations of muscle enzymes. At the present time, it is very seldom possible to bring rigorous proofs to demonstrate the legitimacy of applying to the physiology of living organisms conclusions on tissue metabolism drawn from experiments made with altered systems. These conclusions, in the case of muscle metabolism, make up a developed and logical system, but one which may seem artificial in its unexpected complexity; doubts have been expressed on these points. It is therefore extremely important to establish a correspondence between the well defined points of our chemical and enzymological deductions and what can be established in the living animal. As regards the circulation of phosphorus and the physiological interpretation of its two functions in adenosine triphosphoric acid, it seems that a certain amount of good evidence can now be provided.

This study was done in collaboration with the Institute of Theoretical Physics of the University of Copenhagen. We thank Professor N. Bohr and Professor G. de Hevesy for radioactive $Na_2H^{32}PO_4$, and Miss Hilde Levi for measurements of radioactivity.

PART V

IS PHOSPHORYLATION THE PRIMARY ENERGY-YIELDING REACTION IN CONTRACTION?

In 1929 when ATP was first discovered by Fiske and Subbarow at Harvard it was the "Stella Nova" at the International Congress of Physiology held in Boston. Since then, ATP has retained the center of interest. By 1950 it had even been nominated as the "spark of life" by the *Saturday Evening Post*. It was considered the constituent in muscle bioenergetics responsible for the conversion of chemical energy into mechanical energy.

There were many good reasons to believe that ATP was the key factor. First, ATP occurs in large amounts in skeletal muscle. Second, in the intact muscle, the pyrophosphate phosphorus shows a rapid turnover (exchange with ^{32}P). Third, iodoacetate contractions bring about splitting, first of phosphocreatine, and then of ATP. Apparently phosphocreatine sustains the ATP level. Fourth, the ΔH of splitting of ATP is almost as high as that of phosphocreatine and finally, as will appear in a subsequent chapter, ATP can elicit contraction in muscle fibers and in myosin filaments. The latter also catalyzes specific hydrolysis of ATP to ADP and phosphate. All of these observations gave rise to the optimistic belief that we might at last have arrived at the focus of biological "mechano-chemical" coupling. At this point, A. V. Hill, in a spirited, provocatively humorous birthday message to his old friend, Meyerhof, sparked an international "contest" for the detection of an initial "burst" of ATP splitting during the contraction phase of a single twitch in a muscle held under strict physiological conditions. Hill challenged not only the young biochemists of the fifties but also the eternally young biologist and biochemist, Otto Meyerhof. A proper selection of biological specimens and the development of rapid and exact methods would be mandatory prerequisites for a successful attack on the crucial questions.

Hill discusses his own studies of 1930 on the lowering of vapor pressure during stimulation of muscle(1,2). The depression of vapor pressure and, hence, the increase in osmotic pressure were correlated with known chemical changes. Based on the latter, there was a deficit of 25 to 30 per cent,* as compared with the *observed* depression of vapor pressure. If one assumed that phosphocreatine and ATP were bound to protein and, hence, osmotically inactive, the deficit could be accounted for. Phosphocreatine is not known to be bound to myosin or to other muscle proteins. ATP and ADP, if present in high concentrations, can be shown to bind to myosin(3). Moreover, the contribution of divalent ions, especially magnesium, complexing of ATP should also be taken into account. In any case, Hill's old data, which are quoted in his 1950 article [q.v.], still pose the question of the existence of unknown energy sources for contraction.

Hill's challenge to biochemists gave birth to a number of ingenious projects pertaining to this problem. Curiously enough, neither the initiator of the contest nor the participants were aware that Lundsgaard(4) had, more than fifteen years earlier, subjected the problem to a careful biochemical and physiological examination. Lundsgaard presented his experimental data and his computations in an elaborate article in the 1934 number of the *Biochemische Zeitschrift* dedicated to the seventieth birthday of his teacher, Valdemar Henriques. Since this

*Computations were based on chemical changes like P liberation from phosphocreatine and ATP, and lactate formation.

important paper is rather long and difficult to abridge because of its concentrated argument, I am summarizing it here, garnished with my own comments and I accept responsibility for the paraphrasing.

Lundsgaard was interested in the energetics of muscle contraction at low temperature. His experimental setup, which was geared for that purpose, was constructed as follows:

Two iodoacetate-poisoned frog gastrocnemii were stimulated to perform a tetanus under anaerobic conditions. One was frozen simultaneously with the termination of the tetanus. The other muscle was frozen 1 to 3 minutes after termination of the tetanus. This type of experiment was performed at 20° at 10°, and at 0°. Tension was measured, and mechanical work expressed in caloric equivalents. The phosphocreatine splitting which represented the only metabolic reaction preserved was determined and likewise expressed in calories, using Meyerhof and Lohmann's thermochemical data. At 20° the situation appeared thus: phosphocreatine splitting was the only detectable energy-yielding reaction. Moreover, there was no detectable restitution of phosphocreatine. The splitting of phosphocreatine took place exclusively during the tetanus.

The mechanical work was measured in an isometric myograph in which the length of the muscle remained unaltered during the augmentation of tension. Tension and length were always determined directly. The coefficient Heat/Tension × Length was now expressed in *two* ways.

1. The heat H derived from *chemical* data of phosphocreatine splitting and from Meyerhof's thermochemical data, we shall call *Hchem* (Lundsgaard called it *experimental H*). In our terminology the coefficient would, therefore, become: *Hchem/T × L.*

2. The heat computed, by means of Hill's and Feng's calibration parameters, from *mechanical* data we call *Hmech* (Lundsgaard called it the *theoretical Heat*). According to Hill and Feng, an isometric tetanus at a given temperature shows a constant relation between tension, its duration, rate of stimulation, and heat production. The latter is composed of heat of activation and heat of maintenance. Using a correction factor for different types of muscle (sartorius, gastrocnemius), Hill and Feng were able to correlate heat production with tension, rate and duration of stimulation, and temperature (derived from Hill and Feng's thermomechanical data). The term *Hmech* is, in fact, derived partly from Lundsgaard's experimental mechanical data and partly from Feng's calibration table[5]. The corresponding coefficient would then be *Hmech/T × L.*

The experiments were performed at three different temperatures. The series at 0°–2° comprised ten experiments, whereas five and six experiments were done at 10° and 20°, respectively. The duration of the tetani at the three different temperatures was 25, 10, and 4 sec, respectively. By choosing these durations of the tetani at the different temperatures, the influence of temperature on the heat liberation, according to Feng[5], is compensated, and the heat/tension-length-duration coefficient can be substituted for the heat/tension-length coefficient.

AVERAGE H/T × L (DERIVED FROM LUNDSGAARD, 1934)(4)

	Frozen in tension		Frozen 3 min. after contraction	
	Hchem/T × L (exp.)	*Hmech*/T × L (theor.)	*Hchem*/T × L (exp.)	*Hmech*/T × L (theor.)
0°	**2.7**	3.6	3.7	3.6
10°	3.6	3.4	3.8	3.4
20°	3.1	3.6	3.3	3.6

The only statistically important aberration is the one at 0° (written in bold letters) in the series frozen in contraction. This anomalous figure is attributed to the fact that approximately 30 per cent of the phosphocreatine splitting at 0° takes place *after* the tetanus has terminated. Lundsgaard found such a "delayed splitting" of phosphocreatine even in normal anaerobic muscles, provided that the experiments were performed at 0°–2°. At this low temperature lactic acid formation does not appear in the 3-minute–after period nor is phosphocreatine resynthesized. The main point is that muscles (unpoisoned or poisoned with iodoacetate) at 0° showed a delayed phosphocreatine splitting. During the contraction period, at this low temperature, there is, therefore, a "phosphate" debt which must be balanced by energy-yielding reactions of unknown nature in order to provide energy for part of this contraction process.

I have cited this very long and sophisticated paper in some detail because it seems to have been forgotten in the later literature.

Let us now return to Hill's challenge to the biochemists of the fifties.

If the chemical changes which occur during the contraction phase of a single twitch, are the targets of our studies they must obviously be contrasted with the state of relaxation. As was pointed out, especially by Buchthal, a proper fixing of muscle by freezing or other means becomes a major obstacle if a genuine state of relaxation must be obtained. A sudden freezing of a muscle is usually not instantaneous; hence, stimulation ensues. The contraction brought about by freezing is not particularly important when one is dealing with the chemistry of tetanic contractions. Hence, Fletcher and Hopkins' [q.v.]* original demands remained adequate until the chemistry of the single twitch had to be explored. When attempting to detect chemical changes accompanying one single contraction, it is obviously mandatory to fix a muscle without promoting even one single additional twitch. One way to avoid this "trap" is to try to determine quantitative chemical changes in the intact muscle from the optical changes which may appear during the contraction. This approach resembles the way astronomers observe events taking place in celestial bodies.

Britton Chance selected this approach. He had developed highly sensitive and rapid optical techniques for detecting ADP in cells or cellular particles (the indirect detection was based on the shift in steady state ratios of oxidized to reduced nicotineamid-adenine nucleotide, flavine, and cytochromes). Chance was able to observe ATP splitting during single contractions. Yet, he considered it quantitatively insufficient based on computations from the mechanical energy of the contraction phase. Chance, therefore, thought that a reaction of

*See under authors names for the reprinted material referred to.

unknown nature had to furnish at least part of the primary energy for the contraction. As appears from his paper [q.v.], he had reason to believe that the unknown source was of an appreciable capacity.

Wilkie and his coworkers[6] studied the chemical events during a single twitch with primary emphasis on phosphocreatine breakdown. The Wilkie group as well as Mommaerts and Davies and coworkers were able to detect breakdown of phosphocreatine during a single contraction[6,7,8]. Davies and coworkers exploited the finding of Kuby and Mahowald[9] that the phosphocreatine transphorylation of ADP to ATP (Lohmann reaction) is strongly inhibited by fluorodinitrobenzene. This enabled them to obtain muscles practically free of phosphocreatine but with normal ATP content, to the benefit of a renewed study of ATP changes during a single twitch (contraction peak). Cain and Davies[10] were able to show that ATP breaks down to ADP and AMP (the myokinase reaction is responsible for the latter). The amount broken down in a single contraction of a frog *rectus abdominus* muscle could account for the mechanical energy output if one assumes that 10Kcal are released per Mol P liberated and that this energy is used at about 50 per cent efficiency.

As emphasized by Hill and Howarth (quoted in Davies [q.v.] *et al.*), previous observations indicate strongly that "the chemical reactions which normally occur during contraction can be reversed by a stretch under the influence of the mechanical work supplied." Davies and his coworkers now find, by using the previously mentioned sensitive chemical methods, that stretching inhibits the breakdown of ATP but does not bring about a resynthesis. The authors therefore find themselves unable to make a decisive statement concerning the status of ATP in muscle contraction. They emphasize that, if Hill's conclusions concerning the reversal of chemical processes of muscular activity are applicable to a muscle treated with fluorodinitrobenzene, ATP does not seem to represent the final direct source of energy for muscle contraction. Further experiments coordinating chemical, mechanical and thermic measurements are probably needed to clarify this crucial problem.

References

1. Hill, A. V., and Kupalov, P.: *Proc. Roy. Soc. London B.,* **106** (1930), 445.
2. ———, and Parkinson, J. L.: *Proc. Roy. Soc. London B.,* **108** (1931), 140.
3. Buchthal, F.; Svensmark, O., and P. Rosenfalck,: *Physiol. Rev.,* **36** (1956), 503.
4. Lundsgaard, E.: *Biochem. Z.,* **269** (1934), 308.
5. Feng, T. P.: *Proc. Roy. Soc. London B.,* **108** (1931), 522.
6. Carlson, F. D., Hardy, D. J., and Wilkie, D. R.: *J. Gen. Physiol.,* **46** (1963), 851.
7. Fleckenstein, A., Janke, J., Davies, R. E., and Krebs, H. A.: *Nature,* **174** (1954), 1081.
8. Mommaerts, W. F. H. M.: *Nature,* **174** (1954), 1083.
9. Kuby, S. A., and Mahowald, R. A.: *Fed. Proc.,* **18** (1959), 267.
10. Cain, D. F., and Davies, R. E.: *Biochem. Biophys. Res. Commun.,* **8** (1962), 361.

A CHALLENGE TO BIOCHEMISTS*

A. V. Hill

Biophysics Research Unit, University College,
London, England

(Received for publication March 7, 1949)

Otto Meyerhof has always been betwixt and between: a physiological chemist or a chemical physiologist, perhaps we should call him a "chemiologist". On my shelves are about two hundred of his reprints, his and his colleagues'. The first of these, with its accompanying letter addressing me as "Sehr geehrter Herr Kollege" dated 1911 from Naples, dealt with the heat production of the vital oxidation process in the eggs of marine animals. Next follow papers on the energy exchanges of bacteria, the heat accompanying chemical processes in living cells, the inhibition of enzyme reactions by narcotics (1914). Some time in those apparently peaceful years, before the explosion of 1914, he visited us at Cambridge. Then comes a gap, so far at least as my collection of Otto Meyerhof's reprints is concerned. By 1919 he had moved to Höber's laboratory at Kiel and the long succession of papers began on the respiration, energetics, and chemistry of muscle. And when I say muscle, I mean muscle: living muscle, resting, contracting and recovering from contraction, developing tension and doing work, producing lactic acid and removing it again, using oxygen and glycogen, giving out CO_2 and heat, all things which living muscles are accustomed to do. And since I too was working on living muscle, we were in frequent communication again, after the five years'

*A. V. Hill, "A Challenge to Biochemists," in *Metabolism and Function.* Amsterdam: Elsevier Publishing Company (1950), 4–11.

Reprinted from *Metabolism and Function* appearing as a special issue of *Biochem. Biophys. Acta,* **4**, 1–348 (1950), dedicated to Otto Meyerhof on the occasion of his 65th birthday. Edited by D. Nachmansohn. The article and photograph reprint appears with the permission of the author, editor, and of Elsevier Publishing Company.

A. V. Hill and Otto Meyerhof.

gap. In the summer of 1922, following a suggestion to Hopkins, he visited Cambridge and gave lectures there. I remember "Hoppy" expressing concern lest some anti-German demonstration might take place, but appearing to be satisfied by the comment that if so I should be proud to remove the demonstrator: nothing of course happened. Later, he stayed with me at Manchester and I recall, as an example of his scientific perspicacity, the complete disbelief which he, first of anyone, expressed in experiments he witnessed which six months later were proved to be fraudulent. That was our first reunion after the War, there were many others, in London, Plymouth, Barcelona, Heidelberg, Berlin, Rome and elsewhere. The photograph shows us driving together to Stockholm for the Physiological Congress in 1926.

The results of his researches, and those of his colleagues, are a part of scientific history. They are linked with most that is known of the chemistry of muscle and with much that is established of changes involving phosphate and carbohydrate in the cell. For some years his investigations were concerned mainly with muscle—living muscle: more recently they followed the trend in biochemistry, perhaps even they helped to establish the fashion, of dealing *in vitro* with the enzyme systems of muscle. As late, however, as 1935, he was working on the volume changes of living muscle during contraction and relaxation and relating them to the underlying chemical cause. I read these papers again recently, very carefully, having come to the conclusion that the reversible part of the volume change is attributable mainly or wholly to pressure set up by contraction. The elegance and clarity of Meyerhof's work and its description impressed itself again as it had done in earlier days. One might criticize some of the conclusions, but not the methods or results. To read these papers once more was a sudden pleasure, after so many in which one could not be sure what an author had really done! My last reprint from Heidelberg is dated 1938. Perhaps if Hitler had not driven him from the beautiful Institute and the excellent colleagues and facilities he had there, the succession of papers on muscle—living muscle—might have continued. Alas that they could not! This paper, however, is to challenge him and his disciples to make a few more chemical investigations on living muscle, to see how far the chemistry *in vitro* of muscle extracts can be fitted to the physical facts of muscular contraction.

It is customary for biochemists, (*e.g.,* Baldwin[1], p. 341) to describe "The probable course of events in normal muscular contraction" in some such terms as these:

"On the arrival of a nerve impulse, ATP is broken down, giving rise to ADP and inorganic phosphate, furnishing at the same time the contraction energy. The ADP is promptly converted again into ATP at the expense of phosphagen and no change in the ATP content of the muscle can be detected . . ." Others suppose that contraction is associated with the formation of myosin—ATP and that ATP is broken down in relaxation. By Sandow[2] a slight initial lengthening (in a muscle under tension) after a stimulus ("latency relaxation") is attributed to the formation of a complex between activated myosin and ATP. Most of this is pure speculation, without direct experimental evidence. Unlike Mr. Stalin (Historicus[3]) I have no general theory of revolutions, but I did once write an article (1932), which I think is still worth reading, on "The Revolution in Muscle Physiology"[4]. That was after phosphagen had deposed lactic acid from pride of place as the chief chemical agent in contraction. At that date one could write: "On stimulation, phosphagen breaks down . . . : this is the primary change by which energy is set free". Only four years earlier Ritchie[5] wrote: "On stimulation of a muscle fibre the wave of excitation passes down it; by increasing the permeability of a membrane or by some other means it causes the liberation of lactic acid from a carbohydrate source. The liberated hydrogen ions neutralize the negative charge on a surface protein, of Meyerhof's *Verkürzungsort* . . . and thereby alter the type of structure, the area of surface, and the mechanical constants. This will be the fundamental change." In the lactic acid era the evidence that the formation of lactic acid was the cause and provided the energy for contraction seemed pretty good. In the phosphagen era a similar attribution to phosphagen appeared even better justified. Now, in the adenosinetriphosphate era lactic acid and phosphagen have been relegated to recovery and ATP takes their place. Those of us who have lived through two revolutions are wondering whether and when the third is coming.

It may very well be the case, and none will be happier than I to be quit of revolutions, that the breakdown of ATP really is responsible for contraction or relaxation: but in fact there is no direct evidence that it is. Indeed, no change in the ATP has ever been found in living muscle except in extreme exhaustion, verging on rigor. This is explained by supposing that as soon as ATP is broken down into ADP and phosphate it is promptly restored in the so-called "Lohmann reaction" at the expense of creatine phosphate.

$$\text{ADP} + \text{CP} \longrightarrow \text{ATP} + \text{C}$$

If this happens after each stimulus, then the smallness of the changes involved and their quickness make it extremely difficult to gain any direct evidence on the subject. In a single twitch, for example, the heat set free is about 3 millicalories per gram, which would correspond to the liberation from ATP of $2.5 \cdot 10^{-7}$ g molecule of phosphate per gram of muscle. To measure so small a change, reversed within the duration of a single twitch, might well seem an impossible task.

We should not, however, be so satisfied with the explanation of why no change in ATP is ever found in living muscle that we cease to look for it: for another possibility exists. The total energy available from all sources (lactic acid, phosphagen and ATP) for the anaerobic phase of contraction is about 1 cal/g, corresponding to about 400 twitches. The total energy similarly available after poisoning with iodoacetate (from phosphagen and ATP) is about 0.25 cal/g corresponding to about 100 twitches. From the known amount of ATP present is muscle, the total energy it could provide by breaking off one phosphate is about 0.05 cal/g, corresponding to about 20 twitches. Is it not possible that as stimulation proceeds a balance is reached at some intermediate level between breakdown and restoration? That is the case with phosphagen and lactic acid; in a muscle steadily stimulated (in the presence of oxygen) a certain amount of phosphagen is broken down, a certain amount of lactic is formed, and a steady level is reached between breakdown and recovery. At a still earlier stage one might expect steady stimulation to provide at least a temporary balance between ATP breakdown and restoration.

In frogs' muscles at 20°C, if ATP were the only source of energy a maximal tetanus would lead to its complete breakdown in about 0.5 sec. The suggested balance, if it occurred, would presumably be reached within that time, and when the stimulus ended restoration of the ATP might be completed within another 0.5 sec. The times involved are far too short for chemical manipulation: but biochemists need not be disheartened, frogs' and rabbits' muscles are singularly ill-suited to the enquiry, they are much too quick, why not use muscles which contract more slowly? The muscles of the Mediterranean land tortoise, *Testudo graeca,* commonly imported before the War into England and sold on barrows for 1/– in London streets, take about fifteen times as long to contract as those of a frog and their speed can be further reduced about nine times by lowering the temperature from 20°C to 0°C, or about five times by lowering it to 5°C. This means that the time available for chemical manipulation can be reckoned in large fractions of a minute instead of fractions of a second. Provided, therefore, that the chemical technique is capable of determining a substantial part of the total ATP with reasonable accuracy, the time involved can be made so long that sufficient resolution ought easily to be obtained.

The experiment ought certainly to be made and nobody could make it better than Otto Meyerhof—for he knows how to handle living muscles. The result may not be unequivocal—but it very well may. If no change in ATP is found, but only a change in phosphagen, the *status quo* remains and we can all believe what we like, provided it is consistent with the physical facts described below. But suppose it is found that ATP is broken down at a rate decreasing from the start, reaching a steady concentration after half a minute's stimulation (corresponding to half a second in a frog's muscle at 20°C) and is restored to its original level after (say) a further half minute of rest and recovery. Then at least we can be assured that ATP is really concerned either with the contractile process itself, or with the very early stages of recovery. There are other possibilities and, without trying, it is useless to speculate too much. A German clinician is said to have remarked: "Der Versuch muss gemacht werden und sollte es hundert Bauern kosten". A decision on this important matter is certainly worth a hundred tortoises.

But whatever may be the outcome of this challenge to biochemists, I would invite them also, in their speculations about muscle, to take note of the following facts, all referring to contraction and relaxation, as distinguished from recovery.

1. There is no sign of an endothermic process at any stage of contraction or relaxation. If endothermic processes occur they are balanced, or overbalanced, by exothermic ones.

2. No heat at all is produced during relaxation, apart from that derived from the degradation of work previously performed during contraction (in raising a load, or in stretching elastic material in series with the muscle). When a muscle relaxes without load or tension, no heat is produced after the contractile phase is over.

3. It has been found by quick stretches applied to a muscle shortly after a single shock that the full strength of the contraction, defined as the load which a muscle can just bear without lengthening (and equal to the force of a maximal tetanus) is developed abruptly immediately after the end of the latent period. It is maintained for a time and then declines in "relaxation". If stimulation is continued, each successive shock restores the strength of contraction to its full height.

4. Corresponding to (3) there is a "heat of activation" in a twitch, which is independent of all other factors except the fact of stimulation. The heat of activation starts at its maximum rate before any visible sign of contraction occurs, declining to zero at about the moment when the strength of contraction (see 3 above) begins to fall off, *i.e.,* at the end of the contractile phase.

5. The "heat of maintenance" in a prolonged contraction is the summated effect of the heat of activation following successive elements of the stimulus. It is greater at first corresponding to the more rapid relaxation after a short tetanus, but after a certain duration of stimulus it becomes constant. It is affected only to a minor extent by the length of the muscle. It is greatly increased by a rise of temperature, corresponding to the more rapid relaxation.

6. In twitch and tetanus alike, apart from the heat of activation or the heat of maintenance, energy is given out in two discrete forms, (a) as mechanical work and b) as heat of shortening. The heat of shortening is directly proportional to the change of length over the whole range of shortening, and (for a given change of length) is independent of the work done.

7. Apart from heat of activation or heat of maintenance, the rate at which

total energy, *i.e.*, heat plus work, is given out, is a linear function of the load throughout a contraction:

$$(P + a)\, dx/dt = b(P_0 - P)$$

where x is the amount of shortening up to time t, P is the load, dx is the heat of shortening, P_0 is the maximum isometric tension and b is a constant related to the maximum velocity of shortening under zero load.

8. The constant a in (7) can be obtained either from thermal measurements or from the form of the characteristic relation between load and velocity of shortening. The agreement is good.

9. Relaxation is not an active process. A muscle completely without load or tension does not lengthen again after shortening in response to a stimulus. That its length has really changed and that its fibers or fibrils have not gone into folds is shown by the fact that its latent period is practically the same at a short length as it is at a greater one. If a muscle had to "take up the slack" in fibres or fibrils before its tension could be manifested externally, the latent period would be greatly prolonged.

10. Simultaneous with the earliest sign of mechanical activity after a shock is a change of opacity. This is due to an alteration of light scattering (D. K. Hill[6]). The earliest phase has certain characteristics which distinguish it from a later phase which continues into recovery.

11. If we can assume that excitation occurs at the surface membrane of a muscle fibre, the propagation inwards of the change there started cannot be due to the diffusion inwards of some substance, *e.g.*, Ca ions or acetyl choline, initiating contraction by its arrival at each point. Diffusion is far too slow. Some chain-reaction started at the surface is required.

Nineteen years ago my colleagues and I found (Hill and Kupalov[7]; Hill and Parkinson[8]) in muscles stimulated to exhaustion in nitrogen, a lowering of vapour pressure considerably too large to be accounted for by chemical changes known to occur, if the precursors of the chemical substances produced were themselves osmotically active. In normal muscles complete exhaustion led to a decrease of vapour pressure corresponding to an increased concentration in the free water of a muscle of 0.12 M. The production of 0.35% lactic acid dissolved in the free water, (taken as 0.77 g per g) of the muscle, would lead to a concentration change of 0.050 M. The liberation of creatine and phosphate by the complete breakdown of phosphagen in amounts equivalent to 65 mg. P/100 g would give 0.054 M. The production of phosphate and adenylic acid from ATP in amounts equivalent to 30 mg P/100 g would give 0.012 M. The total, 0.116 M, is not far from that (0.12 M) calculated from the observed change of vapour pressure. We have assumed, however, that the phosphagen and the ATP were not themselves osmotically active; if they had been the increase would have been 0.031 M less, namely 0.085 M instead of 0.12 M. The vapour pressure measurements were certainly not that much wrong.

Again, in muscles poisoned with iodoacetate complete exhaustion led to a mean decrease of vapour pressure corresponding to an increased concentration of 0.050 M. If phosphagen and ATP breakdown are assumed, as above, to be the only chemical reactions involved, the corresponding change of concentration in the free water of the muscle would be 0.066 M. It is impossible, however, in muscles adequately poisoned to ensure that some preliminary breakdown of phosphagen has not occurred: and if the poisoning is not quite sufficient, there is likely to be some formation of lactic acid. Either cause would tend to make the observed change of vapour pressure smaller than that calculated from the assumed breakdowns. Even so, had the phosphagen and ATP originally been osmotically active, the change calculated from the constituents would have been only 0.035 M, considerably less than the 0.050 M observed.

Unless, therefore, some chemical reactions hitherto unknown occur in a muscle stimulated to exhaustion in nitrogen, we are forced to conclude that phosphagen and ATP are not themselves osmotically active in the normal muscle. This would be the case if they were bound to other molecules and their constituents only became free when they broke down. These older experiments are worth recalling now because they are pertinent to the question of how phosphagen and ATP exist in the living muscle. Looking back at them today I see no reason to question their results. If those are correct, ATP and phosphagen exist in a combined form in muscle, exerting no osmotic pressure on their own account until they are broken down.

The work which an isolated muscle of frog or toad can perform under optimal conditions may be as high as 40% of the total energy given out in the initial process, as distinguished from recovery (Hill[9]). This high efficiency is obtained just the same at 0°C as at higher temperatures, and there are no grounds at all for supposing that the nature of contraction is in any way altered, except in speed, by a change of temperature. The muscle twitch is rather stronger at 0°C than at 25°C, and quite as efficient. If theory predicts otherwise, so much the worse for the theory. The highest efficiency is obtained with a comparatively large load and slow shortening; under isotonic conditions, with a load about half the maximum which the muscle can lift. In such a contraction the work done is about twice the heat of shortening: two thirds of the total energy set free, in excess of the heat of activation (or maintenance), is external mechanical work. Under conditions, therefore, of maximum efficiency, the energy is liberated in about the following proportions:

Heat of activation or maintenance	*Work*	*Heat of shortening*
40	40	20

At the other extreme, with zero load and rapid shortening, the situation may be this:

Heat of activation	*Work*	*Heat of shortening*
40	Nil	49

(The heat of activation is the same in both cases.)

The fact that the external work may be so large a fraction of the whole energy liberated in excess of the activation (or maintenance) heat naturally makes one ask whether the heat of shortening may not itself really be work degraded into heat in overcoming some internal resistance to shortening: in that case energy would be liberated in two forms only, heat of activation (or maintenance) and mechanical work. For two reasons, the supposed internal resistance cannot be of a viscous nature: (1) the heat of shortening is independent of the velocity of shortening, and (2) the heat of shortening per cm is the same over the whole range of possible shortening (if it were due to overcoming viscous resistance it would be inversely proportional to the length). The supposed resistance must be constant, and must reside in lines or filaments parallel to the axis of the muscle, it cannot be a volume effect. An obvious objection to the theory of a constant (*e.g.,* frictional) resistance a parallel to and inherent in the contractile elements is that there should then be a constant difference $2a$ between the load at which a muscle just shortened and the load at which it just lengthened: experiment showed (Katz[10]) that no such difference exists. The objection would be valid if a muscle were a single contractile element, with a parallel constant resistance. In fact, however, a muscle fibre is very long relative to its thickness, and its diameter is by no means constant throughout its length. There is no reason to suppose that its maximum force is the same everywhere. If not, in an isometric contraction the stronger regions would tend to shorten at the expense of the weaker regions, and the constant resistance would hinder shortening at one point and lengthening at another (possibly a very convenient arrangement in a system of non-uniform strength). With a large number of such elements in series an increase of load would stretch the weaker elements, a decrease of load would allow the stronger elements to shorten: and the difference of load between observable lengthening and shortening would be small. The objection, therefore, is not really valid.

A stronger objection, raised in 1938[11], is that there are indications that the heat of shortening changes sign when shortening becomes lengthening; and the heat generated in overcoming a frictional resistance does not change sign when the direction of motion is reversed. The difficulty is to get muscles to lengthen reversibly except at very low speeds. Possibly the use of dogfish jaw muscles (Levin and Wyman[12]) which stand stretching well would allow more positive conclusions to be reached. One thing is certain, namely that the work done in making a muscle lengthen does not reappear completely as heat: Some of it is absorbed, presumably, in driving chemical reactions in the endothermic direction. The subject is being investigated afresh by improved methods.

One final word—to continue my challenge to biochemists. Otto Meyerhof's first letter to me, as I wrote at the beginning, came from Naples: all his life he has been ready to vary not only his chemical technique but his biological material. The properties of animals, and of their muscular systems, vary over a very wide range. There is no need to stick to rabbits and frogs. If a problem seems insoluble on one muscle, one should try to define it more precisely to see where the difficulty lies. Discussion with a zoologist, or a visit to a Marine Laboratory, may provide material many times better suited to one's needs. I spent many years trying to measure the heat production of nerve: if I had made the experiment on crabs' nerves instead of frogs' the answer would have come in 1912 instead of 1926. In 1912 it was not possible to define the problem well enough to get a clear direction to non-medullated nerve, but at least one might have taken a chance and not persisted with the frog's sciatic. If one's instruments, or methods, are too slow, one can make them relatively quicker by using slower material—tortoises, toads or even sloths. That means, of course, that biochemists, like biophysicists, must also be biologists (as Meyerhof has always been and as Hopkins was)—but why not?

References

1. E. Baldwin, *Dynamic Aspects of Biochemistry*, Cambridge University Press (1947).
2. A. Sandow, *Ann. N.Y. Acad. Sci.,* **47** (1947) 895.
3. Historicus, *Foreign Affairs,* **27** (1949) 175.
4. A. V. Hill, *Physiol. Revs.,* **12** (1932) 56.
5. A. D. Ritchie, *The Comparative Physiology of Muscular Tissue,* Cambridge University Press (1928).
6. D. K. Hill, *J. Physiol.,* **107** (1948) 40 P.
7. A. V. Hill and P. Kupalov, *Proc. Roy. Soc. B.,* **106** (1930) 445.
8. A. V. Hill and J. L. Parkinson, *Proc. Roy. Soc. B.,* **108** (1931) 148.
9. A. V. Hill, *Proc. Roy. Soc. B.,* **127** (1939) 434.
10. B. Katz, *J. Physiol.,* **96** (1939) 45.
11. A. V. Hill, *Proc. Roy. Soc. B.,* **126** (1938) 136.
12. A. Levin and J. Wyman, *Proc. Roy. Soc. B.,* **101** (1927) 218.

THE RESPONSE OF MITOCHONDRIA TO MUSCULAR CONTRACTION

Britton Chance

Johnson Research Foundation
University of Pennsylvania, Philadelphia

Introduction

In this monograph controversial points hinge, as they always will, upon whether the biological system has changed its physicochemical properties in the transition from its physiological state to its analysis by physical or chemical methods. For example, electron microscopy requires fixation procedures, chemical analyses require cell destruction, and some sensitive electric measurements require rupture of the cell membrane. In spite of the important contributions made by such methods, it is occasionally necessary to employ for control experiments methods that, we may agree, cause no measurable change in the physiological state of the cell. Examples of methods in the "harmless" category are the measurement of temperature and of light absorption, and some measurements of electric potentials. These three methods are, in general, nonspecific, since they do not refer the measurement to a particular location of the cell or to a particular chemical reaction in the cell. However, remarkable progress has been made in microelectrodes for localizing potential. Heat measurement, as used by A. V. Hill[1,2], suffers from a lack of localization and from a confusion as to the source of heat; Hill reviewed at least four kinds of heat that could be responsible for the observed effects[3]. In fact, it is only by time separation that some of these heats can be obtained independent of others. Optical methods would appear to suffer similarly from lack of specificity. For

*B. Chance, "The Response of Mitochondria to Muscular Contraction," *Annals of the New York Academy of Sciences,* 81: Article 2, 477-489 (1959).

example, light-scattering and diffraction measurements carried out by a number of investigators, most recently by William Sleator[4], give chiefly physical evidence of structural changes in the muscle. Recently, however, special spectrophotometric techniques have been developed to minimize the light-scattering and diffraction effects that occur during a muscle twitch and to maximize the small absorbancy changes due to intracellular enzymes[5]. Thus we have a new tool for measuring the metabolic response of muscle to contractions.

Optical methods provide two means for the localization of intracellular events. First, a degree of localization is obtained by spectrophotometric recordings of changes in a cell constituent known to be localized in a regular fashion near a characteristic repetitive structure of the muscle. The location of mitochondria near the I bands of muscle enables us to refer measurements of cytochrome and certain types of pyridine nucleotide changes in excised muscle tissue to this region. Second, localization may be provided by microspectrophotometry[6], which has recently been perfected to the degree that cytochromes in mitochondrial aggregates may be measured fairly accurately[7].

Particularly needed is a method that will indicate changes of intracellular adenosine diphosphate (ADP) and phosphate concentrations within the living, intact muscle a short time after the beginning of the muscular contraction. There is currently a great interest in determining what substances are broken down during the muscle twitch, for these are the data on which any theory on the chemical aspects of the muscular contraction must be built. As explained by Davies *et al.* elsewhere in this monograph, all chemical assays must be done upon the disrupted cell, and any differences between control and experimental muscles involve the assumption that a stimulated muscle behaves in the same way as the control muscle throughout the disintegration procedure. This may or may not be the case. Upon this assumption hang such chemical analyses of muscle.

Respiratory Control

The feasibility of using light-absorption changes in the living muscle to interpret changes of ADP or phosphate in the early phases of muscle twitch depends on a well-known but little-understood phenomenon characteristic of many muscles: namely, that of respiratory control. Respiratory control may be defined as the ability of living tissue to accelerate its respiratory rate during activity and to return to a quiescent state characterized by a slow respiratory rate when metabolic activity has led to a restitution of the chemicals expended during activity. Many aspects of respiratory control have recently been summarized by Lardy.[8] Some quantitative aspects of respiratory control can be determined by studies of isolated mitochondria. These will be reviewed briefly.

Range of respiratory control ratios

For isolated mitochondria it is important to define respiratory control ratio as the quotient of the rate obtained in the presence of added ADP to that obtained following the expenditure of added ADP[9,10]. The reason for setting

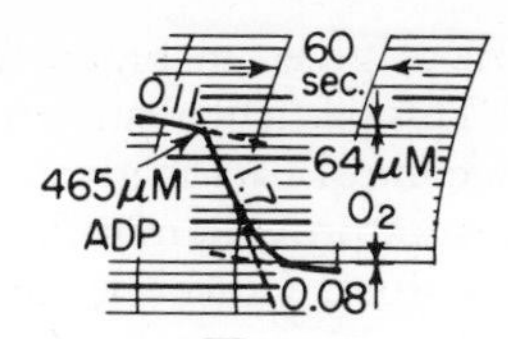

Fig. 1. Example of respiratory control in a heart muscle mitochondrial suspension (experiment 615).

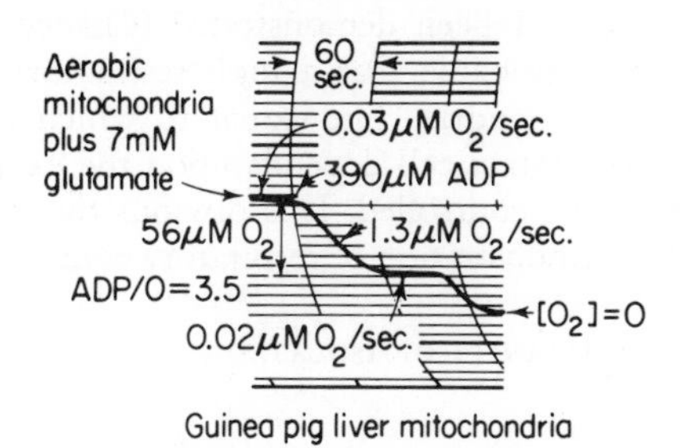

Fig. 2. Example of respiratory control in a guinea pig liver mitochondrial suspension (experiment 372).

up the criterion in this form is that loosely coupled mitochondria contain ATPases that are activated by the ATP (adenosine triphosphate) formed upon the phosphorylation of added ADP.

Some examples of maximal respiratory control ratios are given in Figures 1 and 2, which contain recordings made by means of a vibrating platinum electrode. In heart muscle mitochondria suspended in a sucrose versene medium (Figure 1), an acceleration of respiration upon ADP addition takes place, but upon ADP exhaustion respiration comes to a halt; the ratio of the rates exceeds 20:1. One of the highest values of respiratory control ratios observed

Table 1*

OPTIMAL VALUES OF RESPIRATORY CONTROL RATIOS IN ISOLATED MITOCHONDRIA

Source	Medium	Substrate	Respiratory control ratio	Expt.
Rat liver	Salts	β-Hydroxybutyrate	$>$ 20 (38)	337[12]
Rat heart	Sucrose-versene	α-Ketoglutarate	$>$ 20 (21)	615[9]
Guinea pig liver	Salts	Glutamate	$>$ 20 (65)	372[12]
Guinea pig kidney	Tris-sucrose	Glutamate	17	644†
Turtle liver	Salts	Glutamate	8	641‡
Toad heart	Sucrose-versene	Succinate	5	815§
Ascites tumor cells	Salts	Succinate	6	662‖

*Reproduced by permission of the Ciba Foundation, London, England.[11]
†B. Chance and G. Hollinger (in preparation).
‡Experiments done in collaboration with F. Jöbsis.
§Experiments done in collaboration with J. Ramirez.
‖B. Chance and B. Hess (in preparation).

by us in isolated mitochondria is given in Figure 2, where guinea pig liver mitochondria are represented. This particular curve indicates a ratio of 65:1.

A summary of respiratory control data is given in Table 1[11]. Here it may be seen that many types of tissues provide mitochondria that exhibit respiratory control ratios in excess of 10:1. However, some cells are apparently so resistant to rupture that mitochondria prepared from them may be damaged. This may explain the lower ratios obtained with turtle liver, toad heart, ascites tumor cells and, in particular, mitochondria isolated from yeast cells for which respiratory control has not yet been demonstrated (Chance and Nossal, unpublished data). It is very probable that a high respiratory control ratio is characteristic of intact mitochondria. The extent to which this control ratio can be demonstrated in the intact cell depends upon the extent to which the work done by the cell can be controlled. It is obvious that skeletal muscle is an excellent tissue for the demonstration of respiratory control *in vivo*.

ADP AND PHOSPHATE AFFINITY OF ISOLATED MITOCHONDRIA

Measurements of the stimulation of respiratory rate in response to additions of small concentrations of ADP give, by definition, the ADP affinity of the system. Figure 3 indicates the effect of ADP concentration upon respiratory rate for guinea pig liver mitochondria with succinate as substrate; 20 μM ADP gives half-maximal respiratory stimulation.*

It has been found that spectroscopic responses accompany these respiratory changes and that measurement of the changes in the oxidation-reduction level of any one of several respiratory carriers allows a much more sensitive determination of ADP affinity. Figure 4 gives an example of the addition of small amounts of ADP to azide-treated mitochondria. These mitochondria respond with abrupt changes of oxidation-reduction level (reductions of cytochrome *a*) that are followed by an oxidation when the added ADP has been expended. Larger amounts of ADP cause a greater reduction in the steady state and a longer duration of the cycle, with a correspondingly longer time required to

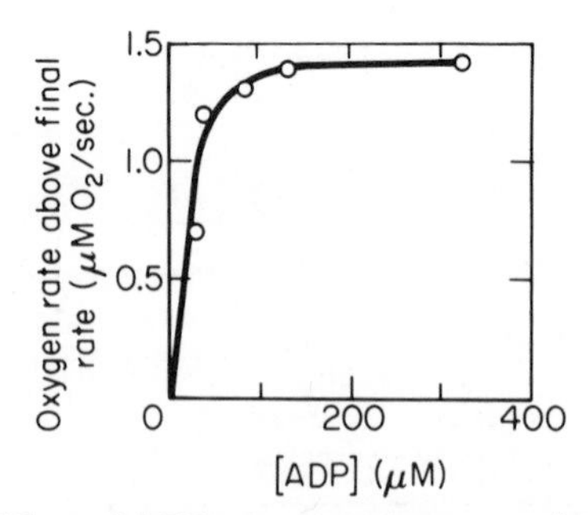

Fig. 3. Effect of ADP concentration upon the respiratory rate of rat liver mitochondrial suspension (experiment 379c).[10] Reprinted by permission from *The Journal of Biological Chemistry*.

*All concentrations are given in moles per liter.

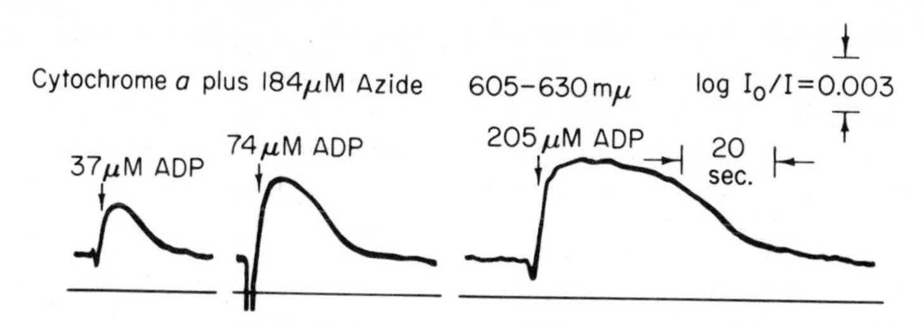

Fig. 4. Spectroscopic response of cytochrome *a* to additions of small concentrations of ADP. Suspension of azide-treated rat liver mitochondria, spectrophotometric recordings with the double-beam apparatus (experiment 388b).[12] Reprinted by permission from *The Journal of Biological Chemistry.*

expend the added ADP. In fact, the area under the curve is proportional to the ADP concentration. These records are examples of the basic principles of the spectroscopic method for the measurement of changes of intracellular ADP concentrations. Four aspects of the indication may be pointed out here:

1. The indication is rapid; at room temperature the steady state is reached in a fraction of a second.
2. The measure is quantitative; the more ADP added, the greater the absorbancy change.
3. The area under the curve also gives a measure of the amount of ADP turned over.
4. The indication is highly specific for ADP; no other known nucleotide causes a similar response.

Similar experiments have been carried out with inorganic phosphate, for which the results are qualitatively similar: phosphate addition causes a change in the steady-state levels of the respiratory carriers. Quantitatively, there are two important differences.

1. The affinity for phosphate is roughly one fiftieth that for ADP at $pH = 7.4$.

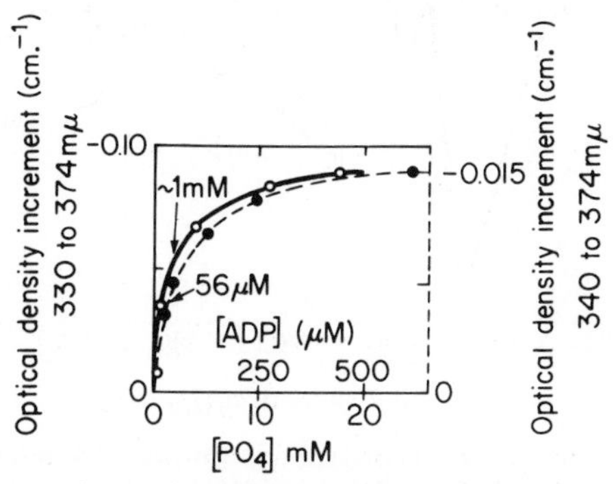

Fig. 5. Comparison of titrations of the response of intramitochondrial pyridine nucleotide to ADP and phosphate. Note different scales of the abscissa (experiments 299b, 463). Reprinted by permission from *Nature*.[5]

2. The nature of the cycle differs: the speed with which the steady state is established is somewhat slower and, since a given amount of phosphate causes considerably less activation of respiration than the same amount of ADP, the cycles are prolonged and do not show the abrupt termination indicated by the ADP cycles. In summary, the response of the mitochondria to phosphate is sluggish and that to ADP rapid.

A comparison of the titration of the pyridine nucleotide of isolated mitochondria with ADP and phosphate is shown in Figure 5.

Methods

Now that we have presented a method that can be applied to the measurement of changes in ADP or phosphate concentrations within living tissues, it is worthwhile to review the experimental methods for such measurements and to indicate their sensitivity.

The wave lengths at which cytochromes can be observed to change their oxidation-reduction state are indicated by the peaks of the absorption bands that appear when an aerobic muscle is treated with nitrogen. The muscle may be stimulated to speed the attainment of anaerobiosis. Such absorption bands are shown in Figure 6 for frog sartorius muscle. This clear recording, obtained in our laboratory by A. Weber, shows absorption bands caused by cytochromes, flavoproteins, and pyridine nucleotides.

The apparatus that we have found most suitable for measuring the changes of intensity of these absorption bands does not measure at one wave length, as is the usual case with the spectrophotometer, but measures instead the differ-

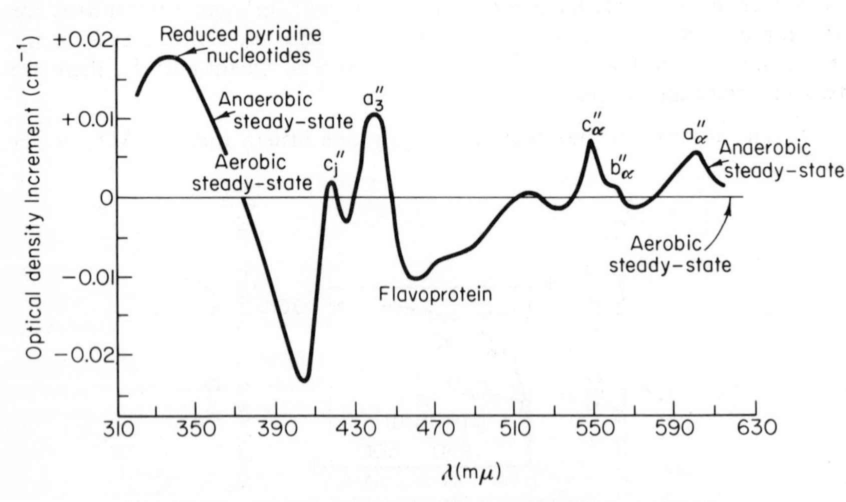

Fig. 6. The spectrum corresponding to the difference between the reduced and oxidized forms of the cytochrome of a 1 mm.-thick frog sartorius muscle. Experimental data by courtesy of A. Weber.

Table 2*

APPROPRIATE WAVE LENGTHS FOR MEASUREMENT OF RESPIRATORY ENZYMES

The Figures in Parentheses Refer to Neutral Reference Wave Lengths Used in Double-Beam Spectrophotometry

O_2 →	cytochrome a_3 →	cytochrome a →	cytochrome c →	cytochrome b
	445 mμ	605 mμ	550 mμ	564 mμ
	(460)	(630)	(541)	(575)

→ flavoprotein →	reduced pyridine nucleotide →	substrate
465 mμ	340 mμ	
(510)	(374)	

*Reproduced by permission from *Science*.[13]

ence of absorption between any two wave lengths. There are a number of advantages to this, the most obvious being that it is possible to select a pair of wave lengths that will measure one cytochrome component largely to the solution of an interfering component. Such wave lengths are indicated in Table 2[13]: the measuring wave lengths appear in the upper line, the reference wave lengths in the lower. A consideration of the difference spectra of Figure 6, together with experimental studies with inhibitors that selectively affect one or two components of the respiratory chain, reveals that differences of intensity recorded at these wave lengths correspond almost exclusively to the component named. The only interference that causes some concern in the intact tissue is that involved in small changes in cytochrome a_3 and large changes of flavoprotein.

Although the autofluorescence of intact tissues has been studied for a number of years and has been attributed to mitochondria[14], we have found the blue fluorescence of the living tissue to be caused chiefly by the reduced pyridine nucleotide of the mitochondria[15,16]. Furthermore, following the results obtained in the laboratories of Theorell[17] and Duysens[18], Baltscheffsky and I have shown that this corresponds to a bound type of pyridine nucleotide with a maximum at 443 mμ[19]. The method appears to be equally as sensitive as the spectrophotometric one, and has the considerable advantage that light transmission through the sample is not required; fluorescence may be excited by oblique illumination and the emitted light may be measured readily with a closely placed photocell[18,20]. The technique would therefore have the advantage of observing changes in the steady state of pyridine nucleotide in the surface layer of a thick muscle. It may also have the further advantage, which I must admit we have not yet investigated, that hemoglobin would probably not interfere. Finally, there is the advantage that the fluorescence signal emitted at 440 mμ is caused primarily by reduced pyridine nucleotide; other components do not have interfering emission bands in this region. However, a difficulty exists in that, with measurements of pyridine nucleotide, the sum of cytoplasmic and mitochondrial pyridine nucleotides is measured. At reduced temperatures, the dominant response to contraction is found to be that of mitochondrial pyridine nucleotide. Microspectrophotometry of intramitochondrial reduced pyridine nucleotide has recently been developed to the point where adequate signals can be obtained from a mitochondrial aggregate[21], but the method has not yet been applied to muscle.

Localization of the Response

It is important to consider whether mitochondria are localized in the proper place in muscle in order to qualify as indicators of the change of ADP concentration caused by muscular contraction. That the cytochrome system is involved in the oxidative recovery process is shown by the cyanide and carbon-monoxide sensitivity of the "activity" respiration. Although the concentration of mitochondria may vary from muscle to muscle, the spectrophotometric method is used in such a way that the response is evaluated in terms of those mitochondria affected by ADP. Thus, mitochondria located near the cell nucleus or in portions of the muscle in which the ADP concentration might not change during the twitch do not adversely affect the recordings. Electron micrographs show clearly that the muscle mitochondria are interwoven with the myofibrils. Not only is there such intimate contact that the mitochondria themselves appear to be deformed, but there probably occurs during muscular contraction a rolling or sliding motion that causes even more intimate contact with the contractile elements and the mitochondria. It is not unreasonable to speculate that the hypothetical pores in the endoplasmic reticulum may provide a very ready exchange of chemicals between the sarcoplasm and the mitochondria. Such an intimacy of contact combined with the possibility of facilitated diffusion due to relative motion and perhaps even pore action suggests that diffusibility of ADP from the myofibril to the mitochondria may not be a limiting factor.

The mitochondria are sharply localized centers of enzyme activity and the actual concentration of cytochrome at the site of optical measurement is very high. At a particular mitochondrion, the preponderance of activity for respiratory resynthesis to other activities is probably very large. By way of analogy, it is useful to consider that the mitochondria act as indicators of the changes of ADP concentrations strategically located in some muscles near the I bands.

Choice of tissue

A muscle approximately 1 mm. thick and 2 to 3 cm. long that is free of myoglobin and that can be readily perfused is optimal for the currently available experimental apparatus. At the present time, experimental data on the response of mitochondria to muscular contraction have been obtained in frog sartorius muscle, toad sartorius muscle, turtle coracohyoideus muscle, and in the heart muscle of the frog, toad and lobster by C. M. Connelly, A. Weber, F. Jöbsis, D. Lubbers, and J. Ramirez.

Muscle holder

One type of muscle holder, developed in collaboration with V. Legallais, has been described elsewhere[13]. In brief, the muscle is held between two thin perforated Lucite plates so that ready access to the oxygenated Ringer's solution is obtained. The optical path in frog sartorius muscle is about 1 mm. long, 4 mm. wide, and 20 mm. or more high. The muscle holder, as used in initial experiments with C. M. Connelly, did not contain methods for measuring tension, but this is now provided in currently used methods developed by F. Jöbsis and V. Legallais. Special muscle holders for use with strips of heart muscle have also been developed (D. Lubbers and J. Ramirez, unpublished data).

Experimental Results

Since reduced pyridine nucleotide has the largest absorption band of the respiratory components, it was logical to begin spectrophotometric experiments on muscle by recording at 340 mμ. In these experiments we used a reference wave length of 386 mμ, since this pair gives a minimal response to oxygenation and deoxygenation of hemoglobin in the few remaining erythrocytes retained by the perfused muscle.

We soon found that lowering the temperature has several advantages. First, the respiratory metabolism is slowed relative to the diffusion of oxygen, and hence the interior of a 1 mm.-thick sartorius muscle can be kept aerobic for a higher number of twitches than at room temperature. As a corollary of this, utilization of ADP or phosphate by the mitochondria is slow, and hence their response to ADP is prolonged. Under these conditions we obtained records of the type shown in Figure 7. In this record no indication of tension is given directly, since it was taken at an early stage in the studies. However, independent controls with tension recordings indicate that the small and downward deflections indicated by the arrows are optical disturbances that accurately mark the time of the twitch. Here we find that single twitches of the oxygenated frog sartorius muscle bathed in oxygenated Ringer's solution at 7°C. and stimulated by silver electrodes give a series of stepwise absorbancy changes. For two reasons, this type of response is significant:

1. It reflects a decrease of absorbancy at 340 mμ with respect to 386 mμ; this is verified by spectra taken at different wave lengths that show the disappearance of a peak at 340 mμ.

2. The time course of the response indicates that the utilization of ADP or phosphate is slow compared to the rate of its production.

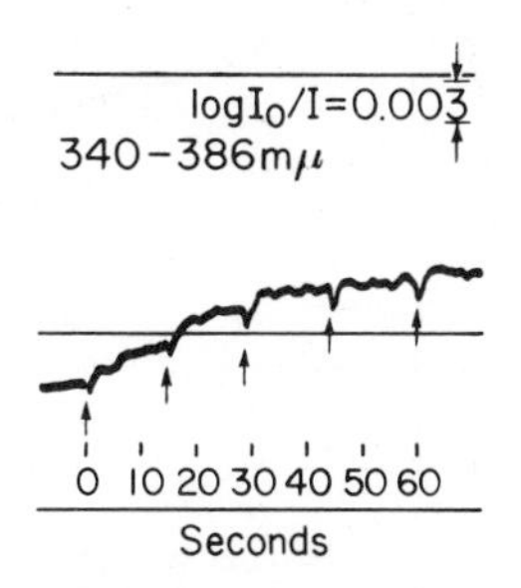

Fig. 7. Response of frog sartorius muscle at 7°C. to isolated twitches (experiment 301-14)[5].

To elaborate on this point, if the mitochondria expend very rapidly the ADP produced in the muscle twitch, then the absorbancy change is small and the "cycle" of oxidation and reduction is very rapid. This does not appear to be the case. If the creatine transphosphorylase or other enzyme systems resynthesize ATP at the same time, there would also be a rapid decay of the ADP level immediately after the twitch. Again, the fact that the optical change is approximately constant for a reasonable time indicates that this reaction is not occurring to an appreciable extent in the time interval recorded. Thus, the kinetics of any process expending ADP other than that under observation must somehow be reflected by the kinetics presented in this record. Within the limits of experimental accuracy there is no such evidence, and mitochondria probably provide the chief process responding to the ADP formed in the muscle twitch under these particular experimental conditions. Further documentation for this possibility is afforded by the high affinity of the respiratory chain for ADP. For example, the mitochondria have at least a tenfold greater affinity for ADP than does the transphosphorylase, although the precise data are not available for the latter under physiological conditions[(22)]. This reasoning also applies to other methods of resynthesis that may be postulated. In conclusion, we believe that the stepwise nature of the optical response obtainable under certain experimental conditions is very significant.

A preliminary calculation suggests that, although the transphosphorylase activity clearly exceeds the respiratory activity at 38°C.[(22)], the relatively greater temperature coefficient of the respiratory activity suggests that the situation would be reversed at the low temperatures (7°C.) at which we usually observe the steady-state changes in cytochromes following a twitch. The fact that rapid cytochrome responses are obtained with both heart and skeletal muscle under appropriate conditions despite their different creatine transphosphorylase activities[(23)] lends further support to this statement.

If a series of twitches at 1.5/sec. is given, the response is much more rapid, as shown in Figure 8. Under these conditions, we can compare the absorbancy change caused by 10 twitches with those obtained by stimulation over longer intervals, as indicated by the arrow (maximum level). From the jogs in the trace of Figure 8, one can readily observe that approximately 6 twitches cause

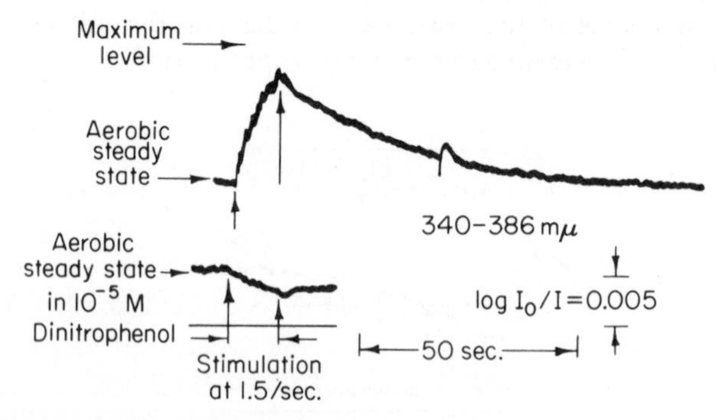

Fig. 8. Response of frog sartorius muscle at 7°C. to a series of twitches. Insert shows effect of dinitrophenol on the response (experiments 301-3, 301-33)[(5)].

a half-maximal absorbancy change. As a control, this record indicates the effect of stimulation of the dinitrophenol-treated muscle in which we obtain not the characteristic disappearance of the 340 mμ band but, instead, a smaller effect in the opposite direction. This control suggests that it is the response of the mitochondrial pyridine nucleotide that dominates under these experimental conditions.

Calculation of the amount of ADP per twitch

Experiments with two different muscles for various stimulation rates and total twitches are summarized in Table 3[5]. It is seen here that, over a stimulation rate of 0.07 to 7.0/sec. and a total number of twitches from 6 to 75, the average number of twitches required to give half-maximal effect is 6.

In studies of mitochondria isolated from rat liver, toad heart, and frog skeletal muscle, we find that about 50 μM ADP causes a half-maximal spectroscopic effect. The ADP released per twitch is then 50/6 μM or 0.009 μmole/gm. The maximum variability of this measurement indicated by Table 1 is about 0.001 μmole/gm. The inherent accuracy for a particular observation is considerably better than this, as may be seen from an inspection of experimental records. We feel that there is now evidence for ADP formation under conditions where the chemical methods have failed to find it.

This small value for the ADP release per twitch supports the negative result obtained by chemical analyses in which the amount of change observed by us is approximately equal to the experimental error of the chemical assay method. In other words, the ADP change under the experimental conditions of Mommaerts[24] and of Fleckenstein *et al.*[25] is too small to measure by the chemical

Table 3*

Changes of Level of Reduced Pyridine Nucleotide Following Muscle Twitches

Frog Sartorius Muscles, 5 to 7°C. in Oxygenated Ringer's Solution Plus Glucose; Muscle Essentially Unloaded

Experiment no	292b	292b	292b	301	301	301	301
Record no	12	13	14	3	4	14	14
Stimulation rate (sec.$^{-1}$)	3	7	1.5	1.5	3	0.07	0.2
Total twitches	75	7	45	15	45	6	27
Maximal oxidation of reduced pyridine nucleotide observed (μM) at given stimulation rate	24	14	23	25	32	11	30
Maximal effect with rapid stimulations (μM)	24	24	24	32	32	32	—
Twitches to give half-maximal effect†	7	6	5	7	7	5	—

*Reproduced by permission from *Nature*[5].

†We have computed the number of stimulations that have occurred in the time required for the optical change to reach half the "maximal effect" with rapid stimulation. In experiment 301–14 this point was reached by linear extrapolation from the initial increments per twitch. More refined calculations were not justified by the accuracy of the observations.

procedure, and the more sensitive spectrophotometric method has yielded positive results.

The question of an alternate interpretation of this assay in terms of phosphate must be considered. However, in the light of the extremely high ratios of phosphate to ADP concentrations found in muscle, we suspect that phosphate does not limit the respiratory rate in the resting muscle, and that the observed responses are caused by ADP rather than by phosphate.

Discussion

In view of the measured heats of activation and shortening and estimates of the work done by a muscle under these experimental conditions, it is reasonable to expect that between 0.3 and 0.5 μmole/gm. of ADP would have been released per twitch if ATP were broken down in the rising phases of the contraction. Such an increased concentration of ADP would be 10 times the half-saturation value and would have caused the absorbancy change to rise immediately to the maximal value, but this is not the case, as is clearly demonstrated by the experimental records. If we interpret the plateaus of the stepwise response to a series of slow twitches, shown in Figure 7, to indicate that ADP utilization during the attainment of the response is negligible, we are faced with the fact that the ADP concentration arriving at the mitochondria is, on a per twitch basis, only a few per cent of that to be expected if ATP were expended directly during the twitch. As an explanation for this discrepancy, it is possible that the muscle is already charged with a high energy compound; this is implicit in the hypotheses of A. F. Huxley and H. H. Weber presented elsewhere in this monograph. According to the theories of these investigators the muscle is charged with a high-energy compound prior to a twitch, so that the first movement of the cross-linked structures does not require the expenditure of ATP. On the other hand, this initial discharge of the crosslinks would lead to a shortening of the sarcomere of only about 400 Å, and hence repetitive recharging and discharging of the crosslinkages would occur in a single contraction, and ADP formation should be observed.

It is now possible to consider both our spectrophotometric observations and Davies' chemical observations in the light of these theories. The important feature of the experimental data presented by Davies elsewhere in these pages is that the store of hypothetical high-energy compound is very nearly completely expended in a single twitch. Inadequate extraction of the compound is unlikely to alter this observation, since it is assumed that the control muscle and the twitched muscle behave similarly in the extraction procedure. It is obvious that, if the muscles behave differently, the assumption of the chemical method (see *Introduction*) is violated, and therefore none of the chemical data is significant. Thus the muscle must depend upon other high-energy stores for twitches subsequent to the first one. This change-over from the endogenous supply to the external supply would be expected to have a marked effect upon the amount of ADP found in the muscle mitochondria, and the increments of absorbancy following a twitch should be small or zero for the first twitch and large for the second and subsequent twitches. As clearly illustrated by

Figure 7, this is not the case; the ADP appears to rise in steps of nearly equal amplitude for single isolated twitches. In fact, if an explanation of our data is based upon the postulated high-energy intermediates, it is apparent that the energy store must be sufficient to tide the muscle over for not one but for several twitches, and that the amount of this store must therefore be in excess of a few micromoles per gram. Thus the optical measurements suggest that any high-energy store postulated as an explanation of the small concentration of ADP arriving at the mitochondria following a twitch must be of considerable magnitude and capable of tiding the muscle over for a few twitches.

Summary

A sensitive optical method has been used to detect specific chemical reactions in the muscle following a twitch. The method is of high sensitivity and, in terms of change of adenosine diphosphate concentration, will record 0.001 μmole/gm. The method has been applied to the question of the amount of ADP formed following the muscle twitch. This is found to be a few per cent of the full amount expected, and various theories involving the activation of crosslinkages between actin and myosin and of tightly bound high-energy reservoirs in the muscle are considered as explanations for this discrepancy.

It is apparent from the optical data that any high-energy store postulated as an explanation of the small amount of ADP formed following the isolated twitches in a frog's sartorius muscle must be sufficient to sustain the muscle for not one but for several twitches.

References

1. Hill, A. V. 1931. Adventures in Biophysics.: 1–162. University Press. Philadelphia, Pa.
2. Hill, A. V. 1949–1950. *Proc. Roy. Soc. London.* **B136**: 195–211.
3. Hill, A. V. 1958. Symposium on Molecular Biology: Elementary Processes of Nerve Conduction and Muscle Contraction. Columbia Univ. and Rockefeller Inst. New York, N. Y.
4. Sleator, W., Jr. 1953. Communications, 19th Intern. Physiol. Congr. Montreal.: 768–769. Thérien Frères, Ltd. Montreal, Canada.
5. Chance, B., and C. M. Connelly. 1957. *Nature.,* **179**: 1235–1237.
6. Caspersson, T.: *J. Roy. Microscop. Soc.,* **60** (1940), 8.
7. Chance, B., R. Perry, B. Thorell, and L. Akerman, 1959. Program & Abstracts, The Biophysical Society. Abstr. **D3**. Pittsburgh, Pa.
8. Lardy, H. A. 1956. Proc. 3rd Intern. Congr. Biochem. Brussels, 1955.: 287. Academic Press. New York, N. Y.
9. Chance, B., and M. Baltscheffsky. 1958. *Biochem. J.* **68**: 283–295.
10. Chance, B., and G. R. Williams. 1955. *J. Biol. Chem.* **217**: 383–393.
11. Chance, B. 1959. Proceedings, Ciba Foundation, London, 1958. Churchill. London, England.
12. Chance, B. and G. R. Williams. 1956. *J. Biol. Chem.* **221**: 477–489.
13. Chance, B. 1954. *Science.* **120**: 767–775.
14. Sjöstrand, F. 1944. *Acta Anat. Suppl.* **1**: 1–163.

15. Chance, B. and H. Conrad. 1958. *Federation Proc.* **17**: 200.
16. Chance, B., and B. Thorell. 1959. *Nature.* In press.
17. Boyer, P. D. and H. Theorell,1956. *Acta Chem. Scand.* **10**: 447–450.
18. Duysens, L. N. M., and J. Amesz. 1957. *Biochim. et Biophys. Acta.* **24**: 19–26.
19. Chance, B., and H. Baltscheffsky. 1958. *J. Biol. Chem.* **233**: 736–739.
20. Chance, B., H. Conrad, and V. Legallais. 1958. Program and Abstracts, The Biophysical Society.: 44–45. Mass. Inst. Technol. Cambridge, Mass.
21. Chance, B., R. Perry, L. Akerman, and B. Thorell. 1959. *Nature.* In press.
22. Kuby, S. A., L. Noda, and H. A. Lardy. 1954. *J. Biol. Chem.* **210**: 65–82.
23. Oliver, I. T. 1955. *Biochem. J.* **61**: 116–122.
24. Mommaerts, W. F. H. M. 1954. *Nature.* **174**: 1083–1084.
25. Fleckenstein, A., J. Janke, R. E. Davies, and H. A. Krebs. 1954. *Nature.* **174**: 1081–1083.

ADENOSINE TRIPHOSPHATE: CHANGES IN MUSCLES DOING NEGATIVE WORK*

Anthony A. Infante, Dzintra Klaupiks and Robert E. Davies

Laboratories of Biochemistry, Department of Animal Biology
School of Veterinary Medicine
University of Pennsylvania, Philadelphia

(Submitted March 30, 1954)

Abstract. *Frog sartorius muscles were isolated, treated with 1-fluoro-2, 4-dinitrobenzene at 0°C, then stimulated tetanically at the length* in situ *and stretched with a Levin-Wyman ergometer during stimulation. The normal adenosine triphosphate breakdown during the tetanus was reduced by about half during the forced stretch. The tension was increased by about 70 percent, but resynthesis of adenosine triphosphate did not occur. Thus, on the basis of A. V. Hill's results, adenosine triphosphate is probably not the direct final energy source for muscular contraction, although it intimately participates in the process. The use of adenosine triphosphate during negative work was less than one-tenth that needed for positive work.*

Perhaps the most dramatic recent discovery concerning the biophysics of muscular activity is that "when a muscle is stretched during a maintained contraction, the whole of the work may disappear"[1]. The findings of Hill, Abbott, Aubert, and Howarth[1-4], based on earlier work by Fenn in Hill's laboratory[5], "leave little doubt that the chemical reactions which normally occur during contraction can be reversed by a stretch under the influence of the mechanical work supplied"[3]. Since the work absorbed in stretching a "contracting" muscle apparently "can reverse the chemical processes of activity"[4], experiments were carried out to find out if adenosine triphosphate (ATP) would be resynthesized under such conditions. Such a close energy coupling would be analogous to the production of ATP during electron

*A. A. Infante, D. Klaupiks, and R. E. Davies, "Adenosine Triphosphate: Changes in Muscles Doing Negative Work," *Science,* **144**: 1577–78 (June 26, 1964).

transport in oxidative phosphorylation compared to the reversal of electron transport by ATP[6].

The methods were similar to those used previously[7-12]. The muscles were treated with 1-fluoro-2,4-dinitrobenzene (FDNB) which can completely inhibit ATP-creatine phosphoryltransferase *in situ* in isolated frog muscle and prevent any regeneration of ATP except through the action of myokinase (ATP-adenosine monophosphate phosphoryltransferase)[8-10]. This made possible the direct demonstration of a breakdown of ATP associated with work during single contractions of isolated intact muscle[8-12], which confirmed the widely held theory of the role of ATP in muscle contraction based on experiments on isolated and reconstituted actomyosin fibers, glycerinated muscle models, and isolated enzymes by Weber, Needham, Szent-Györgyi, Engelhardt, and many others [see review by Huxley[13]].

The sartorius muscles of female frogs (*Rana pipiens*) were dissected in pairs and allowed to rest at room temperature in a physiological bicarbonate-saline solution for 2 to 3 hours. The muscles were then maintained for 35 minutes at 0°C in a similar solution gassed with 5 percent CO_2 + 95 percent N_2, containing 0.38 millimoles of FDNB per liter. They were then mounted at rest length between an immobile support and a jeweler's chain and incubated for an additional 5 minutes at 0°C after which time they were rapidly frozen in a 1:1 mixture of Freons 12 and 13 ($CF_2Cl_2 + CF_3Cl$) at −172°C either without further treatment, or during tetanic isometric contraction, or stretch produced by a Levin-Wyman ergometer. One of the sartorius muscles served as a control for the other pair in each of the series of experiments. The forces involved were recorded from the output of a Grass tension transducer (FT 03) attached to the arm of the ergometer and connected to the muscle by means of the jeweler's chain. The distances moved and the rate of stretching were determined by prior adjustments of the ergometer. The times of stimulation, stretching, and freezing were controlled by a synchronous motor fitted with appropriate switches. The frozen muscle was ground to a fine power at −196°C and extracted and assayed in duplicate[8].

In Table 1, in series B, C, and E, the experimental conditions for the stretching of the muscles were similar to and, in series identical with those used by Hill and Howarth[3] (Fig. 6) for sartorius muscles of English toads (*Bufo bufo*). There was a linear breakdown of ATP with time (series A, C, and H) and there was no significant difference between the numerical results for the changes in ATP and inorganic phosphate. Thus myokinase activity remained slight under these conditions with this muscle[8]. Although the individual changes are small and the variance between the left and right muscles of each pair is relatively high, it is clear from taking together all the various conditions used (Table 1) that stretching an activated muscle caused a reduction in the breakdown of ATP ($P < .001$), even though the tension developed was about 70 percent greater than the isometric tension at rest length. Thus, the changes in series B and D were less than in A and C, respectively, and in series E and G the reduced breakdown of ATP left more ATP and less inorganic phosphate in the stretched muscles than in their pairs which had been stimulated isometrically for the same length of time. The value of P can be calculated both from the distribution of the signs of the observed changes or by combining the probabilities of the individual results by the method of Fisher[14], taking account of the signs of the differences.

The reduction in ATP breakdown was about half, and thus net resynthesis certainly did not occur; otherwise, on the basis of series A, C, and H, the changes in ATP and Pi contents, respectively, would have been +0.3 and −0.3 μmole/g in series F, zero in series B and D, and +0.66 and −0.66 μmole/g in series E and G.

This conclusion that the breakdown of ATP is reduced by stretching an

Table 1

Changes in adenosine triphosphate and inorganic phosphate in isometric tetanic contractions at the rest length *in situ* and in stretches during continued activation of frog sartorius muscles treated with FDNB a 0°C. Tetanic supramaximal electrical stimulation at 12 pulses per second was continuously applied during the stretch. The rate of stretching was 4 mm per 0.5 second in all cases. The average rest length *in situ* was 34 mm for muscles weighing approximately 100 mg. All stretched muscles were frozen just at the end of the stretch. When inorganic phosphate was measured, the muscles were powdered and the powder was extracted by stirring in 0.5*M* perchloric acid at 0°C for 2 minutes. A/portion of the supernatant after centrifugation was immediately assayed for inorganic phosphate by the method of Wahler and Wollenberger[20]. The remaining perchloric acid extract was stirred for 4 minutes at 35°C, neutralized, and assayed for ATP[8, 9]. All values of differences are the means of the difference within each pair ± the standard errors of the means. The numbers in parentheses refer to the number of muscle pairs.

	Treatment of muscle pairs	ATP concn. (μmole/g)		Inorganic phosphate concn. (μmole/g)	
		Av. of means	Diff. in means ± S.E.	Av. of means	Diff. in means ± S.E.
A	Unstimulated control 0.56 second isometric tetanus	2.85 2.52	−0.33 ± 0.10(7)	2.30 2.63	+0.33 ± 0.14(6)
B	Unstimulated control 0.06 second isometric tetanus + 0.50 second stretch	2.62 2.54	−0.08 ± 0.14(8)	1.83 2.02	+0.19 ± 0.11(8)
C	Unstimulated control 0.7 second isometric tetanus	3.40 3.00	−0.40 ± 0.10(6)	2.05 2.53	+0.48 ± 0.13(8)
D	Unstimulated control 0.3 second isometric tetanus + 0.4 second stretch	2.96 2.69	−0.27 ± 0.13(8)	1.85 2.17	+0.32 ± 0.11(9)
E	1.0 second isometric tetanus 0.5 second isometric tetanus + 0.5 second stretch	2.09 2.18	+0.09 ± 0.10(11)	2.59 2.37	−0.22 ± 0.14(5)
F	0.5 second isometric tetanus 0.5 second isometric tetanus + 0.5 second stretch	2.48 2.38	−0.10 ± 0.13(16)	2.50 2.58	+0.08 ± 0.10(11)
G*	1.0 second isometric tetanus 0.5 second isometric tetanus + 0.5 second stretch	2.12 2.28	+0.16 ± 0.12(9)	2.61 2.44	−0.17 ± 0.08(9)
H	Unstimulated control 1.15 second isometric tetanus	2.99 2.17	−0.82 ± 0.08(6)	2.14 2.88	+0.74 ± 0.08(6)

*The muscles were mounted at 90 percent of their rest length *in situ*.

activated muscle is in apparent contradiction to the recent results of Aubert and Maréchal[15] who found an increased breakdown at 0°C of phosphorylcreatine in iodoacetate-treated sartorius muscles of the frog which were stretched several times when activated. However they had used rapid stretches and obtained results quite similar to ours with slower rates of stretch comparable to that used by us[16].

Our results indicate that the net ATP (phosphorylcreatine) breakdown found previously[8,12,17] really is associated with work and not with a function of developed tension × time. The stretched muscles in Table 1 absorbed an average of 400 g-cm of work per gram. A comparison of the ATP requirements for positive work[9] and for this negative work shows that under these conditions the ATP usage during negative work was only about 1/13th that required during positive work. This is in good agreement with the finding that the extra oxygen needed in man for negative work can be less than 1/10th that needed for positive work[18].

These results are relevant to the conclusion of Hill and his colleagues concerning the reversal of the chemical processes of muscular activity during negative work[1-4]. If their conclusion is correct and applicable to sartorius muscles previously treated with FDNB, and if ATP were the direct final energy source for muscle contraction, then it should have been resynthesized. This did not occur; so it seems that ATP is, after all, not the direct final energy source for muscular contraction. The finding of a reduction of ATP breakdown, but not a resynthesis in these experiments was predicted by a recent molecular theory of muscle contraction[19] which is based on the view that ATP is used to extend potential linkages between myosin and actin, which in the presence of calcium, form actual linkages, contract, and do work by the energy of hydrogen and hydrophobic bond formation to make α-helices, this process being repeated cyclically during muscular contraction. Stretching an activated muscle should thus break hydrogen bonds and reduce the breakdown of ATP which has now been observed.

References and Notes

1. B. C. Abbott, X. M. Aubert, and A. V. Hill, *J. Physiol.* (*London*) **111**, 41 P. (1950).
2. ———, *Proc. Roy. Soc. Ser. B.,* **139**, 86 (1951); B. C. Abbott, and X. M. Aubert, *ibid.,* p. 104.
3. A. V. Hill and J. V. Howarth, *ibid.* **151**, 169 (1959).
4. A. V. Hill, *Science* **131**, 897 (1960).
5. W. O. Fenn, *J. Physiol.* **58**, 373 (1924).
6. B. Chance and G. Hollunger, *Federation Proc.,* **16**, 163 (1957); B. Chance, L. Ernster, and M. Klingenberg, in *Biological Structure and Function,* T. W. Goodwin and O. Lindberg, Eds. (Academic Press, New York, 1961), vol. 2.
7. S. A. Kuby and T. A. Mahowald, *Federation Proc.,* **18**, 267 (1959).
8. D. F. Cain and R. E. Davies, *Biochem. Biophys. Res. Commun.,* **8**, 361 (1962).
9. A. A. Infante, Ph. D. dissertation, Univ. of Pennsylvania (1963).
10. A. A. Infante, D. Klaupiks, and R. E. Davies, *Federation Proc.,* **23**, 366 (1964).
11. D. F. Cain, A. A. Infante, and R. E. Davies, *Nature,* **196**, 214 (1962).
12. A. A. Infante and R. E. Davies, *Biochem. Biophys. Res. Commun.,* **9**, 410 (1962).

13. H. E. Huxley, in *The Cell,* J. Brachet and A. E. Mirsky, Eds. (Academic Press, New York, 1960), vol. 4.

14. R. A. Fisher, *Statistical Methods for Research Workers* (Oliver and Boyd, London, ed. 3, 1948).

15. X. Aubert and G. Maréchal, *J. Physiol. (Paris)* **55**, 186 (1963).

16. G. Maréchal, private communication.

17. F. D. Carlson, D. Hardy, and D. R. Wilkie, *Biochemistry of Muscle Contraction,* 23–26 May 1962, J. Gergely, Ed. (Little Brown, Boston in press); R. E. Davies, D. F. Cain, A. Infante, D. Klaupiks, and W. A. Eaton, *ibid.;* W. F. H. M. Mommaerts, *ibid.;* W. F. H. M. Mommaerts, K. Seraydarian, and G. Maréchal, *Biochim. Biophys. Acta,* **57**, 1 (1962); F. D. Carlson, D. J. Hardy, and D. R. Wilkie, *J. Gen. Physiol.* **46**, 851 (1963).

18. E. Asmussen, *Acta Physiol. Scand.* **28**, 364 (1953); ———, in *Muscle as a Tissue,* K. Rodahl, and S. M. Horvath, Eds. (McGraw-Hill, New York, 1962).

19. R. E. Davies, *Nature,* **199**, 1068 (1963).

20. B. E. Wahler and A. Wollenberger, *Biochem. Z.,* **329**, 508 (1958).

21. This work was supported in part by USPHS grants HE-02520 and GPM-12921. One of us (A.A.I.) is a USPHS postdoctoral fellow. We thank Dr. B. Chance for lending us the Levin-Wyman ergometer.

PART VI

ADENOSINE TRIPHOSPHATE (ATP) AND CONTRACTILE PROTEINS

WHAT IS the relationship of ATP and ATP splitting to contraction? It seems appropriate at this point to sketch briefly the development of our present knowledge of myosin, the contractile substance of muscle.

The proteins of muscle were first observed by Kühne and von Fürth. Both described a muscle globulin, myosin (also called *paramyosinogen*) which can be readily coagulated to myosin fibrinogen. During this period, studies of myosin followed much the same line of approach used in studies of proteins involved in coagulation processes like fibrinogen and caseinogen.

Modern knowledge of the molecular basis of muscle contraction and the possible role of phosphorylation in this process began about 40 years ago with the fundamental studies of myosin by Hans Weber[1], John Edsall, and Alexander von Muralt[2,3,4]. These now classic studies provided us with the first exact description of the physicochemical properties of myosin. Moreover, they offered rational methods for extracting and purifying myosin. The peculiar properties of myosin solutions were described in detail. The viscosity was found to be high, typical of a rodlike protein. As Edsall phrased it[2]: "A solution containing 1.5 per cent of muscle globulin is a thick syrup which crawls rather than flows." Edsall and von Muralt succeeded in relating the properties of the anisotropic protein with the double refraction of the intact muscle (see also the reprinted report by Boehm and Weber). The first quantitative observations on the size and shape of the protein were collected in these studies. Native myosin solutions were found to be *monodisperse* with a *uniform size and shape*. Myosin has a large *asymmetry* factor which is also evident from the double refraction of flow. Moreover, the large angle of isocline shown during streaming (orientation under the action of shearing forces) also points to the existence of long and somewhat rigid protein molecules. The double refraction of flow as well as the angular velocity patterns of myosin show characteristic and reproducible curves varying with the protein concentration. Denaturation brings about a loss of the characteristic property of double refraction of flow. In this fundamental series of papers of 1930, von Muralt and Edsall made the important suggestion that the name *myosin* be restricted to the anisotropic monodisperse muscle protein which is responsible for the characteristic double refraction of flow[3,4]. A somewhat abbreviated account of these investigations appeared in the *Transactions of the Faraday Society* in 1930 [q.v.].*

The next big stride forward came in 1932 when Gundo Boehm and Hans Weber, using a very simple device–injection of a myosin solution through a fine needle into distilled water–obtained long myosin filaments which turned out to be not only anisotropic but which showed X-ray diffraction diagrams characteristic of intact muscle (see Boehm and Weber [q.v.] and Weber[5]).

These Boehm-Weber filaments started Engelhardt and Szent-Györgyi on their road to "mechanochemistry." On the other hand, the von Muralt-Edsall monodisperse liquid gels formed the basis for an interesting study by Joseph and Dorothy Needham and their coworkers on the effect of ATP on the asymmetric properties of myosin molecules.

The discovery of actomyosin and actin, which is described in Albert Szent-Györgyi's early paper [q.v.] created a fundamental principle that has influenced our thinking ever since. The relevance of these concepts to intact muscle

*See material under these authors' names in this section.

appeared in an interesting paper by Buchthal and Kahlson, who had found that ATP can elicit a genuine contraction in intact muscle fibers [q.v.]. Later developments are well known and are summarized in several monographs. I have therefore decided to discuss and reprint here only thirteen articles dealing with phosphorylation and muscular contraction.

It was soon discovered that pure myosin could be resolved into subunits. Brief exposure of purified myosin to various proteolytic enzymes brings about a peculiar type of splitting revealing the existence of two types of subunits called *meromyosins* (see reprinted papers): Light meromyosin is a filamentous protein, with a molecular weight of about 160,000[6]; whereas heavy meromyosin, with a molecular weight of about 360,000[6], is a globular protein. Mihalyi and Andrew Szent-Györgyi and John Gergely found that all the ATP^{ase} activity of myosin is confined to the heavy meromyosin.

ATP^{ase} and the Function of Myosin

In his pioneer work of 1940, Engelhardt [q.v.] stressed two points: (1) that myosin is an ATP^{ase} and (2) that the enzymatic splitting of ATP by myosin elicits physical changes in the protein ("mechano-chemical" interaction) if the myosin is in the filamentous form. In a review on phosphorylations in 1942[7], I emphasized another aspect of Engelhardt's discovery, the possibility that myosin might function as a transfer system of phosphoryl groups and that a transitory phosphorylation of myosin might effect an entropy change, or a conformational change (change in the configuration of the protein), which might bring about contraction or relaxation. This idea was independently formulated some years later by Riseman and Kirkwood [q.v.] who elaborated on such a concept by making quantitative estimates. From a thermodynamic point of view, they found a phosphorylation of myosin perfectly acceptable as a basis for mechanical work.

Later, Koshland and his coworkers[8] introduced the use of ^{18}O labeled water and ^{18}O labeled phosphate to examine the question of the existence of a transient phosphorylation of myosin by ATP. An overabundance of ^{18}O from labeled water entering into orthophosphate (as compared with the transfer from ^{18}O labeled phosphate to water) might well indicate that myosin participates in the ATP^{ase} reaction. This is discussed in one of Koshland's papers (see Koshland and Levy). Boyer[9], who had been interested in this type of phenomenon in mitochondria, found indication of the formation of an intermediary compound only if ITP were used. Since ITP does not elicit contraction of myosin filaments, a correlation between contraction and phosphorylation of myosin is not in sight at present, and the high ATP^{ase} activity found in myosin remains a somewhat isolated observation. Most recently Boyer's group (L. Sartorelli, H. Fromm, R. Benson, and P. Boyer: Personal communication, 1966) described an exchange reaction of orthophosphate with ^{18}O labeled water, which takes place if ATP is incubated briefly with myosin in the presence of Mg^{++}. As regards the catalytic site of myosin ATP^{ase}, sulfhydryl (SH) groups occupy a special position.

Titration of the reactive SH groups of heavy meromyosin with mercurials manifests an interesting anomalous response if calcium-activated ATPase is used[10, 11]. Titration of myosin with pCMB in the presence of ATP and Ca^{++} brings about a biphasic reaction feature with respect to the ATPase activity. The activity rises four- to fivefold, reaches a maximum, and then declines[12, Fig. 1].

How Was the Existence of Myosin ATPase Overlooked during the Embden-Meyerhof-Lohmann Era ?

During the period (1929–1939) when Embden, Lohmann and Meyerhof established the phosphoglyceric and phosphopyruvate pathway in glycolysis in muscle, no attention had to be paid to phosphatases. It was simply due to the fact that muscle extracts used for studies on glycolysis were practically free of phosphatases including adenosine triphosphatase (ATPase). The absence of ATPase enabled Lohmann to study the transphorylation of ADP to ATP from phosphocreatine. The discovery of this reaction made it inevitable that ATPases in the insoluble muscle residue would be responsible for the liberation of inorganic phosphate seen during normal contraction. The Lohmann-Meyerhof procedure for preparing active muscle extracts simply prescribed the squeezing of juice from a chilled muscle in a mortar, using distilled water or aqueous potassium chloride extracts. The clear extracts exhibit hardly any detectable ATPase reaction, at least not with the methods available at that time.

A specific ATPase was actually first described by Erik Jacobsen[12] and independently by Barrenscheen as early as 1931. This ATPase was found in liver dispersions. The liver ATPase might be identical with the mitochondrial ATPase. Jacobsen pointed out at that time that muscle extract did not contain any detectable ATPase. This is not in conflict with the report by Engelhardt and Ljubimowa in 1939. In fact, the two authors point out that they made special efforts to extract the insoluble muscle residue left behind in the usual extractions of "soluble" juice. In order to solubilize proteins from the insoluble residue they used buffers and concentrated salt solution as prescribed by Edsall for the extraction of myosin.

In 1947 Meyerhof and his first American graduate student, Wayne Kielley, described a peculiar ATPase in muscle extracts free of myosin[13]. Since myosin ATPase is catalyzed by the divalent ion, calcium[14], and the "new" ATPase was found to be dependent on the presence of magnesium ions, its identity seems clear. Its function may, as later emphasized by Ebashi, be important in the relaxation process (see Section VII).

References

1. Weber, H. H.: *Biochim. Z.,* **158** (1925), 443, 473; *Biochim. Z.,* **189** (1927), 381.
2. Edsall, J. T.: *J. Biol. Chem.,* **89** (1930), 289.
3. von Muralt, A. L., and Edsall, J. T.: *J. Biol. Chem.,* **89** (1930), 315.
4. ———, and ———: *J. Biol. Chem.,* **89** (1930), 351.
5. Weber, H. H.: *Ergeb. Physiol.,* eds. Asher and Spiro, **36** (1934), 109.
6. Kielley, W. W., and Harrington, W. F.: *Biochem. Biophys. Acta,* **41** (1960), 401.

7. Kalckar, H. M.: *Biol. Rev.* Cambridge Phil. Soc., **17** (1942), 28.
8. Levy, H. M., and Koshland, D. E.: *J. Amer. Chem. Soc.,* **80** (1958), 3164.
9. Boyer, P. D.: in *Biochemistry of Muscle Contraction,* ed. J. Gergely. Boston: Little, Brown & Co., 1964, p. 94.
10. Greville, G. D., and Needham, D. M.: *Biochem. Biophys. Acta,* **16** (1955), 284.
11. Kielley, W. W., and Bradley, L. B.: *J. Biol. Chem.,* **218** (1955), 653.
12. ———; *The Enzymes,* 2nd ed., eds. P. D. Boyer, H. Lardy, and K. Myrbäck. New York: Academic Press, Inc., 1961, p. 159.
13. Kielley, W. W., and Meyerhof, O.: *J. Biol. Chem.,* **176** (1948), 591.
14. Bailey, K.: *Biochem. J.,* **36** (1942), 121.

THE X-RAY DIAGRAM OF STRETCHED MYOSIN FILAMENTS*

Gundo Boehm and H. H. Weber

Institut für Physiologie, Freiburg im Breisgau, Germany and
Institut für Physiologie, Münster im Westfalen, Germany

The proteins of the crosstriated skeletal muscle consist–according to more recent analyses (H. H. Weber[1])–of approximately 70% myosin, a globuline like protein and of about 30% myogen, a protein of the albumin type. From analyses by quite diverse methods one might consider myosin as the component which is responsible for the fibrous or fibrillar structure of muscle and which is contractile; solutions of myosin are sols of rods with dissolved particles which deviate considerably from the spherical shape. They show, accordingly, a very high and anomalous viscosity, a tendency for thixothropy, strong birefringence of flow (at a wide angle, which indicates rather long particles) and a positive Gans effect. Earlier attempts by others to resolve solutions, static or flowing, by ultra microscope or by X-ray respectively, were unsuccessful[2]. Therefore it was necessary to investigate whether the solid material oriented in a fibrous type of structure (crystallization of myosin does not materialize[3]) has properties resembling those of the muscle fibril. It is indeed possible by means of a kind of stretch-spin-method to prepare structures from myosin solutions which resemble muscle fibrils mechanically, and optically as well as in their X-ray pattern.

The procedure (H. H. Weber) is as follows: Myosin solutions of greatest purity are injected through a capillary tube into distilled water where they solidify instantaneously into a thread. This thread can be stretched smoothly to about double the original length. It is possible to increase the stretch by

*G. Boehm and H. H. Weber, "Das Röntgendiagramm von gedehnten Myosinfäden," *Kolloid Z.*, **61**, 269–70 (1932). Translated from the German and reprinted with the permission of the authors and of Dr. Dietrich, Steinkopf-Verlag, Darmstadt, Germany. The clear, redrawn X-ray diagrams we owe to Professor G. Boehm.

another 100% after drying and rewetting whereupon the dried threads show the following properties:

The tensile strength is considerable; likewise they do not break readily on folding. The positive birefringence (in reference to longitudinal stretching) showing $n_\gamma - n_\alpha$ of about 0.006 is of the order of magnitude of fresh muscle (0.0025)[4], considering that, in addition to negatively birefringent lipoid, 30% of the dry substance of muscle consists of isotropic myogen. Ultramicroscopically, using the Azimut shutter or the Spierer lens the threads prove optically empty, i.e., elongated particles are not seen.

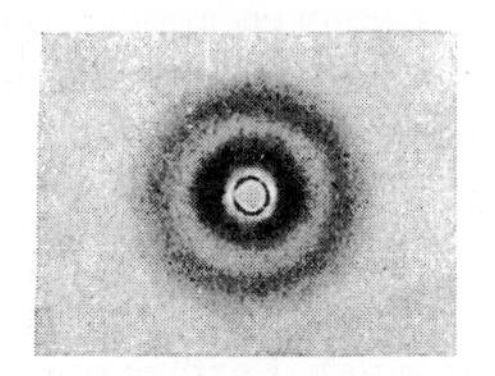

Fig. 1a. Bundled myosin filaments stretched and dried. Exposure 8 min with CuK_α irradiation.

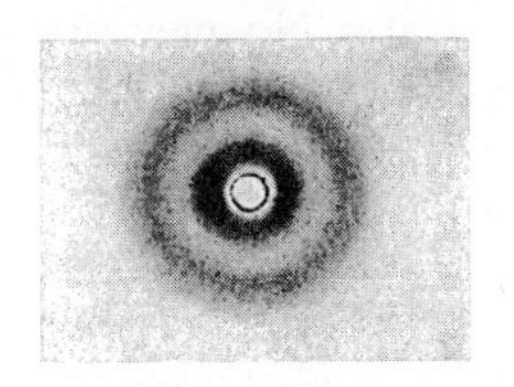

Fig. 1b. Dried sartorius muscle from *Rana esculenta* contracted about 23%. Exposure 8 min with CuK_α irradiation.

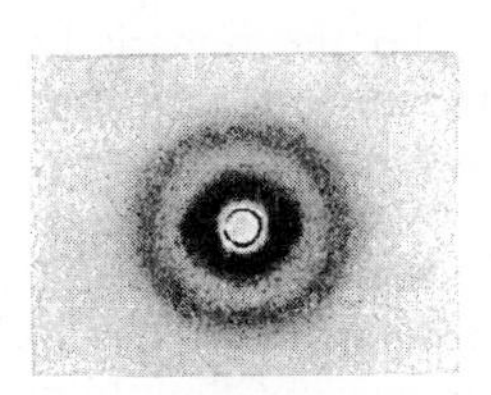

Fig. 1c. Sartorius muscle from *Rana esculenta* dried at maximum tension. Exposure 8 min with CuK_α irradiation.

The photograph of a thread by X-ray, however, reveals a pattern which is indistinguishable from a slightly stretched, dried frog sartorius (Fig. 1a, 1b). These threads, therefore, must contain oriented micelles of myosin.

This finding establishes the following:

1. The X-ray pattern and the birefringence of muscle must be attributed entirely to oriented myosin.

2. The discovery of anisometric structures in stretched gelatine accomplished by X-ray diagram methods of Katz[5] has now been expanded by the demonstration that myosin contains anisometric structures. The structures were shown to have a pronounced tendency to form thread-like aggregates.

These stretched myosin threads represent a type of muscle model which might be suitable for investigation of mechanical properties–perhaps also for thermo-mechanical properties and which may eventually be applied in a study of various types of contractures.

Discussion

[Prepared by W. Haller (London)].

For the interpretation of the birefringence appearing during the stretching of filaments two different possibilities must be considered, the orientation of anisodimensional particles on the one hand, and, on the other, the deformation in the direction of stretch of really isotropic and isodimensional particles. With respect to solutions of myosin, the exact analysis of its birefringence of flow favors the effect of deformation (conf. a theoretical paper in preparation).

The deformation of the colloid particles may be seen as the unwinding and stretching of the loosely coiled, ball-like molecular filaments in a sol. In this connection, we refer to the papers by Haller on the solvation (*Kolloid Z.*, **56** (1931), 257) and by K. H. Meyer, V. Susich, and Valko on the elasticity of lyophilic colloids (*Kolloid Z.*, **59** (1932), 280).

Answer

To the remark by J. R. Katz: No new crystal interferences appear on stretching the myosin threads. One simply obtains a pattern which is completely identical with that of a dried muscle. The pattern is produced by a parallel arrangement of covalent chains as described by K. H. Meyer(6). To the remark by Haller: There can be no doubt that even a static (not flowing) solution of myosin contains anisometric structures. The new theory of birefringence of flow by Werner Kuhn also takes this into account. Furthermore, it is not only the birefringence of flow, but also the anomalous viscosity and to some extent the Gans effect which indicate the presence of elongated particles which do not just originate from the deformation of spherical structures.

Weber–Regardless of the genesis of the birefringence of flow (sol configuration), it must be definitely stated that in the muscle as well as in the myosin thread (gel configuration) the molecular filaments are extended and arranged parallel to each other and remain in this position in the absence of any external deforming forces: completely relaxed muscles and myosin threads floating in a bath have a birefringence ten times higher than the birefringence of flow of myosin solutions if compared on the basis of equal thickness of protein layers.

References

1. Weber, H. H.: *Ber. ges. Physiol.,* (B) **61** (1931), 382.
2. Edsall, J.: *J. Biol. Chem.,* **89** (1930), 294.
3. v. Muralt, A., and J. Edsall: *J. Biol. Chem.,* **89** (1930), 382.
4. v. Ebner, V., Untersuchungen über die Anisotropi organisierter Substanzen (1882).
5. Katz, J. R., *Naturwiss.,* **13** (1925), 900.

Hans H. Weber, 1938. Muscle biochemistry is muscle physiology as well and cannot be separated from it as a teaching topic.

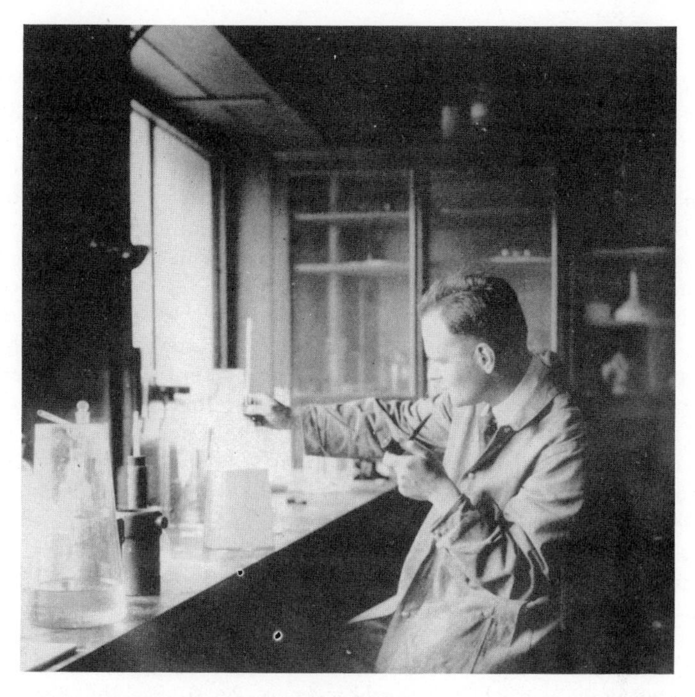

John T. Edsall, Harvard Medical School, 1930. The monodisperse population of elongated protein molecules from skeletal muscle we shall call myosin.

DOUBLE REFRACTION OF MYOSIN AND ITS RELATION TO THE STRUCTURE OF THE MUSCLE FIBRE*

Alexander L. von Muralt and John T. Edsall

Department of Physical Chemistry in the Laboratories of Physiology
Harvard Medical School, Boston, Massachusetts

(Received for publication, July 31, 1930)

Double refraction is an optical property shown by many living tissues. Muscles, bones, tendons and nerves are doubly refractive; the character and magnitude of the effect being however very different for each group. The effect is always the result of an orientation of (optically) anisotropic elements. Since in bones and tendons the normally acting stresses and pressures lie usually in one general direction, it is possible that the high resistance which these tissues show towards mechanical deformation in this direction is due to oriented anisotropic elements. (The anisotropy is accordingly not only an optical, but also a mechanical property.) It is well known that most substances become doubly refractive if they are deformed by external forces; an effect which is generally called accidental double refraction.† It is an interesting fact that those substances which are specially fit to resist large strains (as tendons for instance) possess in the unstrained condition an optical property (double refraction) which, in ordinary substance not showing this resistance, appears as a consequence of the strain. The exact nature of the double refraction of structural substances, as well as the complicated optical behaviour of nerves is at present not known. The underlying cause of the double refraction of the muscle fibres however, seems to become better known, especially in the light of recent experiments, which shall be reviewed in this report.

*A. L. von Muralt and J. T. Edsall, "Double Refraction of Myosin and its Relation to the Structure of the Muscle Fibre," *Trans. Faraday Soc.,* **26**, 837–52 (1930). Reprinted with the permission of the authors and of the publisher.

†The term, "photoelastic effect" which is used in recent publications refers to a special case of accidental double refraction.

Smooth muscle fibres as a whole are doubly refractive whereas the *cross striated* fibres show this effect in the anisotropic disc Q only and to a small extent in the band Z. This appears to be a necessary property connected with contractility. Engelmann[11] who first postulated this relation, based his opinion largely on two important biological observations; namely, that all contractile tissue has been found to be doubly refractive and that this property is lost if the tissue is transformed into functionally different tissue. This disappearance of double refraction is specially striking in the development of the pseudo-electric organ from muscle, which Babuchin has studied. (A more detailed discussion may be found in the excellent book of W. J. Schmidt[34].) It is also in good agreement with the well-known measurements of Hürthle[22]. He found that, during contraction, only the anisotropic disc undergoes definite changes in dimension, which can be interpreted as an indication that contracility is restricted to this part of the muscle fibre. Furthermore it appears that the element producing double refraction is actively involved in the process of contraction. V. v. Ebner[9] observed that the double refraction decreases considerably during contraction, an observation which he defended successfully against objections raised by Bernstein[3]. The conclusive measurements which Stübel and Liang[38] have recently published on the decrease of double refraction in various states of rigor, as well as own observations[27] on the intact fibres of the membrana basihyoidea of the frog, have led us to the belief that the doubly refractive element undergoes considerable changes during contraction. It appears therefore well worth while to discuss the cause of this double refraction and the possible reasons for its disappearance.

Our knowledge of the fine-structure of the living muscle fibre has been enlarged very considerably by the measurements which Stübel[37] has made. They are based on a theory of double refraction of a specialised system, the *Stäbchenmischkörper* which has been developed by Wiener[45]. In view of the importance of this theory, a brief exposition may be given. Wiener calculated the optical properties which would be exerted by a system of small isotropic parallel rods of the refractive index n_1, embedded in an isotropic medium of the refractive index n_2. These rods were to be small compared with the wave-length of light. He found that such a system must be doubly refractive, although the components taken separately are *isotropic,* and that the amount is given by the equation

$$n_e^2 - n_0^2 = \frac{V_1 V_2 (n_1^2 - n_2^2)^2}{(V_1 + 1) n_2^2 + V_2 n_1^2} \tag{1}$$

n_e, n_0 denoting the refractive index of the extraordinary and the ordinary beam respectively, n_1 the refractive index of the rods, V_1 their partial volume, n_2 the refractive index of the medium, V_2 its partial volume.

The three main conclusions which Wiener has drawn from this result of his theory are as follows: (1) Such a system of isotropic rods is always positive doubly refractive, since $(n_1^2 - n_2^2)^2$ is always positive. (2) If the two refractive indices are equal ($n_1 = n_2$), no double refraction occurs. (3) If the double refraction of the system is plotted as ordinate and the refractive index of the medium as abscissa, a parabola results with the vertex at the point $n_2 = n_1$.

If the rod-shaped elements possess a certain amount of double refraction of their own (*Eigendoppelbrechung*) then the vertex of the parabola remains at the same abscissa n_2 but the ordinate is now, instead of 0, equal to the summation of the amounts of double refraction which the rods possess taken separately.

The experimental work of Ambronn and his school[1], who studied the double refraction of many substances (among other variables) as a function of the refractive index n_2 (by imbibition with varying fluids) revealed fine-structures which hitherto have not been accessible. It became even apparent that this new method of investigation, based on Wiener's theory, is more sensitive in certain instances than X-ray spectroscopy. This apparently encouraged Stübel to study the fine-structure of muscle fibres from this point of view and yielded results of importance.

Stübel measured the double refraction of single muscle fibers by changing the fluid of imbibition, and with that the refractive index n_2. If the muscle fibre were to be of a structure similar to the theoretical conception of Wiener's *Stäbchenmischkörper,* then a parabola should result if the measurements were plotted in the indicated way. Stübel himself has not published such a graph, but we have taken his published data and have plotted them, according to Wiener's theory, in Fig. 1.

The points represent measurements at two different wave-lengths, and it seems to us apparent that the experimental data can well be connected by a parabola according to Wiener's theory. The conclusion is, as Stübel pointed out, that the double refraction in the muscle fibre is due: (1) to a structure of parallel rods, small compared with the wave-length of light (*Stäbchenmischkörper*); (2) to a certain amount of double refraction, which is inherent to these rods (*Eigendoppelbrechung*). The vertex of the parabola has an ordinate of 0.0009, which represents the summation of the double refraction of the rods, taken separately (*Eigendoppelbrechung*).

An indication of a third component is given by measurements, represented by the point A in Fig. 1. This value has been obtained by Stübel for muscle fibres which for a long time have been under the influence of lipid-solvents. Lipids, if doubly refractive, are always negative. Their double refraction reduces therefore in the normal muscle fibre the two larger positive components to a lower value. Consequently extraction of these lipids must raise the double refraction in the direction of an increased positive amount (A, Fig. 1), as

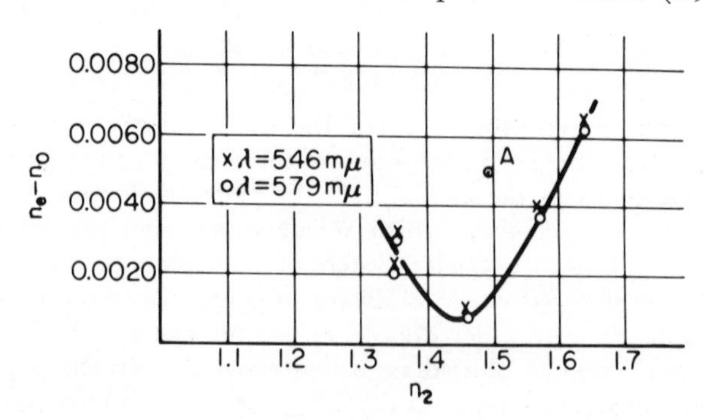

Fig. 1.

observed by Stübel. The exact nature of this negative component, however, is not yet known.

The results of Stübel's measurements can be summarised as follows: The double refraction of the living muscle fibre is due to two positive components, one resulting from a fine-structure of parallel rods, the other from the anisotropy of these rods, and to a negative component produced by the muscle lipids. It may be recalled that this picture of fine-structure is strikingly similar to the hypothetical pictures which O. v. Fürth[16] has given as the underlying basis for his theory of muscular contraction. He assumed that protein particles, in the form of elongated ellipsoids (which are the same as rods) were oriented with their long axes parallel to the axis of the muscle fibre. The size, as well as the formation of these ellipsoids, as assumed by v. Fürth, correspond almost exactly to the fine-structure revealed by Stübel's measurements. It must be mentioned, however, that this experimental evidence for the correctness of v. Fürth's assumptions has no direct bearing on the conclusions which be drew with respect to the mechanism of contraction.

This conception of the muscle fibre as being a system of anisotropic rods or elongated ellipsoids, is not new, although the experimental evidence is only of recent origin. Brücke had already in 1888 advanced a theory to explain the double refraction of muscle fibres which became known as the *Disdiaklastentheorie*. He assumed that small uniaxial crystals, embedded in the muscle fibre were the cause of the anisotropy. They were to be amicroscopically small, all oriented in the long axis of the fibre, possessing a definite amount of positive double refraction. He called them *Disdiaklasten* and they remained hypothetical, although the indications were that they might be protein crystals.

The theory of *Disdiaklasten* was unfortunately in the following years swept away by v. Ebner's theory of internal stresses, which became quite generally accepted. According to this theory, the double refraction of muscle, as well as of other living tissues, was ascribed to the development of internal, asymmetric stresses due to growth. The double refraction was therefore regarded as accidental and was considered to be similar to the effect which appears in most substances as a consequence of strain. Ambronn and his school[1] collected through many years a great amount of evidence against this conception and finally succeeded in showing that v. Ebner's theory can in no way account for the facts. With respect to the muscle fibre it is apparent that such curves as represented in Fig. 1 can never originate from internal stresses. The double refraction must be due to a structure, and it might be permissible to use the old term *Disdiaklasten,* in a slightly altered sense, in designating Stübel's rod-shaped particles. Stübel's measurements have therefore revived Brücke's *Disdiaklastentheorie* in the same way as Ambronn's work in general has placed Nägeli's theory of micellar structure back to its right.

The presence of crystalline micelles in the living muscle fibre has not only been suggested by the optical properties but also by the X-ray spectrograms which have been obtained by Herzog and Jahncke[19] and J. H. Clark[7]. It is usually assumed that the element producing double refraction is identical with the one producing these X-ray patterns. There is, however, at present no indication that this assumption is justified. The evidence that the element producing double refraction (to be exact, the two positive components!) is the myosin molecule, or a particle built up by myosin molecules, is accumulating. Herzog and Jahncke, as well as Clark calculated from their X-ray patterns a period of identity which is of the magnitude 10 Å. U. Such a period of identity appears at present too small for any protein (their molecular weights ranging from 34,000 upwards), unless it is interpreted as J. H. Clark suggested, as being the width of the molecule. This appears, however, to be rather an artificial procedure and it is safer, as O. Meyerhof[26] has pointed out, not to give too much weight at present to the interpretation of these X-ray patterns. It is possible that the further development of the method of Böhm and Schotzky[5] will bring information with respect to the nature of the crystalline element producing the X-ray spectrograms.

The claim that the *Disdiaklasten* have been extracted in form of a muscle protein has been advanced twice in previous years, in both instances without obtaining general recognition. Schipiloff and Danilewsky[33], reported that a muscle protein (myosin), upon drying on a cover slide showed double refraction. This observation was interpreted as being due to the double refraction of the protein particles, which therefore were to be the *Disdiaklasten*. The rather crude method of extraction, as well as the very qualitative character of the observation has provoked criticism and their claim never became recognised. The main objection was based on the fact that orientation of the anisotropic elements is a necessary condition for double refraction, and it was not understood how this could have been brought about by the process of drying. Ambronn and his collaborators, however, have shown in the meantime that freezing and drying can produce orientations, resulting in double refraction. At present it is very hard to give justice to this early claim of Schipiloff and Danilewsky, since their observations were so incomplete; all the same we are rather inclined, on the basis of our own experiments, to believe that the effect which they described was actually due to the anisotropy of myosin.

Thirty years later, Botazzi and Quagliariello[6], suggested that the myosin granula, which they could see under the ultramicroscope in the muscle press-juice preparations were the *Disdiaklasten*. These granules showed, however, no signs of double refraction under the polarising microscope, an observation which recently has been confirmed by Stübel[37]. The suggestion lacks, therefore, in conclusiveness, since any extraction which claims to contain the *Disdiaklasten* must have doubly refractive particles. The *p*H of the solutions which Botazzi and Quagliariello obtained from muscle press-juice is such, that myosin (provided their protein is the same as the one which we studied) is present in the precipitated form. The granules might, therefore, be isotropic precipitates of the myosin, which we found to be anisotropic if in solution. It is well possible that the myosin granules actually contain the *Disdiaklasten,* but that their anisotropy is hidden by the fact that in the granules no specific orientation of the elements prevails.

Muscle globulin solutions which we have prepared and studied, show double refraction of flow or fluxional birefringence (*Strömungsdoppelbrechung,* Freundlich), similar to the V_2O_5 sol. This interesting property has been measured quantitatively and led us to certain conclusions with respect to the nature of the *Disdiaklasten,* as well as to the physico-chemical state of the protein in solution. The measurements and conclusions are briefly reported in the following treatment and thus presented for discussion.

The globulin of muscle was studied in the late nineteenth century by a number of investigators, especially Danilewsky[8], Halliburton (paramyosinogen)[17], and Fürth ("myosin")[16]. In recent years Weber[44], has made important contributions to the chemistry of this protein (myosin) and its quantitative characterisation. The method of preparation which we have employed is in principle the same as that employed by Howe[21] and the earlier observers, the fresh finely ground muscle being extracted by cold, faintly alkaline salt solution (potassium phosphate or chloride, or a mixture of the two: ionic strength 1.3, *p*H 7-8.5). The filtered extract (*p*H near 7) is precipitated by increase or decrease of the ionic strength; the precipitate can be readily concentrated by centrifugation in the cold, and then re-dissolved. The muscle globulin solutions which were used for the study of double refraction were obtained from material reprecipitated from two to six times. Two different sources of material were used, the cheek and jaw muscles of cows (the preparations obtained from this source will be referred to as cow muscle globulin CMG) and the hind leg muscle of perfused rabbits (referred to as rabbit muscle globulin RMG). A detailed description of the method of preparation and the chemical characteristics has been published by one of us (Edsall)[10].

The muscle globulin is a colourless, viscous fluid with great light scattering power. Under the ultramicroscope, as well as under the polarising microscope, the fluid appears optically void. If the solution is allowed to flow through a fine capillary or if it is subjected to shearing stresses between two concentric cylinders, it shows a considerable amount of double refraction. This effect can be observed qualitatively very easily by an arrangement similar to the one described by Zocher[46], known as *Wirbelmethode.* The double refraction of flow which appears in any such arrangement is positive with respect to the direction of orientation of the invisible muscle globulin particles.

For the quantitative study of the double refraction in muscle globulin solutions, an apparatus based on the principle of the concentric cylinders has been designed and built. A detailed description of this apparatus, the method of measurement and the data obtained have been published[28].

The apparatus consists of the rotating concentric cylinders, the termostat and the optical bench with nicols, condensers and compensators mounted on riders.

The concentric cylinders are mounted on a heavy cast-iron stand. The outer cylinder which is of glass (10 cm. high and 4 cm. inside diameter) is rotatable, since it is held in a ball-bearing of 6 cm. diameter by a metal mounting. The inner cylinder is a chromium plated brass cylinder, concentric to the outer cylinder and held fixed by a holder which slides up and down on the cast-iron stand. (Two different sizes of inner cylinders are employed: No. I., $r = 0.91$ cm. and No. II., $r = 1.30$ cm., both of the same height, 9.9 cm) A plate of plane glass, free of internal stresses is sealed to the lower edge of the glass cylinder, thus serving as bottom. The solutions can be filled into the annular space up to the upper edge of the glass cylinder. A plane glass ring, held by a special holder, can then be screwed down to the top of the glass cylinder, thus enclosing the fluid in the annular space between two parallel plates of plane glass. The rotation of the outer cylinder is regulated by a motor, in connection with reducing pulleys.

Parallel to the axis of rotation a beam of parallel, plane polarised light passes with its optic axis at a given radius through the fluid. The light is produced by a powerful arc-lamp, collected by a condenser, polarised by a nicol prism and rendered nearly parallel by a second condenser. All these parts are mounted above the concentric cylinders on a large vertical optical bench carrying all the optical equipment. After having passed the annular space the light passes a quarter wavelength compensator after Sénarmont[35], a condenser, the half-shadow apparatus (after Macé de Lépinay, Leitz Wetzlar), the objective ($f = 80$ mm.) and the analysing nicol, which is provided with scale and vernier, to be read directly to 0.02°. For the convenience of the observer the image of the annular space, as well as of the half-shadow is then observed by a reversion ocular, which consists of a 90° prism and a periplan ocular. With respect to the annular space, the arrangement of the objective and ocular is rendered telecentric by two diaphragmes of 0.15 mm. diameter. Thus only beams which run parallel or almost parallel to the axis of rotation can enter the ocular. By special adjustments care is taken that the optic axis of the lens system is always parallel to the axis of rotation and at a given radius in the annular space. This radius is for cylinder I., $r_1 = 1.44$ cm., for cylinder II., $r_2 = 1.70$ cm. The measured double refraction refers therefore in all the measurements to the double refraction which occurs at these respective radii.

The position of the optic axis in the solutions during rotation is determined by synchronous rotation of the polariser and analyser. Then the compensator and the planes of polarisation are set at 45° to this direction and the amount of double refraction can be measured according to the theory of Sénarmont[35] (see also recently Szivessy[41]), by rotation of the analyser alone, which is necessary in order to extinguish the linear polarised light coming out of the quarter wavelength plate. If this rotation is $\Delta°$ then the phase-difference Δp is given by the relation:

$$\Delta p = \frac{\Delta°}{180} \tag{2}$$

and the double refraction, provided the thickness l of the doubly refractive material is known

$$n_e - n_0 = \Delta p \cdot \lambda \tag{3}$$

Muscle globulin solutions show no dichroism, therefore the orientation of the compensator is comparatively simple. The wave-length λ of light employed in our measurements was kept between 540–550$m\mu$ by the use of Wratten filters 44 and 62.

If the muscle globulin solution in the annular space is viewed as a whole between crossed nicols a characteristic black cross, the cross of isocline or *Wirbelkreuz* (Zocher) appears, while the outer cylinder is rotating. The black arms of this cross indicate those azimuths, where the optic axes of the anisotropic particles coincide with one of the planes of polarisation. The larger of the two angles which the cross of isocline forms with the planes of polarisation is called, by definition, *angle of isocline* (*Kreuzwinkel*) ψ, according to Zocher[46]. This angle is at the same time the smaller of the two angles between the optic axis of the oriented particle and the direction of the velocity gradient between adjoining layers, and also the larger of the angles which the axes of the index ellipsoid form with the radius drawn to the particle.

Theoretically angles of isocline ψ between 45° and 90° are possible. If double refraction of flow were simply a matter of orientation of long needle-shaped particles, orientation in the direction of the stream-lines, or $\psi = 90°$, would have to be expected. The experiments, however, showed that such an assumption is too simple. In pure liquids, for instance, where molecular double refraction of flow occurs, so extensively studied by Vorländer and Walter[43], angles of isocline of 45° are found. On the basis of a theory of Stokes[36], Raman and Krishnan[31] recently showed that this could be explained by assuming orientation of anisotropic molecules under the action of the shearing forces. They found that a good agreement between theory and experiment can be obtained if it is assumed that the molecule is optically and geometrically asymmetric. For most of the long organic molecules involved (fatty acids and higher alcohols for instance) this assumption appears to be reasonable. The prime factor, determining whether a pure fluid produces double refraction of flow or not, is therefore the anisotropy of the molecule. A certain amount of viscosity and strength of the shearing forces is in addition necessary, in order to overcome the disorienting tendency of the temperature motion. Such a highly viscous fluid as glycerin, however, gives hardly any double refraction at all, which in the light of the new conception is interpreted as being due to the lack of anisotropy of the molecule. This assumption appears reasonable, considering the chemical structure of the glycerin molecule.*

Further complications with respect to the mechanism of double refraction and the angle of isocline are encountered in the study of colloids. The classical example of this phenomenon is now the V_2O_5 sol, so extensively studied by Freundlich, Stapelfeld, and Zocher[15]. For this reason the double refraction of muscle globulin will be compared mainly with the behaviour of the V_2O_5 sol, which has been so admirably described in quantitative terms. In these sols the angle of isocline increases from 45° in very young sols towards 90° in old sols. This increase has been ascribed to the growing size of the needle-shaped particles, which become longer with age. Measurements on the increase of the highly abnormal dielectric constants of these sols with age by Errera[12] have independently confirmed this view. It appears, therefore, that, among other factors, the angle of isocline increases with increasing length of the anisotropic particle.

In muscle globulin solutions of intermediate and high concentration, we have found a constant angle of isocline between 76°–78°. For each preparation the angle of isocline was found to have a constant value over the period during which the preparation could be kept free from denaturation and bacteria (2–4 weeks). For smaller concentrations and for very low angular velocities of the outer cylinder smaller values for ψ were found which however were

*This is in marked contrast to the conceptions introduced by the early experiments of Maxwell, Kundt, de Metz[25] and their pupils.[42] From their point of view double refraction of flow in liquids was considered as being an effect similar to the photoelastic effect (*Spannungsdoppelbrechung*) in strained solids, and the experiments were undertaken with the aim of establishing this relation. Theory and experiment, however, could never be brought to agree (Bjoernstahl)[4] and the attempts of Natanson[29] and others on this basis were unsuccessful. The general notion, however, has strangely enough, survived and it may be well to point out that the present experimental evidence can hardly be understood on the basis of this old conception.

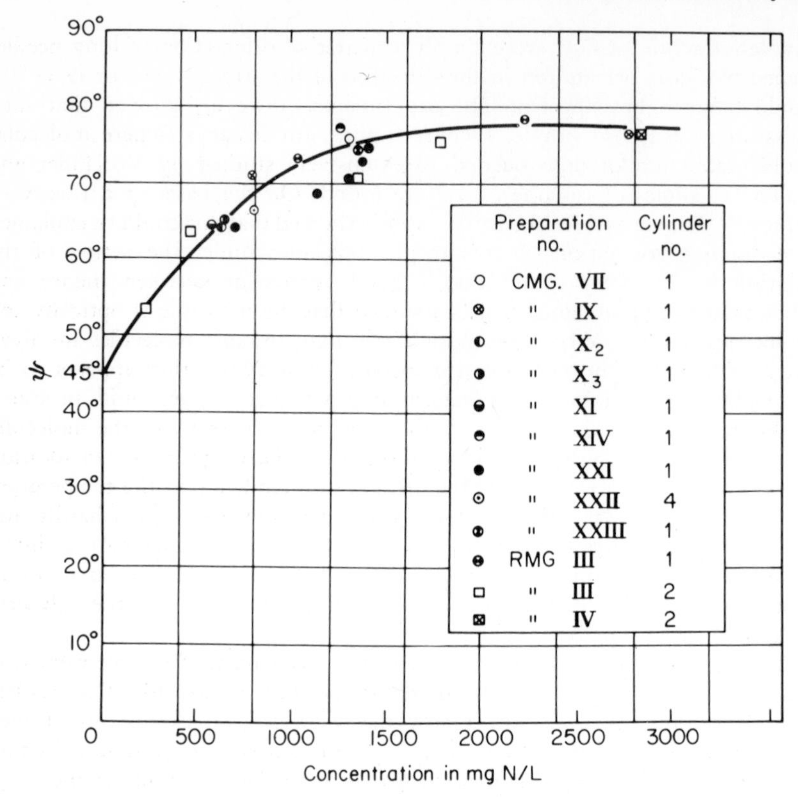

Fig. 2.

reproducible and did not show any of the "ageing" effects, so commonly observed in V_2O_5 sols. The data for ψ in 12 different muscle globulin preparations (taken at the same angular velocity) are represented in Fig. 2. Most of them are between 76°–78°, although for low concentrations a drop towards 45° for concentration 0 was observed. The data for ψ are plotted against concentration as determined by Kjeldahl nitrogen analysis, and a curve of saturation character has tentatively been drawn through them, reaching a limiting value of 78°. The measured points scatter however more than would be expected from an estimation of the usual experimental errors. Hence this scattering must be due to variations between one preparation of globulin and another, not to errors in the measurement of ψ. The error in the Kjeldahl nitrogen analysis is estimated at 2 per cent. and is insufficient to account for the deviations. The Kjeldahl analysis however gives the total protein nitrogen; that of the optically anisotropic together with that of the inert proteins which might be present as impurities. Double refraction of flow is a property not found in many proteins and it is conceivable that this property is restricted to one definite protein within the muscle globulin preparation.

It might also be argued that the muscle globulin actually contains but one definite protein. In this case the fluctuations of ψ would be due to the presence of denatured material formed during the preparation in quantities at present unknown. That this denatured material must have been formed during preparation, and not afterwards, follows from the observation that one given muscle globulin solution always gives the same angle of isocline during its "lifetime." This explanation gains further support through the observation that all reagents which according to Anson and Mirsky[2], bring about denaturation, cause disappearance of the double refraction of flow. Any foreign nitrogenous substance, or any denatured portion of the muscle globulin solution apparently does not participate in the production of the angle of isocline whereas the figures of the Kjeldahl analysis yield the total nitrogen, and by inference the total protein concentration. From this point of view it follows that if muscle globulin could be satisfactorily purified, then the relation between the angle of isocline and the protein concentration would be reproducible.

The angle of isocline which is found more or less consistently for muscle globulin solutions corresponds to the angle ψ which is found in V_2O_5 sols of intermediate age. The behaviour of the latter with respect to the ψ-concentration relation is however different. For small concentrations (less than 0.015 per cent.) the angle of isocline is constant, for higher concentrations it increases irregularly from 57° to about 73°, as Neukircher[14] found. It must be remembered however that the age of the V_2O_5 sol exerts a strong influence on ψ and that it is almost impossible to eliminate this factor in a study of the ψ-concentration relation in such a system. The behaviour of muscle globulin is in this respect in marked contrast to that of the V_2O_5 sol.

The study of the ψ-concentration relation is however simplified, if the measurements are restricted to varying concentrations obtained by dilution from one given stock muscle globulin solution. In that case the percentage error in the Kjeldahl nitrogen analysis, due to the factors discussed above, remains the same for all dilutions, and reproducible values are obtained. In Fig. 3, a series of such measurements on preparation RMG III is represented. ψ is taken as the ordinate, the velocity of revolution of the outer cylinder as abscissa and the concentration as parameter in curves 1–5.*

The measurements have been carried down to the lowest values of concentration and angular velocity which would still allow a definite reading. The extrapolation of the curve of the most dilute sample seems to point to an angle $\psi = 45°$ for the angular velocity 0, and it might be that the other curves have the same origin.

It is apparent that these curves are largely due to a variation of a physical property in the muscle globulin solutions which must be closely related to the viscosity. If the same measurements are taken by increasing the temperature, at constant concentration, a similar set of characteristic curves is obtained. From the work of Freundlich, Neukircher, and Zocher[14], it is known that the angle of isocline is connected with an interesting physicochemical property of certain fluids, described as "rigidity." (An excellent review of these phenomena may be found in Hatschek's book on viscosity[18], Wo. Ostwald has given a comprehensive survey of all the formulæ which have been developed[30] and a mathematical theory has recently been advanced by Reiner.[32]) The

*The viscosities given in Fig. 3 are apparent viscosities η'_{rel} since the muscle globulin solutions possess "rigidity."

muscle globulin solutions were found to be "rigid"; a property which, however, could not be formulated in a quantitative way. In the solution represented by curve 5 in Fig. 3, no change of viscosity with pressure in an Ostwald viscosimeter can be observed, whereas for increasing concentrations this anomaly becomes more and more pronounced. The most dilute solution comes therefore nearest to a fluid with normal viscosity. It has been mentioned that in pure liquids, where the viscosity is normal throughout, the angle of isocline is found to be 45°. From this point of view it appears reasonable that in the solution represented by curve 5 in Fig. 3 this value is

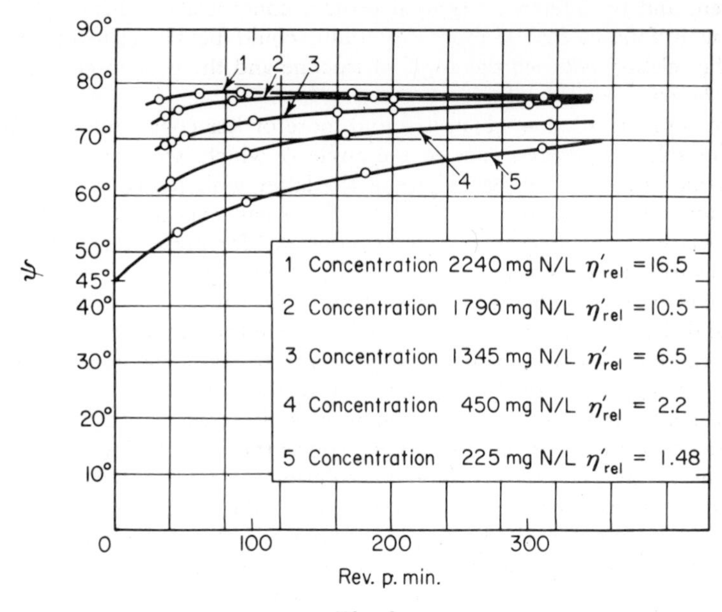

Fig. 3.

approached. In the very young V_2O_5 sol also, the "rigidity" is minimal and correspondingly an angle of isocline near 45° has been found[15]. For the higher values of ψ, which are found in the more rigid solutions Freundlich, Neukircher, and Zocher assume elastic deformation of the oriented particle to such an extent that angles of ψ almost up to 90° result. It must be remembered, however, that the length of the particle increases at the same time with the "rigidity," which complicates the relations considerably. To which extent these two phenomena are the consequence of a common cause, is a fascinating problem which awaits solution.

Without going into the detail of the mechanism of the development of "rigidity" and the increase in length of the particles, we have argued that the appearance of "ageing" effect can always be interpreted as manifestation of such changes. The absence of this effect, as observed in muscle globulin solutions, not only in measurements of ψ, but also in the reproducibility of the apparent viscosity (*i.e.*, "rigidity") and the double refraction of flow (see below), has been interpreted therefore as the expression of constant size and shape of the anisotropic muscle globulin particle. We concluded from this that these particles, present in muscle globulin solutions, are monodisperse and proposed the use of the old term "myosin" in the future for this anisotropic protein alone[28].

The assumption that myosin, according to our definition, is present in muscle globulin solutions, is in agreement with the general evidence that by suitable preparation certain proteins can be obtained in monodisperse form. This evidence has been accumulating in recent years and has been confirmed in the last few years by Svedberg and his collaborators[39], on the basis of their measurements with the ultra-centrifuge. Svedberg[39] has given a list of the molecular weights of proteins, which so far have been obtained in monodisperse form. Muscle globulin is not mentioned in this list but was found together with fourteen other proteins as being polydisperse. Svedberg, however, ascribes this rather to a deficiency in the method of preparation than to a property inherent to these proteins. From Fig. 2 we have concluded that our method of preparation has not yet succeeded in obtaining the myosin in pure form. If our definition of myosin, however, is correct, it should appear to be monodisperse in the ultracentrifuge, provided that the sedimentation velocity of the myosin particles alone can be measured.

The interesting problem immediately arises if it is quantitatively possible to correlate the anisotropy of the myosin particle (it might be the molecule!) with the double refraction in the living muscle fibre. This would be a conclusive proof that the myosin particles are the long sought *Disdiaklasten.*

In Fig. 4 the results of quantitative measurements of the phase difference, expressed in $\Delta°$ (see equation (2) and (3)) on preparation CMG, XIV. are graphically represented.

The phase difference of the solutions was measured with the Sénarmont compensator and the half shadow wedge at varying velocities of revolution (rev. p. min.) of the outer cylinder, using the inner cylinder No. I. The various

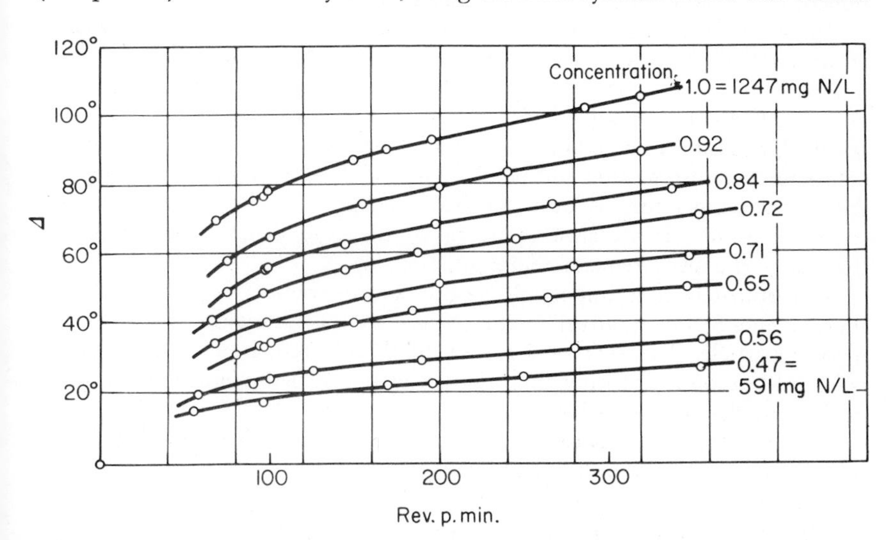

Fig. 4.

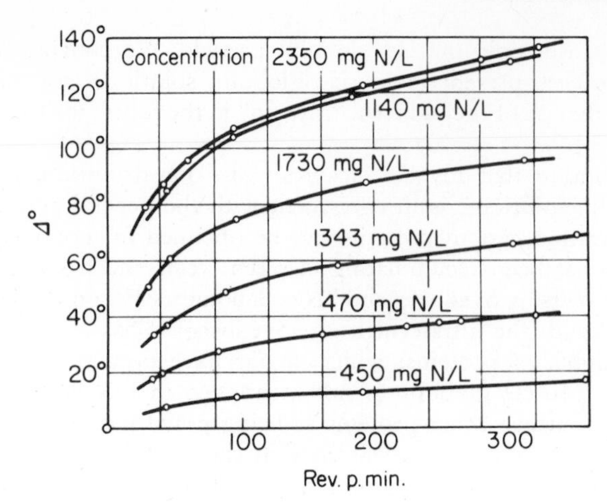

Fig. 5.

concentrations were obtained by dilution and the absolute amount was determined by a number of Kjeldahl analyses on the most dilute and the most concentrated sample.

The data obtained on a preparation of different origin, RMG III., with cylinder No. II., but otherwise identical conditions, are plotted in Fig. 5. While the slope in these curves is slightly steeper (due to the higher velocity-gradient*) it appears that the general character of the curve is in both cases the same, and it may be mentioned that this holds for all the preparations which we have studied. The measured points were, as far as we could observe them, reproducible with respect to "age," which is in a marked contrast to the behaviour of the V_2O_5 sol. This observation has led us, as already mentioned, to the assumption of monodisperse myosin.

In varying the concentration two cardinal factors are changed, the number of anisotropic particles and the viscosity. To which extent the viscosity influences the curves can best be illustrated by Fig. 6, representing two curves taken at different temperature. The variation in this case is largely due to the changed viscosity, to a certain extent also to the increased disorienting action of the temperature motion and possibly to a change in rigidity. The curve of the lower temperature is only reproducible, after having warmed up the solutions to 27°, if precautions are taken to avoid denaturation. This was the case in the measurements represented in Fig. 6.

Unfortunately it is impossible at present to evaluate such curves, as in Figs. 4 and 5, quantitatively with respect to the anisotropy of the single myosin particle. The lack of a theory of double refraction of flow, which takes into account not only the orientation but also the elastic deformation of the particles is responsible.

Three possible explanations for the characteristic shape of the double refraction curves have been mentioned[28].

*If the velocity gradient at the point of observation for the measurements of Figs. 4 and 5 is calculated on the basis of normal viscosity, it is found to be exactly twice as large for Fig. 5 as for Fig. 4.

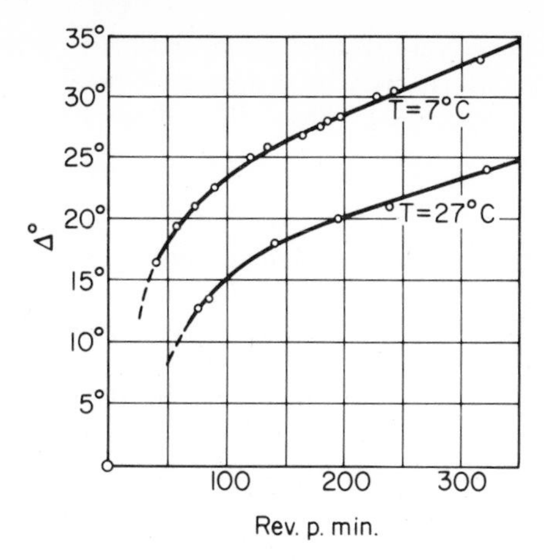

Fig. 6.

1. The curve might be the superposition of a simple saturation curve and a linear term (photoelastic effect) increasing proportionally to the angular velocity of the outer cylinder. The saturation point could in this case be determined by linear extrapolation of the upper (almost) straight part of the curve, being the intersection of this extrapolation with the ordinate.

2. The myosin particle might be an ellipsoid with three different axes. The steep part of the curve might represent the first orientation, regarding the most excentric axis, the smooth slope the second orientation, regarding the second axis.

3. The curve resembles to a certain extent the curve of orientation of dipols (*Langevin function*) in an electric field. It might be that somewhat similar relations obtain for the orientation of the myosin particles in a mechanical field of force.

The anisotropy of the myosin particle is easily lost. If the solutions are not handled very carefully in the cold during preparation, no double refraction of flow occurs, although such physical properties as viscosity remain unaltered (if not increased). All reagents, which according to Anson and Mirsky[2] produce denaturation, produce immediate and total loss of double refraction (within the limits given by our apparatus). In strongly acid and alkaline solutions the effect disappears as irreversibly, as under the influence of denaturing agents. In a moderate range of *p*H (6–9), so far no striking changes have been observed; the observations are, however, scarce and of no definite value.

The Thixotropic Gel

After repeated washing of the muscle globulin solution, the preparation becomes practically salt-free and forms a clear gel, even at an extremely low protein concentration (about 0.03 per cent.) (Edsall[10]). Several of these gels were studied with respect to their angle of isocline in our apparatus. A small displacement of the outer cylinder, not exceeding the limit of elasticity of the gel, produced at once a very distinct cross of isocline at an angle $\psi = 45°$. The shearing forces which arise in this case lie in directions, according to Stokes[36], as to produce such an effect. If the outer cylinder is released the effect disappears at once. If the apparatus, however, is set into permanent rotation, entirely different conditions seemed to prevail. The cross of isocline as it appeared under these conditions, is sketched in Fig. 7.

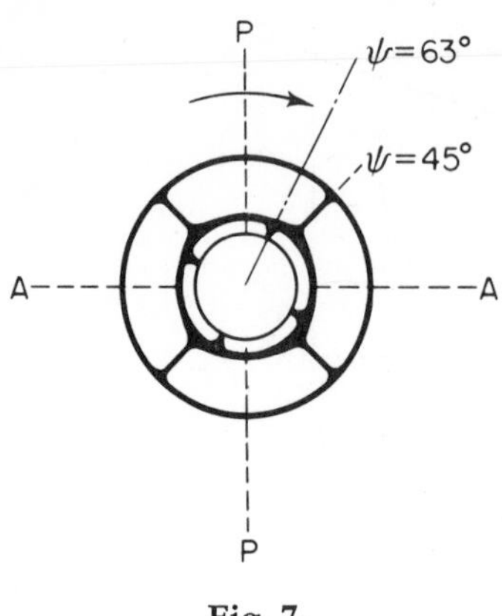

Fig. 7.

The most striking phenomenon is a dark ring separating the two different crosses of isocline. The outer cross has an angle of isocline of 45°, the same as the slightly deformed gel. The inner cross, however, has an angle of about 61°–63° not very clearly measurable, but distinctly not 45°. The black ring separating the two crosses of isocline proves to be isotropic.

The gel has other interesting properties. By mere shaking it is transformed to the fluid state and sets again to a gel after a few minutes. Gels which exert this property have been called "thixotropic"[40], and quite a number of them are known. The "thixotropy" of this gel gives a clue to its behaviour in the concentric cylinder apparatus. The shearing forces which arise in the apparatus in rotation have a tendency to destroy the gel structure permanently. This destruction is most effective near the inner cylinder where the shearing forces are largest. The neighbouring layer around the inner cylinder is therefore in the fluid state and the isotropic ring indicates where the gel "breaks off." From there on the gel moves as a solid block and gives an angle of isocline according to the small elastic stress which is apparently still exerted on it and is a result of the frictional forces at the interface gel-fluid. The fluid layer, however, gives an angle of isocline such as is found in muscle globulin *solutions* (of a similar small protein concentration), which is partly due to orientation of the anisotropic myosin particles. At the interface sol-gel neither definite elastic deformation nor orientation occurs and isotropy must prevail.

This explanation was confirmed by casual observations that small particles in the outer solid block moved at the same angular velocity as the glass cylinder.*

A certain significance is attached to these observations with respect to the behaviour of the living muscle cell. Kühne[23], reported that he observed a nematode wandering through a muscle cell and stated that the nematode had no difficulty in passing through a cross-striated muscle fibre. Where the animal touched the striations with its head, Kühne stated that they disappeared at once, and reappeared after the animal had passed.

*Quartz particles suspended in the gel showed no Brownian movement and did not change their position. The cataphoretic velocity of these particles, however, was that of quartz particles in a fluid. These observations which Dr. H. A. Abramson very kindly made, confirm the thixotropic character of this gel.

Freundlich[13], has suggested that the interior of the muscle cell behaves as a thixotropic gel, which would explain this observation quoted as a curiosity in many text-books. The thixotropic muscle globulin gel is of course far from representing the natural state of the protein in the muscle but the fact that such a gel may be prepared from extracted muscle protein has a certain interest.

The proof that the myosin particles are identical with the *Disdiaklasten* has not been given, but it appears probable from the experiments reported here that this relation exists. In this connection it is of interest that the anisotropy of the myosin particle is such a delicate property, especially in relation to the disappearance of double refraction in the living muscle fibre during contraction. To which extent the "rigid" properties of the muscle globulin solutions and the "thixotropic" properties of the gels are related to the interesting viscous-elastic effects in muscle, reported by Hill and his co-workers[20], is another problem which might bring us nearer to an understanding of the processes involved.

Bibliography

1. Ambronn, H., and Frey, A., *Das Polarisationsmikroskop u.s. Anwendungen,* Leipzig, 1926.
2. Anson, M. L., and Mirsky, A. E., *J. gen. Physiol.,* **13**, 121, 1929.
3. Bernstein, J., *Pfl. Arch.,* **162**, 1, 1916.
4. Bjoernstahl, Y., *The accidental double refraction in colloids,* Uppsala, 1924.
5. Boehm, G., and Schotzky, K. F., *Naturw.,* **18**, 282, 1930.
6. Botazzi, F., and Quagliariello, G., *Arch. int. physiol.,* **12**, 234, 289, 409, 1912.
7. Clark, J. H., *Am. J. Physiol.,* **82**, 181, 1927.
8. Danilewsky, A., *Z. physik. Chem.,* **5**, 158, 1881.
9. Ebner, V. von, *Pfl. Arch.,* **163**, 179, 1916.
10. Edsall, J. T., *J. biol. Chem.,* 1930.
11. Engelmann, Th. W., *Pfl. Arch.,* **7**, 33, 1873; **11**, 432, 1875.
12. Errera, J., *Koll. Z.,* **31**, 59, 1922; **32**, 157, 1923.
13. Freundlich, H., *Protoplasma,* **2**, 278, 1927.
14. Freundlich, H., Neukircher, and Zocher, H., *Koll. Z.,* **38**, 43, 48, 1925.
15. Freundlich, H., Stapelfeld, F., and Zocher, H., *Z. physik. Chem.,* **114**, 161, 1924.
16. Fürth, O. von, *Ergebn, Physiol.,* **17**, 363, 1919.
17. Halliburton, W. D., *J. Physiol.,* **8**, 133, 1887.
18. Hatschek, E., *The viscosity of liquids*, New York, 1928.
19. Herzog, R., and Jahncke, H., *Naturw.,* **14**, 1223, 1926.
20. Hill, A. V., *Muscular activity in man,* New York, 1927.
21. Howe, P. E., *J. biol. Chem.,* **61**, 493, 1924.
22. Hürthle, K., *Pfl. Arch.,* **126**, 1, 1909.
23. Kühne, A., *Virchows Arch.,* **26**, 222, 1863.
24. Kundt, A.: *Wied. Ann.,* **13**, 110, 1881.
25. Metz, G. de, *Wied. Ann.,* **35**, 497, 1888.
26. Meyerhof, O., *Die chemischen Vorgänge im Muskel,* Berlin, 1930.
27. Muralt, A. L. von and Edsall, J. T., *Am. J. Physiol.,* **90**, 457, 1929.
28. Muralt, A. L. von and Edsall, J. T., *J. biol. Chem.,* 1930.

29. Natanson, L., *Phil. Mag.* (6), **2**, 342, 190.
30. Ostwald, Wo., *Koll. Z.,* **47**, 176, 1929.
31. Raman, C. V., and Krishnan, K. S., *Phil. Mag.* (7), **5**, 769, 1928.
32. Reiner, M., *Koll. Z.,* **50**, 199, 1930.
33. Schipiloff and Danilewsky, A., *Z. physiol. Chem.,* **5**, 349, 1881.
34. Schmidt, W. J., *Die Bausteine d. Tierkörpers i. pol. Licht,* Bonn, 1924.
35. Sénarmont, H. de, *Pogg. Ann. Erg.,* **1**, 451, 1842.
36. Stokes, H., *Trans. Cambr. Phil. Soc.,* **8**, 1845.
37. Stübel, H., *Pfl. Arch.,* **20**, 629, 1923.
38. Stübel, H., and Liang,Tse-Yeh, *Chinese F. Physiol.,* **2**, 139, 1928.
39. Svedberg, The, *Koll. Z.,* **51**, 10, 1930.
40. Szegvari, A., and Schalek: *Koll. Z.,* **32**, 318, 1923; **33**, 326, 1923.
41. Szivessy, G., and Münster, Cl., *Z. Physik,* **53**, 13, 1929.
42. Umlauf, K., *Wied. Ann.,* **45**, 306, 1892.
43. Vorländer, D., and Walter, R., *Z. physik. Chem.,* **118**, 1, 1925.
44. Weber, H. H., *Biochem. Z.,* **158**, 443, 1925; **189**, 381, 407, 1927.
45. Wiener, O., *Abh. d. Sächs. Ges. d. Wiss, math. phys. Kl.,* **32**, 1912.
46. Zocher, H., *Z. physik. Chem.,* **98**, 293, 1921.

MYOSINE AND ADENOSINETRIPHOSPHATASE*

W. A. Engelhardt and M. N. Ljubimowa

Institute of Biochemistry, Academy of Sciences of the U.S.S.R.
Moscow, U.S.S.R.
(Received for publication, August 7, 1939)

Ordinary aqueous or potassium chloride extracts of muscle exhibit but a slight capacity to mineralize adenosinetriphosphate. Even this slight liberation of phosphate is mainly due, not to direct hydrolysis of adenosinetriphosphate, but to a process of secondary, indirect mineralization, accompanying the transfer of phosphate from the adenylic system to creatine, the corresponding enzymes (for which the name 'phosphopherases' is suggested) being readily soluble.

In contrast to this lack of adenosinetriphosphatase in the soluble fraction, a high adenosinetriphosphatase activity is associated with the water-insoluble proteins of muscle. This enzymatic activity is easily brought into solution by all the buffer and concentrated salt solutions usually employed for the extraction of myosine. On precipitation of myosine from such extracts, the adenosinetriphosphatase activity is always found in the myosine fraction, whichever mode of precipitation be used: dialysis dilution, cautious acidification, salting out. On repeated reprecipitations of myosine, the activity per mgm. nitrogen attains a fairly constant level, unless denaturation of myosine takes place. Under the conditions of our experiments (optimal conditions have not been determined) the activity of myosine preparations ranged in different experiments from 350 to 600 microgram phosphorus liberated per mgm. nitrogen in 5 min. at 37°. Expressed as

*W. A. Engelhardt and M. N. Ljubimowa, "Myosine and Adenosinetriphosphatase," *Nature,* **144**, 668–69 (1939). Reprinted with the permission of the authors and of the editors of *Nature.*

$$\mathcal{Q}_P\left(=\frac{\mu\text{gm. P}/31\times 22\cdot 4}{\text{mgm. N}\times 6\cdot 25\times \text{hour}}\right),$$

this gives values of 500–850.

Acidification to *p*H below 4, which is known to bring about the denaturation of myosine[1], rapidly destroys the adenosinetriphosphatase activity. Most remarkable is the extreme thermolability of the adenosinetriphosphatase of muscle: the enzymatic activity shown by myosine solutions is completely lost after 10 min. exposure to 37°. This corresponds with the well-known thermolability of myosine[2]. In respect of its high thermolability adenosinetriphosphatase resembles the protein of the yellow enzyme, which when separated from its prosthetic group is also rapidly inactivated at 38° (Theorell[3]). Evidently in the intact tissue of the warm-blooded animal (all experiments were performed on rabbit muscles), some conditions must exist which stabilize the myosine against the action of temperature. A marked stabilizing effect on the adenosinetriphosphatase activity seems to be produced by the adenylic nucleotide itself. As can be seen from the accompanying graph, in the presence of adenosinetriphosphate the liberation of phosphate proceeds at 37° over a considerable period (Curves I, I*a* and I*b*), whereas the same myosine solution warmed alone to 37° for 10–15 min. shows on subsequent addition of adenosinetriphosphate an insignificant or no mineralization whatever.

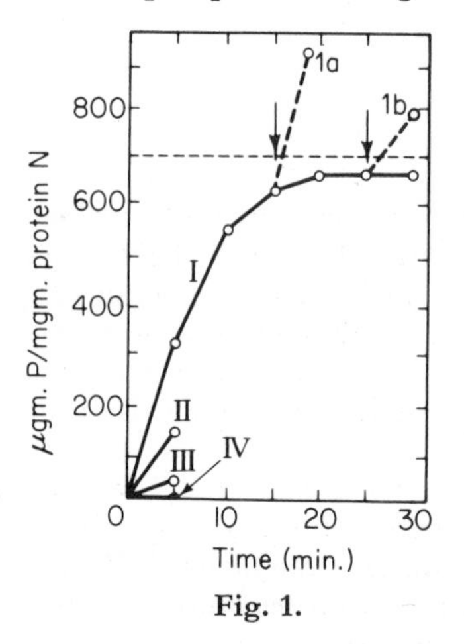

Fig. 1.

Crude buffer extracts accomplish a quantitative hydrolysis of the labile phosphate groups of adenosinetriphosphate; myosine, reprecipitated three times, liberates but 50 per cent of the theoretical amount of phosphorus (see figure). It acts as true adenosine-*tri*-phosphatase and yields adenosinediphosphate, which is not further dephosphorylated and has been isolated in substance. This may serve as a convenient way of preparing adenosinediphosphate, instead of using crayfish muscle[4]. The adenosinediphosphatase is thus associated with the more soluble proteins, occupying an intermediate position between adenosinetriphosphatase and the most readily soluble phosphopherases.

Under no conditions tested could we obtain a separation of adenosinetriphosphatase from myosine. Either the activity was found in the myosine precipitate or else it was absent from the precipitates and from the remaining solution. This disappearance of the enzymatic activity we regard as the result of the start of denaturation of the very unstable myosine.

We are led to conclude that the adenosinetriphosphatase activity is to be ascribed to myosine or, at least, to a protein very closely related to and at

[1] v. Muralt, A., and Edsall, J. T., *J. Biol. Chem.*, **89**, 351, 1930.
[2] Bate Smith, E. C., *Proc. Roy. Soc.*, B, **124**, 136, 1937.
[3] Theorell, H., *Biochem. Z.*, **272**, 155, 1934; **278**, 263, 1935.
[4] Lohmann, K., *Biochem. Z.*, **282**, 109, 1935.

present not distinguishable from myosine. Thus the mineralization of adenosinetriphosphate, often regarded as the primary exothermic reaction in muscle contraction, proceeds under the influence and with the direct participation of the protein considered to form the main basis of the contractile mechanism of the muscle fibre.

ENZYMATIC AND MECHANICAL PROPERTIES OF MUSCLE PROTEINS*

W. A. Engelhardt

Institute of Biochemistry, Academy of Sciences of the U.S.S.R., Moscow, U.S.S.R.

In his Presidential Address at the opening of the International Physiological Congress in Boston, Krogh stated that nature has indeed cared for the interests of scientists, and in a number of cases has created such experimental material and such types of organisms as are especially suited to the study of one or another specific biological problem.

One might say that nature has gone even further. Among her remarkable creations there are such specimens as have been created not only for the solution of one or another specific biological process, but also for the study of the chemical laws of living matter. Muscle must be ranked as the foremost of these.

It is no exaggeration to say that all our understanding of the fundamental biochemical processes which find their basis in the physiological activity of any structure or organ has been wholly based on the results obtained in the study of the chemical dynamics of the muscle. In the study of the chemistry of any specialized structure or organ—whether it be nerve cell, developing embryo, malignant growth, endocrine gland, or young growing bone—the investigator invariably starts from those concepts formed in the study of muscle structure. It is these concepts that form that common basis which

*W. A. Engelhardt, "Enzymatic and Mechanical Properties of Muscle Proteins," *Yale J. Biol. and Med.,* **15**, 21–38 (1942–43). Reprinted with the permission of the authors and of the editors of *Yale J. of Biol. and Med.*

From the Institute of Biochemistry, Academy of Sciences of the U.S.S.R., Moscow. Translation of a review article originally published in the Russian journal, "Advances in Contemporary Biology," **14**, 177–90 (1941). This translation was prepared by Mr. Paul Talalay of the Massachusetts Institute of Technology. Publication of this translation has been approved by the author.

makes possible an understanding of the specific characteristics and peculiarities of the chemical dynamics of one or another biological specimen.

The broad general significance of the specific and individual phenomena which are revealed in the study of the biochemistry of muscle leads us to present here some results obtained recently in our work on the enzymology of muscle tissue. The study of the enzymic properties of muscle proteins has led us, on the one hand, to the question of the relation between the chemistry and mechanics of muscles, and, on the other hand, to some general conclusions regarding the catalytic functions of the proteins of protoplasm, i.e., those which we call "living proteins."

Among the substances whose transformation lies at the basis of the most important metabolic processes of the muscle is adenosine-triphosphoric acid.

```
      NH2
      |
      C
  N /  \ C—N           ┌──────O──────┐          O       O
   |    |     \ CH     |  OH  OH     |          ‖       ‖          OH
  HC    C—N /  ———— C—C—C—C—CH2—O—P—O—P—O—P=O
    \  /                  H  H  H  H              |       |          OH
      N                                        OH      OH
```

This is a most important specific compound, regarded by most investigators as being directly related to the physiological function of the muscle and its contraction and relaxation.

This substance fulfills two functions of primary importance in the energetics of the muscle. On the one hand it acts as a co-ferment in the decomposition of carbohydrates. It transfers phosphoric acid to carbohydrate and thereby prepares the carbohydrate molecule for subsequent decomposition reactions which are the source of energy in muscle activity. On the other hand, adenosine-triphosphoric acid is capable of hydrolytic decomposition with the splitting off of mineral phosphoric acid. We shall chiefly concern ourselves with this latter aspect.

This reaction is strongly exothermic, yielding as much energy per molecule of decomposed adenosine-triphosphoric acid as was previously computed for the reaction of greatest importance for muscle processes—the decomposition of the sugar molecule to lactic acid.

The majority of contemporary investigators visualize the decomposition reaction of adenosine-triphosphoric acid to be that primary exothermic process which serves as the *immediate* source of energy for muscle contraction. The enzyme called adenosine-triphosphatase, which causes this decomposition reaction and which furnishes the chemical basis of the muscle mechanism, is clearly of great interest.

It is strange that this enzyme has been almost completely ignored in contemporary biochemistry and has remained entirely uninvestigated from the enzymatic point of view. Considering the importance of this enzyme and the vital part it plays in muscle metabolism, we[13] undertook a more detailed investigation, which gave results of broader and more widespread significance than at first expected.

In our attempts to isolate adenosine-triphosphatase we found it invariably associated with that fraction of the muscle tissue proteins which is considered to be the contractile component of the muscle proteins, that is, *myosin*.

We give here a brief account of the characteristics of this remarkable protein.

Myosin is a protein the molecules of which show a well-defined elongation. The changes in the internal structural configuration of these molecules, in the sense of their contraction or expansion as shown schematically in the formulas presented below, are the cause of the contraction and relaxation of the muscle fiber.

Changes in the mechanical state of the muscle are the result of changes in the mechanical properties of the molecules and micelles of the myosin. The chief problem of the biochemistry and physiology of muscle is the question of what chemical stimulus brings about molecular changes in the myosin and results in changes in its mechanical state.

In comparison with other proteins myosin is extraordinarily labile. When extracted from the muscle it decomposes at a temperature of 40°C. when treated with acid or alkali. As Edsall[3, 15] has shown, the first sign of the decomposition of the myosin is the loss of one of its characteristic optical properties—stream anisotropy.

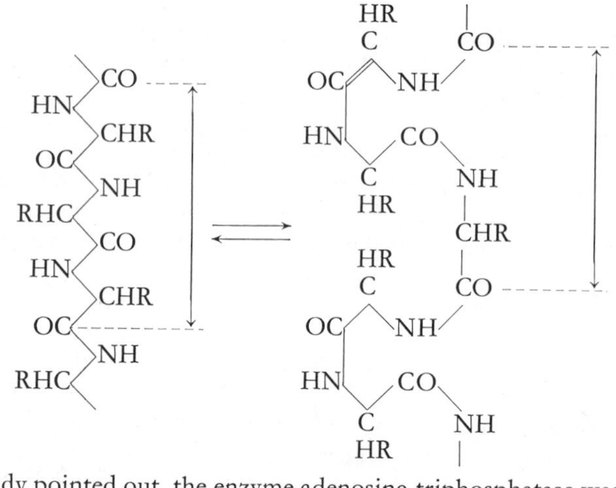

As already pointed out, the enzyme adenosine-triphosphatase was invariably found in the myosin fraction. All attempts at isolation of the adenosine-triphosphatase activity from the myosin by physical means were unsuccessful. For some proteins and enzymes we may effect a separation by means of crystallization. For myosin this is impossible. The customary method of isolating and purifying myosin physically is to allow it to precipitate by lowering the concentration of salts, carefully acidifying it, and salting it out. All these steps were investigated with the one invariable result: in all cases the adenosine-triphosphatase activity accompanied the myosin. Methods for determining the identity of preparations by means of fractionation of the solution in the presence of excess of the undissolved phase, as proposed by Sorensen and widely used by Northrop in his studies of protein enzymes, were difficult to apply to myosin on account of its being extremely hydrophylic. In so far as this method could be successfully used it gave no indication of any lack of identity in our compounds. We therefore came to the conclusion that the adenosine-triphosphatase activity cannot be separated from myosin by physical means.

Not only in its physical relation, but also in all its other properties, adenosine-triphosphatase exhibits such a complete similarity to myosin as can hardly be considered accidental. Just as myosin stands out from all other proteins by its extreme instability, so adenosine-triphosphatase exhibits extreme instability and is thus distinct from the vast majority of other enzymes. The adenosine-triphosphatase activity of specimens of myosin disappears under precisely the same conditions as favor the beginning of the decomposition of myosin, as manifested in the loss of its optical characteristics (anisotropy of flow). Weak acidification to pH 4 causes the decomposition of myosin and the loss of the enzymatic activity. The same occurs when solutions are exposed to a slightly elevated temperature (about 40°). It is known that urea promotes disaggregation of the myosin molecules, with a loss of optical activity. It was found that adenosine-triphosphatase activity also disappears under the action of urea.

It might seem that the results of these purely preparative experiments are insufficient evidence to establish the identity of myosin and adenosine-triphosphatase. The same might be said about the correspondence of the properties of myosin and adenosine-triphosphatase. Nevertheless, the complete agreement of the results obtained along these entirely independent lines of investigation, and the complete lack of data not in agreement with these results, give us the right to maintain that myosin itself, the contractile constituent of the muscle, is the bearer of the properties of adenosine-triphosphatase. We therefore come to the conclusion that the protein, which is the contractile constituent of the muscle, is at the same time the catalytic agent which promotes that chemical reaction which provides the direct source of energy for muscle activity. *Myosin is not only the passive substrate, but is also the active promoter of muscle dynamics.*

As a result of the elongated, rod-like or even thread-like, structure of its molecule, myosin has one extraordinary property. As Weber[21] showed, solutions of myosin injected into water form fibers of considerable strength. These fibers reproduce in great detail the optical properties and molecular structure of the contractile substance of the muscle fibril. They exhibit the double refraction characteristic for the contractile components of the muscle, and they give the typical X-ray diagram of muscle fibers.

It was found that myosin in such fiber form also preserves its enzymic properties—those of adenosine-triphosphatase. We chose myosin threads for the investigation of the relationship between the chemical processes and the mechanical phenomena of muscle. These experiments were the natural outcome and development of results obtained[4] during the study of the enzymatic properties of myosin.

As mentioned above, the activities of muscles, their contraction and relaxation, are the result of changes in the molecular or micellar state of the myosin, caused by chemical agents which appear or disappear in muscle metabolism.

Accordingly, three problems confront muscle physiology and biochemistry. The first is to determine the molecular structure of the contractile component of the muscle. The second is to study as fully as possible the metabolic processes of the muscle and the compounds appearing and disappearing therein. The third is to establish what substances act (and in what manner) upon the molecular structure of the contractile material, changing its mechanical state and thus producing contraction and relaxation.

The first of these problems—the determination of molecular structure—is solved by structural X-ray analysis. Most important for this investigation is the fact that in the myosin thread we have a complete model of the molecular structure of the muscle fiber. As Weber,[20] one of the greatest authorities in this field and to whom we are especially indebted for the study of the physics of muscle proteins and particularly of myosin, says: "the X-ray diagram of the myosin thread is the X-ray diagram of the muscle fiber."

The second problem—the study of muscle metabolism—has during the past twenty years been most extensively investigated. These studies have led to greater depth and precision in our understanding of the chemical processes occurring in the muscle and of the various substances which take part in these processes.

The third problem—the determination of the relation between the chemistry and the mechanical properties of the contractile component of the muscle—has remained completely untouched. We wished to determine whether the myosin thread, being a complete model of the molecular structure of the muscle fibril, could not be used as a mechanochemical model.

For the investigation of the mechanical properties of myosin threads the apparatus shown in Fig. 1 was used.

The apparatus consists of a torsion balance, to the lever of which the myosin thread is attached. When a load is applied to the thread a change in the balance reading results. The change in length is measured by a lightbeam reflected on a scale, calibrated in millimeters, by means of a mirror attached to the balance lever. The extensibility of the thread serves as an index of its mechanical properties. Investigating in this way the effect of various substances on the mechanical properties of the myosin thread, we discovered a phenomenon that is doubtless of the greatest interest.

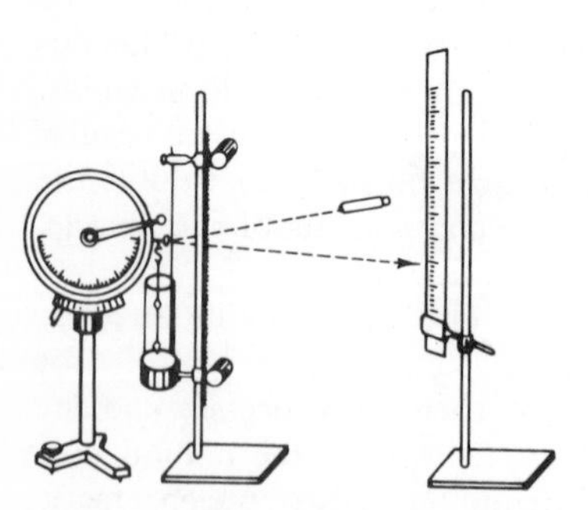

Fig. 1. Arrangement of apparatus for testing the mechanical properties of myosin threads.

Study of a number of compounds which might play a part in muscle metabolism showed that they produced no noticeable effect upon the properties of myosin threads. But one substance—thus far the only one of all those investigated—produced a marked effect, increasing the extensibility of the myosin threads by from one and one-half to two times. This substance was adenosine-triphosphoric acid.

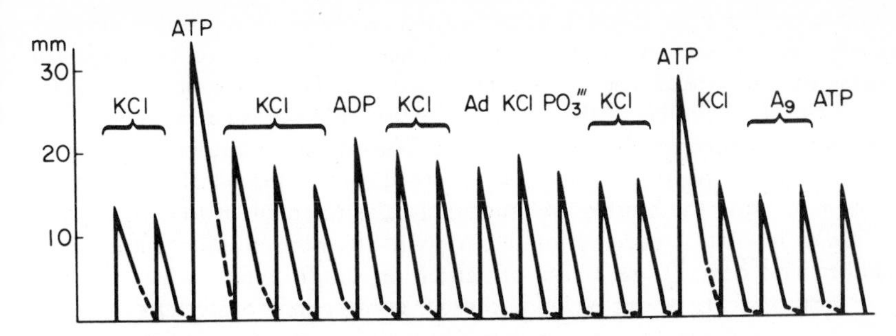

Fig. 2. Load constant, –200 mg. ATP = adenosine-triphosphoric acid; ADP = adenosine-diphosphoric acid; Ad = adenylic acid; PO_3''' = phosphate solution; Ag = $AgNO_3$.

The effect of adenosine-triphosphoric acid is graphically presented in Fig. 2.

Only threads which are relatively fresh have the ability to respond, by changes in extensibility, to the action of adenosine-triphosphoric acid. Threads which have been kept for several days may retain their normal extensibility without any marked change, but their response to the action of adenosine-triphosphoric acid gradually disappears. As a rule this occurs simultaneously with the loss of the enzymatic properties of the fiber.

We found that the adenosine-triphosphatase properties of myosin are completely destroyed by silver salts in extremely small concentrations—of the order of 1: 100,000 molar. In the presence of or after preliminary treatment with silver the myosin threads lose their ability to change their mechanical properties under the action of adenosine-triphosphoric acid. We found the action of silver to be specific: the silver in no way affects the normal extensibility of the myosin threads, but only eliminates the possibility of evoking an additional extensibility by the action of adenosine-triphosphate.

Thermal treatment of the threads that destroys their enzymatic properties also eliminates the effects of adenosine-triphosphoric acid. However, Weber has shown that in this case the normal extensibility of the threads is slightly changed. The general impression (more detailed evidence is required for definite proof) is that myosin threads retain their ability to change their mechanical properties under the action of adenosine-triphosphoric acid only so long as their enzymatic properties remain intact. The evidence at present is insufficient to conclude that there is a causal relationship between these two properties, although this conclusion might seem very tempting. It might be more cautious at present to regard the enzymatic properties as criteria of the native of "paranative" state of the myosin. We consider the loss of enzymatic properties as a sign of the beginning of the decomposition of the myosin. The loss of the ability to respond to the action of adenosine-triphosphoric acid with a change in mechanical properties must also be regarded as a symptom of the beginning of the decomposition of the myosin.

We must emphasize that adenosine-triphosphoric acid affects the myosin thread not in the sense of contraction but of relaxation. However, this fact does not detract from the significance of the results for the problem of muslce activity. Muscle activity consists of two phases which are mutually inseparable: contraction and relaxation. The view that muscle contraction is a purely physical process similar to the discharge of a condenser is supported by much data and by many reliable investigators. The whole of the chemical energy is expended in the process of relaxation—the charging of the condenser—in order to bring the muscle into such a state as to enable it to contract.

Direct uncritical observation has drawn attention to the phase of contraction, and has left the phase of relaxation unobserved. It has, therefore, been customary to speak only of muscle contraction. Systematically this is undoubtedly erroneous; a correctly planned investigation must devote its attention equally to both phases of muscle activity, which together act in indivisible unison. The discovery of the factors responsible for muscle relaxation, that is, the stretching of the contractile elements of the muscle, is just as important as is the discovery of the causes of contraction.

Summarizing the above one may say that the postulate forming the basis of all contemporary theories on the physiological activity of muscles supposes that muscle function depends on the interaction between the contractile substance (myosin) and the chemical compounds which are the substrate or the products of muscle metabolism. Although this postulate is accepted by all, it has never yet been proved.

Experiments of this type were performed, but they either were of a speculative nature or were unsuccessful. As an example we may cite the hypothesis proposed by M. Fischer, and later many times repeated, that changes in the state of muscle proteins are caused by changes in the hydrogen ion concentration. This view is reflected in the theory of Embden and his school of workers. Direct experiments to confirm these theories were not mentioned by the authors. It is necessary to emphasize that in our experiments, contrary to our own expectations, we were unable to find any characteristic changes in the mechanical properties of the myosin threads within the limits of pH 5.0 to 7.5, i.e., within the physiological zone. Outside of this zone, the threads disintegrate, tear, and break.

In another recent investigation, Weber[22] attempted to discover the effect, on the solubility of myosin, of a number of active substances occurring in muscle metabolism. Weber used substances such as adenyl derivatives, creatine, etc., which were also investigated by us, but he could not disclose any influence on solubility.

There is reason to believe that the data presented above, demonstrating, on the one hand, the reaction of the contractile component of the muscle (myosin) with adenosine-triphosphoric acid, which is one of the chief substances playing a part in the energetics of the muscle, and, on the other hand, the change in the mechanical properties of myosin under the influence of this very substance, offer the first case of direct demonstration of the mutual interaction between the bearer of the mechanical properties of the muscle and the bearer of the chemical energy of the muscle. *This is the first demonstration of the relation between the mechanics and the chemistry of muscle.*

We therefore regard three aspects of myosin as being of fundamental importance.

1. Myosin is the bearer of the contractile, mechanical properties of muscle.
2. Myosin is the bearer of catalytic properties, i.e., the promoter of chemical reactions.
3. Myosin takes the form of a chemical transformer which converts chemical energy into mechanical action.

It has been the dream and goal of those working in the field of the physiology and biochemistry of muscle to discover the relation between the chemical processes occurring in the muscle and the changes in the state of its contractile substance–myosin. This dream remained beyond experimental realization. We believe that the experiments described are the first step toward the realization of this dream and its formulation in concrete form. It is clear that this is but a small and modest step, but satisfaction may be derived from the fact that it has been possible to find the way which may lead to further effort.

The results obtained indicate the direction in which we may expect further interesting developments and more detailed investigations.

The first question is the investigation of as many substances as possible which participate in the chemical dynamics of muscle. One might think of intermediate compounds, such as acetylcholine, although one might suppose *a priori* that they function in other regions of muscle dynamics; in our experiments acetylcholine showed no effects. It would be interesting to investigate various enzyme systems and their substrates, more particularly those effecting oxidations. Parallel with the investigation of factors causing the relaxation of the muscle, there should be a vigorous search for the agents of contraction. Perhaps it will be possible to reproduce the action of such typical poisons as caffeine, veratrine, etc., which cause contraction.

Of particular interest would be the study of the refractive properties of myosin threads subjected to certain reactions, since these reflect changes in the micellar structure of the thread. X-ray diagrams of the threads under the influence of "active" substances will doubtless be of interest. One might expect to find changes in the periodicity in the presence of substances changing the mechanical properties of myosin threads.

It is difficult to enumerate all the paths along which further investigations could develop. The significance of the experiments here presented does not rest upon the modest results obtained at present, but in the opening of a new perspective and approach to the study of what might be called the *mechanochemistry* of the muscle.

In addition to myosin, we[(5)] subjected another characteristic muscle protein–myogen–to investigation. As Weber has stated, "muscle is basically composed of myosin and myogen." If myosin is the contractile component of the muscle and the structural basis of the fibril, then myogen makes up the main bulk of the liquid portion of the muscle tissue, the sarcoplasm. Myosin and myogen make up respectively 40 and 20 per cent of the total protein content of the muscle.

Myogen presented a particularly tempting subject for investigation, since Baranowski has succeeded in obtaining it in crystalline form. We undertook to ascertain whether the crystalline myogen of Baranowski is or is not endowed with enzymatic properties.

In this case our path of investigation was the reverse of that used in the case of myosin. In our experiments with myosin we started from a specific enzyme, adenosine-triphosphatase, and concluded that it was indentical with one of the muscle proteins, myosin. With regard to myogen, we began with the examination of one of the muscle proteins and also came to the conclusion that it had enzymatic properties.

Analysis of the crystalline preparations of myogen showed that these crystals possess a high enzyme activity. The crystals are the centers of the activity of aldolase, one of the important enzymes of carbohydrate metabolism. This enzyme causes the splitting of the six-carbon chain of glucose (in the form of its phosphate ester) into two three-carbon molecules which serve as the origin of lactic acid, the final product of the decomposition (anaerobic) of the muscle carbohydrates.

This enzyme action was retained after recrystallization for as many as four times. Thus, once again, it was found that a protein easily obtainable from the muscle in considerable quantity and in a high degree of purity and which is one of the basic proteins of the muscle plasma is a bearer of enzymatic properties.

In the light of these results obtained with the two most important proteins of muscle tissue, it is appropriate to consider some of the general problems connecting enzymology with the biological functions of proteins.

The concept that enzymes are substances present in the cell in minute, almost indeterminable quantities has been commonly accepted. The affirmation that a protein which represents a large and sometimes very considerable part of the cell protoplasm possesses fermentative properties, sounds somewhat unusual and, to many people, improbable.

It is certainly useful to subject some of our most firmly established views and concepts to periodic, critical review. To accept them dogmatically only slows up our scientific progress and hinders our penetration into new fields of knowledge. It is sufficient to quote a single example. When, fifteen years ago, it was reported that enzymes could be obtained in crystalline form, and, indeed, in large quantities, it was regarded with skepticism by all of the most prominent enzymologists of the time. However, in less than one and one-half decades, there occurred a radical change in our point of view: it is no exaggeration to say that the entire study of enzymes, as chemical entities, is at present being developed entirely on the basis of their preparation in crystalline form.

It may be pertinent at the present time to ask how well founded is our concept of enzymes as substances present in minute, barely detectable quantities in the cell. Related to this is another question, namely, whether we are correct in assuming that the protein which constitutes the main bulk of living protoplasm is chiefly merely a colloidal chemical—one which is admittedly very motile but nevertheless a rather passive substance—which, composing the substrate, responds to one or another biochemical reaction or whether it is not the active force which initiates reactions. There is evidence that a review of these two generally accepted concepts is timely, well-founded, and necessary.

As a result of the work on the preparation of crystalline enzymes, the fact that they are substances of a protein nature has become commonly accepted and must be considered well established. We may now postulate *protoplasmic proteins as substances of catalytic nature.*

In the case of myosin, that protein which makes up such an important portion of the muscle tissue, we saw that it was not possible to attribute to it merely the rôle of a passive substrate subjected to chemical reactions. Myosin was found to take an active part in muscle chemistry, to be the catalyst of an important biochemical reaction.

In addition to our own results mentioned above, we may cite a number of results from the literature of the past few years. There is sufficient evidence that in many cases the amount of one or another enzyme is by no means insignificantly small, and may markedly exceed the amount of many substances well known to us, such as phosphates, cholesterin, or extraction products. The content of many enzymes in cells, it is found, is to be measured in tenths of a per cent, it often reaches one or more per cent, and in some cases it even constitutes 10 or more per cent of the cell. Relevant data are recorded in the following table.

CATALYTICALLY ACTIVE PROTEINS

Proteins	Author	Content in % calculated on	Activity Moles of substrate / Moles of enzyme × min.
Urease	Sumner	0.12 dry weight (beans)	—
Flavin-enzyme (liver) (total) flavoprotein	Warburg	1.0 dry weight (liver)	—
Aldehydoxidase	Green	—	550
Diaphorase	Straub	—	7,000(?)
Triosophosphate-dehydrase* (yeast)	Warburg	0.25 dry weight (yeast)	20,000
Lacticodehydrase*	Straub	2.00 water-soluble proteins (muscle)	3,000(?)
Carboxylase	Green	(2.0) (?) dry weight (yeast)	850
Aldolase	Herbert, Green	Water-soluble 5.0 (muscle)	6,600–8,800
Crystalline "myogen"	Engelhardt, Smirnov, & Sakov		1,000
Pepsinogen	Herriot	10 dry weight (gland tissue)	1,020
Myosin	Engelhardt Ljubimowa	40 all muscle proteins	200–2,000
Hemoglobin		90 dry weight (erythrocyte)	6,000

[*Triosephosphate; Lactic dehydrogenase.]

We see that urease, the first enzyme produced in crystalline form, makes up 0.12 per cent of the dry weight of the bean, according to Sumner. Considering the fact that the bean is to a large extent made up of inert reserve food materials, we realize that the urease constitutes several tenths per cent (approaching one per cent) of the active protoplasm. This quantity is not insignificantly small; it is readily measurable.

In liver the group of respiratory enzymes that are derivatives of vitamin B_2, the so-called flavin enzymes calculated on the flavin content, make up about 1 per cent of the dry substance, that is to say, a hundredth part of the solid substance of the liver is composed of this enzyme.

The important enzyme of gaseous exchange–carbonic anhydrase, which controls the discharge of CO_2 from H_2CO_3 in the lungs—is, according to Keilin and Mann[(11)], present in the erythrocytes to the extent of 0.21 per cent of the fresh weight of the cells, that is, about 0.5 per cent of the dry weight. If we consider that 90 per cent of the dry weight of the erythrocyte is hemoglobin, then carbonic anhydrase makes up 5 per cent of all the other solids of the red blood corpuscle.

According to Warburg, the content of oxidoreductase, which takes part in the formation of alcohol and is one of the most important enzymes of fermentation, reaches 0.25 per cent of the dry weight of yeast.

The analogous enzyme of muscle metabolism–lactodehydrase, recently isolated in crystalline form by Straub,–makes up 2 per cent of the total quantity of the water-soluble proteins of the muscle, that is to say, one-fiftieth is lactodehydrase.

Carboxylase, another of the most important enzymes of the fermentation system, which has lately been purified by Green, composes something like 2 per cent of the dry weight of yeast.

Aldolase, which we identify with our crystalline preparations of myogen, composes about 4 or 5 per cent of the total water-soluble proteins of muscle, according to Herbert, Green, and collaborators. This is certainly not an insignificantly small quantity.

Still more remarkable are the results obtained with digestive enzymes. According to Herriot, collaborator of Northrop, the pepsinogen content of the mucous layer of the stomach of a pig is about one gram. Of this quantity 90 per cent, i.e., 0.9 grams is contained in the mucous coat of the fundal part of the stomach, which part composes one-third of the surface area of the stomach. The dry weight of this third is in the best of cases several tenths of a gram; moreover it undoubtedly contains much connective tissue. Therefore the pepsinogen content is several per cent of the dry weight of the tissue. If it were possible to calculate the pepsinogen content of the glandular cells themselves, we would undoubtedly find that it makes up a highly significant portion of the plasmatic proteins of these cells. Probably in these cells, specialized for the elaboration of the given enzyme, the amount of enzyme in the plasma of the cell reaches ten, or even several tens, per cent of the dry weight.

Finally, we shall cite an example which may require some stretch of the imagination, but which is in our opinion fully permissible and appropriate–hemoglobin.

The physiological function of hemoglobin is a typically catalytic one. At the same time the content of this catalytically active protein in the erythrocyte is 90 per cent of the dry weight of the cell. We have become accustomed to this fact—that the content of the catalytic protein in the cell is so high—and it no longer surprises us.

On the basis of these examples it no longer seems completely improbable or untruthful to assume that, inasmuch as the content of catalytic proteins in the highly specialized glandular cell or in the red blood cell can make up a large bulk of the whole of the living material, the protein myosin found in the less specialized muscle cell is endowed with catalytic properties.

Accordingly, it is permissible to conclude that it is possible, and even probable, that the main mass of the living protoplasmic proteins is none other than proteins endowed with one or another catalytic (enzymatic) property. In the light of this concept, protoplasm must not be considered as a mass of undifferentiated protein which serves as the intermediary or substrate of biochemical processes, and which contains minute quantities of enzymes; it is more correctly described as *a conglomerate of catalytically active proteins* which promote the chemistry of the whole of the metabolism of the cell.

Naturally, if each of the enzymes composes such a significant amount—one or more per cent—of the mass of the protoplasm, the question arises as to whether all of the enzymes necessary to the cell can be contained in the 100 per cent of the protoplasmic proteins?

The first answer to such a question is that the figures given above refer in the majority of cases (except perhaps for urease) to enzymes or catalytic proteins which are the basis of processes and functions whose fulfillment is the predominant activity of the cell or tissue in question.

Muscles are entirely specialized for a contractile function. It is, therefore, not surprising to find that the enzyme which lies at the basis of this function composes a predominant part of the protoplasmic protein of the muscle. In the same way the erythrocyte is completely specialized for the transfer of oxygen. We are not surprised that practically the entire substance of the red blood cell is nothing more than a protein adapted for this specialized function.

The processes of specialized predominant metabolism completely overshadow quantitatively those other metabolic processes whose measurement is required for statistical purposes: elementary processes of cell nutrition, growth, etc. The speed of reaction of these latter processes is immeasurably slow, and the quantities involved are extremely small as compared with the processes of functional specialization of metabolism. It is obvious that in order to assure these slowly occurring reactions, it suffices to have present quite insignificant quantities of the corresponding enzymes. And it will not be surprising to find many of these enzymes in very small, almost insignificant, quantities in the cell.

A second point that arises in regard to the possible lack of sufficient proteins to assure all the catalytic requirements of the cell is the assumption that there is a possible polyvalence of the proteins in a catalytic sense, i.e., that possibly a single protein may discharge different catalytic functions as the result of different specific groupings in its structure. As A. E. Braunstein has well said, we shall be dealing with the plurality of the catalytic functions of proteins. Some data have already been collected to substantiate this theory.

We know that hemoglobin has at least two, and probably three, catalytic functions. Aside from the extremely powerful catalytic function in the transfer of oxygen, it also has the properties of peroxidase—admittedly weak, yet wholly perceptible. Apparently it also has the very weak but existing characteristics of catalase.

K. G. Stern[16], in his work on oxidative enzymes, presents the following interesting idea. There may exist in the cell large protein molecules or micelles with a molecular weight of the order of millions, which contain on their surface a network of catalytically active groups, oriented in such a way that they assure a smooth flow of complicated enzymatic processes such as cell respiration.

In one of his papers Euler[6] says that if we have hitherto failed to discover reproducible and well-defined substances with multiple enzymatic properties, this is caused principally by the fact that not only have we failed to look for them, but we have, on the contrary, tried to escape them in order to find unity of function. Perhaps the time has come, says Euler, to direct our effort to the search for such substances with several enzyme functions. We may hope that such investigations will lead to the discovery of such entities as are the building-stones of living protein, of viruses, or of protoplasm.

Two interesting problems now confront the investigator: First, to try to obtain a sufficiently large number of well-defined cell or tissue proteins in a form approaching chemical purity. Second, to examine the products obtained with regard to as many as possible of their enzymic functions.

Investigators have already embarked upon these problems. Dounce and Sumner[2] have already obtained a crystalline protein of globulin nature from liver. Keilin and Mann[11] obtained cuproproteids such as hemocuprein and hepatocuprein. Kuhn[12] obtained ferritin, a crystalline iron-containing protein. In all these cases the authors proceeded in the same manner as we did when we obtained crystalline myogen according to Baranowski's process: they examined their products for enzymatic properties. It is true that the results obtained up to the present have been negative, but this should not be discouraging, because the scope and character of the enzyme actions examined can not be considered as sufficiently comprehensive.

Perhaps such a proposal may appear too vague, but there exist many well-founded reasons for believing that it is highly improbable that it will be possible to obtain proteins of a completely inert nature devoid of catalytic properties from the living and active plasma of the cell. We have the right to assume that *the living protein of the plasma is primarily a catalytically active protein, a protein with the properties of an enzyme.*

References

1. Corran, H. S., Green, D. E., and Straub, F. B.: *Biochem. J.,* 1939, **33**, 793.
2. Dounce, A., and Summer, J. B.: *J. Biol. Chem.,* 1938, **124**, 415.
3. Edsall, J. T., and Mehl, S. W.: *J. Biol. Chem.,* 1940, **133**, 409.
4. Engelhardt, W. A., Ljubimowa, M. N., and Meitina, R. A.: *Trans. Acad. Sci. U.S.S.R.,* 1941 (in press).
5. Engelhardt, W. A., Smirnov, M., and Sakov, N. E.: 1941, Unpublished.
6. Euler, H.: *Ergebn. Vitamin u. Hormonforsch.,* 1938, **1**, 159.
7. Green, D. E., Gordon, A. H., and Subrahmanyan, V.: *Biochem. J.,* 1940, **34**, 764.
8. Green, D. E., Herbert, D. H., and Subrahmanyan, V.: *J. Biol. Chem.,* 1940, **135**, 795.
9. Herbert, D. H., Gordon, A. H., Subrahmanyan, V., and Green, D. E.: *Biochem. J.,* 1940, **34**, 1108.

10. Herriot, R.: *J. Gen. Physiol.,* 1938, **21**, 501.
11. Keilin, D., and Mann, T.: *Biochem. J.,* 1940, **34**, 1163.
12. Kuhn, R.: *Ber. deutsch. chem. Ges.,* 1940, **73**, 823.
13. Ljubimowa, M. N., and Engelhardt, W. A.: *Biochimia,* 1939, **4**, 716; *Nature,* **144**, 668.
14. Mann, T., and Keilin, D.: *Proc. Roy. Soc. B,* 1938, **125**, 303.
15. Muralt, A., and Edsall, J. T.: *J. Biol. Chem.,* 1930, **99**, 315.
16. Stern, K. G.: *Ann. N.Y. Acad. Sci.,* 1939, **39**, 147; Cold Spring Harbor Symposia, **7**, 312.
17. Straub, F. B.: *Biochem. J.,* 1939, **33**, 787; 1940, **34**, 483.
18. Sumner, J. B.: *Ergebn. Enzymforsch.,* 1932, **1**, 295.
19. Warburg, O., and Christian, W.: *Biochem. Z.,* 1939, **303**, 40.
20. Weber, H. H.: *Ergebn. Physiol.,* 1934, **36**, 109.
21. Weber, H. H.: *Pflüger's Arch. f. d. ges. Physiol.,* 1935, **235**, 205.
22. Weber, H. H.: *Naturwiss.,* 1939, **27**, 33.
23. Weber, H. H., and Meyer, K.: *Biochem. Z.,* 1933, **266**, 137.

The editor's guess: Engelhardts exchange views on "bioengineering" with Matthew Meselson. Moscow, 1961.

Joseph Needham brought up the concept of "micromorphology of proteins in structure" in his 1937 birthday essay to Frederick Gowland Hopkins. Both he and his wife worked actively on myosin.

MYOSIN BIREFRINGENCE AND ADENYLPYROPHOSPHATE*

Joseph Needham, Shih-Chang Shen,
Dorothy M. Needham and A.S.C. Lawrence
Biochemical Laboratory, University of Cambridge, England

Our knowledge of the nature of muscular motion has hitherto been divided into two main fields of successful analysis. First there was the discovery of the phosphorylation cycles whereby energy is transferred from carbohydrate breakdown to the muscle fibre; here the earliest landmarks were the investigations of Fletcher and Hopkins(1) and of Harden and Young(2) (see the papers of Parnas(3) and D. M. Needham(4) for up-to-date reviews). Secondly, there was the discovery of the elongated or anisometric character of the particles of muscle-globulin (myosin), demonstrated in the classical paper of v. Muralt and Edsall(5), leading to the application of X-ray techniques to the problem, and the establishment by Astbury(6,7), that muscular contractility is essentially a molecular contractility of protein chains. The exact connexion between these two orders of fact, however, still remains obscure.

In order to bridge this gap, the most promising point of departure seemed to be the important finding of Engelhardt and Ljubimova(8) in 1939 that the enzyme adenylpyrophosphatase is either myosin itself or some protein very closely associated with it. This was confirmed by one of us (D. M. N.) in the following year (unpublished) and recently by Szent-Györgyi and Banga(9). Its importance lies in the facts: (1) that the breakdown of adenylpyrophosphate is, among the processes of intermediary metabolism in the muscle, that nearest in time to the contraction of the fibrils, and (2) that although some adenylpyrophosphate is broken down by transfer of phosphate to hexosemonophos-

*J. Needham, Shih-Chang Shen, D. M. Needham, and A. S. C. Lawrence, "Myosin Birefringence and Adenylpyrophosphate," *Nature*, **147**, 766–68 (1941). Reprinted with the permission of the authors and of the editors of *Nature*.

phate (by an enzyme not present in the myosin fraction), the quantitatively largest part of its breakdown probably occurs by splitting off of free phosphate under the influence of the enzyme adenylpyrophosphatase. It was thought of interest, therefore, to study the effect of adenylpyrophosphate upon the flow birefringence of myosin. The myosin sols were contained in a small annular cell holding just less than 1 c.c., mounted on the fixed stage of a Swift-Dick polarizing microscope, and having the outer co-axial cylinder driven at speeds variable up to 500 rev./min. (shear-rate 4.08 cm./cm./radian). Myosin preparations were also examined at higher dilutions in the co-axial viscosimeter referred to in a recent communication in NATURE by Lawrence, Needham and Shen[10] (shear-rates of 0.73 and 4.2 cm./cm./radian according to the diameter of the inner cylinder used).

Myosin preparations showing strong flow birefringence (Δ50–80°; ψ 50–60° for 1–2 per cent sols at 20° C.) were obtained from rabbit muscle by the usual method. Until this year, myosin sols prepared in Cambridge had nearly always been non-birefringent, and although the reason for this is being made the subject of a special study by Dr. K. Bailey, it may be said here that, provided the rabbit has been starved for twenty-four hours before death, the myosin never fails to show flow birefringence.* In the light of observations such as those reported below on the viscosimetry of myosin, it is likely that these differences are simply due to the size and shape which myosin particles assume when extracted from muscle cells under different conditions. Some preparations of myosin, however, which do not show flow birefringence, may show anomalous flow at low shear-rates, and are then analogous to the preparations of globulin from amphibian embryos described by Lawrence, Needham and Shen[10].

If now a myosin sol containing about 3 per cent of the protein is mixed with adenylpyrophosphate at *p*H 7.0, a considerable decrease in the flow birefringence intensity is immediately observed, for example, 65°–45° (see Fig. 1). This fall occurs in less than a minute at room temperature, but at 0° C. is so lengthened as to allow of its being plotted on a curve. With adenylic acid or inorganic pyrophosphate, the fall is either absent or slight. An important feature of the effect is that it is reversible; after it has occurred, the flow birefringence intensity of the sol rises, reaching its original level in about 2 hr. at 37° C. or overnight at room temperature, and then rising beyond it. It may be added that strict control of the salt concentration is essential, since if increasing amounts of solid potassium chloride are added to a white, almost salt-free, gel of myosin, the flow birefringence intensity continually decreases to reach a minimum at molar concentration, when the sol so formed is most transparent. After that point, with increasing amounts of the salt, the flow birefringence intensity increases, apparently because of the increasing amounts of transparent gel formed by salting-out. Hence in the above-mentioned experiments the ionic strength of the adenylpyrophosphate or other substances added was compensated for in such a way as to bring all samples to a final concentration equivalent to the ionic strength of 0.75 *M*KCl. In order to avoid the effects described by Edsall and Mehl[11], potassium was the only cation present. Finally, viscosimetric measurements indicate that the relative viscosity decreases, but not the anomaly of flow (see Fig. 2*a*).

*The suggestion that the state of nutrition might be a factor was due to Dr. H. Lehmann.

Interpreting the effect, we consider that the first change is a moderate shortening of the myosin particles, with or without a certain amount of disaggregation. This seems not quite analogous with the changes produced by increasing potassium chloride concentrations below molar, for here, although the flow birefringence intensity also falls, and correspondingly the relative viscosity, the anomaly of flow decreases and eventually disappears. Hence the particles must become drastically shorter as well as possibly smaller. After the molar point is passed, though flow birefringence and relative viscosity increase, the viscous anomaly does not again become observable, probably because although the particles lengthen, their axial ratio decreases owing to some form of aggregation. In all such explanations we provisionally ignore the role of intermicellar forces (*cf.* Langmuir[12]) which may well be of importance in the relatively dilute solutions we have used, and must almost certainly be so in more concentrated mixtures such as occur in the muscle fibril itself.

The subsequent rise in flow birefringence intensity during the action of the enzyme, however, is more difficult to interpret, since that of myosin sols in potassium chloride alone will, if placed at 37° C., steadily increase until they set to a thixotropic gel. We have frequently observed in this respect what may be a protective action of adenylpyrophosphate, the sol returning to approximately its original birefringence intensity, and then remaining there for some hours, only in the end to rise. In such circumstances, the parallel increases in relative viscosity and viscous anomaly are also inhibited. This effect on the

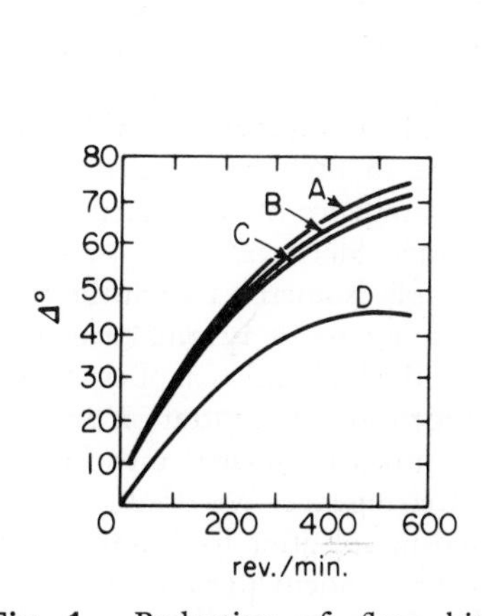

Fig. 1. Reduction of flow birefringence of myosin by adenylpyrophosphate. *A*, potassium adenylate; *B*, control; *C*, potassium pyrophosphate; *D*, potassium adenylpyrophosphate.

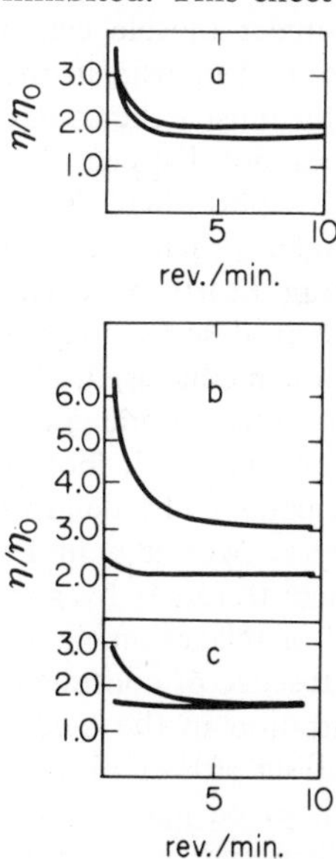

Fig. 2. Viscosity of myosin. *a*, Effect of potassium adenylpyrophosphate; *b*, increase of anomalous viscosity in a dilute birefringent myosin KCl sol held at 37°C. for 2 hr.; *c*, restoration of anomalous viscosity to a nonbirefringent myosin KCl sol by treatment at 37°C. for 2 hr.

particle shape recalls the protective effect of adenylpyrophosphate on the enzymic activity itself at 37° C. discovered by Engelhardt and Ljubimowa[8], and confirmed by us.

The reversibility of the change in flow birefringence intensity of myosin is of much interest, since according to Edsall and Mehl[11], no treatment has ever hitherto been found to restore flow birefringence when this property has once been lost. We find, however, that adenylpyrophosphate is not unique in causing a reversible fall in flow birefringence intensity. After the flow birefringence has been reduced or abolished by ions such as Li′, Ca″, Mg″, or NH_4', or by urea, all that is required to restore it is a short period at physiological temperature (or under some conditions even at room temperature), and its original value is reached long before any signs of gelation appear. It is even possible in this way to restore flow birefringence to a myosin sol which on long standing at 0° C. has lost the property. During such a restoration, or during the rise of birefringence from its original value in, for example, a 0.5 *M* KCl myosin sol subjected to 37° C., relative viscosity increases and flow anomaly appears, first slightly, then markedly (see Fig. 2 *b, c*). The particles must therefore become very much longer.*

Thus the physiological significance of the adenylpyrophosphate effect, which might at first sight seem to be lessened by experiments on simple salts, regains its force in that these themselves give reversible changes in flow birefringence and viscosity, which should throw light on the reversible changes in the intact muscle cell. One cannot, of course, compare too closely the behaviour of myosin in the muscle cell and myosin isolated in the form of a sol. In the muscle the contracting particles do work against resistance; in the sol this does not happen. It should be made clear, however, that potassium adenylpyrophosphate is more effective in causing immediate fall of flow birefringence than any of the other salts so far studied, 1–2 *M* concentrations not being nearly as effective as 0.1 *M* adenylpyrophosphate. None of the effects here described can be due to differences in the refractive index of the dispersion media used, since these are very small (within the limits 1.3336–1.3437). What connexion the reversible changes in flow birefringence here described have with those occurring in the birefringence of the intact muscle fibre remains to be determined. It is known from older work, put on a firm basis by v. Muralt[13] in 1932, that during a single isometric twitch the birefringence decreases by some 35 per cent, returning to its original value by the end of relaxation. It has also been found by Buchtal and Knappeis[14] that in the absence of contraction, changes in hydrogen ion concentration and salt concentration in the medium surrounding a contractile isolated muscle fibre bring about slower reversible changes in its birefringence. We doubt if there is much to be gained by referring to the myosin sol after its birefringence has fallen and it has begun to dephosphorylate adenylpyrophosphate as 'denatured,' unless we are careful to say that the 'denaturation' (which involves neither changes of solubility, nor, as has been shown by Greenstein and

*The restored flow birefringence intensity cannot be ascribed to the photo-elastic effect in a gel under strain because (a) anomalous viscosity appears before the relative viscosity has much increased, and (b) the angle of isocline remains about 70° showing that the particles are orienting freely.

Edsall[15] in the case of the salt effects, changes in titratable –SH linkages) has not gone beyond an early reversible stage at which it is compatible with active enzymic breakdown of the adenylpyrophosphate.

How exactly the adenylpyrophosphate interacts with the myosin in normal muscle remains unclear. But there is a suggestion in the facts already known (entertained for some time past by one of us (D. M. N.) and now put forward in a rather different form by Kalckar)[16] that myosin might be phosphorylated and dephosphorylated, occupying thus the last link in the chain of simultaneous transfers of phosphate ions and energy. As is well known, adenosine-diphosphate can be phosphorylated according to circumstances by creatine-phosphate, phosphopyruvic acid, or diphosphoglyceric acid. We suggest that while it acts as a phosphate acceptor towards these substances, adenosinetriphosphate (adenylpyrophosphate) acts as a phosphate donator towards the contractile protein itself. Contracted myosin would thus have adenylpyrophosphatase activity and phosphate would be transferred from the adenylpyrophosphate to some part of the protein molecule, which would simultaneously extend. Extended myosin would then be in phosphorylated form and charged with energy. When the physical changes touched off by the nerve stimulus occur, inorganic phosphate would be set free, and the energy available from this dephosphorylation used in contraction. It might even be, in view of the effects described in this note, that the contraction of the particles in the fibril would be occasioned by a sudden contact (brought about by changed permeability) with the resynthesized adenylpyrophosphate, the substrate, itself. Such possibilities obviously invite further study.

These experiments, which indicate that adenylpyrophosphate has a direct influence on the relative optical anisometry and on the shape of myosin particles, and possibly on their intermicellar forces, will be continued and extended as better optical and other apparatus becomes available.* They form part of a general investigation of the shape of protein particles arising out of an interest in the proteins of developing embryonic cells. Although they primarily concern the connexion between energy-transfer and myosin-fibril contraction, they are perhaps not without a relevance to inductor-reactor systems in embryos undergoing morphogenesis. The possibility should now be borne in mind that an inductor might be, not a co-enzyme or an auto-catalytically active protein, but a substrate, and that the changes in protein configuration which will lead to a specific histogenesis might be the result of the unavoidable action of an enzyme protein in the competent cell upon the inductor itself.

References

1. Fletcher, W. M., and Hopkins, F. G., *J. Physiol.,* **35**, 247, (1907).
2. Harden, A., and Young, W. J., *Proc. Roy. Soc.,* B, **82**, 321, (1910).
3. Parnas, J. K., *Ergebn. d. Enzymforsch.,* **6**, 57, (1937).
4. Needham, D. M., *Enzymologia,* **5**, 158, (1938).

*The authors wish to take this opportunity of thanking Prof. J. D. Bernal and Dr. J. F. Danielli for the stimulus and benefits of their conversation, and Prof. Tilley for the loan of valuable apparatus.

5. v. Muralt, A., and Edsall, J. T., *J. Biol. Chem.,* **89**, 315 and 351, (1930).
6. Astbury, W. T., "Fundamentals of Fibre Structure" (Oxford, 1933); *Proc. Roy. Soc.,* B, **127**, 30, (1939); *Ann. Rev. Biochem.,* **8**, 113, (1939).
7. Astbury, W. T., and Bell, F. O., Cold Spring Harbor Symp. Quant. Biol., **6**, 109, (1938); Astbury, W. T., and Dickinson, S., *Proc. Roy. Soc.,* B, **129**, 307, (1940).
8. Engelhardt, W. A., and Ljubimowa, M. N., *Nature,* **144**, 668, (1939); *Biochemia,* **4**, 716, (1939).
9. v. Szent-Györgyi, A., and Banga, I., *Science,* **93**, 158, (1941).
10. Lawrence, A. S. C., Needham, J., and Shen, S. C., *Nature,* **146**, 104, (1940).
11. Edsall, J. T., and Mehl, J. W., *J. Biol. Chem.,* **133**, 409, (1940).
12. Langmuir, I., *J. Chem. Phys.,* **6**, 873, (1938).
13. v. Muralt, A., *Archiv f. d. g. Physiol.,* **230**, 299, (1932).
14. Buchtal, F., and Knappeis, G. G., *Skand. Archiv f. Physiol.,* **78**, 97, (1938).
15. Greenstein, J. P., and Edsall, J. T., *J. Biol. Chem.,* **133**, 397, (1940).
16. Kalckar, H., *Chem. Rev.,* **28**, 71, (1941), *Biol. Rev.,* in the Press.

THE CONTRACTION OF MYOSIN THREADS*

A. Szent-Györgyi
Institute of Medical Chemistry, University Szeged, Hungary
(Received for publication, July 6, 1942)

It has been shown by H. H. Weber[1], that a myosin solution, if squirted in a thin jet into water, solidifies in the form of a thread. In this way the myosin can be brought into a form which resembles the muscle fibril in some respects. The myosin thread is an elongated piece of myosin gel.

It has been shown in the preceding paper by Banga and myself that myosin can be obtained from muscle in two different forms which were called myosin A and B. Threads can be prepared from both. For the sake of convenience I will call the threads prepared from the 20 min. extract (see Banga and Sz.) "myosin A threads" while the threads prepared from the 24 h. extract will be called "myosin B threads."

The threads used by previous investigators correspond, in all probability, to our myosin A threads.

The technique of the preparation and observation of threads will be described by M. Gerendás.

A watery extract of muscle was made in the following way: the rabbit's muscle was cut out and minced (as described in the previous paper), suspended in water (1 ml per g of muscle), stirred for 5 minutes at 0°C and squeezed through a cloth. The fluid was then filtered through paper at 0°C.

If a myosin B thread is suspended in this extract and observed under the microscope, a violent contraction will be seen. The thread contracts within 30 seconds to less than half of its length and within 2–3 minutes it reaches a

*I. Banga, T. Erdös, "The Contraction of Myosin Threads": in *Myosin and Muscular Contraction,* ed. A. Szent-Györgyi, **1** (1941–42), 563–573. Reprinted with the permission of the author and of the publisher.

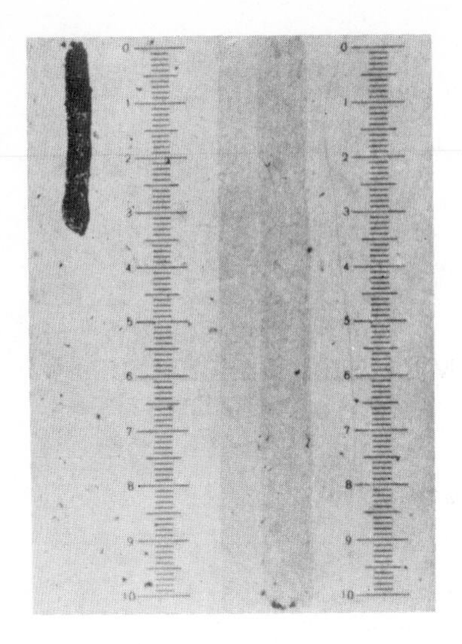

Fig. 1.

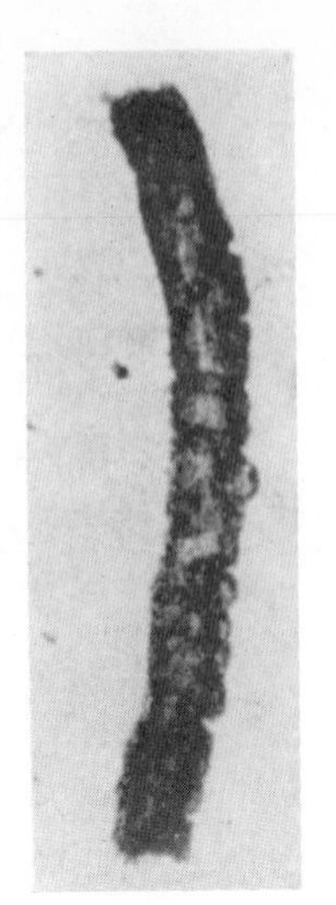

Fig. 2.

maximum contraction of 66%, $\frac{1}{8}$ of its original length. (Fig. 1.) At the same time the thread becomes proportionately thinner, and is seen to become quite dark.* Watched in lateral illumination the transparent thread is seen to turn white, opaque. The rate of contraction depends on the diameter of the thread. Fig. 3 curve a gives the time-curve of the contraction of a relatively thick thread of 0.3 mm diameter. Thinner threads, like those of 0.1 diameter, contract faster but their rate of contraction cannot be measured by the same method because it is too fast. Furthermore thin threads mostly curl up and stretch out again only after they have reached maximum contraction.

If a myosin A thread is suspended in the same fluid no striking change will be observed. (Fig. 4.) Measurement by the ocular micrometer will reveal a

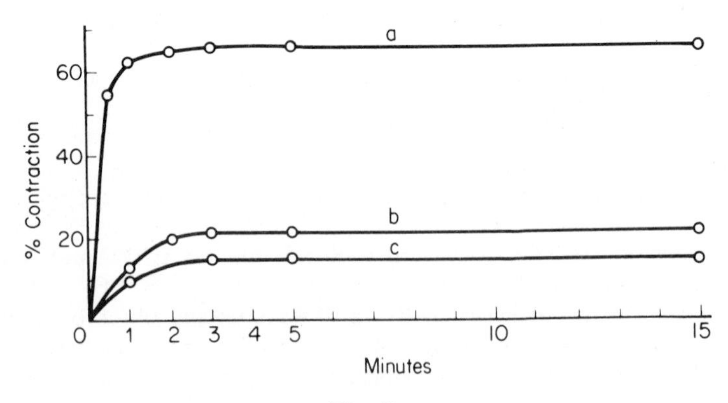

Fig. 3.

*Frequently, the contraction is so violent that the interior of the thread cannot keep pace with the contraction of the outer sheets, so that these latter break up, like a crocodile's skin. Fig. 2. For the same reason the whole thread sometimes breaks up into dark, solid lumps of myosin instead of giving a contraction.

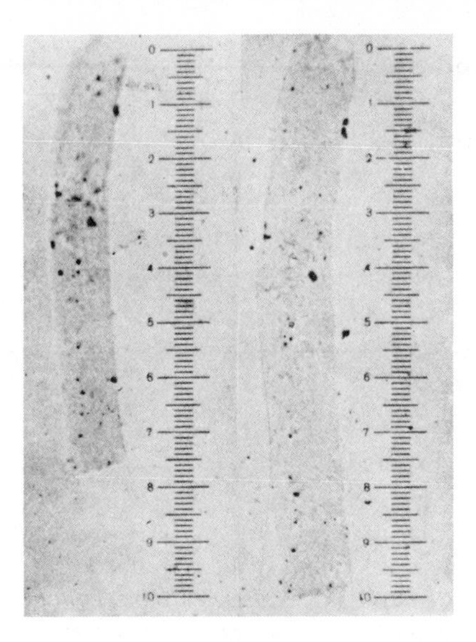

Fig. 4.

weak and slow contraction. (Fig. 3 curve b.) If the myosin solution is filtered twice through a Seitz K filter before the thread has been pulled, the contraction becomes still weaker (Fig. 3 curve c). As shown in the previous paper, the Seitz filter retains the myosin B present in our myosin A preparations as an impurity.

The fresh, watery extract of the muscle contains thus something which causes a violent contraction in myosin B threads but has little influence on myosin A. The active agent seems to be present in excess for the extract can be diluted to 1: 4 with water and will still give the same contraction with thinner threads.

If the muscle suspension is stored overnight at 0°C and filtered only the next day, the extract obtained will found to be entirely inactive. The myosin B thread, suspended in this extract, will show no change at all.

Rabbit muscle contains on the average 3.5 mg adenyltriphosphate (ATP) per g., thus the fresh extract contains ATP in about $\frac{1}{2}$ of this concentration. This ATP is split during storage by the phosphatase present. If the original ATP concentration is restored to the inactivated extract, again the same violent contraction will be obtained as in the fresh extract. Even half of this ATP concentration (0.9 mg per ml) is sufficient to give a maximum effect. This shows that ATP is involved in the observed contraction.

If the same quantity of ATP (0.09%) is dissolved in water and the myosin B thread suspended herein, no contraction will occur and the thread remains entirely unchanged. If we dissolve our ATP in the boiled extract instead of water we obtain a violent contraction again. Even incinerated juice will produce contraction with ATP. This makes it evident that, apart from ATP, inorganic constituents of the extract are also involved in the reaction.

If we use a 0.1 mol KCl solution as solvent for our ATP instead of water the contraction will be much slower. (see Fig. 5 curve b.)

Muscle contains 0.01 mol. Mg. If we add 0.01–0.001 mol. $MgCl_2$ to our

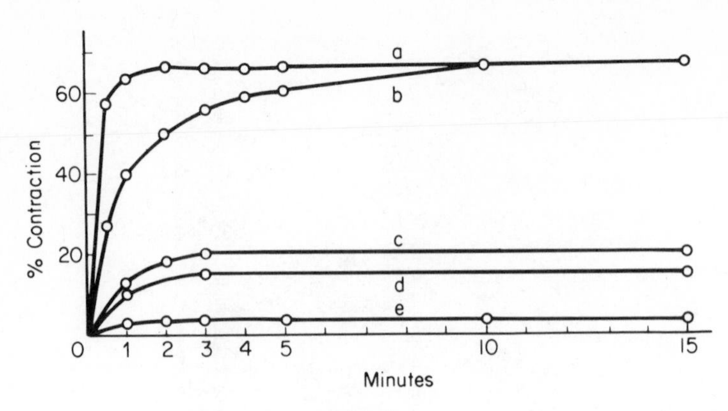

Fig. 5.

0.1 mol KCl and dissolve the ATP in this, the myosin thread suspended in this fluid will give the same violent contraction as in the fresh, watery extract (Fig. 5 curve a). It is evident thus that three factors were involved in the production of the contraction of our myosin B thread: ATP, K and Mg.

The myosin A thread will give the same weak and sluggish contraction (Fig. 5 curve c) with ATP in pure KCl or KCl plus $MgCl_2$. If the myosin solution is filtered twice through a Seitz K filter, the thread prepared from this solution will give a somewhat weaker contraction in KCl (Fig. 5. curve d). If, in addition to the 0.1 mol. KCl, 0.001 Mg is also present there will be practically no contraction at all (Fig. 5 curve e). The contraction of myosin A is not only not enhanced but is almost completely inhibited by $MgCl_2$. If the thread, prepared from unfiltered myosin, gave the same contraction in KCl and $MgCl_2$ (curve c) this was due to the myosin B present as an impurity: the contraction of this myosin B was enhanced and thus compensated the inhibition caused by Mg in the contraction of myosin A.

Myosin B gives thus a strong contraction with 0.1 mol. KCl and ATP and the contraction is greatly enhanced by Mg. Myosin A gives a weak and sluggish contraction with KCl and ATP and the contraction is suppressed by Mg.

If the concentration of KCl is increased, at 0.2 mol. the same reactions are still obtained. But if the concentration is raised to 0.04 mol., the thread, instead of giving a contraction, will dissolve. The action of our ATP—KCl—$MgCl_2$ mixture will depend on the concentration of the KCl present and at higher concentrations the action will be reverted; instead of contraction we will obtain dissolution and instead of aggregation, disaggregation. KCl without ATP will not dissolve the myosin B thread not even in a molar concentration.

Threads prepared from the precipitated, washed and redissolved myosin of these extracts gave identical results.

Adenylic acid, if employed instead of its pyrophosphate ester, ATP, was found to be entirely inactive. It does not give contraction or dissolution.

Neither the effect of KCl, nor that of $MgCl_2$ is specific and can be reproduced by other ions.

In Fig. 6 the effect of KCl is compared with the effect of other halogen salts of K. The abscissa gives the log of the molar concentration of the salt, the ordinate the % of shortening. By bringing the curve under the abscissa I wanted to express dissolution. The broken line means that the dissolution already takes place without the addition of ATP. The threads were prepared from precipitated and washed myosin B and were placed for 5 min. into the salt solution before the addition of ATP (0.018%). No value should be attached to the relative height of the curves, which, in this respect, are not strictly comparable because the results were obtained with different myosin preparations. Readings were made 5 min. after the addition of ATP.

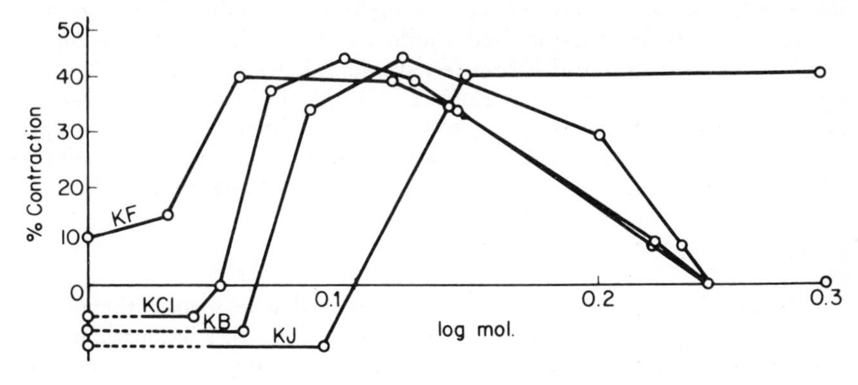

Fig. 6.

It will be seen that the effect of KCl is not specific and that the effect of a salt is not dependent on the kation only but depends on the anion too. With the increasing weight of the anion the curves are shifted more and more to the right. KJ has a very strong effect at relatively high dilutions.

In Fig. 7 the anion (Cl) is kept constant and the kation varied. The differences are not very marked but there is a tendency to the opposite effect, a shift to the right with the decreasing weight of the kation. With Li the effect is distinct. In this curve NH_4Cl (neutralised with NH_4OH) and potassium phosphate (pH 7) are also given. This latter is effective at very high dilutions.

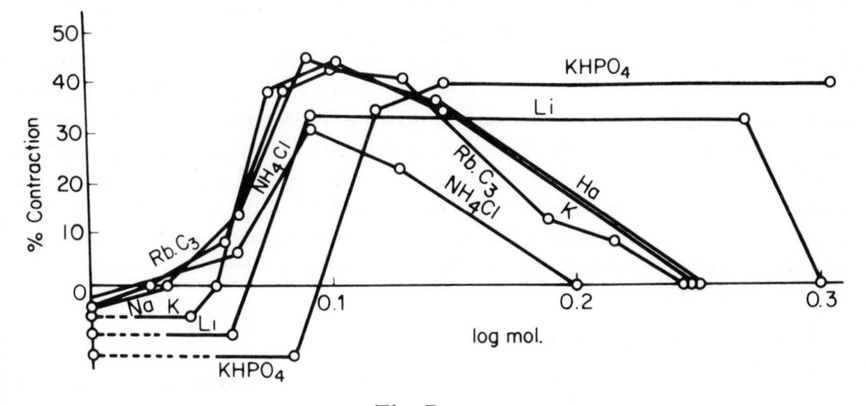

Fig. 7.

The enhancing effect of Mg can be reproduced by Co and Mn as will be shown by M. Gerendás. A detailed study on the effect of varied concentrations and pH will be given by T. Erdös.

Reversibility

The question presents itself whether the contraction observed is reversible or whether it is connected with an irreversible change of the myosin.

If the contracted myosin B thread is transferred into pure water it will remain contracted. This naturally does not mean that the change is irreversible, for the contracted muscle has no reason to relax. Conditions in the muscle, where every myosin micell is fixed within a certain pattern, are different.

If the contracted thread is transferred into Edsall's salt solution, it swells up again. When the thread has reached its original dimensions the swelling can be stopped by transferring it into 0.1 mol. KCl. Such a thread has a perfectly normal appearance and does not contract spontaneously. If ATP is added it contracts again in the same way, as it did the first time. This shows that the contraction is reversible.

Experiments with myosin in situ

We may ask whether the contraction obtained with myosin threads can have any bearing on muscular contraction at all and whether in muscle myosin may behave also as myosin B. It has been shown that myosin gives a contraction even after it has been dissolved and precipitated, once the necessary ions and ATP are present. The contractility of muscle is a very sensitive process and seems to be thus the expression of a subtle organisation. Naturally it is just as possible that a higher organisation is not needed for the contraction itself, but for those changes that elicit this contraction or bring the muscle back to rest again.

To obtain some information on this question, I tried to destroy the finer structure of the muscle as far as possible without destroying the myosin. It is known that the excitability of the muscle is lost in distilled water and also by feezing. Neither of these destroy myosin.

The broad neck muscle of the rabbit was cut into 2 mm. wide strips parallel to the muscle fibres. The strips were placed into distilled water. After one to several hours the strips, if contracted, were stretched to their original length, frozen in solid CO_2 and cut into slices on the freezing microtome. The slices were made parallel to the muscle fibres and were about one fibre thick. Thus they contained one sheet of muscle fibres running through the whole length of the preparation. The slices were put into distilled water and transferred after one to several hours into 0.1 mol. KCl, then placed on a slide under the microscope. After their length had been measured a drop of 0.14% ATP was dropped on them. Immediately a strong contraction began, which reached a maximum within 15–120 seconds and shortened the fibres by 50–60%. The myosin behaved thus as myosin B.

Experiments with myosin suspensions

The last question I want to touch in this paper is whether myosin suspensions give changes which are analogous to the contraction of threads.

The muscle extract containing myosin was neutralised and diluted till the

KCl concentration went down to 0.1 mol. The myosin precipitate was centrifuged, washed and redissolved in Edsall's fluid, precipitated and washed thoroughly again.

The myosin obtained in this way is a fairly stable suspension which settles slowly. Salts in smaller concentration cause precipitation and the suspension will settle somewhat faster. Salt in higher concentration will tend to dissolve the myosin. There is great difference in the behaviour of myosin A and B. The former is much less turbid and has a greater tendency for dissolution.

If, in addition to 0.1 KCl, a small quantity, say 14 mg % of ATP is also added to the myosin B suspension, the precipitation will be greatly intensified. The precipitate immediately becomes roughly granular and settles quickly leaving a clear fluid behind. The effect is very striking. We may call it a "superprecipitation," contrary to the precipitation caused by KCl alone. Mg still enhances the reaction.

4 mol. KCl has no appreciable dissolving action on the myosin B suspension. If ATP is added in addition to this KCl, the myosin dissolves. We can thus say that ATP greatly enhances the effect of salts, bringing about dissolution at concentrations at which the salt by itself is inactive and it also greatly intensifies precipitation.

The phenomena seen in the myosin B suspension are analogous to the phenomena observed on myosin B threads. ATP and higher KCl concentrations dissolve both. ATP and lower salt concentrations, which produce a contraction in the thread, cause a superprecipitation in the suspension.

The analogy is lacking in one point. While smaller salt-concentrations cause by themselves a precipitation in the suspension, salts without ATP never give a contraction in threads. There seems to be a qualitative difference between the precipitating action of salts alone and the precipitation observed in the presence of ATP. Salts alone seem only to cause an aggregation of the myosin micells, while in the presence of ATP they seem to cause some deeper change within the single units, which change expresses itself in the superprecipitation of suspensions and the contraction of threads. Gerendás has found that while salts by themselves have no influence on the double refraction of oriented threads, salts + ATP cause, besides contraction, a complete disappearance of double refraction. Double refraction disappears in muscular contraction also.

Myosin A suspensions behave in an analogous way to myosin A threads. They dissolve without ATP at lower salt concentrations (0.4 mol. KCl) and ATP has only a very slight precipitating action which is not enhanced by Mg.

Summary

It is shown that a myosin B thread, if suspended in a fresh, watery extract of muscle, gives a violent contraction. Myosin A is relatively inactive.

It is shown that three factors are involved in the contraction of myosin: ATP, K and Mg.

At higher salt concentrations, in the presence of ATP, dissolution is obtained.

The action of ions is not specific.

Under the same conditions which cause the myosin thread to contract, the myosin suspensions give a precipitate.

Literature

1. H. H. Weber: *Arch. ges. Physiol.* (Pflüger) **235**, 205, 1934.

Albert Szent-Györgyi sharing his thoughts with his brain trust at Woods Hole, Massachusetts in 1948. "In order to approach the central problems of biology, we have to extend our thinking in two specific directions, into both the sub- and the supramolecular."

ACTIN*

F. B. Straub

Institute of Medical Chemistry, University Szeged, Hungary

It has been shown by Banga and Szent-Györgyi[1] that myosin can be extracted from rabbit's muscle in two different forms.

Myosin A is obtained by extracting the muscle tissue with three volumes of Weber's solution† for 20 minutes in the cold and centrifuged immediately thereafter. The solution of myosin A obtained in this way is viscous and threads may be prepared from it. Neither the viscosity of the solution, nor the threads prepared from it, show any significant change on adding adenyltriphosphate (ATP).

If the muscle is extracted in a similar way for 20 minutes with the same solution, but left to stand for 24 hours in the cold and centrifuged only thereafter, a turbid and very viscous solution is obtained. On addition of ATP the viscosity of such a myosin B solution is decreased to a great extent. Threads prepared from myosin B show a vigorous contraction on addition of ATP and in presence of definite amounts of salts such as KCl and $MgCl_2$.

There was another difference between the two myosin modifications. Whereas both of them would split adenyltriphosphate with the appearence of free posphate, this enzyme action was increased by Mg ions in the case of myosin B, and not increased but rather inhibited in the case of myosin A.

*F. B. Straub, "Actin," *Studies from the Institute of Medical Chemistry, University Szeged,* **2**, 3–15 (1942). Reprinted with the permission of the author and of the publisher.

†This solution contains 0.6 M KCl, 0.04 M $NaHCO_3$ and 0.01 M Na_2CO_3. In previous communications from this Institute (Studies from the Institute of Medical Chemistry, Szeged, vol. **1**) this solution was referred to as "*Edsall*'s solution." Since then it was brought to our attention that this solution was first used by *Weber* and *Meyer*[2].

The problem of understanding the difference between these two modifications of myosin was taken up by studying the factors bringing about the transformation of myosin A into myosin B. These investigations led to the discovery of a new protein present in the muscle stroma. The name *actin* was given to this protein. In combination with myosin it gives the contractile protein of the muscle. As shown later in this paper, myosin B is formed if a certain amount of myosin A and actin are mixed. It follows that myosin A is what was termed by earlier investigators as myosin, myosin B on the other hand is a mixture of a definite amount of actin and myosin.

The ability of a myosin preparation to react with a decrease of viscosity on addition of ATP, was termed by Banga and Szent-Györgyi the "activity" of the respective myosin. It will be shown in this paper that apart from myosin B other mixtures of myosin and actin can be prepared, which show varying degrees of activity. Myosin B is only one of the possible combinations, its significance being only that if the muscle is extracted in the way described above, myosin B will be invariably extracted. But muscle does not contain myosin B, instead it contains an actin-myosin complex with a higher activity than myosin B. We therefore think it advisable to modify the nomenclature put forward by Szent-Györgyi[3] in such a way that *actomyosin* is generally a mixture or compound of actin and myosin, there being many possible actomyosins. One of them is myosin B.

Methods

The viscosimeters used in this work had the following measurements: capillary diameter 0.060 cm, length of capillary 210 mm, diameter of the cylindrical reservoir tube 1.65 cm, amount of outflowing fluid 1.2 = 1.7 ml. The viscosimeters have been placed in an icebath, which was vigorously stirred. The time of outflow of the solution was referred to the time of outflow of the solvent. No correction was taken for the change of specific weight by the presence of the proteins, as the protein content was maximally 3 mg/ml. All measurements have been performed using a buffered KCl solution of pH 7 as solvent. (Its composition see in the following paper of Balenovič and Straub.) 4 ml of myosin solution were placed in the viscosimeter. ATP was added in the form of its K salt, 0.1 ml of a 1.4% solution were added to 4 ml solution.

Determination of Actin in Solution

Any specific property of actin which is in any way proportional to its quantity, may be utilized for its quantitative determination. The combination of actin with myosin to form an actomyosin is such a specific property. The activity of the resulting actomyosin is the higher, the more actin is added to the myosin. It remains to define the measure of the activity of actomyosin. This is complicated by the fact that the decrease of viscosity on addition of ATP depends not only on the actin content but also on the viscosity of the

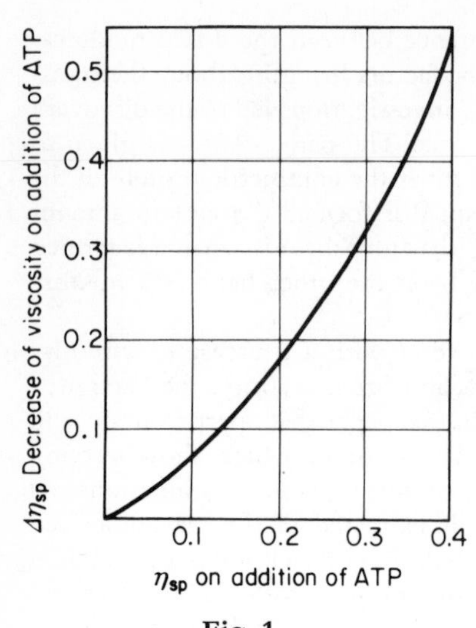

Fig. 1.

actomyosin. This is clearly brought out in Fig. 1. It shows the decrease of the specific viscosity ($\Delta\eta_{sp}$) on addition of ATP as the function of the specific viscosity in presence of ATP (η_{ATP}). The curve is valid for myosin B and it was constructed from the data of Balenovič and Straub (Fig. 1 of the following paper).

The activity of an unknown myosin solution is defined as the relation of its $\Delta\eta_{sp}$ to the $\Delta\eta_{sp}$ of a myosin B solution having the same η_{ATP} value.

As example let us take a myosin solution, which has the specific viscosity (η) = 0.625, in presence of ATP (η_{ATP}) = 0.35. Fig. 1 shows that in the case of myosin B to this value of η_{ATP} corresponds a $\Delta\eta_{sp}$ of 0.41. After the above definition the activity is

$$\frac{0.625 - 0.35}{0.41} 100 = 67\%$$

To a myosin A solution,* which contains 6 mg of myosin, and has no activity (no decrease of viscosity on addition of ATP) different amounts of an actin solution are added, the solution is made up to 6 ml so as to have 0.6 M KCl concentration and pH 7, and the activity of the resulting mixture is determined in every case. The results are plotted against the amount of actin as in Fig. 2. From the curve it is easy to obtain the amount of actin which is necessary to transform the 6 mg of myosin present in the experiment into myosin B (100% activity). From the purest actin so far obtained 1 mg protein is needed to transform 6 mg of myosin into a 100% active myosin. The actin content of an unknown actin solution can be therefore evaluated if we determine the amount of the solution necessary to activate 6 mg myosin from 0 to 100%. In this way the element of arbitrariness introduced by the arbitrary measure of activity, is again eliminated.

There are good reasons to believe that the actin preparations, of which 1 mg activates 6 mg myosin to a 100% active actomyosin, represent the pure actin. This means that myosin B is a compound of 6 mg myosin and 1 mg actin.

That myosin B is a compound of myosin and actin, is supported by the following experiment, in which actin was prepared from myosin B. In the usual way a myosin B was obtained, dissolved in Weber's solution. It was diluted with 5 volumes of distilled water and centrifuged. The precipitate was then treated first with 4 and then with one volume of acetone and left to dry at room temperature. By extracting the dried myosin with distilled water a viscous solution of actin is obtained, which does not contain any

*All experiments described in this paper have been performed using crude myosin A solution, obtained by extracting the muscle tissue with the Weber solution.

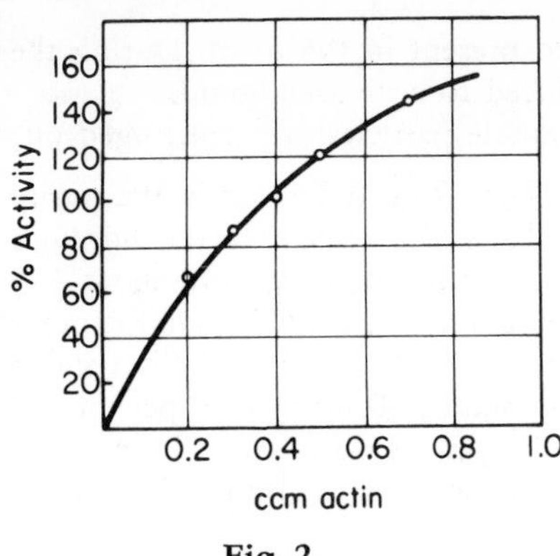

Fig. 2.

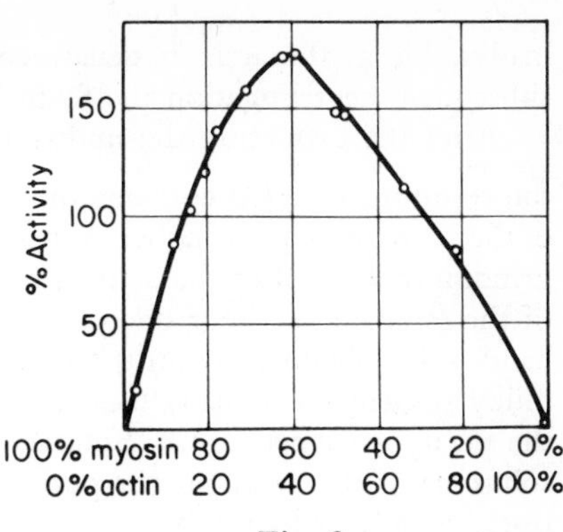

Fig. 3.

myosin. $\frac{1}{4}$–$\frac{1}{3}$ of the estimated actin contents of myosin B can be extracted in this way.

It is seen from Fig. 2. that on addition of more actin than necessary to bring about 100% activation, actomyosins may be obtained with more than 100% activity. Fig. 3. shows the extension of these studies. It is seen that addition of more and more actin, i. e. increasing the ratio actin: myosin, results first in an increase of activity and later in a decrease of it. Finally, actin alone like myosin alone does not show any change on addition of ATP.

The existence of actomyosins more active than myosin B is rather important. It will be described in the following paper that muscle contains an actomyosin with 170% activity. It is seen from Fig. 3. that this is the maximal activity to be achieved at pH 7.

Preparation of Actin

Rabbit's muscle from the legs and of the back are minced in a cooled Latapie mincer. To every 100 g are added 300 ml of Weber's solution and stirred mechanically for 20 minutes at 0°. The mixture is centrifuged and the supernatant solution discarded. The residue is left to stand in the cold room for one or two days. It is then stirred up with 5 times its volume of distilled water and centrifuged again. The solution is again discarded and the residue mixed with 4 volumes of acetone. After 10 minutes the acetone is sucked off and replaced by a fresh lot of 1 volume of acetone. After 10 minutes this is sucked off again and the residue is spread over a filter paper and left to dry at room temperature overnight.

The acetone dried muscle is extracted with 20 volumes of distilled water. The extraction can be carried out in either of the two following ways:

a. The dried muscle is mixed with the distilled water and then left to stand in itself without stirring for 4 hours at room temperature.

b. The dried muscle is ground with the distilled water in a mechanical mortar. In about $\frac{1}{2}$–1 hour, depending on the amount to be extracted, the mixture will become a rather rigid mass of foam. Further grinding is then inadvisable as the actin is denatured when present in the foam. During the subsequent centrifugation the foam is reduced to a viscous liquid.

After the extraction the undissolved muscle particles are centrifuged off; the resulting solution contains the actin. There is no difference in the purity of the actin obtained by either of these procedures. But whereas by the thorough grinding nearly all of the actin is extracted, by procedure (a) only about $\frac{1}{6}$–$\frac{1}{4}$ of the estimated actin of the muscle can be obtained in solution. On the other hand, actin obtained by (a) is a clear solution, that obtained by (b) is a stable milky suspension. Usually these solutions contain 3–4 mg protein per ml and the purity is such that 1.5–2 mg activate 6 mg myosin to 100%. From these solutions of the actin further purification can be achieved by precipitating it isoelectrically or by Ca ions.

The actin solution, obtained by procedure (a) is diluted 8 times by distilled water and then an acetate buffer of pH 4.8 is added. The buffer concentration will be 0.01 M. The precipitate is centrifuged off at room temperature and dissolved by the addition of bicarbonate to neutral reaction.

Actin can be precipitated by adding $CaCl_2$ solution to the actin solution. The amount of Ca necessary to precipitate the actin however depends largely on the concentration of univalent ions already present. Thus about 0.02 M $CaCl_2$ is needed to precipitate the actin of the first extract (which still contains some of the KCl of the Weber's solution), but 0.002 M or less is sufficient to give complete precipitation from isoelectrically purified actin solutions, which are poor in salt.

The Ca-precipitate of actin is inactive. Its activity will however return if KCl is added to the solution. At the same time it is observed that the Ca precipitate will form a stable suspension if KCl is added to it. To ensure this dissolution and reactivation, 40–50 times as much K^+ ion should be added as there was Ca^{++} ion present.

Obviously, the facts recorded here, can be described as a Ca: K antagonism. As the action of Ca is very pronounced and is observed with small, physiological concentrations, moreover the ratio Ca: K, where the effect of Ca is abolished is close to the physiological values, further, considering that actin is a part of the contractile element of the muscle, we may suppose that these observations are related to the Ca: K antagonism in muscle physiology. Further work is planned along this line.

It should be mentioned here that not only Ca^{++}, but also Mg^{++}, Mn^{++} or Sr^{++} ions can precipitate the actin in the same way. On the other hand the effect of Ca^{++} can be overcome by Na^+ ions just as well as by K^+.

It appears that the colloidal state of actin does not materially influence its activity. It was mentioned before that a clear solution obtained by extraction (a) activates myosin to the same extent as the suspension obtained by extraction (b), if both have an equal protein content. For this reason we prefer to prepare actin by the first method and such preparations have been used throughout this study. If the clear actin solution is treated with acid to precipitate it at pH 5–6, when dissolved at pH 7 the actin does not form a clear solution any more, but is more or less cloudy. When precipitated with Ca and redissolved with KCl, it is always a suspension.

The isoelectric precipitation results in removing 10–15% of the protein and the activity of the preparation is raised sometimes by more than 50%. Whether this is due to the removal of inhibitors or some other factor could not be decided. An actin, purified through the Ca-salt, has the maximal activity and all of its protein is precipitated at pH 6 or by new addition of Ca.

As other purification procedures, like precipitation with alcohol or salting out with KCl, did not lead to further purification and because all of the protein was precipitated by minimal Ca concentrations, we think that such preparations are homogeneous and do not contain any significant amount of impurities.

Actin is easily destroyed by heat over 50°. It is stable in a narrow pH zone between pH 7–7.5 but is rapidly destroyed even at 0° by more acid and alkaline reactions. It is precipitated from the solution at pH 6. Dialysis at pH 7 does not diminish its activity. From the solution it can be precipitated by cold alcohol but it is completely destroyed if treated with cold acetone. (This is the more interesting as it is very resistent to acetone when still in the muscle tissue.) If precipitated without loss of activity, the precipitate is voluminous. After centrifuging for 20 minutes at 3000 r.p.m. it still contains more than 99.5% of water.

Thixotropy

The most remarkable property of actin solutions is their very strong thixotropy. A dilute solution of neutral actin containing as little as 3–5 mg protein per ml sets to a gel if left alone at 0° for a few hours. A slight shaking however breaks up the gel immediately. As common with thixotropic gels, the viscosity of actin solutions cannot be determined with accuracy. So much can be said only that it is roughly the same as that of myosin. The time of outflow is determined largely by the shearing forces present in the capillary viscosimeter. For this reason successive determinations of the time of outflow show strongly decreasing values. If the solution is left to stand for a while before the next determination, again a higher value of viscosity will be found. A quantitative study of these phenomena is planned in a system, in which the pressure can be taken into consideration.

Strictly speaking it would be better to use the term resistance (Strömungswiederstand) instead of viscosity. The resistance of a solution results from the true viscosity, which can be measured at high shearing forces and from the elasticity of the solution. That indeed elasticity plays an important part is shown by the fact that the viscosity increases if the time of outflow is longer, i. e., the shearing forces are smaller. The resistance of actin solutions decreases assymptotically with higher shearing forces. It is interesting to point out two parallelisms between actin and thixotropic inorganic sols. One is the permanent double refraction to be dealt with later. The other is the great influence, which inorganic salts exercise on the resistance (apparent viscosity) of actin solutions. It is known from the experiment of Freundlich[4] and the theoretical treatment by Szegvári[5] that the elasticity but not the viscosity is influenced by the salt concentration.

Not only the viscosity of actin, but also the viscosity of actomyosin is influenced by the strong thixotropy of actin. Whereas myosin A solutions show no variation in viscosity during successive determinations, myosin B shows already such an effect, a second run in the viscosimeter gives always lower value than the first one. This results in some ambiguity in the determination of viscosities. The more active an actomyosin is, the less reliable is the determination of viscosity. Therefore we have accepted a certain routine, which would give comparable results. The actin solution was kept at 0° for at least 1 hour prior to use, then it was gently shaken up and the desired amount mixed with the myosin, the mixture immediately put into the icebath and its viscosity determined 5 minutes later. Immediately thereafter two more readings were taken. After the addition of the ATP, two more determinations were made. For the calculation of the activity the first readings were always used. The others served only to see the extent of the thixotropic effect. The solutions are to some extent thixotropic even after the addition of ATP.

Whereas the difficulties caused by the thixotropic effect make the viscosity values uncertain, the determination of activity by this routine procedure can be carried out with satisfying reproducibility, except for actomyosins containing more actin than myosin.

A thixotropic effect with salt free myosin has been observed already by Muralt and Edsall(6). It is clear that these authors dealt with a myosin A preparation, that is an actomyosin of a few per cent activity. This is shown by the very small dependence of the viscosity of their solutions on the shear rate. Freundlich(7) has suggested that the interior of the muscle cell is a thixotropic gel, as evidenced by the observation of Kühne(8). Knowing that the myosin in the muscle fibril is not the myosin studied by Muralt and Edsall, but an actomyosin of about 170% activity, which is a rather strongly thixotropic substance, this assumption is now experimentally supported.

Thixotropic behaviour is an expression of certain properties of actomyosin, which might be of considerable significance in understanding the architecture and the mechanical properties of muscle.

If we assume that the most active actin, described in this paper, is the pure actin, it was calculated (see Balenovič and Straub) that 1 gramm of muscle contains 25–30 mg actin. It is impossible to make such a concentrated solution of actin, which requires a very close packing. Actin reveals by its thixotropy and permanent double refraction (see later) very strong intermolecular forces. At the close packing present in muscle, these intermolecular forces must be very strong and must play an important rôle in determining the mechanical properties of muscle.

High viscosity and thixotropy are two phenomena each pointing to the fact that actin has elongated, rod-shaped molecules of very great assymetry. Double refraction of flow is another typical characteristic of actin solutions, bearing evidence to the same point. The sign of double refraction is identical with that of myosin. Myosin already has a strong double refraction of flow, which appears at very low shearing forces. But its double refraction is far weaker than that of actin solutions. A slight movement of the actin solution gives rise to the appearence of doubly refracting patches, which persist long after the fluid had ceased to move. If there is so much actin in solution that it turns into a thixotropic gel, the double refraction is permanent. It is interesting to remember that according to Freundlich and Schalek(4), thixotropy is observed among those inorganic sols (like aged V_2O_5 sols) which show a permanent double refraction.

Of the physicochemical properties touched upon in this paper, viscosity and double refraction of flow show a parallelism with the actin content of the solution.

The Process of Activation

If actin is mixed with myosin in solution, an actomyosin will be formed instantaneously. When in contact with muscle, myosin will be activated at 0° only in the course of many hours. It has been pointed out by Banga and Szent-Györgyi[1] that there is no activation at all, until the ATP of the muscle is split. That the disappearance of ATP is not the only factor involved in determining the rate of activation, is demonstrated by the following experiment. Rabbit's muscle is extracted at 0° several times in succession with Weber's salt solution. By 3–4 extractions, practically all of the myosin is extracted. On adding an ATP-free myosin solution to the residue, which is by now likewise free of ATP, activation of myosin takes again nearly 24 hours at 0°. The rate of activation can be, however, strongly increased by grinding the muscle residue with sand, prior to the incubation with myosin. In this case activation will be complete within several minutes. From these experiments it follows that the rate determining process of the activation in muscle is the extraction of the actin from the muscle residue.

In absence of myosin, the Weber solution will extract no actin from the muscle residue. The presence of myosin is therefore essential for the extraction of actin. In absence of ATP, as shown by Banga and Szent-Györgyi[1], neither myosin, nor actin are extracted.

To account for these phenomena, we must assume that the actomyosin present in the muscle, dissociates into its components in presence of ATP and the salts of the Weber solution. The myosin is dissolved but the actin is retained by the strong intermolecular forces. After the ATP has disappeared, the myosin already in solution forms a compound again with the actin. By forming this compound, the forces binding actin in its place are overcome and the dissolved myosin dissolves the actin.

The only other way, by which I succeeded to liberate actin is the acetone treatment described above. But even acetone is capable of breaking up the structure only if the latter has been loosened up by a prolonged treatment with the alkaline salt solution.

Actin is not extracted by the usual salt solutions which have been used to study the distribution of muscle proteins. Therefore it is clear that only a fraction of it goes into solution (this gives the few % activity of myosin A), whereas the greater bulk of it remains in the so called muscle stroma. This insoluble protein fraction has been estimated as 15–20% of the muscle proteins. We find (see Balenovič and Straub) that actin represents about 12–15% of the muscle protein. Making an allowance for some connective tissue and the nucleoproteids of the nuclei there is not much protein left unaccounted. We therefore find that there is no other protein left in the muscle stroma, to which the rôle of a structure-protein could be assigned. The possibility that actin is the main structure-protein, receives strong support by its properties, discussed above.

The Formation of the Myosin-ATP Complex

It has been shown by Mommaerts[9] that about one gramm molecule of ATP is needed to give a maximal effect with 100.000 gramm of myosin. From the experiments of the author[10] on myosin A at acid reaction, it became obvious that ATP is bound to the myosin part of the actomyosin complex. This is supported also by the fact that myosin is the carrier of the adenyltriphosphatase activity.

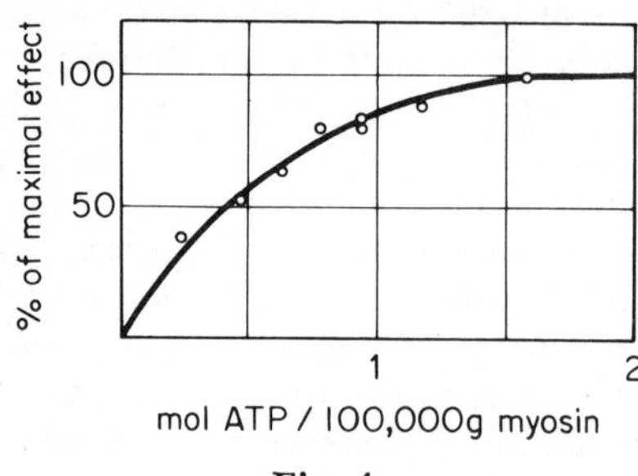

Fig. 4.

On a 100% active actomyosin, prepared from myosin A and actin, I have reinvestigated this problem. As Mommaerts' experiments were performed at the alkaline reaction of the Weber solution, it was important to know the conditions at physiological pH. The viscosities have been determined at pH 7.0 in a solution containing 0.6 M K ions and veronal acetate buffer. The decrease of viscosity on addition of ATP was measured and the results are expressed as % maximal effect and plotted against the amount of ATP in Fig. 4. 0.1 ml ATP solution was added in each case to the 6 mg of actomyosin present in 4 ml salt solution. From four successive determinations after the addition of ATP, the viscosity of the solution at the moment of adding the ATP was determined by linear extrapolation. It is evident, from the nature of the curve that the bond between myosin and ATP is very strong, less than 2 g molecules of ATP being needed for 100.000 g of myosin B to give a maximal effect. There is some dissociation of the myosin-ATP complex, but the results are essentially in agreement with the assumption that 1 g molecule of ATP reacts with 100.000 weight of myosin.

It is quite convenient to work with 6 mg of myosin in the 4 ml of the solution placed in the viscosimeter. In this case about 0.015 mg ATP would give a 50% effect. In our experiments such an amount of ATP caused the viscosity of the myosin to decrease from 1.83 to 1.61. Such a change in viscosity can be determined with satisfactory precision. Considering the absolute specificity of ATP in decreasing the viscosity of myosin, these facts may be utilized for determining very small amounts of ATP.

Summary

A method for extracting a new protein, actin, from muscle tissue is described. Actin forms, together with myosin, actomyosin, the contractile protein of muscle. The most conspicuous properties of actin are its high viscosity, thixotropy and strong double refraction, all proving a great molecular asymmetry.

The antagonistic effect of Ca and K ions on actin is described.

References

1. I. Banga and A. Szent-Györgyi, *Studies from the Institute of Medical Chemistry, Szeged,* **1**, 5 (1942).
2. H. H. Weber and K. Meyer, *Biochem. Z.,* **266**, 137 (1933).
3. A. Szent-Györgyi, *Studies from the Institute of Medical Chemistry, Szeged,* **1**, 67 (1942).
4. H. Freundlich and E. Schalek, *Z. physik, Chem.,* **108**, 153 (1924).
5. A. Szegvári, *Z. physik. Chem.,* **108**, 175 (1924).
6. A. L. v. Muralt and J. T. Edsall, *J. biol. Chem.,* **89**, 315 (1930).
7. H. Freundlich, *Protoplasma,* **2**, 278 (1927).
8. A. Kühne, *Virchow Arch. path. Anat.,* **26**, 222 (1863).
9. W. F. H. M. Mommaerts, *Studies from the Institute of Medical Chemistry, Szeged,* **1**, 37 (1942).
10. F. B. Straub, *ibid.,* **1**, 43 (1942).

ON THE SPECIFICITY OF THE ATP-EFFECT*

F. B. Straub

Institute of Medical Chemistry
University Szeged, Hungary

Actomyosin, in 0.6 *M* KCl, is split into actin and myosin by the addition of small amounts of ATP. This is evidenced by the fall of the viscosity, the original high viscosity returning if the ATP is removed. We find that the same effect is also produced by inorganic pyrophosphate. This is shown in Table I.

Table I

Added substance	Relative viscosity of actomyosin in 0,6 *M* KCl at 0°
None	1.70
0.0007 Mol/lit ATP	1.28
0.00002 " "	1.28
0.0006 " Na-pyrophosphate	1.28
0.00006 " "	1.29
0.00001 " "	1.375
0.000003 " "	1.425

Adenylic acid, ortophosphate and metaphosphate were found to be without any effect. ADP did not show any effect if pure myosin and actin were used for the preparation of actomyosin; it was, however, active if impure myosin

*F. B. Straub, "On the Specificity of the ATP-effect," *Studies from the Institute of Medical Chemistry, University Szeged,* **3**, 38–39 (1943). Reprinted with the permission of the author and of the publisher.

This work was aided by a grant from the Duke of Esterházy.

and actin were used. This is most likely due to contamination of myosin in the latter case by the enzyme system, which brings about the transformation of ADP into some other compound with a pyrophosphate group.

There is some difference between the effect of ATP and inorganic pyrophosphate. Whereas the effect of ATP is almost instantaneous, the effect of pyrophosphate takes some time to develop, especially so at smaller pyrophosphate concentrations. As pyrophosphate is not split by the myosin, its effect is permanent, in contrast to the effect of ATP, which is soon abolished, due to the splitting of ATP by the myosin. If, however, the actomyosin, to which pyrophosphate was added, is precipitated and washed with dilute saline solution, pyrophosphate can be removed and the actomyosin shows again a high viscosity. Thus the effect of inorganic pyrophosphate is reversible.

The action of pyrophosphate depends on the temperature. At 0° and at 6.5° it acts like ATP in very dilute solutions. At 23° however, inorganic pyrophosphate has no effect at all, even in 3.10^{-3} *M* concentration. ATP is fully active not only at this temperature, but at 37.5° also.

If 6.10^{-4} *M* inorganic pyrophosphate is added to a solution of actomyosin at room temperature (22°), there will be no change in its viscosity. If this solution is cooled to 0° its viscosity will become low, just as if ATP would be present. Bringing the solution back to 22°, its viscosity will rise again and can be lowered by the addition of ATP.

From these data it appears that it is the pyrophosphate group of the ATP which is responsible for the viscosity effect. It might therefore be concluded that adenosinediphosphate, prepared through dephosphorylation of ATP by myosin, does not contain a pyrophosphate residue.

It may also be concluded that the splitting of ATP is not involved in its viscosity decreasing effect, since pyrophosphate has the same effect and is not split by myosin.

Dr. T. Erdös has found in this laboratory (unpublished) that on addition of inorganic pyrophosphate, actomyosin threads do not contract at 1.3°, whereas addition of ATP brings about their contraction.

THE ATP MOLECULE*

Albert Szent-Györgyi

Institute for Muscle Research
at the Marine Biological Laboratory
Woods Hole, Massachusetts

Looking at the conventional structure formula of ATP (Fig. 1), one's first impression is that of great complexity. Nature does not indulge in luxuries, so one may wonder why the cell uses such a complex molecule if a P—O—P link is all that is needed. A much simpler inorganic polyphosphate should do just as well.

The molecule has two ends: a phosphate-end and a purine-end. The phosphate-end represents the energy store, (E); one may ask whether the purine-end may not represent E^*, thus providing the molecule with the essential parts needed for the $(E) \longrightarrow E^*$ transformation. The purine contains an extensive system of conjugated double bonds with its nonlocalized π electrons and five N-atoms, each with its lone pair of electrons. As will be discussed later, under conditions, this end of the molecule may also become strongly fluorescent and thus conform to our demands of an E^* transmitter. The purine may thus be instrumental in transforming the (E) of $\sim$P into E^*, when this (E) has to go into biological action and drive the living machine. The whole ATP molecule could thus be not only a storage battery but also a transformer.

What is difficult to see in Fig. 1 is how the energy accepted by the purine-

*A. Szent-Györgyi, "The ATP Molecule:" in *Bioenergetics*. New York: Academic Press, Inc., (1957), pp. 64–73 (Chapter 10). Reprinted with the permission of the author and of the publisher.

This material was presented at the International Enzyme Symposium, Ford Foundation, Detroit, November 1–3, 1955, and published in "Enzymes: Units of Biological Structure and Function," Oliver H. Gaebler, Ed., Academic Press, New York, N.Y.

end could be transmitted to the phosphate-end, and *vice versa,* since the two are separated by the pentose, which has no conjugated double bonds and no π electrons.

But does the ATP molecule really have this structure? Are we not misled by the visual impression made by a structural formula, the correctness of which we fail to question only because we have seen it too often?[1] The ATP molecule has one C—N, one C—C, and one C—O link (marked in Fig. 1 with arrows), which allow a free rotation so that the molecule has a limited freedom to change its shape and curl up. In fact, the outstretched linear form

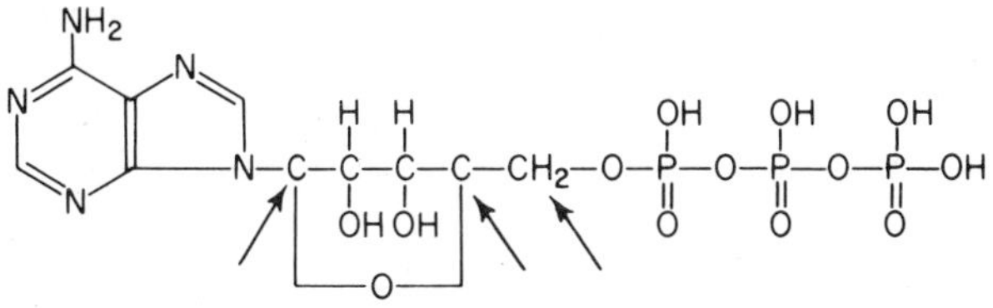

Fig. 1. ATP.

[1] Possibly, chlorophyll also offers an example for such an influence of visual pictures which inhibit our thinking in certain directions. Relations between chlorophyll and carotenes have been sought for a long time without definite results.

The Mg-porphyrin part of chlorophyll is reproduced in Fig. 2a, while the chain of beta carotene is shown in Fig. 3a. The two structures seem to show no relation, the first consisting of pyrrols, coordinated by a metal, while the latter is built of isoprenes. However,

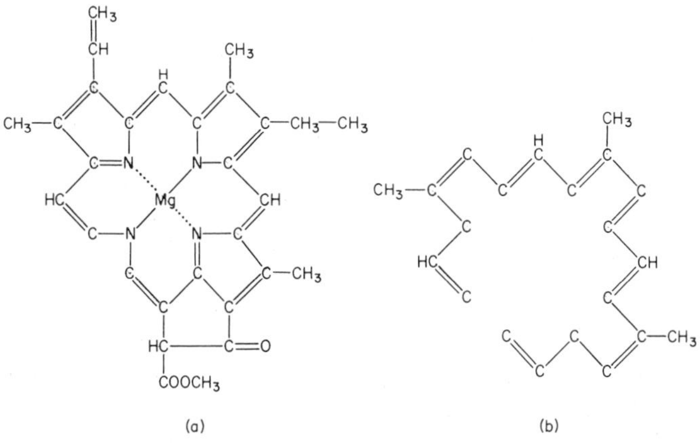

(a) (b)

Fig. 2. (a) Mg-porphyrin part of chlorophyll without the phytol side chain. (b) Same as (a) with Mg, N's, and longer side chains eliminated.

of the molecule in Fig. 1 is an improbable one and the possibility is open that by rotation and the consecutive folding the phosphate and the purine-end of the molecule may approach one another. Naturally, for an energy transfer to take place between the two ends, an "approach" is not enough, since the E of the $\sim$P which has to be transferred is an (E), that is, a bond energy which has no outward action. For the transference of (E) there must be a close fit, point counter point. The atoms which have to touch one another are two O^-'s of the dissociated OH groups of the phosphate chain, and two N's of the purine. The two O^-'s are those lying on either side of the terminal $\sim$P, which has to be split and give up its (E), while the two N's of the purine would be, in all probability, the N of the NH_2 group at position 6 (since this N seems to be most involved in the reaction occurring in muscle contraction), and its second neighbor at position 7.

However, free rotation does not mean free motion. Such a C—C bond does not permit the molecule to bend in any way it pleases, for the valency angles

---

if we eliminate the Mg with the four N's, as well as the longer side chains from the porphyrin, we are left with a chain (Fig. 2b) which is almost identical with the chain of carotene, curled up (Fig. 3b).

If these relations were hitherto not considered, this is probably due to the fact that we have seen the structure formula of pyrrol too often.

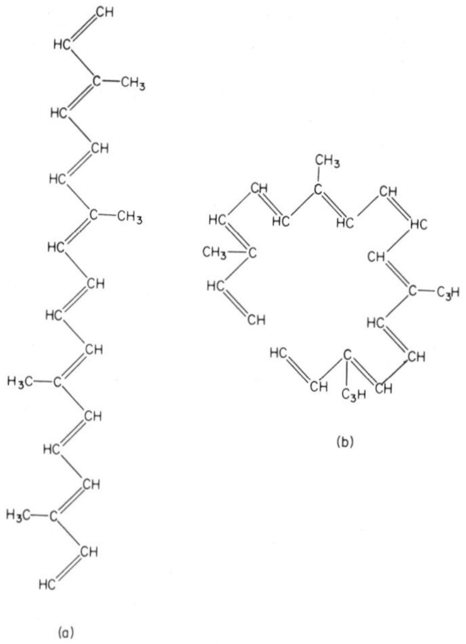

(a)

(b)

Fig. 3. (a) Central chain of carotene without the ionone rings. (b) Same curled up.

have to be kept constant. The motion is not freer than that of the relative motion of two wheels mounted on the same axis. So even with three rotating links the freedom of the ATP molecule is a very limited one and, statistically, the chances that the NH_2 and N_7 would be able to meet the two O^-'s and make a close fit, are very remote. So if the four could meet and make a close fit, it would be probable that this is not mere chance, but has a functional meaning and that the ATP molecule is *made that way and does not happen to be that way*. Whether such a meeting is possible can be decided by building up the molecule of an atomic model which keeps account of atomic radii and limitations of freedom, the rigidity and small flexibility of bond angles. Such a model is the Courtauld atomic model.[2] Figure 4 shows an ATP molecule built up of this model, in its conventional linear form corresponding to Fig. 1. (The molecule is here in its dissociated form, with O^-'s instead of OH's on the phosphates.)

If the molecule is now rotated around the C—C, C—N, and C—O bonds, then the phosphate-end can be folded back so that the O's of the terminal and middle phosphate just touch the N's mentioned. This situation is shown in Fig. 5. If some sort of a link is formed now between the O^-'s and N's, then the P—O—P link, which has to be broken in contraction and supply the energy for it, has formed a ring with the purine. Naturally, for energy to pass from the phosphate to the purine, the contact would have to be an intimate one. H-bridges, possibly, formed between the O's and N's might do, having been shown by Gergely and Evans that H-bonding can establish relations in which π orbitals overlap. It is thus possible that the ATP molecule, activated by myosin, connects its two ends, thus opening the way for an energy transmission from one to the other.

This, however, does not explain the role of the bivalent ions, Ca and Mg, both of which can accelerate the ATP-ase activity of myosin. The possible answer to this problem was given by a chance observation. As discussed in Chapter 4, we may expect energy transmitters to be fluorescent. Since there are violent shifts in energy during contraction, the author expected to find in muscle a fluorescent energy transmitter in high concentration, and prepared alcoholic extracts of muscle, expecting them to show strong fluorescence under the UV lamp. They showed none. However, if a bivalent metal, as Mg, Ca, or Zn, was added (as chloride), an intense blue fluorescence appeared. The fluorescent substance was isolated, identified by McLaughlin, Schiffman, and the author and found to be the metal complex of inosinediphosphate, IDP, the substance produced from ATP by the loss of its terminal phosphate and its hydrolytic deamination.

The probable structure of the metal complex is elucidated by the close analogy of inosine and oxyquinoline (Fig. 6a and b). The latter is known to form with Mg in alcoholic solution a very stable, strongly fluorescent chelate (Fig. 6b). Evidently, an analogous chelate was formed by the inosine (Fig. 6a).

Mg and Ca are known to form very stable, coordinative complexes with

[2] Produced by Griffin and Tatlock, London, available in U.S. through the Anglo-American Scientific Import-Export Co., 185 Devonshire Street, Boston 10, Mass.

Fig. 4. (*above*). ATP of Fig. 1 built of the Courtauld atomic model.

Fig. 5. (*below*). Same as Fig. 2, folded.

polyphosphates, so the possibility is given that the metals form with their four coordinative valencies a quadridentate chelate, connecting the two ends of the ATP molecule. This structure is illustrated by Fig. 7. The model in Fig. 5 shows that there is just enough room left for an Mg between the two N's and two O's. The arrow in Fig. 6b indicates that the metal attracts an electron from the quinolinol (Leverenz). The same can be expected to happen also in the complex of inosine, and if the metal forms a quadridentate chelate linking up also with the phosphates, it may attract electrons also from the latter. The metal may thus serve as a bridge, across which electrons can pass from phosphate to purine. So the Mg actually not only may connect the two ends of a molecule; it may make one single, unique electronic system of the

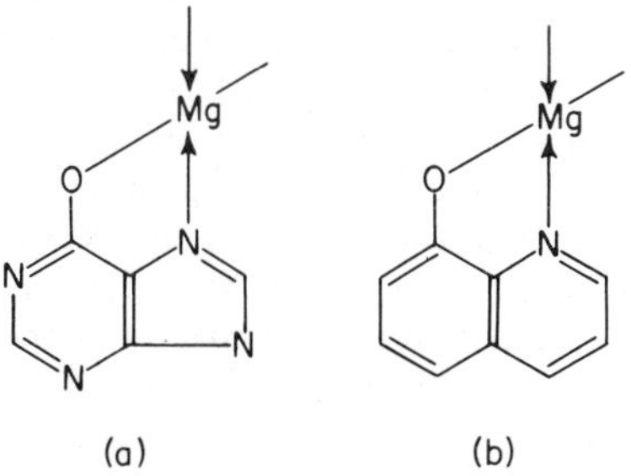

Fig. 6. (a) Mg complex of oxypurine. (b) Mg complex of oxyquinolinol.

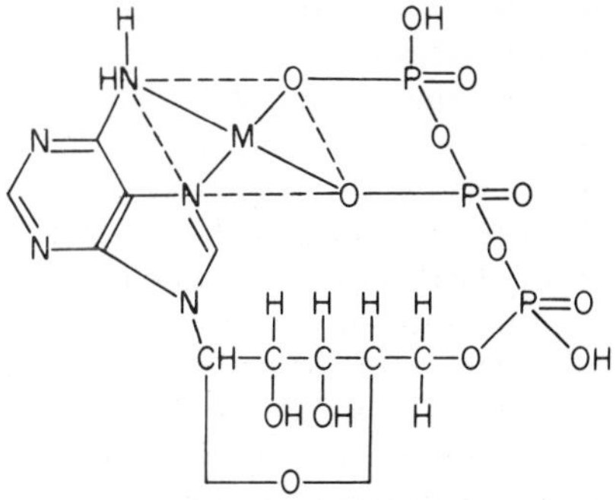

Fig. 7. Possible structure of Mg complex of ATP (interatomic distances are arbitrary).

phosphate chain and the purine with common nonlocalized electrons which could transport energy, the purine having its system of conjugated double bonds and the phosphate the O's with their nonbonded lone pairs of electrons. The P—O—P, which represents the (E), could merge thus with the adenine in one extensive system of mobile electrons.

This opens the possibility that when such a double chelate is formed and the metal attracts electrons from the phosphates, it decreases the energy and strength of the P—O—P bond, which then falls prey to hydrolytic splitting while its energy appears in the purine ring as E^*, completing the $(E) \longrightarrow E^*$ transformation.

That Mg can actually facilitate the passage of electrons from one substance to another with which this metal forms complexes can be demonstrated by mixing an alcoholic solution of 1,2-naphthoquinone and *o*-phenylenediamine (Fig. 8). In this system the quinone oxidizes the diamine very slowly, electrons passing from the latter to the former. This reaction is greatly speeded up by Mg, in analogy to the ATP-ase activity of myosin, which also can occur without Mg, but is greatly accelerated by the metal. The reaction between quinone and diamine is indicated by the darkening of the solution (which can readily be reverted by reducing agents such as ascorbic acid).

One attractive feature of this theory of ATP-ase activity is that it is analogous to E. L. Smith's theory of peptidase activity.

A serious objection may be raised against this theory: the stability constant of the metal chelate of Mg and the adenine or isonine, in water, is very low; the energy of this binding is not great enough to hold the whole quadridentate chelate together against the forces of heat agitation. The Mg-phosphate complex is very stable so that we can expect the ATP to be present in muscle as a Mg complex, there being twice as much Mg in muscle

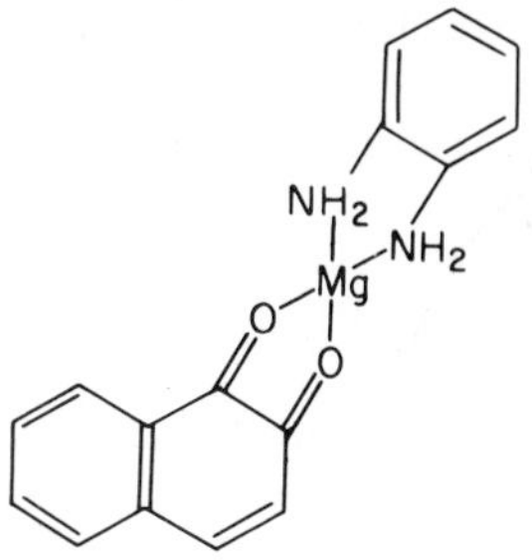

Fig. 8. Mg chelate of *o*-phenylenediamine and naphthoquinone.

as there is ATP, but links formed by Mg with the adenine or purine can be expected to break up in water. Our model in Fig. 5 only says that such a complex *can* be formed, not that it *is* formed. However, we have to remember that the ATP, when undergoing splitting in muscle, is not free but is linked to the protein, myosin, and is "activated" by it. We do not know what "activation" means. Probably it means a binding of the substrate with its consecutive deformation. So it may be that the myosin holds the ATP molecule in the position required for the formation of this bridge between phosphate and purine and this is actually what we mean by activation. One could go even one step further with the argument and say that if the stability constant of the Mg-purine complex were high enough to bring the two ends of the molecule together and make the energy transmission and the splitting of the $\sim$P occur, then this reaction would be of no use to the muscle, because then the $\sim$P would be split and its energy dissipated senselessly. It is one of the basic principles of nature not to use spontaneous reactions which occur by themselves and cannot be kept in hand. If the energetics of the cell consisted of spontaneous reactions, the whole mechanism would have to run down senselessly as a watch does if released from its regulators. We have to demand from any theory of the ATP-ase activity that it should make the splitting of $\sim$P possible only in ATP molecules, bound and activated by the myosin, when the energy of the $\sim$P can be transmitted to the protein and applied usefully. It is thus just the low stability constant of the hypothetic Mg-purine complex which makes the presented theory acceptable.

A word may be said about the third member of the ATP molecule, the pentose which has to connect the adenine and the phosphates in such a way that they come together exactly in the right position. A hexose would not do. But this may not be all there is to it. Nature often kills more than one bird with the same stone. It may be worth noting in this connection that the United States Patent Office has granted two patents for the hardening of gelatine by pentose, a reaction not shared by hexose.[3] This suggests that the ribose can enter into an intimate reaction with the protein which may very well play an important role in the "activation" of ATP.

One last remark may be made about Ca. It greatly promotes the "ATP-ase activity" without promoting contraction or analogous reactions, such as "superprecipitation." It even inhibits them. One tentative explanation which may be given for this behavior could be that Ca, similarly to Mg, forms chelates and weakens the $\sim$P but is unable to transmit its energy to the purine, owing to the great differences in the energy terms of Ca and N. Accordingly, Ca is also unable to promote the oxidoreduction between 1,2-naphthoquinone and *o*-phenylenediamine.

In Chapter 12 we will return once more to a possible additional role of pentose.

While reading the proof sheets of this book my attention was kindly called to a paper of B. H. Levendahl and T. W. James (*Biochim. et Biophys. Acta* **21**, 298, 1956). Its summary may be quoted without comment: "The rotatory

[3] USA Patent No. 2,059,817 (Nov. 3, 1956) and Patent No. 2,180,335 (Nov. 21, 1939), the first granted to the Eastman Kodak Co., Jersey City, N.J., the second to the Agfa Ansco Corporation, Binghamton, N.Y.

dispersion of adenine, adenosine, AMP, ADP, and ATP have been determined. The data have been interpreted to show that the ATP molecule is folded back on itself in such a manner that bonding is permitted between the last two phosphate groups and the amino group of adenine. This stabilized structure is then proposed as necessary for the action of ATP. Parallel 'ATP-like' action of CTP and UTP and other triphosphonucleotides could be explained by possession of a similar configuration."

REMARKS ON THE PHYSICO-CHEMICAL MECHANISM OF MUSCULAR CONTRACTION AND RELAXATION*

Jacob Riseman and John G. Kirkwood

Institute for High Polymer Research, Brooklyn Polytechnic Institute, Brooklyn, New York

Gates and Crellin Laboratories of Chemistry, California Institute of Technology, Pasadena, California

(Received for publication, April 3, 1948)

The physico-chemical processes underlying the contraction and relaxation of muscle have been the subject of much speculation. Recently significant analogies between the elastic behavior of muscle and that of rubber and synthetic elastomers have been investigated by Bull[1] and others. That the structure of striated muscle, and possibly smooth muscle, is much more ordered than those of elastomeric cross-linked high polymers is clearly demonstrated by the studies of Schmitt and his collaborators.[2] Nevertheless, important aspects of structural similarity exist. The basic structural unit of the muscle fibril is considered to be the linear polypeptide chain of the myosin or actomyosin molecule. Like the segments of a polymer network, the long polypeptide chain possesses many internal rotational degrees of freedom which allow it to gain configurational entropy on contraction. Therefore it is probable that the elastic modulus of a structure composed of such elements is in part determined by the dependence of their configurational entropy upon elongation.

The contraction or relaxation of a muscle segment under constant stress is the consequence of a change in the elastic modulus arising from alteration of its structural units. Following certain ideas suggested by the work of K. H. Meyer,[3] we wish to examine the hypothesis that the essential alteration of the structural unit, leading to a change in modulus, is a change in its net electric

*J. Riseman and J. Kirkwood, "Remarks on the Physico-Chemical Mechanism of Muscular Contraction and Relaxation," *J. Amer. Chem. Soc.,* **70**, 2820–21 (1948). Reprinted with the permission of the authors and of the American Chemical Society.

[1] H. B. Bull, *This Journal,* **67**, 2047 (1945).

[2] F. O. Schmitt, M. A. Jakus, and C. E. Hall, *Biol. Bull.,* **90**, 32 (1946); M. A. Jakus and C. E. Hall, *J. Biol. Chem.,* **167**, 705 (1947).

[3] K. H. Meyer, *Biochem. Z.,* **214**, 1 (1929).

charge. A structural unit consisting of a long polypeptide chain can gain or lose electric charge in several ways in response to changes in its physicochemical environment. If the absolute magnitude of the charge, whatever the sign, is increased, electrostatic repulsion between the elementary charges comprising the increment will decrease the elastic modulus by destroying the balance between the external stress and the contractile force arising from configurational entropy. Conversely, if the magnitude of the charge is decreased, the modulus will increase. According to this view, the relaxed state of muscle is an electrically charged state and the contracted state one in which the polypeptide chains are in an uncharged or "isoelectric" condition. We shall presently make some rough estimates of the change in elastic modulus produced by electrostatic repulsion between charges distributed at intervals along a polypeptide chain, after discussing possible mechanisms by which it might gain or lose charge.

Since a polypeptide chain is an ampholyte, it can gain or lose charge in response to a change in the pH of its environment. This mechanism was proposed by K. H. Meyer[3] as a basis for the analysis of the energetics of muscular relaxation and contraction. Meyer's proposal was rejected by Weber[4] on grounds which still must be regarded as inconclusive. Adsorption of cations, for example potassium, by actomyosin segments provides a second method of charging, which the work of Szent-Györgyi suggests may play a role.[5] Nevertheless, both of these charging mechanisms leave obscure the manner in which the chain of carbohydrate oxidation reactions supplies free energy for the muscular processes.

We are inclined to the view that phosphorylation of the hydroxy amino acid residues of the myosin or actomyosin molecule by adenosine triphosphate provides a charging mechanism in best accord with known facts. From its amino acid analysis, myosin is known to contain a large proportion of hydroxy amino acid residues; serine, 3.9%; and threonine, 4.95%.[6] Myosin is also considered to be one of the enzymes involved in the dephosphorylation of ATP to ADP and inorganic phosphate. Assuming that the first step in the dephosphorylation of ATP consists in the phosphorylation of the —OH groups of the hydroxy amino acid residues of the myosin molecule, we conclude that at pH of 7, the approximate pH of the sarcoplasm, each $—H_2PO_4$ group will be approximately singly ionized to $—HPO_4^-$. Thus the phosphorylation process would impart to the neutral sites originally occupied by —OH, approximately one unit of negative charge. The extension of the myosin chain and that of the structure of which it is the unit, resulting from electrostatic repulsion between the negatively charged $—HPO_4^-$ groups, stores up the free energy, released in the degradation of the high energy phosphate bond of ATP, in the form of negative configurational entropy of extension. Subsequent dephosphorylation of the myosin molecule with release of inorganic phosphate ion to the sarcoplasm would remove negative charge from the molecule and release the stored free energy as mechanical work in contraction

[4] H. H. Weber, *ibid.,* **217**, 430 (1930).

[5] A. Szent-Györgyi, "Chemistry of Muscular Contraction," Academic Press, Inc., N. Y., 30 (1947).

[6] M. L. Anson and J. T. Edsall, "Advances in Protein Chemistry," Vol. I, Academic Press, Inc., New York, N. Y., 1944, p. 310.

of the structure. The coupling between the carbohydrate oxidation process and the mechanical processes of muscle activity is thus clarified by the proposed mechanism. ATP, regenerated in the chain of oxidation reactions, serves as a carrier of free energy released in these reactions to the myosin units of the muscle structure.

In order to examine the quantitative implications of our hypothesis, we will make some rough estimates of the change in elastic modulus E of a flexible linear molecule produced by attaching electric charges of equal magnitude at equal intervals along its length. If the terminal groups of the molecule are separated by a distance L and n charges of the same sign and magnitude e are attached to the chain at points separated by equal numbers of bonds, the random coil model of a flexible linear molecule leads to the following approximate estimate of the increment in elastic modulus ΔE, produced by the charge increment ne

$$\Delta E = -\frac{8}{3}\frac{N\rho}{M}\frac{n^2e^2}{D_e L}$$

where D_e is an effective dielectric constant, N is Avogadro's number, M the molecular weight of the linear molecular unit of the structure and ρ is the density of the structure, regarded as a three-dimensional network with L the average distance between net points. Since electrostatic interactions between the charges of neighboring chain segments are neglected and other gross approximations have been employed in its derivation, Eq. (1) is intended to provide no more than an estimate of the order of magnitude of ΔE.

The elastic modulus of muscle[1] is of the order of magnitude 10^5 dynes/sq. cm. In order to produce a change of this order of magnitude by charging the structural units, we estimate from Eq. (1) a magnitude of n of the order of 100, with the rough assignments of value $\rho = 1$, $M = 10^6$, $L = 10^4$A, $D_e = 100$ to the other parameters. The values of L and M are those determined for the myosin or actomyosin molecule in solution, and D_e is assumed to be of the order of magnitude of the dielectric constant of water. The estimated number of charges would require phosphorylation sites situated at intervals of 100 Å. along the myosin or actomyosin molecule. This value is not inconsistent with the hydroxy amino acid content of myosin.

The observation of Needham[7] that the flow birefringence of myosin solutions is diminished by the addition of ATP seems at first to be in contradiction with our hypothesis. However, it seems that the effect was observed under conditions leading to the dissociation of actomyosin into actin and myosin, according to Szent-Györgyi.[5]

We have deliberately avoided placing undue emphasis on hypothetical structural details of muscle and on the detailed analogy between the elastic properties of muscle and elastomers. The essential qualitative aspects of our suggestions, (a) change in the elastic modulus of the structure due to alteration of the electric charge of the structural unit, considered to be a polypeptide chain rich in hydroxy amino acid residues; (b) charging of the structural unit through phosphorylation of the hydroxyl groups by ATP, are to a large extent independent of assumed structural details.

[7] J. Needham, Shih-Chang Shen, D. Needham, and A. S. C. Lawrence, *Nature*, **147**, 766 (1941).

TRYPSIN DIGESTION OF MUSCLE PROTEINS III. ADENOSINETRIPHOSPHATASE ACTIVITY AND ACTIN-BINDING CAPACITY OF THE DIGESTED MYOSIN*

E. Mihályi† *and Andrew G. Szent-Györgyi*

Institute for Muscle Research at the Marine Biological Laboratory
Woods Hole, Massachusetts

(Received for publication, July 1, 1952)

Myosin has the specific property of combining with actin to form actomyosin. This complex is reversibly dissociated into its components by adenosinetriphosphate (ATP)[1]. Myosin also seems to be associated with several enzymic activities[2-4]. Numerous attempts to separate the enzymatically active component from the myosin itself have failed[5]. Myosin is a very delicate protein and undergoes denaturation even under very mild conditions[6]. The first sign of this, when solubility and other physicochemical properties are hardly affected, is the loss of the specific activities such as actomyosin formation and the ATPase activity[7]. Enzymatically inactive myosin preparations have always been obtained by procedures connected, very probably, with the alteration of the myosin molecule[8]. On the other hand, water-soluble enzymes with ATPase activity have been described[9,10], but their specificity, activation, etc., are very different from those of myosin, which makes it improbable that the enzymic activity of myosin is due to the adsorption of these enzymes.

Myosin has been shown to split into two well defined components by a short digestion with trypsin[11]. By prolonging the digestion period these

*E. Mihályi and A. G. Szent-Györgyi, "Trypsin Digestion of Muscle Proteins. III. Adenosinetriphosphatase Activity and Actin-Binding Capacity of the Digested Myosin," *J. Biol. Chem.*, **201**, 211–219 (1953). Reprinted with the permission of the authors and of the American Society of Biological Chemists, Inc.

Sponsored by Armour and Company, Chicago, Illinois, and the American Heart Association.

†Present address, Harrison Department of Surgical Research, University of Pennsylvania, Philadelphia, Pennsylvania.

components are further degraded to smaller fragments not precipitated by trichloroacetic acid[12].

Gergely[13,14] has reported that trypsin digestion leaves the ATPase and deamidase activity of myosin unimpaired. Perry[15] similarly found that the ATPase activity of myosin remained intact long after its viscosity and the drop in viscosity on the addition of ATP in the presence of actin decreased. He reported that the anastomosed network characteristic of actomyosin also disappeared. However, Gergely and Perry did not differentiate in their investigations between the first rapid splitting of myosin and the following gradual breaking down of the split products. The investigations reported here were undertaken in order to see how far the specific activities of myosin are affected by the first, rapid, and second, slow, phase of the trypsin digestion, and whether both, or only one, of the primary split products possess the specific properties.

Experimental

Materials

Myosin, actomyosin, and actin were prepared as described in Paper I[11].

Commercial trypsin crystallized twice (with 50 per cent $MgSO_4$) and preparations of crystallized soy bean trypsin inhibitor were obtained from the Worthington Biochemical Laboratory.

An ATP preparation from the Armour Laboratories was purified by precipitation with $BaCl_2$ at pH 4 and converted into the sodium salt.

Methods

ATPase activity of digested myosin

Myosin was incubated with trypsin at 23° for 7 hours. The experimental procedure was as follows: 100 ml. of myosin solution of 5.8 mg. per ml. were mixed with 10 ml. of 0.5 M borate buffer of pH 8.5, and 10 ml. of trypsin solution of 2.5 mg. per ml. of protein were added. Samples of 10 ml. were withdrawn at given time intervals and mixed with 1 ml. of trypsin inhibitor solution of 5 mg. per ml. in order to stop digestion. Aliquots of the samples were used to determine the ATPase activity and the amount of non-protein nitrogen. A blank mixture was also prepared in which trypsin was previously mixed with the trypsin inhibitor and then added to the myosin solution. The blank was also kept at 23°; samples were taken at different time intervals and then kept at 0° until the activity determinations.

The ATPase activity of the samples was determined in reaction mixtures of 3 ml., containing 1 mg. of ATP and 0.22 to 0.44 mg. of native or digested myosin. Salt and buffer concentrations were 0.4 M KCl, 0.004 M $CaCl_2$, and 0.033 M borate buffer of pH 9.2. The samples were incubated for 5 minutes at 35°, then deproteinized, and the free phosphate in the filtrate was determined by the Fiske-Subbarow colorimetric method[16]. A myosin-free bank determination was made in all instances and its value subtracted from those obtained with myosin.

The amount of the digestion products not precipitated in 0.30 M trichloroacetic acid was determined by means of the optical density at 278 mμ of the filtrates. The details of the procedure were described in Paper II.

Fig. 1 summarizes the results, which show the ATPase activity and the amount of protein left in solution by 0.30 M trichloroacetic acid plotted against the time of digestion. The activity of the blank did not change appreciably during the course of the experiment. On the contrary, the activity of the digested samples decreased markedly, and, after 7 hours of incubation with trypsin, only approximately 10 per cent of the original activity was left. The formation of non-protein nitrogen proceeded at a much slower rate, and even at the end of the reaction period more than 60 per cent of the protein was still precipitated by 0.30 M trichloroacetic acid.

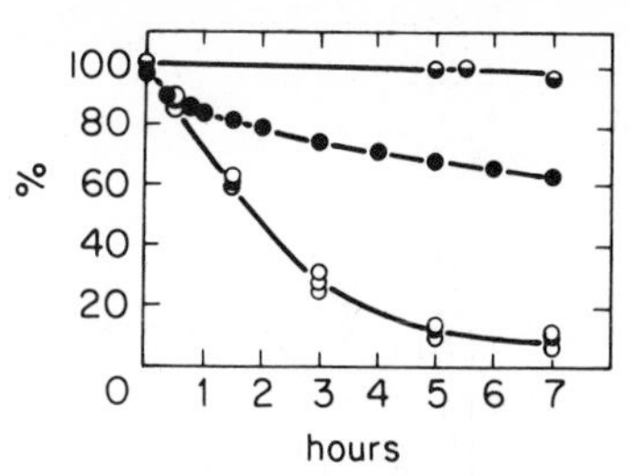

Fig. 1. ATPase activity and the amount of protein precipitated by trichloroacetic acid during the trypsin digestion of myosin. Ordinate, per cent of the original activity and myosin quantity; abscissa, incubation time in hours. ○, enzymic activity of the digested samples; ◓, enzymic activity of the trypsin-free blank; ●, amount of protein precipitated by 0.30 trichloracetic acid.

Under the conditions indicated above the first phase of tryptic digestion, *i.e.*, the splitting of the native myosin into two components, was complete in about 2 minutes[12]. As Fig. 1 shows, the ATPase activity only decreases by about 13 per cent even after 30 minutes digestion. This indicates that the first phase of tryptic digestion leaves the ATPase activity of myosin unimpaired.

Combination of digested myosin with actin

Myosin was digested up to a point at which, according to the ultracentrifugal analysis, no native myosin was left, and the amount of non-protein nitrogen formed was still negligible[11]. To 10 ml. of myosin solution of 15 mg. per ml., 1 ml. of 0.5 M phosphate buffer, pH 7.3, and 1 ml. of trypsin solution of 0.5 mg. per ml. of protein (dissolved in 0.0025 N HCl) were added. After an incubation of 10 minutes at 23° the digestion was stopped by adding 1 ml. of trypsin inhibitor solution of 1.25 mg. per ml. A blank mixture was also prepared with trypsin and trypsin inhibitor mixed before being added to the myosin solution.

The combination of myosin with actin was determined by the method described earlier by one of us[17]. To a constant amount of myosin increasing amounts of F-actin were added and the viscosity response of the samples upon ATP addition determined. In Fig. 2 the difference of the logarithm of viscosities before and after ATP addition (designated as $\Delta \log \eta_{rel.}$) is plotted against the amount of actin added. With native myosin $\Delta \log \eta_{rel.}$ increases linearly with the actin amount up to a certain value at which a marked break occurs. This point was considered as the point of saturation of myosin with actin. The digested myosin behaves similarly. As was expected, the rise in viscosity on the addition of actin and the viscosity drop caused by ATP are much smaller than with native myosin, but it is very significant that the break

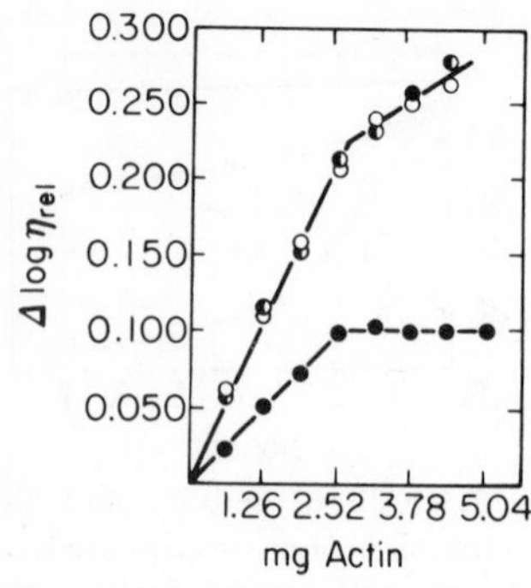

Fig. 2. Viscosity response of mixtures of actin and native or trypsin-digested myosin upon ATP addition. Ordinate, difference in the logarithm of viscosities before and after ATP addition; abscissa, mg. of actin added to 14.74 mg. of myosin in a total volume of 6 ml. ○ and ◐, trypsin-free and trypsin plus trypsin inhibitor-containing blanks; ●, trypsin-digested myosin.

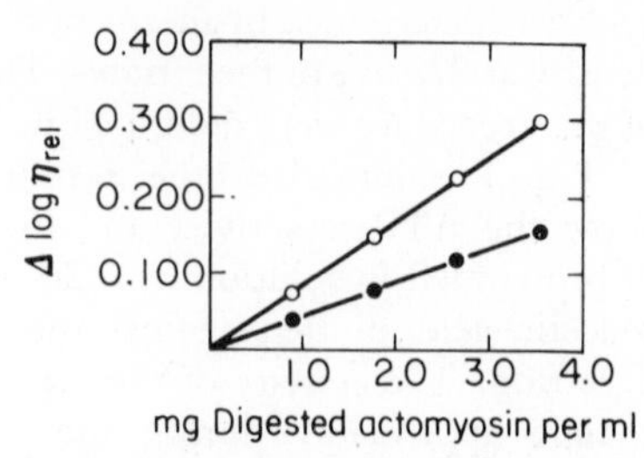

Fig. 3. Logarithm of the relative viscosity of digested actomyosin plotted against protein concentration in the absence (○) and in the presence (●) of ATP. Ordinate, log of the relative viscosity; abscissa, mg. of digested actomyosin per ml.

of the curves with both the native and the digested myosin occurs when approximately the same amount of actin has been added.

The actin-myosin ratio at the point of equivalence was smaller in these experiments than the value reported earlier[17]. Repeated experiments showed that the actin myosin ratio at the point of equivalence was 1:5.0 to 5.3 per weight instead of the value 1:4.3.

Reaction of digested actomyosin with ATP

Actin is very slowly degraded by trypsin[12]; therefore, with the digestion of actomyosin, the myosin component should be chiefly affected.

After actomyosin was digested with trypsin to the extent that, according to the ultracentrifugal determination, no native actomyosin was left, it was dialyzed free of the non-protein digestion products[11]. The viscosity of the digested actomyosin preparations was determined at different concentrations in the presence and in the absence of ATP. In Fig. 3 the logarithm of the relative viscosities is plotted against concentration. Both with and without ATP the points fall on a straight line. The activity of the native actomyosin, calculated with the logarithmic formula of Portzehl, Schramm, and Weber[18], corresponded to the stoichiometric actinmyosin ratio. Therefore, at equal protein concentrations, the digested actomyosin and a stoichiometric mixture of digested myosin and actin should have the same value for Δ log $\eta_{rel.}$. The protein concentration at the point of equivalence in Fig. 2 is 2.92 mg. per ml. with Δ log $\eta_{rel.}$ equal to 0.101. In Fig. 3 for the same protein concentration there is a corresponding difference of 0.113. The two figures are in reasonably good agreement.

Relation of components of first phase of digestion to specific activities of myosin

In the previous sections it was demonstrated that the split products of myosin combined with actin and retained the original ATPase activity. To decide whether both myosin fragments obtained by the short digestion possess the specific activities, or whether these are linked to only one of them, ultracentrifugal experiments were performed. Actin purified by the method of Mommaerts[19] was added to the digested myosin (1 mg. of actin to 4.3 mg. of myosin) and the sedimentation of this solution compared with that of the original myosin digest. The proteins were dissolved in solutions of 0.6 M KCl and 0.0067 M phosphate buffer of pH 7. Fig. 4 shows the sedimentation patterns obtained. After actin addition the sedimentation of the originally slower sedimenting component remained the same within the experimental errors, indicating no interaction between this fraction and action. On the other hand, the sedimentation of the originally faster sedimenting component increased by approximately 20 times. In other experiments when less actin was added, the characteristic peak of the fast fraction appeared. These findings indicate that it is the fast fraction which is responsible for the actin-combining capacity of the myosin molecule. At the same time they suggest an easy method of separation of the two components by differential ultracentrifugation. This separation was achieved as follows: To the digested myosin a stoichiometric amount of actin was added, both proteins being dissolved in 0.6 M KCl, then centrifuged in the preparative ultracentrifuge for 120 minutes at 40,000 r.p.m. The supernatant, containing the light fraction, was recentrifuged under the same conditions. The second supernatant, as shown by the sedimentation in the analytical ultracentrifuge, contained only the light fraction. 42 per cent of the original digest was present in this fraction. The sedimented fast fraction-actin complex of the first centrifugation was dissolved in 0.6 M KCl and

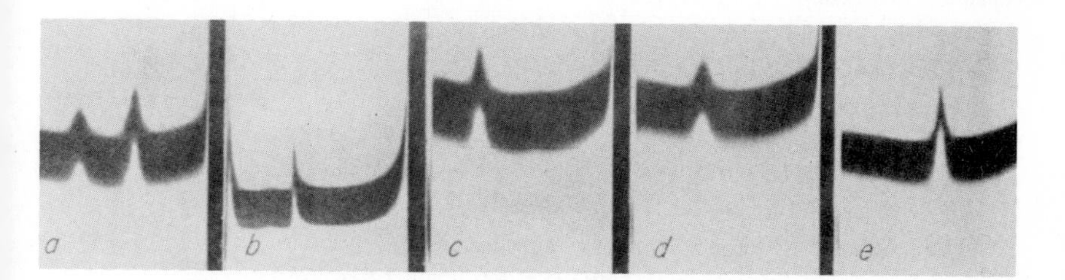

Fig. 4. Sedimentation patterns of digested myosin with and without the addition of actin and that of the isolated components. Temperature, 17–20°; speed, 59,780 r.p.m. (a) Digested myosin in the absence of actin 105 minutes after reaching full speed. Bar angle, 45°. (b) Digested myosin in the presence of actin 3 minutes after reaching full speed. Bar angle, 65°. (c) The same as (b) 95 minutes after reaching full speed. Bar angle, 34°. (d) Isolated slow component 130 minutes after reaching full speed. Bar angle, 35°. (e) Isolated fast component 84 minutes after reaching full speed. Bar angle, 45°.

ultracentrifuged again, then redissolved in 0.6 M KCl. The actin was removed from this complex by adding 80 mg. of ATP to 140 mg. of total protein, and 0.005 M $MgCl_2$, and centrifuging the mixture for 120 minutes at 40,000 r.p.m. ATP dissociated the complex, and actin sedimented while the fast fraction was left in the supernatant. Some of the ATP was split during this procedure, but special measurements showed that enough remained intact to keep the complex dissociated. The supernatant was dialyzed against 0.6 M KCl, 0.0067 M neutral phosphate buffer for the ultracentrifugal analysis, and against 0.6 M KCl for the enzyme activity determinations. Every step was made in the cold as far as possible.

The sedimentation constants of the components of the digest, with and without actin, as well as those of the isolated fractions are given in Table I. They were not extrapolated to zero protein concentration. The slow fraction sediments in all the samples with the same velocity, whereas the fast fraction sediments somewhat faster in the isolated state. The small difference might be due to concentration effects, or to interaction between the slow and fast

Table I

SEDIMENTATION CONSTANTS OF FRACTIONS OF TRYPSIN-DIGESTED MYOSIN

Material	Protein concentration	s_{20}*	
		Slow fraction	Fast fraction
	mg. per ml.		
Digested myosin	5.14	2.62	5.37
" " + actin	5.14†	2.70	80.0 *Ca*
Isolated slow fraction	2.04	2.67	
" fast "	3.66		5.70

*s_{20} expressed in S (Svedberg units) and calculated to 0.6 M KCl and 0.0067 M neutral phosphate buffer.
†The figure gives only the concentration of the digested myosin.

Table II

COMPARISON OF ATPASE ACTIVITY OF MYOSIN, DIGESTED MYOSIN, AND ISOLATED FRACTIONS

Material	Phosphate liberated per mg. protein in 5 min. at 35°
	γ
Myosin	83.7
Digested myosin	80.7
Fast fraction + actin*	130.0
Isolated fast fraction	131.6
" slow "	11.5

*The protein introduced with actin was omitted in the calculations.

components in the unfractionated mixture. Both fractions in the isolated state appeared homogeneous and sedimented with a symmetrical peak.

The results of the ATPase determinations of the native myosin, unfractionated digest, and isolated fractions are shown in Table II. The ATPase activity of the fast fraction calculated per mg. of protein is higher than that of the native or digested myosin. The increase corresponds to the assumption that only the fast fraction has ATPase activity, since the latter constitutes approximately 55 to 60 per cent of the myosin. The small activity of the slow fraction is probably due to contamination with the fast component. It seems that the ATPase experiments were more sensitive for the determinations of impurity, since no fast fraction could be seen in the ultracentrifuge at the low protein concentration (2.04 mg. per ml.) used.

Discussion

The experiments described clearly indicate that the first phase of trypsin digestion does not affect the ATPase activity of myosin, while the second, slow phase is accompanied by the progressive loss of the enzymic properties with the appearance of non-protein nitrogen. The rate of the decrease of the fraction precipitated by trichloroacetic acid is much slower than that of the decrease of the enzymic activity, because the secondary degradation of myosin is not an *all or none* process. Complete parallelism between activity and precipitable protein could only be expected with a specific method in which the intact split products of the first phase can be selectively precipitated, but not the further degraded ones. Trichloroacetic acid is not suitable for this purpose, since it even precipitates split products with a molecular weight of 4000[20], *i.e.*, well below what is considered the lower molecular weight limit of proteins.

The viscosity data suggest that the first phase of the trypsin digestion does not affect the capacity of myosin to bind actin either. Myosin and split myosin bind the same amount of F-actin at the point of equivalence. This was also corroborated by the ultracentrifugal experiments.

Trypsin does not affect the linkage between actin and myosin. In the digested actomyosin, actin and the myosin split products are still linked together, as the effect of ATP on the viscosity of digested actomyosin solutions shows.

The two components obtained during the first phase of tryptic digestion are not equivalent in respect to ATPase activity and actin-binding capacity. It seems that the active centers necessary for these specific activities are linked only to the faster sedimenting component. It was this difference in the behavior of the two components which was the basis for their separation. Since there is no loss in the specific activities of myosin during the first phase of the digestion, the inactivity of the slower sedimenting component cannot be due to denaturation. The specific activities of myosin seem to be linked to one of the subunits and there seems to be no difference, as far as activities are concerned, whether these units are free or built into a long native myosin molecule.

Myosin strongly enhances the polymerization of globular actin[21]. The complex which myosin forms with F-actin at low ionic strength superprecipitates on ATP addition[1]. Trypsin-digested myosin did not give superprecipitation in similar conditions, nor did it have any effect on the rate of polymerization of actin, indicating that these properties are connected with the intactness of the myosin molecule.

It is of interest to mention that proteolytic split products of immune globulins also retained their full biological activity[22-24]. Pappenheimer and Robinson[25] have reported that partially digested diphtheria antitoxin binds more toxin than the same amount of the native one. Grabar[26] made a similar observation with pepsin-digested pneumococcus antibody. In the specific precipitate formed with the latter and the pneumococcus polysaccharide the polysaccharide-protein ratio was twice as large as that obtained with the native antibody. Apparently an inactive part of the molecule was split off by the proteolytic enzymes. This assumption was proved by Petermann and Pappenheimer[27] in their ultracentrifugal analysis showing that the diphtheria antitoxic pseudoglobulin is split by proteolytic enzymes into two halves, only one of them being able to combine with the toxin.

Summary

The specific activities of myosin (ATPase activity and actin-binding capacity) are unaffected by splitting of the molecules in the first rapid phase of tryptic digestion.

The activities belong to the faster sedimenting one of the two components obtained in the first phase of tryptic digestion. Separation of the two components was achieved.

The secondary slow degradation by trypsin of these split products causes the disappearance of the specific activities.

References

1. Szent-Györgyi, A., *Acta physiol. Scand.*, **9**, suppl. 25, 41 (1945).
2. Lyubimova, M. N., and Engelhardt, V. A., *Biokhimiya*, **4**, 716 (1939).
3. Hermann, S. V., and Josepovits, G., *Nature*, **164**, 845 (1949).
4. Humphrey, G. F., and Webster, H. L., *Australian J. Exp. Biol. and Med. Sc.*, **29**, 17 (1951).
5. Engelhardt, V. A., *Advances in Enzymol.*, **6**, 147 (1946).
6. Mirsky, A. E., *Cold Spring Harbor Symposia Quant. Biol.*, **6**, 150 (1938).
7. Bailey, K., *Advances in Protein Chem.*, **1**, 289 (1944).
8. Singher, H. O., and Meister, A., *J. Biol. Chem.*, **159**, 491 (1945).
9. Needham, J., Shen, S. C., Needham, D. M., and Lawrence, A. S. C., *Nature*, **147**, 766, (1941).
10. Kielley, W. W., and Meyerhof, O., *J. Biol. Chem.*, **174**, 387 (1948); **176**, 591 (1948); **183**, 391 (1950).
11. Mihályi, E., and Szent-Györgyi, A. G., *Biol. Chem.*, **201**, 189 (1953).
12. Mihályi, E., *J. Biol. Chem.*, **201**, 197 (1953).
13. Gergely, J., *Federation Proc.*, **9**, 176 (1950).
14. Gergely, J., *Federation Proc.*, **10**, 188 (1951).

15. Perry S. V., *Biochem. J.,* **48**, 257 (1951).
16. Fiske, C. H., and Subbarow, Y., *J. Biol. Chem.,* **66**, 375 (1925).
17. Szent-Györgyi, A. G., *J. Biol. Chem.,* **192**, 361 (1951).
18. Portzehl, H., Schramm, G., and Weber, H. H., *Z. Naturforsch.,* **5b**, 61 (1950).
19. Mommaerts, W. F. H. M., *J. Biol. Chem.,* **188**, 553 (1951).
20. Butler, J. A. V., Dodds, E. C., Phillips, D. M. P., and Stephen, J. M. L., *Biochem. J.,* **42**, 116 (1948).
21. Laki, K., and Clark, A. M., *Arch. Biochem.,* **30**, 187 (1951).
22. Pope, C. G., *Brit. J. Exp. Path.,* **20**, 132 (1939).
23. Petermann, M. L., *J. Biol. Chem.,* **144**, 607 (1942).
24. Petermann, M. L., and Pappenheimer, A. M., Jr., *Science,* **93**, 458 (1941).
25. Pappenheimer, A. M., Jr., and Robinson, E. S., *J. Immunol.,* **32**, 291 (1937).
26. Grabar, P., *Compt. rend. Acad.,* **207**, 807 (1938).
27. Petermann, M. L., and Pappenheimer, A. M., Jr., *J. Phys. Chem.,* **45**, 1 (1941).

STUDIES ON MYOSIN-ADENOSINETRIPHOSPHATASE*

J. Gergely†

National Institute of Arthritis and Metabolic Diseases,
National Institutes of Health, Bethesda, Maryland

(Received for publication, July 25, 1952)

The ATPase[1] activity of myosin and actomyosin preparations has been established by previous studies[2,3]. There has been some speculation whether myosin itself is the ATP-splitting enzyme or the ATPase proper is attached in a secondary manner to the myosin molecule[4]. Polis and Meyerhof[5] succeeded in preparing, by means of a La salt precipitation, a myosin fraction that had 2- to 3-fold increased ATPase activity. The question, however, remained open whether this fraction of higher activity was still to be considered myosin or represented ATPase separated from the contractile protein.

To obtain further insight into the relationship of ATPase and myosin we have treated myosin with trypsin and investigated the ATPase activity, the viscosity of this preparation, and some of its reactions with actin.

*J. Gergely, "Studies on Myosin-Adenosinetriphosphatase," *J. Biol. Chem.*, **200**, 543–550 (1953). Reprinted with the permission of the author and of the American Society of Biological Chemists, Inc.

A preliminary report of this work was presented before the American Society of Biological Chemists at Atlantic City, April, 1950. Since the publication of this report some of the results have been independently confirmed by Perry[1].

This work was carried out during the tenure of a Special Research Fellowship of the Experimental Biology and Medicine Institute, United States Public Health Service.

†Present address, Cardiovascular Research Laboratory, Massachusetts General Hospital, and Department of Medicine, Harvard Medical School, Boston, Massachusetts.

[1] The following abbreviations are used in this paper: ATP, adenosinetriphosphate; ADP, adenosinediphosphate; ITP, inosinetriphosphate; ATPase, adenosinetriphosphatase; ITPase, inosinetriphosphate.

Methods

Enzyme assay

Liberation of inorganic phosphate from ATP was measured at 39°. If not otherwise stated, the incubation mixture had the following composition: 200 μM of glycine buffer, pH 9.2, 200 μM of KCl, 20 μM of $CaCl_2$, and 2 μM of ATP. The total volume was 2 ml. In order that specific activities (units per mg. of protein) may be expressed in terms of the customary Q_P[(6)], 1 unit was defined as the amount of enzyme necessary to liberate 1/22.41 μM of inorganic phosphate per hour. The amount of enzyme was so chosen that not more than 40 per cent of the initial ATP was hydrolyzed during the incubation time of 5 minutes. Under these conditions the amount of phosphate liberated was proportional to the amount of enzyme present in the incubation mixture.

Determinations

Inorganic phosphate was determined according to Fiske and Subbarow[(7)] with a Coleman model 14 photoelectric colorimeter. Viscosity measurements were carried out in Ostwald type viscometers, outflow time being about 100 seconds. The purity of ATP preparations was determined on the basis of their labile (inorganic phosphate liberated in 9 minutes at 100° in 1 N HCl) phosphate content, optical density at 265 mμ, and differential spectrophotometric measurements according to Kalckar[(8)]. A Beckman model DU spectrophotometer was used for ultraviolet absorption measurements. Protein concentration was determined from the light absorption at 280 and 260 mμ[(9)]. Occasional N determinations[2] were carried out to check this procedure, the factor 6.2 being used for converting N values to protein.

Preparations

Myosin was prepared essentially according to Szent-Györgyi[(3)], by extraction of minced rabbit muscle with 0.6 M KCl-phosphate buffer, pH 6.5, and crystallization on diluting to a final concentration of 0.06 M KCl. Actin contamination, if present, was removed as actomyosin by adding ATP[(10)]. Actin was prepared according to Feuer *et al.*[(11)]. Apyrase and myokinase were prepared according to Kalckar[(12)], and 5-adenylic acid deaminase according to S. P. Colowick.[3]

ATP was obtained as the Ba salt from the Sigma Chemical Company, St. Louis, and converted into the Na salt for use. ITP was prepared according to Kleinzeller[(13)]. Crystalline trypsin was obtained from the Worthington Biochemical Laboratory.

Results

Trypsin digestion of myosin

1 ml. of a myosin solution, 0.6 M KCl, and 10 to 15 mg. of protein per ml., pH 8, were incubated at 25° with 10 γ of trypsin. At the end of the incubation, the mixture was diluted 10-fold with ice-cold distilled water and stored at 0°.

[2] Kindly performed by Dr. W. C. Alford.

[3] Personal communication.

The trypsin-treated myosin showed a decreased viscosity and a changed behavior towards actin. A solution containing actin and untreated myosin at an ionic strength of about 0.6 shows a drop in viscosity on the addition of ATP. At ionic strength of about 0.1 the addition of ATP brings about superprecipitation. A 15 minute incubation of myosin with trypsin was sufficient to cause the superprecipitation reaction to disappear. The change in the viscosity decrease on addition of ATP in the presence of actin is shown in Fig. 1, together with the viscosity changes of myosin. In contrast to the effect of trypsin on the viscosity of myosin and on its ability to react with actin in a typical way, the ATPase activity remained practically unchanged (Fig. 1). In fact, in some experiments there was a slight increase in activity after digestion.

Some information about the change in the shape of the myosin molecule on trypsin digestion may be obtained by determining the specific viscosity at various protein concentrations. Assuming the myosin molecule has the shape of a rotational ellipsoid, the axial ratio can be calculated[14]. In Fig. 2, the specific viscosity both of native and of trypsin-treated myosin is plotted against the volume fraction of the protein. The apparent axial ratio decreases from 25 to 10, indicating a rather profound change in molecular structure, even under the conditions that leave the enzymatic activity towards ATP unchanged.

Purification of the ATPase from trypsin-treated myosin preparations has been attempted. While various methods, including ammonium sulfate, acetone,

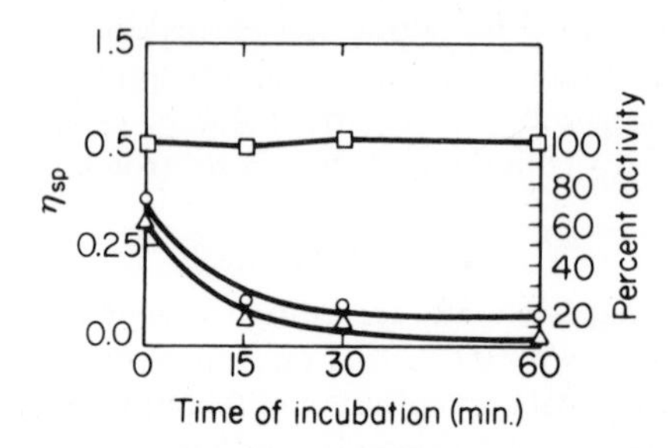

Fig. 1. Effect of trypsin on ATPase activity, specific viscosity, and actin-combining ability of myosin. Trypsin was added at 0 time to a myosin solution in 0.6 M KCl, pH 8, at 25°. Samples were taken after various periods of incubation. □, ATPase activity expressed as percentage of activity at 0 time. Details of the assay are given in the text. ○, specific viscosity of myosin. The mixture in the viscosimeter consisted of 0.3 ml. of the myosin-trypsin digest, 2.4 ml. of 0.6 M KCl, 0.6 ml. of 0.6 M K-phosphate buffer, pH 7.0. The viscosity was measured at 0°. △, drop in specific viscosity on addition of ATP to a solution containing actin and samples from the myosin-trypsin digest. Composition of the solution in the viscosimeter: 0.3 ml. of the myosin-trypsin digest (10 mg. of myosin per ml.), 0.15 ml. of F-actin (4 mg. of protein per ml.), 2.4 ml of 0.6 M KCl, 0.6 ml. of 0.6 M K-phosphate buffer, pH 7.0. The viscosity was measured before and after addition of 0.1 ml. of 0.01 M ATP-Na salt at 0°.

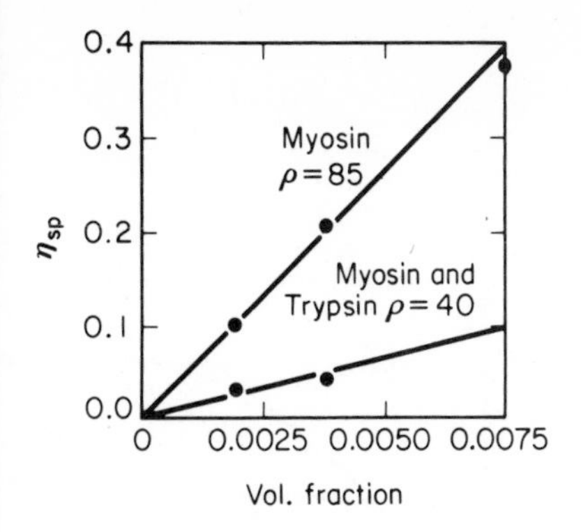

Fig. 2. Effect of protein concentration on the specific viscosity of native and trypsin-treated myosin. 0°, 0.01 M phosphate buffer, pH 7, 0.6 M KCl. The volume fraction was calculated, assuming a partial specific volume of 0.75 for myosin ρ = axial ratio.

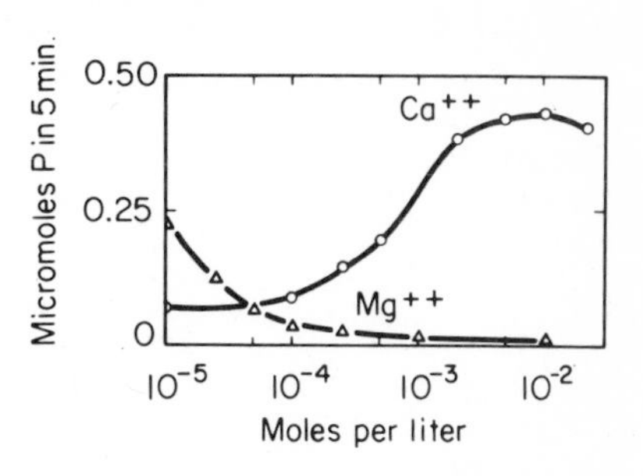

Fig. 3. Effect of Ca^{++} and Mg^{++} ions on ATPase obtained after trypsin treatment of myosin, pH 9.2; 0.1 M glycine buffer. Initial ATP concentration 0.001 M, protein concentration 5 γ per ml., total volume 2 ml. In the experiments with Mg^{++}, $Ca^{++} = 5 \times 10^{-3}$ M.

and alcohol fractionation and adsorption on aluminum Cγ gel, have failed, it was found that a pH precipitation step involving the removal of an inactive fraction yielded a preparation of 2- to 3-fold increased activity. Myosin was incubated with trypsin for 45 minutes; after 10-fold dilution the pH was adjusted to 6.3 to 6.5 (ionic strength 0.06) and the resulting precipitate centrifuged at 10,000 × *g*. The supernatant, which showed increased specific activity, contained 50 to 80 per cent of the original total activity. The precipitate was redissolved in 0.1 M glycine buffer, pH 9.2, and the pH of the supernatant was adjusted to the same value. Table I illustrates this procedure.

Table I

PURIFICATION OF MYOSIN-ATPASE

	Q_P	Protein	Total units
		mg.	
Myosin	6,100	10.0	61,000
〃 + trypsin	6,100	10.0	61,000
pH 6.5 ppt	2,000	6.5	13,000
〃 6.5 supernatant	15,000	3.5	52,500

For details see the text. The figures in the second and third columns refer to 1 ml. of the original myosin solution.

Effect of ions

The purified enzyme obtained after trypsin digestion showed activation by Ca^{++} and inhibition by Mg^{++} ions (Fig. 3). The pH dependence is shown in Fig. 4.

Splitting of ITP

It has been shown that myosin also catalyzes the hydrolysis of ITP[13]. We have investigated the splitting of ITP by the trypsin-treated purified preparation. Fig. 5 shows the effect of substrate concentration on the initial rates of hydrolysis, together with data on ATP. With ATP the half saturation concentration could not be obtained, since it seemed to be at such a low level that the measured rates could not properly be considered as initial rates. Its value must certainly be less than 10^{-4} M per liter. For ITP the corresponding value is about 3.5×10^{-4}. Inspection of Fig. 5 shows that, while the K_M (Michaelis-Menten constant) value for ATP is smaller than that for ITP, the rate

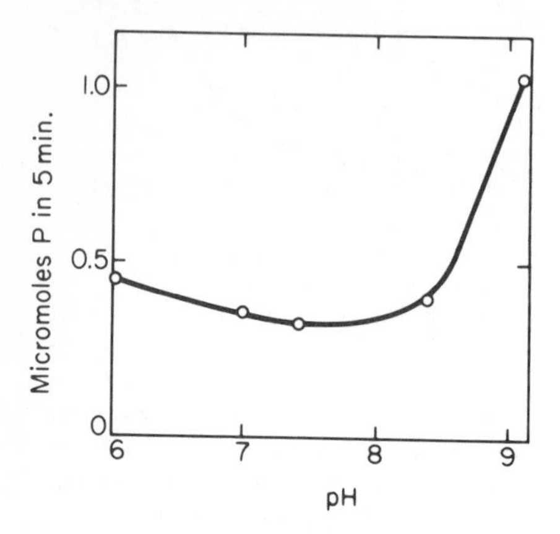

Fig. 4. Effect of pH on ATPase obtained after trypsin treatment of myosin; 0.1 M glycine buffer. Initial ATP concentration 0.001 M, Ca^{++} 0.005 M, KCl 0.1 M, protein concentration 10γ per ml., total volume 2 ml.

Fig. 5. Effect of ATP and ITP concentration on the initial rate of inorganic phosphate liberation. pH 9.2, 0.1 M glycine buffer, 0.005 M Ca^{++}, 0.1 M KCl, protein concentration 5γ per ml., total volume 2 ml.

Table II

Rate of Phosphate Liberation and ADP Formed and Calculated

Added	P liberated	ADP formed	ADP calculated
	μM	μM	μM
2 μM ATP	0.76	0.35	0.38
4 〃 ITP	1.20		
2 〃 ATP + 4 μM ITP	0.66	0.26	0.33

0.1 M glycine buffer, *p*H 9.2, 0.1 M KCl, 0.005 M $CaCl_2$, total volume 2.0 ml., incubation at 39°, time 5 minutes. The reaction was stopped by addition of 0.2 ml. of 20 per cent perchloric acid. After neutralization with $KHCO_3$ and filtration, inorganic phosphate was determined in one aliquot and ADP according to Kalckar in another. The last column shows the amount of ADP that would correspond to the inorganic phosphate liberated, assuming that only ATP was split.

of the enzymatic reaction, with the enzyme saturated, is higher in the case of ITP than in the case of ATP. In view of this fact, a competitive inhibition by ATP of phosphate liberation from ITP would be expected. A detailed analysis of the competition could not be carried out, since the value of K_M for ATP could not be determined. It can, however, be shown that the rate of phosphate liberation is indeed reduced in a mixture of ATP and ITP, compared with the rate of ITP splitting alone (Table II). That this effect is due to competition is supported by enzymatic analysis of the products. Most of the inorganic phosphate appearing in an ATP-ITP mixture can be accounted for as ADP, showing that it is the ATP moiety that is predominantly attacked, although at a slower rate than ITP.

Discussion

The data presented here show that, by trypsin treatment of myosin, an ATPase preparation can be obtained that is devoid of the characteristic actin-combining ability of the original material. Kielley and Meyerhof(15) have recently described a water-extractable ATPase in muscle that is activated by Mg^{++} and inhibited by Ca^{++} ions. The enzyme described in this paper behaves like native myosin in so far as it is activated by Ca^{++} and inhibited by Mg^{++} ions, and also with respect to the effect of pH on its activity. It has been shown(13, 16, 17) that myosin splits both ATP and ITP and that the affinity for ATP is greater than that for ITP(16, 17). Furthermore, ITP is split at a higher rate than ATP. It seems that a similar situation exists in the case of our preparation and that there is indication of competition between the two substrates.

Viscosity measurements suggest that trypsin treatment of myosin leads to the formation of fragments of lower dissymmetry than that of the original molecule. It would appear that not all of the fragments possess uniform enzymatic activity, since part of the protein can be removed, leaving a preparation of 2- to 3-fold increased activity.[4]

As far as the relationship of ATPase and myosin is concerned, either of the following pictures is consistent with the results. It is conceivable that myosin, one of the contractile proteins of muscle, may form a rather tight complex with ATPase proper, and that this enzyme is liberated by trypsin. Indeed, such seems to be the case with respect to 5-adenylic acid deaminase.(19) Or, it is possible that ATPase activity resides in the contractile-protein molecule, but, while splitting this molecule into smaller units destroys most of its properties, ATPase emerges intact. This second picture implies that a simpler molecular organization is sufficient for ATPase activity than is necessary for interaction with actin, as manifested by the superprecipitation reaction and by the effect of ATP on the viscosity of an actomyosin solution.

[4] A. G. Szent-Györgyi has recently reported(18) that the two fractions obtained by the pH precipitation appear to be homogeneous as judged by electrophoretic and ultracentrifugal criteria. He also obtained a crystalline preparation of the inactive fraction. On the basis of ultracentrifugal evidence it appears that the active fraction combines with actin and can be dissociated from it by ATP. The existence of two, *ultracentrifugally* distinct, fractions in a myosin-trypsin digest was first shown by E. Mihályi (unpublished).

Summary

1. It has been shown that trypsin-treated myosin no longer reacts typically with actin. There is no superprecipitation reaction and addition of ATP at $\Gamma/2 = 0.6$ does not significantly decrease the viscosity. ATPase activity remains unchanged.

2. Partial purification of ATPase from a trypsin digest of myosin has been described.

3. The effect of H^+, Mg^{++}, and Ca^{++} ions on such a preparation has been studied. The behavior of the trypsin-treated preparation with respect to these parameters is the same as that of native myosin.

4. Viscosity measurements indicate a breaking up of myosin into fragments of decreased asymmetry.

5. The effect of substrate concentration on the ATPase and ITPase activity of trypsin-treated myosin has been studied. It has been shown that these two substrates compete with each other.

6. The relationship of myosin and ATPase has been discussed.

References

1. Perry, S. V., *Biochem. J.,* **48**, 257 (1951).
2. Engelhardt, V. A., and Ljubimowa, M. N., *Nature,* **144**, 668 (1939).
3. Szent-Györgyi, A., *Acta physiol. Scand.,* **2**, suppl., 25 (1945).
4. Engelhardt, V. A., *Advances in Enzymol.,* **6**, 147 (1946).
5. Polis, B. D., and Meyerhof, O., *J. Biol. Chem.,* **169**, 389 (1947).
6. Bailey, K., *Biochem. J.,* **36**, 121 (1942).
7. Fiske, C. H., and Subbarow, Y., *J. Biol. Chem.,* **66**, 375 (1925).
8. Kalckar, H. M., *J. Biol. Chem.,* **167**, 445 (1947).
9. Warburg, O., and Christian, W., *Biochem. Z.,* **303**, 40 (1939).
10. Spicer, S. S., and Gergely, J., *J. Biol. Chem.,* **188**, 179 (1951).
11. Feuer, G., Molnar, F., Pettko, E., and Straub, F. B., *Hung. acta physiol.,* **1**, 150 (1948).
12. Kalckar, H. M., *J. Biol. Chem.,* **167**, 461 (1947).
13. Kleinzeller, A., *Biochem. J.,* **36**, 729 (1942).
14. Mehl, J. W., Oncley, J. L., and Simha, R., *Science,* **92**, 2380 (1940).
15. Kielley, W. W., and Meyerhof, O., *J. Biol. Chem.,* **176**, 591 (1948).
16. Mommaerts, W. F. H. M., and Seraidarian, K., *J. Gen. Physiol.,* **30**, 401 (1947).
17. Spicer, S. S., and Bowen, W. J., *J. Biol. Chem.,* **188**, 741 (1951).
18. Szent-Györgyi, A. G., *Federation Proc.,* **11**, 296 (1952).
19. Gergely, J., *Federation Proc.,* **10**, 188 (1951).

THE RELATIONSHIP BETWEEN SULFHYDRYL GROUPS AND THE ACTIVATION OF MYOSIN ADENOSINETRIPHOSPHATASE*

W. Wayne Kielley and Louise B. Bradley

Laboratory of Cellular Physiology, National Heart Institute,
National Institutes of Health, Bethesda, Maryland

(Received for publication, May 27, 1955)

Greenstein and Edsall[1] in early observations on the quantitative estimation of sulfhydryl groups in the muscle protein myosin classified the SH groups of the native protein into "free" and unavailable groups on the basis of porphyrindine titration with nitroprusside as an external indicator. In subsequent experiments, Singer and Barron[2] observed that combination of the "free" groups of myosin with *p*-chloromercuribenzoate (PCMB) resulted in little alteration of the adenosinetriphosphatase (ATPase) activity of myosin, while addition of PCMB equal to the total SH led to complete inhibition. It has been observed by one of us[3] that low concentrations of PCMB resulted in some increase in ATPase activity and this was also observed occasionally by Polis and Meyerhof[4].

On reexamination of the influence of PCMB titration of myosin SH, we observed that a marked increase in ATPase activity occurred when approximately one-half the sulfhydryl groups had been titrated, when Ca^{++} was employed as ATPase activator. On the other hand, only inhibition occurred when ethylenediaminetetraacetic acid (EDTA) was employed as activator (*cf.* [5,6] for EDTA activation). The reaction of myosin with *N*-ethylmaleimide (NEM) gave similar results.

After the work presented here was completed, two reports[7,8] appeared

*W. W. Kielley and L. B. Bradley, "The Relationship between Sulfhydryl Groups and the Activation of Myosin Adenosinetriphosphatase," *J. Biol. Chem.*, **218**, 653–659 (1956). Reprinted with the permission of the authors and of the American Society of Biological Chemists, Inc.

A preliminary report of this work was presented at the 127th meeting of the American Chemical Society, Cincinnati, Ohio, April 2, 1955.

on the influence of phenylmercuric acetate and dinitrophenol on myosin ATPase. In these the authors appear to be dealing with similar phenomena.

Experimental

MATERIALS AND METHODS

Myosin

Myosin was prepared by either of two procedures. One is a slight modification of that of Tsao[9], the other a modification of the procedure of Weber and Portzehl[10]. The products obtained were indistinguishable in so far as the work presented here is concerned.

The only modification in the procedure of Tsao was the inclusion of 0.01 M EDTA in the saturated ammonium sulfate. No attempt was made to ascertain that this was essential. However, Tsao observed that his preparations were rather low in ATPase activity, whereas our preparations were of consistently high activity. In the modified Weber and Portzehl procedure, the muscle was extracted with 3 volumes of 0.5 M KCl, 0.1 M K_2HPO_4 for 20 or 30 minutes, and then treated in the Waring blendor for $\frac{1}{2}$ minute. The material was then centrifuged, and the myosins A and B were precipitated from the supernatant solution by dilution with 10 volumes of water. The redissolved precipitate (0.5 M KCl) was then adjusted to 0.28 M KCl at pH 6.7 to 6.8 and the precipitate (myosin B) removed. The myosin A of the supernatant solution was obtained by further dilution to 0.04 M KCl. The latter fraction, after resolution in 0.5 M KCl, was then subjected to $(NH_4)_2SO_4$ fractionation as in the first procedure.[1] This consisted of addition of a saturated solution of $(NH_4)_2SO_4$ (adjusted to pH 6.5 to 7.0 and containing 0.01 M EDTA) to the myosin solution to 40 per cent saturation. The precipitate formed was removed and discarded. The supernatant solution was then brought to 50 per cent saturation by further addition of the $(NH_4)_2SO_4$ solution. The precipitate was collected by centrifugation, redissolved in 0.5 M KCl, and dialyzed against 0.5 M KCl overnight. Protein was estimated from micro-Kjeldahl nitrogen determinations with correction for NH_4^+ remaining after dialysis.

Adenosine triphosphate was obtained commercially as the crystalline sodium salt.

Ethylenediaminetetraacetic acid (EDTA) was obtained commercially and used as the potassium salt (pH 7.0).

The method of Whitmore and Woodward[11] was used to prepare PCMB or to purify the material obtained commercially.

NEM was obtained commercially and purified by sublimation.

ATPase activity

Enzymatic activity was measured in a system containing 0.001 M ATP, 0.02 M histidine (pH 7.6), either 0.005 M $CaCl_2$ or 0.001 M EDTA, and either 0.05 M KCl (when $CaCl_2$ is present) or 0.4 M KCl (when EDTA is present).

[1] Both myosin B and myosin A without the $(NH_4)_2SO_4$ fractionation exhibit a behavior identical to that described here.

Treatment of myosin with sulfhydryl reagents

PCMB

PCMB was added to myosin dissolved in 0.5 M KCl in quantities indicated in Fig. 1 and Table I. The tubes containing the treated protein were kept chilled and assayed immediately by adding aliquots containing 20 to 30 γ of protein to the medium given above (final volume 1.0 ml.).

The spectrophotometric procedure of Boyer[12] was employed to follow the reaction of protein SH with PCMB.

NEM

N-Ethylmaleimide in the quantities indicated in Fig. 2 was added to myosin dissolved in 0.5 M KCl and adjusted to pH 7.2, 7.6, or 8.0 with 0.02 M histidine. The tubes containing the treated protein were allowed to stand at 0° for 2 and 20 hours before assaying aliquots containing 20 to 30 γ of protein for ATPase activity in the medium given above with volume of 1.0 ml.

RESULTS

The influence of PCMB on myosin ATPase is presented in Fig. 1. Curve C represents the spectrophotometric titration of myosin SH by PCMB. The

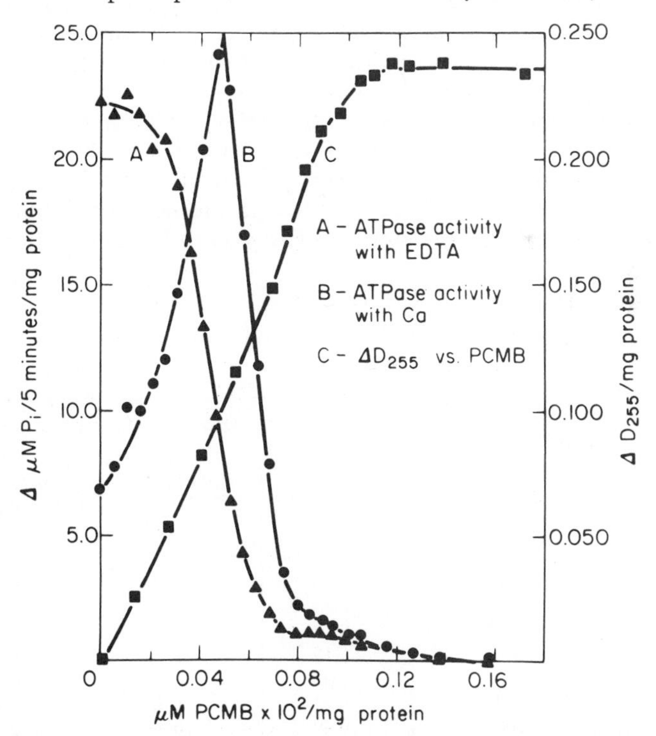

Fig. 1. Titration of myosin with PCMB. Curve A, ATPase activity in the presence of EDTA: Curve B, ATPase activity in the presence of Ca^{++}; and Curve C, spectrophotometric titration of myosin with PCMB. For other conditions see the text.

total SH is calculated to be 0.11 μmole of SH per mg. of protein. Curve B presents the behavior of ATPase activity with Ca^{++} as activator. Titration of about one-half the groups led to about a 4-fold increase in activity in this experiment (between 3- and 4-fold generally) with inhibition at higher concentrations of PCMB, complete inhibition corresponding to complete titration of the SH. With EDTA as activator (Curve A) only inhibition was observed, the process being complete when only about three-fourths of the groups was titrated and the form of the curve suggests that only one-half of the groups was involved in this inactivation, and that about one-quarter of the groups is not essential for enzymatic activity. The inhibition by PCMB was reversed to some extent by other sulfhydryl compounds, as was indicated by Singer and Barron[(2)].

When assayed in the presence of Ca^{++} it can be seen (Table I) that the effect of an amount of PCMB (0.04 μmole per mg. of protein) approaching one-half the total SH was readily reversed by β-mercaptoethanol. However, for a higher concentration of PCMB, approaching complete inhibition, no reversal was observed. In contrast, the activity measured in the presence of EDTA exhibits some reversal of the effect of both low and high PCMB concentrations by β-mercaptoethanol. The pattern of these results suggests that under the conditions (time, etc.) removal of the mercurial from the protein by even the large excesses of β-mercaptoethanol is incomplete and that those mercaptide groups reacting most readily with the β-mercaptoethanol may be sulfhydryl groups reacting first with PCMB. In the case of Ca^{++} activation, these latter SH groups appear to be inhibiting groups when free.

The results of incubating myosin with NEM at pH 7.2 and 8.0 (both at 0°) are shown in Fig. 2. Tsao and Bailey[(13)] observed that only about 50 per cent of the sulfhydryl groups of native myosin will react with this reagent. Comparison of the results for 2 and 20 hour incubations of myosin with NEM at 0° indicates that the reaction is slow. It is also evident from Fig. 2 that the rate

Table I

REVERSAL OF EFFECT OF PCMB ON MYOSIN ATPASE BY β-MERCAPTOETHANOL

Activator	μmoles β-mercapto-ethanol per μmole PCMB	μmole inorganic phosphate per 5 min.		
		No PCMB	0.04 μmole PCMB per mg. protein	0.07 μmole PCMB per mg. protein
Ca^{++}	0	0.13	0.38	0.09
"	10	0.13	0.12	0.06
"	100	0.14	0.11	0.06
"	1000	0.13	0.11	0.09
EDTA	0	0.45	0.18	0.01
"	10	0.44	0.40	0.15
"	100	0.49	0.38	0.15
"	1000	0.50	0.42	0.29

and extent of the reaction are influenced by pH. Though only a fraction of the SH groups react with NEM, comparison with PCMB indicates that a fractional reaction with either reagent may not involve the same groups and that "availability" of any particular SH group depends on the binding agent. Thus for an apparent reaction about one-half the SH with NEM (on the basis of reagent concentration no more than one-half could have reacted) the same phenomenon of increased enzymatic activity occurs as with PCMB in the presence of Ca. However, the magnitude of this increase and the influence of reagent concentration on its development are different from that observed with PCMB. The difference in behavior relative to the SH reagent is particularly noticeable with EDTA as activator when, with NEM, complete inhibition occurs when no more than one-quarter of the SH could have combined with the reagent, in contrast to the reaction of three-quarters of the SH groups with PCMB before complete inhibition occurs in EDTA activation.

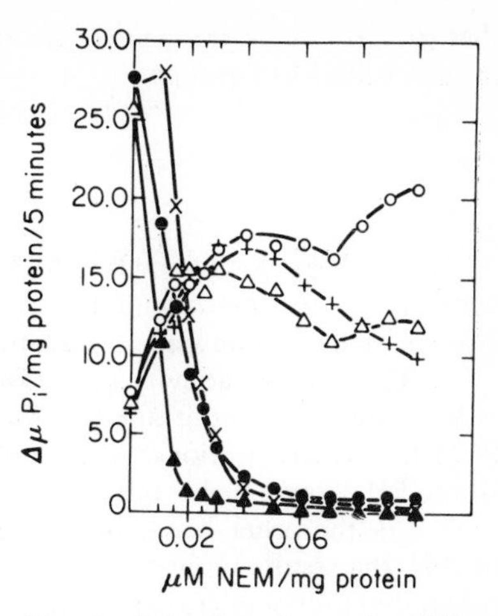

Fig. 2. Behavior of myosin ATPase treated with NEM. ○, at 2 hours, pH 7.2 activity measured with Ca^{++}; ●, at 2 hours, pH 7.2 activity measured with EDTA: △, at 20 hours, pH 7.2 activity measured with Ca^{++}; ▲, at 20 hours, pH 7.2 activity measured with EDTA: +, at 2 hours, pH 8.0 activity measured with Ca^{++}; ×, at 2 hours, pH 8.0 activity measured with EDTA. For other conditions see the text.

Discussion

From results with the two sulfhydryl-binding agents it appears possible to conclude that myosin possesses more than one sulfhydryl group per active center. It also appears likely that not all the SH groups are directly concerned with the enzymatic activity. Of the SH groups occurring at the active center one or more appear to participate in an interaction with some unidentified group, and this interaction prevents the enzyme from exhibiting its potential maximal activity in the presence of Ca as activator.

The behavior of the enzyme in the presence of EDTA as contrasted to that in the presence of Ca suggests that EDTA may interfere with this sulfhydryl interaction which results in suppression of activity when Ca^{++} or K^{+} is present as activator. The characteristic properties of EDTA further suggest that the unidentified group interacting with the SH groups is a metal. In a recent note Friess *et al.*[14] have concluded that, if a metal is involved in the EDTA activation, this metal must be very firmly bound and that it is not removed by treatment with EDTA. Analysis of EDTA-treated myosin leads them to suggest that this metal may be Mg. The observations presented here further suggest that the groups on the protein involved in this binding of metal are, at least in part, sulfhydryl groups.

Summary

1. The behavior of myosin ATPase after treatment with *p*-chloromercuribenzoate or *N*-ethylmaleimide has been studied with either Ca^{++} or ethylenediaminetetraacetic acid as activator.

2. The ATPase activity of myosin exhibited a marked increase with Ca^{++} as activator when about one-half the sulfhydryl groups had been titrated with PCMB. Further titrations led to inhibition, the process being complete when all the SH groups had reacted.

3. On the other hand, only inhibition was observed with EDTA as activator, the results indicating that removal of only one-half of the SH groups is required for complete inactivation when EDTA is used as activator.

4. The results with NEM were qualitatively similar. With EDTA as activator, binding of no more than one-quarter of the SH is required for complete inactivation.

5. In attempting to reverse the effects of PCMB with β-mercaptoethanol, only the activity as measured in the presence of EDTA appeared to follow a course which is the reverse of PCMB combination, starting at any degree of PCMB combination.

References

1. Greenstein, J. P., and Edsall, J. T., *J. Biol. Chem.,* **133**, 397 (1940).
2. Singer, T. P., and Barron, E. S. G., *Proc. Soc. Exp. Biol. and Med.,* **56**, 129 (1944).
3. Kielley, W. W., Thesis, University of Minnesota (1946).
4. Polis, B. D., and Meyerhof, O., *J. Biol. Chem.,* **169**, 389 (1947).
5. Friess, E. T., *Arch. Biochem. and Biophys.,* **51**, 17 (1954).
6. Bowen, W. J., and Kerwin, T. D., *J. Biol. Chem.,* **211**, 237 (1954).
7. Greville, G. D., and Needham, D. M., *Biochim. et biophys. acta,* **16**, 284 (1955).
8. Chappell, J. B., and Perry, S. V., *Biochim. et biophys. acta,* **16**, 285 (1955).
9. Tsao, T. C., *Biochim. et biophys. acta,* **11**, 368 (1953).
10. Weber, H. H., and Portzehl, H., *Advances in Protein Chemistry,* **7**, 162 (1952).
11. Whitmore, F. C., and Woodward, G. E., *Org. Syntheses,* coll. **1**, 159 (1941).
12. Boyer, P. D., *J. Am. Chem. Soc.,* **76**, 4331 (1954).
13. Tsao, T. C., and Bailey, K., *Biochim. et biophys. acta,* **11**, 102 (1953).
14. Friess, E. T., Morales, M. F., and Bowen, W. J., *Arch. Biochem. and Biophys.,* **53**, 311 (1954).

EVIDENCE FOR AN INTERMEDIATE IN ATP HYDROLYSIS BY MYOSIN*

Daniel E. Koshland, Jr. and Harvey M. Levy

Brookhaven National Laboratory
at New York University, New York

Ever since the hydrolysis of ATP by myosin was discovered the mechanism of this reaction has been a subject of intensive research. In some way the energy of the phosphate bond must be transferred to muscle protein so that chemical energy can be transformed into mechanical work. Formation of a phosphoryl-myosin intermediate has been suggested by many workers, but evidence for such an intermediate has been elusive. Exchanges between ADP labeled with ^{32}P and ATP have given negative results (Koshland, Budenstein and Kowalsky, 1954) and no ^{32}P-labeled myosin, other than that absorbed by non-covalent forces (Gergely and Maruyama, 1960), has yet been identified with the hydrolysis reaction.

If one examines the possible ways in which the ATP molecule might split, two possibilities exist: cleavage could occur at the terminal phosphorus or at the middle phosphorus atom. To resolve this problem, ^{18}O studies were initiated. Under all circumstances cleavage occurred between the terminal phosphorus and its oxygen atom (Table 1) as shown by the absence of ^{18}O in the ADP produced (Koshland and Clarke, 1953; Levy and Koshland, 1959). Under certain circumstances—for example, in the presence of calcium ion—

*Daniel E. Koshland, Jr. and Harvey M. Levy, "Evidence for an Intermediate in ATP Hydrolysis by Myosin," *Biochemistry of Muscle Contraction*. ed. J. Gergely. Boston: Little, Brown & Co., 1964, pp. 87–93. Reprinted with the permission of the authors and of the publisher.

Research carried out at Brookhaven National Laboratory under the auspices of the United States Atomic Energy Commission and at the New York University through the support of the United States Public Health Service, Research Grant 6276.

Table 1

Oxygen Exchange Catalyzed by Myosin and Actomyosin

Conditions: 0.125 M Tris, pH 7.4, 0.10 M KCl, 0.01 M $MgCl_2$, 0.01 M ATP; 2–6 mg of protein per ml; 25°C.

$$\text{Ad—O—}\underset{\text{O}^-}{\overset{\text{O}}{\overset{\|}{\underset{|}{\text{P}}}}}\text{—O—}\underset{\text{O}^-}{\overset{\text{O}}{\overset{\|}{\underset{|}{\text{P}}}}}\text{—O} \;\wr\; \underset{\text{O}}{\overset{\text{O}}{\overset{\|}{\underset{|}{\text{P}}}}}\text{—O}^- \longrightarrow \text{ADP} + {}^{18}\text{O}^-\text{—}\underset{\text{O}^-}{\overset{\text{O}}{\overset{\|}{\underset{|}{\text{P}}}}}\text{—O}^-$$
$$\text{H} \;\wr\; {}^{18}\text{OH}$$

System	^{18}O Atoms per Phosphate Group	
	ADP	P_i
Myosin	None	3.4
Actomyosin	None	2.8

the amount of oxygen observed in the terminal phosphorus introduced by ^{18}O from the water was the theoretical number, that is, one out of the four of the phosphate ion (Koshland, Budenstein and Kowalsky, 1954). However, in other cases—for example, in the presence of magnesium ion—this number was found to be much higher than theoretical, for example, three out of the four oxygen atoms (Levy and Koshland, 1959). The extra oxygens must have been introduced by an exchange reaction.

In studying the nature of this exchange reaction, it was soon found that the results could be explained by postulating an intermediate step in the myosin-catalyzed hydrolysis (Levy and Koshland, 1959). Myosin did not catalyze exchange of oxygen between H_2O and P_i, nor did it cause exchange between H_2O and unhydrolyzed ATP. Since exchange did not occur with the initial reactant, ATP, nor the final product, P_i, we deduced that it occurred at some intermediate stage in the reaction. Although the precise nature of the intermediate was not delineated from this experiment, it was the first direct evidence for an intermediate stage in myosin hydrolysis, and the ^{18}O was used as a probe for determining some of the properties of this intermediate.

A year or so ago Dr. Boyer was examining this system from a different angle and observed that there was a lack of rigor in the reasoning leading to an intermediate stage. To test this he and Dr. Dempsey (Dempsey and Boyer, 1961) incubated ^{18}O-labeled P_i in a medium of $H_2{}^{16}O$ and found that an exchange between the ^{18}O of the P_i and the ^{16}O of the water occurred during the hydrolysis of ATP. In the absence of ATP, this exchange was not detectable. Thus, Boyer and Dempsey confirmed our findings that no exchange occurred with the P_i of the medium in the absence of ATP; but they went further and established that in the presence of ATP an exchange with the P_i did occur. They then performed kinetic analyses that indicated that all of the exchange occurring in myosin hydrolysis could be accounted for by this ATP-induced exchange with the P_i of the medium. If true, this would essentially eliminate the evidence for the intermediate stage that we had postulated.

Table 2

OXYGEN EXCHANGE DURING ATP HYDROLYSIS

Conditions: 0.125 M Tris, pH 7.4, 0.10 M KCl, 0.01 M $MgCl_2$, 0.005 M P_i initially, 0.005 M ATP initially, 2 to 5 mg of protein per ml, 25°C.

Initial Position of ^{18}O	Per Cent Hydrolysis	Oxygen Atoms Exchanged/Phosphate Ion Designated		
		Initially Added P_i	Total P_i Present*	Calculated for Exchange at Intermediate Stage
P_i	27	0.28		
P_i	108	0.56		
P_i	130	0.50		
Water	127		1.44	2.00
Water	130		1.40	1.85

*P_i added initially plus P_i produced by hydrolysis.

Table 3

OXYGEN EXCHANGE DURING ITP HYDROLYSIS

Conditions: 0.125 M Tris, pH 7.4, 0.10 M KCl, 0.01 M $MgCl_2$, 0.005 M Pi, 0.005 M ITP, 2 to 5 mg of protein per ml, 25°C.

Initial Position of ^{18}O	Per Cent Hydrolysis	Oxygen Atoms Exchanged/Phosphate Ion Designated		
		Initially Added P_i	Total P_i Present*	Calculated for Exchange at Intermediate Stage
P_i	27	0.10		
P_i	95	0.02		
Water	71		0.76	1.92
Water	82		0.76	1.78

*P_i added initially plus P_i produced by hydrolysis.

To clarify this discrepancy we performed further experiments (Levy *et al.*, 1962) of the types reported in Tables 2 and 3. Parallel experiments were set up in which the conditions were identical except for the initial position of the ^{18}O label. The parallel experiments used the same buffer (magnesium chloride) and initial amounts of P_i and ATP. In one case, the water contained ^{18}O and the P_i was unlabeled, whereas in the other case the P_i contained ^{18}O and the water was unlabeled. The latter measures exchange occurring only via P_i of the medium. The former measures exchange occurring both at the medium level and at the intermediate stage. If the two values were the same, the evidence for an intermediate stage would be excluded, but if the latter were lower than the former, evidence for such an intermediate stage would be established.

In the ^{18}O-labeled P_i experiments shown in Table 2 it can be seen that approximately 0.5 oxygen atoms were exchanged per phosphate group during ATP hydrolysis. In the parallel experiment with ^{18}O initially in the water the P_i of the final product contains approximately 1.4 atoms per phosphate ion. This is an average value for the total P_i present at the end of the experiment, i.e., P_i produced by hydrolysis and P_i initially present. Thus many more atoms are exchanged in the experiment which is sensitive to an intermediate as well as a medium exchange than in an experiment which is sensitive only to a medium exchange. Thus, both types of exchange must exist.

If one assumes that the exchange is made up of two components, an exchange reaction with the P_i of the medium, as postulated by Boyer, and another exchange occurring at an intermediate level, as originally postulated by us, it is possible to correct these values so that one may calculate the number of atoms exchanged at the intermediate stage. When this is done it is found that two atoms of oxygen exchanged per phosphate group at the intermediate stage (last column of Table 2). It should be emphasized that the two atoms introduced by exchange are over and above the one atom introduced due to cleavage of the phosphorus-oxygen bond by hydrolysis. The figure 2.0 for intermediate exchange is greater than the 1.4 observed for total P_i because the latter is averaged over all the phosphate present including the 0.005 M P_i added initially. The 2.0 figure is calculated for the P_i produced by hydrolysis, since the initially added P_i should undergo only a medium exchange.

Although the difference between the exchange values in the ATP experiments were outside of experimental error, it was still desirable to establish this conclusion with even greater certainty. If the medium exchange was a time-dependent exchange, as pointed out by Boyer, then an accentuation of the difference between the medium-exchange reaction and the intermediate exchange might be obtained by studying ITP. ITP is hydrolyzed by myosin in the presence of magnesium at approximately ten times the rate of ATP. Nevertheless, it was found to exchange about the same amount of oxygen overall. Therefore, the medium-exchange reaction should be drastically reduced for the same per cent hydrolysis since much less time is required. In Table 3 such evidence is presented. In this case the time required for 70 per cent hydrolysis was very short and during this time interval essentially no exchange occurred between ^{18}O-labeled P_i and $H_2{}^{16}O$. That is, the amount of exchange at the medium level was not detectable. On the other hand, in the experiment with ^{18}O-labeled H_2O where exchange at the intermediate level is detectable a figure of 0.76 atoms of ^{18}O per total phosphate present was obtained. Using the same type of calculation used in Table 2, the exchange per phosphate group produced by hydrolysis of ITP is 1.9 atoms or essentially two out of the four atoms. These results therefore confirm the conclusion that exchange does, indeed, occur at the intermediate level.

To confirm this result further, another method of accentuating the intermediate exchange was performed. Dinitrophenol is known to activate the magnesium-catalyzed hydrolysis of ATP by myosin. Therefore, the ATP experiments were repeated, but this time dinitrophenol was added so that the hydrolysis was essentially complete in 1.5 hours. The amount of exchange between ^{18}O-P_i and $H_2{}^{16}O$ was negligible (0.027) in the short interval (Table 4).

Table 4

OXYGEN EXCHANGE DURING ATP HYDROLYSIS ACTIVATED FIVE-FOLD WITH 0.03 M DNP

Initial Position of ^{18}O	Per Cent Hydrolysis	Oxygen Atoms Exchanged/Phosphate Ion Designated		
		Initially Added P_i	Total P_i Present*	Calculated for Exchange at Intermediate Stage
P_i	84 (1.5 hr)	0.027		
P_i	105 (20 hr)	0.45		
Water	90 (1.5 hr)		2.46	2.44
Water	120 (20 hr)		2.90	2.23

*P_i added initially plus P_i produced by hydrolysis.

During the same interval when ^{18}O-labeled H_2O and $P_i{}^{16}O$ were used, an exchange of 2.46 atoms per phosphate occurred. Again correcting for the small medium exchange, it is possible to calculate that under these conditions 2.44 atoms of oxygen are exchanged per phosphate group at the intermediate stage in the reaction.

This system also allowed us to check our conclusions in regard to ATP which stood for longer intervals. Even though the hydrolysis of ATP was essentially complete in 1.5 hours we let the solution stand for 20 hours and the $P_i{}^{18}O$-$H_2{}^{16}O$ medium exchange increased, as to be expected for a time-dependent process.

Our conclusion from these studies, therefore, is that we have been able to confirm Dempsey and Boyer's discovery of a medium exchange between the P_i of the medium and the oxygen in the water catalyzed by myosin during ATP hydrolysis. However, we disagree in the finding that this exchange can account for all of the ^{18}O exchange occurring during hydrolysis of ATP or ITP by myosin. In our routine experiments the amount of exchange occurring at the intermediate stage is appreciably higher than that occurring at the medium level and under special circumstances—for example, with a rapidly hydrolyzing nucleotide such as ITP or in the presence of an activated ATP hydrolysis, i.e., with dinitrophenol—essentially no medium exchange at all is observed in an interval when intermediate exchange occurs readily.

The reason for the discrepancy in these two laboratories is not yet understood. It has been our pleasure to have a most cooperative and friendly exchange with Dr. Boyer. From his data above we would certainly be led to the same conclusions which he has obtained. It seems therefore that several possibilities exist: (1) an inhibitor is present in his preparation; (2) an activator is present in our preparation; or (3) some basic modification of the protein has occurred in one laboratory or the other to lead to these rather different results. We had found under other circumstances, for example, in the presence of barium, that no ^{18}O exchange occurs (Levy and Koshland, 1959). Thus the ^{18}O-exchange reaction must depend on some general features of the active site, and these features do not necessarily parallel nucleotide triphosphate hydrolysis. Thus it is not unreasonable that an inhibitor or some modification of the protein itself would change the conformation at the active site in such a way that the ATPase activity would be retained but the ^{18}O-exchange reaction would be decreased.

Once convinced that an ^{18}O-exchange reaction did indeed occur at an intermediate stage, we used this property to study some of the properties of the protein. Since these studies have been reported elsewhere (Levy *et al.*, 1960; Yount and Koshland, 1963), I shall only summarize them very briefly. The most striking conclusions are that the properties of the ^{18}O-exchange reaction do not parallel the properties of the hydrolysis rate. The metal-ion dependence, the *p*H dependence, the nucleotide specificity, and the temperature dependence of the two processes are different.

On the other hand, there are some striking similarities between the ^{18}O-exchange specificity and the specificity for contraction. For example, the nucleotides which show the highest ^{18}O exchange are the most effective nucleotides for contraction. The metal ions showing the highest ^{18}O exchange are the most effective metal ions for contraction. It seems therefore that there are some properties of the active site which are essential for the ^{18}O-exchange reaction which are not essential for the overall rate of hydrolysis. These properties have certain strong similarities to the requirements for contraction. One can conclude that some features of the active site which are essential for contraction can be probed by this ^{18}O-exchange reaction and can lead to a greater understanding of the relation of the hydrolysis to the mechanism of muscular contraction.

PART VII

RELAXATION AND CONTRACTION

OUR APPROACH to the physiology of muscle contraction was greatly influenced, as has been mentioned, by the demonstrations of the mechanical effect of ATP on myosin. The concept of special regulators which promote relaxation was indicated in 1942 in Bailey's[1] extensive paper from the Low Temperature Station in Cambridge, England. Bailey pointed out the physiological importance of calcium ions as a stimulating agent of myosin ATP^ase; yet in order for the ions to exert a regulating effect on contraction and relaxation, they, as well as ATP, must be confined to certain areas.

One might mention that there was good evidence that ATP was confined to the isotropic (single refractive) dark bands of muscle. This was first demonstrated by Caspersson and Thorell [q.v.]* at the Karolinska Institutet in Stockholm by means of their scanning methods of absorption in the ultraviolet, using a specially constructed quartz microscope. Subsequent observations by Hoagland and coworkers[2] showed likewise that isotropic zones revealed in photomicrographs in polarized light correspond exactly to the deeply absorbing striae revealed in the ultraviolet. These observations were made by a method in which half of a field of a given muscle section, photographed in polarized light, could be compared with the remaining half photographed in ultraviolet light.

More recently D. K. Hill[3] studied the localization of adenine nucleotides in the myofibril using tritium labeling and autoradiography after precipitation of nucleotides with uranyl acetate. Hill's observations confirmed the early work by Caspersson and Thorell that the isotropic bands (I bands) are the seat of adenine nucleotides. Their detailed electronmicrographs also made it possible to state that while most of the nucleotides are found between the fibrils, they are found also in a component of the reticulum.

Meanwhile, when the glycerinated muscle preparation became a common biological test preparation, Marsh [q.v.] and Bendall[4], who were also at the Low Temperature Station in Cambridge, reported that they were able to convert a glycerinated muscle preparation from a contracted to a relaxed state by means of a protein fraction from muscle. This fraction was called the *relaxing factor*. Important further solutions to this problem originated particularly in Ebashi's laboratory in Tokyo and in Weber's group in Heidelberg. This collection includes a paper by Ebashi [q.v.] in which he correlates Kielley and Meyerhof's[5] early observations on a magnesium-dependent ATP^ase in the supernatant with the presence of relaxing factors in the same fractions. The test system actually contains a soluble enzyme, submicroscopical granules which are rich in calcium ions and myosin or actomyosin filaments.

The question regarding the nature of conformational changes occurring in myosin or actin during contraction might eventually be elucidated by wide-angle X-ray crystallography [cf. H. Huxley[6]].

References

1. Bailey, K.: *Biochem. J.,* (1942).
2. Hoagland, C. L., Shank, R. E., and Lavin, G. I.: *J. Exp. Med.,* **80** (1944), 9.

*See material in list of reprinted papers.

3. Hill, D. K.: *J. Physiol.,* **153** (1960), 433. ———: *J. Cell Biol.,* **20** (1964), 435.
4. Bendall, J. R.: *Proc. Roy. Soc. London B,* **142** (1954), 409.
5. Kielley, W. W., and Meyerhof, O.: *J. Biol. Chem.,* **176** (1940), 591.
6. Huxley, H.: *Biophys. Symp.,* Vienna 1966.

THE LOCALIZATION OF THE ADENYLIC ACID IN STRIATED MUSCLE-FIBRES*

T. Caspersson and Bo Thorell

Chemical Department, Karolinska Institutet,
Stockholm, Sweden

(Received for publication, April 15, 1942)

Of the groups of compounds that go to build up the main part of the musculature, only two have a marked selective light absorption in ultra-violet† between 2,000 and 3,000 Å, namely the purine derivatives and the proteins. Of the first-mentioned, the adenylic acid and the adenyl pyrophosphoric acid constitute more than 95%. The pyrimidine nucleus in these latter causes a pronounced absorption-band at 2,600 Å. The selective absorption of the albuminous substances is conditioned by tyrosin and tryptophan. These amino-acids both have absorption-bands around 2,800 Å. The average ultra-violet absorption of the muscle-tissue will consequently be predominantly determined by the sum of the absorptions of the adenine derivatives, the tyrosin and the tryptophan. Owing to the high specific absorption of the adenylic acid—at 2,600 Å it is more than 100 times higher than that of myosin—its absorption-band will have approximately the same height as that of the muscle-protein in spite of the low concentration. (Cf. below, Fig. 1.).

Investigations carried out by different authors on material from widely differing animal-species show an extensive agreement in the adenine content as well as in the tyrosin and tryptophan content of the muscle-protein (Bailey 1937), so that there seems to be justification for the assumption that the main features of the structure are the same for large groups of animals.

*Excerpt from T. Caspersson and B. Thorell, "The Localization of the Adenylic Acid in Striated Muscle-Fibres," *Acta Physiol. Scandinav.*, **4**, 97–117 (1942). Reprinted with the permission of the authors and of the publisher.

†The unorthodox spelling of biochemical terms has been left untouched since it does not interfere with the clarity of the text. The European spelling of amino acids, such as tyrosine and tryptophane, has also been preserved.

Figure I shows an absorption-spectrum of muscle-fibres from Drosophila funebris, taken with the technique described below (cf. page 531). The absorption-curve shows a pronounced maximum at 2,600 Å. Above this is

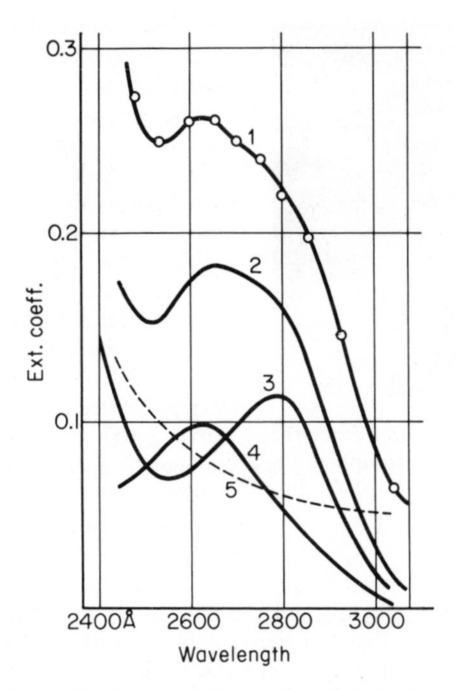

Fig. 1. Curve 1, absorption-spectrum of living muscle-fibre from Drosophila funebris, taken with so little resolving power that the striation has no influence on the course of the curve (cf. text, page 528). 2, the same absorption-spectrum after correction has been made for light-dispersion (curve 5 has been subtracted). 3 and 4, the absorption-curves for myosin and adenylic acid respectively, in which curve 2 can be resolved. 5, the course of the light-dispersion, calculated on the assumption that at 3,100 Å this conditions the whole absorption and is inversely proportional to the fourth power of the wavelength.

superposed a maximum at 2,800 Å, appearing chiefly as an increase in breadth of the first-mentioned max. The course of the absorption-spectrum of the adenylic acids is known (Dhéré 1906, Gulland and Holiday 1934, Heyroth and Loofburouw 1931) and the myosin can be calculated (cf. Caspersson 1940). The measured curve can then easily be resolved into the two component curves (cf. Caspersson loc. cit.), as has been carried out on Fig. 1. In order to make such a resolution into two components one needs only two points. The fact that the agreement in all curve-points is good shows that no substances with other absorption-character than those of the adenine and the protein markedly affect the absorption of the muscle-tissue. The concentration-relation adenylic acid to protein of the absorption-type of the myosin, which may be calculated from the curves, is very near 1:100 in good agreement with the chemical investigations of musculature from different animals which give as an average 10—20% protein and 0.1—0.2% adenylic acid.

Fig. 2. Living muscle-fibres from the thigh-musculature of Drosophila funebris photographed at 2,570 Å. Enlarged 720 times.

At the adenine absorption-maximum the ultra-violet microphotograph of the living muscle-fibre (Drosophila) clearly shows the striated structure, in that certain segments absorb more than others (Fig. 2). It is not possible to decide off-hand whether we have here to do with a refraction phenomenon or a true absorption. If photographs of the same living muscle-fibre are taken with a *wide aperture for the illuminating bunch of rays* one observes the following: the striated structure appears between 2,400 Å and 2,950 Å—most clearly between 2,500 and 2,850 Å. Under 2,400 Å the absorption in both segments is so high that the structure appears only as an indication; over 2,950 Å the absorption in both segments is so low that the picture of the structure begins to disappear. Such series have been performed on musculature from frog, fresh-water crayfish and the fruit-fly, with identical results. They thus show preliminarily that the striation in the ultra-violet picture must be at least partly conditioned by real light-absorption, and not only by refraction-phenomena as in visible parts of the spectrum. It appears, moreover, that the absorptions in the different muscle-segments change in different ways with the wave-length. In order, however, to decide which substances condition the absorption, *complete absorption-spectra must be measured for the different structural elements.*

A method for the measurement of absorption-spectra of small objects has been worked out by one of the present authors and has been described elsewhere (Caspersson 1936, 1940). The lower limit for the magnitude of the element that can be measured is determined by the wave-length of the light employed. In order to obtain a detailed absorption-spectrum the magnitude should not be less than four times the wave-length of the light, as the light-diffraction will otherwise introduce a factor that is difficult to calculate. The measuring is considerably facilitated and accuracy increased if the dimensions are somewhat greater. As striated musculature in general has such short segments that the above-mentioned value is approached, it is most suitable to look for material with especially long segments.

Investigations on Drosophila Funebris

For these investigations muscle-fibres from coxa, trochanter and femur were used. This musculature has segments of up to 10 to 15 μ. The appearance at 2,570 Å of the living muscle-fibre may be seen in Fig. 2. Certain segments absorb strongly. The nuclei, which lie within an axial zone, contain isolated grains of strongly absorbing substance (nucleic acids).

As the measuring takes a comparatively long time, during which the preparation must be quite motionless and unchanged, it is not possible to measure living material; the preparation must first be treated with an appropriate fixing fluid. This must fulfil the following requirements: (1) The distribution of the absorbing substances that are to be investigated must not be noticeably changed. (2) The protein substances must not be precipitated so that surfaces entailing an abrupt alteration in the refractive index can arise. Owing to the impossibility of defining the physico-chemical conditions in the muscle-fibre of the substances in question, model experiments are of but slight value. Instead, a direct comparison was made in ultraviolet of the living muscle-fibre with that which had been treated with fixative. A series of different combinations were tried, the best results being obtained with 25% acetic acid in absolute alcohol saturated with lanthane acetate. In this fixative a muscle-fibre is not noticeably changed in the aspects referred to in the course of twenty-four hours.

When measuring, a region with an approximate diameter of 1 μ was projected into the photo-electric cell. Such measuring-points have been marked in Fig. 3. By means of a special arrangement it is possible to move the preparation

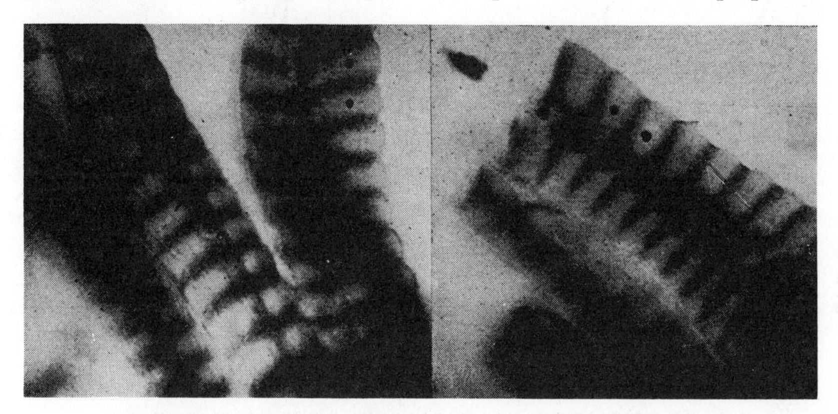

Fig. 3. Muscle-fibres from Drosophila. The dots mark the position and approximate size of the fields that in connection with the measurement were projected into the photo-cell. 2,570 Å 720 times. A pair of curves from each fibre are to be found in Figure 4.

and afterwards return it to its original position with a smaller error than corresponds to the magnitude of the smallest element that the objective can resolve, i.e. 0.05 μ. The selected measuring regions are projected into the photo-cell in turn. Measuring takes place in wave-length after wave-length, as a rule from 2,480 Å to 3,100 Å.

Little-working musculature

The preparation was made from young flies from flask cultures. The flies were anaesthetized with ether and dissected, after which the preparation was immediately transferred to the fixative or was examined at once in the body-fluid.

Figure 4 shows absorption-curves from three different muscle-fibres. In

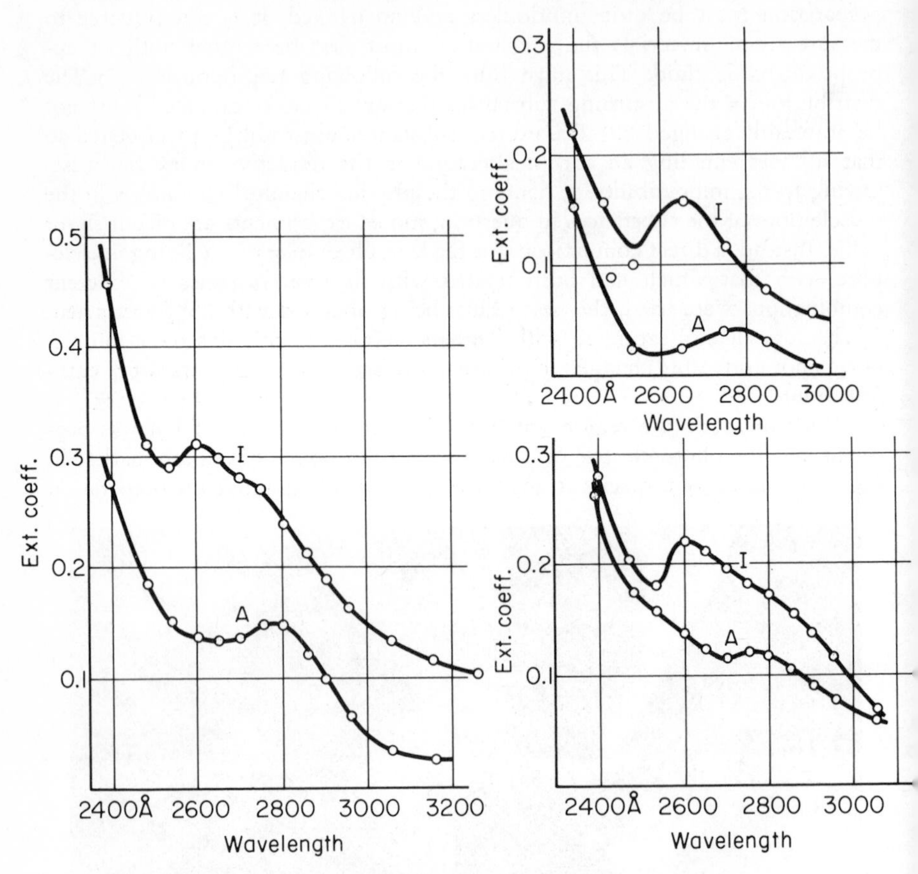

Fig. 4. Absorption spectra from strongly ultra-violet absorbing (isotropic) segments and weakly ultra-violet absorbing (anisotropic) segments lying adjacent to one another (cf. Fig. 3). Three different muscle-fibres. The segments are marked I and A respectively according to the double-refractive conditions.

each part-figure are drawn 2 curves (I and A respectively), deriving from two adjacent segments. All show a strong adenine band in the one segment, while this is only weakly represented or entirely missing in the other segment. In the more weakly absorbing segment the protein band appears clearly. There is a protein-band of the same magnitude also in the other segment; but this appears only as an increase in breadth or bulging of the adenine maximum. In some muscle-fibres also the more weakly absorbing segment shows an adenine absorption of varying strength (cf. Fig. 4c). This is, however, always considerably lower than in the adjacent segment.

TABLE OF QUOTIENTS BETWEEN THE EXTINCTION-COEFFICIENTS FOR THE WAVELENGTHS 2,650 Å AND 2,850 Å FOR POINTS IN DIFFERENT SEGMENTS

Isotropic segment	Adjacent anisotropic segment
1.48	1.03
1.44	1.27
1.47	1.22
1.49	0.96
1.59	1.17
1.45	1.37
1.42	1.14
2.00	1.42
1.83	1.24
1.39	1.17
1.33	0.97
1.92	1.15
1.61	1.11
1.66	1.18
1.85	1.00
1.80	1.28

In order to demonstrate the magnitude of the variations in the ratio archive the table gives a series of quotients between the extinction coefficients at 2,650 and 2,850 Å for pairs of adjacent segments from different muscle-fibres in different preparations. The value of this quotient for pure adenylic acid is 2.3 and for a protein of the composition of myosin 0.9 (cf. Fig. 1). These quotients have throughout a higher value for the segments that in the table are referred to under I, which shows that these are richer in adenine nucleotides.

From these quotients we can calculate the approximate relation between adenylic acids and protein of the myosin type. These relations are illustrated in Fig. 5. (Cf., however, the paragraph below.)

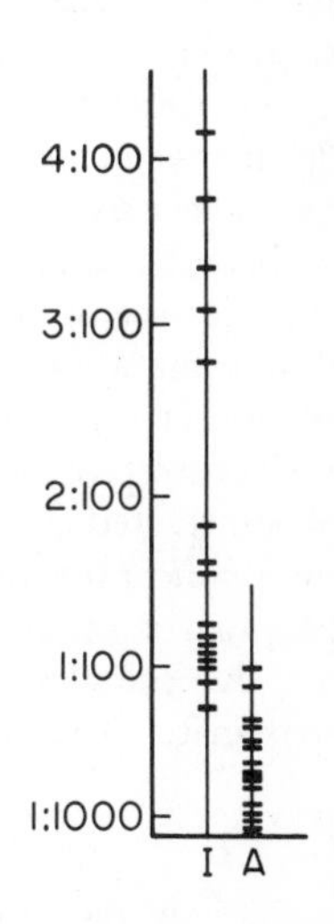

Fig. 5. The relation of adenylic acid to protein in different segments, corresponding to the quotients in the table, calculated without correction for the light-dispersion. Although the light-dispersion has a strong tendency apparently to equalize the differences (cf. text), the big differences may be clearly seen.

In calculating the absolute quantities of adenylic acid and proteins from the measuring data one must also observe the loss of light arising through light-dispersion. In the complete absorption-spectra it is possible to introduce quite a good correction for this, as this loss of light is approximately inversely proportional to the fourth power of the wave-length. If there were no such dispersion of light the ultra-violet absorption would correspond to that of solutions of the substances in question and thus over 3,000 Å fall to a low value. In the muscle-fibres, however, a weak absorption still appears even at 3,200 Å, which must be due to light-dispersion. As the wave-length relation of the latter is known one can introduce a correction for this (see curve 5, Fig. 1). This correction cannot, of course, be introduced for the quotients given above, so that the value of the latter is limited to affording a support to the general validity of the data given above for a large material and a demonstration of the range of their variation. The error that is introduced in the quotients through dispersion of light entails in the quotients a stronger rise of the extinction coefficients at the shorter wave-length and affects most the A-segment. It will thus considerably affect the result in *a direction contrary to the expected effect,* so that it cannot falsify the result.

The spiral arrangement in the muscle-fibres also complicates the measuring. This arrangement can be observed in at least large parts of the muscle-fibres, and it often makes it difficult to get muscle-partitions freely projected from adjacent segments. The error that may be conceived to arise goes in the same direction as the previous one, i. e. it tends erroneously to equalize the concentration differences.

Absorption-spectra and the quotients given in the tables thus indicate that the greater part of the adenine derivatives are localized to the segments with the strongest ultra-violet absorption. In the majority of the muscle-fibres with especially long segments that were measured, the concentration-relation was considerably below 1 to 10 (calculated after correction for light-dispersion). The protein concentrations cannot be compared with any high degree of accuracy. Certain, there are reasons for assuming that the distribution of amino-acids in myosin is fairly homogeneous in the animal kingdom, but we have not sufficient experimental data for the other protein substances. If we assume that these have approximately the same tyrosin and tryptophan content as myosin (as seems to be the case for rabbit musculature), the protein concentration in the weaker absorbing segments of Drosophila muscle would be somewhat higher than in the other segments. The greatest difference that was observed corresponds to approximately 3:2. When the adenine absorption has been subtracted also the absorption under 2,500 Å gives a measure of the protein concentration that however is not alone determined by the tyrosin and tryptophan content. Owing to the non-selective character of this part of the curve, this calculation, too, will include considerable errors. It gives however, approximately the same values as the estimation of protein concentration given above.

Comparison between Ultra-Violet Absorption and Double Refraction

In order to decide what structural elements observable with other methods correspond to the segments rich in adenylic acid, living muscle-fibres were photographed in both ultra-violet and polarized light. Comparison of the plates showed that *the strongly ultra-violet absorbing substances were localized in the isotropic segments.*

Working musculature

The muscle-material has in all the cases so far discussed derived from flies in narrow culture-flasks. Cultivation took place at room-temperature.

Under these conditions the flies alternately hop and fly short distances. The best material for experimental purposes was obtained from young individuals whose skin had just commenced to harden. These do not move about much. Before dissection they were anaesthetized with ether or chloroform. The muscles may thus be regarded as working relatively little. In order to make a comparison with especially strongly working musculature the flies were made to perform as much work as possible with their leg muscles. Faradic stimulus was applied so that the flies, deprived of their wings, had to move about on a plate with two comb-shaped electrodes, whose teeth interlocked and which were connected up with a source of current. The stimulus was so strong that the flies sprang several mm. into the air when they contacted it. A stimulus of a more physiological kind was obtained by letting the flies move about on a plane disc whose slope towards the horizontal plane was continuously changed. In this way they could be kept in continuous movement for several minutes.

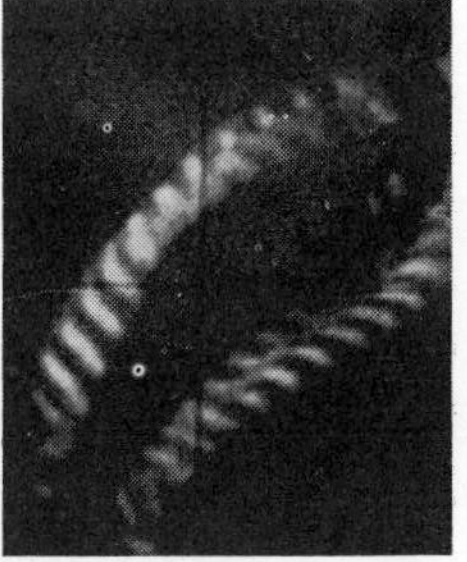

Fig. 6. Living muscle-fibre from Astacus between crossed Nicol's prisms. With the aid of the fine details in the original, which appear too weakly on reproduction, comparisons may be made with the corresponding ultra-violet picture, when it will be seen that the strongly absorbing segments correspond to the isotropic segments. 720 times.

The muscle-preparations were made in the same way as described above, and were examined both in the living and in the fixed states. The structure showed a constant difference as compared with the resting muscle-fibres in that the sharp differentiation of strongly and weakly absorbing bands was more or less smudged out (Fig. 7). The strongly absorbing segments at lower magnification appeared to be increased in height and diffusely delimited. The structure is never quite wiped out. At high magnitudes the absorbing material in the resting muscle gives the appearance of rows of small irregular heaps of grains lying close together. The size of the single grains is near the limit of the resolving power of the ultra-violet microscope. In the fatigued muscle the anisotropic segments are extremely finely longitudinally striated. The striation appears most clearly nearest to I and diminishes in clarity towards the middle of A. The structure is conditioned by extremely fine streaks of strongly ultra-violet absorbing substance, which are connected with the heaps of grains in I. Probably the only explanation of how such a picture could arise is that strongly absorbing substances from I have wandered down towards the middle of A through a system of very fine longitudinal fibrils. As the absorption-character of the substances agrees on the ultra-violet photographs with that of the substance of the I-segments, and despite their fineness the striae contrast sharply with the surrounding protein, the specific extinctions must lie much higher than for the protein-substance in A, and be comparable with those of the adenine derivatives in I. It must thus be the adenine derivatives in I that in the course of muscle-activity partly wander down towards A. This probably also explains the circumstance that certain of the above-mentioned quotients and absorption-spectra from only slightly working musculature show a varying though low content of these substances in A.

In experiments of this type it is difficult to avoid artefacts. In order to exclude such we investigated unfixed muscle-fibres that had been left to lie in body-fluid or in Ringer's solution for varying periods. When muscle-fibres had been thus left for a longer period, pictures of an extension of absorbing substances from I to A also appeared. The streaks that appear in this connection are, however, coarse and crude, and at the same time one observes changes in the fine structure of I. In connection with attempts to find suitable fixing media (page 531) these last-mentioned pictures were also observed. There is thus a very definite difference between the pictures of fine striae that appear in the fatigued muscle and are lacking in the resting muscle, and the pictures that are observed in both resting and fatigued muscle-fibres when the

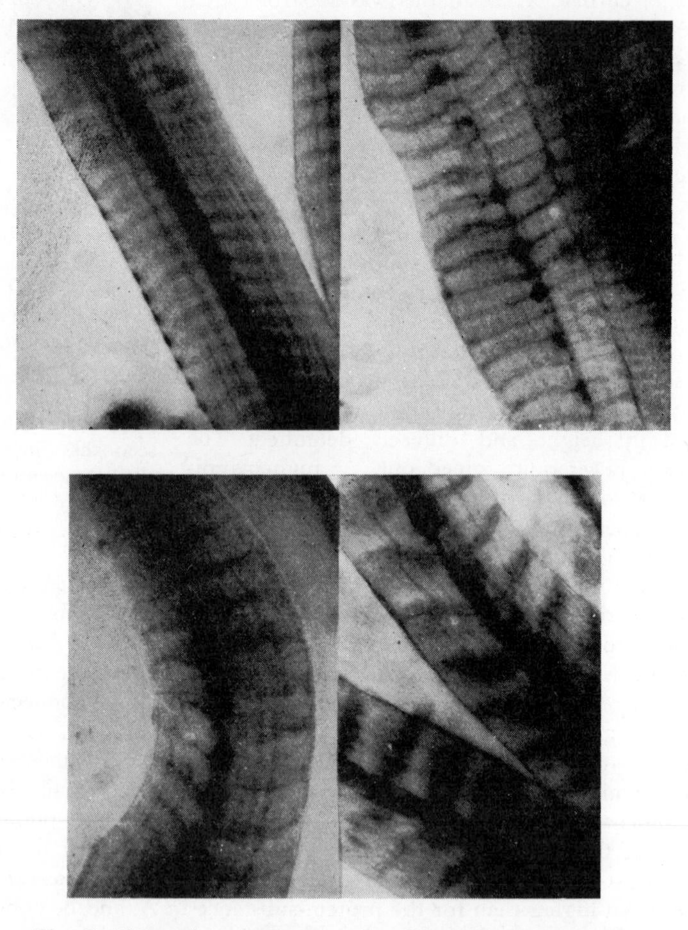

Fig. 7. Two pairs of pictures taken from living muscle-fibres from Drosophila at 2,570 Å and showing in low and high enlargement respectively the difference between strongly working (left pictures) and feebly working (right pictures) muscle-fibres.

preparation is left to lie for a long time or is treated with unphysiological solutions. It thus seems justified to assume that what has been described above cannot be referred to artefacts, but that in working musculature the adenine derivatives wander down towards A.

Investigations of Other Musculature

In order to decide in how far the observations made with reference to Drosophila funebris possess general validity certain other musculature was also examined.

Astacus fluviatilis

In parts of its musculature the fresh-water crayfish has muscle-segments of magnitudes comparable with those found in Drosophila. Muscle-fibres from the dorsal hypodermal muscle-layer were examined. Ultra-violet microphotographs showed the same typical arrangement as has been described above for Drosophila. A series of quotients were taken, also in the same way as above-described and with the same result as for Drosophila. The polarization investigation showed that the absorbing substance lay collected in the isotropic segments. (Fig. 6.)

Ascaris megalocephala

In the extremely large sub-epidermal muscle-cells of Ascaris only a smaller part of the cytoplasm is differentiated as myofibrillae (cf. Fig. 8). Here then the possibility presents itself of examining in one and the same cell equally thick layers of differentiated and non-differentiated cytoplasm and to measure absorption-spectra with moderate enlargements. The fibrilla-part has a strong ultra-violet absorption. Absorption-spectra therein show a high concentration of substances with purine absorption as well as a high protein concentration, fully comparable with the musculature of the insects and vertebrates. In the fibrilla-free part, on the other hand, there are only extremely small quantities of substances with the 2,600-band, so that the protein-band appears almost pure (Fig. 9). The protein concentration is also rather lower in these parts.

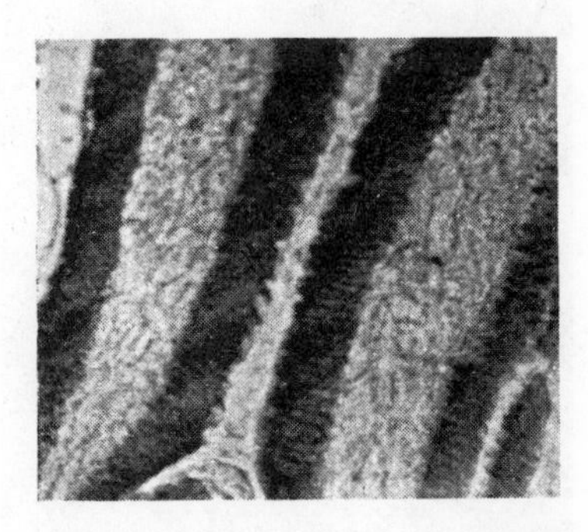

Fig. 8. Parts of two muscle-cells from the sub-epidermal muscle layers of Ascaris. The peripheral layers differentiated. Wavelength 2,570 Å. Enlarged 300 times.

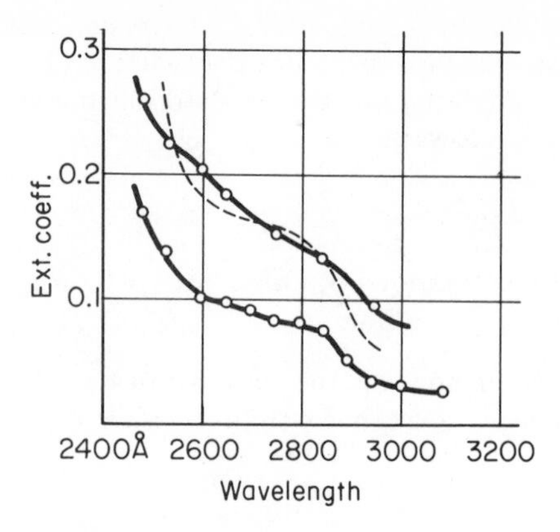

Fig. 9. Absorption-spectra of points in the differentiated (upper continuous curve) and undifferentiated parts (lower continuous curve) of muscle-cells from Ascaris. The dotted curve gives the values of the lower curve, multiplied by an arbitrary factor. Comparison with the curve from the differentiated part shows, even without correction for light-dispersion, that the absorption of this part is proportionally higher than that of the undifferentiated part around 2,600 Å (the adenylic acid band) but lower around 2,800 Å (the protein band) and below 2,500 Å (where the protein absorption rapidly rises while the adenylic acid absorption sinks). From this it appears that the differentiated part contains much more adenylic acid than the other part.

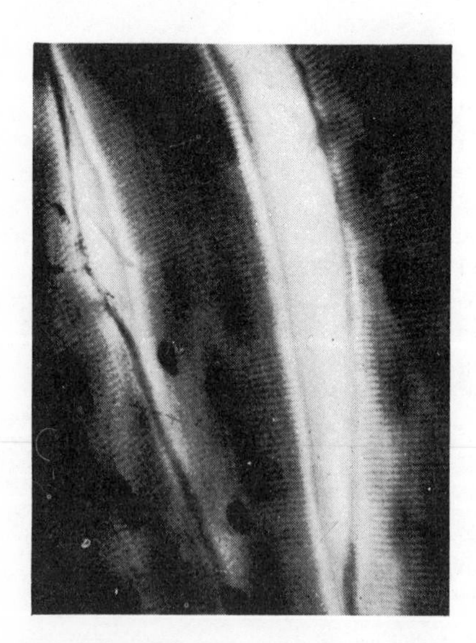

Fig. 10. Living muscle-fibres from frog, photographed in body fluid at 2,570 Å. Enlarged 720 times.

Vertebrates

Striated musculature from frog, rat and dog was examined with ultra-violet micro-photography. The arrangement of alternately strongly and weakly absorbing segments appears clearly. Fig. 10 shows a living muscle-fibre from a frog. That here, also, despite the smallness of the structure, we have to do with a true absorption and not only with light-refraction appears from serial photographs in different wave-lengths, which show the same conditions as have been described above for Drosophila musculature. Owing to the shortness of the muscle-segments and the great thickness of the muscle-fibres in comparison therewith no attempts were made to work out detailed absorption spectra.

In order to compare the relation

between double-refracting and ultra-violet absorbing segments finely pulverized asbestos was sprinkled over the preparation of the living muscle. On the ultra-violet photographs and the plates that were taken in polarized light it was then possible to identify isolated segments with the guidance of the sharp-edged, double-refracting grains of asbestos. *Here, too, the vaguely double-refracting and the strongly ultra-violet absorbing segments were in agreement.* (In certain material also a very fine strongly absorbing line goes through the central part of the strongly double refracting segment.)

Cardiac musculature also shows the typical arrangement of strongly and weakly absorbing bands. In the fibres of Purkinje, however, there is striation only in the outermost layers and there merely as an indication. Absorption-spectra were measured from the central region of Purkinje's fibres(calf) and from the adjacent musculature. The preparation was 5 μ in thickness and the diameter of the region that was projected into the aperture of the photocell about 5 μ. The muscle-segments were considerably lower than that value and were passed obliquely by the light-rays. The muscle-structure was thus not "resolved" by the photo-cell. As the light-absorption is dependent solely on the *amount* of substance that the ray of light passes through and is not affected by the distribution of the absorbing substance, the *absorption-curve in this case will be approximately the same as it would be if the substances were in molecular solution.* The only condition that must be fulfilled if the structure-factor is to be eliminated in this fashion is that the structure-elements shall be numerous and so small in comparison with the magnitude of the measured region that the different parts of the photo-cell receive more or less equal amounts of light. Figure 11 shows absorption-curves. They show that Purkinje's fibres lack the high concentrations of adenine derivative that are characteristic of the contractile muscle-cells.

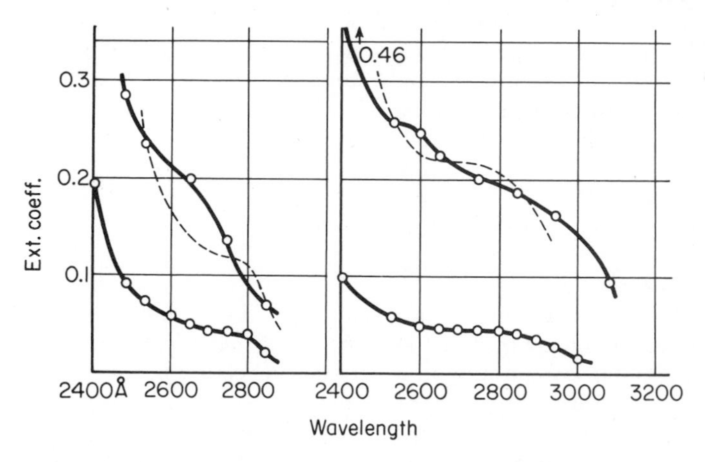

Fig. 11. Two pairs of curves from points in Purkinje's fibres (the lower continuous curves) and from adjacent cardiac musculature (the upper continuous curves). The dotted curves have been plotted in the same way as in Figure 8 and show the same conditions as there.

Survey of the Results

The ultra-violet absorption exercised by the adenine derivatives at 2,600 Å is as compared with protein substances of the muscle-protein type exceedingly high—the order of magnitude of the extinction-coefficient relation is about 100:1; in consequence, elements that are rich in adenine derivative will appear clearly in ultra-violet pictures taken at this wave-length even if large quantities of protein are in the immediate vicinity. Owing to the dispersion of light in the muscle-structure, that is composed of elements with considerably different refractive index, such pictures are, however, far from unambiguous.

As compared with other tissues the musculature is distinguished by great quantities of adenylic acids. In order to demonstrate that these are for the most part combined with the contractile elements, absorption-spectra of the epithelial muscle-cells from *Ascaris* were measured, where only a part of the cytoplasm is differentiated to form myofibrillae. The fibril-part showed a high concentration of substances with the adenine-band (as well as fairly high protein concentrations), while the fibril-free cytoplasm showed a pure protein spectrum. A similar comparison is offered by Purkinje's fibres in the mammalian heart. In the calf they possess only a thin layer superficially differentiated to myofibrillae. Absorption-spectra of the part poor in fibrillae show a very low content of substances with selective absorption about 2,600 Å, while the cells of the adjacent cardiac muscle-tissue with the same layer-thickness show a high band in this place.

By measuring complete ultra-violet absorption-spectra of the isotropic and the anisotropic parts in the musculature of arthropods with especially long segment it was possible to show that the main part of the adenine derivatives in non-working musculature lie collected in I. The partitions of the vertebrate musculature are too small to allow of an exact absorption-measurement; but the photographic series taken at various wave-lengths show with very great probability that the case is the same also here.

Attempts to compare fatigued and unfatigued musculature were carried out on Drosophila material, which has especially long segments. In the fatigued musculature substances containing adenylic acid were seen to wander down from the isotropic to the anisotropic segments.

Discussion

A number of different circumstances speak strongly in favour of the assumption that myosin, first investigated in detail by Edsall and Muralt, is the prime factor in muscle-contraction. It has been found in all examined muscle-substance, and its composition in widely varying classes of animals is strikingly homogeneous. (See Bailey, 1937). That the long polypeptide chain of the myosin molecule, like that of keratin, can be stretched and contracted by folding of the molecule-chain has been demonstrated by Astbury and Dickinson (1935). These workers were in one case even able to show such a folding in living musculature in connection with contraction.

Comparisons between the polarization-optical and Roentgen-optical qualities of myosin fibres and muscle (see Weber) show that the double-refractive muscular part consists of myosin-rods arranged with complete axial parallelism. These rods have a diameter of about 45 Å and a length of at least 500 Å. The protein content in the single rod amounts to about 70%. The rods are built up of at the most 20 fibre-molecules the involuted state of whose molecular chains can be changed. The double-refractive muscle-segment thus constitutes an organ that is extremely highly specialized for the performance of a special task and in which for mechanical reasons there is but little room for other substances than the contraction-conditioning myosin. I is not in the strict sense isotropic; for Schmidt (1934) has shown that it has a double refraction that in the material investigated amounts to about 10% of A's. It is probably conditioned by small quantities of myosin. Other protein substances must, however, be present, as is shown both by desiccation experiments (Noll and Weber 1934) and the absorption-measurements given in the foregoing. The main part of the contraction takes place in A, while I's share has been discussed. Studnitz found that on a weak contraction I's height was not changed, though it diminished on strong contraction. Whether we have here to do with an active actual contraction or with a passive shortening brought about by A's change of form cannot be decided. The decrease in height is, even on strong contraction, less than that of the adjacent A. The rôle played by the isotropic segments in the mechanical process of contraction seems thus to be a subordinate one, or in any case less immediate than that of the anisotropic segments.

The glycolytic processes that constitute the energetic basis for the contraction seem in different types of muscle-tissue to follow a course similar to that found in the type of musculature that has been most carefully investigated, namely, skeletal muscle. (See regarding smooth musculature: Dworaczek and Baarenscheen, 1937; Meeraus and Lorbeer, 1937. Re cardiac musculature: Ochoa 1937). The amount of creatine phosphate varies. In certain muscles its task seems to be performed to a greater or lesser extent by arginin phosphate. Adenylic acid and adenyl-pyrophosphoric acid are invariably found.

Unambiguous data from various quarters show that the breaking up of the adenyl-pyrophosphoric acid, which reaction is exothermic and rapid, precedes the other processes in the breaking down of the carbohydrates. It probably takes place in the very first stage of the contraction (see e. g. Lundsgaard, 1938). The subsequent reactions: the breaking down of phospho-creatine and phospho-arginin etc. as well as the actual glycolysis, serve to restore the adenyl-pyrophosphoric acid and to maintain it at a constant phosphorylation level, which in the resting muscle seems to be high.

The demonstration in the foregoing that the adenine derivatives in the resting muscle are chiefly localized in the isotropic segments shows that the segmentation of the striated musculature implies a chemical differentiation covering more than the differences in myosin-concentration that the polarization-optical investigations render probable. The most probable explanation is that the formation of the adenylpyrophosphoric acid (by far the most predominant adenine derivative in resting muscle) takes place in these. The striation of the muscle would thus be conditioned by a chemical differentiation in a contractile part and a part where the formation of substances that supply energy for the contraction takes place. There is also a certain support for such a view in earlier observations. That the staining conditions in I and A change in connection with the contraction has been known for long, but on account of the complicated physico-chemical conditions it seems at present impossible to draw any conclusions whatsoever regarding the course of the chemical processes. With glycogen-staining in resting muscle Studnitz (1934) found glycogen both in I and A. After work the stainable granules disappeared, and on restitution they reappeared first in I. This speaks in favour of a particularly important rôle for I in the glycogen conversion. The observations alone that during strenuous muscular work the adenine derivatives wander along the fine fibrillar structure into A must, since artefacts can in all probability be excluded, be interpreted as a direct transport of adenyl-pyrophosphoric acid, from I direct to the contractile elements, as this acid constitutes the main part, possibly all, of the adenine derivatives during rest. The energy from the glycolysis in I is thus transported direct by adenylic acids to the contractile protein chains and there used for the contraction. Whether this takes place as an introduction to a contraction or as a restitution after a contraction is irrelevant in the present discussion.

As regards the chemical process in connection with the transferance of energy from the adenyl-pyrophosphoric acid that has flowed down in Q and is in contact with the oriented myosin-rods, absorption-data cannot, of course, give any help. Certain chemical investigations seem, however, to show that such a direct transferance as has been postulated as most probable in the light of the above data, is chemically quite conceivable. Engelhardt and Ljubimowa (later confirmed by Szent-Györgyi and Banga 1940) and D. M. Needham[1] found that *the ferment that splits the adenyl-pyrophosphoric acid cannot function except in the presence of myosin or else is myosin itself;* and they maintain that "the mineralization of adeninetriphosphate, often regarded as the primary exothermic reaction in muscle contraction, proceeds under the influence and with direct participation of the protein considered to form the main basis of the contractile mechanism of the muscle fibre."

J. Needham, Shen, D. M. Needham and Lawrence (1941) have further confirmed this in that they have shown that adenyl-pyrophosphate has a markedly strong capacity to decrease the double refraction of flow in a myosin solution. This effect, that is referred to a change in the degree of dispersity and the particle-length is, curiously enough, reversible. As a possible mechanism for the interaction between adenyl-pyrophosphate and myosin in the muscle these authors suggest the hypothesis that the myosin itself is phosphorylized and is the last link in the chain of simultaneous energy and phosphate transferance.

These observations, that show that a transferance of energy direct from adenyl-pyrophosphate to myosin is in a high degree probable, supplement well what has been shown above. If the myosin, or a therewith closely associated protein is responsible for the dephosphorylation and thus also the liberation of energy—and according to Needham, Shen, Needham and Lawrence's hypothesis also responsible for the transferance of the energy by means of phosphorylation to the actual myosin molecules—then this process must take place above all where the main part of the myosin is localized, i.e. in the anisotropic segments. This constitutes the continuation of the process that

[1] Not published, cited in J. Needham et collab. (1941).

has been postulated above on the strength of the cyto-chemical data, and it thus seems probable that the energy with adenyl-pyrophosphate is transported not only to but also on to the contractile myosin.

On account of the short transport-roads the diffusion after the existing concentration-falls is probably sufficient to explain the wandering of the released adenyl-pyrophosphate from I and of the adenylic acid in the opposite direction, which may also be a part of the explanation of the circumstance that musculature with short segments is capable of quicker action than is musculature with long segments.

Whether the muscle-contraction is released by the diffusing of the adenyl-pyrophosphates to the contractile myosin-chains (from which they are spatially separated) either by permeability changes or some similar process, or whether this process is a restitution-process after contraction of a myosin-chain bearing potential energy can only be decided by kinetic investigations. The strict way in which the substances in the resting muscle are separated from one another makes the former process seem at present fairly probable.

The characteristic structure of the striated muscle thus seems to be conditioned by a differentiation in parts with a chiefly contractile function and parts whose chief functions are in the service of the chemical conversion. In connection with the contraction, energy is transported from the latter to the former by means of adenyl-pyrophosphate of the corresponding phosphatase, which may possibly be the myosin itself. This arrangement, with parts only a few μ in height, just enables an extraordinary effectivity and rapidity of function, which qualities are among the most prominent of the striated muscle-fibre.

Summary

Ultra-violet absorption-spectra have been measured of individual isotropic and anisotropic segments in striated muscle-fibres from insects. By means of comparisons with ultra-violet microphotographs it is shown to be probable that the results obtained in this way are applicable to striated musculature in general. The absorption-curves show that the main part of the adenine derivatives (adenylic acid and adenyl-pyrophosphoric acid) in the resting muscle are localized in I. These segments seem to have their main function as the seat of chemical energetic processes while A is the seat of the contractile elements. From observations on fatigued muscle-fibres one deduces that energy is transferred from I to A by adenyl-pyrophosphate whose energy is released by the phosphatase that is bound to the myosin.

The background to the striation of the muscle-fibres thus seems to be a differentiation of contractile parts and chemically-working parts respectively. The transport of energy from the latter to the former is performed by adenyl-pyrophosphoric acid.

The work has been supported by funds from the Rockefeller Foundation and the foundation "Thérèse och Johan Anderssons minne."

References

Astbury, W., and S. Dickinson, *Nature,* 1935. **135**. 95, 765.
———, and S. Dickinson, *Ibidem,* 1936. **137**. 909.
Bailey, K., *Biochem. J.* 1937. **31**. 1406.
Caspersson, T., *Skand. Arch. Physiol.* 1936. **73**. Suppl 8.
———, *Chromosoma,* 1940. **1**. 562.
———, *J. micr. Soc.* 1940. **60**. 8.
———, and B. Thorell, *Naturwissenschaften,* 1941. **29**. 363.
Dhéré, Ch., C. R. *Soc. Biol.,* Paris 1906. **60**. 34.
Dworaczek, E., and H. Barrenscheen, *Biochem. Z.,* 1937. **292**. 388.
Edsall, J., and A. v. Muralt, *J. Biol. Chem.,* 1930. **89**. 289. 315.
Engelhardt, W., and M. Ljubimowa, *Nature* 1939. **144**. 668.
Gulland, H., and E. Holiday, *J. Chem. Soc.* 1934. 1639.
Heyroth, F., and J. Loofburouw, *J. Amer. Chem. Soc.,* 1931. **53**. 3441.
———, and J. Loofburouw, *Ibidem,* 1934. **56**. 1728.
Lundsgaard, E., *Ann. Rev. Biochem.* 1938. **7**. 377.
Meeraus, W., and G. Lorbeer, *Biochem. Z.,* 1937. **292**. 397.
Needham, J., S. Shen, D. M. Needham, and A. Lawrence, *Nature* 1941. **147**. 766.
Noll, D., and H. Weber, *Pflüg. Arch. ges. Physiol.,* 1934. **235**. 234.
Ochoa, S., *Biochem. Z.,* 1937. **290**. 62.
Schmidt, W. J., *Z. Zellforsch.,* 1934. **21**. 224.
Studnitz, G., *Ibidem.,* 1934. **23**. 1. 270.
Szent-Györgyi, A., and I. Banga, *Nature,* 1940. **145**.
Weber, H., *Naturwissenschaften* 1939. **27**. 33.

APPLICATION OF ADENOSINE TRIPHOSPHATE AND RELATED COMPOUNDS TO MAMMALIAN STRIATED AND SMOOTH MUSCLE*

Fritz Buchthal† *and Georg Kahlson*
Department of Physiology
University of Lund, Sweden
(Received for publication, June 6, 1944)

In a former note[1] an account was given of the stimulating effect of adenosine triphosphate and related substances on the isolated striated frog muscle fibre. When adenosine triphosphate is applied to *striated mammalian muscle* (m. tib. ant. of the decerebrated cat) by close arterial injection[2] in amounts of 0.05–0.53 mgm. per gm. muscle ($1.46–14.6 \times 10^{-6}$ mol./ml. $= 0.1–1.0 \times 10^{-6}$ mol./gm. muscle) a rapid, tetanic contraction is released which is

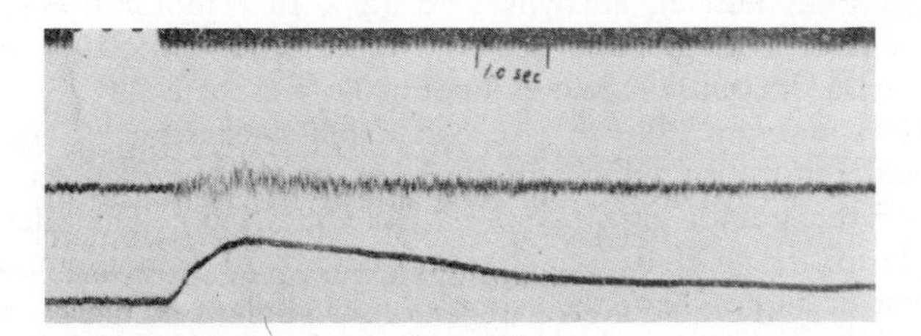

Fig. 1. Mechanical response and action potentials of M. Tibialis Ant. (cat) after close arterial injection of 0.6×10^{-6} Mol. adenosine triphosphate per gm. muscle.

*Fritz Buchthal and Georg Kahlson, "Application of Adenosine Triphosphate and Related Compounds to Mammalian Striated and Smooth Muscle," *Nature,* **154**, 178–79 (1944). Reprinted with the permission of the authors and of the editors of *Nature.*

†[On leave during the German occupation of Denmark from the Laboratory of Zoophysiology, University of Copenhagen.]

[1] Buchthal, F., Deutsch, A., and Knappeis, G. G.: *Nature,* **153** (1944), 774.

[2] Brown, G. L.: *J. Physiol.,* **92** (1938), 22 P.

accompanied by interfering electrical activity (see accompanying record). Threshold dose and mechanical response are identical in non-curarized and curarized preparations, the effect of total curarization being insured by inexcitability of the sciatic nerve towards maximal stimuli and by insensitiveness of the muscle to intra-arterial injection of 50 μgm. acetylcholine. Intra-arterial injection of 5 μgm. acetylcholine after previous treatment of the non-curarized preparation with adenosine triphosphate releases a mechanical response with a considerably longer duration and higher tension than the same dose of acetylcholine does to a muscle without previous application of adenosine triphosphate.

Adenosine diphosphate ($1.7–7.0 \times 10^{-6}$ mol./ml. = $0.12–0.5 \times 10^{-6}$ mol./gm. muscle) likewise initiates tetanic contractions. Adenylic acid ($2.6–8.1 \times 10^{-6}$ mol./ml. = $0.2–0.6 \times 10^{-6}$ mol./gm. muscle) has stimulating effects on the non-curarized preparation, while the curarized muscle only reacts slightly or not at all. Inorganic sodium triphosphate ($2–12 \times 10^{-6}$ mol./ml. = $0.15–0.9 \times 10^{-6}$ mol./gm. muscle) and sodium pyrophosphate ($2.5–17 \times 10^{-6}$ mol./gm. muscle) release tetanic contractions in curarized and non-curarized muscle, while inorganic sodium orthophosphate in amounts of 20×10^{-6} mol./ml. (= 1.7×10^{-6} mol./gm. muscle) is ineffective.

All chemical stimuli were applied iso-osmotically by replacing a corresponding amount of sodium chloride in the Thyrode solution with the substances in question; *p*H of the solution was 7.3.

When applied to *smooth muscle* (stomach and bladder of the cat by intra-arterial injection, and small intestines of the guinea pig by adding the substance to the surrounding Ringer bath) *only adenosine triphosphate initiates strong activity*. The threshold dose is approximately 0.3×10^{-6} mol./ml. (= 0.04×10^{-6} mol./gm. muscle). The atropinized preparation which is insensitive to strong doses of acetylcholine still reacts to adenosine triphosphate. Adenosine diphosphate (2.5×10^{-6} mol./ml. = 0.35×10^{-6} mol./gm. muscle), adenosine diphosphate plus orthophosphate ($5.0 + 5.0 \times 10^{-6}$ mol./ml. = $0.7 + 0.7 \times 10^{-6}$ mol./gm. muscle), adenylic acid (2×10^{-6} mol./ml. = 0.26×10^{-6} mol./gm. muscle), inorganic sodium pyrophosphate (7×10^{-6} mol./ml. = 1.0×10^{-6} mol./gm. muscle) and sodium orthophosphate (14×10^{-6} mol./ml. = 2.0×10^{-6} mol./gm. muscle) are completely ineffective on the stomach and the bladder of the cat, while pyrophosphate causes contractions in the small intestines of the guinea pig ($0.22–0.7 \times 10^{-6}$ mol./ml.). The effect of inorganic triphosphate (6×10^{-6} mol./ml. = 0.83×10^{-6} mol./gm. muscle) is either absent or slight. Thus the action of adenosine triphosphate on smooth muscle is highly specific and corresponds to its specific effect on the flow birefringence of purified myosin solutions.[3]

These experiments support the view that adenosine triphosphate is an essential agent in the release of normal muscular contration.

[3] Needham, J., Shih-Chang Shen, Needham, D. M., and Lawrence, A. S. C.: *Nature*, **147** (1941), 766. Needham, J., Kleinzeller, A., Miall, M., Dainty, M., Needham, D. M., and Lawrence, A. S. C.: *Nature,* **150** (1942), 46.

MORPHOLOGY IN MUSCLE AND NERVE PHYSIOLOGY*

Francis O. Schmitt

Department of Biology, Massachusetts Institute of Technology, Cambridge, Massachusetts

(Received for publication, May 19, 1949)

As applied to biology, morphology embraces the study of the structure of cell and tissue constituents from gross and microscopic anatomy through the colloidal range and even to the molecular and atomic levels. With the introduction of electron microscopy it is now possible to visualize directly the structure of objects throughout the colloidal range. It is not unrealistic to expect that technical development will make possible direct visualization of such biologically important objects as the smaller protein molecules and possibly even the polypeptide chains. Simultaneously the theory and techniques of X-ray diffraction are also progressing. This method is already able to deal effectively with the analysis of the internal architecture of certain crystalline proteins; a major hurdle appears to be the development of suitable computing methods—a matter chiefly of technology and patience. Progress is also being made in the analysis of the less regularly constructed, but no less important biologically, fibrous proteins and complexes of proteins with lipids, nucleic acids and polysaccharides. This, too, is a matter of painstaking, patient development of techniques, experimental and theoretical.

Morphology is a science in its own right. The analysis of the detailed structure of the molecules and complexes which occur in tissues is largely the task of the physicist who, in turn, must depend upon the chemist to identify, isolate, purify and characterize the individual constituents. In the normal course, as physicists and chemists become interested in such substances, one may

*Excerpt from Francis O. Schmitt, "Morphology in Muscle and Nerve Physiology," Meyerhof Dedicatory Vol., *Biochem. Biophys. Acta,* **4**, 68–71, 76 (1950). Reprinted with the permission of the author, the editor, and the Elsevier Publishing Company.

expect knowledge in this branch of crystallography slowly to unfold. Slowly because such complex, frequently imperfectly structured materials are not attractive to most crystallographers who are likely to regard them as "sick crystals," as one colleague expresses it. Actually, some of the most important protein crystals are far from "sick" structurally; upon the regularity of the internal structure of their molecules depend such fundamental vital properties as are manifested in the phenomena of immunology, genetics, and the ordered processes of growth and development. Relatively minute changes in the structure of certain protein molecules may make the organism sick (Pauling *et al.*[1], recently referred to sickle cell anemia as a "molecular disease"!). The great biological significance of structural studies has stimulated many physicists and chemists to devote their efforts to the problem. Hopefully their numbers will grow.

The detailed analysis of biomolecular structure is a long term task. The analysis starts with a rough characterization of the main structural features of a particular tissue entity. With the aid of the electron microscope the biologist relatively untrained in the discipline of crystallography can and must take an active interest in this phase. As the analysis becomes more detailed, eventually leading to the localization of the constituent atoms, the task becomes more that of the crystallographer. The physiologist and biochemist must make use of the information available at the moment in attempting to account for biological phenomena.

To what extent has structure analysis been of assistance in solving major physiological problems and what is the outlook for further advance in this field? In seeking a perspective regarding such a question a consideration of muscle contraction and nerve conduction may be instructive because of the contrast which these problems present in respect of inherent susceptibility to morphological investigation and to progress already accomplished. The following account is necessarily brief and attempts merely to indicate the trend of research in this field.

Muscle Contraction

Contractility is particularly favourable for morphological study because it involves structural changes at all levels of observation. The voluminous literature of muscle histology, devoted largely to striated muscle, led to few important physiological clues. Perhaps the "reversal of striation"[2] on contraction was among the most suggestive. Even observations in polarized light were difficult to interpret. The positive form birefringence indicated that the fibrous proteins have widths small with respect to the wavelength of light. The relative isotropy of the *I* bands was long misinterpreted as indicating disorientation in these regions. Muralt and Edsall's demonstration of the positive birefringence of myosin focused attention on this protein as the contractile substance of muscle. Astbury's identification of myosin as the source of the wideangle X-ray pattern of muscle, together with his hypothesis of intramolecular folding during contraction, helped to seek in myosin the substratum of contraction[3].

In the short time since electron microscopy has been applied to the problem, important advances have been made. The view that myosin is localized in the *A* bands, already discredited by quantitative considerations, was disproven by electron microscopy, which showed that the protein filaments extend as parallel bundles continuously through both *A* and *I* bands[4]. The relative isotropy of the *I* bands is therefore not due to disorientation. Recently the view has been taken that the isotropy results from the presence of negatively birefringent substances in the *I* bands which compensate the positive birefringence of the myosin; this material has been variously reported as nucleotides[5,6], lipids[7] and phosphoproteins (*N* material)[8].

In contraction the protein filaments remain relatively straight and parallel, indicating that the contractile unit is thinner than the filaments (ca 150 Å). The distribution of the dense material in the *A* bands and on the *Z* membrane changes in agreement with the histological picture of reversal of striation.

Morphological studies were greatly stimulated by advances in our concepts of mechano-chemical coupling mediated by high-energy phosphate bonds and by the discovery by the Szeged group that myosin is composed of two proteins, a water-soluble myosin and actin, the actomyosin complex being sensitive to the action of adenosine-triphosphate (ATP). The general morphological features of the water-soluble myosin and the globular and fibrous actin were soon demonstrated with the electron microscope[9], together with the dissociating effect of ATP on the actomyosin threads[10].

Of great significance in the morphological approach to the contractile mechanism is the axial periodicity demonstrated both by small-angle X-ray diffraction[11] and by electron microscopy[4]. This period has a value of about 400 Å in uncontracted fibres and appears to be characteristic of muscle generally, for Bear has observed it not only in vertebrate striated muscle but also in various invertebrate muscles which are generally regarded as being of the smooth type. In electron micrographs the filaments have a beaded appearance which gives rise to a fine banding of the myofibril, the distance between bands being about 400 Å. Draper and Hodge[12] have shown the period very strikingly in electron micrographs of platinum-shadowed preparations. In their preliminary note they state that the axial period varies inversely with the degree of shortening of the muscle. Variations in the 400 Å period with fibre length were also noted by Bennett[13] who believes to have shown that the filaments have a helical structure. If these points are satisfactorily documented and confirmed we shall have visual evidence of the contractile phenomenon at the near-molecular level.

Actually the relation between the 400 Å axial period demonstrated by X-ray diffraction and the pseudo-period of about the same value seen in electron micrographs is not clear. The largest meridional spacing observed in the X-ray patterns is about 27 Å which is an order of the larger period. If the situation is similar to that of paramyosin[14,15] one might expect that the period which might be observable as cross bands in the electron microscope, would have a value of about 27 Å; the larger period of about 400 Å would be manifested as a geometric pattern of discontinuities within the bands. However, depending on the type of geometry of the intraperiod structure, discontinuities at a spacing larger than 27 Å may appear in electron micrographs. The solution of this problem will have to await a more detailed X-ray analysis and attainment of very considerably increased electron microscope resolution of the structure of the filaments.

Astbury, Perry, Reed, and Spark[16] have observed a spacing of 54 Å in fibrous actin. At large angles the pattern is not that of an alpha protein. This led the authors to the conclusion that the large-angle pattern of muscle is due to myosin while the small-angle pattern is due to actin; the full muscle pattern derives from actomyosin which exists as a complex in muscle. While this may prove to be the case, the diffraction evidence is not yet sufficiently detailed to require this conclusion.

The electron microscope investigation of contractility might be facilitated by examination of *in vitro* models such as the actomyosin-ATP system described by Szent-Györgyi[17]. This would be true if such systems permitted higher resolution than could be achieved in the myofibril and, particularly, if the essential properties of such a system faithfully portray those of muscle. Recently Szent-Györgyi[18] has found that muscles thoroughly extracted with 50% glycerol at low temperatures are capable of contraction when treated with ATP and produce the same tension as the intact muscle when maximally excited. Differences in the behaviour of this model as compared with intact muscle are attributed to the fact that the model may lack some of the proteins, lipids and other substances with which the actomyosin is normally associated in muscle. From studies of this model, as from the previous one of Varga[19], the conclusion was reached that contractile substance is composed of functional units, "autones", and that contraction represents an all-or-none equilibrium reaction of these units; contraction and relaxation are two distinct allotropic states of the autones.

Unfortunately, as admitted by Szent-Györgyi[18] and as amplified by Sandow[20] none of the partial systems and models thus far proposed fully displays the essential properties of muscle. So far as the morphological evidence is concerned, Perry, Reed, Astbury, and Spark[21] have shown by X-ray and electron microscope studies that the changes which occur when ATP is added to actomyosin is an intermolecular syneresis, the contraction occurring in a direction normal to that which characterizes muscle contraction. Moreover, there is no evidence from X-ray results for the existence of two distinct states of the "auxones". Upon contraction the large-angle pattern indicates a change from an alpha to a poorly defined, disoriented beta configuration. Efforts to obtain a characteristic small-angle pattern from contracted muscle have thus far met with failure. What little electron microscope evidence bears on this point suggests that the 400 Å axial period shows a continuous change in value with change in fibre length rather than two distinct states.

However valuable partial systems and models may be from the biochemical viewpoint, it is evident that, in the investigation of structural mechanism which is characteristic of muscle, final answers will be obtained by observation of nothing less complex than the muscle fibre itself.

There is no reason to doubt that the combination of X-ray diffraction and electron microscopy will be equal to the task of disclosing the molecular changes which occur in contraction. The small-angle X-ray analysis is particularly promising and may be expected in the near future to portray the main features of the lattice of Bear's Type II protein. The more detailed structure at smaller separations, involving the configurations of polypeptide chains in relaxes and contracted muscle seems more difficult of unique solution unless more diffraction data can be obtained at large angles.

An electron microscope investigation of the extra-filamentous structures of the striated myofibril, including the materials concerned in the "reversal of striation", the *Z* membranes and the binding material which connects filaments to each other and to the sarcolemma laterally, offers much promise. However, primary interest attaches to the detailed structure within the filament and the changes of this structure with contraction. As compared with paramyosin the task of the electron microscopist will be considerably more exacting because of the smaller spacings involved. Obviously, at this level of size the most critical judgement of image quality and of optical artifacts will be required.

* * *

References

1. L. Pauling, H. A. Itano, S. J. Singer, and I. C. Wells, *Science,* **109** (1949), 443.
2. H. E. Jordan, *Physiol. Rev.,* **13** (1933), 301.
3. W. T. Astbury, *Proc. Roy. Soc.* (*London*) B, **134** (1947), 303.
4. C. E. Hall, M. A. Jakus, and F. O. Schmitt, *Biol. Bull.,* **90** (1946), 32.
5. R., Caspersson, and B. Thorell, *Acta Physiol. Scand.,* **4** (1942), 97.
6. C. L. Hoagland, *Currents in Biochemical Research,* D. Green, Ed., Interscience, New York 1946, 413.
7. E. W. Dempsey, G. B. Wislocki, and M. Singer, *Anat. Record.,* **96** (1946), 221.
8. A. G. Malotsy, and M. Gerendás, *Nature,* **159** (1947), 502.
9. M. A. Jakus, and C. E. Hall, *J. Biol. Chem.,* **167** (1947), 705.
10. S. V. Perry, R. Reed, W. T. Astbury, and L. C. Spark, *Biochem. Biophys. Acta,* **2** (1948), 674.
11. R. S. Bear, *J. Am. Chem. Soc.,* **67** (1945), 1625.
12. M. H. Draper, and A. J. Hodge, *Nature,* **163** (1949), 576.
13. H. S. Bennett, *Anat. Record,* **103** (1949), 7.
14. R. S. Bear, *J. Am. Chem. Soc.,* **66** (1944), 2043.
15. C. E. Hall, M. A. Jakus, and F. O. Schmitt, *J. Applied Phys.,* **6** (1945), 459.
16. W. T. Astbury, S. V. Perry, R. Reed, and L. C. Spark, *Biochim. Biophys. Acta,* **1** (1947), 379.
17. A. Szent-Györgyi, *Muscular Contraction,* Academic Press, New York 1947.
18. A. Szent-Györgyi, *Biol. Bull.,* **96** (1949), 140.
19. L. Varga, *Hung. Acta Physiol.,* **1** (1946), 1, 138.
20. A. Sandow, *Ann. Rev. Physiol.,* **11** (1949), 297.
21. S. V. Perry, R. Reed, W. T. Astbury, and L. C. Spark, *Biochim. Biophys. Acta,* **2** (1948), 674.

THE DOUBLE ARRAY OF FILAMENTS IN CROSS-STRIATED MUSCLE*

H. E. Huxley

Medical Research Council, Department of Biophysics,
University College, London, England

(Received for publication, February 27, 1957)

Introduction

Until about five years ago, the view was generally held that the contractile structure in striated muscle consisted of a single set of longitudinal filaments which extended continuously through each sarcomere (*i.e.*, from one Z line to the next) and which were, perhaps, even continuous between successive sarcomeres. In the A bands, some extra material, the A substance, was believed to be present, lying either on or between the filaments. It was believed that some internal folding or coiling of the filaments was responsible for contraction.

Since then, studies on striated muscle by x-ray diffraction[1], electron microscopy[2], and phase contrast light microscopy[3], have suggested that a rather different type of structure is present. The essence of this structure is that it consists of two separate sets of longitudinal filaments in each sarcomere, which overlap in certain regions and thereby produce the characteristic band pattern of striated muscle. In this model, changes in the length of the muscle are brought about by a process in which the two sets of filaments slide past each other. The changes in the visible band pattern of muscle which occur as a result of stretch or contraction can be explained in a very straightforward way by the model, as can a number of other aspects of the behaviour of muscles[4–6].

Despite the apparent straightforwardness of the observations and arguments which led to this model, its validity has been questioned by some workers

*H. E. Huxley, "The Double Array of Filaments in Cross-Striated Muscle," *J. Biophysic. and Biochem. Cytol.*, **3**, 631–647 (1957). Reprinted with the permission of the author and of The Rockefeller University Press.

recently[7–9] primarily on the basis of their own electron microscope observations; and attempts have been made to revive the so called "classical" model. It is the author's view that the force of the original arguments, particularly those deriving from the light microscope observations, has not been fully appreciated in these attempts; and that in consequence, certain electron microscope observations have been accepted at their face value, and theories constructed from them (which conflict with other observations on muscle) perhaps a little too readily. However, electron microscope observations should be able to stand on their own; and so it is important to establish what the facts of the situation really are, and to find out just at what point the error in argument or interpretation has crept in.

The issue revolves primarily around the appearance of longitudinal sections of muscle in the electron microscope. At first sight, one would expect the double array of filaments to be readily visible in such sections. However, it is usually found that the sections show what appears to be only a single set of filaments. It is this observation which has led some people to call into question the original interpretation of the cross-sectional views of muscle, which did appear to show a double set of filaments. The purpose of the present paper is to discuss the conditions under which one might expect to see the double array of filaments, if it exists, in longitudinal sections of striated muscle, and to describe the appearance of such sections when the requisite conditions are fulfilled.

Section thickness and visibility of filaments

The separation between the filaments which form the primary array in the A bands of striated muscle is, in fixed and embedded material, about 300 to 350 A. The separation between adjacent layers of filaments in this lattice will depend on the plane in which the particular set of layers is drawn, relative to the hexagonal axes of the lattice. The maximum separation of layers will be found in the direction indicated in Text-fig. 1 (the $10\bar{1}0$ crystallographic direction), and will be approximately 250 A; the distance between primary filaments in these layers will be the fundamental lattice spacing, 300 to 350 A. The next largest layer separation will be found in the direction indicated in Fig. 2 (the $11\bar{2}0$ crystallographic direction); the separation of these layers will be about 150 A,* and the distance between primary filaments in the layers will be 520 to 600 A. The lattice can of course be divided up into layers in an infinite number of other directions, but these will all give layer separations smaller than the ones already described, and the filaments in any given layer will be further apart.

Thus the appearance of longitudinal sections through this structure will depend rather critically both on the plane of sectioning and on the section thickness. If the section thickness exceeds the layer separation in that particular direction, then obviously the section will contain two or more superimposed layers of filaments. The appearance of the section in the electron

*[Although the letter "A" is used to indicate both A bands and Angström units, we have not found any passage in which the clarity of the paper is jeopardized. Hence, we have left the terminology unaltered.]

microscope (whose depth of focus is of course much greater than the layer separation), will then be governed by the way in which the filaments in successive layers are stacked above each other. When two or more filaments lie vertically above each other in the sections, they appear as a single dense filament. In general, however, this will not happen, and a wide variety of different patterns will result. This is indeed what one observes in the electron microscope with the usual type of longitudinal sections of muscle, but of course one tends to select those areas in which the "filaments" show up most clearly. The possibility of seeing secondary filaments in sections which contain more than one layer of primary filaments will be governed by the extent to which, in projection, they overlap with each other and with primary filaments. Examination of Text-fig. 1 will show that even in the most favourable case (layers parallel to the $11\bar{2}0$ direction), and even with idealized filament profiles, the secondary

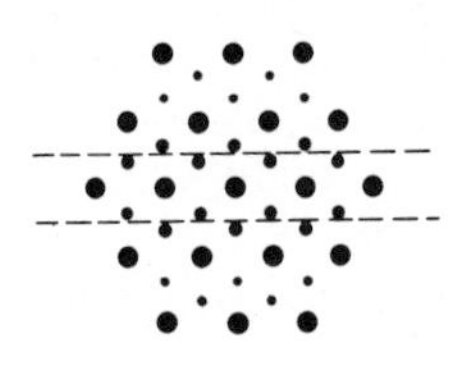

Text-Fig. 1a. Diagram showing end-on view of a double hexagonal array of filaments; the two dotted lines indicate the outline of a longitudinal section, about 250 A in thickness parallel to the $10\bar{1}0$ lattice planes, at the appropriate level to include one layer of primary filaments, and two layers of secondary filaments.

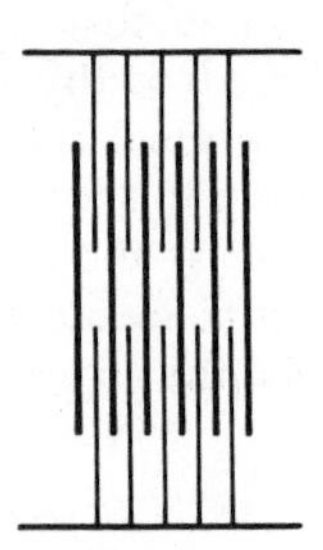

Text-Fig. 1b. Diagram showing expected appearance of longitudinal section, cut as indicated in Text-fig. 1*a*. Note simple alteration of primary and secondary filaments. The latter will represent two filaments lying vertically above each other in the section.

filaments in the projected view of sections of thickness 300 A and upwards would almost touch each other and the primary filaments alongside them. And imperfections in the sharpness of the outline of the filaments, and in their alignment, will aggravate this effect. The result would be that the space between the primary filaments would appear to be filled with material of more or less uniform density. In other directions, the overlapping will be worse, and in the case of layers parallel to the $10\bar{1}0$ planes, primary and secondary filaments would lie in the same vertical register. This argument will be illustrated later by the superposition of photographs of single layers of the structure.

We see, then, that although these sections may in some instances give some sort of a picture of the main array of filaments, they will be of little use for showing details of the structure between these primary filaments. In order to allow any reasonable possibility of seeing secondary filaments (if they exist), the section must contain only a single layer of primary filaments. This can be achieved if sections of the type shown by the dotted lines in Text-figs. 1 and 2 are employed. These represent section thicknesses of 250 A and 150 A respectively. And although it is frequently stated or implied that sections of this thickness have been used, we will see in a moment that there is reason to believe that the real thickness has often been substantially greater.

For any given filament to remain within a 200 A thick section for a distance of, say, 1 to 2 microns, its orientation must be correct to better than one degree, and it must not diverge from a straight course by more than 100 A. Now these are rather exacting requirements, and it is surprising, to say the least, that what appear to be perfectly oriented layers of filaments can so often be seen all over the field when longitudinal sections of muscle, cut without taking any particular precautions to obtain perfect orientation, are examined in the electron microscope.

The explanation lies in the fact that the "bright silver" sections often used in electron microscope studies represent layers of the structure which, before sectioning, were considerably more than 300 A in thickness. The actual thickness of such sections, as measured in the interference microscope, is in the range 600 to 1000 A. However, these sections may be reduced in length by as much as a factor of 30 per cent in the cutting direction, as compared with the block face, without any increase in their width; it follows that an increase in thickness must have taken place, and that the thickness of the sectioned layer was 400 to 700 A. Thus the usual silver sections will contain more than one layer of filaments, and in those areas where the filaments can be seen at all clearly, *i.e.* where they are in vertical register, the fact that they may be tilted about a horizontal axis with respect to the plane of the section will not be detectable for there will always be *some* filaments in the section, to give the appearance of a continuous "filament," and the density variation as one filament leaves the section and another one enters will not be very great. When the sections contain only one layer of filaments, however, they exhibit quite a different appearance, and even with the most carefully oriented material, it can immediately be seen that most of the filaments are not accurately parallel to the plane of the section and remain in it for only a relatively short distance. This provides quite a good test of section thickness. And it will be clear from what has gone before that sections which do not pass this test will be too thick for secondary filaments to be separately resolved.

The best conditions for seeing the structure in between the primary filaments are obtained with sections of the kind illustrated in Text-fig. 2. These contain only one primary and one secondary filament in the thickness of the section, which needs to be about 150 A. The primary filaments in the layers will be about 500 A apart, and the appearance of an array with this rather large spacing, therefore, provides another indication of section thickness. If secondary filaments exist, we should expect to find two of them between each pair of primary filaments, as shown in Text-fig. 2 *b*. The sections appear dark grey in colour when floating on a water surface and viewed by reflected light. The muscle shows the same kind of "pseudo striations" (as layers of filaments enter and leave the section), as have previously been observed[10] in insect flight muscle, where the spacing of the array is greater and the pseudo striations easier to obtain. With a little care, it is usually possible to find examples of layers of filaments which stay in the section for the whole length of an A band.

Sections of the kind shown in Text-fig. 1 also contain only a single layer of

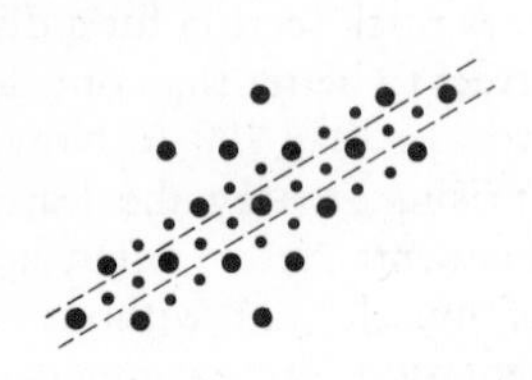

Text-Fig. 2a. As in Text-fig. 1, but with dotted lines showing a longitudinal section about 150 A in thickness, cut parallel to the $11\bar{2}0$ planes of the lattice; primary and secondary filaments in this all lie in the same layer, and two secondary filaments occur between each pair of primary ones.

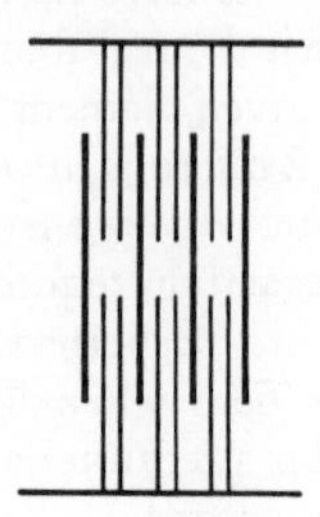

Text-Fig. 2b. Diagram showing expected appearance of longitudinal section, cut as in *2a*. Note characteristic appearance of two secondary filaments between each pair of primary filaments.

primary filaments, but would contain two secondary filaments in vertical register; these should, however, still be seen separately from the primary filaments, as indicated in Text-fig. 1 *b*; the sections in this case are about 250 A in thickness, and they also appear dark grey in colour.

The reduction of the dimension of the section in the cutting direction has already been remarked upon. This, of course, will produce changes in the embedded tissue, and if the filaments are parallel to the knife edge, they will be moved towards each other and the whole appearance obscured. In order to see the arrangement of the filaments most clearly, they should therefore be sectioned perpendicular to the knife edge, a fact already remarked upon by Spiro[7].

Thus the longitudinal sections of striated muscle which are likely to show the secondary filaments (if they exist), are ones which have been cut with the fibre axis perpendicular to the knife edge, and which are so thin that they appear dark grey when examined in reflected light, and show muscle with pseudo striations when examined in the electron microscope. It is this type of section which has been used in the present studies.

Materials and methods

The main source of material was rabbit psoas muscle, both fresh and glycerinated[11]; fresh muscle from frog sartorius and frog musculus extensor longus digiti IV was found to give results almost indistinguishable from those obtained with fresh psoas muscle. Specimens were fixed for varying lengths of time in buffered osmium tetroxide[12]. Optimum results were obtained using a fixation time of $^1/_2$ hour at room temperature. The material was dehydrated in an ethyl alcohol series and embedded in methacrylate in the usual way in No. 4 gelatin capsules. A polymerization temperature of 60°C. was employed; no improvement in the appearance of the tissue was noted when prepolymerized plastic was employed. Additional staining was usually provided by dissolving 1 per cent phosphotungstic acid (PTA) in the absolute alcohol used in the dehydration. This rendered the tissue very hard and as a result it was more difficult to cut good sections. Thick, damaged sections of tissue stained with PTA can often give the impression that the PTA has itself caused structural damage. But when thin, smooth sections are obtained from the same block, they show that the tissue is preserved perfectly satisfactorily. It was often found advantageous to use a harder plastic to compensate for the hardness of the tissue; mixtures containing up to 50 per cent methyl methacrylate were quite practicable. Sections were cut on an extensively modified Hodge-Huxley-Spiro microtome[13, 14], using a glass knife. They were collected on Smethurst-Highlight copper grids coated with carbon films[15].

The sections were examined in a Siemens Elmiskop I. A beam current of 13 μa., an accelerating voltage of 80 kv., and a condenser aperture of 200 microns were used. Condenser 1 was set at the fifth click position from zero on the coarse control; this gave a reasonable compromise between specimen contamination and overheating. A molybdenum objective aperture 35 μ in diameter was employed; the astigmatism was kept below about $\Delta f = 0.2\ \mu$ (3 clicks on the fine focus control). An instrumental magnification of 40,000 was the one most commonly employed. Photographs were taken on Ilford special contrasty lantern plates, and were developed in Ilford PQ universal developer.

Results

1. The arrangement of the filaments

The appearance of striated muscle in longitudinal sections of the required thinness and orientation is shown in Fig. 1. The pattern seen varies from one fibril to the next, and from one sarcomere to the next in any given fibril, depending on the orientation of the axes of the hexagonal lattice of myofilaments with respect to the plane of sectioning. Sections through the lattice in the $11\bar{2}0$ direction are shown in Figs. 2 and 3. (In the latter figure, the orientation is inaccurate by about one degree and the section passes from one layer of the lattice to the next.) These electron micrographs should be compared with the theoretical diagrams shown in Text-fig. 2. Two thin filaments, lying in between each pair of thick ones, are invariably seen in such sections. The thick filaments are continuous from end to end of the A bands, and their distance apart is about 500 to 550 A, which agrees with the theoretical value. In fibrils at about rest-length, the thin filaments terminate before they reach the centre of the A band, leaving a gap in the middle, the H zone. This gap may be seen in the centre of each sarcomere over a wide field (*e.g.* Fig. 1 and also Fig. 9) and therefore cannot be due to a chance displacement of the filaments. In slightly stretched muscle (Fig. 4), the ends of the secondary filaments are farther apart and a wider H zone appears; in slightly contracted muscle (Fig. 7), the ends of the thin filaments come together in the centre of the A bands and the H zone disappears. Intermediate stages are shown in Figs. 5 and 6.

The appearance of sections cut in the $10\bar{1}0$ direction is illustrated in Fig. 8. This should be compared with the theoretical diagram shown in Text-fig. 1. In this type of section too, a second set of filaments, terminating at the edges of the H zone, can be seen lying between the primary filaments which extend continuously from end to end of the A band. This time, however, a simple alternation of primary and secondary filaments is observed; the secondary "filaments" will in general really consist of two superposed secondary filaments, as reference to Text-fig. 1 will make clear. The different extent to which the secondary filaments penetrate into the A band at different muscle lengths is illustrated by Figs. 11 and 24 which show a stretched muscle with its wide H zone. Cross-bridges between primary and secondary filaments are prominent features of all these pictures.

If a slightly oblique section is cut through the structure, the appearance shown in Fig. 9 is obtained. This pattern may at first look very complicated, but longer inspection of it will show that it results simply from the section passing through successive layers of the double hexagonal lattice of filaments. The absence of the secondary filaments from the H zones is well demonstrated in this picture. Many of the sarcomeres have the appearance shown at higher magnification in Fig. 10. Reference to Text-fig. 2 will show how this has arisen. As the section passes out of one layer of filaments and enters the next, the thick filaments in this layer appear in a position midway between the thick filaments of the preceding layer; and between the thick filaments of the new set, pairs of thin filaments again appear. The ends of the thick and the thin filaments, *i.e.*, the points where they leave the section, lie on the same straight line, and the two types of filaments in a given layer must therefore lie in the same plane. It will be apparent from all this that the secondary filaments cannot be "shaved-off edges" of the primary filaments in the layers immediately above and below.

If the muscle is fixed "unrestrained" (as opposed to being held at a fixed length), the characteristic appearance of primary and secondary filaments is still seen perfectly normally in longitudinal sections. Fig. 12 shows an example of a fibril fixed in this way. It will be seen that the only difference between this fibril and the one shown in Fig. 2 (which was held at a fixed length), is that the general orientation of the filaments is poorer, particularly in the I bands, where thin filaments enter and leave the section in many places. In the lower part of Fig. 12, the section, which must have been extremely thin, has passed through four secondary filaments between a very widely spaced pair of primary ones. This would correspond to a section parallel to the $21\bar{3}0$ crystallographic direction.

The filament arrangement can be illustrated in a slightly different way, by Fig. 15. This represents a transverse section through a fibril in the region of the H zone. The section is not quite accurately transverse, and it intersects the boundaries between the H zone and the rest of the A band. Thus in the center of the field there is a simple array of filaments and on either side, a compound array of filaments; and the continuity of the rows of thick, primary filaments between the two regions, and the way in which the set of six secondary filaments appear around each of the primary filaments at the boundary can be readily seen.

The continuity of the secondary filaments in the A band with the I band filaments is difficult to demonstrate with diagrammatic clarity over a whole fibril, although it can be seen in restricted area (Figs. 2, 13, and 14). The difficulty arises because the degree of order of the filaments in the I band is substantially lower than in the A band, and in so far as we are concerned with the separation of overlapping structures, we are already right on the limit of the present technique. The alignment of the I band filaments is considerably improved by holding the muscle at a fixed length, or under tension, during fixation, dehydration, and infiltration, but unfortunately, not to the extent that departures from linearity by about 100 A no longer occur; and this is sufficient to take a filament out of the section, or to bring it into the section if it lay just above or below. Thus the I band filaments rarely lie completely in the section over their whole course. And it will not help matters to use thicker sections, for then thick and thin filaments will overlap in the A bands. Furthermore, the Z lines seem to shrink sideways slightly during fixation and embedding, with the result that the array of I band filaments is constricted there, leading to further non-parallelism at the A-I boundary. The addition of these complications to the initial requirements of section thickness and orientation has, so far, not allowed electron micrographs of the A-I boundary to provide a conclusive demonstration that the primary filaments *always* terminate there completely, and that the secondary filaments *always* continue on into the I band. The real evidence for this lies in the observations of the changes in the degree of overlap of primary and secondary filaments at different muscle lengths; it is very difficult indeed to see how such changes can be brought about unless the secondary filaments are connected to the Z lines. The electron micrographs referred to do, however, show examples of the termination of the primary filaments and the continuity of the secondary ones.

2. Appearance of thicker sections

It is of some interest now to consider what the appearance of these structures would be in thicker sections. It has already been pointed out that the best chance of seeing secondary filaments in sections containing more than one layer of primary filaments would be given by sections cut in the $11\bar{2}0$ direction. And it was argued that even in this case, the secondary filaments would almost certainly be obscured by overlap. This effect can now be illustrated. Fig. 16 shows a photograph of a well oriented single layer $11\bar{2}0$ section. Fig. 17 shows the same photograph printed twice, with the second printing shifted so as to place the second layer of filaments in the position they would occupy in the projected view of two layers of the lattice. It will be seen that the secondary filaments are concealed very effectively, although the bridges still appear. Fig. 18 shows a typical electron micrograph of a thicker section (bright silver); again, the secondary filaments are completely obscured.

3. Fine details of structure

Bridges are observed between primary and secondary filaments. These bridges form part of the primary filaments, and in stretched muscles can still be seen projecting into the space from which the secondary filaments have been withdrawn. The secondary filaments, on the other hand, can, in the I bands, be seen to have no projections on them.

Between any adjacent primary and secondary filament there seems to be a bridge about every 400 A. This distance is arrived at by counting the number of bridges[15-20] in one half of an A band, and dividing this number into the half-length of the band (0.75 μ) measured in the light microscope; because of a variable amount of dimensional change produced in the sections during cutting, it is not possible to measure the spacing directly—it usually appears much less than 400 A. The bridges between a secondary filament and the three primary filaments around it do not occur in register, but seem to be spaced out fairly evenly along the secondary filaments, as though they were attached about every 133 A. The bridges on any given primary filament do not occur in register either; one often gets the impression that they are disposed in a helical fashion (*e.g.* Fig. 27), one turn of the helix occupying 400 A. The bridges do not seem to have any precisely fixed form; they are more or less at right angles to the filaments, and there does not appear to be any preferential direction of tilt, even in muscle fixed under continuous tension. But displacement of structures may occur during sectioning.

The ends of the primary filaments appear to be tapered, down to a point, for about the last 1,000 to 2,000 A of their course; this is seen too consistently for it to be an effect of section orientation.

The bridges, as has been remarked before[2] may also be seen in cross-sections through the double array of filaments (Figs. 19 and 20). The angular form of the cross-section of the filaments themselves is also apparent, a fact already commented on by Sjöstrand and Andersson[9]. In the H zone, this effect is particularly noticeable, and all sorts of bizarre shapes appear when the section is sufficiently thin to give a projection of only a short length of filament (Figs. 21 to 23).

4. Appearance in unextracted muscle

The observations as they have been described were made on glycerinated psoas muscle. However, no essential difference was found when fresh muscle was fixed and examined, and the secondary filaments can be seen perfectly well in suitably thin longitudinal sections (Figs. 25 and 26). The contrast, though, is very much lower than in glycerinated material, and it is evident that a good deal of diffuse and granular background material is present. It seems likely that this represents some of the soluble muscle proteins (which make up 30 to 40 per cent of the total), which are partly removed during glycerol extraction and the subsequent washing—a procedure which leaves the contractile structure in almost full working order[6,11].

Discussion

1. Previous work

The results which have been described above give full support to the "sliding filament model" of striated muscle, and they are not compatible with the other types of structure which have been proposed by Spiro[7], by Hodge[8], and by Sjöstrand and Andersson[9]. The latter are all essentially single array, coiling filament models. As we have already observed, it was the misleading appearance of muscle in longitudinal section which gave rise to the doubts about the sliding filament model, and now that that difficulty has been resolved, the primary motive for considering other types of structure has been removed. However, the existence of these alternative models has created a certain amount of confusion; particularly as reasons other than the invisibility of the secondary filaments in longitudinal sections have been put forward in their support. It appears from the present electron microscope evidence that these models are wrong, but in order to clear up a little of this confusion, it is desirable to discuss some of the other evidence as well.

In the first place, these three hypotheses are incompatible with a large part of the original evidence on which the idea of a sliding filament model was based, and in the main, this evidence was simply set aside during their elaboration. Certain parts of it were challenged on *ad hoc* grounds[1] which it may not be essential to discuss now. The evidence has been described in detail in previous papers[1–6] and it is unnecessary to repeat it here. The observations, and the conclusions drawn from them, still all hold good. And the identification of myosin as the A substance has been confirmed recently by comparative interference microscope and biochemical measurements[16,17]. However, it does seem desirable to draw particular attention to one of the original arguments in favour of the sliding filament model, one based on simple observations in the light microscope, which has either been overlooked or whose force has been greatly underestimated. It concerns the changes in the length of the H zone which accompany changes in sarcomere length.

During stretch and contraction, the A bands remain at approximately constant length, and changes in the length of the muscle are accounted for by changes in the length of the I bands. Most significant of all, however, is the fact that the H zones (the lighter regions in the middle of the A bands) are longer in stretched muscle than in muscle at rest-length, and that they shorten and disappear as the muscle contracts. Over the range of lengths from the stretched condition down to the point at which the H zones close up completely, the distance between the Z lines and the edges of the H zones remains approximately constant. If myosin extraction is carried out at any particular sarcomere length in this range, then it is found that the "ghost" fibril, left after the removal of the A substance, consists of the Z lines and material extending from the Z lines up to the position of the edge of the H zones

[1] *E.g.* that the "secondary filaments" seen in the end-on sections were really obliquely sectioned bridges; it was not clear how this effect could be produced with sections of conventional thickness.

before extraction[4]. These large, easily recognised differences in the length of the H zones, which are unaccompanied by any appreciable changes in the over-all length of the A bands, and which are very easy to observe either in isolated myofibrils or in sectioned whole muscle, provide, in the author's view, perhaps the most direct and compelling argument of all for the sliding filament model. It is very difficult to see how these observations, coupled with the present ones on the changes in the extent of overlap of primary and secondary filaments, can be explained in any other way. The alternative theories that have been put forward depend not on contesting this evidence, but on presuming that their incompatibility with it can in some unspecified way be removed; they take as their starting point the belief that there is no such thing as a system of two overlapping sets of filaments. Let us consider the positive evidence advanced in their support.

Spiro observed that the number of filaments present in the A bands, on either side of the H zone was, in rest-length muscle, three times as great as in the H zone itself. This is in agreement with the previous observations[2]. In shortened muscle, however, only a single set of filaments was observed.

This latter observation has not been confirmed in the present studies, nor, contrary to the suggestion, has it been found that any difference, other than a straightening out of the I band filaments, results from holding a muscle at a fixed length rather than leaving it unrestrained during the preparative procedures. The evidence presented in support of a progressive transformation of thin into thick filaments is incomplete, for no intermediate stages were shown; the argument rests on the picture in which a complete "transformation" into thick filaments has occurred. It seems possible that the appearance of the structures has been affected by the freezing and thawing process used to produce the contraction; certainly considerable destruction of the thin filaments in the residual I bands has taken place.

Hodge presents a graph, apparently showing a linear relationship between sarcomere length and axial period, and suggests that it is this change in period which produces contraction. It is also suggested that the I band filaments may be incorporated into the A band filaments in some way. It is to be doubted, however, whether variations of $\pm$ 20 per cent (*i.e.* in the range 320 A to 480 A) are significant, for the measurements were made on mechanically fragmented material which had been allowed to dry down onto the electron microscope grid. During this process, variable amounts of shrinkage and distortion will affect both axial period and sarcomere length simultaneously. The two points on the graph at a sarcomere length of 1.5 μ and an axial period of 250 A seem to lie outside the range of scatter, but their significance is questionable until the effect is shown to be a reproducibile one correlated with changes in band pattern. One must point out that the most striking feature of very many of the published electron micrographs which show the axial period of muscle is that the number of periods in the A bands always seems to be constant (at about 38), while the number of periods in the I band varies with the degree of contraction. Specific attention is drawn to this fact by Edwards *et al.*[18], and their Fig. 6 shows the phenomenon with great clarity. It was also commented on earlier by A. F. Huxley (personal communication, 1953). Such a change in the number of periods in the I bands is incompatible with a linear relationship between sarcomere length and axial period. Moreover, the approximate constancy of the length of the A band has been well documented by observations in the light microscope[4–6], and if the number of periods there remains constant, then so must the length of each period, as x-ray measurements had indeed indicated[1].

The second argument used by Hodge derives from the apparent continuity, in many pictures, of the I band filaments with the main filaments in the A band. However, as we have already seen, it is almost impossible, even using the very thinnest sections, to prove which set of filaments in the A band is continuous with the I band filaments; in the pictures which have appeared previously in the literature, several layers of filaments are included in the sections, and no valid argument, either way, about filament continuity can be based on them. Indeed, in sections which show filaments in the A band most clearly, the secondary and primary filaments will lie in vertical register with each other.

The evidence from insect flight muscle will be considered in a separate paper, but a preliminary report (Huxley and Hanson[19]) has shown that in this type of muscle too, the secondary filaments can be seen clearly in thin, suitably oriented longitudinal sections as well as in cross-sections; in this case, however, the secondary filaments are located midway between *two* primary ones.

It is observed that a clear appearance of secondary filaments is difficult to obtain in cross-sections of muscle, and it has been suggested that this is due to a variability of the structure. But it is the present author's experience that the main difficulty with sections of conventional thickness is simply one of orientation. The secondary filaments are rather thin (about 50 A) and lie quite close to the primary filaments, and unless they are absolutely perpendicular to the plane of sectioning, the picture they give can be quite confused and misleading. Such a picture can be transformed into one which shows the secondary filaments, in transverse section, perfectly well all over the field, and, incidentally, all simultaneously in a given fibril (*e.g.* Fig. 19), simply by taking great care over the orientation of the block face.

Sjöstrand and Andersson, also, have expressed the view that only a single set of continuous filaments is present in striated muscle. But the picture of a longitudinal section which they have published[9] in support of this is not a convincing one, and, judging by the side spacing between filaments, contains more than one layer of them. It is also claimed that the axial distance between the cross-bridges which they have seen varies with the length of the sarcomere. However, the observations could equally well be accounted for by the reduction of the sarcomere length—and, of course, of any other axial periodicities—which takes place during sectioning unless the cutting edge is parallel to the fibers; this effect can reduce the dimensions by a factor of two or more in very thin sections. This criticism also applies to their claim that the A bands decrease in length when sarcomeres shorten.

It would seem, then, that the additional arguments advanced for the various alternative hypotheses are all rather insecurely founded. The hypotheses were neither designed to, nor are able to, account for much of the other evidence available concerning the structure of muscle. They were put forward essentially because of the failure to observe a double array of filaments in longitudinal sections. This failure occurred because the sections used, though thin, were not thin enough for the purpose of studying a close-packed array of filaments.

2. The present results

There are many features of this type of structure which might profitably be discussed in relation to the mechanism of muscular contraction. However, most of them have already been considered at length in previous papers[1–6], and we will comment here on only a few special aspects.

One simple feature of the structures seen in these longitudinal sections is that, no matter whether the muscle is stretched, at rest-length, or shortened, the cross-bridges between primary and secondary filaments have the same form, *i.e.* they are, more or less, at right angles to the filaments. The only difference between the muscles at different lengths is that the two sets of filaments overlap to different extents. This must mean that as the filaments slide past each other during contraction, the bridges between them remain attached for a short distance only, and they must then detach from the secondary filaments and reattach at a point a little further along. We will refer to the separation between these points of attachment on the secondary filaments as the "step distance." Thus if contraction is brought about by some movement of, or interaction at, the bridges, then one complete contraction of the muscle will represent a number of cycles of operation of the contraction mechanism associated with the bridges. This idea is by no means a new one, but it is as well to draw attention to it again now that more direct evidence is available to support it. If the bridges represent sites of actin-myosin interaction and ATPase activity, if they represent the sites at which chemical energy is transformed into a mechanical deformation and hence into external work, then each site will be able to operate a number of times during a single contraction of the muscle. This possibility offers obvious advantages over a mechanism in which the active sites can function only once during each contraction.

In an earlier paper[4], the increase in the length of the I bands during stretch of muscle *in rigor* (*i.e.* when the contractile system was "locked"), was commented on. Probably the latter part of the rather large extensions used was non-physiological (although it was reversible); but the present observations of the difference in the degree of orientation of the I band filaments, as between muscle fixed when unrestrained, and when held at a fixed length, does direct attention again to the possibility that part of the series elastic component[21] might be accounted for by a tendency of the I band filaments, in normal circumstances, to depart slightly from perfect linearity. After all, the filaments are very thin, they are a long way apart, and they must be undergoing continuous Brownian movement. If this were the case, one would expect the effective length of the series elastic component to be different at different muscle lengths, *i.e.* to depend on the length of the thin filaments "exposed" in the I bands. This should be susceptible to experimental test.

The low birefringence of the I bands compared with that of the A bands is, of course, in part due to there being less oriented structural protein there (about $\frac{1}{3}$ the amount). And the greater space available between the filaments might allow the concentration of the molecules of the soluble proteins to be higher there than in the A bands, thereby increasing the effective refractive index of the solution, and decreasing the form birefringence of the array. But the low birefringence may also be contributed to by an inherently poorer orientation of the array of I band filaments themselves.

The details of the arrangement of the cross-bridges between the filaments have several interesting features. Each secondary filament is connected to each of the three primary filaments around it once every 400 A. The three bridges involved are not in register but appear to be spaced out evenly along the secondary filaments. Thus one would expect these filaments to possess a threefold screw axis. A structure for action has recently been proposed by Selby and Bear[(20)] on the basis of x-ray diffraction data. This structure is described in terms of a two-dimensional net, but the authors draw attention to, and indeed favour, an equally valid interpretation in terms of a two-chain helix with a 406 A axial period. It is interesting to note that when one builds this helical structure, it turns out to possess a threefold screw axis. The axial distance between nodes in the Selby and Bear structure is about 54 A. If these nodes represent actin monomers, to which bridges from the myosin may attach, then the minimum step distance would be 54 A; and as the nodes occur on a helix, the secondary filaments would have to spin about their long axes as the muscle contracted. A step distance of 54 A is compatible with a mechanism in which one molecule of ATP is split at each of the bridges for each step of the contraction when the muscle is exerting a maximum tension of about 4 kg./cm^2.

In the micrographs so far examined, the primary filaments also give the impression of having a helical structure, with the bridges protruding from them along a sixfold screw axis, with six bridges in 400 A. It is, incidentally, perfectly possible to build a structure from such filaments with the bridges in the right places for them to lie on a threefold screw axis on the secondary filaments. The tapered ends of the primary filaments would be a natural consequence of their being built from a helical array of long molecules, with their axes parallel to the helix, spaced out along it with a period less than the length of the molecules. The observation by Szent-Györgyi[(22)] that it is the H meromyosin subunit of myosin which combines with actin and which functions as an ATPase would suggest that the bridges or protuberances seen on the primary filaments represent or contain this part of the myosin molecule. The total number of these bridges in one gram of muscle is easily computed, and is found to be approximately 5×10^{16}. The number of molecules of myosin present in one gram of muscle may be calculated from the molecular weight of myosin[(23)] and the known myosin content of muscle[(16, 17, 24, 25)], and is found to be in the range 4 to 8×10^{16}. Thus each of the bridges seen probably represents the active part of one molecule of enzyme.

Thus the details of the structure seem to be capable of interpretation in terms of the known properties of actin and myosin; but as yet, they do not reveal the nature of the basic mechanism involved in contraction.

Summary and Conclusions

The conditions under which one might expect to see the secondary filaments (if they exist) in longitudinal sections of striated muscle, are discussed. It is shown that these conditions were not satisfied in previously published works for the sections were too thick. When suitably thin sections are examined, the secondary filaments can be seen perfectly easily. It is also possible to see clearly other details of the structure, notably the cross-bridges between primary and secondary filaments, and the tapering of the primary filaments at their ends. The arrangement of the filaments and the changes associated with contraction and with stretch are identical to those already deduced from previous observations and described in terms of the interdigitating filament model in previous papers. There are therefore excellent grounds for believing that this model is correct. The alternative models which have been proposed appear to be incompatible both with the present observations and with much of the other available evidence.

I am indebted to Professor Bernard Katz for the encouragement that he has given to this work, to the Medical Research Council for their support, and to the Wellcome Trust for the provision of an electron microscope.

References

1. Huxley, H. E., *Proc. Roy. Soc. London Series B,* 1953, **141** 59.
2. Huxley, H. E., *Biochim. et Biophysica Acta,* 1953, **12** 387.
3. Hanson, J., and Huxley, H. E., *Nature,* 1953, **172** 530.
4. Huxley, H. E., and Hanson, J., *Nature,* 1954, **173** 973.
5. Huxley, A. F., and Niedergerke, R., *Nature,* 1954, **173** 971.
6. Hanson, J., and Huxley, H. E., *Symp. Soc. Exp. Biol.,* 1955, **9** 228.
7. Spiro, D., *Exp. Cell Research,* 1956, **10** 562.
8. Hodge, A. J., *J. Biophysic. and Biochem. Cytol.,* 1956, **2** No. 4, suppl., 131.
9. Sjöstrand, F. S., and Andersson, E., *Exp. Cell Research,* 1956, **11** 493.
10. Hodge, A. J., Huxley, H. E., and Spiro, D., *J. Exp. Med.,* 1954, **99** 201.
11. Szent-Györgyi, A., Chemistry of Muscular Contraction, 1951, New York, Academic Press Inc., 144.
12. Palade, G. E., *J. Exp. Med.,* 1952, **95** 285.
13. Hodge, A. J., Huxley, H. E., and Spiro, D., *J. Histochem. and Cytochem.,* 1954, **2** 54.
14. Huxley, H. E., *Proc. Internat. Conf. Elect. Micr.,* London, 1954, in press.
15. Watson, M. L., *J. Biophysic. and Biochem. Cytol.,* 1955, **1** 183.
16. Huxley, H. E., and Hanson, J., *Biochim. et Biophysica Acta,* 1957, **23** 229.
17. Hanson, J., and Huxley, H. E., *Biochim. et Biophysica Acta,* 1957, **23** 250.
18. Edwards, G. A., Ruska, H., de Souza Santos, P., and Vallejo-Freire, A., *J. Biophysic. and Biochem. Cytol.,* 1956, **2** No. 4, suppl., 143.
19. Huxley, H. E., and Hanson, J., *Proc. 1st European Regional Conf. Elect. Micr.,* Stockholm, Almqvist & Wiksell, (1956), 202.
20. Selby, C. C., and Bear, R. S., *J. Biophysic. and Biochem. Cytol.,* 1956, **2** 71.
21. Hill, A. V., *Proc. Roy. Soc. London Series B.,* 1953, **141** 104.
22. Szent-Györgyi, A. G., *Arch. Biochem. and Biophysics,* 1953, **42** 305.
23. Holtzer, A., *Arch. Biochem. and Biophysics,* 1956, **64** 507.
24. Hasselbach, W., and Schneider, G., *Biochem. Z.,* 1951, **321** 461.
25. Szent-Györgyi, A. G., Mazia, D., and Szent-Györgyi, A., *Biochim. et Biophysica Acta,* 1955, **16** 339.

Explanation of legends to Figures 1–27 beginning on page 567 and ending on page 579 of this volume, for the article by H. E. Huxley, *"The Double Array of Filaments in Cross-Striated Muscle."* Reprinted from *J. Biophysic. and Biochem. Cytol.*, **3**, 631–647 (1957). (Pages 522–579 in this volume.)

General Notes: (1) All photographs shown are of material fixed in buffered osmium Tetroxide for ½ hour.

(2) Due to the change in the dimensions of the sections during the cutting process, an effect which is very marked when extremely thin sections are involved, the scale of the structure along the axis of the muscle is considerably foreshortened. Dimensions at right angles to the filaments are, however, largely unaffected.

(3) The lower magnification photographs were taken at an initial magnification of 8,000, the high magnification ones at 40,000, and the very high magnification photograph (Fig. 27) at an initial magnification of 160,000.

(4) Except where otherwise stated (Figs. 25 and 26), the material used was glycerinated psoas muscle.

LEGENDS

Fig. 1. Low power general view of fairly well oriented thin section through a number of striated myofibrils. Examples of sections parallel to the $10\bar{1}0$ lattice planes are marked a, and of sections parallel to the $11\bar{2}0$ lattice planes, b. Magnification, 60,000.

Fig. 2. Section through one sarcomere, parallel to $11\bar{2}0$ planes, showing primary filaments, with large interfilament spacing (about 500 A), and pairs of secondary filaments in between them. The interruption of the secondary filaments at the edges of the H zone is readily visible. The characteristic tapering of the primary filaments at the ends of the A bands is seen quite generally in this and other pictures, and seems to occur independently of effects caused by the filaments passing out of the section. The cross-bridges between primary and secondary filaments can also be seen and counted. The tapering of the primary filaments seems to occur over the length occupied by the last four to six bridges, *i.e.* the last 1600 to 2400 A of the filament. The primary filaments are somewhat thickened in the H zone, elsewhere their diameter is 110 to 120 A; the diameter of the secondary filaments is 50 to 60 A. Magnification, 175,000.

Fig. 3. Section through one sarcomere, almost parallel to $11\bar{2}0$ plane. The filaments run at an angle of about 1° to the plane of the section, and near the centre of the sarcomere, the section passes from one layer of the lattice to the next. The substantially poorer orientation of the filaments in the I bands, as compared with the A bands, is apparent. However, the order is sufficiently good for one to see that the secondary filaments in the I bands do not have projections on them. Magnification, 195,000.

Fig. 4 through 7. Sections through sarcomeres in $11\bar{2}0$ direction, showing variation in length of gap between ends of secondary filaments in centre of A bands at different sarcomere lengths. Fig. 4 shows a slightly stretched sarcomere; Fig. 7 a slightly contracted one. Magnification, 100,000.

Fig. 8. Section through one sarcomere, parallel to $10\bar{1}0$ planes. A simple alternation of primary and secondary filaments is seen, the secondary filaments terminating at the edges of the H zone. The secondary filaments appear thicker than in Figs. 2 and 3 because in general they will really consist of two superposed secondary filaments in the plane of the section, as shown in Text-fig. 1. Magnification, 150,000.

Fig. 9. Low power picture of oblique section through a number of myofibrils. The filaments run at an angle of about 5 to 10° to the plane of the section, and an interesting variety of patterns is seen; the hexagonal lattice of filaments in any given sarcomere may be rotated or twisted slightly relative to that in the adjoining one. The absence of secondary filaments from the H zones may be verified in this type of section. Magnification, 25,000.

Fig. 10. Oblique section through one sarcomere. The thicker primary filaments in successive layers occur midway between the filaments in the preceding layer. The pair of secondary filaments between each pair of primary filaments are located in the same plane as the primary filaments, and leave the section simultaneously; in the H zone, only the primary filaments are visible. Magnification, 150,000.

Fig. 11. Stretched muscle, showing long H zones. Section approximately parallel to $10\bar{1}0$ planes. The secondary filaments can be seen to extend only a short distance into the A bands. The regularity in the arrangement of the filaments in the A bands of stretched specimens is always found to be lower than in rest-length fibrils; possibly the absence of secondary filaments from a long region in the center of the A band allows the primary filaments there, no longer cross-linked via the secondary filaments, to become disarranged. The specimen shown in this section was not held at a fixed length during fixation, and the I bands too are disorganised and somewhat shortened. Magnification, 45,000.

Fig. 12. Section through fibrils which were not held at a fixed length during fixation. Over most of the picture, the section is parallel to the $11\bar{2}0$ planes and shows the characteristic appearance of primary and secondary filaments. The secondary filaments in the I bands are considerably disoriented. In the lower part of the picture, the section passes through about four secondary filaments in between very widely spaced pairs of primary filaments. One secondary filament can be seen running along the outside of this lower fibril. Magnification, 150,000.

Figs. 13 and 14. Sections showing primary and secondary filaments at the A–I boundary and at the H–A boundary. The tapering of the thick filaments at the A–I boundary is readily seen, and in places they may be observed to terminate, while the secondary filaments continue on into the I bands. Because of the inherent lack of order at this boundary, however, this appearance cannot be seen in every case. Magnification, 150,000.

Fig. 15. Cross-sectional view of fibril in neighborhood of the H zone. The section has passed through the A band proper on the left and right hand sides of the picture, and through the H zone in the center of the picture. The rows of primary filaments in the H zone can be followed through into the A bands (particularly on the right hand side of the picture); at the H–A boundary, the array of secondary filaments around each of the primary ones makes its appearance. Magnification, 150,000

Fig. 16. Typical section through fibril parallel to $11\bar{2}0$ planes, section thickness probably about 100 to 150 A. Magnification, 150,000.

Fig. 17. Same photograph, but doubly printed to imitate appearance of similarly oriented section about 300 A thick (*i.e.* containing two layers of primary filaments). It will be seen that the secondary filaments are now obscured.

Fig. 18. Thicker section (probably 600 to 700 A in thickness) through muscle fibrils. This is a section of conventional thickness (silver in color when viewed in reflected light); the secondary filaments cannot be distinguished. Magnification, 50,000.

Figs. 19 and 20. Cross-sections through A band region, showing double hexagonal array of primary and secondary filaments; and the cross-bridges between them. Magnification, 150,000.

Fig. 21. Similar section, but printed with higher contrast to bring out the angular form of many of the filaments in cross-section. Magnification, 150,000.

Fig. 22. Very thin cross-section through an H zone, showing the pronounced noncircular appearance of the filaments in this region. Magnification, 100,000.

Fig. 23. Highly magnified view of an H zone, showing the triangular and other geometric forms assumed by the filaments in cross-section. Magnification, 200,000

Fig. 24. Longitudinal section through adjacent contracted and stretched sarcomeres. Note the longer I bands and H zone in the stretched sarcomere; the withdrawal of the secondary filaments from the H region is well marked. Magnification, 35,000.

Fig. 25. Low power view of longitudinal section of rabbit psoas muscle fixed while in the living state (*i.e.* no glycerol treatment). The characteristic patterns of primary and secondary filaments can be seen just as in glycerinated materials, although the contrast and clarity of the picture is lower. Magnification, 20,000.

Fig. 26. Longitudinal section of fresh muscle (as in Fig. 25), cut parallel to $11\bar{2}0$ planes, showing pairs of secondary filaments in between the widely spaced pairs of primary filaments, in the same manner as in glycerinated material. Magnification, 100,000.

Fig. 27. Highly magnified view of central region of an A band, sectioned parallel to $11\bar{2}0$ planes, showing primary and secondary filaments and bridges between them. The termination of the secondary filaments at the H zone is readily seen. The initial magnification of this photograph was 150,000, and the instrumental resolution is substantially better than 10 A, judging from measurements on the background granularity due to phase-contrast effects very close to focus. The axial spacing between the bridges is greatly foreshortened by the change in section dimensions produced during the cutting of this very thin section. The appearance of bridges between pairs of secondary filaments is probably due to those bridges which extend out of the plane of the section to the primary filaments in the layers above and below. In many places, there are indications that the filaments have a helical structure. Magnification, 600,000.

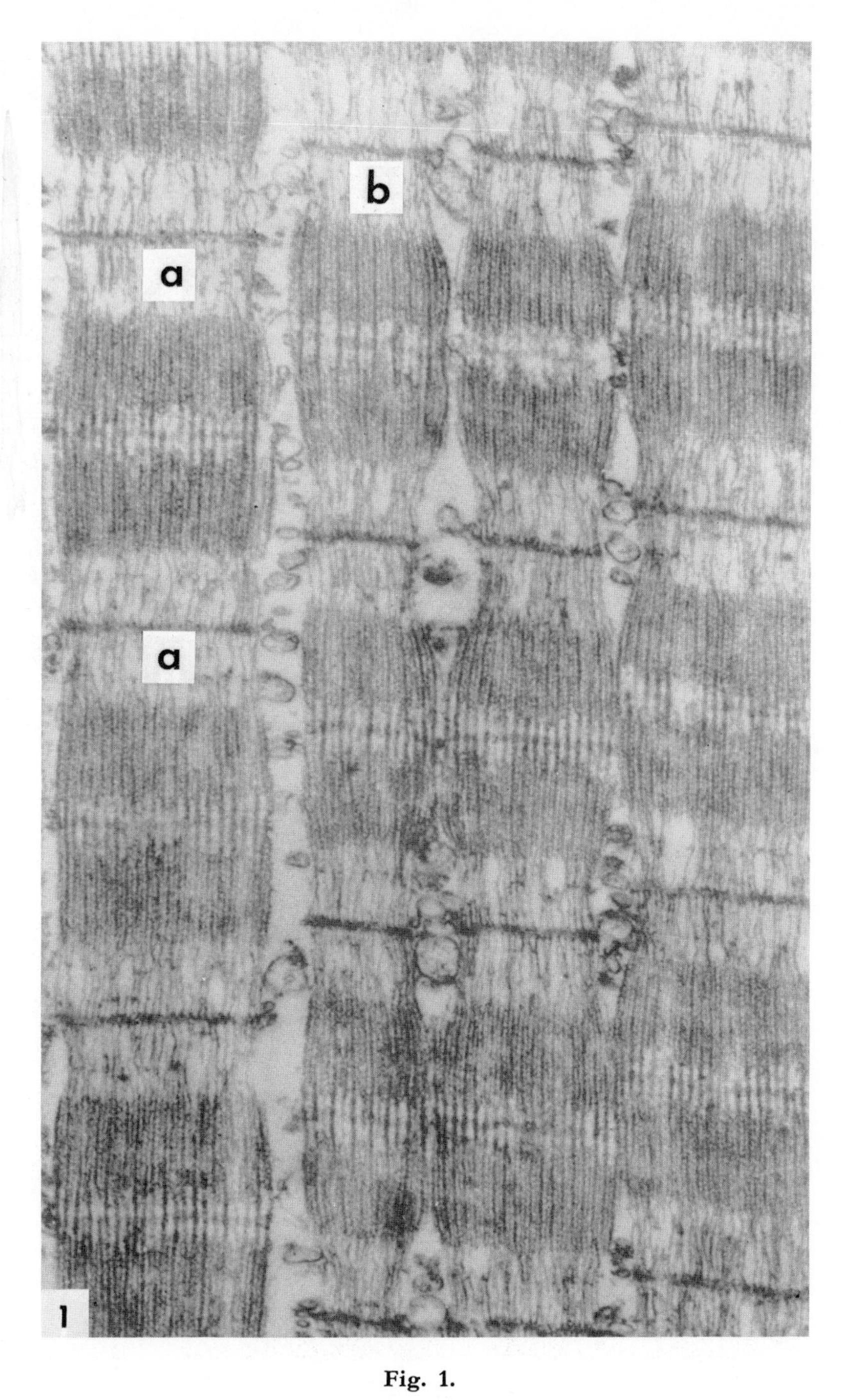

Fig. 1.

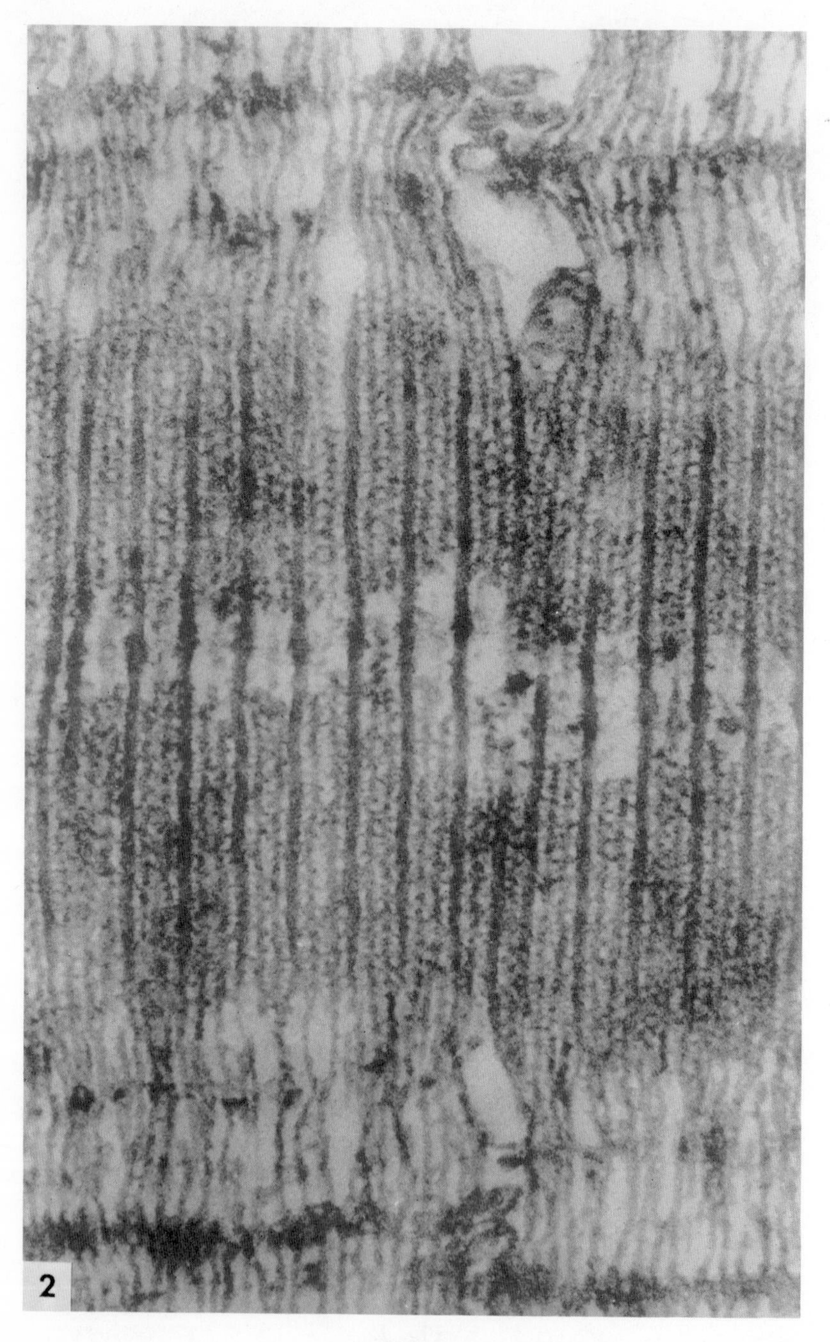

Fig. 2

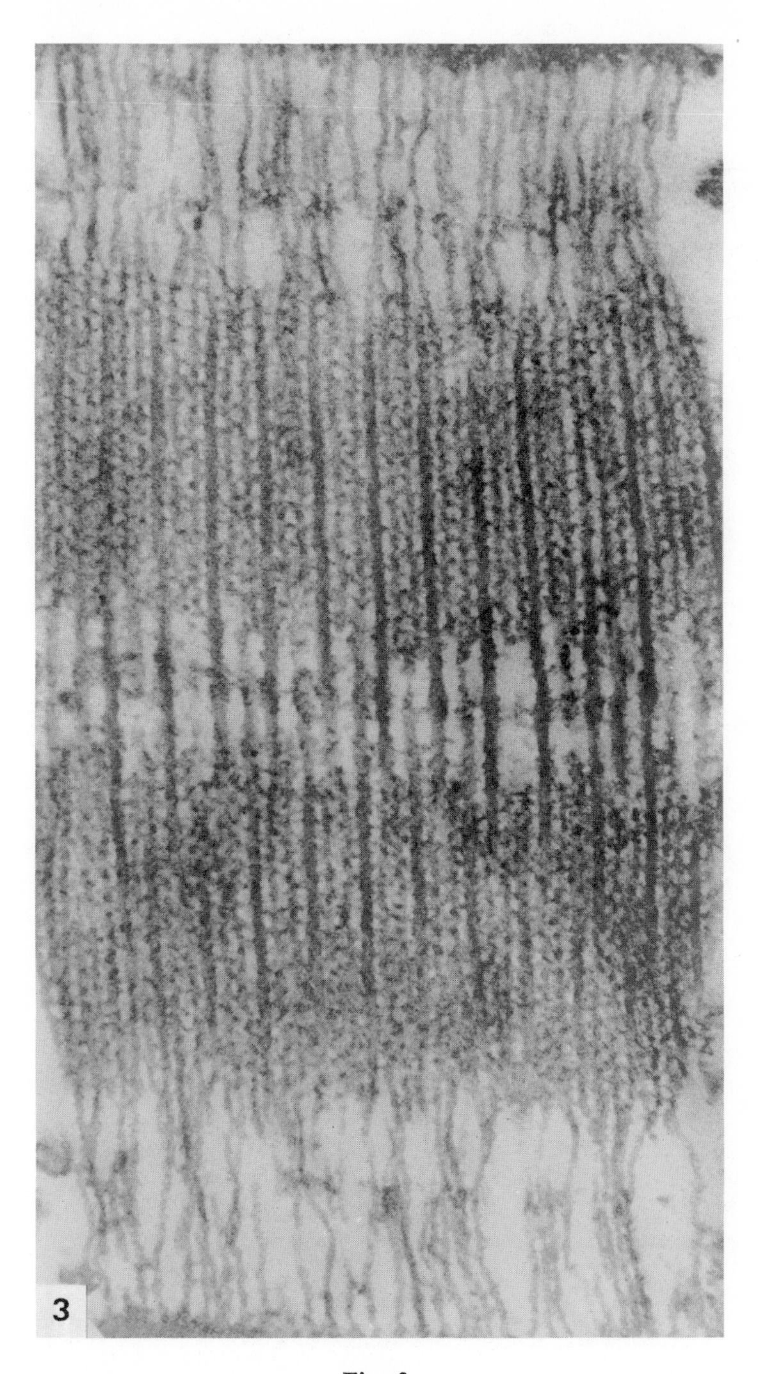

Fig. 3.

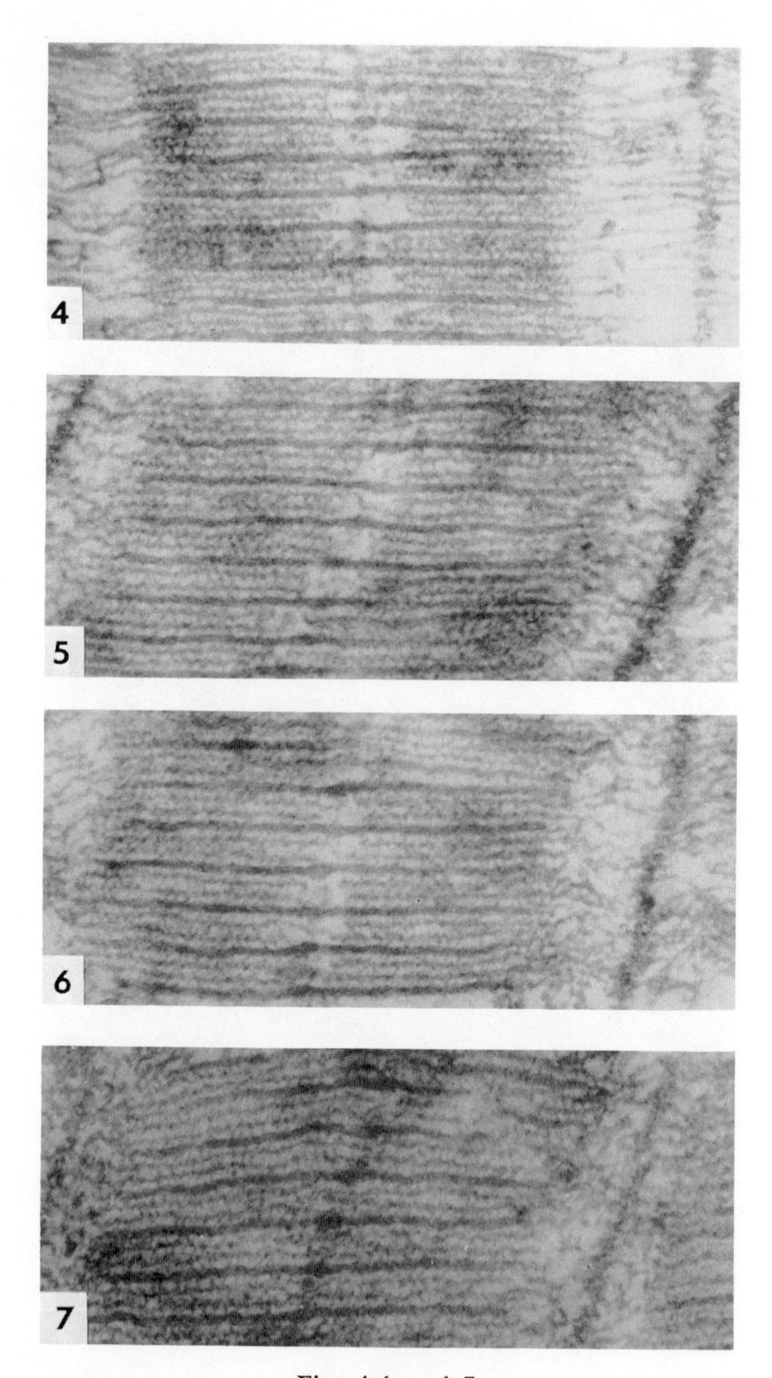

Figs. 4 through **7.**

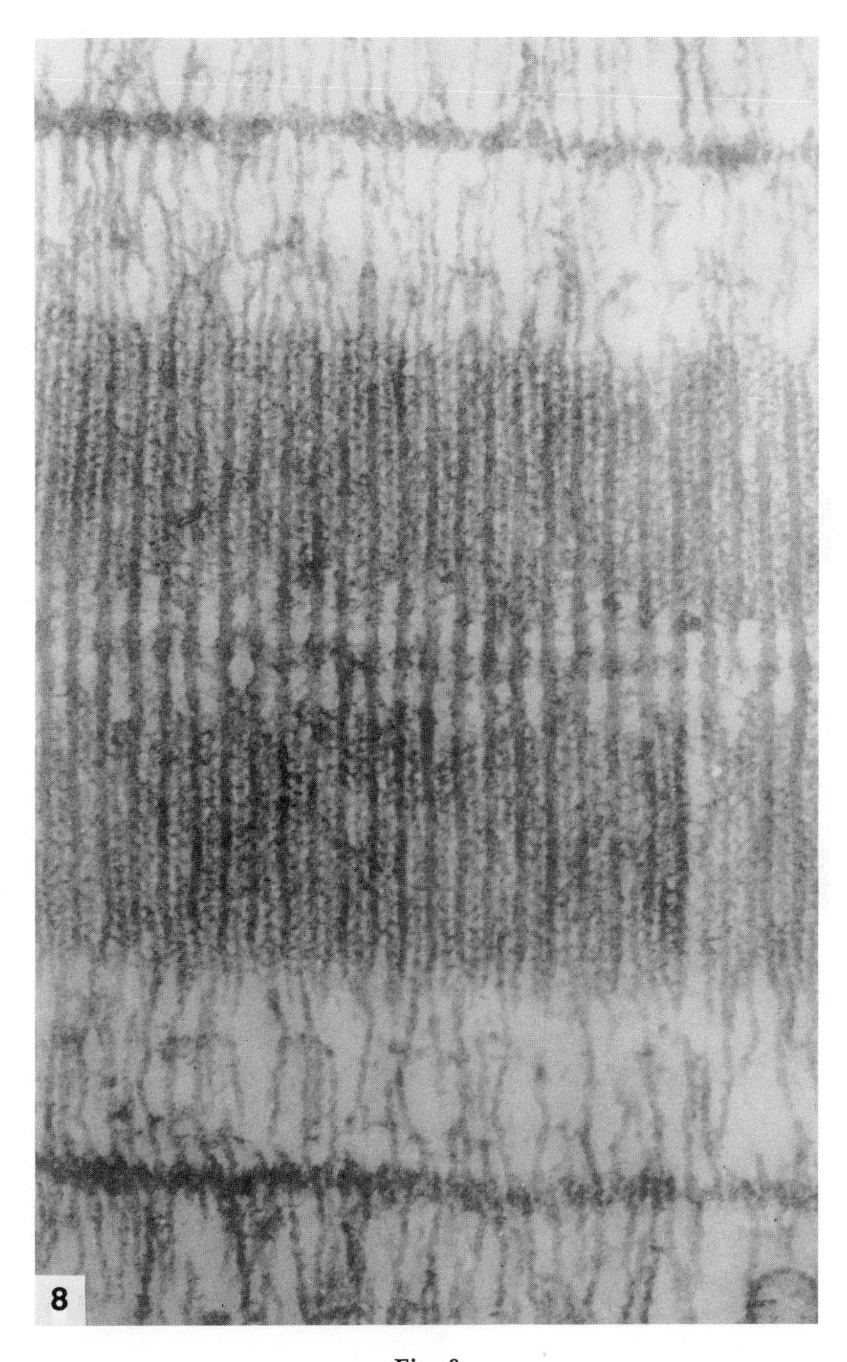

Fig. 8.

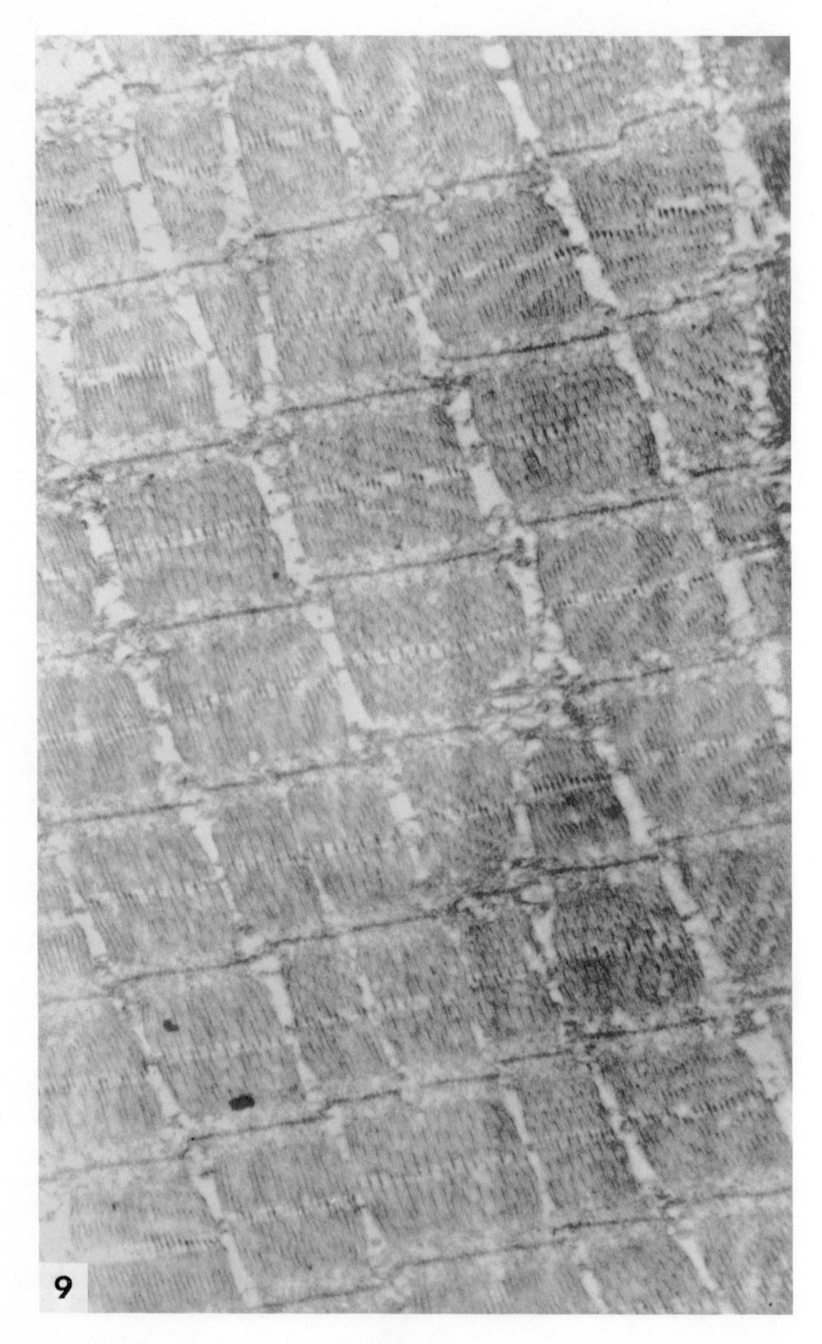

Fig. 9.

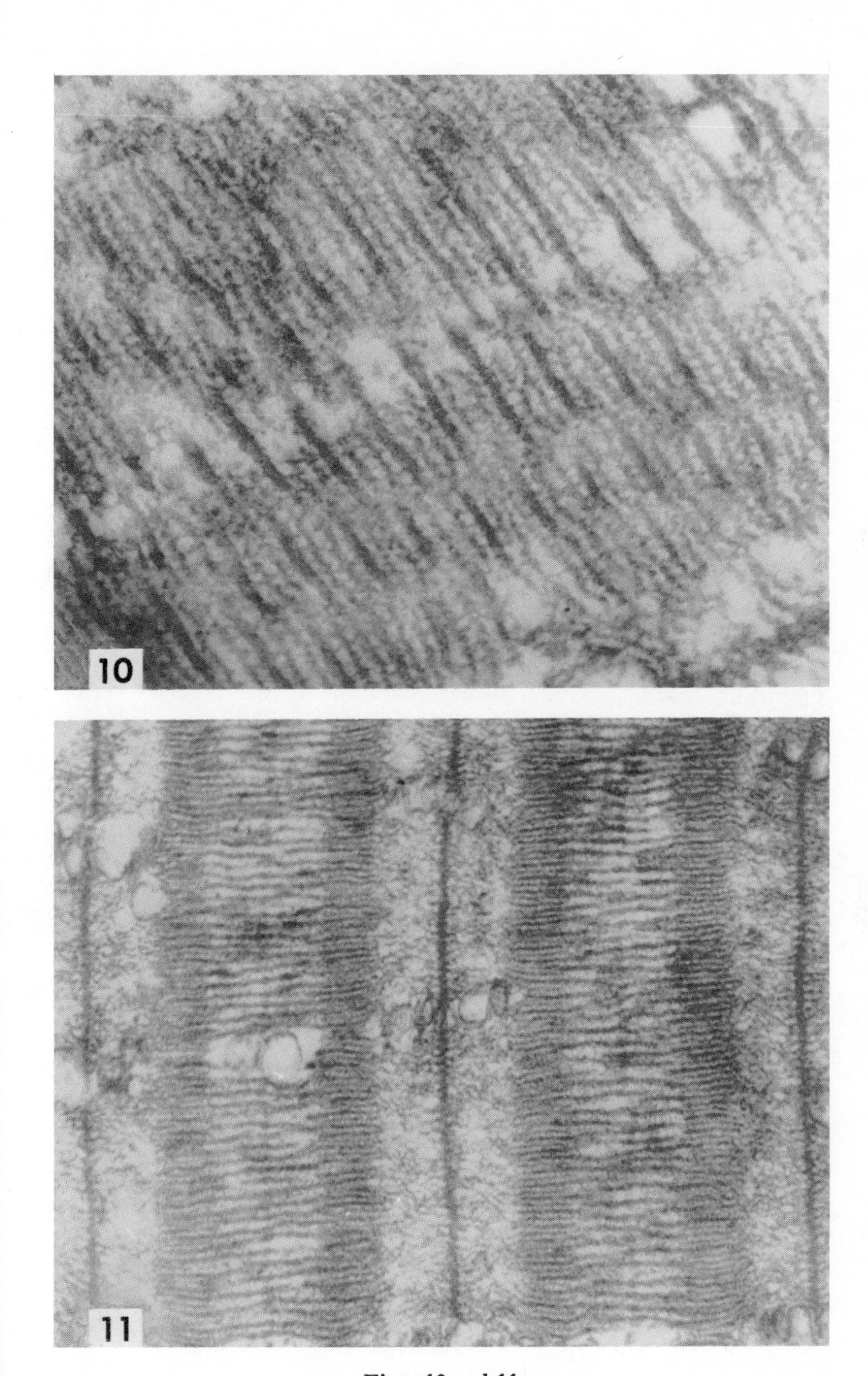

Figs. 10 and **11.**

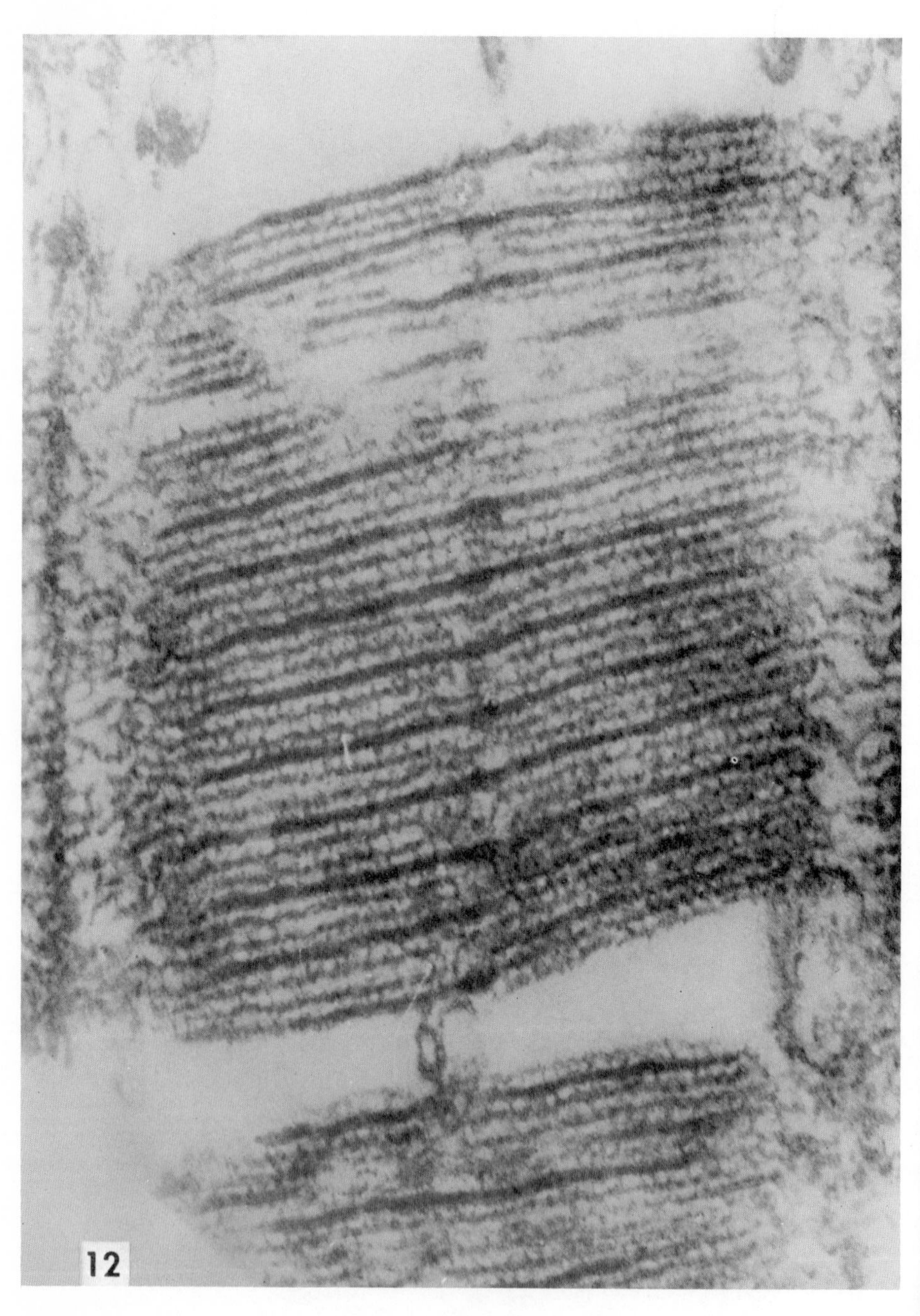

Fig. 12.

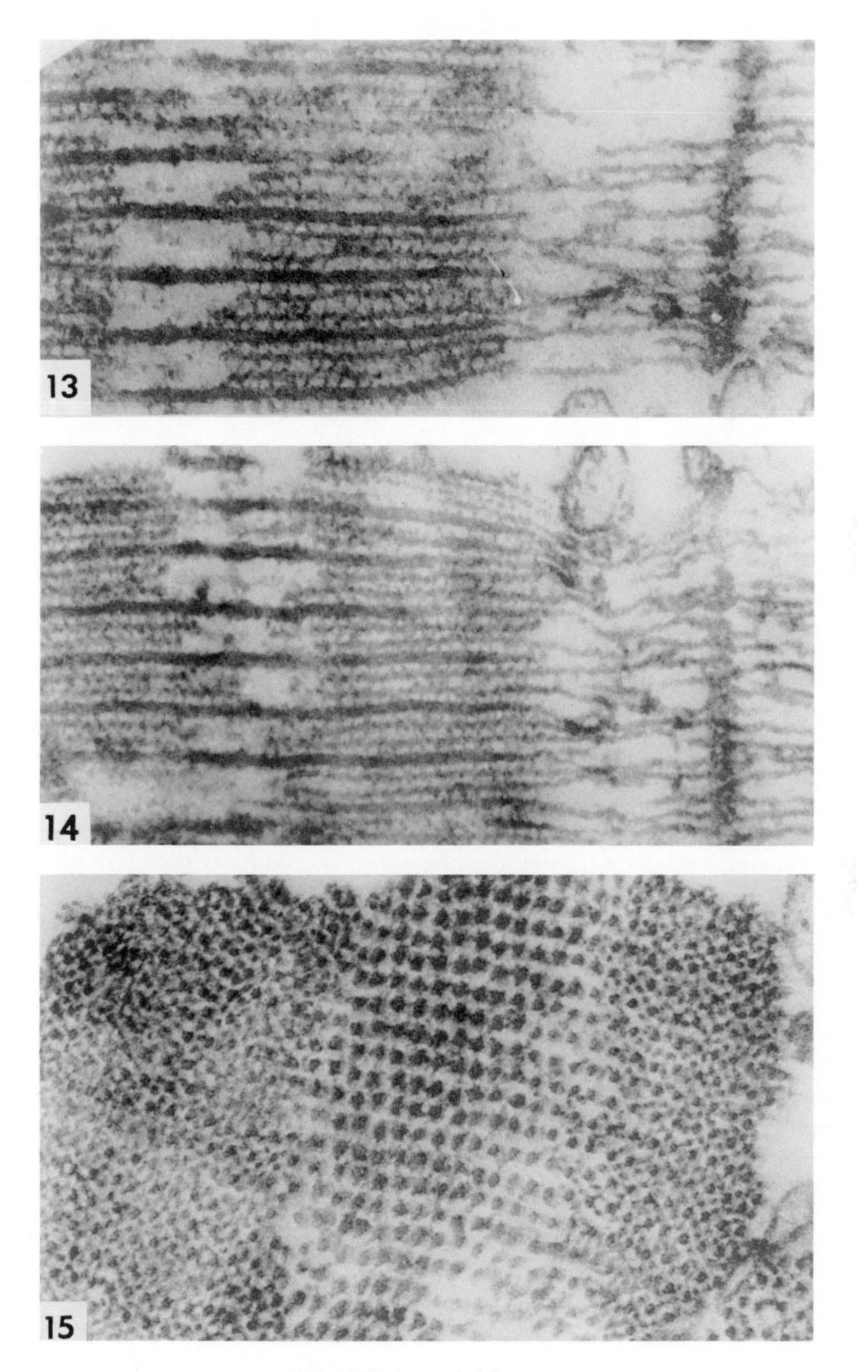

Figs. 13 through **15.**

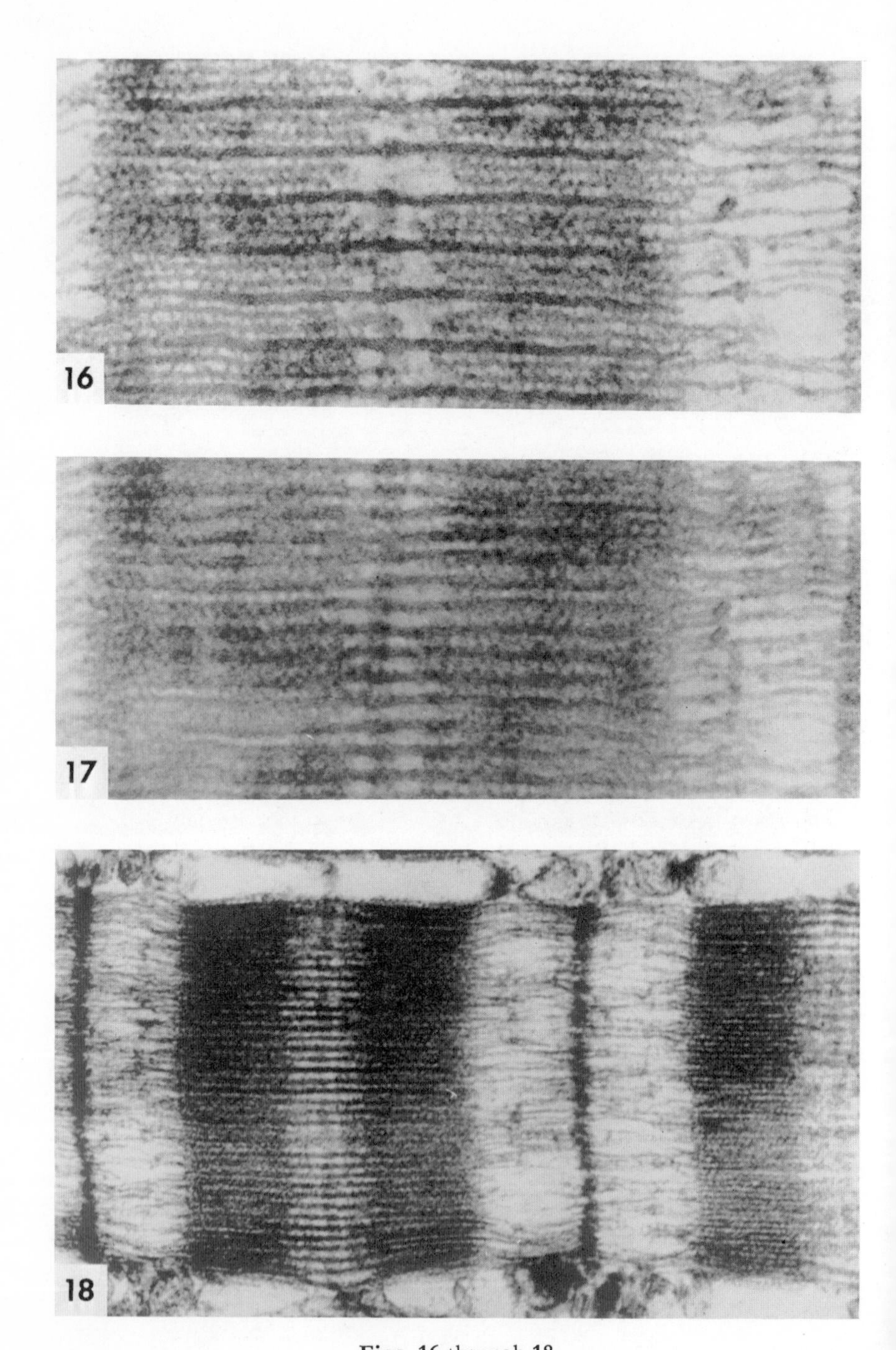

Figs. 16 through **18.**

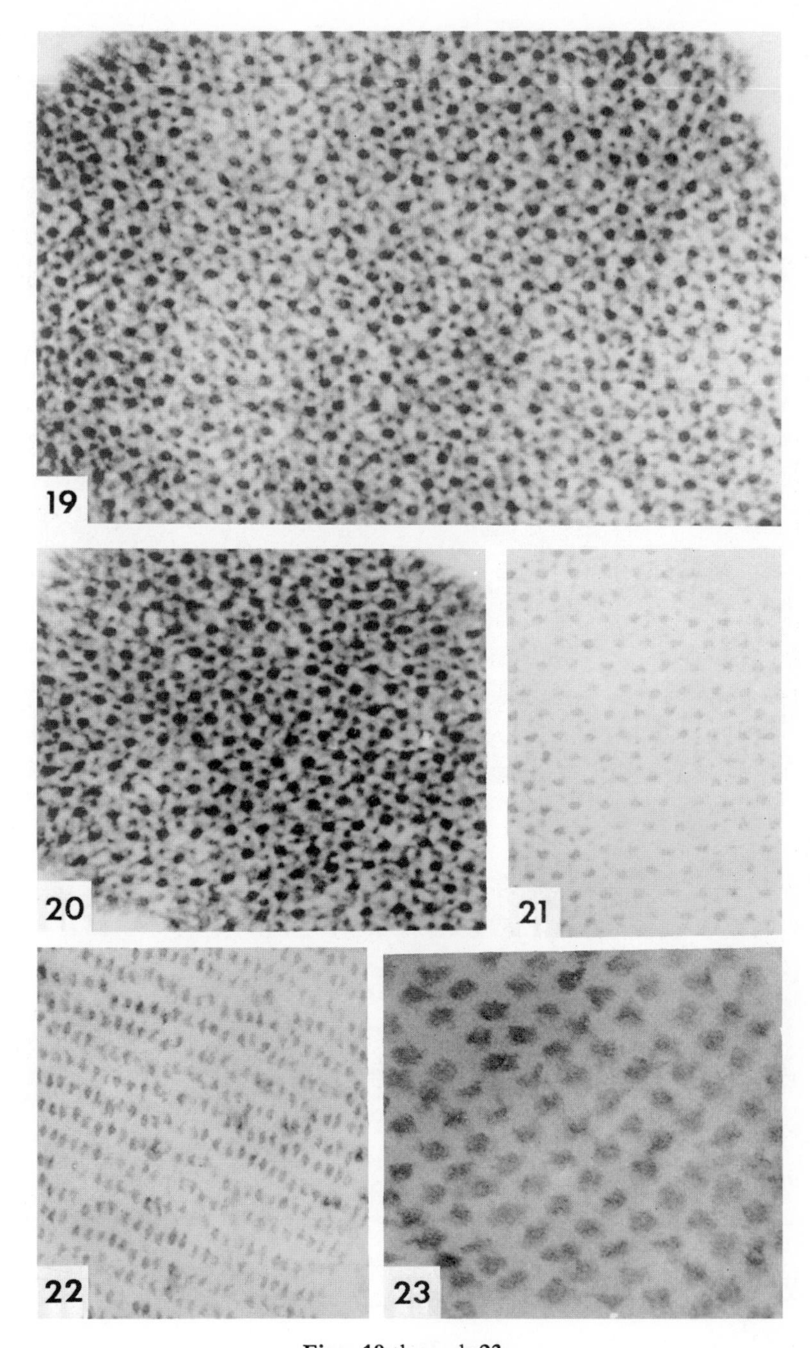

Figs. 19 through **23.**

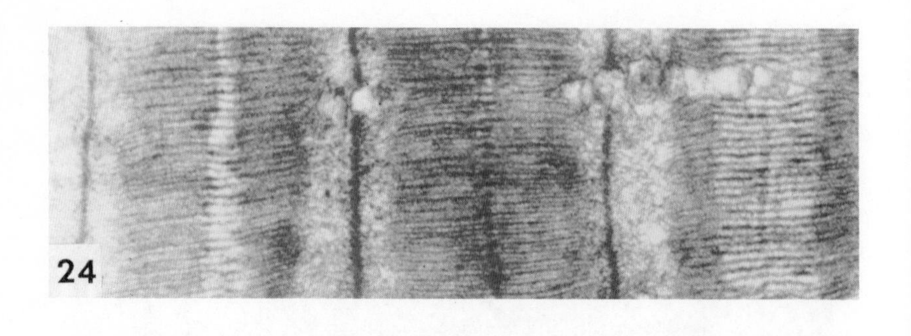

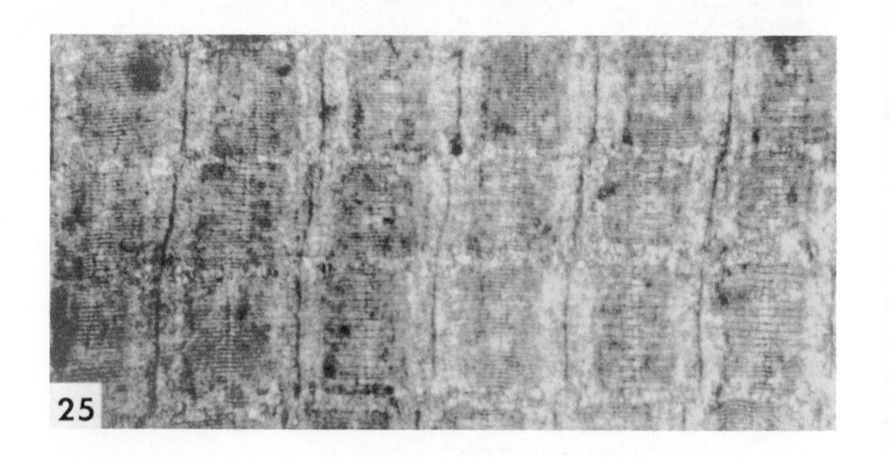

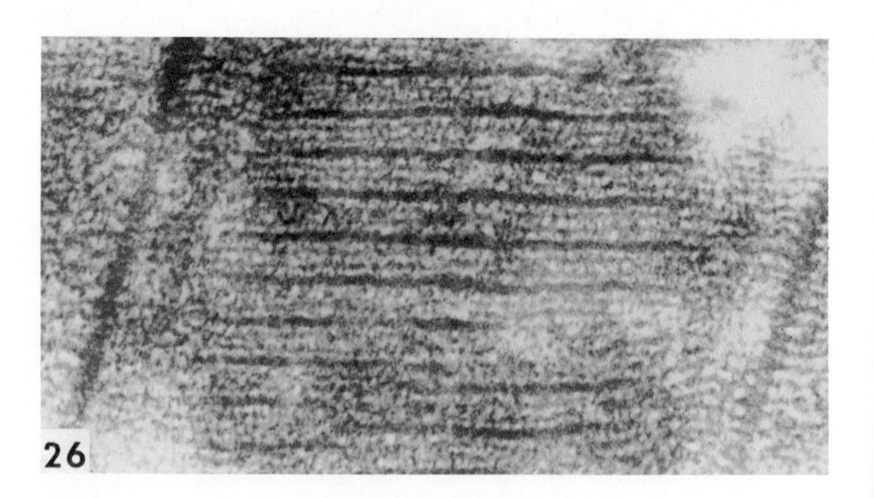

Figs. 24 through **26.**

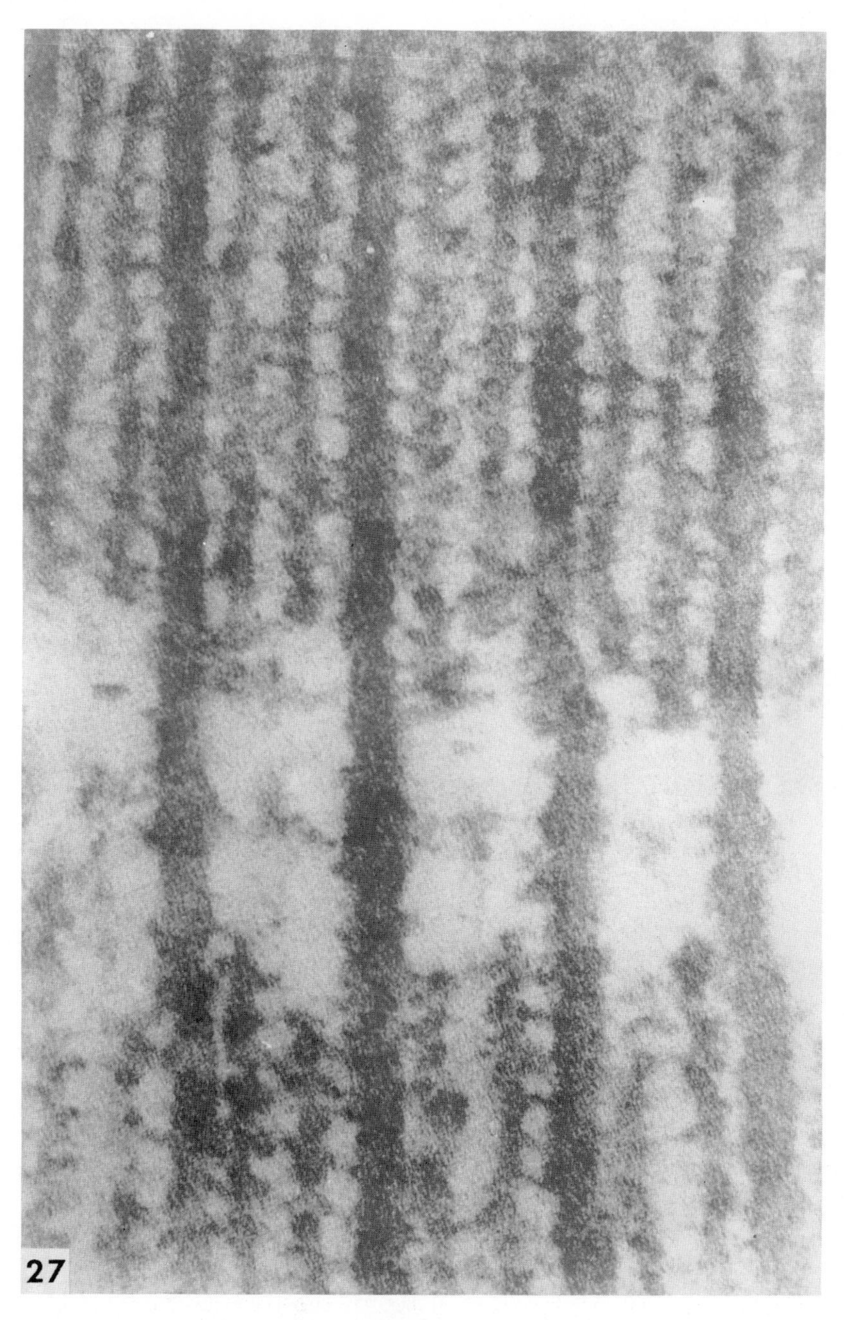

Fig. 27.

CHEMICAL REACTIONS DURING CONTRACTION AND RELAXATION*

Hans H. Weber

Institut für Physiologie im Max Planck Institut für
Medizinische Forschung, Heidelberg, Germany

Since this symposium is chiefly intended for discussion and since the introductory lectures are to serve as a basis for it, I feel that it is not useful to present a complete and detailed picture of the biochemistry of the muscle. Instead, I will confine myself to a sketch in which I shall emphasize the gaps in our present knowledge as a basis for general consideration.

We know from Emden and Parnas and particularly from the eminent biochemist, Meyerhof, and his well-known pupil, Lohmann, that lactic acid fermentation has no other purpose than the formation of adenosine triphosphate (ATP) from adenosine diphosphate (ADP) and inorganic phosphate (P). In addition, we know from a large number of scientists, some of whom are present in this room, that the same is true of the oxidative energy-yielding metabolism.

The experiments of A. V. Hill, especially those carried out during the historically significant period of his collaboration with Meyerhof, have shown us that all the processes leading to the demonstrated ATP formation take place during the anaerobic and aerobic *restitution* phase. Consequently, the energy transformed during the *working* phase can originate only from the *splitting* of the formed or restituted ATP. We ask whether this energy is transferred directly to the contractile structures; for it is conceivable that the energy

*Hans H. Weber, "Chemical Reactions during Contraction and Relaxation," *Molecular Biology,* New York: Academic Press, Inc., 1960, 25–36. Reprinted with the permission of the author and the publisher.

is transferred from the ATP to another intermediate, and only then to the contractile structure.

In experiments with living muscle, attempts have been made to answer this question by measuring the ATP level during and directly after the performance of moderate work at low temperature. It was reasoned that the ATP level would decrease temporarily during the working phase if the energy liberated by ATP splitting were transferred directly to the contractile structure. A significant decrease of the ATP level is never found in the case of performance of moderate work. In general there is only a reduction of the phosphocreatine level, because the restoration of ATP by transphosphorylation from phosphocreatine occurs just as rapidly as the splitting of ATP. Recently Britton Chance(1) with his elegant method corrected this result a little, showing that the ATP level does fall somewhat even with single twitches of the muscle, but so little that the ATP splitting thus evidenced is far from being sufficient to meet the energy requirement of the twitch. In very recent and systematic work carried out by Krebs, Fleckenstein, Davies and co-workers(2) on the one hand, and by Mommaerts(3) on the other, it has been shown in addition that in the case of the performance of very little work at near 0°C not even the phosphocreatine level is measurably reduced.

All these observations seem to indicate that the energy liberated by ATP splitting is *not* transferred *directly* to the contractile system. Principally, however, it may be objected to this interpretation that the ATP splitting during the working phase can be much greater than the reduction of the ATP level, because the ATP level may be the result of both processes—ATP splitting and ATP resynthesis occurring at the same time (i.e., the change in the ATP turnover can be much greater than in the ATP level). The experiments of Krebs, Fleckenstein, and coworkers(2), as well as those of Mommaerts(3), approach this problem from an entirely different premise than those of Britton Chance(1). Therefore I do not want to anticipate the discussion of this point in my lecture.

The situation becomes much simpler when the analysis of the working phase is carried out with isolated contractile systems rather than with living muscle; for with the isolation of the contractile systems, the multiplicity of the ATP restoring processes is excluded. Isolated contractile systems are capable of splitting ATP, and thus of contracting with performance of work(4–6). On the other hand, they are not capable of restoring the split ATP.

The washed muscle fibrils which Dr. H. E. Huxley has just described are examples of such isolated systems taken from cross-striated muscles. They have been introduced into scientific research in Chicago(7), and especially in Cambridge, England, by Perry(8). It is possible to leave the fibrils and contractile elements of movable cells in the tissue. If the cell and muscle membranes are destroyed and all soluble proteins, enzymes, and substrates of metabolism are extracted, the entire muscle fiber(4–6) and cell(9) behave as if they were composed only of the isolated contractile structures. This method was developed for muscle fibers by Albert Szent-Györgyi(4) and for cells by Hoffmann-Berling(9).

Finally the contractile protein can be extracted from muscle and cells and purified by the usual methods of protein chemistry(10, 11). From the solution of these highly purified proteins, well-oriented threads can then be produced, which likewise contract with the performance of work upon the addition of appropriate chemical substances[10, 11]. This so-called actomyosin thread was produced for the first time in my laboratory 25 years ago[12].

In a physiological ionic environment all these isolated contractile systems split added ATP and thereby contract. If the ATP splitting is reversibly inhibited by poisons[13] or by a physiological factor[14, 15] originating from the muscle granules[16], then a reversible relaxation occurs.

We know from countless experiments during the last 10 years that the contraction of isolated contractile structures is more or less the same as the contraction of living muscles and cells[5]. Maximal tension and its dependence on temperature, maximal shortening, speed of shortening, and efficiency are quantitatively the same[5, 6]. The identity extends so far, that the differences existing among the various muscle and cell types with respect to these phenomena can also be found in the isolated contractile systems prepared from these different muscles and cells[17].

Thus, it is well established that the isolated contractile systems of all types of muscles as well as of many movable cells require ATP as an operative substance for the contraction cycle. In its function as an operative substance, ATP can be replaced by any one of the nucleoside triphosphates (NTP) so far investigated.[18] The mechanical power of the contractile systems is greatest, however, with ATP. A particularly significant reduction in mechanical power is observed when the six $-NH_2$ nucleoside triphosphates are replaced by six $-OH$ nucleoside triphosphates[18]. None of the other known substances participating in ATP restitution has any effect. It makes no difference whether or not these other substances contain energy-rich phosphate bonds. Thus, we see that the reaction between ATP and the contractile structure is quite a specific one. This is a strong argument in favor of the physiological significance of ATP.

The conclusion that ATP splitting is a prerequisite for contraction, and that its inhibition results in relaxation and renders contraction impossible, still does not account completely for the interaction between ATP and contractile structure; for, if the ATP splitting in an isolated contractile system is interrupted not by inhibition but by the removal of ATP, the system does not relax but becomes rigid[13]. The presence of ATP, i.e., the binding of ATP to the contractile structure, without simultaneous ATP splitting, has a plasticizing effect[13], and in its function as a plasticizer, ATP can be replaced not only by other nucleoside triphosphates, but also by any other inorganic polyphosphate[13].

From the results reported so far, it may be concluded that ATP functions as a contracting substance in all muscles and in many cells providing that it is split by the contractile structures. If it is merely bound by these structures, however, and not split, then its effect is that of a relaxing or plasticizing substance. ATP splitting, however, by any ATPase other than the contractile protein system does not result in contraction.

Since living muscle always contains ATP, the above conclusion means that the relaxation phase of muscle contraction is not an active process, but simply the end of the ATP splitting and of the contraction phase of the contraction cycle thus induced. This interpretation is in perfect agreement with the mechanical and thermodynamic results obtained by A. V. Hill in experiments with *living* muscle.

We have heard from H. E. Huxley that the contraction of the muscle fibril is not caused by a *shortening* of the fine filaments within the fibril, but by a *sliding* of the actin filaments alongside the L-myosin filaments. This brings us to the question of conceiving a mechanism through which ATP splitting can result in a sliding of actin filaments alongside L-myosin filaments. In order to find such a mechanism, we have only to modify, in one single point, the well-known metabolic reactions in which the energy liberated by ATP splitting is required for chemical synthesis.

Let us select as a model of such a metabolic reaction the synthesis of fatty acid chains from acetate by coenzyme A in the presence of ATP and Mg.

(1) acetate$^-$ + A—P ~ P ~ P $\rightleftharpoons$ A—P ~ acetate + P ~ P
Activation of the COOH group of the acetate.

(2) A—P ~ acetate + HS—CoA $\rightleftharpoons$ acetate ~ SCoA + AMP
First reaction of the activated COOH group with the SH group of coenzyme A.

(3) acetate ~ SCoA + acetate ~ SCoA $\rightleftharpoons$ acetoacetyl ~ SCoA + HSCoA
Second reaction of the activated COOH group with the CH_3 group of another CoA—S acetate.

(4) Further reaction of similar type forming longer carbon chains.

Such a series of reactions leads to a sliding of filaments if we make the following additional assumptions: (1) The group activated by transphosphorylation is attached to the one filament. (2) The various functional groups reacting successively with this active group are arranged in a linear periodicity alongside the other filament.

"A sequence of reactions during the course of which the active group of one filament successively reacts with several linearly arranged groups of the other filament has the inevitable result that the active group travels alongside the other filament while being shifted from the first to the last of the groups involved" (from H. H. Weber, see footnote, page 25) (Fig. 1).

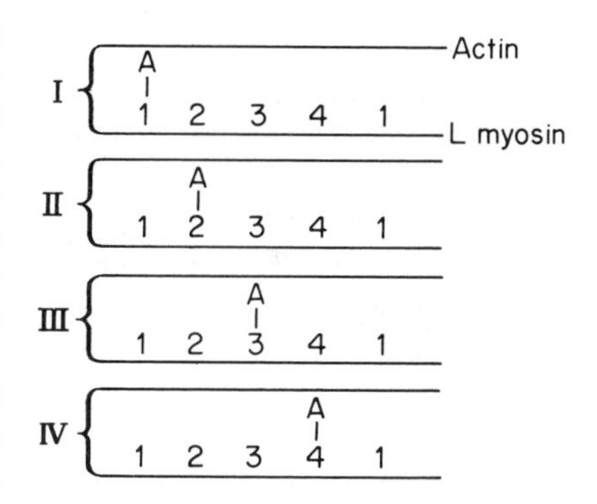

Fig. 1. Shifting of one filament alongside another filament produced by a sequence of chemical reactions. A= active group of the actin filament. 1, 2, 3, 4 = groups of the L-myosin filament successively reacting with A.

Consequently we suppose that the terminal phosphate residue of the ATP is transferred, for example, to an acid group of the actin filament activating this group (Fig. 2); that is, the activation of the acid group. We assume that this phosphate residue is exchanged with a sulfhydryl group of the L-myosin filament. This corresponds to the conversion of acetyl-adenosine-5′-phosphate (AC-AMP) into acetyl-coenzyme A, the second step of fatty acid synthesis.

In neither of these reactions is there a liberation of energy; for the free energy of the terminal phosphate bond in the ATP, like the binding of phosphate to an acid group of the actin filament, and like the sulfhydryl-ester bond between actin and L-myosin filament has a standard energy of about 8000 cal. However, the succeeding reactions of the activated group of the actin filament with further functional groups of the L-myosin filament must liberate energy; for the shift of the actin filament alongside the L-myosin filament during contraction implies performance of work. The energy contained in the sulfhydryl-ester bond is almost completely liberated if at the conclusion of the sequence of reactions the filaments are united by means of a normal ester bond. For such an ester bond contains only about 2000–3000 cal. There may be one or two bonds containing a medium amount of energy between the first and the last types of bonds.

Consequently it is supposed that the active group of actin is passed along from the bond of the SH group by the bond of a phenolic OH group to a bond of an alcoholic OH group. Subsequently this normal ester bond is hydrolyzed, and the acid group of actin is rephosphorylated. Afterward the sequence of reactions represented begins anew.

"Since the rephosphorylated actin group is now close to the sulfhydryl group of the next period and remote from the sulfyhdryl group of the first period, the active actin group is shifted alongside the *next* longitudinal period of the L-myosin filament if the muscle is unloaded. The single shifting steps

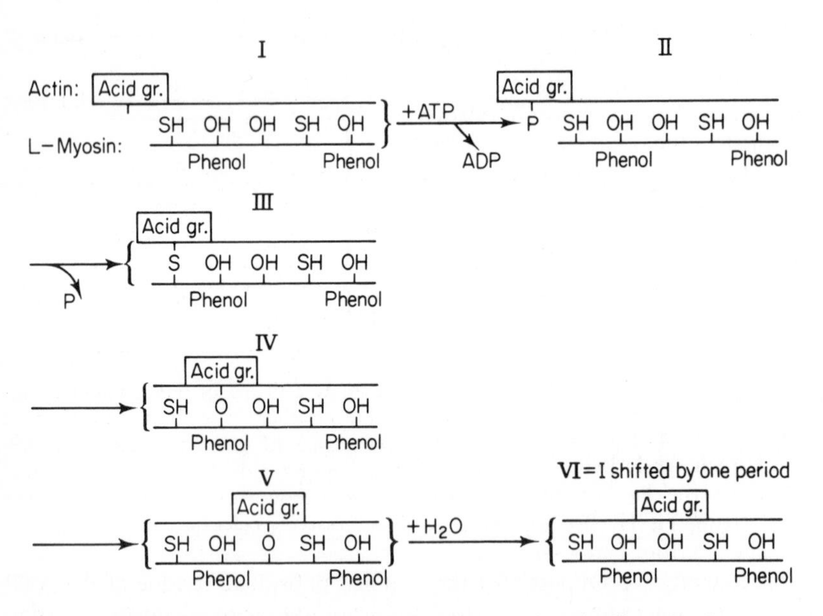

Fig. 2. Diagram of a series of chemical reactions leading simultaneously to ATP hydrolysis and to a shifting of the molecules of the actin filaments alongside the molecules of L-myosin filaments. The figures I to VI denote successive chemical conditions and stages of shifting.

without interruption amount to the visible external shortening which can be measured.

"The more heavily the muscle is loaded, however, the more frequently during rephosphorylation the actin filament with its active group is withdrawn by the load into the initial position. This increases the number of shifting steps necessary for the fibril to shorten by a definite amount. As a result, the amount of energy released per centimeter of shortening increases the more heavily the muscle is loaded" (from H. H. Weber, see footnote, page 580).

On the other hand, the loading of the muscle counteracting the shortening can be interpreted as an apparent increase of the activation energy of reactions III to V of Fig. 2 which induce the shortening. Therefore, the more heavily the muscle is loaded the more slowly these reaction steps succeed one another. This means, however, that the liberation of energy decreases with increasing load if it is calculated in terms of unit of time, and not in terms of centimeters of shortening. Both consequences of the suggested reaction scheme—increasing liberation of energy per centimeter of shortening, decreasing liberation per unit of time with increasing load—have been established in the classic work of A. V. Hill on the thermodynamics of the *living* muscle.

This principle of the proposed sequence of chemical reactions may require modifications in detail. I should like to present several arguments in favor of this principle for our discussion of this reaction scheme:

1. In contrast to the earlier theories of contraction the proposed reaction sequence leads to thermodynamic results, at least qualitatively, that agree with the above-mentioned thermodynamic observations made by A. V. Hill on living muscle.
2. It is certain that not only the contraction of the actomyosin system but also the Mg-activated ATP *splitting* is due to an interaction between actin and L-myosin; for under physiological conditions, the rate of Mg-activated ATP splitting by actomyosin is about ten times as great as the splitting rate of the components actin and L-myosin when separated. On the other hand, Hasselbach[19] has shown that ATP splitting in the living muscle is activated by Mg.
3. It is quite likely that the transfer of energy from ATP to the working system takes place by means of transphosphorylation. This is true of all the cases investigated so far in which the energy of ATP splitting is required for chemical processes. There are a number of experimental indications that the ATP splitting by the contractile structures also begins with transphosphorylation[20, 21].
4. We have known for a long time that the contraction of the isolated contractile systems and the ATP splitting by these systems are completely inhibited, if the SH groups of the actomyosin are blocked or destroyed[5].
5. Some recent unpublished observations concerning the mechanism of the physiological *relaxation* phase can be explained easily by the principle suggested. This will be illustrated in greater detail.

It is understandable that in the presence of ATP relaxation always occurs when ATP splitting is inhibited. But relaxation is also induced by means of the *physiological relaxing factor,* although this factor does not inhibit ATP splitting completely even in the highest concentrations (Fig. 3). This relaxing factor

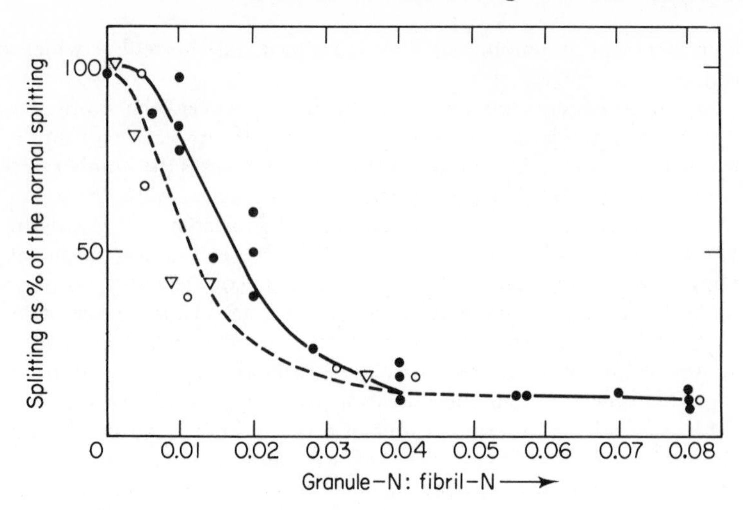

Fig. 3. Dependence of ATP splitting on the proportion of granules to fibrils. Key: ●, 0.4 mg protein/cc; o, 4 mg protein/cc; inverted triangle, 9 mg protein/cc.

was discovered by Marsh and Bendall and shown by Portzehl to be a product of certain muscle granules. This factor brings about total relaxation if the Mg-activated ATP splitting is reduced by the relaxing factor to approximately 10% of the normal ATP splitting (Fig. 3). The speed of this residual splitting, however, which cannot be suppressed by the factor is *exactly the same* as the speed of the ATP splitting occurring in the actomyosin system when the actin and the L-myosin are completely dissociated from each other.

That the factor dissociates actin and L-myosin in the presence of ATP can also be shown in another way. We know from H. E. Huxley that the slipping of the actin filaments from the L-myosin filaments is responsible for the fact that muscle fibrils can be easily stretched under ATP. Therefore, the degree of association of both types of filaments can also be determined from the resistance against stretch. Table I shows that this resistance, that is to say, the association of actin and L-myosin filaments, is greatest in the rigid state (in rigor) when no plasticizing polyphosphates are present. Furthermore, this table shows that the association of the filaments is still great when ATP is present and is split at the same time, that is, during contraction. This is understandable from the reaction scheme which was described.

During relaxation the resistance to stretch is by far the smallest, since the interaction between actin and L-myosin, which depends upon ATP splitting, is absent. There is, however, a very considerable difference between the relaxation produced by mersalyl and the relaxation brought about by the factor. In the case of factor-induced relaxation the static resistance is zero (Table I). Both the kinetics of the ATP splitting and the complete disappearance of the resistance to stretch under the influence of the relaxing factor indicate that the factor has no relaxing effect, because it inhibits ATP splitting incompletely. On the contrary, the difference between ATPase poisons and the factor strongly suggests that the factor renders contraction impossible and diminishes splitting by preventing any interaction between actin and L-myosin filaments. For this interaction is the prerequisite not only for contraction but also for Mg-activated ATP splitting at a normal rate.

Table I

Static Resistance Against Stretch (Extracted Muscle Fibers)

	As $g \times cm^{-2} \times L \times \Delta L^{-1}$
Rigor	~ ·10,000
Contraction	~ 5000
Relaxation	
Produced by mersalyl	~ 250[a]
Produced by factor	~ 0[b]

[a] Portzehl(13).
[b] Hasselbach and Weber(22).

References

1. Chance, B., and Connelly, C. M. *Nature* **179**, 1235 (1957).
2. Fleckenstein, A., Janke, J., Davies, R. E., and Krebs, H. A. *Nature* **174**, 1081 (1954).
3. Mommaerts, W. F. H. M. *Am. J. Physiol.* **182**, 585 (1955).
4. Szent-Györgyi, A. *Biol. Bull.* **96**, 140 (1949).
5. Weber, H. H., and Portzehl, H. *Progr. in Biophys. and Biophys. Chem.* **4**, 60 (1954).
6. Ulbrecht, G., and Ulbrecht, M. *Biochim. et Biophys. Acta* **11**, 138 (1953); **13**, 319 (1954).
7. Schick, A. S., and Hass, G. M. *Science* **109**, 487 (1949).
8. Perry, S. V. *Biochem. J.* **48**, 257 (1951).
9. Hoffmann-Berling, H. *Fortschr. Zool.* **11**, 142 (1958).
10. Portzehl, H. *Z. Naturforsch.* **6b**, 355 (1951).
11. Dörr, D., and Portzehl, H. *Z. Naturforsch.* **9b**, 550 (1954).
12. Weber, H. H. *Arch. ges. Physiol., Pflüger's* **235**, 206 (1934).
13. Portzehl, H. *Z. Naturforsch.* **7b**, 1 (1952).
14. Marsh, B. B. *Biochim. et Biophys. Acta* **9**, 247 (1952).
15. Bendall, J. R. *Nature* **170**, 1058 (1952).
16. Portzehl, H. *Biochim. et Biophys. Acta* **26** 373, (1957); Weber, H. H. *Ann. Rev. Biochem.* **26**, 667 (1957).
17. Weber, H. H. "The Motility of Muscle and Cells," 69 pp. Harvard Univ. Press, Cambridge, Massachusetts, 1958.
18. Hasselbach, W. *Biochim. et Biophys. Acta* **20**, 355 (1956).
19. Hasselbach, W. *Biochim. et Biophys. Acta* **25**, 562 (1957).
20. Ulbrecht, G., and Ulbrecht, M. *Biochim. et Biophys. Acta* **25**, 100 (1957).
21. Ulbrecht, G., Ulbrecht, M., and Wustrow, H. J. *Biochim. et Biophys. Acta* **25**, 110 (1957).
22. Hasselbach, W., and Weber, H. H. *Biochim. et Biophys. Acta* **11**, 160 (1953).

THE PROTEINS OF ADDUCTOR MUSCLES*

K. Bailey†
Stazione Zoologica, Naples, Italy
(Received for publication, August 4, 1955)

Although the physiological behaviour of adductor muscles has frequently been studied, very little is known of their biochemistry. The present research, carried out during the Spring and Summer of 1955, is devoted especially to those protein components which are presumed to take part in contraction.

Adductor muscles usually consist of two readily distinguishable parts, the one white and opaque, the other somewhat yellow and translucent. A very detailed description is to be found in the early paper of Marceau (1909), who confirmed earlier observations that the vitreous portion is a fast muscle, enabling the mollusc to close the shell, whilst the opaque part contracts more slowly and maintains its contraction over very long periods if need be to keep the valves closed. The slow muscle is entirely smooth, but the fast muscle after fixation may show a simple transverse striation, or an appearance which Marceau interpreted as a helix, whose pitch varied with the degree of contraction. Plenck (1924), however, from his work on the striated muscles of the snail, considered that oblique or spiral striations were artifacts of fixation or of strains caused by uneven contractions, and infers that this may be true of Marceau's findings for lamellibranch muscle.

Hall, Jakus & Schmitt (1945) first isolated from many molluscan muscles

*K. Bailey, "The Proteins of Adductor Muscles," *Pubbl. Staz. Zool.,* **29**, 96–108 (1955). Reprinted with the permission of the publisher.

The author wishes to express his gratitude for the kindness accorded to him by Dr. R. Dohrn and Dr. P. Dohrn during his stay in the Stazione Zoologica; also to Drs. F. and A. Ghiretti who so readily acquainted him with all the facilities of the laboratory.

†The late Dr. Bailey was a Fellow of Trinity College, Cambridge University, on sabbatical leave to the Stazione Zoologica.

(foot, neck and adductor) a type of fibril which took up an electron stain to give in the electron microscope a precise geometrical distribution of spots with a full repeating period of 725A. In a parallel investigation, Bear (1944) showed that the same fibre period could be deduced from small-angle diffraction of these muscles, but that in the case of the coloured, striated portion of *Pecten* muscles, the diffraction pattern was closer to that given by mammalian skeletal muscle. However, it is implied in the paper of Hall, Jakus & Schmitt (1945) that these special fibrils, to which they gave the name paramyosin, could be obtained from both fast and slow muscles, though they advocate the use of the slow. Paramyosin has not been isolated as a chemical entity: from the fact that no change of spacing can be observed in fibrils isolated from relaxed or contracted muscles, Schmitt (1947) has suggested they may play some purely mechanical function unassociated with contraction.

By analogy with mammalian muscle, the long spacings of paramyosin fibrils could be due to an actin component, since the large spacings given by intact muscle arise from the actin component and not from the myosin, which *in situ* or extracted contributes only a wide-angle α-pattern. [For a full discussion of X-ray data, see Bailey (1954) and Huxley (1954)]. Indeed, the identification of paramyosin by purely physical methods has set some difficult biochemical problems which can be enumerated as follows: (1) Is the paramyosin fibril only part of the muscle structure, or does it constitute the whole? Since the spacings of the isolated fibril can be obtained from the whole muscle, it must necessarily constitute a considerable part of the muscle. (2) Is it a single component or a heterogeneous structure? In addition to the small-angle pattern, whose origin is unknown, it also gives the typical wide-angle α-pattern, which must arise from a myosin-like protein. The larger spacings could be derived therefore from this protein or from some other component.

The present study is a preliminary account of a more protracted investigation on the biochemical behaviour of these muscles. It has not been possible to prepare in reasonable quantity by maceration of the muscle the very fine fibrils of paramyosin, and work has been confined to whole muscle. Quantitative aspects of the protein components have been investigated, using both fast and slow muscles and various kinds of extracting solution. A myosin-like protein has been isolated in crystalline form and its properties examined, but so far, its relation to paramyosin is obscure.

Material and Methods

Adductors of the oyster and of *Pinna nobilis* only were used. The former were excised with the valves closed except in some later experiments in relation to the sensitivity of extracts towards adenosine triphosphate (ATP) and inorganic pyrophosphate (PP). In these cases, relaxation could be effected by immersion for 15 min. in sea water containing 5% ether. The *Pinna* specimens were narcotised in 2% urethane-sea water. After excision, the muscle was cooled in ice, and homogenisation effected in a small Waring blender fitted with submerged plunger to prevent frothing. All extractions were carried out between 0 and 5°C, and the centrifuged residues extracted at least three times.

Nitrogen determination was made with the microkjeldahl in duplicate. The protein was precipitated from extracts with trichloroacetic acid (final concentration 5%) and washed free of non-protein nitrogen. Viscosity tests were carried out in the Ostwald viscosimeter, flow time 60 sec. (1.5 ml.) for water at 20°C.

Results

1. Nitrogen partition of oyster muscle

The results given in Table 1 were obtained in experiments carried out at different times using a pool of at least six adductors. The following points are worthy of mention. The connective tissue as estimated by the residue remaining after homogenising in N/100 NaOH is very small (~3% of the total protein); a higher figure is obtained (7–8%) if the muscle is first extracted with M-KCl followed by exhaustive washing in urea (6.6 M containing phosphate pH 7).

Table I

N Partition in the Adductor of the Oyster

	White (slow) portion	Yellow (fast) portion
Total protein as % fresh wt.	15.2[a]	14.5[a]
	13.8, 14.3 } 14.0[b]	12.6, 12.4 } 12.5[b]
As % of the total protein:		
Connective tissue	2.7[c]	2.9[c]
	7 [d]	8 [d]
Water soluble	13.9	21
	13.9	
Soluble $I = 0.2$	17.2	30
	16.9	
Trichloroacetic-soluble N as % total N	18	24
	20	

a. muscle dissolved in N/100 NaOH; b. direct determination on minced muscle (considered more accurate); c. residue after dissolution in N/100 NaOH; d. residue after dissolution in 6 M-urea buffered with phosphate at pH 7.

The water-soluble protein (14–20%), which by analogy with skeletal muscle contains mainly the enzymes of glycolysis, is extremely low, indicating—if we allow also for connective tissue—that 70–80% of the muscle is fibrillar. For comparison, the sarcoplasmic protein of rabbit skeletal muscle amounts to 35–37% of the total intracellular protein (Hasselbach & Schneider, 1951), rat diaphragm 32% (Stewart, 1955) and fish (cod) about 20% (Dyer, French & Snow, 1949). There appear to be significant quantitative differences between slow and fast muscles, a fact emphasised also in the response to different kinds of extracting solution (see below).

2. Extraction of oyster adductors by various salt solutions

In Table II are recorded the amounts of protein extracted with KCl solutions of differing ionic strength, buffered with phosphate at pH 6.8. In both muscles there is a significant rise in extractable protein between I = 0.2 and 0.3, but in the case of the white portion, little more is extracted until I = 0.93. The yellow part, on the other hand, is more soluble at all ionic strengths and the total extracted at I = 0.93 is 65% as against 43% for the white. The response of the white portion seemed to suggest that only one globulin component was being extracted at neutral pH, a surmise having some support from the greatly increased extraction if the ionic strength is lowered but the pH raised (Weber's solution). Further experiments (see below) soon showed that M-KCl extracts both of the slow oyster muscle and of the entire adductor of *Pinna* which does not possess a striated portion contained predominantly a globulin capable of crystallisation when dialysed to 1 = 0.2, pH 6.5–7.0.

Table II

Solubility of the Proteins of Oyster Adductor as a Function of Ionic Strength[a]

Ionic strength	White portion	Yellow portion
0.2	17	21
0.3	25	30
0.43	29	32
0.63	25	48
0.93	43	65
Weber's solution[b]	76	80

a. KCl solutions containing phosphate 1 = 0.03, pH 6.8; b. 0.6 M-KCl, 0.04 M-$NaHCO_3$, 0.01 M-Na_2CO_3, pH 9.1. pH of extracts themselves 8.5.

It must be admitted that the different physiological state of the two parts of the muscle at closure may account for the differences of response to extractants, but superficially, the yellow portion behaves more like mammalian striated muscle. Time has not permitted a study of the action of the reagents which dissociate actin and myosin (ATP and PP) when included in the extracting solution.

3. Crystallisation of a myosin-like component

It had been noticed that in the application of Bailey's (1948) method for the preparation of tropomyosin from *Pinna* muscle that the ethanol-dried fibre gave a viscous extract on mixing with M-KCl, and that the protein contained in it, unlike tropomyosin, denatured entirely on acidification to pH 4.6. When, however, the KCl extract was dialysed against changes of progressively dilute KCl, a gel-like precipitate appeared consisting entirely of crystalline needles.

Further experiments soon showed that the same protein could be obtained from M-KCl extracts of wet muscle, thus avoiding the use of organic reagents. The two methods will be called "wet" and "alcohol" respectively, and are outlined below.

Wet method: *Pinna* muscle is minced with 2 vols. of water and the whole diluted to 25 vols. with KCl.-phosphate (I = 0.05, pH 6.8). After centrifuging, the residue is mixed with solid KCl to give a molar solution and diluted to 10 vols. (on the original wt. basis) with M-KCl-phosphate. After standing overnight at 0°, the residue is centrifuged down, washed three times with a little KCl, and then discarded. The combined extracts are dialysed against progressively diluted KCl until a slight precipitate forms, at about 0.3 M. This is usually amorphous and is centrifuged off. Further dialysis of the clear supernatant against 0.15 M-KCl usually gives crystals without amorphous material.

Alcohol method: The muscle is minced as before, washed with 10 vols. of cold water, and 3 times with 10 vols. of cold ethanol. After finally drying in ether, the residue, while ether-damp, is mixed with 5 vols. (original wt. basis) of M-KCl and left overnight at 0°. The method then follows that above, except that the separation of any amorphous material is usually unnecessary. The supernatant from the first crystallisation contains a water-soluble viscous protein which will be further investigated.

Recrystallisation

Method 1: After dissolving by adding an equal volume of M-KCl pH 6.8, the solution is centrifuged to remove any haze, or filtered through a small pad of paper pulp. The dialysis procedure is then repeated.

Method 2: At pH 6.5 and 25° the crystals slowly dissolve at I = 0.4; on cooling to 0°, beautiful needles, or rosettes of needles, are deposited.

4. Properties of the crystalline protein

General: The crystals are deposited in the pH range 6.5–7 at an ionic strength of 0.15–0.2. At the more acid pH, they dissolve slowly in 0.6 M-KCl and more readily in 0.5 M-KCl at pH 7. The needles first obtained from *Pinna* muscle have been isolated also from both slow and fast muscles of the oyster (see Fig. 1) and it must be assumed that the protein is of general distribution in all adductors. In the polarising microscope, the crystals are intensely birefringent. When washed free of salt, the crystals dissolve at pH 8.6 and below pH 4, the acid medium causing complete denaturation; surprisingly, they remain largely undenatured if rapidly dried in cold ethanol or freeze-dried, procedures which denature mammalian myosin completely.

From solution, the protein precipitates in neutral or slightly alkaline ammonium sulphate between 23 and 29% saturation, and gives a positive nitroprusside test. At pH 6.5, precipitation by dilution begins at I = 0.33 and is complete at I = 0.25.

Viscosity and birefringence

Solutions of the protein are extremely viscous. In M-KCl pH 6.6 the intrinsic viscosity (Table III) is 2.56 compared with a value of ~ 2 for rabbit myosin (Mommaerts, 1945) and 0.52 for tropomyosin (Tsao, Bailey & Adair,

Fig. 1. Typical crystals of the myosin-like component of adductor muscles. This specimen was obtained from the fast oyster adductor by the alcohol method.

Table III

VISCOSITY OF TWICE-CRYSTALLISED *PINNA* PROTEIN AT 20°

Medium	Protein concentration (c, in g./100 ml.)	η_{sp}/c
M-KCl, pH 6.6	0.434	3.71
	0.260	2.98
	0.173	2.87
	0.130	2.77
	0.065	2.36
		Extrapolated 2.56
Urea 6.6M; phosphate 0.044M; pH 7	0.217	2.0
	0.145	1.9
	0.108	2.0
		Extrapolated 2.0

Intrinsic viscosity $[\eta] = \lim_{c \to 0} \eta_{sp}/c$; i. e. 2.56 and 2 respectively for the two media.

1951). In strong urea solutions the particles appear to become more symmetrical, the intrinsic viscosity falling to 2.0.

Prof. S. Ranzi has kindly carried out measurements of flow birefringence on solutions of concentration 0.35% in 0.5 M-KCl at pH 7. The solutions were strongly flow birefringent and by conventional calculation showed a greater asymmetry than rabbit myosin; e. g., the latter, prepared by the method of Green, Brown & Mommaerts (1953) is calculated to be ~ 3000A long, end *Pinna* protein 4,500A. These results support the inference derived from viscosity, that the protein is one of the most asymmetric yet found.

5. Amount of crystalline protein in adductors

The method adopted for the estimation of the protein is necessarily empirical and the figures derived are therefore minimal. The muscle was minced in cold water and diluted with 25 parts of KCl-phosphate (I = 0.05, pH 7) to remove sarcoplasmic protein. The residue was centrifuged down and allowed to stand overnight at 0° in 10 parts of M-KCl. After centrifuging and washing twice, the extracts were dialysed to a final ionic strength of 0.15. The precipitate of crystals and some amorphous material was centrifuged down, dissolved in 0.5 M-KCl at pH 7, and dialysed against 0.4 M-KCl when a slight precipitate of amorphous material was centrifuged off. The supernatant on dialysis against 0.2 M-KCl deposited only crystals.

Table IV

Estimation of Crystalline Globulin in Adductor Muscles as % of the Total Protein

Exp.	Animal	Amount protein
1	Pinna (grey part)	31
1	Pinna (white part)	33
2	Pinna (whole muscle)	28
3	Oyster (white part.)	23
3	Oyster (yellow part.)	26.5

It will be seen from Table IV that in a parallel experiment, both white and grey portions of *Pinna* adductor, which histologically are unstriated and indistinguishable (private communication from Dr. H. Reichel; see also Bandmann & Reichel, 1954) gave values representing 30% of the total protein. The white and yellow portions of the oyster gave ~ 25%, but in this case, the precipitate obtained from the white only was crystalline. (Whilst crystals can always be obtained from white and yellow by the ethanol method, their appearance is erratic in the latter case, and always accompanied by non-crystalline material.) The yield of crystals by the ethanol method, assayed only in the case of *Pinna*, is considerably less, ~ 12%.

From these results it would seem that the protein is a major component of adductor muscles.

6. The nature of the crystalline globulin

The general properties already outlined suggest that the protein resembles myosin rather than actin. The latter protein has in fact been prepared from the fast muscle of the oyster by the method of Tsao & Bailey (1953), which involves butanol- and acetone-treatment of the fibres, and subsequent extraction with water. Under these conditions, the globulin is unextracted, but a protein is obtained which after precipitation at pH 4.6 and re-solution combines with rabbit myosin to give a complex dissociable by ATP and PP. Myosin can be characterised by its ATP-ase activity, or by its ability to combine with actin. Under conventional conditions (pH 8 in glycylglycine buffer, Ca activation) the ATP-ase activity of a homogenate was very low, and it is possible that conditions of assay need to be explored. For this reason, and also for its simplicity, the viscometric response of extracts or crystals towards ATP and PP was investigated.

Table V

SENSITIVITY OF EXTRACTS OF OYSTER ADDUCTOR (WHITE)

(Extracts obtained by grinding the muscle with quartz in presence of 0.95 M-KCl-phosphate, pH 7)

Experiment	Materials	Sensitivity to PP
1a	Rabbit myosin + rabbit actin	50
1b	〃 〃 + oyster actin	77
2a	Fresh oyster extract	32
2b	Same + rabbit actin	55
2c	Dissolved crystals + actin	23
3a	Fresh oyster extract	33
3b	Stored in ice 12 hr.	13
3c	Dissolved crystals	0
3d	Same + actin	15
3e	Residue of crystals	0
3f	Same + actin	40
4a	Oyster extract stored 12 hr.	11
4b	Same + actin	13
4c	Dissolved crystals + actin	13
4d	Dissolved crystals + residue + actin (*p*H 7.7)	33
4e	Dissolved residue alone	0
4f	Same + actin	24

Viscosity was measured in a medium of ice and water before and after addition of PP (final concentration 0.006 M). As a measure of the diminution of viscosity caused by the latter, the results can be expressed in terms of sensitivity (S_{PP}), defined (Weber & Portzehl, 1952).

$$S_{PP} = \frac{\log \eta - \log \eta_{PP}}{\log \eta_{PP}} \times 100$$

where η is the original relative viscosity and η_{PP} the final after addition of ATP (Weber) or PP. Extracts were fortified by addition of rabbit actin, shown to be active by its response to rabbit myosin. Only the slow part of the oyster has been investigated.

The results and observations can be enumerated as follows:

1. Freshly prepared extracts in 0.95 M KCl-phosphate pH 7 show an erratic response to PP which is enhanced by addition of actin (Table V. exp. 2 a and b). After standing in ice overnight, the sensitivity diminishes quite markedly (exp. 3 a and b).

2. On dialysis against 0.2 M-KCl, it is possible to obtain crystalline preparations almost free of amorphous material. After dissolving, the sensitivity of the solution is virtually zero, but a small response can be elicited after addition of actin (exp. 3 c and d).

3. On dissolution in 0.6 M-KCl, the crystals leave behind an amorphous residue which dissolves at a more alkaline pH. Inactive in itself, this solution after addition of actin has a fairly high sensitivity (exp. 3 e and f; 4 e and f), but only in one experiment of several have crystals again been obtained on re-dialysis.

From the foregoing discussion and the representative data of Table V it is not possible to derive any conclusion. On the one hand, it could be argued that the crystalline globulin in its native state combines with actin, but that its activity is sufficiently labile (as found, for example, in whole extracts) to have largely disappeared by the time the crystals are isolated.

On the other hand, the less soluble residue, collected after dissolving the crystals, shows a relatively high sensitivity after addition of actin, and it is possible that this could contain a more conventional type of myosin component, the low activity of the crystals being due to contamination with this latter. Moreover, if the assay of the crystalline component is correct, there remains some 55% of the total protein to be accounted for as fibrillar protein.

Discussion

Lajtha (1949) has compared the proteins of the slow adductor of the oyster with those of rabbit muscle. He prepared myosin B (i. e. actomyosin) by extracting with Weber's solution, and studied the synaeresis behaviour of gels made by squirting the extracts into dilute salt solution. Although the synaeresis observed on addition of ATP shows that the gels contained an "active" actomyosin, it should be noted that the forces required to elicit synaeresis need be extremely small, and do not reflect the activity of the gel in a quantitative manner. For this reason, viscosity studies are to be preferred. He notes, however, that the ATP-ase activity of the extracts is rather feeble.

Lajtha also reports the preparations of actin and of myosin A (i. e. actin-poor myosin). Without giving details, he considers both these proteins are similar to those of rabbit, but it is likely that the extracts which he supposed to contain myosin A actually contained the globulin component which has been crystallised in the present work, and which certainly differs from conventional myosin. He assumes that paramyosin is a component additional to those he studied.

Future work must decide whether the crystalline component is in fact the myosin component of adductor muscles whose actin-combining capacity is easily lost, or whether there exists a more conventional myosin which sometimes contaminates it. In any case, the high viscosity of the protein may help to explain the peculiar plastic properties of these smooth muscles.

Some preliminary X-ray investigations have been carried out by Prof. Astbury and Dr. Beighton of Leeds University. Oriented films of the wet crystals show a strong α-pattern, and also a meridional spot of spacing 70 A, which coincides with one of the two strongest reflections obtained by Bear (1944)—the other being at 147 A.—for muscles containing paramyosin. On present evidence, therefore, it is not possible to decide whether the component is identical with paramyosin until other reflections are observed.

NOTE ADDED IN PROOF

Subsequent work has shown that the globulin in its amino acid pattern is a tropomyosin (see Bailey, K.—1956—Invertebrate tropomyosin.—*Proc. Biochem. Soc., Biochem. J. Vol. 64, No. 1*). In addition, there exists in lamellibranch adductors (Rüegg, J. C. Unpublished) and in cephalopod muscle smaller amounts of a water-soluble tropomyosin, which survives the method advocated by Bailey (1948) for the preparation of mammalian tropomyosin (cf. Yoshimura, K.—1955—Studies on the tropomyosin of squid.—*Mem. Faculty of Fisheries, Hokkaido Univ. Vol. 3, No. 2*).

Summary

1. The partition of nitrogen amongst the protein components of fast and slow portions of oyster adductor has been determined.

2. In solutions of increasing ionic strength the amount of protein extracted at pH 7 increases progressively, and is not complete at an ionic strength of 1. The globulins of the fast muscle are more soluble than those of the slow.

3. From the oyster adductor (fast and slow portions) and from *Pinna nobilis*, a crystalline globulin of myosin type has been isolated. Viscosity and birefringence studies show it to be more asymmetric than rabbit myosin. It appears to comprise about 25–30% at least of the total protein.

4. Since the actin-combining capacity of the crystals is low, the function of the protein in contraction is still obscure. Likewise, its possible relation to the paramyosin component of Schmitt and Bear has to be demonstrated.

References

Bailey, K.—1948—Tropomyosin: a new asymmetric protein component of the muscle fibril. —*Biochem. J., Vol. 43, p. 271.*

——— & Huxley, H. E.—1954—Structure Proteins. I. Muscle.—In *The Proteins. Editors H. Neurath & K. Bailey, Vol. II B, p. 1024.*

Bandmann, H. J. & Reichel, H.—1954—Struktur und Mechanik des glatten Schliessmuskels von Pinna nobilis.—*Zeits. f. Biol., Vol. 107, p. 67.*

Bear, R. S.—1944—X-ray diffraction on protein fibres. II. Feather rachis, porcupine quill tip and clam muscle.—*J. Amer. Chem. Soc., Vol. 66, p. 2043.*

Dyer, W. J., French, H. V. & Snow, J. M.—1950—Proteins in fish muscle. I. Extraction of protein fractions in fresh fish.—*J. Fish. Res. Bd. Canada, Vol. 7, p. 585.*

Green, I., Brown, J. R. C. & Mommaerts, W. F. H. M.—1953—Adenosinetriphosphate systems of muscle.—*J. Biol. Chem., Vol. 205, p. 493.*

Hall, C. E., Jakus, M. A. & Schmitt, F. O.—1945—The structure of certain muscle fibrils as revealed by the use of electron stains.—*J. Applied Phys., Vol. 16, p. 459.*

Hasselbach, W. & Schneider, G.—1951—L-myosin and actin content of rabbit muscle.—*Biochem. Z., Vol. 321, p. 462.*

Lajtha, A.—1949—The muscle proteins of invertebrates.—*Pubbl. Staz. Zool. Napoli, Vol. 21, p. 226.*

Marceau, F.—1909—Morphologie l'histologie et la physiologie comparées des muscles adducteurs des mollusques acéphales.—*Arch. de Zool. Exp. Gén., (5), Vol. 2, p. 295.*

Mommaerts, W. F. H. M.—1945—On the shape and dimensions of myosin particles in solution.—*Arkiv. Kemi Mineral. Geol. Vol. 19, No. 17, p. 1.*

Plenck, H.—1924—Die Muskelfasern der Schnecken.—*Zeit. f. wiss. Zoologie, Vol. 122, p. 68.*

Schmitt, F. O.—1947—Electron microscope and X-ray diffraction studies of muscle structure. —*Ann. New York Acad. Sci., Vol. 47, p. 799.*

Stewart, D. M.—1955—Changes in the protein composition of muscles of the rat in hypertrophy and atrophy.—*Biochem. J., Vol. 59, p. 553.*

Tsao, T. C. & Bailey, K.—1953—The extraction, purification and some chemical properties of actin.—*Biochem. Biophys. Acta, Vol. 11, p. 102.*

——, Bailey, K. & Adair, G. S.—1951—The size, shape and aggregation of tropomyosin particles.—*Biochem. J., Vol. 49, p. 27.*

Weber, H. H. & Portzehl, H.—1952—Muscle contraction and fibrous muscle proteins.—*Adv. Prot. Chem., Vol. 7, p. 161.*

THE EFFECTS OF ADENOSINE TRIPHOSPHATE ON THE FIBRE VOLUME OF A MUSCLE HOMOGENATE*

B. B. Marsh

Low Temperature Station for Research in Biochemistry and Biophysics
University of Cambridge and Department of Scientific and Industrial Research
Cambridge, England

Observations in many laboratories during recent years leave little doubt, even if they provide no final proof, that actin, myosin, and adenosine triphosphate (ATP) are the essential components for muscular contraction. The striking contraction-like synaeresis caused by addition of ATP to actomyosin gels or threads at low salt concentration (Szent-Györgyi[1]), and, on a more organised system, the rapid shortening of glycerol-treated muscle fibres in presence of ATP (Korey[2]), have been taken to support those theories which assume that relaxation of muscle coincides with energy release to the contractile machine by the enzymic dephosphorylation of ATP. On the other hand, the heat measurements of Hill (summarised by Hill[3]) cannot easily be reconciled with an active relaxation theory, while several observations on model systems would appear to find a more ready interpretation in terms of active, rather than passive contraction; in particular there may be noted the prevention of shortening of washed fibres (Korey[2]), and of "contraction" of actomyosin gels (Kuschinsky and Turba[4]), by sulphydryl reagents, and the lengthening of loaded actomyosin threads caused by inorganic pyrophosphate and other pyrophosphate compounds which are not enzymically split (Engelhardt and Ljubimowa[5]). The relation between contraction-like effects and the presence or decomposition of ATP is thus obscure, and indeed the supposedly intimate connection between synaeresis of actomyosin and contraction of muscle is by no means firmly established (Perry, Reed, Astbury and Spark[6]).

*B. B. Marsh, "The Effects of Adenosine Triphosphate on the Fibre Volume of a Muscle Homogenate," *Biochem. Biophys. Acta,* 9, 247–260 (1952). Reprinted with the permission of the author and of the publisher.

The present investigation was commenced to study the effect of ATP on the water retention of a muscle homogenate, and was suggested by the familiar observation that, when fresh muscle is minced, the sarcoplasmic fraction becomes more readily extractable as glycolysis proceeds, while the extraction of fibrillar protein becomes more difficult. Although fluid retention is greatly diminished by a fall in pH (Empey[7]), this does not appear to be the whole explanation since the expression of muscle juice is facilitated by post-mortem glycolysis even when acid production is minimised; furthermore, the striking effect of ATP on actomyosin suggests that hydrolysis of ATP might alter the water relations of muscle during the onset of rigor mortis. The investigation has proved unexpectedly fruitful in revealing the presence in muscle of a labile factor which profoundly affects the volume response of muscle fibres to added ATP. A brief account of this work has already appeared[8].

Experimental

Fig. 1. The modified centrifuge. Casing removed to illustrate position of light source and path of light beam.

The modified centrifuge

It was suggested by Dr. K. Bailey that a laboratory centrifuge might be modified in some way to permit continuous reading of the volume of the fibre layer of a muscle brei during centrifugation. This was achieved quite simply in the following manner.

A 60 watt lamp was placed below a slit in the base of the centrifuge, vertically below the horizontal position of the spinning tube, and a corresponding slit (for viewing) was cut in the lid of

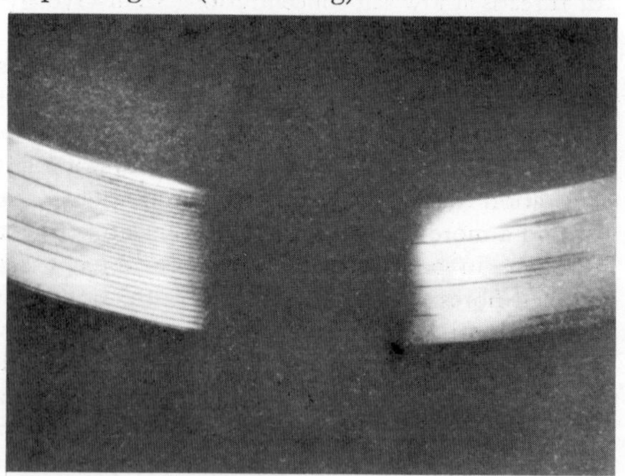

Fig. 2. (a) The centrifuge cup, showing position of the slits relative to the graduations on the centrifuge tube; (b) Appearance of the same cup when spinning at 1000 r.p.m. The direct beam from the light source has been removed by a shield (central black area) during photographic exposure.

the machine, vertically above the light source. Two slits were made in a metal centrifuge cup, each about 5 cm $\times$ 1 cm, diametrically opposite and in line with the axis of the cup, and in such positions that, when the cup was spinning horizontally, light passed from the source, through both slits and out of the viewing slit in the lid, once during each revolution. The counter-balancing cup was not modified.

When a graduated centrifuge tube containing liquid was spun in this apparatus, not only the graduation lines, but also the etched numbers were clearly visible through the fluid. When the tube contained both solid and liquid layers, the heavier material appeared as a uniform black area while the supernatant fluid allowed the transmission of light to the observer. There was a very sharp boundary between the two phases, and the volume of the more solid phase could be read continuously by reference to the lowest graduation line visible through the liquid layer. The effect was independent of centrifuging speed, and was as clear when direct current replaced A.C. mains to supply the light source. The apparatus bears no resemblance to the stroboscopic microscope-centrifuge of Harvey[(9)].

The centrifuge is illustrated in Fig. 1 and the appearance of a tube during and after spinning is shown in Fig. 2.

Materials

The psoas and adductor muscles of rabbits were used throughout the investigations. Most of the animals were well fed and were killed by decapitation after thirty minutes' complete relaxation produced by intraperitoneal or intravenous injections of myanesin (Bate-Smith and Bendall[(10)]).

Adenosine triphosphate (ATP) was prepared by the method of Lohmann as described by Needham[(11)].

Technique

Following the example of the Szeged school[(1)], potassium chloride (0.16 *M*) was the medium generally used in the present work. About 2 g muscle were homogenized in 8 ml of this solution for $\frac{1}{2}$–3 minutes, using the apparatus of Marsh and Snow[(12)], and the brei was quickly transferred to a graduated centrifuge tube (10 ml, 0.1 ml divisions) in the modified centrifuge, which was rapidly accelerated to about 2,800 r.p.m. (relative centrifugal force at tip of tube about 1750). Taking zero time at the commencement of centrifuging, the volume of the heavier fibre layer was recorded against time as its surface level reached each graduation mark.

A considerable decrease in fibre volume occurred due merely to centrifugal packing, but this could be practically eliminated in one of two ways, depending on the initial state of the muscle. (a) When fresh, pre-rigor muscle was examined, the complex curve relating volume to time of centrifuging was "corrected" for the simple packing effect by dividing the observed volume by that of a brei of rigor muscle at the same time of spinning. Trials showed that the fibre volume of a homogenate of muscle in full rigor decreased smoothly with time of centrifuging, no further change being detected after about 80 minutes, and the volume at any other time was expressed as a multiple of the 80-minute volume

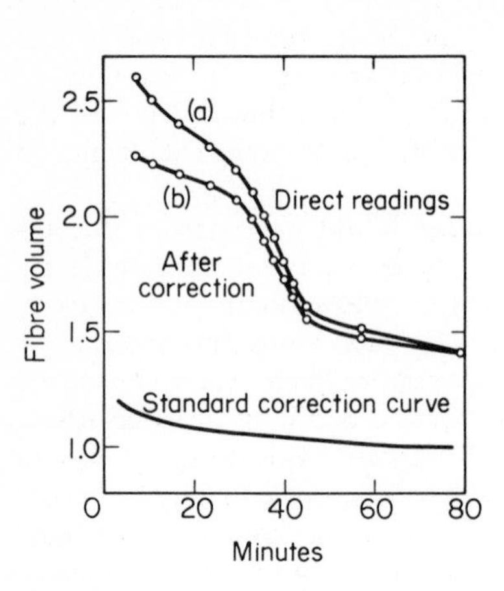

Fig. 3. The fibre volume of a homogenate of fresh muscle as a function of time of centrifuging. (a) Curve obtained by direct observation; (b) the same after dividing each observed value by a correction factor (derived from the standard correction curve) to eliminate simple centrifugal packing.

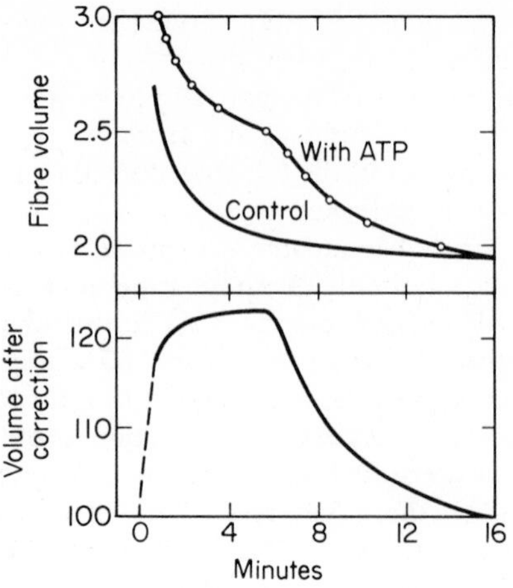

Fig. 4. The swelling effect on fibre volume of added ATP. Upper: Fibre volume before and after addition of ATP to a brei of rigor muscle. Lower: Volume behaviour after correcting for simple centrifugal packing. 1 mg ATP-P added at time 0.

(*e.g.* 3-minute volume = 1.20; 20-minute volume = 1.09, etc.). This series of values was used to "correct" other curves for simple centrifugal packing. The method is illustrated in Fig. 3. (b) When the effect of added reagents, notably ATP, was examined on muscle homogenized after the development of rigor mortis, it was found simpler to obtain first a "control curve" for that particular brei. The small fibre plug was then re-incorporated with the supernatant fluid by stirring, the reagent was added to the suspension of fibre pieces, and the fibre volume was again observed in the centrifuge. The second curve obtained was then corrected for simple packing by dividing the observed values by those of the control curve at corresponding times of spinning. This is illustrated in Fig. 4. It will be observed that, while method (a) provides a final curve expressing volume in ml, method (b) gives a curve in which volume is represented as a percentage of that in absence of the added reagent.

Results

Fibre volume and rigor mortis

When the fibre volume of a fresh actively-glycolysing muscle brei was examined as a function of time of centrifuging, a curve resembling (a) in Fig. 3 was obtained, and on correction for packing this was transformed to curve (b) in that diagram. Three more corrected curves are illustrated in Fig. 5, selected at random from over fifty such observations. In all these experiments each curve consisted of three distinct phases which for convenience will be described as pre-rapid, rapid, and post-rapid, respectively. The fibre volume at commencement of the rapid phase may be considered as unity to permit calculation of rate and extent of volume diminution, and to aid these determinations, the three linear phases may be extrapolated to give precise values at their points of intersection for volume and time at the beginning and end of the rapid phase.

The pre-rapid phase

This occupied up to 50 minutes, its duration being markedly affected by glycogen reserves (regulated by control of feeding) and by the degree of exhaustion of the animal at death (controlled by myanesin). In these respects it may be compared with the delay period preceding the elasticity changes which accompany the onset of rigor mortis in whole muscle (Bate-Smith and Bendall[13]). Thus, well-fed animals treated before death with myanesin almost invariably had muscles which, taken soon after death, showed a pre-rapid phase of at least 20 minutes, and occasionally over 40 minutes, duration, while at the other extreme the muscles of two insulin-treated animals with negligible glycogen content (ultimate pH 7.16, 7.18) showed in one case a prerapid phase of three minutes and in the other this phase was complete before observations could be commenced. There was little or no volume decrease during the pre-rapid phase, the observed diminution being almost exactly accounted for by the packing effect, and except in two cases the rate of decrease did not exceed 0.5% per minute, calculated on the volume at commencement of the rapid phase.

The rapid phase

This commenced quite suddenly, the rate of volume diminution sometimes increasing ten-fold within five minutes after a pre-rapid phase occupying several times this period. The rapid phase showed wide variations in velocity

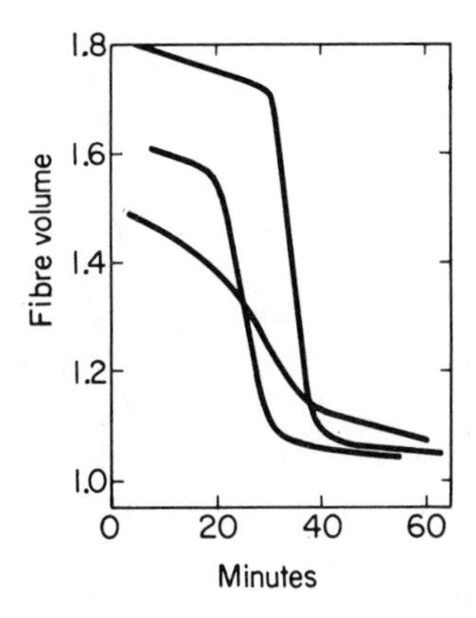

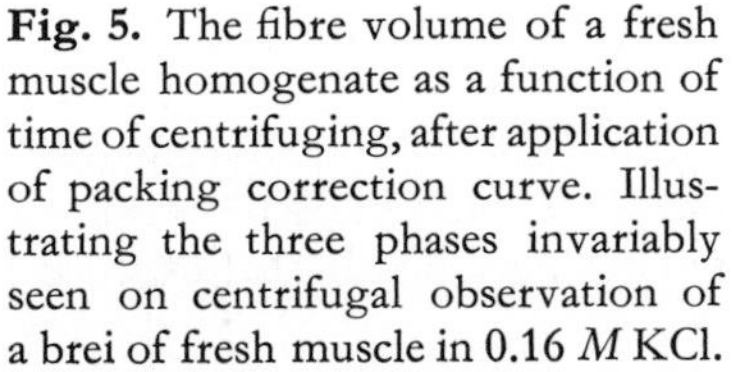
Fig. 5. The fibre volume of a fresh muscle homogenate as a function of time of centrifuging, after application of packing correction curve. Illustrating the three phases invariably seen on centrifugal observation of a brei of fresh muscle in 0.16 *M* KCl.

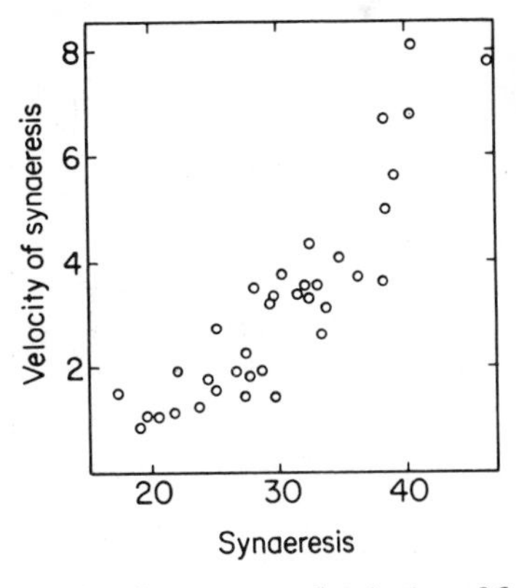

Fig. 6. The extent and velocity of fibre synaeresis in centrifuging homogenates of fresh muscle. Extent: Fibre volume decrease, during rapid phase, as a percentage of initial volume. Velocity: Extent of synaeresis per minute during rapid phase.

and extent, the briefest lasting only $3\frac{1}{2}$ minutes and the longest $21\frac{1}{2}$ minutes, while the volume decrease in the fibres varied from 17 to 46% of the value at commencement of the rapid phase. The rate of volume diminution varied from 1-8% per minute, which, on individual curves, represented an increase over the rate of shrinkage in the pre-rapid phase of 5–10 times.

The post-rapid phase

The decrease of fibre volume ceased quite abruptly at the end of the rapid phase, and the volume thereafter remained constant or decreased only slowly at about the same rate as in the pre-rapid phase. No further change in volume behaviour occurred, this phase being linear for as long as observations were continued.

Relation between velocity and extent of synaeresis

A large volume decrease was associated with a rapid phase of short duration, and some degree of control over these variables was possible in that tissue homogenized for a longer period exhibited greater synaeresis, the rapid phase occupying a shorter time interval. Thus a fibre volume decrease of 20% and occupying 20 minutes was observed in a brei, the preparation of which involved homogenizing for 30 seconds; while muscle reduced by treating for 3 minutes diminished in fibre volume by about 40% in 6 minutes. This is better expressed in Fig. 6, in which each of the 36 points represents the relation between the velocity and extent of synaeresis in the rapid phase of one homogenate. Analysis in this way was sometimes impossible since the pre-rapid phase was occasionally too brief to permit accurate determination of its linearity and slope, while sometimes the post-rapid phase was obscured by a light precipitate, presumably denatured myogen, which descended from the supernatant liquid to blur the boundary and prevent further observation. Nevertheless there are sufficient points to indicate a definite relation.

Effect of varying KCl concentration

The three-phase diagram relating fibre volume to time of centrifuging was observed when homogenates were prepared in water or in KCl solutions of concentrations up to 0.4 *M*. The turbid supernatant liquid and blurred boundary obtained when molarities of more than 0.3 were used made volume observation difficult, but in all cases the same type of behaviour was recognizable. The pre-rapid phase was shortened considerably when the molarity was low; the duration of this phase in water was only $3\frac{1}{2}$ minutes compared with 12 minutes when another sample of the same muscle was examined in 0.16 *M* KCl.

Effect of other ions

Centrifugal examination of a brei prepared in Ringer-Tyrode solution (Na^+, K^+, Ca^{++}, Mg^{++}, Cl^-, $H_2PO_4^-$, HCO_3^-) revealed a volume/time relationship very different from the normal three-phase behaviour. An immediate synaeresis occurred, the fibre volume decreasing very rapidly to only one half the expected value. There was no suggestion of a pre-rapid phase, and within a few seconds of commencement of centrifuging the fibres had packed so tightly that their volume remained almost unaltered despite prolonged spinning. Further investigations, using solutions of Tyrode constituents either alone or in various combinations as homogenizing fluid, showed that calcium was the ion responsible for the new type of volume behaviour. When calcium was omitted from the Tyrode solution the usual three-phase diagram was observed, but the presence of 0.02% calcium chloride, either alone or with other salts, invariably permitted only the sudden synaeresis.

Effect of iodoacetate

Several homogenates were examined after preparation in 0.16 *M* KCl + 0.002 *M* sodium iodoacetate to prevent completely the formation of lactic acid. On centrifuging, the three-phase diagram was obtained in all cases, but the duration of the pre-rapid phase was shortened considerably. Thus in one pair of experiments this phase lasted only 10 minutes in the presence of iodoacetate while the control (without inhibitor) showed a delay of 26 minutes. On another occasion the corresponding times were 7 and 22 minutes respectively. The observed synaeresis is not connected directly, therefore, with post-mortem acid production, but must obviously be intimately associated with a change occurring fairly early post-mortem.

The effect of added ATP on fibre volume

To test the hypothesis that the dephosphorylation of ATP was related to the rapid phase of volume diminution, the ester was added back to the shrunken system in amounts of about 1 mg labile-P (in 1 ml solution) per 10 ml brei (containing about 2 g muscle). One or other of two large and very different effects was caused by this treatment—a freely reversible increase in fibre volume or an irreversible synaeresis.

Reversible volume increase

ATP addition to a muscle brei which had passed through a rapid phase of shrinkage only shortly before almost invariably produced the effect illustrated by Fig. 4. The fibre volume as read, before applying the correction factors for simple centrifugal packing, was observed to remain fairly constant, or to decrease less than that of the fibres of the control brei, for some minutes. This phase was followed by a rapid volume decrease which ceased when the volume was about equal to that of the fibres of the control homogenate after the same time of centrifuging. After applying the control curve to eliminate the packing effect the volume changes resembled the lowest curve of Fig. 4, consisting of a brief period of rapid volume increase of 15–25%, a phase of up to five minutes duration during which this high level was maintained or slightly increased, and another brief period during which the fibre volume decreased rapidly to about its original value. This series of changes usually occurred when ATP was added to a rigor brei prepared on the same day as the animal was killed, and the same effect was observed in about two of every three experiments when this procedure was applied to a brei prepared from rigor muscle after storage for 24 hours at 0°C. Volume reversal could often be repeated several times on the same brei by further additions of ATP, although maxima and minima tended to be slightly lower with each repetition.

The effects of the presence of other salts on this volume behaviour was examined. ATP was added to homogenates prepared in solution of Tyrode constituents, and with one exception, fibre volume increases were observed in all cases where the fibres of a control brei in KCl showed the swelling effect. In the presence of 0.02% $CaCl_2$, however, this was replaced by irreversible synaeresis. The volume response to added ATP therefore paralleled exactly that of homogenates of fresh muscle prepared in various solutions as previously described.

Homogenates of rigor muscle in isotonic KCl were unaffected by the addition of inorganic phosphate (4 mg P) or of inorganic pyrophosphate (4 mg acid-labile P).

Irreversible synaeresis

Addition of ATP to a brei of *rigor* muscle frequently caused an immediate and considerable synaeresis instead of the fibre swelling previously described. This shrinkage was extremely rapid and was almost invariably complete within about 20–30 seconds. Its onset could often be recognised, before centrifugal observation was recommenced, by a characteristic increased whiteness of the fibre particles. The volume behaviour is illustrated in Fig. 7. The apparent slight swelling which followed the large synaeresis, as shown in the lowest curve of that figure after correcting for centrifugal packing, was observed in all cases of considerable shrinkage, and would appear to be due merely to the extremely "superpacked" nature of the particles. The appearance of these pieces was very different from that of the fibres of the control brei, and their density, after losing almost half their water content during synaeresis, must have been greater than before ATP addition, so it is not surprising that the particles failed to follow exactly a course related linearly to that taken by the fibre pieces of the control homogenate. The slight volume increase following synaeresis is considered, therefore, to be of no real significance.

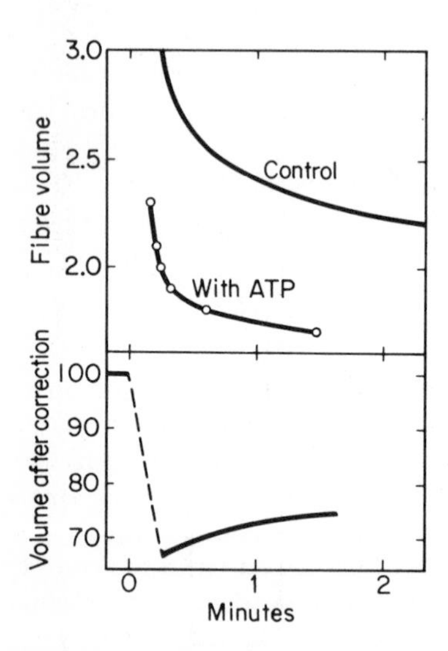

Fig. 7. The shrinking effect on fibre volume of added ATP. Upper: Fibre volume before and after ATP addition to a brei of rigor muscle. Lower: Volume behaviour after correcting for simple centrifugal packing. 1 mg ATP-P added at time 0.

Well over fifty volume shrinkages of this type were observed in the course of the investigation, and by comparing the volumes before and after ATP addition after one minute's centrifuging in each case, a mean volume diminution of 23% was calculated, with extreme limits of 11% and 43%. No method of reversing this shrinkage was found. Washing the fibres to remove ATP or further addition of ATP proved equally ineffective, the volume/time relationship remaining unchanged from that of the first large shrinkage.

Intermediate behaviour

In addition to the very distinct effects described above, many experiments produced volume/time curves of greater complexity. In some, for instance, a volume increase of 5–10% above the control was followed after 2–3 minutes by a rapid phase during which the volume decreased to 3–6% below that of the control. In other cases an immediate synaeresis of about 8% was followed by a swelling which increased the fibre volume almost to that of the control, and a second synaeresis then reduced the volume to a steady value about 8% below that of the control. These curves were at first mistaken for other quite distinct types of volume behaviour, but it later became obvious that the systems were all displaying effects intermediate between reversible swelling and irreversible synaeresis. By graphically compounding these two effects in varying ratios it was possible to reproduce even the most complex curves obtained in centrifuge experiments. All the observed effects were therefore the resultant of two volume changes of similar magnitude but opposite in direction and of different rates. While one was clearly a reversal of the synaeresis observed when the fibre volume of a fresh, actively-glycolysing muscle brei was examined, the other appeared to be directly related to the synaeresis of extracted actomyosin and the shortening of glycerol-treated muscle fibres induced by added ATP.

The factor controlling volume changes

To this point, the only clues to the possible nature of the factors influencing the response of the fibres were the apparent sensitivity to calcium of the volume-swelling mechanism, and the increased tendency toward fibre shrinkage with increasing age of the preparation. Other factors were investigated.

The effect of pH

The volume response to added ATP was found to be unaffected by the presence of 0.002 *M* iodoacetate. In order to prepare homogenates of fixed pH a sample of muscle was homogenized in 0.16 *M* KCl, and when the pH had attained the required value iodoacetate was added. Creatine phosphate and ATP decomposition were complete within a few minutes, after which the centrifuge test could be performed with no possibility of pH shift. In the pH range 7.2–5.8, to which observations were confined, pH had no consistent effect on volume response to ATP, and both swelling and synaeresis were found at any pH within this range. However, the *extent* of volume change was markedly dependent on pH, the total change (maximum increase above control + maximum decrease below control) decreasing from 20–25% at pH 6.7–7.2 to 8–10% at pH 5.8–6.3.

The effect of ATP concentration

Although immediate synaeresis could be produced by adding lesser amounts of ATP, it became clear that this was not the essential difference between the two volume effects since on many occasions immediate synaeresis followed the addition of very appreciable amounts of ATP. Nevertheless the results were of some interest; in those systems which displayed fibre swelling on the addition of appreciable amounts of ATP, the minimum concentration of that ester required to cause a full volume increase was found to be 0.07 mg ATP-P per g muscle (assuming that the distribution of added ATP was uniform throughout both fibre pieces and suspending medium). Maximum immediate synaeresis was brought about by the addition of 0.02 mg labile P per g muscle.

The sarcoplasmic factor

Replacement of the supernatant liquid of a centrifuged homogenate by fresh 0.16 *M* KCl solution altered entirely the volume response to ATP. In those systems which swelled on addition of ATP, the intermediate type of volume behaviour was obtained merely by re-suspending the fibres in fresh KCl solution, and when the fibre pieces were twice washed in KCl solution with alternate centrifuging, they responded to ATP addition by shrinking immediately. Some degree of control over the volume effects was now possible in that synaeresis could be brought about at will by a standard procedure, as illustrated by Fig. 8, but of greater significance was the indication of the existence of a substance, soluble in 0.16 *M* KCl solution, which must be intimately concerned in volume increase effects. This will be referred to as the "factor."

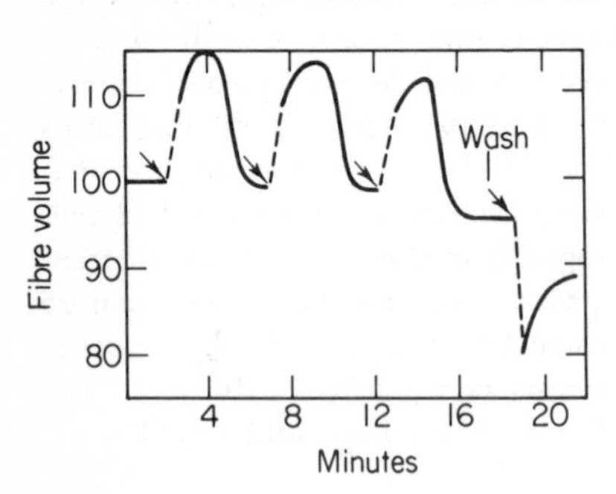

Fig. 8. Reversible volume increase, and the irreversible synaeresis observed after washing the homogenized fibres. Fibre volume expressed as a percentage of the control volume after the same time of centrifuging. 0.76 mg ATP-P added at each arrow.

Nature of the factor

Solutions suspected to contain the factor were examined by observing in the centrifuge the volume response to added ATP of washed fibres suspended in the solution. Absence of the factor was shown by an immediate synaeresis, and presence of the factor by reversible volume swelling, while intermediate behaviour—a slight swelling followed by appreciable synaeresis—was interpreted as an indication of the presence of the factor in diminished quantity.

Evidence for protein nature

The constituents of Tyrode solution, either separately or in combination, permitted only irreversible synaeresis when they were tested with washed fibres. The supernatant fluid from a spun homogenate lost all factor activity after removal of heat-coagulated material (2 minutes at 60°), or by removal of the precipitate formed after neutralization to pH 7 of an acidified aliquot (pH 3.6, HCl). These demonstrations of the lability to heat and acid of the factor suggested that it consists, at least in part, of protein material, and extracts (known by the centrifuge test to contain the factor) were therefore dialysed at 0°C for 16 hours in cellophane sacs against either distilled water or 0.16 *M* KCl. Of six such experiments, four showed that no factor activity remained within the sac, while the other two showed partial activity estimated from the positions of their centrifuging curves to be about 25% and 50%. In no case could any activity be detected in the outer dialysing fluid, and when this was concentrated and added to the inner solution no enhancement of activity was observed. In all these experiments a small amount of precipitated protein failed to redissolve in KCl solution. A rapid dialysis (4 hours) against ice-cold distilled water, with frequent changes of fluid and continual agitation of the sac did not inactivate the volume-reversing factor, full activity remaining within the sac. On this occasion all the precipitated protein readily redissolved in 0.16 *M* KCl solution.

Fractionation with ammonium sulphate failed to provide any information, possibly because of the necessity of prolonged dialysis, and elution by phosphate buffer of proteins adsorbed on calcium phosphate also failed, no fraction showing any factor activity after dialysis.

The magnesium-activated ATP-ase described by Kielley and Meyerhof[14] was prepared, but showed no activity. The same result was obtained when tests were made using several relatively pure glycolytic enzymes (phosphorylase, lactic dehydrogenase, aldolase, enolase, creatine phosphokinase, and triose phosphate dehydrogenase) kindly supplied by Dr. B. A. Askonas[15].

Function of the factor

It was found that the rate of dephosphorylation of ATP by the fibers of a muscle brei varied over a wide range, being considerably higher, in general, when the fibres were washed before incubation with ATP. A closer examination revealed a correlation between presence of the volume-reversing factor and diminution in the ability to split ATP. Those systems, the fibres of which increased in volume on ATP addition, released inorganic phosphate from excess ATP at a rate of only about 40 μg P/minute/g muscle, while in those homogenates which responded to added ATP by immediate fibre synaeresis, inorganic phosphate appeared at the rate of about 300–400 μg P/minute/g muscle. (All measurements were made at 17°C). Since almost 10 μg P were produced per minute per g muscle by soluble enzymes in the former case, the ratio of fibrillar ATP-ase activities in the two systems was about 1:10. No separation of this apparent ATP-ase inhibition from the fibre-swelling factor has been achieved. The former effect was not due to aerobic resynthesis of ATP since the presence of 0.01 *M* cyanide did not alter the rate of production of inorganic phosphate. Neither was it due to a preferential diversion of ATP to other reactions (for instance, the phosphorylation of fructose-6-phosphate), since more complete analysis showed that the amount of ATP removed during incubation agreed with the amount of inorganic phosphate produced, and no accumulation of other phosphate esters was detected.

Changes in micro-structure

Although the volume changes reported above suggest that changes in fibre length were involved, the swelling and synaeresis might conceivably have been caused merely by changes in rigidity of the fibre pieces, or by alterations in fibre diameter at fixed length. Changes in systems displaying either increased or decreased volume immediately on ATP addition were examined microscopically (× 100). A drop of a suspension of fibre particles in either 0.16 *M* KCl solution or in an extract containing the swelling factor was placed within a wax circle of 4 mm diameter mounted on a glass slide. ATP solution was added by pin-point and stirred into the suspension, and measurements were made at intervals.

Those fibres suspended in KCl solution responded very rapidly to added ATP and within a few seconds had attained their ultimate shape. Their length decreased by about 60% and their breadth increased by about 40%, causing an overall volume decrease of approximately 20%. In presence of the volume-increasing factor the effect was very different, the fibre lengthening by about 20% during the first minute, and by a further 20% in the next ten minutes when shortening did not intervene before this time. Shortening then occurred at a rate of 10–20% per minute till the particle had returned to within 10% of its initial length, after which no further change occurred. No alteration in diameter was detected, but as attention was directed towards length changes, the possibility of breadth alterations is not excluded. The duration of the phase of increasing or high volume was long compared with the corresponding phase in the centrifuge tests, possibly because ATP removal in the former case was achieved solely by enzymic decomposition while in the centrifuge this process was augmented by centrifugal transfer of ATP from the fibres to the supernatant fluid. Apart from this, the observations on micro-structure were entirely consistent with the volume behaviour of the fibres during observation in the centrifuge. In particular, fibre swelling corresponded to lengthening, and synaeresis to shortening.

Discussion

It may be argued, with some justification, that the volume and length changes in muscle fibre pieces described in this work are related only to the onset and reversal of rigor mortis. Certainly the slowness of the changes and the fact that fibre shortening and synaeresis occur only at low ATP concentrations might be taken as evidence that the observations have no connection with muscular contraction. Nevertheless the magnitude of these changes and their obvious dependence on an intimate relation between fibre proteins and ATP suggest that the system may be regarded as a model muscle undergoing contraction and relaxation. Indeed, the physiological completeness, high degree of fibrillar organisation, and ready reversibility of volume changes in the present work suggest that the muscle homogenate is more nearly a true model system than actomyosin threads or glycerol-treated fibres.

From the evidence of electron micrographs, Perry *et al.*[6] concluded that the synaeresis of actomyosin under the influence of ATP is not the process underlying the act of muscular contraction. The present work seems in no way incompatible with this conclusion. The partial dehydration of the fibres of the muscle homogenate is regarded, not as the fundamental cause, but rather as one effect, of fibre shortening; the volume changes are merely a useful guide to the extent of those fundamental changes in molecular configuration which affect the state of hydration. That such large volume changes are not observed in living muscle is no argument against the validity of comparing the homogenate with intact muscle for, whereas the fibre pieces have been so extensively damaged as to permit the ready outflow of fluid, this change is paralleled in the intact muscle by an appreciable increase in pressure during contraction (Hill)[16]. Thus, while agreeing with Perry *et al.*[6] that an actomyosin system cannot, "by a process of alternate dehydration and rehydration, reproduce the mechanical properties of the cell", we consider that synaeresis may still provide, in certain circumstances, a valuable index to more deep-seated changes of molecular configuration.

The relaxation-like effect observed in those systems which swelled on ATP addition appears to be a new phenomenon in muscle models, the only comparable effect being a lengthening of actomyosin threads, when loaded, under the influence of the pyrophosphate group. Since fibre swelling has been shown to depend upon the presence of an extremely labile factor, it is not surprising that other model systems show no similar response to ATP; solubility of the factor in water and low salt concentrations would ensure its absence from actomyosin threads or gels, while due to its instability it would be absent from glycerol-treated fibres. As reported here, the factor appears to be accompanied by an ability to suppress the ATP-ase activity of the fibril, and it will be assumed that the fibre-swelling effect is associated with this enzymic inhibition. It must be admitted, however, that the only reason for supposing the identity of the volume-increasing factor and the ATP-ase-inhibiting factor is the simultaneous disappearance of both from an extremely complex system, and the components responsible for the two effects may prove to be different substances, alike only in their instability. Should only one component be responsible for both effects, it is suggested that the rate of ATP decomposition will be a much more sensitive guide to presence or absence of the factor than the centrifuge test, which in the present work provided only a semiquantitative determination of factor activity, and might conceivably have failed to record relatively appreciable amounts of the component.

Assuming that fibre-swelling and enzyme inhibition are intimately related—possibly by a masking of essential groups of myosin, or by formation of a "protective conjugation" between factor and ATP, as suggested by Engelhardt[17]—how are the changes in fibre length and volume to be explained in relation to the presence or decomposition of ATP? In those systems which displayed fibre swelling on addition of ATP, presence of the ester in appreciable concentration was greatly prolonged, and an increase of volume and length is therefore to be associated with the *presence,* rather than with the rapid breakdown, of ATP. On the other hand, in those homogenates which exhibited immediate synaeresis, the relatively high ATP-ase activity ensured that presence of the ester in appreciable amount was maintained for no more than a very brief period of time. If ATP-ase activity exceeded the rate of ATP diffusion into the fibre pieces, ATP could not attain, even momentarily, a significant concentration at the site of its interaction with the contractile proteins, yet energy release proceeded at a high rate. A volume decrease is thus to be associated with the near *absence* of ATP, coinciding with, or preceded by, its rapid breakdown. This explains why the presence of calcium permitted only synaeresis on ATP addition; calcium is a powerful activator of myosin ATP-ase (Bailey)[18] and its presence would be expected, therefore, to exert an effect opposite to that of the labile factor.

Similarly the effect of adding smaller amounts of ATP, in presence of the factor, finds a simple explanation. Addition of the ester in large amount causes fibre swelling since an appreciable concentration of ATP can be maintained. Addition of small amounts of ATP causes only synaeresis because, even at a low rate of dephosphorylation, energy is being supplied to the contractile system while at the same time ATP is present in insufficient amount to maintain saturation of the relevant centres. The conditions for synaeresis—low concentration of ATP and breakdown of the ester—are thus satisfied. Since 0.07 mg ATP-P per g muscle is required to permit a complete reversible cycle of fibre swelling, while 0.02 mg ATP-P causes maximum immediate synaeresis in 1 g muscle, it is justifiable, on the above reasoning, to assume that 0.02 mg ATP-P is that amount which just saturates the ATP-reacting groups of 1 g muscle, while 0.05 mg ATP-P is the minimum amount required to maintain this saturation while maximum increase in volume is being attained. Since ATP decomposition proceeds, in presence of the factor, at a rate of about 40 μg P/minute/g muscle, the duration of the phase of rapidly increasing fibre volume should be about $1\frac{1}{4}$ minutes, and it was indeed found that this phase lasts about 1–2 minutes in all cases.

It is also of interest that, assuming 100 mg myosin/g muscle, 0.02 mg ATP-P per g muscle corresponds to 1 mole of ATP per 310,000 g myosin. This may be compared with the ratio of 1 mole of ATP per 100–300,000 g myosin required to dissociate completely the actomyosin complex in 0.5 M KCl (Mommaerts)[19]. Interpreting these observations on the intermolecular level, it is considered that the contractile proteins of relaxed muscle are maintained in the dissociated (actin + myosin) state because of the greater affinity of the sulphydryl groups of myosin for ATP than for actin (Bailey and Perry[20]). Because of the presence of the labile factor reported above, the rate of ATP breakdown is slow, and consequently a state where ATP is temporarily absent does not occur; hence actomyosin cannot be formed. Contraction involves a transformation to the actomyosin condition, due to (a) an increased rate of energy release from ATP, and (b) a temporary absence of ATP from the vicinity of the —SH centres of myosin—these changes are caused by increased ATP-ase activity, as a result of either Ca^{++} release[18], or a temporary inactivation of the labile factor, or both effects. Relaxation is regarded as a return to a low ATP-ase activity by reversal of factor inactivation or by removal of Ca^{++}, this permitting accumulation of ATP which dissociates actomyosin to its component proteins, and the structure imbibes water. This swelling may be due to the greater water-retaining ability of the dissociated protein, or to the relatively greater freedom of the component proteins, or perhaps to the elastic properties of the sarcolemma which, on disruption of the actomyosin structure by ATP, might attempt to regain its normal resting length.

The above demonstration that ATP-ase activity in muscle fibres can be greatly diminished by the factor directs attention to the relatively high rate at which ATP dephosphorylation proceeds in other model systems. Analysis merely of the medium containing such systems after "contraction" induced by ATP might well indicate that only a small amount of the ester has been decomposed, but it by no means follows that "contraction" is due to the presence (as distinct from the decomposition) of ATP. Provided that, in the model itself, and not in the medium, ATP dephosphorylation can keep pace with inward diffusion of the ester, contraction-like effects will necessarily be observed.

The author is grateful to Dr. K. Bailey, under whose guidance and supervision this work was carried out, and to Dr. E. C. Bate-Smith and Mr. J. R. Bendall for valuable discussions. The latter undertook the experimental treatment of the rabbits used during the investigation. Dr. B. A. Askonas kindly provided the enzymes tested for factor activity. The photographs were taken by Mr. D. P. Gatherum.

Summary

1. By means of a centrifuge simply modified to permit continuous reading of the fibre volume of a spinning muscle brei, a decrease in the fluid-retaining ability of fresh homogenized fibres has been detected. This synaeresis, amounting to 17–46% of the initial volume, occurs after an interval of up to 50 minutes commencing when ATP has virtually disappeared.

2. Addition of ATP to the shrunken system causes either a volume increase of 10–25% followed by a decrease to about the orginal volume, or an immediate irreversible synaeresis of 10–40%. The former effect is accompanied by reversible fibre lengthening, and the latter by irreversible shortening.

3. The essential difference between the two effects is due to the presence in the former system of a labile component, probably a protein, in the absence of which ATP addition invariably causes irreversible synaeresis.

4. The labile factor appears to exert its effect by diminishing greatly the ATP-ase activity of the homogenized fibres.

5. The results are discussed in relation to the behaviour of other model systems and to muscular contraction.

References

1. A. Szent-Györgyi, *Chemistry of Muscular Contraction,* New York, 1947.
2. S. Korey, *Biochim. Biophys. Acta,* **4** (1950) 58.
3. A. V. Hill, *Biochim. Biophys. Acta,* **4** (1950) 4.
4. G. Kuschinsky, and F. Turba, *Biochim. Biophys. Acta,* **6** (1951) 426.
5. W. A. Engelhardt, and M. N. Ljubimowa, *Biokhimiya,* **7** (1942) 205.

6. S. V. Perry, R. Reed, W. T. Astbury, and L. C. Spark, *Biochim. Biophys. Acta,* **2** (1948) 674.
7. W. A. Empey, *J. Soc. Chem. Ind.,* **52** (1933) 230T.
8. B. B. Marsh, *Nature,* **167** (1951) 1065.
9. E. N. Harvey, *J. Franklin Inst.,* **214** (1932) 1.
10. E. C. Bate-Smith, and J. R. Bendall, *J. Physiol.,* **107** (1947) 2P.
11. D. M. Needham, *Biochem. J.,* **36** (1942) 113.
12. B. B. Marsh, and A. Snow, *J. Soc. Food Agric.,* **1** (1950) 190.
13. E. C. Bate-Smith, and J. R. Bendall, *J. Physiol.,* **110** (1949) 47.
14. W. W. Kielley, and O. Meyerhof, *J. Biol. Chem.,* **176** (1948) 591.
15. B. A. Askonas, *Biochem. J.,* **48** (1951) 42.
16. A. V. Hill, *J. Physiol.,* **107** (1948) 518.
17. W. A. Engelhardt, *Advances in Enzymol.,* **6** (1946) 147.
18. K. Bailey, *Biochem., J.,* **36** (1942) 121.
19. W. F. H. M. Mommaerts, *Biochim. Biophys. Acta,* **4** (1950) 50.
20. K. Bailey, and S. V. Perry, *Biochim. Biophys. Acta,* **1** (1947) 506.

KIELLEY-MEYERHOF'S GRANULES AND THE RELAXATION OF GLYCERINATED MUSCLE FIBERS*

Setsuro Ebashi

Department of Pharmacology, Faculty of Medicine
University of Tokyo, Tokyo

Introduction

Bozler[1] has shown that the high concentration of ATP† could produce the relaxation of freshly glycerinated muscle fibers shortened by the low concentration of ATP. Later, Bendall[2] introduced the Marsh factor[3] into the experiment of muscle model and succeeded in demonstrating a contraction-relaxation cycle of glycerinated muscle fibers by using the physiological concentration of ATP. While Bozler[4] as well as Goodall and A. G. Szent-Györgyi[5] discovered independently that PC could develop a remarkable relaxation in relatively fresh fibers, the latter research group suggested that PC-creatine-phosphokinase system might have an essential role in the relaxation of glycerinated fibers. This postulation was later confirmed by Lorand[6]. However, Bendall[7] identified the Marsh factor with an extraordinary stable enzyme, namely myokinase, in spite of the very labile nature of the original factor. Recently Lorand and Moos[8] have demonstrated that the phosphoenolpyruvate-pyruvatekinase system has a relaxing effect on glycerinated fibers.

In view of circumstantial evidence, these enzyme systems appear to act as ADP phosphokinetic enzymes in the process of relaxation, although Bendall

*Setsuro Ebashi, "Kielley-Meyerhof's Granules and the Relaxation of Glycerinated Muscle Fibers," *Conference on the Chemistry of Muscular Contraction*. Tokyo: Igaku Shoin Ltd., 1957, pp. 1–6. Reprinted with the permission of the author and of the publisher.

†The following abbreviations are used in this paper: ATP—adenosine triphosphate; ADP—adenosine diphosphate; AMP—adenosine 5′—phosphate; PC—phosphocreatine; ATPase—adenosinetriphosphatase; Fr. A—fraction A; Fr. B—fraction B. As for Fr. A and Fr. B see the text.

himself has postulated a quite different conception in regard to myokinase-system. Goodall's recent report[9], discovering the relaxing activity of carnosine phosphate, may suggest the presence of another ADP-phosphokinetic enzyme.

These findings, however, do not definitely settle the problem concerned with the relaxation of glycerinated fibers. It is a well known fact that glycerinated fibers preserved for a long time or exhaustively washed in salt solution do not lengthen even in the presence of a sufficient amount of these enzyme systems. In regard to this problem, it has been discovered in our laboratory[10] that the protein factor in muscle extract connected with the relaxation consists of at least two different components, which were tentatively called Fr. A and Fr. B. The latter fraction contains a considerable amount of myokinase, and the relaxing activity of the fraction may be explained by the enzyme to a certain extent[11]. As for Fr. A, its activity has been shown to be inseparable from the granular fraction which is identical with Kielley-Meyerhof's granules[11,12,19].

The role of Kielley-Meyerhof's granules in relaxation is a matter of considerable interest, although yet it is not at all clear. The chief attention has been focused on the problem as to whether the ATPase activity of the granules has an essential part in relaxation or not[12]. It is a fact that all the manipulation depressing the enzymatic activity has been always accompanied by the abolition of the relaxing activity. And, moreover, the enzymatic activity is a good index for the relaxing action so far as the freshly prepared granules are concerned. However, the relaxing activity of the granules has been shown to be far more sensitive to various deteriorating procedures than the enzymatic activity, i.e. the former activity is abolished without forfeiting the latter one due to the application of certain treatments such as: aging, freezing and thawing, precipitation at pH 5.5 and resolubilization, and treatment by certain surface active agents. These facts may indicate that an unknown labile nature or natures of the granules other than ATPase is essential for relaxation. Nevertheless, this may not necessarily mean that the enzymatic activity has nothing to do with the relaxing activity.

Enzymatic Activity of So-Called Relaxing Factors in Fibers

As is well known, the glycerinated fibers tend to lose their lengthening capacity while they are being preserved in glycerol-water. It is not yet clear whether this phenomenon depends on an irreversible change, or denaturation, of contracting system or a gradual decrease of so-called relaxing factors in fibers.

Fig. 1 shows the enzymatic activities of intrinsic factors in fibers such as myokinase, creatine phosphokinase and Kielley-Meyerhof's granules at different preservation periods. Measurement of enzymatic activities was conducted by a method essentially similar to that of Kielley-Meyerhof's[13], Bowen-Kerwin's[14] and Noda-Kuby-Lardy's[15], however, a slight modification was added. Glycerinated fibers were immersed in 20% glycerol-water (salt free) at 0°C for 30 minutes, then homogenized in the solution of 0.1 M KCl containing 0.02 M $NaHCO_3$ and supernatant was used for enzymatic determination.

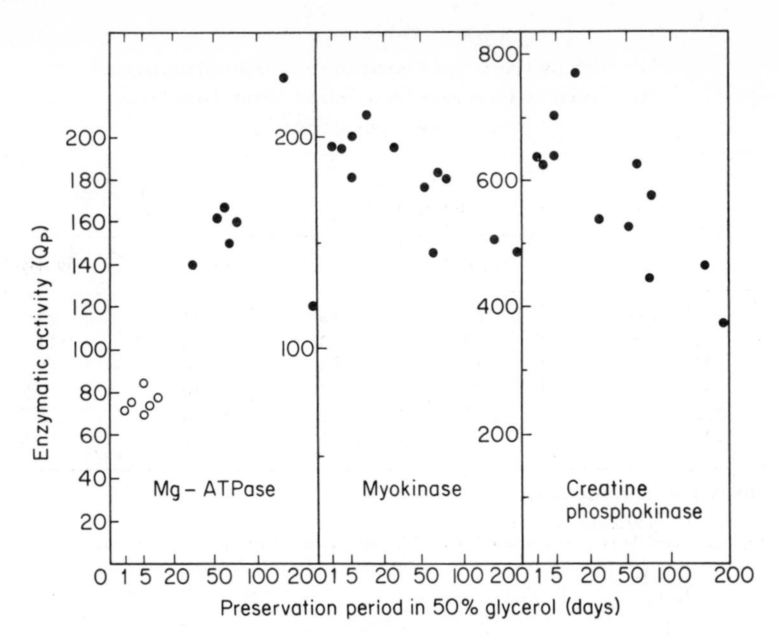

Fig. 1. Enzymatic activity of so-called relaxing factors in the glycerinated fibers at various periods of preparation in 50% glycerol-water.

The content of the soluble enzymes, namely creatine phosphokinase and myokinase, was shown to have remained rather constant during the first several weeks of glycerination. If we consider a longer period of preservation, a definite tendency of decrease was observed, especially in the case of creatine phosphokinase. This decrease seems, however, to be too small to explain the loss of extensibility of glycerinated fibers during preservation.

It is worth noticing that both enzymes, especially creatine phosphokinase, escaped from fibers far more rapidly than in glycerol-water (Table 1).

Mg-ATPase of Kielley-Meyerhof's granules showed a peculiar tendency, i. e. enzymic activity showed no decrease at all, but rather definitely showed an increase around the 10th week. It must be mentioned that the enzymic action of prepared Kielley-Meyerhof's granules persisted longer than its relaxing ability, and the former frequently showed an increase up to twice the initial value after the latter had disappeared. There is no direct evidence, however, that the phenomenon in prepared granules corresponds to that of Mg-ATPase in muscle fibers.

From the above results, it seems likely that the escape of the so-called relaxing factors from the fibers is not the chief reason for the disappearance of extensibility during preservation. Some irreversible changes, or denaturation, of the contracting system in fibers may play a more important role in this problem. In connection with this, the very labile and curious natures of Kielley-Meyerhof's granules appear to have some relation with the phenomena, although yet decisive evidence has not been presented.

Table 1

ACTIVITY OF CREATINE PHOSPHOKINASE AND MYOKINASE REMAINED IN GLYCERINATED FIBERS AFTER IMMERSION IN SALT SOLUTION

	Creatine phosphokinase		Myokinase	
	6 days*	65 days*	6 days*	65 days*
0 min.**	62	54	21	22
10	23	23	19	14
30	17	15	16	11
60	11	12	10	8.5
120	6	6	4.5	6
240	4	5	2	3.5

activity was expressed as $Q_P \times 10^{-1}$.

*preserved time in 50% glycerol-water.

**elapsed time after the immersion of fibers in 0.1 M KCl solution containing 0.01 M $MgCl_2$ at 0°C.

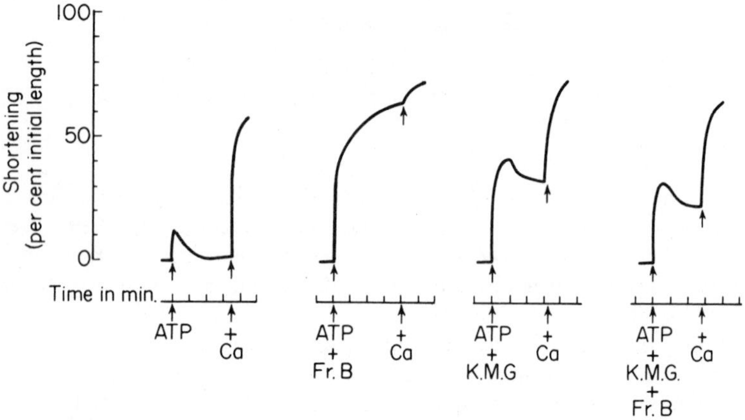

Fig. 2. Effect of Kielley-Meyerhof's granules and Fr. B (myokinase) on fresh fibers treated with an anionic surface active agent.

Fibers, at 3 days of preservation, were used. Before use, they were immersed in 20% glycerol for about 1 hour and washed in 0.15 M KCl solution containing 0.01 M $MgCl_2$ for 5 minutes. The figure on the left shows the brief contraction and spontaneous relaxation of non-treated fiber by ATP alone. Other fibers were treated with 0.01% triethanolamine alkylbenzene sulfonate for 3 minutes before the experiment at room temperature (22°C). ATP: 5 mM. Mg-ATPase of K.M.G. or Kielley-Meyerhof's granules: Ap = 9400. Myokinase of Fr. B: Ap = 8,300. Ap = 22.4 × number of μM P split or transferred per ml. per hour at 38°C. Contamination of K.M.G. by myokinase or creatine phosphokinase was neglected because of its small amount, not more than 500 in Ap. Ca: 8 mM. Bath solution: 0.15 M KCl containing 0.01 M $MgCl_2$. pH: 6.4 ~ 6.6. Tension: about 150 gm/cm². Temperature: 22°C.

Effect of Ionic Surface Active Agents and Phospholipases on Fresh Glycerinated Fibers

Fresh glycerinated fibers can be lengthened by ATP of relatively lower concentration alone. In this case, there is no direct evidence that myokinase as well as Kielley-Meyerhof's granules plays an essential role in the process of relaxation. The two factors, especially the latter, have only been proved to act upon those fibers which are preserved in glycerol water and brought up to a condition where the two factors can exert their influences on the fibers. Of course, fresh fibers might be converted into such fibers by washing them out in salt solution. But this sort of experiment can not always be reproduced.

A certain concentration of various ionic surface active agents, irrespective of cationic or anionic (desoxycholate, taurocholate, glycocholate, cholate, sulfate ester of higher alcohol, alkylbenzene sulfonate, alkylpyridinium and benzalkonium chloride), were able to bring the very fresh fibers into such a condition as mentioned above, i. e., the fibers could not be lengthened by myokinase-system or creatine phosphokinase-PC-system, but the collaboration of Kielley-Meyerhof's granules with these systems resulted in the relaxation of the fibers, as shown in Figs. 2 and 3.

In the above concentration of these agents, Mg-ATPase in the treated fibers did not decrease, but increased by 20 ~ 50%.

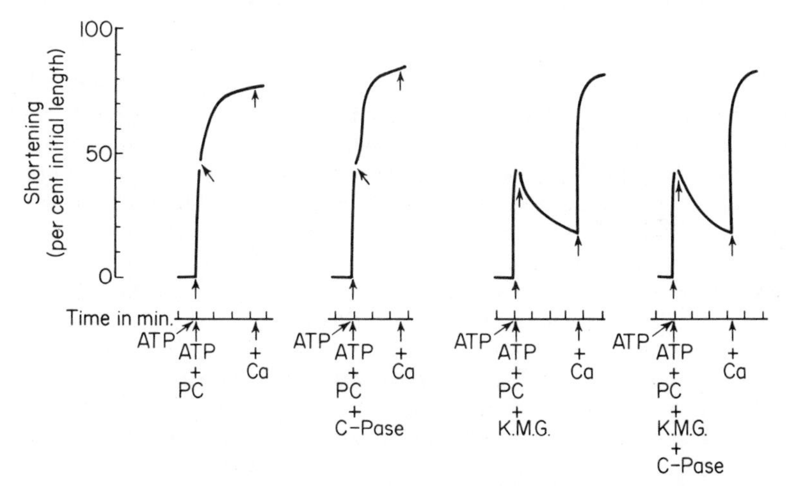

Fig. 3. Effect of Kielley-Meyerhof's granules and creatine phosphokinase on fresh fibers treated with a cationic surface active agent.

Fibers, stored for 2 days, were treated with 0.015% of benzyl alkonium chloride for 3 minutes at room temperature (24°C) and washed for 2 minutes before use. PC: 10 mM. Mg-ATPase of K.M.G.: Ap = 11,500. C-Pase or creatine phosphokinase: Ap = 60,500 or 32 units/ml. (This sample contained 38 units/mg.). Control experiment revealed that fibers treated as above contained creatine phosphokinase of about 130 in Qp or 20,800 in Ap. pH: at 6.0 ~ 6.2. Temperature: 24°C. As for others, see the legend for Fig. 2.

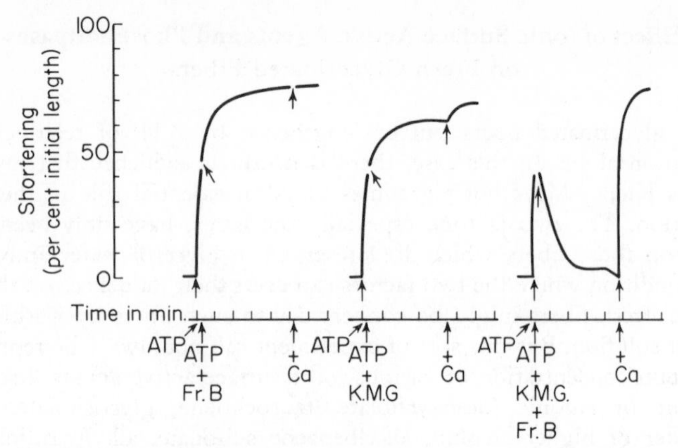

Fig. 4. Effect of Kielley-Meyerhof's granules and Fr. B (myokinase) on fresh fibers treated with phospholipase C.

Fibers, stored for 5 days, were treated with phospholipase C of Clostridium Welchii, 270 units of Macfarlane and Knight (17) per ml., for 5 minutes at room temperature, 24°C, and washed for 3 minutes before use. This treatment decreased the Mg-ATPase in fibers to 7% of the control value. Mg-ATPase of K.M.G.: Ap = 9,200. Temperature: 24°C. As for others, see the legend for Fig. 2.

Phospholipase C was known as a more or less specific destroyer of Mg-ATPase of Kielley-Meyerhof's granules[16]; it was shown also to be the eliminator of the relaxing capacity of the granules[12]. If it was applied to the fresh fibers, the situation was almost the same as in the case of surface active agents (Fig. 4). Phospholipase A from the venom of agkistrodon blomhoffi (Boie) also showed similar results (Fig. 5). Both enzymes by themselves had nothing to do with the contraction and relaxation of fibers if the process was limited to a very short period.

In the case of surface active agents, there is no definite evidence that they act principally on the Kielley-Meyerhof's granules. But in these cases there may be some possibilities that these two enzymes have preferentially affected the granules. In fact, Mg-ATPase in fibers was almost abolished by these treatments. However, these two enzymes are not so highly purified; so the possibility of interference from different contaminated enzymes can not be excluded.

Concluding Remarks

Several experiments referring to the relaxation of glycerinated psoas muscles have been presented with special reference to the relaxing capacity of Kielley-Meyerhof's granules. All of them are not conclusive, but may suggest the important role of the granules in relaxation of glycerinated fibers.

Acknowledgements. I wish to express my cordial thanks to Prof. Kumagai for his constant encouragement and advice, and also to Dr. F. Ebashi for her valuable assistance.

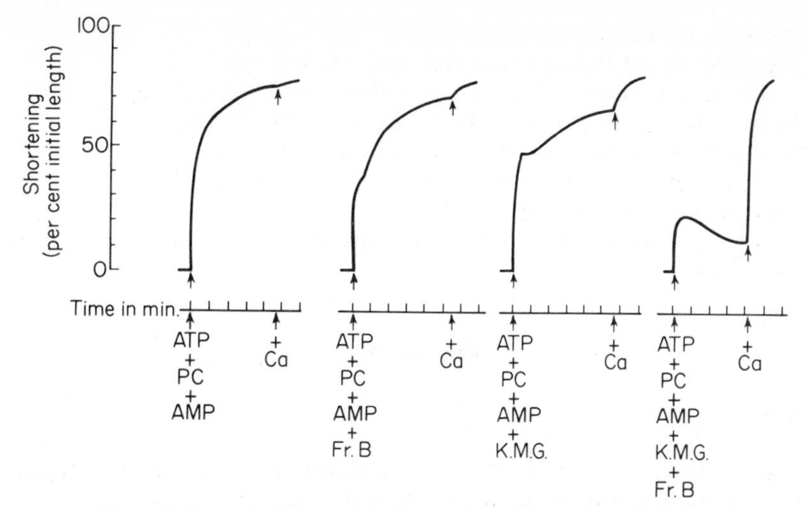

Fig. 5. Effect of Kielley-Meyerhof's granules and Fr. B (creatine phosphokinase) on fresh fibers treated with phospholipase A.

Fibers of 2 days preservation were treated with phospholipase A of venom from agkistrodon blomhoffi (Boie) for 5 minutes at room temperature (20°C) and washed 3 minutes before use. 1 ml. of the phospholipase preparation could liberate 2.4 μM of fatty acid per hour from 300 mg lecithin in 3% solution at 37°C (18). After the treatment, Mg-ATPase of fibers was shown to be less than 3% of the control. Mg-ATPase of K.M.G.: Ap = 12,200. Creatine phosphokinase of Fr. B: 26,000 in Ap. Myokinase of Fr. B: 8,600 in Ap. Myokinase or creatine phosphokinase remained in fibers was estimated from other control experiments; the former was 130 to 230 in Qp and the latter 100 to 190. AMP: 5 mM. AMP was shown to be inhibitory to the relaxation by myokinase-system (19). 0.02 M maleate-buffer (pH 6.1) was used. Temperature: 20°C. As for others, see the legend for Fig. 3.

References

1. Bozler, E., *Am. J. Physiol.* **167**, 276 (1951).
2. Bendall, J. R., *Nature,* **170**, 1058 (1952); *J. Physiol.,* **121**, 232 (1953).
3. Marsh, B. B., *Nature,* **167**, 1065 (1951); *Biochim. et Biophys. Acta,* **9**, 247 (1952).
4. Bozler, E., *J. Gen. Physiol.* **37**, 63 (1953).
5. Goodall, M. C. and Szent-Györgyi, A. G., *Nature,* **172**, 84 (1953).
6. Lorand, L., *Nature,* **172**, 1181 (1953).
7. Bendall, J. R., *Proc. Roy. Soc.* (London), B, **142**, 409 (1954).
8. Lorand, L., and Moos, C., *Federation Proc.* **15**, 121 (1956); *Biochim. et Biophys. Acta,* **24**, 461 (1957).
9. Goodall, M. C., *Nature,* **178**, 539 (1956).
10. Ebashi, S., Takeda, F. and Kumagai, F., *Folia Pharmcologica Japonica,* **51**, 107 (1955).
11. Kumagai, H., Ebashi, S. and Takeda, F., *Nature,* **176**, 166 (1955).

12. Ebashi, S., *Arch. Biochem. and Biophys. in press.*
13. Kielley, W. W. and Meyerhof, O., *J. Biol. Chem.,* **176**, 591 (1948).
14. Bowen, W. J. and Kerwin, T. D., *Biochim. et Biophys. Acta,* **17**, 522 (1955).
15. Noda, L., Kuby, S. A. and Lardy, H. A., *J. Biol. Chem.,* **209**, 191 (1954).
16. Kielley, W. W. and Meyerhof, O., *J. Biol. Chem.,* **183**, 391 (1950).
17. Macfarlane, M. G. and Knight, B. C. J. G., *Biochem. J.,* **35**, 884 (1941).
18. Fairgairn, D., *J. Biol. Chem.,* **157**, 633 (1945).
19. Ebashi, S., Takeda, F., Otsuka, M. and Kumagai, H., *Symposium Enzyme Chem.,* (Japanese), **11**, 11 (1956).

Discussion

Gergely, chairman

Gergely: Briggs and Portzehl suggested that relaxation could be obtained by a preparation, similar to the Kielley-Meyerhof granules, without adding any phosphate-transfer system. What do you think of this view?

Ebashi: I would like to present two possible explanations: either the fiber still contained an effective amount of phosphate-transferring-systems, or the granules themselves contained phosphate-transferring systems, especially myokinase. According to my experiments the granular fraction obtained by only one precipitation contained a sufficient amount of myokinase for relaxation.

Lorand: The actual data of Briggs and Portzehl as published in *Biochim. Biophys. Acta* last July, show that there is a need for creatine phosphate in the relaxation of even "single" fibers. We ourselves, proved that "single" fibers do not relax unless a phosphate donor (in our case phosphoenolpyruvate) is added. Furthermore, Portzehl's report (*Biochim. Biophys. Acta*, July, 1957) clearly states that the Marsh factor used in her experiments contained appreciable amounts of myokinase. How then can one discount the need for an ATP producing step in relaxation from the evidence so far?

W. W. Kielley: Did you study the relaxing activity of liver granules?

Ebashi: Yes, I did; I also tried the heart granules, but they were all ineffective.

PART VIII

PRECURSORS OF POLYMERS

Enzymatic synthesis of *macromolecules* had not been observed *in vitro* to any significant extent until thirty years ago, when Carl and Gerty Cori discovered α-glucose-1-phosphate (G-1-P) and glycogen phosphorylase. The isolation and chemical synthesis of "Cori ester" made it possible to observe *glycogen* synthesis in tissue extracts[1,2]. Since glucosyl groups of G-1-P react with the 4 hydroxyls groups of the *terminal* glucosyl units, the rate depends to a great extent on the glycogen concentration as well as the number of branches, each branch providing a terminal *recipient* point. The equilibrium depends exclusively on the ratio G-1-P/P. If the ratio is close to one, almost 80 per cent of the G-1-P is converted to glycogen.

Nevertheless, the intact cell does not take advantage of this pathway to any detectable extent[3]. Leloir and his coworkers discovered a derivative of G-1-P, UDP-glucose, which was found to be a powerful donor of glucosyl residues for glycogen biosynthesis[4,5] in the presence of the proper enzyme (UDPG-glycogen synthetase).

Leloir and his coworkers discovered a derivative of G-1-P, UDP-glucose, which was found to be a powerful donor of glucosyl residues for glycogen biosynthesis[4,5] in the presence of the proper enzyme (UDPG-glycogen synthetase).

In the biosynthesis of a highly complex polysaccharide like chondroitin sulfate, the sulfate stems from a specific active sulfate donor, which Hilz and Lipmann identified as a 3′ phosphate derivative of 5′ adenylsulfate[6].

Three important precursors involved in the biosynthesis of the *cell membrane* and the *cell wall* were discovered a decade ago. Park[7] described uridine diphospho N-acetyl muramic acid peptides in staphylococci treated with penicillin in the prelytic stage. Later work by Park and Strominger, and by Strange, showed the direct relevance of this type of nucleotide for cell wall biosynthesis. The other component of the rigid cell wall layer of the grampositive bacteria, ribitol, was found by Baddiley and his group, who also described the "ribityl donor", cytidine diphosphoribitol[8]. Various nucleotides important for the biosynthesis of the exposed cell wall surface lipopolysaccharides have been described by Nikaido, Ginsburg[10], and others.

An important precursor of phospholipid biosynthesis, cytidine diphosphocholin, the active form of "phosphorylcholine", was described in 1955 by Kennedy and Weiss[9].

Nucleic Acid Synthesis

The active precursor of *polyribonucleotides* can be either diphospho- or triphosphonucleosides, depending on the organism, the enzyme, and the nature of the polynucleotides. The diphosphonucleoside dependent polynucleotide synthetase from *Azetobacter Vinlandii* came, as is well known, to play an historic role in a great development (see Manago and Ochoa [q.v.]*). No less dramatic was the discovery of DNA polymerase from *E. coli* (see Kornberg [q.v.]). These developments are too fresh in mind to require much comment. Perhaps one aspect of the biosynthesis of nucleic acids should be emphasized. The electro-

*See under author's names in list of material reprinted in this section.

philic reactant is consistently a "nucleotidyl" residue which reacts with the nucleophilic partner, the 3-hydroxyl group of the pentose of a terminal nucleoside unit of the nucleic acid. This general mechanism also determines the polarity of the lengthening of the nucleic acid chain. The order of the bases in DNA and RNA is usually predetermined and is complementary to the bases of the primer strand according to the Watson-Crick model[11]. The weak interatomic forces (hydrophobic forces, hydrogen bonds, and electrostatic forces) are highly important contributors to the "individuality" of macromolecules. This applies especially to nucleic acids and to proteins.

Protein Synthesis

Early suggestions about the mechanism of peptide synthesis centered around reversal of peptide hydrolysis or around transpeptidation. The latter proved to be effective in the synthesis of a number of special low molecular peptides. But the key system to protein synthesis (which we owe largely to the early work of Hoagland, Keller, Loftfield, and Zamecnik[12]) was observed to be a more involved system than had been anticipated. Later, it became clear that the complicated mechanism actually represented a specific and highly sophisticated "translation" mechanism[13].

Comments on this subject, however, will be confined to a brief discussion of the chemical nature of active "amino acids." It is again worth recalling (as Zamecnik does in his Harvey Lecture[12]) Chantrenne's predictions of 1948, which grew out of his previously cited successful chemical synthesis of hippuric acid from *di*-substituted phosphorous compounds (*di*-benzoylphosphate and glycine), and which showed that di-benzoyl-phosphate is indeed a much more effective benzoyl donor than mono-benzoyl-phosphate[14]. These observations foreshadowed developments beginning seven to eight years later, when it was realized that the active amino acids are acyl compounds forming an anhydride with the phosphate group of 5-adenylic acid rather than with phosphate (see Hoagland and Berg). The amino acyl adenylate which is bound to the protein is a donor of amino acyl, or alternatively, of "adenyl," depending on the availability of enzymes and nucleophilic agents.

References

1. Cori., C. F., Colowick, S. F., and Cori, G. T.: *J. Biol. Chem.,* **121** (1937), 465.
2. ———, Schmidt, G., and Cori, G. T.: *Science,* **89** (1939), 464.
3. Robbins, P. W., Traut, R. R., and Lipmann, F.: *Proc. Natl. Acad. Sci.,* **45** (1959), 6.
4. Caputto, R., Leloir, L. F., Cardini, C. E., and Paladini, A. C.: *J. Biol. Chem.,* **184** (1950), 333.
5. Leloir, L. F., and Cardini, C. E.: *J. Amer. Chem. Soc.,* **79** (1957), 6340.
6. Hilz, H., and Lipmann, F.: *Proc. Natl. Acad. Sci.,* **41** (1955), 880.
7. Park, J. T.: *J. Biol. Chem.,* **194** (1952), 877, 885, 897.
8. Baddiley, J., Buchanan, J. G., Cass, B., and Mathias, A. P.: *J. Chem. Soc.,* **78** (1956), 4583.
9. Kennedy, E. P., and Weiss, S. B.: *J. Amer. Chem. Soc.,* **77** (1955), 250.

10. Ginsburg, V.: *Adv. Enzymol.,* ed. Nord, F. F.: **26** (1964), 35. New York: Interscience Publishers, John Wiley and Sons, Inc.
11. Watson, J. D.: *Molecular Biology of the Gene*. New York: Benjamin Company, Inc., 1965.
12. Zamecnik, P. C.: *The Harvey Lectures,* Ser. 54, "Historical and Current Aspects of the Problem of Protein Synthesis." New York: Academic Press, Inc., 1960, pp. 256–81.
13. Hoagland, M. B.: "The Relationship of Nucleic Acid and Protein Synthesis as Revealed by Studies in Cell-Free Systems," in *Nucleic Acids,* eds. Erwin Chargaff and J. N. Davidson, **3**, chap. 37. New York: Academic Press, Inc., 1960.
14. Chantrenne H.: C. R. Lab Carlsberg, *Ser. Chim.,* **26** (1948), 297.

THE ACTIVATING EFFECT OF GLYCOGEN ON THE ENZYMATIC SYNTHESIS OF GLYCOGEN FROM GLUCOSE-1-PHOSPHATE*

Gerty T. Cori and Carl F. Cori

Department of Pharmacology, Washington University,
School of Medicine, St. Louis, Missouri

(Received for publication, October 23, 1939)

The enzyme (glycogen)-phosphorylase which catalyzes the reaction, glycogen + $H_3PO_4 \rightleftharpoons$ glucose-1-phosphoric acid, has been found in mammalian tissues and in yeast. When glucose-1-phosphoric acid is used as substrate and the activity is to the left, a striking difference is noted between enzyme preparations from liver on the one hand and from skeletal, heart muscle, and brain on the other hand. In the case of liver phosphorylase activity starts immediately upon addition of glucose-1-phosphate and coenzyme (adenylic acid), while a lag period is observed with the enzymes prepared from the other tissues.† During the lag period the reaction proceeds at a barely measurable rate until quite suddenly a burst of activity sets in, suggesting an autocatalytic reaction. The lag period which is short (5 to 45 minutes) in comparatively crude enzyme preparations is greatly prolonged with further purification of the enzyme. In some muscle enzyme preparations a lag of 24 hours was observed and others seemed to have lost activity altogether. When these enzymes were tested with glycogen and inorganic phosphate as substrates, so that the reaction proceeded to the right, no lag was observed.

It was noted that liver phosphorylase preparations always included some glycogen (10 to 50 mg. per cent), while no or only doubtful traces of glycogen were found in the enzyme preparations from other tissues. It has been possible

*Gerty T. Cori and Carl F. Cori, "The Activating Effect of Glycogen on the Enzymatic Synthesis of Glycogen from Glucose-1-Phosphate," *J. Biol. Chem.*, **131**, 397–98 (1939). Reprinted with the permission of the authors and of the American Society of Biological Chemists, Inc.

†Cori, G. T., Cori, C. F., and Schmidt, G., *J. Biol. Chem.*, **129**, 629 (1939).

to abolish the lag period and to reactivate seemingly inactive enzymes by adding small amounts of glycogen; a concentration of 0.5 to 10 mg. per cent of glycogen, depending on the enzyme used, was sufficient to start enzyme activity. One may conclude that this enzyme which synthesizes a high molecular compound, glycogen, requires the presence of a minute amount of this compound in order to start activity.

Min.	Per cent of glucose-1-phosphate converted to glycogen	
	A	B
3		38.2
8	1.5	60.8
15	2.2	76.6
25		79.2
40	2.9	79.2

A representative experiment is shown in the accompanying tabulation. The enzyme was prepared from an aqueous extract of muscle by adsorption with aluminum hydroxide, elution with sodium glycerophosphate, concentration of the elution, and repeated precipitation with 0.3 saturated ammonium sulfate. The reaction mixture contained 0.014 M of glucose-1-phosphate and 0.001 M of adenylic acid adjusted to *p*H 6.9. In A no glycogen was added; in B glycogen was added to give a concentration of 10 mg. per cent.

NEUBERG AND CORI

Who is convincing whom?

[The late Professor Carl Neuberg was Director of Kaiser Wilhelm Institut für Biochemie, Berlin-Dahlem from 1925 to 1934, and subsequently, was Visiting Professor at New York University 1936–1950.]

DISCUSSIONS NOT ON THE AGENDA
from
A Symposium on Respiratory Enzymes, University of Wisconsin, 1941. (Courtesy of the University of Wisconsin Press.)

BIOSYNTHESIS OF GLYCOGEN FROM URIDINE DIPHOSPHATE GLUCOSE*

L. F. Leloir, J. M. Olavarria[1]*, Sara H. Goldemberg*[2] *and H. Carminatti*[3]

Instituto de Investigaciones Bioquímicas
Fundación Campomar y Facultad de Ciencias Exactas y Naturales
Buenos Aires, Argentina

(Received for publication, October 14, 1958)

Introduction

The biosynthesis of glycogen from UDPG[4] with a liver enzyme has been reported previously[(1)]. The reaction has been investigated further using a partially purified preparation from rat muscle with which the general properties of the system have been studied.

Methods

Analytical

Glycogen was estimated by the phenol-sulfuric acid method[(2)], after digestion with KOH and ethanol precipitation[(3)]. A sample of glycogen prepared as described by Somogyi[(4)] was used as standard. Its concentration was checked against glucose using the anthrone method[(5)]. UDP was esti-

*L. F. Leloir, J. M. Olavarría, Sara H. Goldemberg, and H. Carminatti, "Biosynthesis of Glycogen from Uridine Diphosphate Glucose," *Archives of Biochemistry and Biophysics,* **81**, 508–520 (1959). Reprinted with permission of the authors and of Academic Press.

This investigation was supported in part by a research grant (No. G-3442) from the National Institutes of Health, U. S. Public Health Service.

[1] Fellow of the Consejo Nacional de Investigaciones Científicas y Técnicas.

[2] Investigator of the Instituto Nacional de Microbiología.

[3] Investigator of the Comisión Nacional de Energía Atómica.

[4] Abbreviations used: UDPG: uridine diphosphate glucose; UDP: uridine diphosphate; G-1-P: glucose 1-phosphate; G-6-P: glucose 6-phosphate; Tris: tris (hydroxymethyl) aminomethane; EDTA: ethylenediamine tetraacetate.

mated as described by Cabib and Leloir[6], but halving the amounts of reagents. UDPG was measured spectrophotometrically with a partially purified UDPG dehydrogenase[7]. Phosphorylase was estimated as described by Cori *et al.*[8]. Protein was measured by the methods of Kunitz and McDonald[9] and of Warburg and Christian[10]. Amylase activity was determined under the same conditions as the glycogen-forming enzyme but without UDPG or G-6-P. After deproteinization with $Ba(OH)_2$ and $ZnSO_4$, the reducing substances were measured according to Park and Johnson[11]. G-6-P was estimated spectrophotometrically[12]. Radioactivity was measured with a gas-flow counter. Radioactive sugars in paper chromatograms were located with a silver reagent (13) and then eluted, plated, and counted. Approximately half of the added counts were detected after this treatment.

Substrates

UDPG was isolated from yeast as described by Pontis *et al.*[14]. UDPG labeled in the glucose moiety was prepared by incubating C^{14}G-6-P with UDPG and a crude *Saccharomyces fragilis* extract[15] and was purified by paper chromatography.

Enzymes

The method of Ballou and Luck[16] was used for the preparation of wheat β-amylase and that of Cori *et al.*[8] for phosphorylase. A crude preparation of maltase was obtained as described by Weidenhagen[17] and dialyzed.

Assay of the glycogen-forming enzyme

The standard reaction mixture contained 0.23 μmole UDPG, 0.5 μmole G-6-P, 0.4 mg. glycogen, 3.75 μmoles Tris-maleate buffer of pH 8.5, 0.25 μmole EDTA, and enzyme, in a final volume of 0.05 ml. Incubations were carried out at 37° for 30 min. The reaction was stopped by heating for 1 min. in boiling water, and the UDP formed was measured. Under the conditions of the test, added UDP did not disappear on incubation with crude or purified muscle extracts.

The initial rate of reaction was calculated from a two- or three-point time curve by applying the equation for first-order reactions and extrapolating the value of k to zero time.

Preparation of the enzyme

Rat muscle was used in most of the experiments. Only a small part of the activity was extracted with water from minced muscle. In order to obtain active extracts, it was necessary to homogenize the tissue thoroughly with a blendor. The extraction was usually carried out with water. With phosphate or pyrophosphate buffers the yield was slightly higher, but more inactive protein was extracted. The enzyme could be precipitated from the aqueous extracts either with 0.41 saturated ammonium sulfate or by adjusting the pH to 5.8–6.0.

The procedure which was used in most of the preparations was as follows. The muscles from two rats (about 55 g.) were cooled, minced, suspended in 3 vol. of cold water, and homogenized for 2–3 min. in a Waring blendor. The homogenate was immediately centrifuged at $12{,}000 \times g$ for 10 min. When this procedure was carried out quickly and at low temperature, the pH did not drop below 6.5.

The supernatant crude extract was divided into two portions. One (about 1/4) was heated at 100° for 5 min., and the precipitate was removed by centrifugation. The rest of the supernatant fluid (its volume is represented as v in the following) was acidified to pH 5.8—6.0 (chlorophenol red as indicator), and after 15 min. at 0° was centrifuged at $24{,}500 \times g$ for 10 min. The precipitate was suspended in water, adjusted to pH 5.8–6.0, and centrifuged. The precipitate was then suspended in $0.2v$ of heated extract and centrifuged again after freezing and thawing. The supernatant fluid, which was usually turbid and contained most of the activity, was acidified, centrifuged, and washed with water as previously. The precipitate ("second precipitate") was suspended in $0.05\, v$ of 0.015 M Tris—maleate buffer of pH 7.4 containing 0.005 M EDTA.

The activity of these preparations decreased about 50% after storage overnight at —15°. More stable preparations were obtained by lyophilizing the "second precipitate" and extracting with $0.05\, v$ of 0.1 M pyrophosphate buffer of pH 8 containing 0.65 mM reduced glutathione. The latter preparations were

Table I

RESULTS OF THE PURIFICATION PROCEDURE

Micromoles/hr./mg. protein

	Crude extract	Purified preparation	Ratio Purified/crude
UDP formation	1.4	17.5	12.5
Phosphorylase	75	34	0.45
Amylase (in glucose equivalents)	0.02	0.018	0.9
Protein concentration (*mg./ml.*)	38.7	8.2	—

completely colorless and transparent and were stable for at least a week when stored at pH 7 and —15°. The results obtained in the purification by the second procedure are shown in Table I. The purification obtained in that experiment was 12-fold with respect to protein, and the yield was about 10%. In other preparations the purification ranged from 20- to 40-fold.

Results

STOICHIOMETRY

The stoichiometry of the reaction catalyzed by the glycogen-forming enzyme is shown in Table II. It may be observed that, in the complete system, for each mole UDPG that was utilized about 1 mole UDP was formed and

1 mole glucose was added to glycogen. In the absence of G-6-P or glycogen, the chemical changes were much smaller.

Effect of increasing enzyme concentration

As shown in Fig. 1, increasing amounts of UDP were formed with increasing amounts of enzyme. It may be noted that for practical reasons the test system contained only 0.23 μmole UDPG, so that a linear relation was not to be expected. Furthermore, inhibition by UDP probably influences the course of the reaction.

Table II

Stoichiometry

Complete system as described in text, but with amounts doubled and 0.072 mg. glycogen.

	UDPG disappearance	UDP formed	Glycogen formed
	μmole	*μmole*	*μmole glucose*
Complete system	0.13	0.13	0.12
No G-6-P	0.05	0.03	0.05
No glycogen	0.015	0.02	—

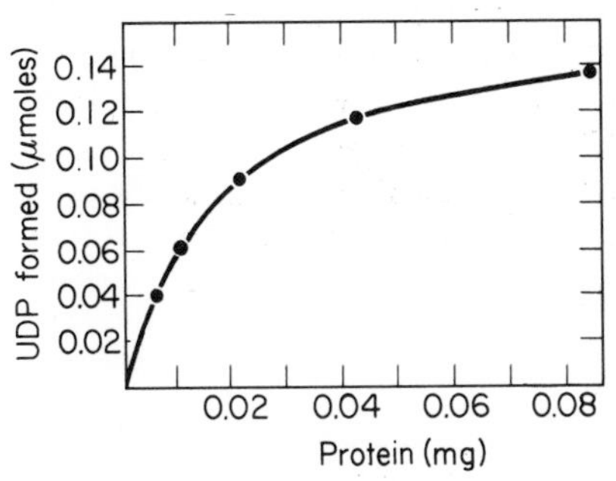

Fig. 1. The effect of enzyme concentration on the formation of UDP. Standard system as described in text.

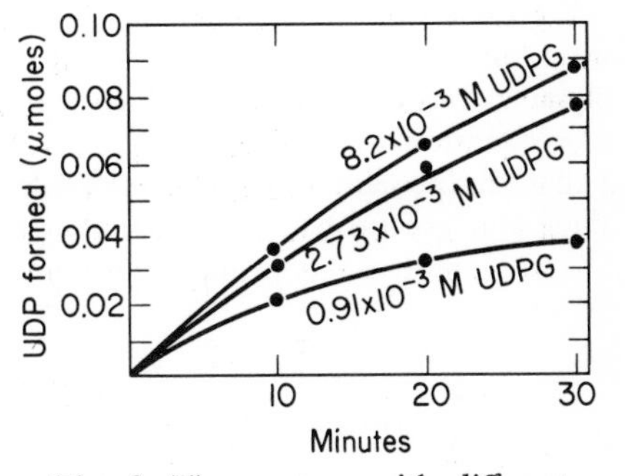

Fig. 2. Time curves with different amounts of UDPG. Standard system as described in text.

Effect of UDPG concentration

The course of the reaction with different UDPG concentrations is shown in Fig. 2. From these values, initial rates were calculated as described under *Methods,* and the method of Lineweaver and Burk[18] was applied. The value obtained for the Michaelis constant, which should be considered only approximate, was 5×10^{-4} *M*. It may be mentioned for comparison that the K_m for G-1-P in the phosphorylase reaction is 5.7×10^{-3} *M*[19].

pH Optimum

As shown in Fig. 3, the pH for maximal activity in Tris–maleate buffer was 8.3.

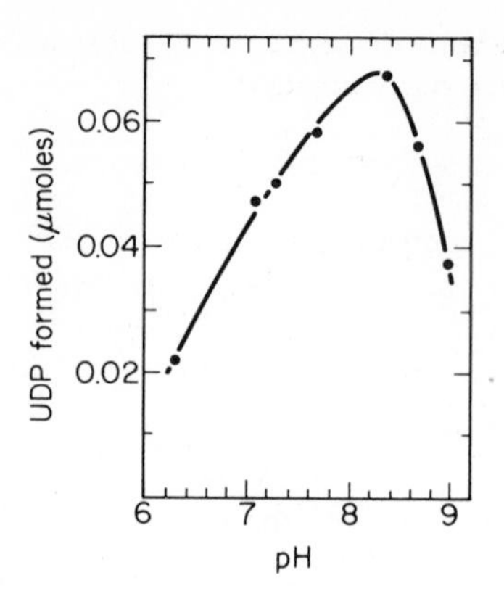

Fig. 3. pH-activity curve. Standard system as described in text, with Trismaleate buffers of different pH. The final pH was checked on aliquots with a glass electrode. The solutions were neutralized before proceeding to the enzymic estimation of UDP.

PRIMER REQUIREMENTS

Figure 4 shows the effect of glycogen concentration on the rate of UDP formation from UDPG. The results of experiments employing different polysaccharides as primers are shown in Table III. Application of a first-order equation to such results gave values which changed greatly with time so that the percentages given in Table III were calculated from single time values and are probably not strictly proportional to priming ability. It may be observed that the primer requirements of the glycogen-forming enzyme

Table III

ACTIVATING EFFECT OF DIFFERENT POLYSACCHARIDES

The results are expressed in per cent activation in relation to glycogen. The final concentration of polysaccharide was 5.4 mg./ml. in every case. Such a concentration of glycogen was just enough to produce maximal activity. All activities were tested with the same enzyme preparation. Numbers in brackets represent a second experiment. Incubation time: 30 min.

Polysaccharide	UDP formation from UDPG	P_i formation from G-1-P
Glycogen	100	100
Phosphorylase limit dextrin from glycogen[a]	59 (52)	28 (20)
β-Amylase limit dextrin from glycogen[a]	30 (31)	15 (20)
Glycogen treated with α-amylase.	0 —	6 —
Commercial soluble starch (blue with iodine)	17 (20)	5 (10)
Soluble starch[b] (red with iodine)	19 (13)	0 (0)
Potato starch (heated in alkali)	59 (48)	9 (20)

[a] Prepared as described by Hestrin (20).
[b] Obtained by treatment with acid according to Lintner (21).

are not very different from those of muscle phosphorylase. That is, the best primer is glycogen and its activity decreases after degradation. A more detailed study will be required to determine whether the differences observable in Table III are in fact real.

Substances which failed to give any detectable stimulation when tested as primers were: glucose, mannose, maltose, lactose, cellobiose, trehalose, raffinose, melibiose, and gentiobiose. Commercial dextrin (Difco) produced a slight activation.

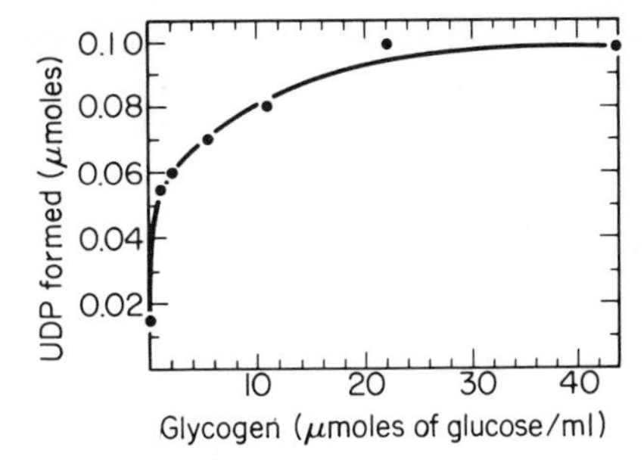

Fig. 4. The effect of varying glycogen concentration on enzyme activity. Standard system as described in text.

Activators and inhibitors

Phloridzin at 3.3×10^{-3} *M* concentration inhibited 48%. At the same concentration the inhibition of phosphorylase was 42%. These values were obtained from the change in initial rates calculated as described under *Methods*. With glucose at 0.05 *M* concentration, the inhibition was 52%, as compared to 72% for phosphorylase. The effect of added UDP was tested both by the standard method and by measuring glycogen formation. At 10^{-3} *M* of added UDP, the inhibition was 31%; and at 5×10^{-4} *M* it was 18%.

The reaction was inhibited 80% by 0.2 *M* potassium borate and 58% by 0.1 *M* KCN. Two antidiabetic substances 1-*p*-tolylsulfonyl-3-butylurea and 1-(*p*-aminobenzenesulfonyl)-3-butylurea (0.2% final concentration) inhibited slightly (15 and 8%, respectively). Cysteine or reduced glutathione produced a slight activation on some enzyme preparations. The following substances did not change the activity: Mg^{++}, G-1-P, adenosine triphosphate, adenosine 5′-monophosphate, galactose 1-phosphate, and insulin.

The action of glucose 6-phosphate

Partially purified preparations of the glycogen-forming enzyme were found to be activated by heated extracts. Heated pigeon muscle extracts were particularly active. The active substance proved to be stable in acid, labile in alkali, and behaved as an acid when treated with anion-exchange resins. Many known substances were tested as possible substitutes for the heated extract. Of these G-6-P and fructose 6-phosphate proved to be active. Since the preparation contained an active phosphoglucoisomerase, it could not be decided which of the two produces the activation. A sample of G-6-P obtained by chemical synthesis was found to be active. Many other substances were tested as possible substitutes for G-6-P. The following were ineffective: glucose, maltose, trehalose, fructose 1-phosphate, trehalose phosphate, sucrose phosphate, lactose, citrate, G-1-P, adenosine triphosphate, adenosine 5′-monophosphate, glutamic acid, inorganic phosphate, and galactose 1-phosphate. Activation was obtained with glucosamine 6-phosphate and with galactose 6-phosphate. The sample of the latter was contaminated with G-6-P, but the amount appeared to be too small to account for the activation.

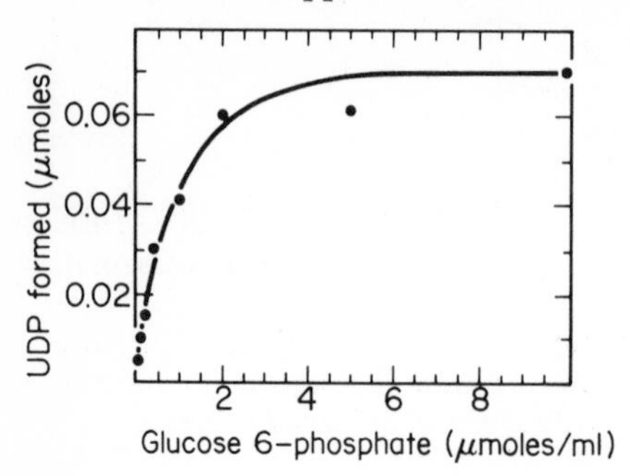

Fig. 5. The effect of changing G-6-P concentration. Standard system as described in text.

A curve showing the formation of UDP with different G-6-P concentrations is shown in Fig. 5. Half-maximal velocity was attained at about 6×10^{-4} *M* concentrations of G-6-P. In order to obtain information on the mechanism by which G-6-P increases the rate of reaction, some experiments were carried out with labeled compounds. As shown in Table IV, while the radioactivity of C^{14} UDPG was recovered in the glycogen, that of C^{14} G-6-P was not. Moreover, estimations of G-6-P + fructose 6-phosphate showed that there was no measurable change in concentration during the reaction (Table V).

Table IV

INCORPORATION OF RADIOACTIVITY INTO THE GLYCOGEN

Complete system as described in text except for the additions indicated and that the amount of glycogen was increased to 1.2 mg.

The reaction was stopped by adding 0.9 ml. of 33% KOH. After heating 20 min. at 100°, the glycogen was precipitated with 1.25 ml. ethanol, boiled, centrifuged, reprecipitated, and plated for counting the radioactivity.

Additions				Counts/min. in glycogen
UDPG		G-6-P		
μmole	*counts/min.*	*μmole*	*counts/min.*	
0.16	4,950	0.80	0	2,550
0.16[a]	4,950[a]	0.80	0	0
0.25	0	0.43	51,500	120
0.25[a]	0[a]	0.43	51,500	30

[a] Added after incubation.

Table V

NONDISAPPEARANCE OF G-6-P

Standard system as described under *Methods* except for the G-6-P concentration.

	Incubation time	G-6-P estimation	UPD formed
	min.	*μmole*	*μmole*
Complete system	0	0.097	0
Complete system	30	0.098	0.057
No glycogen	30	0.099	0.003

The assumption that hexose phosphate might act as primary acceptor of the glucose residue and that a disaccharide phosphate would then serve as donor to the glycogen, was examined by several methods. One was to add phosphopyruvate and pyruvate kinase during the reaction in order to favor the first step by removing the UDP. Even under these conditions the reaction required a polysaccharide primer, showing that the first step would not take place without the second, or that only one step is involved in the overall reaction. Other experiments consisted in adding disaccharide phosphates such as sucrose or trehalose phosphate instead of UDPG and testing for increase in glycogen. The results were negative.

Radioactive glycogen

Incubation of C^{14} UDPG led to the labeling of the glycogen. In a typical experiment 1.2 μmoles of C^{14} UDPG (12,400 counts/min.) were incubated with an eightfold dose of the standard mixture containing 4.8 mg. glycogen. After 30 min. at 37° the mixture was boiled in alkali, precipitated with 1.25 vol. ethanol, and reprecipitated. An aliquot was plated for measuring radioactivity. The number of counts per minute in the glycogen fraction for the total sample was 10,000. That is, the incorporation reached about 80%. It may be mentioned that the C^{14} UDPG preparation contained about 25% UDP-galactose, so that the incorporation obtained was complete within experimental errors. Controls in which the reaction was stopped at $t = 0$ gave no radioactivity.

Reaction products

The method used for the estimation of UDP is not very specific. Other nucleoside diphosphates and also adenosine monophosphate are active in this test system. However, uridine monophosphate is inert. In order to check the formation of UDP in the enzymic reaction, the products were run on paper with an ethanol–ammonium acetate solvent of pH 7.5 (22). A spot with the same mobility as UDP was observed in the zone corresponding to the complete sample, but not in that of controls where the reaction was stopped at zero time or incubated without UDPG. Furthermore, it may be mentioned that no inorganic phosphate was found to be liberated in the reaction, and that no formation of oligosaccharides could be detected by paper chromatography.

As to the characterization of the polysaccharide formed in the reaction, the usual methods could not be used because only small amounts of UDPG were available. Therefore enzymic degradation and paper chromatography were utilized. The procedure consisted in isolating the glycogen formed from C^{14} UDPG as previously described. The radioactive glycogen obtained was treated with β-amylase which is known to hydrolyze alternate $1 \rightarrow 4$ α-linkages yielding maltose. The reaction mixture was then treated with 3 vol. methanol. The soluble portion which contained all the radioactivity was run on paper with the pyridine–butanol–water solvent (6: 4: 3)[23]. As shown in Fig. 6A, the product obtained behaved like maltose with a trace of glucose.

All the radioactivity was found in the disaccharide spot. Since the aforementioned solvent did not separate the disaccharides, a sample which had been chromatographed was wetted with borate buffer and submitted to electrophoresis. As shown in Fig. 6B, the radioactivity was found in a spot having the same mobility as maltose. Electrophoresis was carried out after

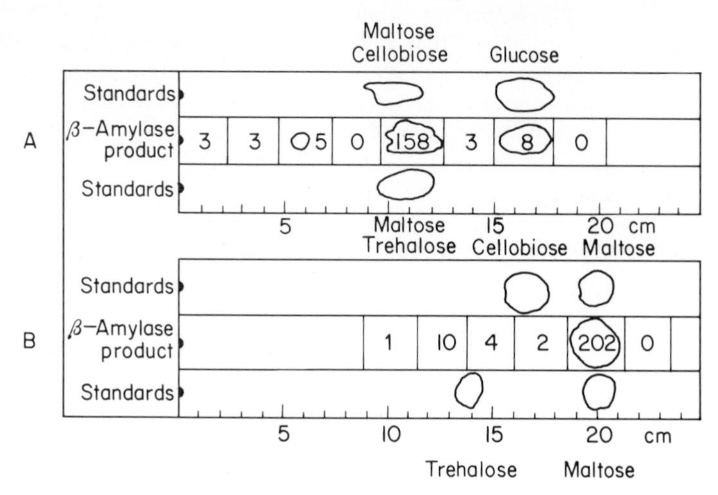

Fig. 6. (A) Paper chromatography of the product obtained by the action of β-amylase on the glycogen formed from radioactive UDPG. The glycogen was obtained as described in text. An aliquot was treated with β-amylase at pH 6 (20) for 20 hr. at 37°. Three volumes of methanol were added. The supernatant which contained all the radioactivity was spotted on paper and run with the butanol-pyridine-water solvent. The radioactivity of the sugar spots was then detected as described in text. The squares in the graph show how the papers were cut, and the numbers represent counts per minute after subtracting background. (b) Same as (A), but after chromatography the paper was wetted with 0.05 *M* borate buffer of pH 9 and submitted to electrophoresis at 600 v. for 6 hr.

Fig. 7. Paper chromatography of the product obtained by the action of β-amylase on radioactive glycogen submitted to the subsequent action of a maltase-containing yeast extract. General procedure as in Fig. 5.

paper chromatography because in that manner the interference of salts and residues of protein was obviated.

As an additional test, the product obtained by the action of β-amylase was treated with a maltase-containing yeast extract. As shown in Fig. 7, the radioactivity appeared in a spot migrating like glucose.

In other experiments the radioactive glycogen was degraded with phosphorylase and excess inorganic phosphate[20]. On precipitation with 3 vol. methanol, no radioactivity remained in the precipitate containing the limit dextrin. The soluble portion was freed from methanol by evaporation and separated into two fractions by $ZnSO_4$ and $Ba(OH)_2$ precipitation. The precipitate containing G-1-P was hydrolyzed with acid[24], and the free sugar formed was chromatographed on paper with the butanol–pyridine–water solvent. Radioactivity was found only in the glucose spot.

Distribution

The activity of extracts of different rat organs is shown in Table VI. It may be observed that more activity is found in muscle, heart, and liver than in the other organs. Positive results were also obtained with pigeon liver and muscle and with rabbit muscle.

Discussion

There are some facts which indicate that the classical scheme for glycogen synthesis may not be correct since it implies the action of phosphorylase, and the ratio of inorganic phosphate to G-1-P as measured in whole tissues is usually too high for net synthesis to take place[25]. Furthermore, agents which cause glycogen degradation produce a concomitant increase in phosphorylase concentration. Thus the glycogenolytic action of epinephrine[26,27] glucagon, and high Na^+ ions[28] has been attributed to an increase in phosphorylase activity. Other facts have also been interpreted as meaning that phosphorylase functions mainly in the degradation of glycogen[25,29]. If these conclusions are correct, then some other mechanism or mechanisms should be involved in glycogen synthesis and the enzyme described in this paper might well serve this purpose. In the intact rat liver the rate of glycogen formation is about 30 μmoles/hr./g.[30] and, as shown in Table VI, liver extracts under optimal conditions can catalyze the transfer of about 190 μmoles

Table VI

Distribution in Different Rat Organs

Results expressed as micromoles UDP formed/hr./g. tissue. Organs were homogenized in 3 vol. water.

Organ	UDP formed	Organ	UDP formed
Muscle	220	Spleen	37
Heart	166	Kidney	31
Liver	187	Lung	36
Brain	32		

glucose from UDPG/hr./g. tissue. The values for muscle are 5–17[31] and 220, respectively. In the whole organ the enzyme is, presumably, not working at saturating concentrations of substrate and activator so that the two values might be in fair agreement.

It may be mentioned that tissues have much more phosphorylase than the glycogen-forming enzyme; thus when the rate of glycogen formation from UDPG and from G-1-P was compared in crude muscle extracts under optimal conditions and in the presence of activators (G-6-P and adenylic acid, respectively), it was found that the latter process (phosphorylase reaction) was 20–50 times faster.

The tests which have been carried out on the radioactive glycogen formed from C^{14} UDPG indicate that the glucose residue becomes attached by $\alpha(1 \rightarrow 4)$ linkage. This is the same type of linkage which is hydrolyzed by phosphorylase.

As to the mechanism by which G-6-P increases the activity of the glycogen-forming enzyme, no clue was obtained. The effect was also obtained with fructose 6-phosphate, glucosamine 6-phosphate, and galactose 6-phosphate. No indication in favor of the intermediary formation of a disaccharide phosphate could be shown.

Summary

An enzyme which leads to the formation of glycogen according to the equation:

$$\text{UDPG} + \text{primer} \longrightarrow \text{UDP} + \text{glucosyl}\ \alpha\,(1 \longrightarrow 4)\ \text{primer}$$

has been studied.

The optimal conditions for activity were determined with a partially purified preparation from rat muscle. The reaction requires the presence of a polysaccharide as primer and is strongly activated by hexose 6-phosphates. Using UDPG labeled in the glucose moiety, it was found that the radioactivity was transferred to the glycogen from which it could be removed as maltose with β-amylase or as G-1-P with phosphorylase. Thus it seems that the glucose residue becomes linked $\alpha(1 \rightarrow 4)$ to the polysaccharide. Several inhibitors were tested as well as the occurrence of the enzyme in different organs.

References

1. Leloir, L. F., and Cardini, C. E., *J. Am. Chem. Soc.* **79**, 6340 (1957).
2. Dubois, M., Gilles, K. A., Hamilton, J. K., Rebers, P. A., and Smith, F., *Anal. Chem.* **28**, 350 (1956).
3. Montgomery, R., *Arch. Biochem. Biophys.* **67**, 378 (1957).
4. Somogyi, M., *in* "Methods in Enzymology" (S. P. Colowick and N. O. Kaplan, eds.), Vol. III, p. 3. Academic Press, New York, 1957.
5. Trevelyan, W. E., and Harrison, J. S., *Biochem. J.,* **50**, 298 (1952).
6. Cabib, E., and Leloir, L. F., *J. Biol. Chem.* **231**, 259 (1958).

7. Strominger, J. L., Maxwell, E. S., Axelrod, J., and Kalckar, H. M., *J. Biol. Chem.* **224**, 79 (1957).
8. Cori, G. T., Illingworth, B., and Keller, J. P., *in* "Methods in Enzymology" (S. P. Colowick and N. O. Kaplan, eds.), Vol. I, p. 200. Academic Press, New York, 1955.
9. Kunitz, M., and McDonald, M. R., *J. Gen. Physiol.* **29**, 393 (1945–46).
10. Warburg, O., and Christian, W., *Biochem. Z.* **310**, 384 (1941).
11. Park, J. T., and Johnson, M. J., *J. Biol. Chem.* **181**, 149 (1949).
12. Kornberg, A., and Horecker, B. L., *in* "Methods in Enzymology" (S. P. Colowick and N. O. Kaplan, eds.), Vol. I, p. 323. Academic Press, New York, 1955.
13. Trevelyan, W. E., Procter, D. P., and Harrison, J. S., *Nature* **166**, 444 (1950).
14. Pontis, H. G., Cabib, E., and Leloir, L. F., *Biochim. et Biophys. Acta* **26**, 146 (1957).
15. Trucco, R. E., *Nature* **174**, 1103 (1954).
16. Ballou, G. A., and Luck, J. M., *J. Biol. Chem.* **139**, 233 (1941).
17. Weidenhagen, R., *in* "Die Methoden der Fermentforschung" (E. Bamann and K. Myrbäck, eds.), Band II, p. 1749. George Thieme Verlag, Leipzig, 1941.
18. Lineweaver, H., and Burk, D., *J. Am. Chem. Soc.* **56**, 658 (1934).
19. Cori, C. F., Cori, G. T., and Green, A. A., *J. Biol. Chem.* **151**, 39 (1943).
20. Hestrin, S., *J. Biol. Chem.* **179**, 943 (1949).
21. Vanino, L., "Handbuch der Präparativen Chemie," Band II, p. 204. Ferdinand Enke, Stuttgart, 1937.
22. Paladini, A. C., and Leloir, L. F., *Biochem. J.* **51**, 426 (1952).
23. Jeanes, A., Wise, C. S., and Dimler, R. J., *Anal. Chem.* **23**, 415 (1951).
24. Cardini, C. E., and Leloir, L. F., *Arch. Biochem. Biophys.* **45**, 55 (1953).
25. Niemeyer, H., "Metabolismo de los hidratos de carbono en el hígado," Universidad de Chile, Santiago, 1955.
26. Sutherland, E. W., and Cori, C. F., *J. Biol. Chem.* **188**, 531 (1951).
27. Sutherland, E. W., *Ann. N.Y. Acad. Sci.* **54**, 693 (1951).
28. Cahill, G. F., Ashmore, J., Zottu, S., and Hastings, A. B., *J. Biol. Chem.* **224**, 237 (1957).
29. Beloff-Chain, A., Catanzaro, R., Chain, E. B., Masi, I., Pocchiari, F., and Rossi, C., *Proc. Roy. Soc.,* (London) **143**, 481 (1955).
30. Catron, L. F., and Lewis, H. B., *J. Biol. Chem.* **84**, 553 (1929).
31. Stadie, W. C., Haugaard, N., and Vaughan, M., *J. Biol. Chem.* **200**, 745 (1953).

ENZYMATIC SYNTHESES OF PYRIMIDINE AND PURINE NUCLEOTIDES. I. FORMATION OF 5-PHOSPHORIBOSYLPYROPHOSPHATE*[1]

Arthur Kornberg, Irving Lieberman and Ernest S. Simms

Department of Microbiology, Washington University
School of Medicine, St. Louis, Missouri

(Received for publication, March 5, 1954)

In studies on the incorporation of orotic acid into pyrimidine nucleotides, we have observed its conversion to uracil by liver preparations and a requirement for adenosine triphosphate (ATP) and ribose-5-phosphate (R5P) for this reaction.[2] With an enzyme preparation purified about 20-fold from extracts of pigeon liver acetone powder, the following reaction has been observed

$$\text{ATP} + \text{R5P} \longrightarrow \text{5-phosphoribosylpyrophosphate (PRPP)} + \text{adenosine-5'-phosphate (A5P)} \qquad (1)$$

With PRPP isolated from reaction (1) and with a partially purified enzyme preparation from yeast, we have been able to show that adenine is converted to A5P and that uridine-5′-phosphate (U5P) is formed from orotic acid. The evidence, to be presented at a later date, suggests the equations

$$\text{Adenine} + \text{PRPP} \rightleftharpoons \text{A5P} + \text{inorganic pyrophosphate (PP)} \qquad (2)$$

$$\text{Orotic acid} + \text{PRPP} \rightleftharpoons \text{orotidine-5'-phosphate} + \text{PP} \qquad (3a)$$

$$\text{Orotidine-5'-phosphate} \longrightarrow \text{U5P} + CO_2 \qquad (3b)$$

*Arthur Kornberg, Irving Lieberman, and Ernest S. Simms, "Enzymatic Synthesis of Pyrimidine and Purine Nucleotides. I. Formation of 5-Phosphoribosylpyrophosphate," *J. Amer. Chem. Soc.*, **76**, 2027–28 (1954). Reprinted with the permission of the authors and of the American Chemical Society.

[1] This investigation was supported by a research grant from the National Institutes of Health, Public Health Service.

[2] I. Lieberman, A. Kornberg, E. S. Simms, and S. R. Kornberg, *Federation Proc.*, in press. B. Karger and C. E. Carter have also obtained evidence for the conversion of orotic acid to uracil by liver extracts with uridylic acid as an intermediate (personal communication).

The further metabolism of U5P leading to the production of uracil and, in the presence of ATP, to the formation of uridine diphosphate (UDP) and uridine triphosphate (UTP) has been demonstrated with enzyme preparations from yeast and liver. These reactions, the mechanisms of which are now under investigation, are summarized in equations (4) and (5).

$$\text{U5P} \longrightarrow \text{uracil} \qquad (4)$$

$$\text{U5P} \longrightarrow \text{UDP} + \text{UTP} \qquad (5)$$

Balance studies in support of equation (1) are given in Table I.

Table I

STOICHIOMETRY OF PRPP SYNTHESIS[a]

	Micromoles		
	0 min.	60 min.	Δ
ATP[b]—Exp.	12.5	1.6(1.8)[e]	−10.9
—Control	13.1	12.5(11.1)[e]	− 0.6
PRPP[c]—Exp.	0.0	10.9(10.0)[e]	+10.9
—Control	0.0	0.0	0.0
A5P[d]—Exp.	0.0	9.8(10.2)[e]	+ 9.8
—Control	0.0	0.0(0.0)[e]	0.0

[a] The experimental (Exp.) incubation mixture (10.0 ml.) contained 0.40 ml. of ATP (0.03 *M*, 2.2 × 10^4 c.p.m./μ mole), 1.00 ml. of R5P (0.025 *M*), 0.20 ml. of reduced glutathione (0.5 *M*), 0.20 ml. of $MgCl_2$ (0.1 *M*), 0.50 ml. of KF (1 *M*), 0.20 ml. of phosphate buffer (1 *M*, *p*H 7.4) and 2.00 ml. of the enzyme preparation (containing 0.72 mg. of protein). The control incubation mixture lacked R5P. Incubation was at 35° for 60 min. [b] Determined spectrophotometrically by the combined action of hexokinase and glucose-6-phosphate dehydrogenase (A. Kornberg, *J. Biol. Chem.*, **182**, 779 (1950)). With added myokinase, the extent of TPN reduction was exactly doubled, indicating the absence of adenosine diphosphate (ADP). [c] Determined spectrophotometrically by the removal of orotic acid (see equation (3)), or the production of A5P (see equation (2)); methods are unpublished. [d] Determined spectrophotometrically by Schmidt's A5P deaminase (H. M. Kalckar, *J. Biol. Chem.*, **167** (1947), 445). [e] Values in parentheses were determined by chromatography on Dowex-1 anion exchange resin; ATP and A5P were estimated by optical density measurement at 260 mμ; PRPP was estimated as indicated in footnote (*c*).

ATP labeled with P^{32} in the two terminal phosphate groups was used. In the control sample (R5P absent), ATP was not removed to any significant extent and the production of PRPP and A5P was not detectable. In the presence of R5P, almost all the ATP added disappeared and was matched by the appearance of equivalent molar quantities of PRPP and A5P. No ADP or inorganic orthophosphate was produced from ATP after 30 min. or 60 min., when 65 or 87%, respectively, of the ATP was consumed.

PRPP was isolated by ion-exchange chromatography as a discrete symmetrical zone and estimated spectrophotometrically by enzymatic condensation with adenine (equation (2)), or with orotic acid (equation (3)). Eight fractions selected from this PRPP zone (representing approximately 80% of the PRPP) contained pentose, enzymatically active PRPP (equations 2 or 3), acid-labile P and total P in molar ratios (within 5% of the average value) of 1.00: 0.94: 2.04: 3.08. The average specific radioactivity (c.p.m./μmole) of these fractions was 2.24 × 10^4 as compared with values of 2.27 × 10^4 and

2.14×10^4, respectively, for the ATP at zero time and that isolated from the control sample.

A solution of PRPP, containing 1.45 μmoles of enzymatically active material was heated at 65° in 0.1 *M* acetate buffer, *p*H 4.0. After 10 and 40 minutes, respectively, 0.51 and 0.00 μmole of enzymatically active PRPP remained; 0.96 and 1.73 μmoles of reducing substance (referred to ribose) appeared. In another experiment involving a 30 min. heating period, the removal of 1.13 μmoles of PRPP was matched by the formation of 1.07 μmoles of reducing substance and 0.99 μmole of PP (determined by ion-exchange analysis); 0.30 μmole of inorganic orthophosphate was also produced.

These observations taken together with evidence, to be presented later, for the quantitative conversion of PRPP to A5P (equation (2)) or to U5P (equation (3)) lead us to propose a provisional structure of 5-phosphoribosyl-pyrophosphate for the activated ribose compound.

ADP did not replace ATP in equation (1) and ribose-1-phosphate (R1P) reacted at only 11% of the rate observed with R5P; this reactivity of R1P is likely due to its conversion to R5P by phosphoribomutase activity in the enzyme preparation. A sample presumed to contain ribose-1,5-diphosphate[3], derived from glucose-1,6-diphosphate[4] and R1P by the action of phosphoglucomutase[5] and glucose-6-phosphate dehydrogenase, was inactive in place of PRPP in equation (3).

It is evident that PRPP may also prove to be the intermediate involved in the synthesis of ribotides of acylic purine precursors,[6] nicotinamide[7] and other nitrogenous compounds, and in the system for A5P synthesis described by Saffran and Scarano.[8] There is the further possibility that a 2-deoxy PRPP will prove to be the active condensing agent in the biosynthesis of deoxynucleotides.

[3] H. Klenow, *Arch. Biochem.*, **46**, 186 (1953).

[4] Kindly furnished by Dr. L. F. Leloir.

[5] Kindly furnished by Dr. D. H. Brown.

[6] Greenberg, G. R., *J. Biol. Chem.*, **190**, 611 (1951); Williams, W. J., and Buchanan, J. M., *ibid.*, **203**, 583 (1953).

[7] I. G. Leder, and P. Handler, *ibid.*, **189**, 889 (1951).

[8] M. Saffran and Scarano, E., *Nature*, **174**, 949 (1953).

ENZYMATIC SYNTHESIS AND BREAKDOWN OF POLYNUCLEOTIDES; POLYNUCLEOTIDE PHOSPHORYLASE*[1]

Marianne Grunberg-Manago and Severo Ochoa

Department of Biochemistry, New York University
College of Medicine, New York

(Received for publication, April 7, 1955)

In the course of experiments on biological phosphorylation mechanisms[2] it was found that extracts of *Azotobacter vinelandii* catalyze a rapid exchange of P^{32}-labelled orthophosphate with the terminal phosphate of ADP,[3] IDP, UDP, CDP and (less rapidly) GDP. There is no reaction with the corresponding nucleoside triphosphates or monophosphates (tried ATP, ITP, AMP, IMP). The exchange is accompanied by the liberation of P_i and requires Mg^{++}. Employing the rate of the ADP-P_i exchange as an assay, the enzyme activity has been purified about 40-fold through ammonium sulfate fractionation and $Ca_3(PO_4)_2$ adsorption steps. The ratio of the rates of ADP-P_i exchange to P_i liberation remained constant.

On incubation of the purified enzyme with IDP, in the presence of Mg^{++},

*Marianne Grunberg-Manago and Severo Ochoa, "Enzymatic Synthesis and Breakdown of Polynucleotides; Polynucleotide Phosphorylase," *J. Amer. Chem. Soc.*, **77**, 3165–66 (1955). Reprinted with the permission of the authors and of the American Chemical Society.

Marianne Grunberg-Manago is Chargée de recherches, Centre National de la Recherche Scientifique, on leave of absence from the Institut de Biologie Physico-Chimique, Paris.

[1] Supported by grants from the U. S. Public Health Service, the American Cancer Society (recommended by the Committee on Growth, National Research Council), the Rockefeller Foundation, and by a contract (N6onr279, T.O. 6) between the Office of Naval Research and New York University College of Medicine. Presented at the April, 1955, meeting of the Federation of American Societies for Experimental Biology in San Francisco.

[2] M. Grunberg-Manago, and S. Ochoa, *Fed. Proc.*, **14**, 221 (1955).

[3] Abbreviations: diphosphates of adenosine, inosine, guanosine, uridine, and cytidine, ADP, IDP, GDP, UDP, and CDP; orthophosphate, P_i; adenosine and inosine monophosphates, AMP, and IMP; inosine-2′- and 3′-monophosphates, 2′-IMP and 3′-IMP; inosine diphosphatase, IDPase; trichloroacetic acid, TCA; tris (hydroxymethyl) aminomethane, Tris; specific activity, SA; micromoles μM.

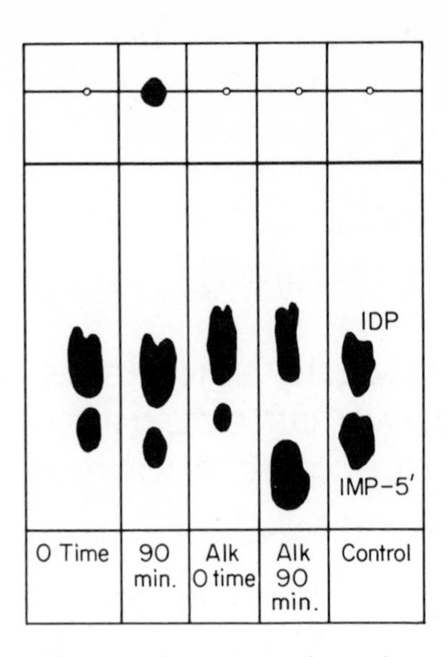

Fig. 1. Identification of product of reaction of IDP with *Azotobacter* enzyme by paper chromatography. Solvent system of Krebs and Hems; spots located by UV absorption. The three degrees of shading indicate strong, medium, and weak absorption respectively; 0.97 mg of enzyme (SA, 12) incubated 90 minutes at 30° with 25 μM. IDP in the presence of 12 μM. $MgCl_2$ and 90 μM. Tris buffer, *p*H 8.1; final volume, 2.5 ml. Mixture deproteinized by heating 1 minute at 100° and equal aliquots (without and with hydrolysis with 0.4 *N* KOH for 22 hours at 37°) used for chromatography. The IDP is contaminated with small amounts of 5′-IMP. Incubation results in decrease of IDP and appearance of an ultraviolet absorbing material which remains at the origin of the chromatogram. After alkaline hydrolysis, this material disappears and is replaced by a product migrating somewhat faster than 5′-IMP. This has been identified as a mixture of 2′- and 3′-IMP. The SA of the enzyme is defined as units/mg. protein. One enzyme unit catalyzes the exchange of 1.0 μM. of P_i^{32} with ADP in fifteen minutes at 30° under standard assay conditions. SA of initial enzyme extract was 0.3.

50–60% of the nucleoside diphosphate disappears with liberation of a stoichiometric amount of P_i. The missing nucleotide is accounted for by a water-soluble, non-dialyzable product which is precipitated by TCA or alcohol. Its solutions are rather viscous and exhibits a typical nucleotide ultraviolet absorption spectrum. Judging from its chromatographic behavior on Dowex anion exchange columns[4] the material is strongly acidic. It yields IMP (Fig. 1) on mild alkaline hydrolysis[5] and thus appears to be an IMP. 2′- and 3′-IMP have been identified as products of hydrolysis of the IMP polymer by alkali and

[4] Based on W. E. Cohn, This Journal, **72**, 1471 (1950).
[5] E. Vischer and E. Chargaff, *J. Biol. Chem.*, **176**, 715 (1948); C. E. Carter, This Journal **72**, 1466 (1950).

Table I

STOICHIOMETRY OF REACTION WITH IDP OR IMP-POLYNUCLEOTIDE

In experiment 1, 1.29 mg. of purified *Azotobacter* enzyme (SA, 10) as incubated with 24.8 μM. IDP, in the presence of 12 μM. $MgCl_2$ and 90 μM. Tris buffer, *p*H 8.1; final volume, 2.5 ml. Reaction was stopped by heating 1 minute at 100°. The IDP remaining in an aliquot of the supernatant was removed by hydrolysis to IMP and P_i with an excess (0.08 mg.) of purified ox liver IDPase[10] for 40 minutes and the enzyme destroyed by heating 1 minute at 100°. In experiment 2, an aliquot of the IDPase supernatant was incubated with 0.65 mg. of the *Azotobacter* enzyme (at *p*H 7.4). In experiment 3, 10.4 μM. (as mononucleotide) of a dialyzed solution of the IMP polynucleotide (isolated by TCA precipitation after incubation of IDP with *Azotobacter* enzyme as in experiment 1) was incubated with 0.65 mg. of enzyme (SA, 9) in the presence of 7 μM. $MgCl_2$ and 80 μM. Tris buffer, *p*H 8.1; final volume, 1.4 ml.; temp., 30° throughout. Values are expressed in μM. per ml. of reaction mixture. IDP was determined as the P_i liberated by IDPase; P_i was determined by the method of Lohmann and Jendrassik[7]; the polynucleotide was precipitated with TCA, dissolved in buffer, and determined from the light absorption at wave length 247 mμ at *p*H 7.0. This was corrected for the absorption ratio mononucleotide/polynucleotide (factor, 1.2) and expressed as mononucleotide. ϵ 247 for IMP was taken to be 13.2 at acid *p*H.[8]

Experiment	Incubation, min.	IDP	P_i	Polynucleotide
1	0	9.76	1.06	
	90	4.30	7.10	4.96[a]
	Δ	−5.46	+6.04	+4.96
2	0	0	14.2[b]	3.19
	60	2.3	12.0	1.20
	Δ	+2.3	−2.2	−1.99
3	0	0.69	8.10	7.50
	90	1.96	6.74	6.21
	Δ	+1.27	−1.36	−1.29

[a] Corrected for losses. [b] Some P_i contributed by *Azotobacter* enzyme solution.

5′-IMP by snake venom phosphodiesterase preparations.[9] This identification is based on (a) paper chromatography with the Krebs and Hems[6] and C80A[10] solvent systems, (b) liberation of P_i on hydrolysis for 20 minutes at 100° with 1.0 HCl,[11] and (c) behavior toward 5′- and 3′-specific nucleotidases.[9] These results suggest that 5′-mononucleotide units are linked to one another either through 2′- or 3′-phosphoribose ester bonds, or both, as in nucleic acid. Similar polymers have been obtained with the other nucleoside diphosphates so far tried (ADP, UDP).

[6] H. A. Krebs and R. Hems, *Biochem. et Biophys. Acta,* **12**, 172 (1953).

[7] K. Lohmann and L. Jendrassik, *Biochem. Z.,* **178**, 419 (1926).

[8] H. H. Kalckar, *J. Biol. Chem.,* **167**, 429 (1947).

[9] We are indebted to Dr. C. E. Carter and Dr. L. A. Heppel for generous gifts of snake venom preparations containing phosphodiesterase and 5′-nucleotidase (J. M. Gulland and E. M. Jackson, *Biochem. J.,* **32**, 590, 597 (1938); R. O. Hurst and G. C. Butler, *J. Biol. Chem.,* **193**, 91 (1951)), and of 3′-nucleotidase (L. Schuster and N. O. Kaplan, *J. Biol. Chem.,* **201**, 535 (1953)). The latter enzyme was a gift of Dr. Kaplan to Dr. Heppel.

[10] L. A. Heppel, personal communication. This solvent consists of 800 ml. saturated ammonium sulfate, 180 ml. sodium acetate, and 20 ml. isopropanol.

[11] C. E. Carter, ref. 5.

The reaction catalyzed by the *Azotobacter* enzyme is readily reversible. In the presence of the enzyme and Mg^{++}, the IMP-polynucleotide undergoes phosphorolysis to IDP. Table I shows the stoichiometry of the reaction with IDP in both directions. Phosphorolysis by the purified enzyme of nucleic acid isolated from *Azotobacter* has been shown through the incorporation of P_i^{32} and chromatographic identification of radioactive GDP, UDP, CDP, and ADP. Further, the labelled GDP and UDP were specifically hydrolyzed by IDPase.[12] The above results indicate that the new enzyme (or enzymes) catalyzes the reaction.

$$n\text{X–R–P–P} \rightleftharpoons (\text{X–R–P})_n + n\text{P}_i$$

where R is ribose and X may be adenine, hypoxanthine, guanine, uracil or cytosine, and suggest that, in analogy with polysaccharides, reversible phosphorolysis may be a major mechanism in the biological breakdown and synthesis of polynucleotide chains. Studies of the reaction with mixtures of several nucleoside diphosphates, the distribution of the enzyme (already known to be present in other microörganisms), and further work on its behavior toward natural nucleic acids, are in progress.

[12] G. W. E. Plaut, *Federation Proc.,* **14**, 263 (1955). We are indebted to Dr. G. W. E. Plaut for a generous gift of this enzyme. It catalyzes the hydrolysis of IDP, GDP, and UDP but is inactive on ADP and CDP.

ENZYMIC SYNTHESIS OF DEOXYRIBONUCLEIC ACID*

Arthur Kornberg, I. R. Lehman, Maurice J. Bessman and E. S. Simms

Department of Microbiology, Washington University School of Medicine, St. Louis, Missouri

(Received for publication, May 2, 1956)

We have reported[1] the conversion of ^{14}C-thymidine via a sequence of discrete enzymic steps to a product with the properties of DNA†.

$$\text{Thymidine} \xrightarrow{\text{ATP}} \text{T5P} \xrightarrow{\text{ATP}} \text{TTP} \xrightarrow{\text{ATP}} \text{“DNA”} \qquad (1)$$

The thymidine product is acid-insoluble, destroyed by DNAase, alkali-stable and resistant to RNAase. We have now extended these studies to include adenine, guanine and cytosine deoxynucleotides, and with partially purified enzymes from *E. coli* we have studied further the nature of the polymerization reaction.

^{32}P-labeled deoxynucleotides were prepared by enzymic digestion of DNA obtained from *E. coli* grown in a ^{32}P-containing medium; the nucleotides were then phosphorylated by a partially purified enzyme. The principal product of T5P phosphorylation was separated as a single component in an ion-exchange chromatogram and identified as TTP. The ratios of thymidine: acid-labile P: total P were 1.00: 2.03: 3.08. Enzymic formation of the di- and triphosphates of deoxyadenosine and the pyrimidine deoxyribonucleosides

*Arthur Kornberg, I. R. Lehman, Maurice J. Bessman, and E. S. Simms, "Enzymic Synthesis of Deoxyribonucleic Acid," *Biochem. Biophys. Acta,* **21**, 197–98 (1956). Reprinted with the permission of the authors and of the publisher.

†Abbreviations used are: DNA, deoxyribonucleic acid; ATP, adenosine triphosphate; T5P, thymidine-5′-phosphate; TDP, thymidine diphosphate; TTP, thymidine triphosphate; DNAase, deoxyribonuclease; RNAase, ribonuclease.

has been observed[2] and the presence of pyrimidine deoxyribonucleoside polyphosphates in thymus extracts has been reported[3].

Polymerization of TTP requires ATP, a heat-stable DNA fragment(s), provisionally regarded as a primer, and two enzyme fractions (called S and P; previously[1] called A and B, respectively) each of which has thus far been purified more than 100-fold (Table I). Preliminary studies suggest that TDP can replace TTP and has the same requirements for incorporation into DNA; a decision as to the more immediate precursor requires further purification of the system.

Table I

REQUIREMENTS OF THE PURIFIED SYSTEM

Extracts of *E. coli B* prepared by sonic disintegration were treated with streptomycin to yield a precipitate (fraction SP) and a supernatant fluid (fraction SS). Ammonium sulfate, gel and acid fractionation procedures applied to fractions SP and SS yielded fractions P and S, respectively, *E. coli* DNA was prepared by heating fraction SP (optical density at 260 $m\mu = 15$) at 70° for 10 minutes. To produce "primer," 0.1 ml of *E. coli* DNA was combined with 40 γ of fraction SP; after 1 hour at 37° in the presence of $5 \cdot 10^{-3} M$ $MgCl_2$, the mixture was heated for 10 minutes at 80°. The complete system contained (in 0.3 ml) 0.014 μmole of TTP ($1.5 \cdot 10^6$ c.p.m./μmole), 0.1 μmole of ATP, 0.10 ml of "primer," 10 γ of fraction S, 1 γ of fraction P, 1 μmole of $MgCl_2$, and 20 μmoles of glycine buffer, pH 9.2. After incubation for 30 minutes at 37°, 0.05 ml of crude *E. coli* extract ("carrier") and 0.3 ml of 7% perchloric acid were added. The precipitate was washed, plated and its radioactivity measured.

	mμmoles DNA-P/hour
Complete system	1.48
No ATP	0.20
No "primer"	0.11
No enzyme fraction S	0.07
No enzyme fraction P	0.04

"Primer" for the *crude* enzyme fraction was obtained (1) by the action of crystalline pancreatic DNAase on *E. coli* DNA or (2) on thymus DNA, or (3) by an *E. coli* enzyme fraction (SP) acting on DNA contained in it. However, "primer" for the *purified* enzyme fraction was obtained only with method (3); the action of pancreatic DNAase on either *E. coli* thymus DNA did not yield "primer". These findings imply the existence of an activity in the crude enzyme fraction responsible for the formation of active "primer". The chemical properties of the unpurified "primer" resemble those of a partial digest of DNA.

Utilization of the polyphosphates (presumably triphosphates) of adenine, guanine and cytosine deoxynucleosides for DNA synthesis occurs at rates approximately equal to those for TTP in crude enzyme fractions, but at appreciably slower rates with the enzyme purified for TTP polymerization (Table II). These changes in ratio suggest the presence of different enzymes for each of the deoxyribonucleoside triphosphates. Mixtures of these triphosphates, each tested at concentrations near enzyme saturation, gave additive or superadditive rates, further suggesting different enzymes for each of the substrates and a facilitation of polymerization by such mixtures.

Table II

CONVERSION OF FOUR DEOXYNUCLEOSIDE TRIPHOSPHATES

The incubation mixtures and assays were as described in Table I except that (I) the concentrations of deoxynucleoside triphosphates were $1.5 \cdot 10^{-5} M$, and (2) the crude enzymes were 60 γ of fraction SP and 240 γ of fraction SS.

Triphosphates	Tested with crude enzymes	Tested with purified enzymes (for TTP)
	mμmoles DNA-P/hour	
Thymidine (T)	0.8	5.48
Deoxyguanosine (G)	0.6	0.98
Deoxycytidine (C)	0.8	1.44
Deoxyadenosine (A)	0.6	1.28
T + G	2.2	14.6
T + G + C	4.4	19.6
T + G + C + A	6.4	22.0*
T + G + C + A (no "primer")	2.0	0.28

*65% conversion of substrate.

Studies are in progress to define the mechanism of the polymerization reaction and the linkages and sequences in the DNA-like produce formed. Further investigations with phage-infected *E. coli*[1] and studies with biologically active DNA may begin to clarify the question of how genetically speicfic DNA is assembled.

Fellowship support of I. R. Lehman by the American Cancer Society, of M. J. Bessman by the Public Health Service, and grants to A. Kornberg by the Public Health Service and the National Science Foundation are gratefully acknowledged.

References

1. A. Kornberg, I. R. Lehman, and E. S. Simms, *Federation Proc.*, **15** (1956) 291.
2. H. Z. Sable, P. B. Wilber, A. E. Cohen, and M. R. Kane, *Biochim. Biophys. Acta,* **13** (1954) 156. L. I. Hecht, V. R. Potter, and E. Herbert, *ibid.*, **15** (1954) 134.
3. R. L. Potter and S. Schlesinger, *J. Am. Chem. Soc.,* **77** (1955) 6714.

AN ENZYMIC MECHANISM FOR AMINO ACID ACTIVATION IN ANIMAL TISSUES*

Mahlon B. Hoagland

Medical Laboratories of the Collis P. Huntington
Memorial Hospital of Harvard University
Massachusetts General Hospital, Boston

(Received for publication, December 4, 1954)

Zamecnik and Keller[1] have shown that rat liver soluble protein fraction is essential for the incorporation of ^{14}C-labeled amino acids into the protein of microsomes in the presence of ATP**. The data presented in the present paper suggest a mechanism for activation of amino acids in this supernatant fraction.

^{32}PP, when incubated with ATP, dialysed 100,000·*g* rat liver supernatant, $MgCl_2$ and KF becomes incorporated into ATP†, ††. This incorporation is enhanced two to three fold by the addition of a complete complement of pure L-amino acids. The incorporation into ATP of ^{32}P and ^{32}PP in microsomes and of ^{32}P in supernatant is unaffected by amino acids.

The enhancement of exchange is dependent both on the total concentration and the number of amino acids added. If the total concentration of amino acids is held constant, at 0.04 *M*, the PP-ATP exchange increases as the number of L-amino acids is increased (Table I). 0.4 ml of a rat liver 105,000·*g* supernatant fraction, prepared by the method of Zamecnik and Keller[1] with minor modifications, dialysed 16 hours against the buffered sucrose medium in which the

*Mahlon B. Hoagland, "An Enzymic Mechanism for Amino Acid Activation in Animal Tissues," *Biochem. Biophys. Acta,* **16**, 288–89 (1955). Reprinted with the permission of the author and of the publisher.

This is publication No. 835 of the Cancer Commission of Harvard University.

**Abbreviations: ATP—adenosinetriphosphate; AMP—adenosinemonophosphate; PP—inorganic pyrophosphate; P—orthophosphate.

†The ^{32}PP was prepared by, and the gift of Dr. M. E. Jones.

††Pabst K_2 ATP was used routinely but similar results were obtained with the Sigma Chemical Company crystalline product.

Table I

The Effect of Increasing Numbers of Amino Acids on PP-ATP Exchange

Number of amino acids added	Counts per minute per μM AVP	% Exchange
0	4038	28
4	5660	40
8	7960	56
12	8270	58
16	8970	63
20	9360	66

homogenate was prepared, was incubated for 7 minutes at 37° in a volume of 1.1 ml with 5 μM ATP, 4 μM $MgCl_2$, 100 μM tris(hydroxymethyl) aminomethane buffer *p*H 7.6, 60 μM KF and 4 μM ^{32}PP, *p*H 7.5 containing 128,000 c.p.m. Every amino acid was chromatographically pure, they were added in groups of four, with a total concentration of 40 μM in each tube. (Changing the order of addition of the amino acids gave similar results.) The reaction was stopped with trichloroacetic acid and the ATP and PP separated and determined by charcoal adsorption according to the method of Crane and Lipmann[2]. Under the above conditions, no ATP or PP was hydrolysed. % exchange = counts per minute per μM ATP divided by total counts per minute per μM ATP plus PP · 100.

More than 2 μM/ml of each amino acid, on the average, does not further increase the exchange. Equivalent numbers and concentrations of D-amino acids are totally ineffective.

AMP is only slightly inhibitory to the system and ^{14}C-uniformly-labeled AMP fails to exchange with ATP*.

PP does not accumulate in the system in measurable quantity even though

Table II

ATP Loss, and Hydroxamate Appearance in the Presence of Amino Acids and Hydroxylamine (in μM/ml)

Addition	Hydroxamate formed	ATP lost	Pi formed
—	0	2.31	4.64
NH_2OH	0.34	1.39	2.78
AA	0	2.25	4.51
AA + NH_2OH	0.69	2.25	4.51
Δ due to AA alone	0	0	0
Δ due to AA in presence of NH_2OH	0.35	0.86	1.73

*Prepared from Schwarz ^{14}C-uniformly-labeled ADP enzymically (hexokinase and myokinase) and isolated by paper electrophoresis. The ^{14}C ADP was generously supplied by Dr. John Gergely.

the PPase activity is completely inhibited by KF, and PP-ATP exchange is proceeding. However, in the absence of KF and in the presence of high concentration of *salt-free* NH_2OH, a mixture of L-amino acids produces a considerably increased "hydrolysis" of ATP and a corresponding appearance of two equivalents of P. Concomitantly there is produced a significant quantity of hydroxamic acid (Table II). Rat liver was homogenized with 1.5 volumes of 0.05 *M* KCl and the 100,000·*g* supernatant from this homogenate was dialysed for 20 hours against 100 times its volume of 0.05 *M* KCl. 1.0 ml of the supernatant was then incubated for 30 minutes at 37°C in a final volume of 2.0 ml with 10 μM MgK_2 ATP, 100 μM tris buffer *p*H 7.6, and 2400 μM of freshly prepared salt-free hydroxylamine. An equimolar mixture of 12 amino acids (each of which alone failed to produce measurable quantities of hydroxamate—phe, val, leu, try, i-leu, lys, thr, ser, alan, his, arg, gly) was used at a total concentration of 12 μM/ml. Analyses:

1. hydroxamate was measured directly on a 1.0 ml aliquot of the reaction mixture by the method of Lipmann and Tuttle[3], using as an internal standard a mixture of the hydroxamates of six amino acids (leu, i-leu, val, lys, gly, ala).*
2. After killing the enzyme in the remaining 1.0 ml by boiling 2 minutes, P was determined by the method of Fiske and Subbarow on a 0.1 ml aliquot and compared to a zero-time control.
3. ATP was similarly determined on a 0.1 ml aliquot hydrolysed for 12 minutes in 1 *N* HCl. (In the absence of KF, all PP formed is split to P during the incubation.)

D-amino acids are inert here also. The appearance of hydroxamates, like the exchange, is dependent on the number of amino acids added, as well as the concentration. A purified enzyme from this supernatant, when incubated with leucine and NH_2OH, has yielded a product indistinguishable from leucine hydroxamate by paper chromatography.

These findings permit the following tentative formulation of the process by which amino acids are activated for protein synthesis. If E_1 is an activation site on an enzyme for a particular amino acid (AA_1), this site binds ATP in such a way as to labilize the AMP-pyrophosphoryl linkage:

$$E_1 \text{—}\ulcorner\qquad\qquad\urcorner\text{—} + \text{ATP} \rightleftharpoons E_1 \text{—}\ulcorner \text{AMP} - \text{PP} \urcorner\text{—} \tag{1}$$

The amino acid then displaces the PP:

$$E_1 \text{—}\ulcorner \text{AMP} - \text{PP} \urcorner\text{—} + \text{AA}_1 \rightleftharpoons E_1 \text{—}\ulcorner \text{AMP} - \text{AA}_1 \urcorner\text{—} + \text{PP} \tag{2}$$

These two equilibria would permit PP-ATP exchange to occur only in the presence of amino acids, would account for the failure of AMP to inhibit the reaction or to exchange with ATP and the failure of PP to accumulate in the absence of an acceptor. The additive effect on exchange of different amino acids suggests a separate activation site for each.

In the presence of NH_2OH, the carboxyl-activated amino acid reacts to form the hydroxamate, while AMP and PP accumulate, since the enzyme is regenerated to cycle again:

$$E_1 \text{—}\ulcorner \text{AMP} - \text{AA}_1 \urcorner\text{—} + \text{NH}_2\text{OH} \longrightarrow E_1 + \text{AA}_1 - \text{NHOH} + \text{AMP} \tag{3}$$

* American Cancer Society Scholar in Cancer Research.

(The possibility of the AMP ~ AA dissociating from the enzyme is not ruled out and is consistent with the results if the dissociation were of small extent.)

The above formulation is similar to that of Maas[4] for the enzymic synthesis of pantothenic acid in a purified *E. coli* extract where detailed stoichiometry of the reaction has been worked out.

The natural intracellular counterpart of NH_2OH in reaction (3) above might be expected to be the amino group of amino acids or peptide chains in the microsomes where arrangement of amino acid sequence and condensation of peptide chains would occur[5]. Examination of the interrelationship of these two fractions is in progress in this laboratory.

The author wishes to thank Dr. F. Lipmann, Dr. P. C. Zamecnik and Dr. E. B. Keller for valuable discussions.

References

1. P. C. Zamecnik and E. B. Keller, *J. Biol. Chem.*, **209** (1954) 337.
2. R. K. Crane and F. Lipmann, *J. Biol. Chem.*, **201** (1953) 235.
3. F. Lipmann and L. C. Tuttle, *J. Biol. Chem.*, **159** (1945) 21.
4. W. K. Maas and G. D. Novelli, *Arch. Biochem. Biophys.*, **43** (1953) 236 and personal communication.
5. E. B. Keller, P. C. Zamecnik, and R. B. Loftfield, *J. Histochem. and Cytochem.*, **2** (1954) 378.

PARTICIPATION OF ADENYL-ACETATE IN THE ACETATE-ACTIVATING SYSTEM*[1]

Paul Berg

Department of Microbiology, Washington University
School of Medicine, St. Louis, Missouri

(Received for publication, April 21, 1955)

The exchange of inorganic pyrophosphate[2] (PP) with the terminal phosphates of ATP may be mediated by reversible pyrophosphorolysis of dinucleotide coenzymes[3] or by recently postulated mechanisms involving enzyme-intermediate compounds.[4,5] In an attempt to isolate and define more precisely the nature of these enzyme-intermediate compounds, an enzyme has been purified from yeast which catalyzes the exchange between $P^{32}P^{32}$ and ATP with acetate as an obligatory requirement (Table I).
The enzyme also carries out the over-all synthesis of acetyl-CoA according to reaction 1.

$$\text{ATP} + \text{acetate} + \text{CoA} \rightleftharpoons \text{acetyl-coA} + \text{A5P} + \text{PP} \quad (1)$$

Acetate appears to act catalytically in exchange reaction; thus 0.04 μM. of acetate catalyzed the incorporation of 0.15 μM. of $P^{32}P^{32}$ into ATP.

*Paul Berg, "Participation of Adenyl-Acetate in the Acetate-Activating System," *J. Amer. Chem. Soc.*, **77**, 3163–64 (1955). Reprinted with the permission of the author and of the American Chemical Society.

[1] This work was carried out under the tenure of a Postdoctoral Fellowship and Scholarship of the American Cancer Society and was supported by funds from the U. S. Public Health Service and the National Science Foundation.

[2] The following abbreviations have been used: PP, inorganic pyrophosphate; ATP, adenosine triphosphate; CoA, coenzyme A; A5P, adenosine-5′-phosphate.

[3] A. Kornberg, *J. Biol. Chem.*, **182**, 779 (1950); A. W. Schrecker, and A. Kornberg, *ibid.*, **182**, 795 (1950); A. Munch-Petersen, H. M. Kalckar, E. Cutolo, and E. E. B. Smith, *Nature,* **172**, 1036 (1953).

[4] M. E. Jones, F. Lipmann, H. Hilz, and F. Lynen, THIS JOURNAL, **75**, 3285 (1953).

[5] M. B. Hoagland, *Biochim. et Biophys. Acta,* **16**, 288 (1955).

Table I

REQUIREMENTS FOR $P^{32}P^{32}$ EXCHANGE WITH ATP

Reaction mixture: 100 μM potassium phosphate buffer, *p*H 7.5; 5 μM. $MgCl_2$; 2 μM. ATP; 2 μM. $P^{32}P^{32}$; 2 μM. potassium acetate; 1 unit purified enzyme; volume, 1.0 ml.; temperature 37° for 20 minutes; ATP separated from $P^{32}P^{32}$ by charcoal adsorption[6]; 1 unit of enzyme forms 1 μM. acetyl-CoA in 20 minutes.

		Minus			
Components	Compl.	ATP	Mg^{++}	Acetate	Enzyme
μM. $P^{32}P^{32}$ inc.	0.55	0.01	0.01	0.02	0.00

That this exchange is associated with the acetyl-coA forming activity is supported by the fact that the ratio of acetyl-coA formation[7] to PP-ATP exchange increased with purification when acetate was omitted from the exchange assay, but remained essentially constant when acetate was present.

Evidence has now been obtained which supports the following mechanism for reaction 1.

$$\text{ATP} + \text{acetate} \rightleftarrows \text{adenyl-acetate} + \text{PP} \quad (2)$$

$$\text{Adenyl-acetate} + \text{coA} \rightleftarrows \text{acetyl-coA} + \text{A5P} \quad (3)$$

A compound with the properties of adenylacetate was prepared by the reaction of di-silver adenylate and acetyl chloride and partially purified by ion-exchange chromatography.[8] This preparation was enzymatically converted to ATP^{32} in the presence of $P^{32}P^{32}$ and the purified acetate-activating enzyme (Table II). In the absence of enzyme, there was no detectable formation of ATP and no significant disappearance of hydroxylamine reacting material. The ATP formed was characterized chromatographically, spectroscopically, and by its phosphorylation of glucose with hexokinase.[3]

Table II

STOICHIOMETRY OF CONVERSION OF ADENYLACETATE TO ATP

Reaction mixture: 50 μM. potassium phosphate buffer, *p*H 7.5; 5 μM. $MgCl_2$; 2.0 μM. $P^{32}P^{32}$; 0.75 μM. adenyl acetate (based on conversion to ATP); 1 unit of enzyme; volume, 1.0 ml.; temperature 37°.

	ATP[a]	PP	Adenyl acetate[b]
Δ 30 minutes	+0.28	....	−0.30
Δ 60 minutes	+0.42	−0.47	−0.43

[a] ATP separated from $P^{32}P^{32}$ by charcoal adsorption. [b] Determined by the method of Lipmann and Tuttle.[9]

Adenyl acetate was converted to ATP at approximately the same rate as the PP exchange with ATP. Under comparable conditions, 0.31 μM. $P^{32}P^{32}$ was incorporated into ATP and 0.23 μM. of adenyl acetate was converted to ATP.

[6] R. K. Crane, and F. Lipmann, *J. Biol. Chem.*, **201**, 235 (1953).
[7] M. E. Jones, S. Black, R. M. Flynn, and F. Lipmann, *Biochim. Biophys. Acta*, **12**, 141 (1953).
[8] I am grateful to Dr. David Lipkin, Washington University, for suggesting the method of preparation of adenyl acetate.
[9] F. Lipmann, and L. C. Tuttle, *J. Biol. Chem.*, **153**, 571 (1944).

The formation of acetyl-coA from adenyl acetate and coA is supported by the following evidence. The addition of 2 μM. of CoA to the reaction mixture (see Table I) caused an 80% inhibition of the exchange reaction and a 95% inhibition of the conversion of adenyl acetate to ATP. This is consistent with a competition of coA with PP for the adenyl acetate. Furthermore, the product, enzymatically formed from adenyl acetate and coA, disappeared in the presence of phosphotransacetylase and arsenate.[10] The formation of acetyl-coA was dependent on the presence of coA and the enzyme, and the arsenolysis was dependent on the phosphotransacetylase. Starting with 0.28 μM. of adenyl acetate, there was a formation of 0.19 μM. of acetyl-coA as measured by the disappearance of hydroxylamine reacting material in the presence of arsenate and phosphotransacetylase. Further experiments on the isolation and characterization of the acetyl-coA are in progress.[10a]

It seems likely, therefore, that the primary reaction in the activation of acetate and perhaps of the higher fatty acids[11] involves an acyl group cleavage of ATP, resulting in the liberation of pyrophosphate and in the formation of an acyl adenylate. This formulation is supported by the recent O^{18} exchange studies of Boyer[12] and by Jencks[13] who obtained a coA-independent activation of octanoic and similar fatty acids by pyrophosphate split of ATP. Definitive proof of this general mechanism must await experiments demonstrating the enzymatic synthesis of adenyl acetate from ATP and acetate (reaction 2), and the formation of adenyl acetate from acetyl-coA and A5P (reaction 3).

During the course of the above work a methionine requiring PP–ATP exchange system was also purified from yeast.[14] It is possible that this and the previously reported pantoic[15] and amino acid activated PP-ATP exchanges[5] may occur by the formation of the corresponding adenyl-acyl group derivatives. Thus the pantothenate peptide bond formation may represent a variation of the acetate activation with the amino group of β-alanine serving as the acyl group acceptor in place of the sulfhydryl group of coA.

[10] E. R. Stadtman, G. D. Novelli, and F. Lipmann, *ibid.,* **191**, 365 (1951).

[10a] *Note added in proof:* Further evidence for reaction (3) was the quantitative enzymatic conversion of adenyl-acetate to acetyl-CoA as measured by the malic dehydrogenase-condensing enzyme system (Stern, Shapiro, Stadtman and Ochoa, *J. Biol. Chem.,* **193**, 703 (1953); I am deeply indebted to Dr. S. Ochoa for a gift of crystalline condensing enzyme). The acetyl-CoA formed was also characterized by the enzymatic acetylation of *p*-nitro aniline with a partially purified acetylating enzyme from pigeon liver (Tabor, Mehler and Stadtman, *ibid.*, **204**, 127 (1953).

[11] A. Kornberg, and W. E. Pricer, Jr., *ibid.*, **204**, 329 (1953); G. R. Drysdale, and H. A. Lardy, *ibid.,* **202**, 119 (1953); H. R. Mahler, S. J. Wakil, and R. M. Bock, *ibid.,* **204**, 454 (1953).

[12] P. D. Boyer, O. J. Koeppe, W. W. Luchsinger, and A. B. Falcone, *Fed. Proc.,* **14** 185 (1955).

[13] W. P. Jencks, *ibid.*, **12**, 703 (1953).

[14] P. Berg, unpublished.

[15] W. K. Maas, quoted by F. Lipmann, *Science,* **120**, 855 (1954).

PART IX

REGULATION OF ENERGY METABOLISM

The Pasteur-Meyerhof Reaction

MORE THAN 100 years ago, Pasteur demonstrated that the fermentation of yeast either recedes or stops completely under aerobic conditions. He did not commit himself, however, to a particular mechanism which would suppress fermentation.

It was Meyerhof who, in 1919, pointed out the close relationship between fermentation in yeast and lactic acid formation in animal tissues. He reported that lactic acid formed in frog muscles under anaerobic conditions seemed, under aerobic conditions, to be resynthesized to glycogen. His chemical data were consistent with Hill's myothermic measurements of the ratios between initial heat and recovery heat.

Meyerhof[1] attributed the lack of accumulation of lactic acid in muscles working aerobically to a continuous resynthesis of lactic acid to glycogen. This resynthesis supposedly kept pace with the generation of lactic acid. During the aerobic recovery period, about one-fifth of the lactic acid which had accumulated during anaerobic work was oxidized, while the other four-fifths was reconverted to glycogen. This cycle has been called the "Meyerhof Cycle" or the "lactic acid cycle."

In 1930 Carl and Gerty Cori studied the cycle of lactic acid in the liver. They found that lactic acid (D-lactic acid) is a particularly efficient source of glycogen in the liver[2]. They formulated a cycle in which the glycogen in working muscles is broken down to lactic acid which, via the blood stream, finds its way to the liver where it is reconverted to glycogen[2], a cycle now known as the "Cori cycle."

At that time it was taken for granted that muscle invariably generates large amounts of lactic acid during a work period. A variety of new questions arose with Lundsgaard's discovery of alactacid muscle contractions. If lactic acid formation is not a prerequisite for mechanical work, what happens under normal conditions? Under conditions of aerobic work when respiration is able to meet the demands of the working load, is lactic acid generated at all? Perhaps aerobic metabolism arrests lactic acid formation. After all, respiration is much more economical with respect to its utilization of carbon than is fermentation. During a prolonged work period demanding increased consumption of fuel, the organism would gain by switching off the expensive emergency power supply, glycolysis, as soon as an increased oxygen supply, adequate for the increased aerobic metabolism, becomes available.

That this type of regulation actually takes place in the intact organism was demonstrated in 1936 by Ole Bang of the Copenhagen Physiological Laboratories in his studies on muscle metabolism in man[3]. Bang made quantitative observations on physically trained subjects performing work of medium intensity over an extended period of time on an ergometer bicycle (the so-called "Krogh ergometer cycle"). A brief initial burst of lactic acid accumulation in the capillary blood invariably occurred, but during continuation of the working period the blood lactic acid levels, surprisingly enough, fell markedly, sometimes even reaching resting values before termination of the working period. These observations are best explained by the assumption that lactic acid is formed only during the first minutes of the working period, and that as soon as the general and local blood circulation has adapted to the increased demand for fuel, thus permitting adequate respiration, further formation of lactic acid ceases (Lipmann, [q.v.]*).

If respiration suppresses or even arrests glycolysis, the question arises whether oxygen per se may exert a direct inhibition on one or more of the enzymes that catalyze the Embden-Meyerhof pathway of lactic acid formation. Biochemical models of possible mechanisms by which aerobic oxidations might directly interfere with fermentative pathways were developed by 1935 by Fritz Lipmann [q.v.]. The approach actually had been outlined a decade earlier by Otto Warburg. It was he who coined the name "the Pasteur reaction" for the arrest of glycolysis by respiration (see the reference to Pasteur's publication in Warburg's article[(4)]). Warburg also found that ethyl carbylamine is a specific and powerful agent for interrupting the Pasteur reaction[(4)].

One of the most interesting examples of interference with the Pasteur reaction remains the old discovery by Warburg[(4)] that this reaction is highly defective in tumor cells. This phenomenon, which remains one of the major puzzles in cell biology, is somewhat beyond the scope of this book.

As appears from the Lipmann review, dyes like dichlorophenol indophenol, which oxidizes sulphydryl compounds to di-sulphides, bring about a strong inhibition of glycolysis in the presence of oxygen. Lipmann [q.v.] quotes Rapkin's observations on the striking inhibition of S-S glutathione of the oxido-reduction between phosphoglyceraldehyde and pyruvate. The review also cites the early work by Bumm and Appel and Quastel and Wheatley in which it was found that cysteine and SH-glutathione unleash aerobic glycolysis in slices of animal tissue. These observations and later ones (especially by Fromageot and Chaix, cited by Lipmann) make it clear that SH compounds interrupt the Pasteur reaction. Conversely, S-S compounds or dyes which may catalyze the formation of S-S compounds work the other way. Thus, Dickens (Lipmann, [q.v.]) found that the typical "Warburgian" glycolysis in sarcoma was inhibited by the oxidizing dye pyocyanine. This inhibition depended on the presence of oxygen. The Lipmann review spells out the interpretation and possible importance of oxidation-reduction processes for the Pasteur reaction and regulation of anaerobic carbohydrate metabolism.

Allosteric Inhibitions as Basis for Regulation of Metabolic Pathways

It later became clear that lactic acid production can be regulated by mechanisms other than oxidation-reduction processes. Allosteric transitions of enzymes like glycogen phosphorylase or phosphofructokinase, brought about by nucleotides or simple sugars, also seem to be important to the regulation of muscle metabolism.

Phosphorylase *a* is active in the absence of adenylic acid, whereas phosphorylase *b* activity depends on the presence of this nucleotide. This observation,

*See under authors' names in list of material reprinted in this section.

made by the Coris almost 30 years ago, later became central to the interpretation of allosteric regulation phenomena.

The action of phosphorylase *a* combined with that of the so-called debranching enzyme (cf. G. T. Cori[5]) can account for the extensive lactic acid formation in anaerobic working muscle in which a large part of the glycogen is broken down.

In 1943 Gerty Cori and Arda Green showed that phosphorylase *a* can be converted to phosphorylase *b* by a muscle enzyme (the "PR" enzyme), or by crystalline trypsin[5]. In 1953, Keller and Cori [q.v.] established that phosphorylase *b* has a molecular weight of only about half the molecular weight of phosphorylase *a*. A reversible equilibrium reaction was later described by Sutherland and his coworkers[6] and by Krebs and Fischer[7].

It was first realized by Krebs and Fischer[7] that the presence of calcium ions, even in minute amounts (as could be encountered, for instance, in experiments using filter paper for filtration), will initiate a conversion in the extracts of the adenylate-dependent phosphorylase *b* to the fully active phosphorylase *a*. The target of this activation by calcium is actually the phosphorylase *b* kinase, an enzyme which catalyzes the phosphorylation reaction of phosphorylase *b*, a step crucial for its conversion to phosphorylase *a*[8].

In order to arrive at a clear decision about the type of alterations which might occur during the transition from rest to contraction, problems additional to those encountered in experiments on muscle extracts had to be considered. The problem of obtaining a muscle in a truly resting state is a tricky one, as was first realized by Fletcher and Hopkins more than 60 years ago. Freezing a resting muscle, so often used in the study of the biochemistry of resting muscle, only contributes to perturb the resting muscle, even to the point of eliciting strong contractions. A special technique of a rapid deep freeze is a prerequisite for arresting the muscle in the resting state. This was accomplished particularly successfully in the experiments described by Danforth *et al.* [q.v.]. The outcome clearly confirmed the observations by Krebs and Fischer[7]; resting muscle contains largely the *b* form of phosphorylase which is practically inactive since it depends on the availability of 5-adenylate, a nucleotide which is not normally present in either resting or working muscle. In the active, contracting state the transition of *b* to *a* by the activation of the phosphorylating mechanism presumably is brought about primarily by a mobilization of calcium ions, which act as activators of phosphorylase *b* kinase. Altogether this ingenious type of modulation mechanism dramatically regulates the availability of fuel (i.e., muscle glycogen) during work. The biological role of phosphorylase in rapidly making glycogen available for energy metabolism therefore seems quite clear.

It should be pointed out that Krebs, Kent, and Fischer[8] described essentially a new type of function of ATP: Phosphorylation of an enzyme in the service of regulation. The reaction can be formulated as follows:

$$\text{2 phosphorylase } b\text{(serine)} + \text{2 ATP} \xrightarrow[\text{kinase}]{\text{phosphorylase}}$$

$$\text{1 phosphorylase } a\text{(2P-serine)} + \text{2 ADP}$$

In the equation "serine" refers to a serine of the protein molecule, phosphorylase *a* having 2 phosphoserines. Reversible conversion of phosphorylase *a* to phosphorylase *b* is catalyzed by a highly specific phosphatase[9] (alias Coris' PR enzyme).

It was mentioned that UDP glucose-glycogen synthetase serves in glycogen synthesis, phosphorylase *a* in breakdown[10]. Sutherland [q.v.] subsequently discovered a new nucleotide, 3′,5′ cyclic adenylic acid, which is crucial to the activation of phosphorylase. Sutherland and coworkers[11] as well as Krebs and Fischer[7], found that the conversion of phosphorylase *a* to phosphorylase *b* requires ATP, a specific kinase as well as 3′,5′ cyclic adenylate. The latter apparently activates the phosphorylase kinase (Sutherland and Rall[11]). These recent developments are well known; it seems, therefore, unnecessary to elaborate more on this topic.

Another enzyme of the Cori-Embden-Meyerhof pathway, phosphofructokinase, first described by Zacharias Dische in 1935[12] has recently been found to be subject to complicated metabolic regulation. In this connection the reviewer recalls his discussions of 25 years ago with Dr. Carl Cori on the peculiar phenomenon of accumulation of hexose monophosphate in iodoacetate-poisoned muscles. At that time it had been established that iodoacetate specifically inhibits glyceraldehydephosphate dehydrogenase[13]. Accordingly, a muscle extract poisoned with iodoacetate should, upon addition of glycogen, accumulate hexose diphosphate and triosephosphate, but not hexosemonophosphate. This is actually the case. Yet, in the original experiments by Lundsgaard in iodoacetate-treated intact muscle performing anaerobic work before the onset of rigor, hexosemonophosphate was found to accumulate in considerable amounts. The mono-ester formation was dominant at the time ATP levels were sustained by transphosphorylation from phosphocreatine. Not until the onset of rigor did hexosediphosphate accumulation become dominant; by this time the ATP has been spent[14].

The two steps which govern the conversion of hexosemonophosphate to hexosediphosphate are the conversion of glucose-6-phosphate to fructose-6-phosphate and the phosphorylation of the latter by ATP to fructose-1,6-diphosphate. Since iodoacetate in intact muscle most effectively practically completely blocks the dehydrogenation of triosephosphate (a fission product of fructose-1,6-diphosphate), the two above-mentioned steps clearly could not be rate limiting. We therefore reasoned that the accumulation of hexosemonophosphate in an iodoacetate-treated muscle performing mechanical work must be due to a regulatory inhibition of the phosphorylation step to diester. The enzyme phosphofructokinase catalyzes this phosphorylation step which can be formulated as follows:

$$\text{ATP} + \text{fructose-6-phosphate} \longrightarrow \text{ADP} + \text{fructose-1,6-diphosphate}.$$

Prior to their discovery of the phosphorolysis of glycogen to hexosemonophosphate (glucose-1-phosphate and glucose-6-phosphate), the Coris had advanced ideas about hexosemonophosphate accumulation in intact muscle—ideas which were forerunners of the later development of regulatory interactions of hormones.

At the International Congress of Physiology in Leningrad in 1935, the Coris presented some interesting observations on accumulation of hexosemonophosphate in resting muscle under anaerobic conditions[15], cf. also[2]). They showed that the hormone epinephrine causes the levels of hexosemonophosphate and lactic acid in the anaerobic muscle to rise at the expense of glycogen, with no drop in the levels of phosphocreatine or ATP. This increase is augmented by a combination of dinitrophenol and epinephrine. In 1943 Engelhardt and Sakov[16] suggested an inhibition of phosphofructokinase as a possible basis for the suppression of glycolysis seen in the Pasteur reaction.

About 20 years later, Sutherland ([q.v.] and [6]) showed that epinephrine stimulates glycogen phosphorylase, and Lardy and Parks and others[17, 18] found that excess ATP, especially on a background of low levels of inorganic phosphate, strongly inhibits phosphofructokinase. The recent work by Lowry and his coworkers ([q.v.] and [19]) illustrates the ingenious complexity of feedback inhibitions. There seem to be two sites for ATP, an active site and an inhibitory site. Moreover, inorganic phosphase, 5′-adenylic acid, fructose-6-phosphate and fructose-1,6-diphosphate, which do not exert any inhibition, can compete with ATP for the second site. When they occupy the second site, ATP inhibition subsides. Later developments from the same laboratory[19] point to the existence of many more sites. For instance, citrate inhibits phosphofructokinase from brain[21]. "Phosphofructokinase seems to be responsive on the one hand to the balance between a high-energy phosphate expenditure and high-energy phosphate formation, and on the other, to the metabolic mixture offered to the cell."[20]

Lynen has discussed problems of the phosphate cycle in a review[21] dealing largely with alcohol fermentation. He reminds us of the effect of polymetaphosphate formation from inorganic phosphate and of the abolition of the Pasteur effect after uncoupling of oxidative phosphorylation by 2,4-dinitrophenol.

Regulation of Some Respiratory Cycles

Respiration depends strikingly on the availability of phosphate acceptors, especially in animal tissues. The general importance of the availability of phosphate acceptors like glucose or ADP for the rate of respiration in mitochondria has been clearly demonstrated by Britton Chance, as illustrated in the graph taken from one of his publications[22]. It can be seen that addition of ADP brings about a rapid reoxidation of DPNH and an increase in oxygen consumption.

Although the respiratory cycles of the cell seem so perfect, and especially the Krebs cycle, H. L. Kornberg came to realize that they may very well be "leaky" and hence would be in need of replenishment by independent routes. This brings us to a brief discussion of the so-called Cooper-Kornberg reaction. The existence of an efficient pathway for synthesis of phosphopyruvate from pyruvate[23] may be of major importance for keeping a metabolic cycle like the Krebs cycle operating in cells, especially in growing cells. It was found[23] that *E. coli* grown on lactate contain an enzyme that catalyzes the reaction

$$\text{pyruvate} + \text{ATP} \longrightarrow \text{phosphoenolpyruvate} + \text{AMP} + \text{P}$$

This reaction proceeds practically to completion in the direction of phosphopyruvate, 5′ AMP and inorganic phosphate. Phosphopyruvate generated can be

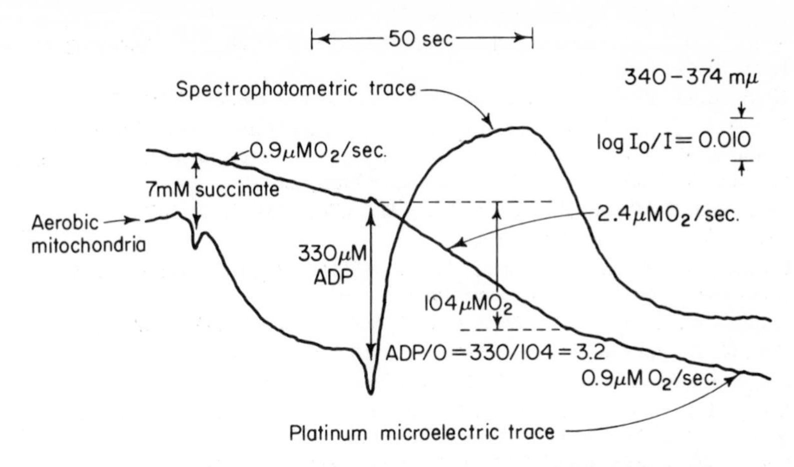

Fig. 1. Changes in the steady-state oxidation-reduction levels of DPN caused by the addition and subsequent exhaustion of ADP. The reaction kinetics are recorded by the double-beam spectrophotometer and the oxygen utilization is recorded by a vibrating platinum electrode. Note that the decrease of respiration and the increased DPN reduction are closely synchronized. DPN reduction indicated by a downward deflection.

subsequently converted to oxaloacetate. The Cooper-Kornberg reaction provided, therefore, an independent inlet to the Krebs cycle.

The Cooper-Kornberg reaction may also be crucial to an important phosphorylation system discovered recently by Roseman and his group[24]. In this phosphorylation system phospho-enol-pyruvate acts as the phosphoryl donor, and the histidine group of a heat-stable protein functions as a provisional acceptor. This reaction is catalyzed by a specific enzyme (so-called "enzyme I"). The phosphoryl histidine enzyme catalyzes subsequent transphosphorylation. This system, which seems to be critical to active sugar transport in *E. coli,* is actually driven exclusively by phospho-enol-pyruvate. The level of phospho-enol-pyruvate may, however, depend on the ATP level by means of the above-mentioned Cooper-Kornberg reaction.

The phosphorylation of phosphoglucomutase by glucose-1, 6-diphosphate[25] may also serve regulatory purposes.

References

1. Meyerhof, O.: *Pflügers Arch. gesant. Physiol.,* **175** (1919), 20; *Pflügers Arch. gesant. Physiol.,* **182** (1920), 284.
2. Cori, C. F.: *Physiol. Rev.,* **11** (1931), 143.
3. Bang, O.: *Skand. Arch. Physiol.,* **74**, Suppl. 10 (1936), 51–82; Lundsgaard, E.: *The Harvey Lectures,* **32** (1937), 65–88.
4. Warburg, O.: *Über den Stoffwechsel der Tumoren.* Berlin: Springer-Verlag, 1926; Warburg, O.: *Biochem. Z.,* **172** (1926), 432.
5. Cori, C. F. and Cori, G. T.: *Les Prix Nobel in 1947.* Stockholm: Norstedt Sönner, 1949, p. 126.

6. Sutherland, E. W., Jr., and Wosilait, W. D.: *Nature,* **175** (1955), 168.
7. Krebs, E. G., and Fischer, E. H.: *J. Biol. Chem.,* **216** (1955), 113; Fischer, E. H. and Krebs, E. G., *ibid.,* p. 121.
8. Krebs, E. G., Kent, A. B., and Fischer, E. H., *J. Biol. Chem.,* **231** (1958), 73.
9. Graves, D. J., Fischer, E. H., and Krebs, E. G., *J. Biol. Chem.,* **235** (1960), 805.
10. Leloir, L. F., and Cardini, C. E.: in *The Enzymes,* eds. P. D. Boyer, H. Lardy, and K. Myrbäck, **6** (1962), 317. New York: Academic Press, Inc.
11. Sutherland, E. W., Jr. and Rall, T. W.: *J. Biol. Chem.,* **232** (1958), 1077.
12. Dische, A.: *Biochem. Z.,* **280** (1935), 248.
13. Parnas, J. K., Ostern, P., and Mann, T.: *Nature,* **134** (1934), 1007; Racker, E.: *Mechanisms in Bioenergetics.* New York: Academic Press, Inc., 1965.
14. Lundsgaard, E.: *Biochem. Z.,* **227** (1930), 51; *Biochem. Z.,* **233** (1931), 322.
15. Cori, G. T., and Cori, C. F.: *J. Biol. Chem.,* **116** (1936), 119, 129.
16. Engelhardt, V. A., and Sakov, N. E.: Biokhimiya, **8** (1943), 9.
17. Lardy, H. A., and Parks, R. E., Jr.: in *Enzymes: Units of Biological Structure and Function,* ed. O. Gaebler. New York: Academic Press, Inc., 1956, p. 584.
18. Bueding, E., and Mansour, J. M.: *J. Biol. Chem.,* **237** (1962), 629.
19. Passonneau, J., and Lowry, O. H.: *Biochem. Biophys. Res. Commun.,* **13** (1963), 372.
20. Parmeggiani, A., and Bowman, R. H.: *Biochem. Biophys. Res. Commun.,* **12** (1963), 268.
21. Lynen, F.: "Phosphatkreislauf und Pasteur Effekt." *Proc. Internat. Sympos. Enz. Chem.,* Maruzen, Tokyo, 1957, p. 25.
22. Chance, Britton: "Interaction of ADP with Respiratory Chain," in *Enzymes: Units of Structure and Function,* ed. H. Lardy. New York: Academic Press, Inc., 1956, Fig. 8, p. 454.
23. Cooper, R. A., and Kornberg, H. L.: *Biochem. Biophys. Acta,* **104** (1965), 618–20.
24. Kundig, W., Ghosh, S., and Roseman, S.: *Proc. Natl. Acad. Sci.,* **52** (1964), 1067; Kundig, W., Kundig, F. D., Anderson, B. and Roseman, S.: *J. Biol. Chem.,* **241** (1966), 3243.
25. Najjar, V. A. and Pullman, M. E., *Science,* **119** (1954), 631.

LIPMANN AND MEYERHOF

Why is the Pasteur effect?

[The late Professor Otto Meyerhof, Nobel Laureate, 1925, was Director, Kaiser Wilhelm Institut für Biologie, Berlin-Dahlem, 1923–1930; Director, Kaiser Wilhelm Institut für Medizinische Forshung, Physiologie in Heidelberg, 1931–1938; Visiting Professor, University of Pennsylvania, 1940–1950.]

DISCUSSIONS NOT ON THE AGENDA
from
A symposium on Respiratory Enzymes, University of Wisconsin, 1941.
(Courtesy of the University of Wisconsin Press.)

PASTEUR EFFECT*

Fritz Lipmann

Department of Biological Chemistry, Massachusetts
General Hospital and the Department of Biological Chemistry,
Harvard University School of Medicine, Boston

With respect to their dependence on oxygen supply, organisms may be classified into[1] strict aerobes, equipped only with respiratory metabolic systems,[2] strict anaerobes, equipped only with anaerobic fermentative metabolic systems, and[3] facultative organisms, equipped with both respiratory and fermentative systems. This commonly used classification should not be followed too rigidly, however, for intermediate states between the main classes are common in nature, and adaptive interconversion has been widely observed.

The organisms in each of the first two groups rely exclusively on one form of energy supply, respiratory or fermentative, respectively. The third group, however, has developed the two mechanisms side by side. It is with this latter group that we shall deal, and more specifically with the interrelation between their respiratory and fermentative mechanisms.

Most doubly equipped organisms possess in the Pasteur effect a regulatory device that enables them to use, as occasion demands, either their aerobic or their anaerobic systems. By the operation of this effect their fermentative apparatus is blocked in the presence of sufficient oxygen, and energy is furnished almost exclusively by the far more efficient and powerful respiratory apparatus. When oxygen is lacking, however, the fermentation system is brought into operation.

The following example may serve to illustrate the energetic structure of a facultative anaerobic organism. A power plant uses as a source of energy

*Fritz Lipmann, "Pasteur Effect," *A Symposium on Respiratory Enzymes*. Madison: Univ. of Wisconsin Press (1941–42), pp. 48–73. Reprinted with the permission of the author and of the publisher.

cheap water power; this may be compared to the "cheap" respiratory energy. But because of seasonal variations of flow the water power may not be entirely reliable and hence as a safeguard against a deficiency in the supply of power a more expensively operating steam engine is built into the plant; this may be compared to "expensive" fermentation. For obvious reasons the plant will be equipped with a switch mechanism—its "Pasteur effect"—which keeps the steam engine from functioning so long as the water flow supplies sufficient energy but throws it into operation when water power is lacking.

An impressive physiological example of a mechanism utilizing both respiratory and fermentative energy supply is the muscle. Figures 1 and 2, after experiments by Bang[1], reproduce measurements of oxygen consumption and lactic acid formation (blood lactate) on human beings during physical work. During a prolonged period of not too hard work (Figure 1) blood lactate at first increases moderately but returns, during the first quarter of the period, almost to the resting level. An anaerobic energy supply is observed only at the beginning, when an adequate oxygen supply is lacking, and until respiration climbs to the equilibrium level. For excessive short-term work the picture is different, as shown in Figure 2. An excessive and long-continued increase of blood lactate signifies a large expenditure of anaerobic energy. In such a situation the adaptation is much too slow to supply oxygen in time, and the muscle has to rely almost entirely on the anaerobic energy of glycolysis.

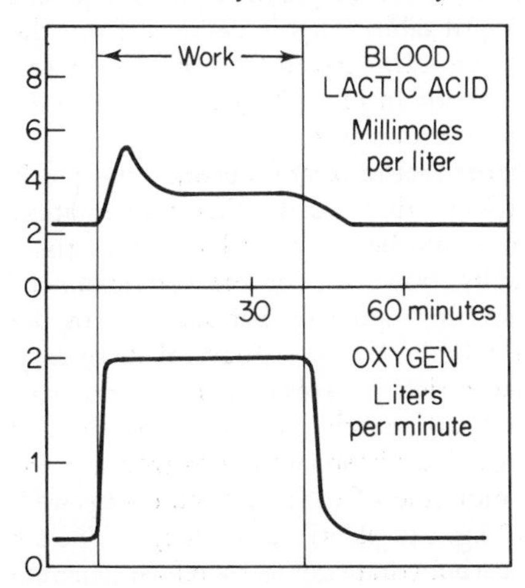

Fig. 1. Oxygen consumption and blood lactate with moderate work. The slight initial rise of lactate output coincides with the period of adaptation, before the oxygen consumption rises to the equilibrium level. (After O. Bang, ref. 1.) [Both abscissae signify identical time scales related to work or rest periods; the denotation "Oxygen liters per minute" simply refers to rates of oxygen consumption at various periods.]

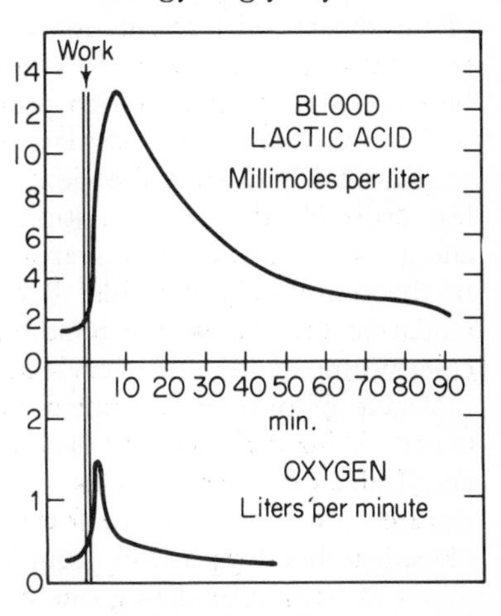

Fig. 2. Oxygen consumption and blood lactate with strenuous work. The excessive lactate formation signifies predominantly anaerobic energy supply. A moderate rise of oxygen consumption occurs first after completion of the work in the period of restitution. The slow fall of lactate in the blood indicates a relatively slow removal of lactate by resynthesis or oxidation. (After O. Bang, ref. 1.) [See editor's comments to Fig. 1.]

The Efficiency of Aerobic and Anaerobic Metabolism

Fermentations are energy-yielding rearrangements of the atoms constituting the glucose molecule. These are oxidation-reduction reactions in which, after cleavage, one part of the molecule is oxidized at the expense of the other part, which accordingly is reduced. Energetically probably the most efficient reaction is one such as the propionic acid fermentation: $3\,C_6H_{12}O_6 = 4\,C_2H_5COOH + 2\,CH_3COOH + 2\,CO_2 + 2\,H_2O$. The energetic yield is 61 kg.-cal. of heat per mole of glucose. According to Burk[2], the change in free energy is approximately 18 kg.-cal. higher per mole of glucose than the heat exchange. The probable maximum for a fermentative breakdown of carbohydrate thus amounts to 79 kg.-cal, per mole. This is 11.5 per cent of the 686 kg.-cal. to be obtained by respiratory breakdown. The more common lactic acid and alcoholic fermentations do not reach this maximum but yield only 54 kg.-cal. (36 kg.-cal. plus 18 kg.-cal, entropy change), or 7.9 per cent of the heat of combustion.

These values represent the theoretical maximum for fermentation and respiration. To compare the efficiencies of the two reactions in the cell we must know how much of the energy of each reaction is actually available to the cell. In a recent paper[3] the author has pointed out that from 40 to 70 per cent of the theoretical fermentation energy is utilizable. This is deduced from the fact that in the muscle up to 40 of the 54 kg.-cal. derived from glycolysis can be stored as four energy-rich phosphate bonds in phosphagen, the energy of which is utilizable for muscular work and other purposes.

Until fairly recently the view was favored that respiration energy was much less utilizable than fermentation energy—in other words, that fermentation energy was relatively more valuable than would be indicated by a comparison of theoretical caloric yields. Recent results, however, for the conversion of oxidation into phosphate bond-energy strongly indicate that such is not the case. With oxidation of pyruvic acid in the brain, Ochoa[4] found that for each molecule of oxygen consumed four energy-rich phosphate bonds were generated. With carbohydrate oxidation in heart muscle, according to Belitzer and Tzibakova[5] as many as seven energy-rich phosphate bonds (for nomenclature cf. ref. 3) might be formed per molecule of oxygen. One energy-rich phosphate bond represents from 10 to 12 kg.-cal. of utilizable energy. The six moles of oxygen oxidizing one mole of carbohydrate could therefore generate from $11 + (4 \text{ to } 7) \times 6 = 260$ to 460 kg.-cal. of utilizable energy, or 40 to 68 per cent of the theoretical.

The unexpectedly high yield obtained in these recent experiments shows that there is probably no great difference in utilizability between fermentation and respiration. Therefore it seems permissible to take the theoretical caloric value of 54 and 686 kg.-cal., respectively, as a basis for comparing their efficiencies, and to conclude that only one-twelfth to one-ninth of the total possible energy is made available to the cell by the anaerobic fermentation of the glucose molecule.

From this calculation the superior economy of respiratory metabolism becomes evident. To draw the same amount of energy from fermentation as from respiration the cell must use from nine to twelve times as much substrate. In reality the anaerobic energy is rarely equal to the aerobic. As a rule a fully developed facultative anaerobe uses anaerobically only four to eight times as much substrate as aerobically, thus reaching on the average half the energy level of the aerobic state. From these considerations the economical and regulatory aspect of the Pasteur effect becomes evident. Through its operation the voluminous fermentative metabolism is allowed to proceed only in anaerobiosis, as is indicated in the following scheme, which represents the increase in glucose utilization following change from aerobic to anaerobic conditions:

Oxygen	Nitrogen
($Q_{O_2} = 6$; $Q_G^{O_2} = 0$)	($Q_G^{N_2} = 10$)
1. Glucose + 6 $O_2 \rightarrow$ 6 CO_2 + 6 H_2O	1. Glucose $\rightarrow$ 2 Lactate
	2. Glucose $\rightarrow$ 2 Lactate
	3. Glucose $\rightarrow$ 2 Lactate
	4. Glucose $\rightarrow$ 2 Lactate
	5. Glucose $\rightarrow$ 2 Lactate

Some examples of "ideal" facultative anaerobic cells are given in Table 1. From the Q values the corresponding glucose consumption and the caloric yields are calculated. One cubic millimeter of respiration oxygen corresponds to the utilization of 1.34 micrograms of glucose and the yield of 5.2×10^{-3} cal.; one cubic millimeter of fermentation carbon dioxide corresponds to the utilization of 4.03 micrograms of glucose and the yield of 1.2×10^{-3} cal.

In the cases cited the large fermentative metabolism disappears completely in aerobiosis. In the experiment with the fish retina almost three-fourths of the aerobic energy is made available through anaerobic metabolism and corre-

Table 1

EFFECT OF OXYGEN ON GLYCOLYSIS

Organism or tissue	Q_{O_2}	$Q_F^{O_2}$	$Q_F^{N_2}$	Substrate consumption: rate*	Substrate consumption: anaerobic/aerobic	Caloric yield: rate†	Caloric yield: anaerobic/aerobic	Reference
Torula								
anaerobic	——	—	260	1.04		0.31		6
aerobic	−180	18	—	0.31	3.4	0.94	0.33	
Embryonic heart								
anaerobic	——	—	28	0.11		0.034		7
aerobic	−13.6	0	—	0.018	6.0	0.071	0.48	
Pigeon brain								
anaerobic	——	—	28	0.11		0.034		8
aerobic	−16	0	—	0.022	5.0	0.083	0.41	
Fish retina (30° C.)								
anaerobic	——	—	29	0.12		0.035		9
aerobic	− 9.6	1	—	0.017	7.0	0.050	0.70	

*mg. glucose per mg. dry weight per hour.
†calories per mg. dry weight per hour.

spondingly a three- to sevenfold anaerobic increase of glucose consumption occurs.

The Metabolic Structure of Cells

In the middle of the metabolic type scale are placed the organisms that alternate between anaerobic and aerobic metabolism of similar efficiency. In the upper part are predominantly aerobic types, e.g., kidney and liver and many plant cells with relatively small fermentative capacity, and at the top of the scale is a strict aerobe, azotobacter, with Q_{O_2} of 2000 and no trace of fermentation. On the anaerobic side, below the middle, are a variety of types representing gradations down to exclusively anaerobic life. Here fermentation is partly or wholly persistent in the presence of oxygen, and respiration becomes a more or less residual function.

Bacteria

The anaerobic type of life is most common among the bacteria, but it occurs frequently in the animal kingdom, especially among invertebrates. In all stages of phylogenetic development, life either chooses or is forced to adapt to anaerobic conditions, and similar metabolic arrangements correspond to similar environmental conditions. Transition from alternative to exclusive anaerobiosis is well illustrated by a tabulation of the metabolism of the common yeasts (Table 2), taken from Meyerhof's classical paper[6].

The almost exclusively anaerobic cultured yeasts used in the manufacture of alcoholic beverages presumably developed from the alternating torula or wild yeast type. Baker's yeast is intermediate, having a fair respiration and partially persistent aerobic fermentation. The metabolic type is not rigidly fixed. With aeration there is adaptation to the respiratory type, and with the exclusion of air the reverse is easily achieved (Table 2). It may be noted that the reappearance of respiration is accompanied by aerobic repression of fermentation. The history of the manufacture of baker's yeast is an impressive illustration of the economic superiority of aerobic metabolism[10]. The earlier "Vienna" procedure of growing yeast without agitation has now been almost entirely replaced by the aeration procedure, for it has been found that through aeration the yield can be greatly increased with the same amount of culture

Table 2

Yeast Metabolism

Type	Q_{O_2}	$Q_F^{O_2}$	$Q_F^{N_2}$	$\frac{(Q_F^{N_2} - Q_F^{O_2}) \times 3}{Q_{O_2}}$	Inhibition per cent
Wild yeast	−180	18	260	4	93
Baker's yeast	−87	95	274	6.2	65
Brewer's yeast	−8	213	233	7.5	8
Same after 15 hours aeration	−73	113	193	3.3	42

fluid. In metabolic terms, the same amount of metabolized substrate yields a larger amount of yeast material with economical respiration than with uneconomical fermentation.

As a measure of the Pasteur effect two differently derived units are recorded in the last two columns of Table 2. In the first of the two the Meyerhof Oxidation Quotient is calculated. This relates the disappearance of fermentation to the magnitude of respiration. When three times the difference between fermentation in nitrogen and fermentation in oxygen is divided by the respiration in oxygen, the quotient represents the relation between the glucose equivalent of fermentation and that of respiration. Disregarding underlying theoretical implications, it states how much fermentation glucose is replaced by oxidized glucose when respiration is allowed to occur. Stressing the economical significance of the quotient, we have proposed to call it a replacement quotient[3]. In the next column the percentage of inhibition is calculated. From the recorded figures one would suspect a relationship of some kind between the magnitude of respiration and the Pasteur effect. This observation originally led Meyerhof to a universal application of his resynthesis theory as an explanation of the Pasteur effect.

Although a broader discussion of the theory of the Pasteur effect is reserved for a later paragraph, some interesting experiments relating to the constancy of the Meyerhof Quotient may be mentioned here. These experiments on the effect on the oxidation quotient of the Z-factor of v. Euler, a factor stimulating only fermentation, are taken from a paper by Meyerhof and Iwasaki[11]. See Table 3.

Table 3

Influence of Medium of Fermentation and Oxidation Quotient (Reference 11)

Yeast	Medium	Q_{O_2}	$Q_F^{O_2}$	$Q_F^{N_2}$	O.Q.*	Inhibition per cent
Baker's yeast I	glucose in phosphate	−37.5	44.5	119	6	62
	plus wort	−37	96	212	9.4	55
Baker's yeast II	glucose in phosphate	−40.5	105	199	7	47
	plus molasses	−45.6	228	374	9.6	39
Strain XII	sugar in phosphate	−10	49	57	2.3	14
	plus yeast extract	−10.5	137	167	8.5	18

*Meyerhof oxidation quotient.

For the same yeast, *Q* values, the oxidation quotient, and the percentage of inhibition are listed with and without the stimulating factor. Here with the same respiration but variations in anaerobic and aerobic fermentation, the relative influence of respiration on the disappearance of fermentation increases, whereas the percentage of inhibition remains constant. These experiments are more consistent with an explanation of the Pasteur effect as an inhibition of some kind resulting from the presence of oxygen but independent of the magnitude of respiration. Similar results were recently reported by Burk,

Winzler, and du Vigneaud[12], who studied the metabolism of a biotin-deficient yeast. They found oxidation quotients up to 20, whereas 12 is the theoretical maximum for a resynthesis theory.

The results of these experiments are discussed in some detail in order to draw attention to two lines of approach to an explanation of the phenomenon. The first set of data, showing parallel rates of respiration and fermentation disappearance suggests an interaction between the two lines of reaction which leads to the establishment of a *dynamic equilibrium* of some kind. The second set of data shows conditions where the aerobic inhibition of increased fermentation is independent of the rate of respiration, which remains practically constant. This suggests a *direct inhibition* through oxygen.

Table 4 surveys lactic and propionic acid bacteria and chlorophyll-bearing plant tissue. The three types of lactic acid bacteria are representative examples which duplicate in every respect the metabolic types of the yeasts shown in Table 2. *Lactobacillus delbrückii** deserves special comment. It shows a well-developed Pasteur effect without possessing any catalysts of the hemin type. The oxygen-producing plant tissues show only a small fermentative capacity.

Table 4

BACTERIAL AND PLANT METABOLISM IN GLUCOSE

Material	Organism	Q_{O_2}	$Q_F^{O_2}$	$Q_F^{N_2}$	O.Q.	Aerobic inhibition per cent
Lactic acid bacteria (13)	*Bacterium cereale*	−189	49	305	3.9	84
	Bacterium Delbrückii (hemin-free)	−109	79	188	3	58
	Lactobacillus casei	0	287	316		9
		−3	255	277	22	8
Propionic acid bacteria (14)	*Propionibacterium pentosaceum*	−15	4	20	3	80
Lathyrus plant (15)	Sprout	−3.5	0	1		
	Leaf	−1.4	0	0.8		
Algae (15)	*Chlorella pyrenoidea*	−5	0	1.6		
	Coelastrum proboscideum	−12.7	0	2.7		

ANIMAL TISSUES

The fairly frequent occurrence among invertebrates of organisms adapted to anaerobic life has been mentioned. Particularly among the worms is found a great variety of partly or wholly anaerobic forms. Here, on a higher phylogenetic level, appears a metabolic stratification similar to that in the yeasts and bacteria. There seems to be a gradual transition from the facultative anaerobic free-living worms to the obligate anaerobic parasitic forms[16]. As early as 1909 Lesser[16a] made important quantitative experiments on the interrelation between respiration and fermentation in earthworm metabolism. He found

*This organism has been referred to by a variety of names in the biochemical literature, including: *Bacterium Delbrückii, Lactobacillus Delbrükii, Bacterium cereale, Bacillus acidificans longissimus,* etc.

that fermentation resulted in the consumption of from four to six times as much glycogen. These same ratios were later found by Meyerhof to hold likewise for frog muscle[17]. As end products in worm fermentation Lesser found, in addition to lactic acid, large amounts of higher fatty acids, especially valeric acid (cf. also 18 and 19). An increase of glycogen utilization in anaerobiosis is described for many other types of worms[19]. The parasitic worms living in the practically oxygen-free intestinal fluids show predominantly anaerobic metabolism. Although they are able to respire aerobically, their fermentation does not seem to be inhibited nor their glycogen consumption diminished in oxygen[19].

The higher vertebrates and especially the warm-blooded animals must be considered as essentially aerobic organisms. This is not true for all their parts, however, nor under all conditions. The experiment shown in Figure 2 is an example of partial anaerobiosis: the muscle suddenly put under high strain must rely predominantly on a supply of anaerobic energy. In Table 5 metabolic figures for representative tissues are assembled. During recent years interesting examples of adult tissues with predominantly anaerobic metabolism have been described. Relatively large and aerobically persistent glycolysis has been found, for example, in the medulla of the kidney and in cartilage[20, 21]. Dickens correlates the metabolic pattern with the relatively poor blood supply of these tissues. Large and aerobically persistent glycolysis in the intestinal mucosa recently reported by Dickens and Weil-Malherbe[22] might likewise be correlated with the previously mentioned lack of oxygen in the intestinal fluid surrounding it.

Table 5

METABOLISM OF ANIMAL TISSUES

Types	Tissue	RQ	Q_{O_2}	$Q_G^{O_2}$	$Q_G^{N_2}$	O.Q.	Reference
Largely aerobic	Kidney	0.8	−20	0	0–8		80
	Liver	0.6–0.9	−12	0–2	0–12		80
Facultative anaerobic	Brain	1	−16	0	26	4.9	36
	Pituitary	—	−12	0	13	3.3	81
	Testis	0.7–1	−14	8	14	1.3	80
	Medulla of kidneys	0.97	− 9	16	23	3.3	20
	Mucosa jejuni	0.86	−16	25	23	0	22
Largely anaerobic	Sperm, human	—	− 1	6.5	8		82

These observations show that the pronounced anaerobic metabolism of embryonic tissue and of malignant growth (Table 6) is not an isolated phenomenon. Here are tissues, as has been pointed out in an earlier paper[23], which through environmental conditions, such as insufficient circulation, are forced to rely partly on a supply of anaerobic energy (cf. 20). Recent measurements by Philip[24] on respiration of the early developmental stages of the chick embryo give evidence that this must be the case. This corroborates our earlier findings based on less conclusive experimental data. Philip's remarks may be quoted here: "The study of the early blastomere," he says, "has revealed that oxygen diffusion limits the oxygen consumption in oxygen tensions of the air.

Table 6

METABOLSIM OF TUMOR TISSUE*

Tumor	Q_{O_2}	$Q_G^{O_2}$	$Q_G^{N_2}$	O.Q.	Inhibition per cent	Reference
Flexner-Jobling carcinoma, rat	−55	23	23	0	0	54
	− 7	16	24	3	33	
Adenocarcinoma, human male	− 9	16	29	4.2	42	75
	− 1.2	5	12	15	58	
	0	9	22	∞	59	
Jensen sarcoma	−14	16	34	3.9	53	39
Walker sarcoma 256	−22	25	46	2.9	46	39

*The average R.Q. for all the tissues was 0.85

This indicates that the early blastomere may actually be in a state of partial anaerobiosis." And later in the same discussion: "The considerations presented suggest that some of the energy used during early periods of growth can be probably associated with the rapidly increasing size of the embryo during early periods before the circulation system can function as an adequate oxygenating mechanism." The relatively poor vascularization of most tumors is evidence that the last statement holds likewise for malignant growth. We are led then to the conclusion that the high capacity of the anaerobic metabolism present in normal and in malignant growing tissues should be attributed to their partly anaerobic state of life rather than to an unlikely special growth function of glycolysis.

A curious phenomenon of hyperfunction of the Pasteur effect is observed in human beings at high altitudes(25). The relatively high lactic acid level of the blood which would be expected at low oxygen pressure is observed only before adaptation occurs(25). After adaptation to the new environment the lactate level of the blood becomes normal. Even with exhausting work the lactic acid concentration remains very low, 2 to 3 millimolar, as compared with a blood level of 13 millimolar reached with exhausting work at sea level (Figure 2). Apparently a special mechanism prevents the muscle from utilizing too much of the anaerobic energy supply even at low oxygen pressure. Dill in his book on life at high altitudes(26) makes interesting comments on this phenomenon: "It is as though the body, realizing the delicacy of its situation with regard to oxygen supply, sets up an automatic control over anaerobic work which renders impossible the severe acid-base disturbances which can be voluntarily induced at sea level."

Interpretation of the Pasteur Effect

During recent years discussion has centered more or less around the question whether the effect depends upon respiratory activity as such, or upon an inhibition produced by the action of oxygen. In the first case the rate of respiration with its output of energy would be a determining factor and a state of dynamic equilibrium would result; part or even all of the respiration energy would be spent or fixed to revert or repress glycolytic breakdown. If, however, the effect is brought about through oxygen, or more specifically by oxidative inhibition of an essential part of the glycolytic enzyme system, then the reaction may be independent of the rate of respiration and involve no transfer of energy.

Equilibrium schemes

A fuller understanding of the partial reactions involved in fermentation and respiration has given new impulse to the discussion of their interrelationship as manifested in the Pasteur effect. The fact that cozymase and adenylic acid, the two transmitter substances in fermentation, are likewise participants in respiration has given rise to some interesting suggestions.

Ball[(27)] has pointed out that, aerobically, respiratory oxidants such as flavin may compete with pyruvic acid for the reduced cozymase. Pyruvic acid would disappear largely through oxidation rather than through fermentative reduction. Adler and Calvet[(28)], however, comparing the ratio of oxidized to reduced cozymase in aerobic and in anaerobic baker's yeast found no significant difference, but a ratio of nearly one to one in both cases.

Adenylic acid, the other common transfer system, was first linked with the Pasteur effect through work done by Ostern and Mann[(29)]. They found that the addition of adenosine triphosphate (ATP) to mashed muscle depressed aerobic glycolysis and raised the Meyerhof Quotient from 2.2 to 4. Later Lennerstrand[(30)] discussed the possibility that with aerobic over-phosphorylation of adenylic acid (Ad), the ratio ATP:Ad might become too high to permit adenylic acid to function effectively as a transmitter in fermentation.

A scheme based on the recent development of the biochemistry of phosphate turnover has been presented and discussed in detail elsewhere[(3)]. Figure 3 is taken from this paper. The upper cycle, revolving clockwise, represents anaerobic glycolysis; the lower cycle, revolving counter-clockwise, represents aerobic resynthesis. It appears that the clockwise run of the glycolytic cycle depends on the outflow of the energy-rich phosphate created in the reaction. An actual reversal of the cycle back to the aldehyde stage may occur when through the aerobic influx of new energy-rich phosphate the carboxyl-bound phosphate in 1,3-diphosphoglyceric acid (Ph-glyceryl-Ph) cannot be removed.

In the Meyerhof-Warburg equilibrium reaction, as the diagram shows, inorganic phosphate is bound when the reaction proceeds to the right and is set free when it proceeds to the left. Therefore inorganic phosphate concentration can become a rate-determining factor. Meyerhof *et al.*[(31)] and Belitzer[(32)] have pointed out that in muscle the increased concentration of inorganic phosphate through creatinephosphate breakdown should be regarded as the cause of the release of metabolic activity due to stimulation. Along similar lines, Johnson[(33)] recently suggested that the lowering of inorganic phosphate concentration might be a possible cause of aerobic inhibition of glycolysis. Inorganic phosphate concentration seems, however, to be high in most cells except in resting muscle, where most of the phosphate is bound to creatine. Phosphate is generally considered to be the intracellular anion. How much of this is really free phosphate and how much is labile phosphate broken down by chemical manipulations remains to be determined[(34)].

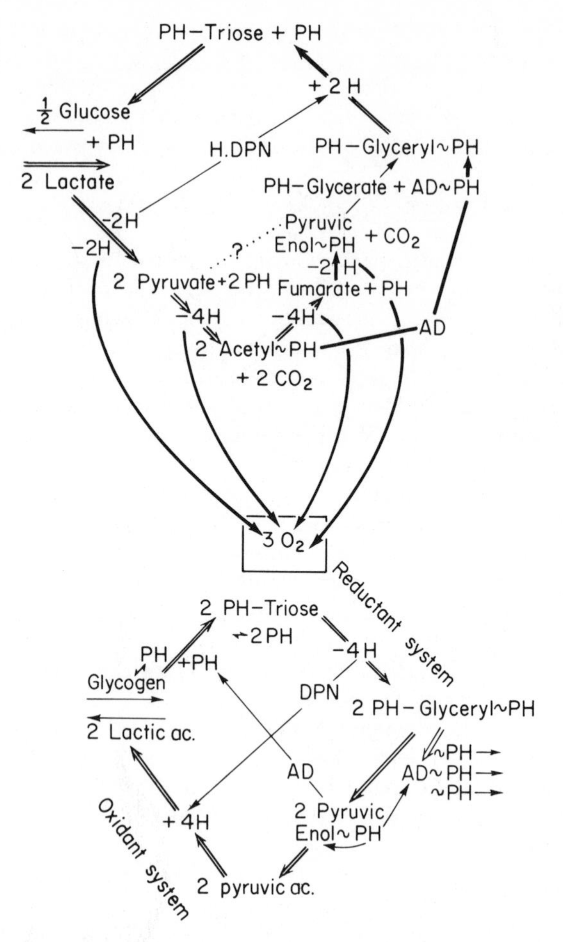

Fig. 3. Schemes for anaerobic breakdown (upper cycle) and aerobic resynthesis (lower cycle) of carbohydrate. (From F. Lipmann, *Advances in Enzymology,* Vol. 1, 1941).

In our opinion, the value of such schemes is limited because they disregard controlling factors in the cell which undoubtedly must regulate the routes of phosphate turnover—that is, the synthesis as well as the breakdown of intermediates. The Pasteur effect cannot be due merely to an "open" equilibrium; it must be due to specific transmitter systems. Evidence for this is the fact that the "linkage" between respiration and glycolysis may be interrupted without impairing the reactions proper.

Inhibition of the Pasteur Effect

A considerable variety of compounds are able to interrupt the Pasteur effect, or the Pasteur reaction, as Warburg[35], after discovering the specific action of ethyl carbylamine, first called the phenomenon. Table 7 presents

Table 7

AEROBIC RELEASE OF GLYCOLYSIS

Inhibitor	Concentration	Q_{O_2}	Q_G^O	R.Q.	Tissue	Reference
Ethyl carbylamine	0	− 13	19		Jensen sarcoma	35
	10^{-3} M	− 14	32			
Carbon monoxide, light	88%	− 11	6	0.72	Allantois	37
Carbon monoxide, dark	88%	− 11	11			
Oxygen pressure	95%	− 25.5	0	1.05	Chorion	38
	5%	− 22	11	0.7		
Phenosafranine	0	− 13	1	1.03	Brain	39
	10^{-5} M	− 13	21	0.98		
Glutathione	0	−230	74		Yeast	40
	2.5×10^{-3} M	−205	255			

a survey of the agents which have been given the most study and which have proved most effective. Similar effects were found with phenylhydrazine by Dickens[36], with dinitrocresol and dinitrophenol by Dodds and Greville[41], and with HCN on certain plant cells by Genevois[42].

The common effect is the release of aerobic glycolytic action up to an anaerobic level, while respiration remains quantitatively unchanged. In Laser's experiments[37,38] with carbon monoxide and low oxygen pressure the respiratory quotient was lowered, indicating qualitative changes of respiration. With phenosafranine, however, Dickens[39] found that the respiratory quotient of the brain remained unity and the manometric and chemical determinations of lactic acid were in excellent agreement. In this case, at least, it seems very probable that the interruption of the Pasteur reaction occurred without a qualitative change of respiration.

The action of metal-specific inhibitors has been of great interest. Work in this field has revived discussion of the question whether the effect is dependent on, or independent of, respiratory activity. The old observation that in most tissues cyanide released aerobic fermentation by inhibiting respiration was taken as indisputable proof of the dependence of the Pasteur effect on the intactness of respiration. Consequently ethyl carbylamine action, affecting only the Pasteur effect, but leaving primary respiration intact, was interpreted as inhibition of a reaction linking respiration to glycolysis.

A differential inhibition of respiration and Pasteur reaction by carbon monoxide was observed by Warburg[43a] in yeast experiments. Mainly interested in the respiratory effect of carbon monoxide, he remarked only incidentally upon the relatively higher sensitivity of the Pasteur reaction. Later, Laser[38] showed that in animal tissues the differences in sensitivity were pronounced. Frequently, he found, carbon monoxide had little or no effect on respiration but did cause aerobic glycolysis to appear. The release of aerobic glycolysis in animal tissues had been observed by Warburg and Negelein[43], but had been considered as a secondary effect due to inhibited respiration. From some preliminary measurements of the effect of light on aerobic glycolysis in retina in the presence of carbon monoxide, the spectrum of the *Atmungsferment* was charted. Since Laser[38] had found respiration in retina to be uninfluenced by carbon monoxide, these measurements, as Stern and Melnick[44] recognized, had to be reinterpreted as preliminary measurements of the spectrum of the Pasteur agent-carbon monoxide compound. Stern and Melnick then measured carefully with the Warburg illumination technique the relative absorption spectrum of the Pasteur agent-carbon monoxide compound. The decrease in aerobic fermentation on irradiation was plotted against wave length. This decrease may be assumed to be due to the decomposition of the Pasteur agent-carbon monoxide complex. The resulting spectrum was very similar to that of the respiratory enzyme. Such measurements were made on retina[44] and yeast[45]. Recently Melnick[45a] charted the spectrum of the respiratory enzyme of animal tissue by using heart muscle extracts in which, in contrast to the intact tissue, respiration is sensitive to carbon monoxide[46]. The bands developed from these measurements are reproduced in Figure 4. The spectra of the respiratory enzymes in yeast and in animal tissue, respectively, differ greatly, as do those of the Pasteur enzymes. In each case, however, the spectrum of the Pasteur enzyme follows closely that of the respiratory enzyme, deviating only in the absorption at longer wave lengths. The consistent, although small, differences are considered as evidence of the existence of two definitely different enzymes, one catalyzing the final step in the oxidation of metabolites, the other catalyzing the oxidative inactivation of a part of the glycolytic system.

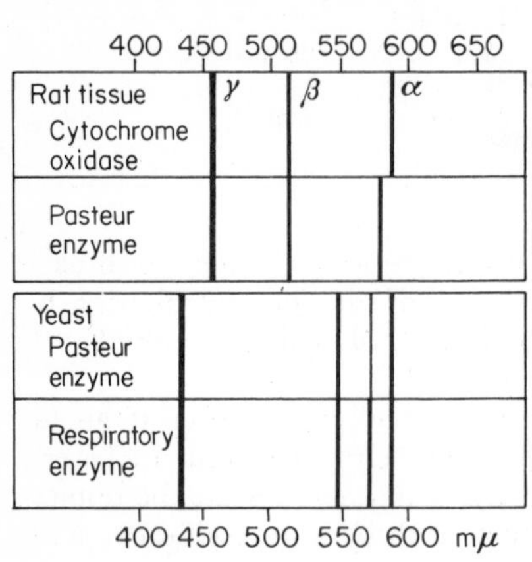

Fig. 4. Spectra of the respiratory and Pasteur enzymes in animal tissue and in yeast.

This analysis seems a very promising approach at least to an elucidation of the events taking place on the oxygen side. The similarity between the respiratory and Pasteur enzymes suggests a direct reaction between oxygen and the transmitter. This is further indicated by the difference in the affinity of the two enzymes for oxygen[37,47,76]. The peculiarities of the carbon monoxide effect on respiration and a change of respiratory quotient at low oxygen tension[37] and with carbon monoxide[38], however, seem to indicate that the present interpretation may not represent the final solution.

In spite of the interest that attaches to the metal-specific anticatalysts, it should not be overlooked that such inhibitors as phenosafranine, dinitrocresol[41], and glutathione can hardly be considered metal-specific. In an extensive study of the action of phenosafranine and other phenazine derivatives, Dickens[39,48] presents evidence which suggests that a flavin enzyme may participate in the transmission of the aerobic inhibition. The relationship between flavin and the Pasteur effect is indicated also by its occurrence in the hemin-free *Lactobacillus delbrückii*[13], where flavin is the only respiratory catalyst.

The disturbance of the Pasteur effect in brain that attends a lack of ionic balance represents a phenomenon of great complexity. Ashford and Dixon[49] observed a profound metabolic change in brain slices suspended in tenth molar potassium chloride. Aerobically they found increased respiration and appearance of glycolysis; and anaerobically, gradual and irreversible disappearance of glycolysis. They correlated the metabolic changes with the well-known increase in cell permeability through potassium ion[50]. Dickens and Greville[51] showed subsequently that the potassium effect is specific for brain and is not found in other tissues, and that omission of calcium had a similar effect. Continuing on similar lines, Weil-Malherbe[52] observed definite effects of potassium and also ammonium ions at much lower concentrations than those used by Ashford and Dixon.

This effect of electrolyte on brain metabolism signifies a great lability of the Pasteur mechanism. Warburg[54] has emphasized that the Pasteur mechanism is universally very sensitive to unphysiological surroundings. For example, in rat embryo aerobic glycolysis is high in Ringer solution but low or absent in serum or amniotic fluid. Effects of this type must be taken as an indication that aerobic disappearance of glycolysis is the result of an easily disturbed balance of reactions.

Reversible oxidative inhibition of glycolysis in extracts

In order to approach experimentally the possibility that oxidative inhibition might be the cause of aerobic disappearance of fermentation, I studied some time ago the effect of oxidizing agents on glycolysis and fermentation in extracts[55, 56]. It was shown that the fermenting system was inactivated by small amounts of iodine and quinone. By adding indophenols as oxidants inhibition in oxygen was provoked, which disappeared in its absence, when the oxidizing dye was reduced by constituents of the extract and through its enzymatic activity. Two experiments of this type with muscle extract and yeast juice, respectively, are summarized in Table 8. It appears that addition of the dye reproduces a Pasteur effect, which occurs, however, with negligible respiratory activity, demonstrating the possibility of a reversible oxidative inactivation.

Table 8

Induced Pasteur Effect in Cell Extracts

Extract	Addition	O_2	CO_2 in O_2	CO_2 in N_2	O.Q.	Aerobic inhibition per cent	Ref.
		Cubic millimeters per hour					
Yeast	6 × 10^{-3} M naphthol-sulfonate indophenol	−49	106	710	37	85	56
	None	—	1140	1120	—	0	
Muscle	10^{-3} M dichloro-phenol indophenol	− 4	20	380	270	95	55
	None	—	440	425	—	0	

The inhibition of glycolysis with iodine was subsequently studied more carefully by Gemmill and Hellerman(57). With concentrations just high enough to obtain fairly complete inhibition they were able to recover the activity by adding glutathione or cysteine. This suggested strongly that the oxidative inhibition was due to reversible oxidation of enzyme SH-groups. Rapkine(58, 59) then showed that the oxidoreduction between phosphoglyceraldehyde and pyruvic acid was at least one of the partial reactions being blocked by oxidation, presumably of enzyme-SH. This reaction system could be inactivated by S-S-glutathione and reactivated by SH-glutathione. More recently Rapkine found the same reaction reversibly inactivated by dichlorophenolindophenol (personal communication), which might explain my earlier results with the complete glycolytic system. With these experiments the possibility of a reversible oxidative inactivation has become firmly established. It is therefore of little significance for the question at issue that, as was shown by Michaelis and Smythe(60), many dyes, irrespective of oxidation-reduction potential, inhibit irreversibly by various mechanisms, or that naphtholsulfonate indophenol with different yeast preparations leads earlier to irreversible inactivation than in our experience.

Besides the system studied by Rapkine, a number of partial enzymes of glycolysis were found to undergo oxidative inactivation followed by reactivation with glutathione. These reactions are as follows:

phosphoglyceraldehyde + pyruvate $\longrightarrow$ phosphoglycerate + lactate(58)

glycogen + phosphate $\rightleftarrows$ glucose-1-phosphate(61, 62)

glucose-1-phosphate $\rightleftarrows$ glucose-6-phosphate(62)

adenosinediphosphate + glucose $\longrightarrow$ adenylic acid + glucose-6-phosphate(63)

The activation by thiol compounds of glycolysis in extracts, described by Geiger and Magnes(64) and Michaelis and Runnström(65) thus becomes easily understadable.

Thiol influence on fermentation and glycolysis in intact cells

The function of glutathione is not yet well understood. It is present in practically every cell in fairly large amounts. Frequently it has been suggested that it performs the role of an oxidation-reduction buffer. The very complexity of intracellular metabolism prevents us from making more than vague statements of that type. The protection against oxygen injury which thiol compounds give to strict anaerobes, first observed by Quastel and Stephenson(53), lends support to the assumption that their function is one of stabilization.

Observations on intact cells as well as on cell-free enzyme systems suggest a regulatory effect of thiol compounds on glycolysis and fermentation. As yet it is impossible to correlate definitely the action on intact fermenting cells and on fermenting enzyme extracts or partial systems, but a promising approach seems to be opened which is worth very careful consideration.

Release of aerobic glycolysis with glutathione was first observed by Bumm and Appel[66] with sliced animal tissues. Soon afterward Quastel and Wheatley[40] made an interesting study of the effects of glutathione and cysteine on the metabolism of baker's yeast. One of their experiments is included in Table 7 above. Glutathione interrupts the Pasteur effect without affecting respiration. They noted that an extract of brewer's yeast had much the same effect as glutathione, which they ascribe to the large content of thiol compounds in brewer's yeast. With cysteine the effect on aerobic glycolysis was the same, but respiration was markedly inhibited. The respiratory inhibition was specific for glucose and absent when glycerol was used as substrate. More recently Runnström and Sperber[67] undertook a study of the cysteine effect. Accompanying the release of aerobic fermentation they found an inhibition of the synthesis of higher carbohydrates from glucose. This interesting observation would be still more significant if respiration were not inhibited at the same time. The alternative between glucose fermentation and synthesis to glycogen suggests that thiol compounds are able to upset the normal aerobic reaction course from glucose-6-phosphate over Cori-ester to glycogen, forcing the glucose monoester into the fermentation cycle.

With *Propionibacterium pentosaceum* a number of interesting observations were made by Fromageot and Chaix[14]. Dilute suspensions of repeatedly washed bacteria did not ferment in the presence of minute amounts of oxygen. This inhibition was counteracted by very small concentrations of cysteine or hydrogen sulfide. With unwashed and concentrated suspensions small amounts of oxygen did not affect the fermentation, but aerobically fermentation disappeared (normal Pasteur effect, Table 4) and was released by thiol compounds. They concluded that normally a substance is present in bacteria which protects the fermentation system against the action of small amounts of oxygen. Since with impoverished organisms protection can be restored with cysteine, they assumed the protecting substance to be a thiol compound. With high oxygen pressure the physiological concentration of the protective system is not high enough to counteract the oxidative inhibition and aerobic disappearance of fermentation; that is, the Pasteur effect occurs. When the concentration of thiol compound is increased, the oxidative inhibition is blocked again, and aerobic fermentation appears. In other words, the occurrence of fermentation depends on the relative concentrations of SH-compound and oxygen, respectively.

An observation reported by Dickens[68] with pyocyanine should be mentioned here. In the presence of this dye "anaerobic" glycolysis of sarcoma was inhibited when measured in unpurified nitrogen containing 0.3 per cent oxygen. At the same time a slight color remained, indicating slight reoxidation of the dye. The color and the inhibition disappeared when chromous chloride was used to absorb the traces of oxygen. The parallel between this phenomenon and our dye-induced Pasteur effect in extracts, as well as Fromageot's effect of low oxygen pressure on propionic acid bacteria, is obvious.

Despite the complexity of dye effects on living cells[68,69], Dickens came to the conclusion that in general there is a tendency for dyes with high oxidation-reduction potential to increase the Pasteur effect. In harmony with this generalization is the increase of the Pasteur effect in tumors by ferricyanide[70] and in yeast by indophenols[71]. The complexity of the dye effects, however, is illustrated by the action of methylene blue, which, according to early observations by Gerard[72], releases aerobic glycolysis in muscle, while it was found by Barron[73] to inhibit aerobic glycolysis in erythrocytes. Nevertheless there seems to be a parallelism between dye action in extracts and in cells and a correlation between thiol and dye effects.

Pasteur effect with very low respiration

That aerobic inhibition of fermentative metabolism is independent of respiration can be most clearly demonstrated through the occurrence of the Pasteur effect with very low respiration. Aside from the dye-induced Pasteur effect in extracts, some examples of such phenomena in living cells have already been discussed, such as inhibition by traces of oxygen or inhibition in the presence of relatively low respiration in yeast[11,12]. As a rule parallelism between the appearance of anaerobic metabolism and the disappearance of respiration is to be expected by the very nature of the phenomenon. The fact that the most-used inhibitors of respiration are metal-specific and are likewise more or less pronounced inhibitors of the Pasteur reaction, has greatly complicated the analysis. Inhibitors of respiration which interrupt the chain, not at the end where the iron catalysts are operating but at an earlier stage, might be expected under favorable conditions to interrupt respiration without affecting the Pasteur reaction. Malonate and maleate, which block the Szent-György cycle, might react in this way. Weil-Malherbe[52] has indeed found with maleate poisoning that there is no appreciable aerobic glycolysis in brain when respiration has already declined to very low levels. I found, with malonate, similar effects on embryonic heart[74]. In Table 9 a survey is given of these and other experiments, where with animal cells a Pasteur effect was found with low respiration. This was observed by Rosenthal and Lasnitzki[75] with some human cancer without inhibitors and by Kempner and Gaffron[76] with myeloblasts at 6 per cent oxygen pressure. It should be remembered that in Kempner's

Table 9

Pronounced Pasteur Effect with Very Low Respiration

Tissue	Addition		Q_{O_2}	$Q_G^{O_2}$	$Q_G^{N_2}$	O.Q.	Aerobic inhibition per cent	Reference
Adenocarcinoma,	none		−1.2	5	12	15	58	75
human male	none		0	9	22	∞	59	
Brain of rat	10^{-2} M maleate	1 hr.	−10	2	19	5	90	52
		2 hr.	−5	0.5	16	9	97	
		3 hr.	−2	3	16	19.5	82	
Embryonic	2.5 × 10^{-2} M	(1)	−1	4	35	72	89	74
chicken heart	malonate	(2)	−4	4	32	20	87	
		(3)	−11	6	31	6.8	80	
		(4)						
		1 hr.	−11	8		6.5	75	
		2 hr.	−6	6	32	13	83	
	in serum-malonate	(5)	−9.5	5	16	3.5	69	

experiments with myeloblasts, while the Pasteur reaction was unaffected, respiration declined with falling oxygen pressure: the Q_{O_2} in 95 per cent oxygen was 8; in 6 per cent oxygen, 3.2. Laser[37] found the reverse with chorion, retina, and mouse liver; that is, little influence of low oxygen tension on respiration but inhibition of the Pasteur reaction.

Malonate does not have the effect described above on all tissues. I found with pigeon brain a decrease of respiration accompanied by a large increase of aerobic glycolysis[74]. Similar results were reported by Kutscher and Sarreither[77] with skeletal muscle.

Conclusion and Outlook

It has not been my purpose to give a complete survey of the work in this field. The recent reviews by Burk[83, 84] constitute a competent discussion of the problem as a whole, especially with regard to earlier work and thoughts. Our purpose here has been to summarize mainly the facts that indicate the occurrence of an oxidative inhibition. In general the evidence may be considered indicative but not conclusive, except in a few instances. The cell may choose to eliminate unneeded anaerobic metabolism by an inhibitory mechanism rather than by a counterforce, but there are indications that such inhibition acts upon reactions directing the internal flow of energy. Substances interrupting the Pasteur linkage likewise interrupt synthetic reactions, as has been shown in the case of cysteine[67] and especially dinitrophenol[41]. Clifton[78], while in Kluyver's laboratory, made the discovery that dinitrophenol inhibits completely the synthetic processes in microorganisms. His study was based on the work of Barker[85], who demonstrated that with "resting" organisms only part of the disappearing non-nitrogenous metabolite could be accounted for by oxidation, while a large part was converted into cell material, presumably carbohydrate. This conversion was completely interrupted in the presence of dinitrophenol, in which case catabolic breakdown continued until all material was oxidized[78, 79].

Dinitrophenol has therefore become an important tool for the study of the relation between anabolic and catabolic processes, which must be determined by the flow of energy-carrying reactions. In a recent paper[3] where I have discussed the generation and transfer of energy-rich phosphate bonds it is stated that a major part of metabolically yielded energy is converted primarily into phosphate bond energy. An understanding of the means by which the cell directs the flow of energy-rich phosphate bonds into predetermined reactions should lead to a more precise understanding of the mechanism of regulative cell reactions such as the Pasteur effect and the probable related action of the hormones.

References

1. Bang, O., Dissertation, Copenhagen, 1935.
2. Burk, D., *Proc. Royal Soc.* (London), **B 104**, 153 (1929).
3. Lipmann, F., Advances in Enzymology, **1**, 99 (New York, 1941).

4. Ochoa, S., *J. Biol. Chem.,* **138**, 751 (1941).
5. Belitzer, V. A., and Tzibakova, E. T., *Biokimia,* **4**, 516 (1939).
6. Meyerhof, O., *Biochem. Z.,* **162**, 43 (1925).
7. Warburg, O., and Kubowitz, F., *Biochem. Z.,* **189**, 242 (1927).
8. Lipmann, F., *Skand. Arch.,* **76**, 255 (1937).
9. Nakashima, M., *Biochem. Z.,* **204**, 479 (1928).
10. Warburg, O., *Biochem. Z.,* **189**, 350 (1927).
11. Meyerhof, O., and Iwasaki, K., *Biochem. Z.,* **226**, 16 (1930).
12. Burk, D., Winzler, R. T., and du Vigneaud, V., *J. Biol. Chem.,* **140** Proc., xxi, (1941).
13. Davis, J. G., *Biochem. Z.,* **265**, 90; **267**, 357 (1933).
14. Fromageot, G., and Chaix, P., *Enzymologia,* **3**, 288 (1937).
15. Genevois, L., *Biochem. Z.,* **186**, 461 (1927).
16. v. Bunge, G., *J. Physiol. Chem.,* **12**, 565 (1888).
16a. Lesser, E. J., *Ergebnisse d. Physiol.,* **8**, 742 (1909).
17. Meyerhof, O., *Pflügers Archiv.,* **185**, 11 (1920).
18. Slater, W. K., *Biochem. J.,* **19**, 604 (1926).
19. v. Brand, T., *Ergebnisse Biol.,* **10**, 37 (1934).
20. Dickens, F., and Weil-Malherbe, H., *Biochem. J.,* **30**, 659 (1936).
21. Dickens, F., and Weil-Malherbe, H., *Nature,* **138**, 125 (1936).
22. Dickens, F., and Weil-Malherbe, H., *Biochem. J.,* **35**, 7 (1941).
23. Lipmann, F., *Biochem. Z.,* **261**, 157 (1933).
24. Philips, F. S., *J. Exp. Zoology,* **86**, 257 (1941).
25. Edwards, H. T., *Am. J. Physiol.,* **116**, 367 (1936).
26. Dill, D. B., Life, Heat and Altitude, p. 173 (Harvard University Press, 1938).
27. Ball, E. G., *Bull. Johns Hopkins Hosp.,* **65**, 253 (1939).
28. Adler, E., and Galvet, F., *Arkiv Kemi, Mineral. Geol.,* **12B**, no. 32 (1936).
29. Ostern, P., and Mann, T., *Biochem. Z.,* **276**, 408 (1935).
30. Lennerstrand, A., *Naturwissenschaften,* **25**, 347 (1937).
31. Meyerhof, O., Schulz, W., and Schuster, P., *Biochem. Z.,* **293**, 309 (1937).
32. Belitzer, V. A., *Enzymologia,* **6**, 1 (1939).
33. Johnson, M. J., *Science,* **94**, 200 (1941).
34. Lipmann, F., *J. Biol. Chem.,* **134**, 463 (1940).
35. Warburg, O., *Biochem. Z.,* **172**, 432 (1926).
36. Dickens, F., *Biochem. J.,* **28**, 537 (1934).
37. Laser, H., *Biochem. J.,* **31**, 1671 (1937).
38. Laser, H., *Biochem. J.,* **31**, 1677 (1937).
39. Quastel, J. H., and Wheatley, A. H. M., *Biochem. J.,* **26**, 2169 (1932).
40. Quastel, J. H., and Wheatley, A. H. M., *Biochem. J.,* **26**, 2169 (1932).
41. Dodds, E. C., and Greville, G. D., *Lancet,* **112**: **I**, 398 (1934).
42. Genevois, L., *Biochem. Z.,* **191**, 147 (1927).
43. Warburg, O., and Negelein, E., *Biochem. Z.,* **214**, 64 (1929).
43a. Warburg, O., *Biochem. Z.,* **177**, 471 (1925).
44. Stern, K. G., and Melnick, J. L., *J. Biol. Chem.,* **139**, 301 (1941).
45. Melnick, J. L., *J. Biol. Chem.,* **140**, Proc., xc (1941).
45a. Melnick, J. L., *Science,* **94**, 118 (1941).
46. Keilin, D., and Hartree, E. F., *Proc. Roy. Soc.* (London), **B 127**, 167 (1939).
47. Bumm, E., Appel, H., and Fehrenbach, K., *Z. physiol. Chem.,* **223**, 207 (1934).
48. Dickens, F., and McIlwain, H., *Biochem. J.,* **32**, 1615 (1938).

49. Ashford, C. A., and Dixon, K. C., *Biochem. J.*, **29**, 157 (1935).
50. Dixon, K. C., *Nature*, **137**, 742 (1937); *Biol. Rev.*, **12**, 431 (1937).
51. Dickens, F., and Greville, G. D., *Biochem. J.*, **29**, 1468 (1935).
52. Weil-Malherbe, H., *Biochem. J.*, **32**, 2257 (1938).
53. Quastel, J. H., and Stephenson, M., *Biochem. J.*, **20**, 1125 (1926).
54. Warburg, O., Stoffwechsel der Tumoren (Berlin, 1926).
55. Lipmann, F., *Biochem. Z.*, **265**, 133 (1933).
56. Lipmann, F., *Biochem. Z.*, **268**, 205 (1934).
57. Gemmill, C. L., and Hellerman, L., *Am. J. Physiol.*, **120**, 522 (1937).
58. Rapkine, L., *Biochem. J.*, **32**, 1729 (1938).
59. Rapkine, L., and Trpinac, P., *Compt. rend. soc. biol.*, **130**, 1516 (1939).
60. Michaelis, L., and Smythe, C. V., *J. Biol. Chem.*, **113**, 717 (1936).
61. Gill, D. M., and Lehmann, H., *Biochem. J.*, **33**, 1151 (1939).
62. Cori, G. T., and Cori, C. F., *J. Biol. Chem.*, **135**, 733 (1940).
63. Colowick, S. P., and Kalckar, H. M., *J. Biol. Chem.*, **137**, 789 (1941).
64. Geiger, A., and Magnes, J., *Biochem. J.*, **33**, 866 (1939).
65. Michaelis, L., and Runnström, J., *Proc. Soc. Exp. Biol. Med.*, **32**, 343 (1935).
66. Bumm, E., and Appel, H., *Z. physiol. Chem.*, **210**, 79 (1932).
67. Runnström, J., and Sperber, E., *Nature*, **141**, 689 (1938).
68. Dickens, F., *Biochem J.*, **30**, 1064 (1936).
69. Elliott, K. A. C., and Baker, Z., *Biochem. J.*, **29**, 2396 (1935).
70. Mendel, B., and Strelitz, F., *Nature*, **140**, 771 (1937).
71. Hoogerheide, J. C., *Dissertation, Leiden*, 1935.
72. Gerard, R. W., *Am. J. Physiol.*, **97**, 523 (1931).
73. Barron, E. S. G., *J. Biol. Chem.*, **81**, 445 (1929).
74. Lipmann, F., unpublished.
75. Rosenthal, O., and Lasnitzki, A., *Biochem. J.*, **196**, 340 (1928).
76. Kempner, W., and Gaffron, M., *Am. J. Physiol.*, **126**, 553 (1939).
77. Kutscher, W., and Sarreither, W., *Z. physiol. Chem.*, **265**, 152 (1940).
78. Clifton, C. E., *Enzymologia*, **4**, 246 (1937).
79. Doudoroff, M., *Enzymologia*, **9**, 59 (1940).
80. Elliott, K. A. C., Greig, M. E., and Benoy, M. P., *Biochem. J.*, **31**, 1003 (1937).
81. Fujita, A., *Biochem. Z.*, **197**, 75 (1928).
82. MacLeod, J., *Am. J. Physiol.*, **132**, 190 (1941).
83. Burk, D., Occasional Publications, *Am. Assoc. Adv. Sci.*, **4**, 121 (1937).
84. Burk, D., *Cold Spring Harbor Symp.*, **7**, 420 (1939).
85. Barker, H. A., *J. Cell. Comp. Physiol.*, **8**, 231 (1936).

ENZYMIC CONVERSION OF PHOSPHORYLASE *a* TO PHOSPHORYLASE *b**

Patricia J. Keller† *and Gerty T. Cori*

Department of Biological Chemistry, Washington University
School of Medicine, St. Louis Missouri

(Received for publication, April 23, 1953)

Two forms of rabbit muscle phosphorylase were described by Cori and Green in 1943: phosphorylase *a* which exhibits 60 to 70% of maximal activity without the addition of adenylic acid (adenosine-5′-phosphate), and phosphorylase *b* which has no or minimal activity unless adenylic acid is added to the reaction mixture[(1)]. In the presence of adenylic acid, both forms are equally active. Extracts of muscle and other tissues were shown to contain an enzyme called PR (prosthetic group-removing) which converts phosphorylase *a* to the *b* form[1]. It was at that time believed that phosphorylase *a* contained firmly bound adenylic acid and that the PR enzyme removed it from the protein. However, in 1944 it was stated that all attempts to demonstrate free adenylic acid (or pentose) after PR action gave negative results[(2)]. Subsequent attempts to show the presence of adenine in phosphorylase *a* were also unsuccessful[(3)].

Molecular changes of greater magnitude than removal of a prosthetic group have recently been found to accompany the conversion of phosphorylase *a* to phosphorylase *b*. A smaller protein, which can be identified as phosphorylase *b*, is formed in the reaction catalyzed by PR. Sedimentation patterns in the ultracentrifuge show this second molecular species to be absent from solutions of crystalline phosphorylase *a* but present in all reaction mixtures in which conversion to phosphorylase *b* has occurred.

*Patricia J. Keller and Gerty T. Cori, "Enzymic Conversion of Phosphorylase *a* to Phosphorylase *b*," *Biochem. Biophys. Acta,* **12**, 235–238 (1953). Reprinted with the permission of the authors, the editor, and of the Elsevier Publishing Company.

†Fellow of the National Institutes of Health, United States Public Health Service. This report has been taken from a dissertation to be submitted by P. J. Keller in partial fulfilment of the requirements for the degree of Doctor of Philosophy in Biological Chemistry, Washington University.

The extent of conversion of phosphorylase *a* to phosphorylase *b* during PR action is determined by assaying enzymic activity in the presence and in the absence of adenylic acid. Results are expressed as % of specific activities (activity without/activity with adenylic acid × 100). From these data the relative proportions of the *a* and *b* forms of phosphorylase in reaction mixtures have been calculated[1].

Fig. 1A shows the sedimentation pattern of a 0.6% solution of phosphorylase *a*, which is homogeneous in the ultracentrifuge. This preparation was 66% as active in the absence as in the presence of adenylic acid. The activation of phosphorylase *a* by adenylic acid would seem from this to be real and not due to contaminating phosphorylase *b*. The sedimentation constant ($s_{20,w}$) of phosphorylase *a* is 13.2 Svedberg units.

Concomitant changes in ultracentrifugal and enzymic behavior when phosphorylase *a* is incubated with the PR enzyme have been followed. Figs. 1B, 1C, 1D, and 1E show the sedimentation patterns of reaction mixtures in which increasing amounts of phosphorylase *a* have been converted to the *b* form. The purified PR enzyme used in these experiments represented no more than three % of the total protein present, and the concentration did not exceed 0.015%. At this concentration no protein peak for the PR enzyme would be visible. A second protein component, with a sedimentation constant of 8.2, can be seen to arise at the expense of phosphorylase *a*. In Fig. 1E the conversion is almost complete. The area of the smaller protein in % of the total protein is at all times equivalent to the enzymically determined proportion of phosphorylase *b*. Addition of adenylic acid (0.001 M) to the enzyme solution during ultracentrifugation had no effect on the rate of sedimentation nor on the relative proportions of the two proteins.

The protein product of the PR-catalyzed reaction behaves as a single molecular species throughout the course of centrifugation. This suggested that phosphorylase *a* is split by PR into equal or nearly equal parts. Molecular weight determinations have shown this to be the case. The molecular weights of phosphorylases *a* and *b* have been calculated using the constants and formula listed in Table I. The partial specific volumes ($\overline{V}_{20}$) were determined, using the Linderstrom-Lang gradient tube as described by Taylor[6].

The sedimentation constant of phosphorylase *a* reported here, namely 13.2, is of the same order as that found by Oncley in 1943 which was 13.7. Phosphorylase *b* was not centrifuged at that time.

Table I

MOLECULAR CONSTANTS OF PHOSPHORYLASES *a* AND *b*

Constant	Phosphorylase *a*	Phosphorylase *b*
$S_{20,w}$*	13.2	8.2
$D_{20,w} \times 10^7$	2.6	3.3**
V_{20}	0.751	0.751
M.W.***	495,000	242,000

*expressed as Svedberg units
**determined by Green[5]
***calculated from $\text{M.W.} = \dfrac{RTs}{D(1-\overline{V}_{20})}$

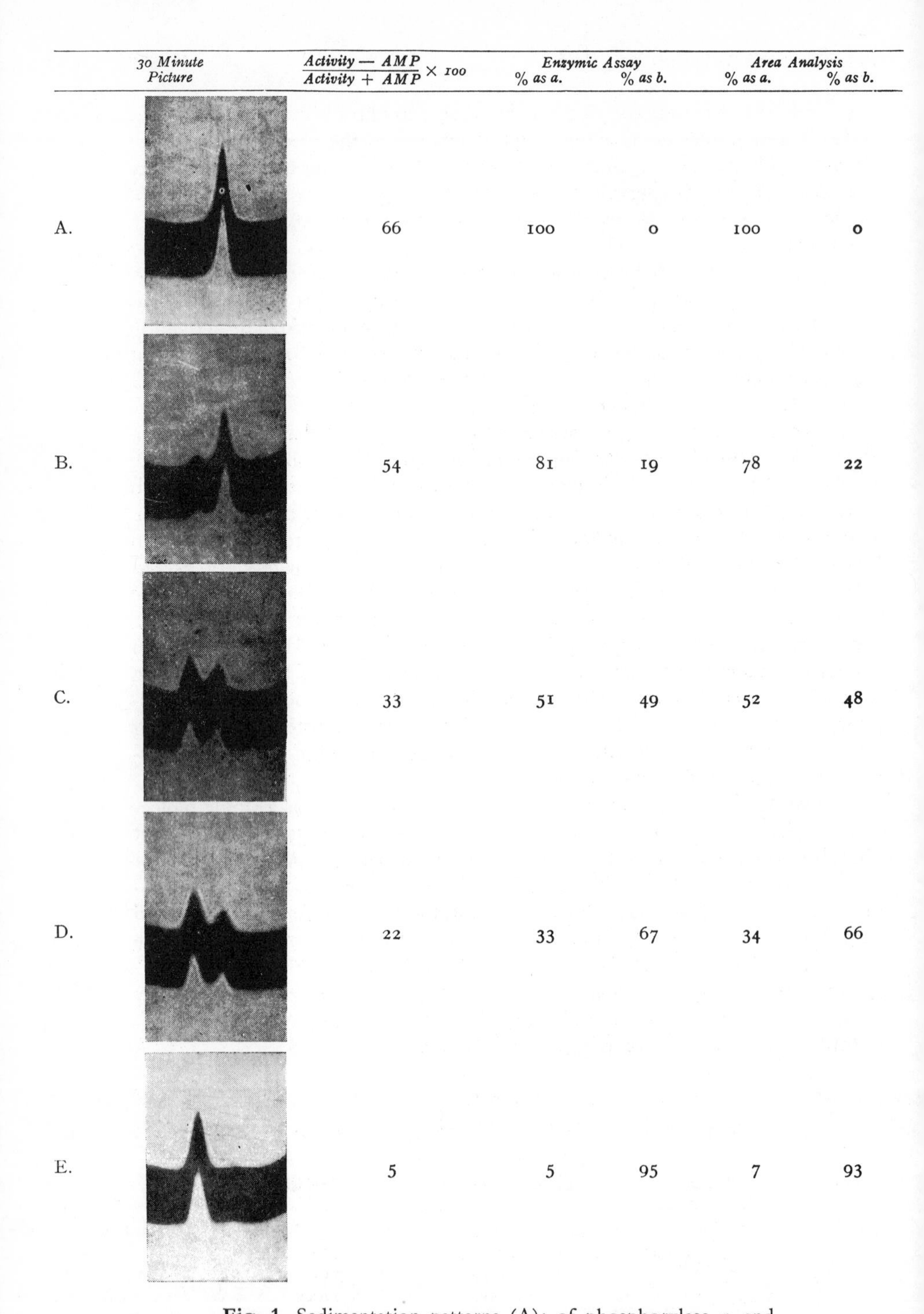

	30 Minute Picture	$\frac{\text{Activity} - AMP}{\text{Activity} + AMP} \times 100$	Enzymic Assay		Area Analysis	
			% as *a*.	% as *b*.	% as *a*.	% as *b*.
A.		66	100	0	100	0
B.		54	81	19	78	22
C.		33	51	49	52	48
D.		22	33	67	34	66
E.		5	5	95	7	93

Fig. 1. Sedimentation patterns (A): of phosphorylase *a*, and (B), (C), (D) and (E): of reaction mixtures of phosphorylase *a* and PR in which increasing amounts of phosphorylase *a* have been converted to phosphorylase *b*. Centrifugations were carried out in the Spinco ultracentrifuge operating at a speed of 59,780 r.p.m.

The diffusion constants for enzymes *a* and *b* call for some comment. Oncley[4] determined the diffusion constant for phosphorylase *a* in a neutralized cysteine-glycerophosphate buffer. The experiment was done at 25°C. Pictures were taken between 625 and 2036 minutes. Oncley reports formation of insoluble cystine as well as the appearance of skewed curves during the course of the experiment. Despite these difficulties a value between 3.2* and 3.8* seemed reasonable. The molecular weight calculated by Oncley was set between 340,000 and 400,000.

Green, in 1944, determined the diffusion constant of phosphorylase *b* and found it to be 3.3[5]*. The apparent similarity of diffusion constants for phosphorylase *a* and *b* pointed to molecular weights of the same order of magnitude.

In the present experiment, versene (ethylenediaminetetraacetic acid) was used instead of cysteine to solubilize phosphorylase *a*, and a five-time recrystallized preparation of enzyme was used. The experiment was run at 2°C. The same value (2.6) was found for ascending and descending limbs of the diffusion cell and the curves were symmetrical between 138 and 4000 minutes.

In retrospect it would seem that during Oncley's measurements phosphorylase *a* was being converted to *b* by traces of PR which may adhere to phosphorylase even through several recrystallizations.

How the PR enzyme effects the halving of phosphorylase *a* is the subject of further investigation. While prosthetic group-removing enzyme is a misnomer, it is proposed to retain the name PR (phosphorylase-rupturing enzyme). In chromatographic experiments neither peptide fragments nor free amino acids were found to be released in the *a* to *b* conversion. It may be recalled that crystalline trypsin will convert phosphorylase *a* to a form enzymically identical with phosphorylase *b*[2]. At *p*H 6 this reaction proceeds preferentially to general proteolysis. However, the PR enzyme shows neither proteolytic nor esteratic activity when tested with substrates which are acted upon by trypsin. The identity or non-identity of trypsin-formed phosphorylase *b* and the PR product is also being investigated.

Miss Carmelita Lowry's technical assistance is gratefully acknowledged.

Summary

Phosphorylase *b*, the product of the action of a muscle enzyme (PR) on crystalline rabbit muscle phosphorylase *a*, has one half the molecular weight of the *a* form. The molecular weights are 242,000 and 495,000 respectively.

References

1. G. T. Cori, and A. A. Green, *J. Biol. Chem.,* **151** (1943) 31.
2. G. T. Cori, and C. F. Cori, *J. Biol. Chem.,* **158** (1944) 321.
3. S. F. Velick, and L. F. Wicks, *J. Biol. Chem.,* **190** (1951) 741.
4. J. L. Oncley, *J. Biol. Chem.,* **151** (1943) 21.
5. A. A. Green, *J. Biol. Chem.,* **158** (1944) 315.
6. J. F. Taylor, *Federation Proc.,* **9** (1950) No. 1.

*$D_{20, w} \times 10^7$

THE RELATION OF ADENOSINE-3′, 5′-PHOSPHATE AND PHOSPHORYLASE TO THE ACTIONS OF CATECHOLAMINES AND OTHER HORMONES*

Earl W. Sutherland and T. W. Rall

Department of Pharmacology, Western Reserve University
School of Medicine, Cleveland, Ohio

Table of Contents

*Excerpt from Earl W. Sutherland and T. W. Rall, "The Relation of Adenosine-3′,5′-Phosphate and Phosphorylase to the Actions of Catecholamines and other Hormones," *Pharmacol. Rev.*, **12**, 265–299 (1960). Reprinted with the permission of the authors and of The Williams & Wilkins Co., Baltimore, Maryland.

Observations reported by the authors and co-workers as in preparation were supported in part by grants from the U. S. Public Health Service (H-2745 and HTS-5228) and from Eli Lilly and Company.

I. Introduction

Catecholamines are observed to produce effects in many different organs and tissues, these effects being manifest in a number of ways. Thus, the heart will beat more rapidly and more forcefully in response to catecholamines, while smooth muscle may contract or relax depending on its anatomical location, or on its physiological or pharmacological status, or both. The central nervous system may respond with evidences of heightened or decreased activity depending on experimental conditions. Many tissues display alterations in metabolism after exposure to catecholamines; an especially prominent metabolic effect is the increased glycogenolysis observed in liver and skeletal muscle. Other effects of catecholamines include changes in the secretory activity of the salivary glands, where a thick but sparse secretion occurs.

The cataloguing, classification, and quantitation of these diverse effects has led to extensive use of the concept of receptors in order to supply a unifying framework upon which to build an analysis of the action of sympathomimetic amines. Usually in these conceptual pictures, the receptor is viewed as a substance in or on a cell, which, when in combination with an active agent, may initiate or modify intracellular events resulting eventually in the observable response. For the most part the intervening events between receptor-agent interaction and observable response have remained obscure, due probably in part to the lack of knowledge of the basic biochemistry and biophysics involved in the observable response, such as muscular contraction, rhythmicity, or secretion. Also hampering such investigations has been the necessity of working with intact cells in order to have an observable response. However, in the case of the glycogenolytic action of epinephrine and related amines, the knowledge of some of the biochemical reactions involved in the control of glycogen breakdown and the ability to obtain observable effects in broken cell systems has led recently to the discovery of adenosine-3′, 5′-phosphate (cyclic 3,5-AMP). The formation of this compound appears to be a very early result of the interaction of active agent and receptor.

It will not be a purpose of the authors to review in general the actions of catecholamines. The metabolic effects of epinephrine and related amines were reviewed in detail by Ellis in this Journal in 1956[1]; the effects on the central nervous system have been reviewed even more recently by Rothballer[2]. The subject of sympathomimetic amine receptors and their blockage has been reviewed recently in this Journal by Furchgott[3], Ahlquist[4], Nickerson[5], and Slater and Powell[6]. Instead, an attempt will be made to survey the available information regarding the formation and properties of cyclic 3,5-AMP, with emphasis upon the relation of this compound to the action of catecholamines. In order to aid in the evaluation of the role of cyclic 3,5-AMP, an attempt will be made to survey the current status of the relation of phosphorylase activation to the glycogenolytic and other actions of catecholamines. Finally, an attempt will be made to discuss some of the implications which arise from consideration of the role of cyclic 3,5-AMP in the action of catecholamines and other hormones.

II. Adenosine-3',5'-Phosphate

A. History and chemical properties

Adenosine-3',5'-phosphate (cyclic 3,5-AMP) was discovered in 1957 by two separate groups of investigators. Its formula is shown in Figure 1. Cook, Lipkin and Markham were investigating the hydrolysis of adenosine triphosphate (ATP) in the presence of barium hydroxide and found several products besides the major products, adenylic acid and inorganic pyrophosphate, one of these being the compound under discussion[7][8][9]. During the same period of time, a heat stable factor, the accumulation of which was increased by epinephrine or glucagon[10][11][12] was isolated from hepatic tissue and was crystallized[12][13]. Both groups determined that the compound had an adenine, ribose, phosphate ratio of 1:1:1 and contained no monoesterified phosphate, and wrote separately to Dr. Leon Heppel for a purified enzyme which might help in the elucidation of its structure. Dr. Heppel informed the groups that the tentative structures proposed were identical, and, as a consequence of this information, samples were exchanged and found to be identical before the final structure was assigned. In an early report[7] it was thought that the compound was a dinucleotide, but subsequent studies showed that the compound was a mononucleotide of adenylic acid with the phosphate attached to the 5-position of the ribose and also to the 3-position[8], as shown in Figure 1.

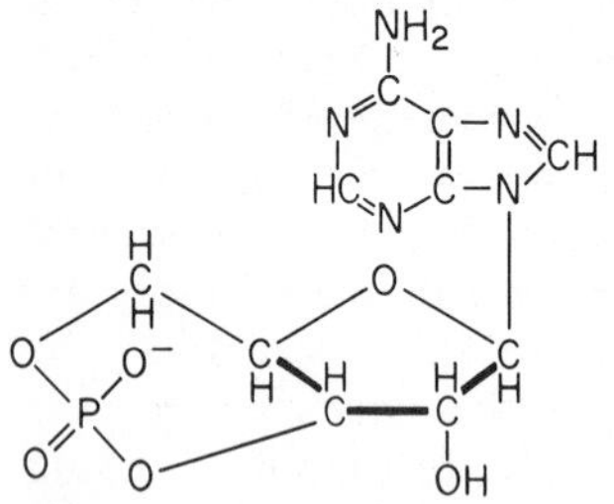

Fig. 1. Adenosine-3',5'-phosphate.

The ultraviolet spectrum and the molar extinction coefficient of cyclic 3,5-AMP are essentially the same as those of adenosine-5-phosphate. The compound is very stable chemically (*e.g.*, when compared to adenosine 2-', 3'-phosphate) and may be boiled at neutral, or slightly acid or alkaline *p*H for one-half hour or more with no appreciable loss of activity[12]. It has been synthesized by two relatively simple procedures, one by hydrolysis of ATP in the presence of barium hydroxide as mentioned[9], the other by dehydration of adenylic acid in the presence of dicyclohexylcarbodiimide, a procedure used by Khorana *et al.*[14][15] and by Lipkin *et al.*[8].

B. Formation of cyclic 3,5-AMP by tissue preparations

1. *Preparation and properties of the cyclizing enzyme*

Formation of cyclic 3,5-AMP was demonstrated first when particles from liver were incubated with ATP, magnesium ions, and other additives[11][13]. This finding stemmed from the observation that the effect of epinephrine in homogenates, leading to increased levels of active phosphorylase, was lost if a particulate fraction was removed by centrifugation, even with low gravitational forces (see III, A-1-c).

The material which catalyzes the formation of cyclic 3,5-AMP from ATP will be referred to as the cyclizing enzyme in this discussion, even though more than one enzyme may be involved. A more appropriate name, such as adenyl cyclase, may be used in the future, depending on the outcome of investigations on specificity and on the mechanism of the reaction. Although studied for approximately two years, little published information regarding its properties is available. Preparation of particles containing this enzyme has been described, as well as conditions of incubation for formation of cyclic 3,5-AMP[10]. More recent experiments[16] have shown that particles from liver or muscle may be washed in hypotonic solutions, then in hypertonic solutions, and subsequently can be dispersed or solubilized in the presence of 1.8% Triton-X-100. The concentrations of certain other enzymes may be reduced by fractionation on cellulose columns. The cyclizing enzyme, however, is relatively labile, behaves much as a lipoprotein, and has been quite difficult to study because of these characteristics. In addition, a considerable amount of other protein with similar characteristics is present in the Triton solution. All the protein including the cyclizing enzyme is difficult to separate from the detergent. Several other enzymes are known to be present; the only one studied in any detail is an ATPase, which, like the cyclizing enzyme, is readily inactivated by freezing at —20°C, but which may be separated in large part from the cyclizing enzyme with use of cellulose column chromatography.

The usual system employed for production of cyclic 3,5-AMP by tissue preparations contains buffer, ATP, magnesium ions, caffeine, and fluoride ions. The addition of fluoride appears to stimulate maximally the production of cyclic 3,5-AMP and in most cases, little additional effect of epinephrine is observed in the presence of 0.01 M NaF. Frequently, 0.02% to 0.5% of the ATP (0.002 M) is converted to cyclic 3,5-AMP.

The partially purified cyclizing enzyme is often inactivated by slow freezing, or storage at —20°C, even after previous rapid freezing to —70°C or lower. Alcohol-ether extractions of active precipitates cause inactivation, as does heating or incubation with the proteolytic enzymes, trypsin or chyomotrypsin. Partially purified preparations may be frozen rapidly and stored at —70°C for days or weeks with little loss of activity.

The mechanism of the cyclization reaction has been explored to some extent, using particulate preparations from dog liver and Triton extracts of particles from dog skeletal muscle. Liver particles catalyzed the formation of C^{14} cyclic 3,5-AMP from 8-C^{14}-ATP, with no decrease in specific activity during the process, thus indicating that the purine portion was derived from the added ATP, rather than from tissue sources[10]. In other experiments, using extracts of skeletal muscle particles as catalyst, cyclic 3,5-AMP was synthesized from α-labeled ATP (*i.e.*, adenine-ribose-P^{32}-O-P^{31}-O-P^{31}), and also from β- and γ-labeled ATP (*i.e.*, adenine-ribose-P^{31}-O-P^{32}-O-P^{32})[17]. Cyclic 3,5-AMP, derived from α-labeled ATP, contained the isotope in good yield, while cyclic 3,5-AMP prepared from the terminally labeled ATP contained only trace amounts of isotope. These findings suggest that the cyclization process may involve a relatively simple mechanism, perhaps analogous to the chemical synthesis from ATP catalyzed by barium hydroxide.

Broken cell preparations have been employed in almost all studies of the formation of cyclic 3,5-AMP. Haynes has reported a study of its formation in slices from beef adrenals and the effect of ACTH on the formation within the slice[18]. Column chromatography was employed at times to concentrate the compound, since the concentration in extracts of slices was often below that required for assay. Studies of this type involving slices or whole organs may supply more valuable information in the future, even though the assay and characterization of small amounts of cyclic adenylic acid in tissues appears to be difficult[19].

2. *Location of cyclizing enzyme*

a. Distribution in tissues: The cyclizing enzyme appears to be present in every animal tissue that has been examined with the possible exception of dog blood cells. No systematic and thorough study of distribution has been made, however, and in most cases biological assay alone has been used as evidence for the formation of cyclic 3,5-AMP by tissue preparations. The biological assay is based on the ability of cyclic 3,5-AMP to increase the rate of formation of phosphorylase from dephospho-phosphorylase in tissue extracts[10][11][20].

Using not only biological assay, but also isolation and other characterization procedures, formation of cyclic 3,5-AMP has been clearly demonstrated with preparations from liver, heart, skeletal muscle, and brain of the dog[12], with preparations from beef adrenal cortex[18] and with preparations from the flatworm, *Fasciola hepatica*[21]. As judged by biological assay, the cyclizing enzyme was present in the spinal cord[22], kidney, intestinal mucosa, intestinal muscle, aorta, uterus, testis, lung, spleen, and omental fat of the dog[16]. Active preparations were obtained from the liver of cat, rat and chicken, from the brain of cat, pigeon, ox, pig and sheep, and from blood cells of chickens and pigeons. Some apparent activity was noted in chub minnows and in fly larvae. The annelid, *Lumbricus terrestis,* was quite active in forming cyclic 3,5-AMP, again as judged by biological assay[16]. To date, the crude preparations most active in forming the cyclic nucleotide have been from *F. hepatica* and from mammalian brain cortex.

The cyclizing enzyme, therefore, is widely distributed in animal tissue and appears to be present in all four of the phyla that have been investigated. Its presence in the widely separated phyla, Chordata and Platyhelminthes, has been well documented.

b. Cellular location: In general, most of the cyclic 3,5-AMP-forming activity of broken cell preparations is found in the particulate fraction collected with low gravitational forces; for example, 600 × *g* (or 100 × *g*) for 17 minutes has been used for collection of active particles from liver. (A variable amount of activity, 20% to 50% or occasionally more, may remain in the supernatant fractions above this precipitate.) The precipitate collected at 600 × *g* or 1000 × *g* may be washed several times and the activity may be recollected several times at the same low forces with high recovery of activity. Fractions of this type have been called "nuclear" fractions by many workers, and obviously attention was called to the possibility that the cyclizing enzyme might be associated with the nuclei.

Support for this possibility comes from two sources. Chicken and pigeon erythrocytes contain nuclei and also contain readily demonstrable amounts of cyclizing enzyme, while the erythrocytes of dog are anucleate and contain little or no cyclizing enzyme when tested under the same conditions(16). In addition, a study was made of the specific activities of the "nuclear" fraction, of the mitochondrial fraction, and the mitochondria and microsomes trapped in the first "nuclear" precipitate(23)(16). Guides to fractionation were gravitational forces, microscopic examination, and determination of an enzyme activity that is localized in mitochondria (cytochrome oxidase) and one that is concentrated in microsomes (inorganic pyrophosphatase). Fractionation followed by repeated refractionation of the nuclear fraction showed clearly that cyclic 3,5-AMP-forming activity was not associated with the mitochondria nor with the microsomes, but with some portion of the "nuclear" fraction. Although an early report showed that particulate fractions free of intact nuclei were active(11), it could be proposed that cyclizing activity was still associated with broken nuclei, and that this might even explain the losses of activity into supernatant fractions on initial centrifugation.

While the cyclizing enzyme may be closely associated with the nucleus, other possibilities exist, with the cell membrane being a leading contender for consideration. The chicken erythrocyte preparations contained cell membranes, as well as the nuclei. Although the dog erythrocyte preparations contained cell membranes, one could speculate that the membranes from these anucleate cells were relatively inactive at this advanced stage of life. Furthermore, some cell membranes have been noted in liver "nuclear" preparations for some time, although one might expect membranes to sediment with difficulty because of their high lipid content. The recent report of Rajam and Jackson is of considerable interest in relation to this question. They studied the behavior of cell membranes from ascites tumor cells during centrifugation, using antigen-antibody techniques, as well as microscopy. They report that the cell membranes sediment readily at low forces and appear at the very bottom of the tubes after centrifugation(24). Finally, the dispersal of rather large amounts of protein of apparent lipoprotein nature (including the cyclizing enzyme) with high concentrations of Triton-X-100 lead one to suspect that the particle has considerable amounts of lipid material in association.

In summary the particles associated with the cyclizing enzyme sediment readily and are not mitochondria or microsomes. The two most likely candidates for specific association with cyclizing activity are nuclei and cell membranes. Further studies with avian erythrocytes may clarify this perplexing question.

* * *

5. *Possible mechanism of action of catecholamines on particulate preparations*

The exact mechanism of action of the catecholamines remains unknown. Several possibilities, however, are apparent at this early stage. The cyclizing enzyme itself may be stimulated, or an inhibitor of the enzyme may be neutralized in some fashion. If the actual reaction involves more than one enzyme, the same considerations would apply to one or more of the participating enzymes. Another possibility is that some intermediate is formed and that this intermediate is protected by the hormone. The cyclizing enzyme might catalyze the formation of some compound other than cyclic 3,5-AMP as the major product of its enzymatic activity, *e.g.,* formation of a 3,5-bond in a polynucleotide. In this case the major reaction might be diverted by the hormones into the formation of cyclic 3,5-AMP. Or perhaps, the cyclizing enzyme is embedded in a lipoprotein complex in close proximity to another enzyme utilizing ATP. If the activity of the neighbor were to be suppressed by catecholamines, the cyclizing enzyme could have better access to the substrate (ATP) and thus form more cyclic 3,5-AMP. In such a case the "receptor" would be the neighboring enzyme and not the cyclizing enzyme. At the present time, such speculations are of interest, primarily because they indicate possible experimental approaches.

It is unlikely that the stimulation of the accumulation of cyclic 3,5-AMP by catecholamines is due to a protection of cyclic 3,5-AMP itself. Attempts to show effects of catecholamines on phosphodiesterase preparations that inactivate cyclic 3,5-AMP have been uniformly negative. Furthermore, preparations forming cyclic 3,5-AMP and responding to addition of catecholamines have been purified to the stage where essentially all phosphodiesterase activity was absent. In these and cruder preparations no effect of the catecholamines on cyclic 3,5-AMP concentration has been noted, unless ATP was present in the media. Therefore, unless some unknown ATP-dependent inactivating system was present, the effect of the amines would be at some other site. Although the presence of such an inactivating system appears to be an unlikely possibility, it should be noted that an ATP-stimulated adenylic acid deaminase has been described by Mendicino and Muntz(25) and by Waitzman(26).

* * *

F. Action of cyclic 3,5-AMP

1. *Activation of phosphorylase*

It is clear that cyclic 3,5-AMP promotes the accumulation of active phosphorylase from inactive phosphorylase in a number of tissues. This has been demonstrated in homogenates and extracts of liver and in extracts of heart of the dog(10). This effect of cyclic 3,5-AMP has also been observed in extracts of rabbit muscle by Krebs, Graves and Fischer(27) and in extracts of adrenal glands by Riley and Haynes(28). The effects in extracts or homogenates are produced by final concentrations of cyclic 3,5-AMP in the 10^{-7} M range(10). These concentrations are near the concentrations of catecholamines required for the stimulation of active phosphorylase formation in liver homogenates (see III, A-1-c).

The effect of cyclic 3,5-AMP on active phosphorylase concentration has been observed with two preparations that contained intact cells, liver slices[29][12] and adrenal slices[18]. In both cases relatively large amounts of the cyclic compound were required for a consistent effect. With liver slices, concentrations in the 10^{-5} M range were effective, while with adrenal glands even higher concentrations were required (8×10^{-4} M). These findings have been interpreted as indicating that the cyclic nucleotide enters cells very poorly, since much lower concentrations were effective in broken cell preparations. Two experiments carried out by Dr. Roger Jelliffe and the authors with anesthetized dogs showed only a slight rise in blood sugar following the intravenous injection of up to 4 mg/kg of cyclic 3,5-AMP. Part of the injected cyclic 3,5-AMP was found in the urine of the dog.

Experiments with adrenal sections from rats have been reported by Haynes, Koritz and Peron, in which the effect of cyclic 3,5-AMP on steroid production was studied. Clear-cut and large effects were observed using 2×10^{-3} M solutions of cyclic 3,5-AMP[30], but were not observed using five related nucleotides. According to the theory proposed by Haynes[18], the increased steroid production would be related to the increased levels of adrenal phosphorylase on incubation with cyclic 3,5-AMP.

In the studies mentioned above, attempts have been made to evaluate the specificity of the response to cyclic 3,5-AMP. A number of related compounds have been inactive in the various systems. When tested with liver preparations and using phosphorylase activation as the indication of activity, cyclic 3,5-uridylate, synthesized by Khorana *et al.,* and cyclic 3,5-inosinic acid, prepared by Lipkin by deamination of cyclic 3,5-AMP, were found to have about 0.4% and 2% of the activity of cyclic 3,5-AMP, respectively[31]. Thus, the system responding to cyclic 3,5-AMP by activation of phosphorylase seems to be much more sensitive to changes in the base than was the phosphodiesterase.

The exact mechanism of action of cyclic 3,5-AMP on the activation of phosphorylase remains unknown. The activation of phosphorylase is brought about by transfer of phosphate from ATP to inactive phosphorylase (dephosphophosphorylase) catalyzed by a kinase which is discussed in detail in III, A-1-c. The activation of phosphorylase can proceed in the apparent absence of the cyclic nucleotide; ATP, but not added cyclic 3,5-AMP, appears essential. When kinase preparations that catalyze the formation of active phosphorylase from inactive phosphorylase are purified, they continue to require ATP for activity, but are not affected by addition of the cyclic nucleotide. In cruder preparations, such as homogenates or extracts, the kinase appears to be restrained by an unknown factor(s), and when this restraint is operative, effects of cyclic 3,5-AMP can be observed. In liver extracts, prepared by collecting the supernatant fraction of homogenates after centrifugation at $11{,}000 \times g$, effects of cyclic 3,5-AMP are prominent. After centrifugation at $100{,}000 \times g$, the effects of cyclic 3,5-AMP may be less prominent, because the kinase activity without cyclic 3,5-AMP is greater; the absolute increase in kinase activity due to cyclic 3,5-AMP addition, however, may be as great as that noted in extracts prepared by centrifugation at $11{,}000 \times g$ (100). Thus, one restraining factor, removable by centrifugation at 100,000 $\times$ *g*, although not essential for the action of 3,5-AMP, apparently may magnify its action. Krebs and Fischer[27][32] have reported that phosphorylase *b* kinase from skeletal muscle may be extracted in a form which is completely inactive at *p*H 7.0 or below, but active at higher *p*H. Cyclic 3,5-AMP plus ATP or calcium ions activate the inhibited or inactive kinase. In liver extracts[33] zinc ions and, to a lesser extent, copper ions, increased the kinase activity, while trypsin in small amounts is a very active stimulant[10]. Makman *et al.*[34] have noted a factor in plasma that increases phosphorylase activation in liver extracts. It is obvious that a number of agents influence the rate of phosphorylase activation by the kinase and the relation of these to the action of cyclic 3,5-AMP is not clear. Possibly, the kinase occurs in an inactive or inhibited form that can be activated by several agents. Cyclic 3,5-AMP may directly or indirectly displace or neutralize an inhibitor or a group attached to the kinase, or may aid the transfer of phosphate to inactive phosphorylase in another unrecognized manner.

III. The Relation of Phosphorylase Activity to the Action of Catecholamines

A. Individual tissues

1. *Liver*

* * *

c. Broken cell preparations: It was known for some time that the concentration of phosphorylase in homogenates and crude extracts of liver would decline rapidly unless NaF was present[35]. Subsequently, an "inactivating" enzyme inhibited by fluoride was purified several hundred-fold from liver extracts[36]. When highly purified liver phosphorylase became available[37], it was observed that the enzymatic inactivation of phosphorylase was accompanied by the release of small amounts of inorganic phosphate[38]. Both the fully active and maximally inactivated phosphorylase migrated in the ultracentrifuge at the same rate, both proteins having an apparent molecular weight of around 240,000[36]. The amount of inorganic phosphate appearing during the inactivation process was calculated to be two molecules per molecule of phosphorylase. The "inactivating" enzyme, apparently belonging to the category of protein phosphatases, was named phosphorylase phosphatase.

The product of the phosphatase reaction displays about 2 to 3% of the enzymatic activity of the original phosphorylase when assayed in the absence of adenosine-5′-phosphate and only about 10 to 15% of the original activity when assayed in the presence of adenosine-5′-phosphate. While deserving of the name "inactive phosphorylase," the product of the phosphorylase phosphatase reaction has been named dephospho-phosphorylase to call attention to the chemical nature of the change involved and to provide a name that could also be applied to the product of the muscle phosphorylase *a* to *b* conversion. It was hoped that this would avoid the confusion that arises from the fact that muscle phosphorylase *b* displays nearly the same enzymatic activity as the *a*

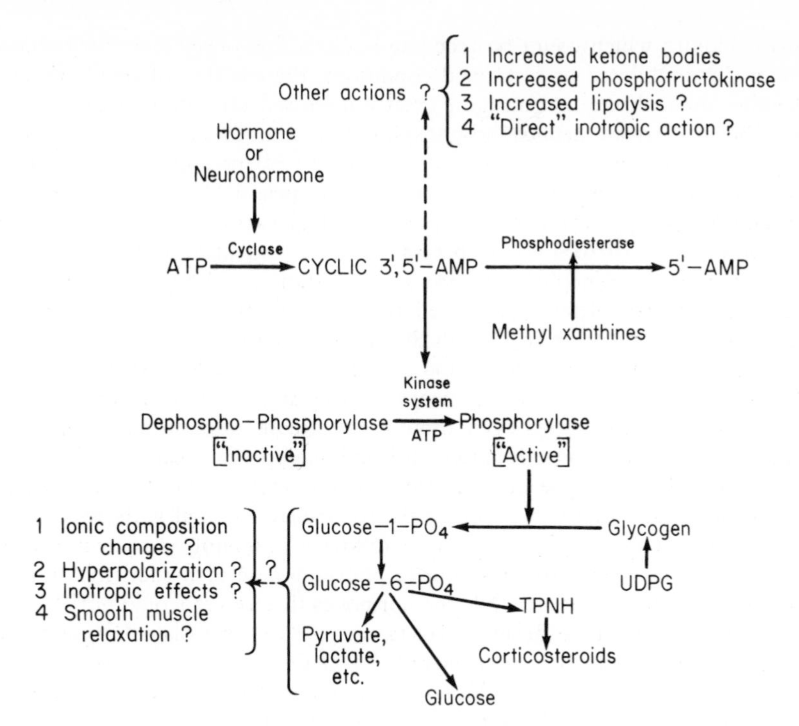

Fig. 2. Summary of some interrelationships involving adenosine-3′,5′-phosphate.

form when assayed in the presence of sufficient quantities of adenosine-5′-phosphate. Since dephospho-phosphorylase still contains perhaps as many as six molecules of TCA-insoluble, esterified phosphate per molecule of protein, the prefix "dephospho-" should not be construed to indicate the complete absence of phosphate, in the same sense that "dehydro" does not indicate the complete absence of hydrogen atoms in a molecule.

Radioactive phosphorylase was recovered from slices of dog liver that had been incubated so as to contain a high proportion of dephospho-phosphorylase and then had been exposed to epinephrine or glucagon and P^{32} orthophosphate. More than 70% of the radioactivity could be recovered as orthophosphate after treatment of the labeled enzyme with phosphorylase phosphatase. Thus, it appeared that the reactivation process involved phosphorylation of the inactive enzyme, as would be predicted from the fact that inactivation involved dephosphorylation[39, 38, 36]. The radioactivity in the reactivated phosphorylase appeared in the phosphate esterified to serine residues[40].

An enzyme, dephospho-phosphorylase kinase, was then detected and was partially purified from extracts of liver which would transform dephospho-phosphorylase into phosphorylase in the presence of Mg^{++} ions and the proper phosphorylating agent, ATP[39]. In crude, relatively concentrated homogenates of liver, the transformation or reactivation process progressed more slowly than in more dilute homogenates, or than would be expected from the amount of kinase activity which could be recovered after a few purification procedures. Under these inhibited or restrained conditions, the addition of small amounts of either epinephrine or glucagon greatly increased the rate of formation of phosphorylase from dephospho-phosphorylase, the amount of the increase being related to the amount of hormones added[11]. Under optimal conditions, the half-maximal effect was usually observed at near 1×10^{-7} M epinephrine (around 0.02 μg per ml) and near 1×10^{-8} M glucagon[42]; *l*-norepinephrine was only about 10 to 15% as potent[42, 11] and *dl*-isopropylarterenol about twice as potent as epinephrine[41]. At a concentration of 100 μg per ml, ergotamine tartrate completely suppressed the effect of up to 2 μg per ml of epinephrine, while not interfering with the effect of glucagon[42].

It was found that epinephrine and glucagon were unable to stimulate phosphorylase formation in broken cell preparations that had been centrifuged at low speeds (1200 to 2000 $\times$ *g*) even though the kinase reaction also proceeded slowly in such supernatant fractions. Recombination of a small amount of the low-speed pellet with the supernatant fraction restored the hormone effect. It was then found that incubation of the particulate material with Mg^{++} ions, ATP, and the hormones for a short time followed by boiling and centrifugation yielded extracts which would stimulate phosphorylase formation in the low-speed supernatant fractions, while the hormones themselves would not[11]. The active principle in these heated extracts was eventually isolated[13], and was identified as adenosine-3′,5′-phosphate[8, 12]. As discussed in previous sections, the effect of epinephrine and glucagon was more directly concerned with the conversion of ATP to adenosine-3′,5′-phosphate, and the ability to influence phosphorylase concentration was thus a property of the cyclic nucleotide.

The stimulatory influence of adenosine-3′,5′-phosphate on the dephospho-phosphorylase kinase reaction depends both upon the inhibited or partially active state of the kinase in crude preparations and upon the response of some system to the cyclic nucleotide acting to overcome or relieve the restraint. While the opposition of phosphorylase phosphatase is a very important restraint upon the accumulation of the product of the kinase reaction, the effect of adenosine-3′,5′-phosphate is undiminished in aged preparations exhibiting little or no phosphorylase phosphatase activity and is even augmented in the presence of NaF[33], an inhibitor of the phosphatase. The reader is referred to a previous section (see II, F) for further discussion of the relationship of cyclic 3,5-AMP and dephospho-phosphorylase kinase activity.

Broken cell preparations of dog liver, from whole homogenates to 100,000 $\times$ *g* supernatant fractions, respond to graded concentrations of cyclic 3,5-AMP of the order of 10^{-7} M with graded increments of phosphorylase formation[10, 11]. Using 11,000 $\times$ *g* supernatant fractions of dog liver homogenates, supplemented with partially purified dephospho-phosphorylase, an assay system was devised to estimate the content of cyclic 3,5-AMP in various biological materials and in experimental samples. The various non-cyclic adenine mononucleotides, as well as adenosine-2′,3′-phosphate, had no detectable effect in this system[10]. The inosine and uridine analogues of adenosine-3′,5′-phosphate, prepared and kindly supplied by Dr. D. Lipkin and Dr. H. G. Khorana, respectively, were relatively inactive (see II, F). It will be of great interest to determine the effect of other analogues, prepared as potential mimetic or blocking agents on broken cell systems, as well as on intact cell preparations in which cyclic 3,5-AMP is inactive or acts only at high concentrations.

2. *Skeletal muscle*

Working primarily with skeletal muscle, the Cori's carried out their classical work with phosphorylase over a period of years, which led to crystallization of the enzyme from rabbit muscle extracts and detailed studies of the properties of the enzyme[43-51]. The enzyme catalyzed the readily reversible reaction: glycogen + inorganic phosphate $\rightleftarrows$ glycogenglucose residue + glucose-1-phosphate, with the equilibrium of the reaction in favor of glycogen synthesis. Phosphorylase *a* could be converted to phosphorylase *b* which was inactive when assayed in the absence of adenosine-5′-phosphate, but fully active in the presence of adenosine-5′-phosphate; the molecular weight of phosphorylase *b* was approximately one-half that of the *a* form. (In these two respects the inactive phosphorylase (dephospho-phosphorylase) from liver differs from that from muscle, *i.e.,* the inactive form from liver shows little or no activity when assayed with adenosine-5′-phosphate, and the molecular weights of the active and inactive forms are the same as discussed in the previous section.) It was assumed that phosphorylase catalyzed both the synthesis and degradation of glycogen *in vivo* as it did *in vitro.*

The concept of the roles of phosphorylase *a* and *b* in relation to the function of skeletal muscle has been revised in recent years. In 1945, it was concluded[45] that resting muscle of rabbits contains mainly phosphorylase *a*, while during strong contractions, phosphorylase *a* is converted to the *b* form. Later, however, it was shown that epinephrine increased the relative amount of phosphorylase *a* in rat diaphragm[52]. These results with rat diaphragms did not bear directly on the problem of resting *versus* exercised muscle, but were puzzling since both epinephrine and muscular contractions led to glycogen breakdown and lactic acid production. Krebs and Fischer[53] restudied the problem of phosphorylase activity in rabbit muscle and found that essentially all the phosphorylase of resting rabbit muscle was in the *b* form. At the same time, Fischer and Krebs reported that phosphorylase *b* can be converted to phosphorylase *a* in muscle extracts in the presence of ATP and a divalent ion[54]. In subsequent reports they have pursued the studies of the interconversions of rabbit muscle phosphorylase and conclude that ATP donates phosphate to the *b* form, the phosphate being esterified to serine residues in the enzyme[55,56]. The conversion of phosphorylase *a* to *b* represents a phosphatase action[57], and the overall process would be basically the same as that occurring in the interconversions of the enzyme from liver and heart. Krebs, Graves and Fischer[27][32] have reported that calcium ions may increase the rate of formation of active phosphorylase in extracts of rabbit muscle.

In 1956, Dr. Carl Cori reported results from a restudy of the problem of phosphorylase activity in resting and active muscle[58]. Using both frogs and rats, he reported confirmation of the observations of Krebs and Fischer[53], *i.e.,* the active form, phosphorylase *a*, is relatively low during rest and increases rapidly during contraction. With fatigue, the *a* form decreases. Epinephrine injection into rats caused a marked increase in the phosphorylase *a* content of skeletal muscles. The increase occurred within one minute after intravenous injection; subcutaneous injection caused a sustained increase of the phosphorylase *a* content of muscle. Epinephrine also caused an increase in the active phosphorylase content of isolated frog muscle. These findings were also reported by Cori and Illingworth[59].

As a result of these more recent investigations, the following concept has gained acceptance. Resting skeletal muscle contains largely inactive phosphorylase, *i.e.*, phosphorylase *b* or dephospho-phosphorylase. On stimulation the active phosphorylase content increases rapidly. The factors involved in this rapid activation of phosphorylase have not been elucidated. The active phosphorylase of skeletal muscle decreases on prolonged stimulation leading to fatigue. Addition of epinephrine to muscle *in vivo* and *in vitro* produces an increased content of active phosphorylase.

Ellis has studied the biochemical effects of epinephrine in relation to its effects on muscle and has reviewed this area recently in this Journal[60]. Readers are referred to this article for references and interpretations of work in this area. In brief summary, activation of glycogenolysis may be important for the muscular effects of epinephrine, but the picture is not at all clear. Perhaps hexosephosphates are implicated in the control of contractility, as was proposed by Ellis previously[61]. Muscular contraction itself is complex and the sequence of events occurring in contraction has not been clarified; therefore, one might anticipate some difficulty in attempting to relate the complex events of glycogenolysis to the unknown sequence of events occurring during contraction. In any event, Ellis has shown that the relative potencies of the sympathomimetic amines in producing inotropic effects in rat diaphragm were correlated with the relative potencies in producing glycogenolysis[62].

* * *

References*

1. Ellis, S.: The metabolic effects of epinephrine and related amines, *Pharmacol. Rev.,* **8**: 485–562, 1956.
2. Rothballer, A. B.: The effects of catecholamines on the central nervous system, *Pharmacol. Rev.,* **11**: 494–547, 1959.
3. Furchgott, R. F., The receptors for epinephrine and norepinephrine, *Pharmacol. Rev.,* **11**: 429–441, 1959.
4. Ahlquist, R. P.: The receptors for epinephrine and norepinephrine, *Pharmacol. Rev.,* **11**: 441–442, 1959.
5. Nickerson, M.: Blockade of the actions of adrenaline and noradrenaline, *Pharmacol. Rev.,* **11**: 443–461, 1959.
6. Slater, I. H., and Powell, C. E.: Some aspects of blockade of inhibitory adrenergic receptors or adrenoceptive sites, *Pharmacol. Rev.,* **11**: 462–463, 1959.
7. Cook, W. H., Lipkin, D., and Markham, R.: The formation of a cyclic dianhydrodiadenylic acid by the alkaline degradation of adenosine-5′-triphosphoric acid., *J. Amer. Chem. Soc.,* **79**: 3607–3608, 1957.
8. Lipkin, D., Cook, W. H., and Markham, R.: Adenosine-3′: 5′-phosphoric acid: a proof of structure, *J. Amer. Chem. Soc.,* **81**: 6198–6203, 1959.
9. Lipkin, D., Markham, R., and Cook, W. H.: The degradation of adenosine-5-triphos-

*Only those references that accompany the reprinted portion of this article have been reproduced here. They have been numbered consecutively.

phoric acid by means of aqueous barium hydroxide. *J. Amer. Chem. Soc.,* **81**: 6075–6080, 1959.

10. Rall, T. W., and Sutherland, E. W.: Formation of a cyclic adenine ribonucleotide by tissue particles, *J. Biol. Chem.,* **232**: 1065–1076, 1958.
11. Rall, T. W., Sutherland, E. W., and Berthet, J.: The relationship of epinephrine and glucagon to liver phosphorylase. IV. The effect of epinephrine and glucagon on the reactivation of phosphorylase in liver homogenates, *J. Biol. Chem.,* **224**: 463–475, 1957.
12. Sutherland, E. W., and Rall, T. W.: Fractionation and characterization of a cyclic adenine ribonucleotide formed by tissue particles, *J. Biol. Chem.,* **232**: 1077–1091, 1958.
13. Sutherland, E. W., and Rall, T. W.: The properties of an adenine ribonucleotide produced with cellular particles, ATP Mg^{++} and epinephrine or glucagon, *J. Amer. Chem. Soc.,* **79**: 3608, 1957.
14. Khorana, H. G., Tener, G. M., Wright, R. S., and Moffatt, J. G.: Cyclic phosphates III. Some general observations on the formation and properties of 5-, 6- and 7-membered cyclic phosphate esters, *J. Amer. Chem. Soc.,* **79**: 430–436, 1957.
15. Tener, G. M., Khorana, H. G., Markham, R., and Pol, E. H.: Study on polynucleotides. II. The synthesis and characterization of linear and cyclic thymidine oligonucleotides. *J. Amer. Chem. Soc.,* **80**: 6223–6230, 1958.
16. Sutherland, E. W., Rall, T. W., and Menon, T.: (with the technical assistance of James Davis and Arleen Maxwell), Distribution, preparation and properties of the cyclizing enzyme forming adenosine-3′,5′-phosphate. In preparation.
17. Rall, T. W., and Sutherland, E. W.: (with the technical assistance of Arleen Maxwell and James Davis), Mechanism of the cyclizing reaction forming adenosine-3′,5′-phosphate. In preparation.
18. Haynes, R. C., Jr.: The activation of adrenal phosphorylase by the adrenocorticotropic hormone, *J. Biol. Chem.,* **233**: 1220–1222, 1958.
19. Butcher, R. W., Jr., Sutherland, E. W., and Rall, T. W.: Measurement of adenosine-3′, 5′-phosphate in heart and other tissues, *The Pharmacologist* **2**: No. 2, 66, 1960.
20. Rall, T. W., and Sutherland, E. W.: Dephospho-phosphorylase kinase, In: Methods in Enzymology, Vol. V, edited by S. P. Colowick and N. O. Kaplan, New York: Academic Press. In press, 1960.
21. Mansour, T. E., Sutherland, E. W., Rall, T. W., and Bueding, E.: The effect of 5-hydroxytryptamine (serotonin) on the formation of adenosine-3′,5′-phosphate by tissue particles from the liver fluke, Fasciola, hepatica, *J. Biol. Chem.,* **235**: 466–470, 1960.
23. Sutherland, E. W., Rall, T. W., and Menon, T.: Production of a cyclic adenine ribonucleotide produced with cellular particles. In Abstract Amer. Chem. Soc. Div. of Biol. Chem., p. 8C, April 13, 1958.
24. Rajam, P. C., and Jackson, A.: A cytoplasmic membrane-like fraction from cells of the Ehrlich mouse ascites carcinoma, *Nature,* **181**: 1670–1671, 1958.
25. Mendicino, J., and Muntz, J. A.: The activating effect of adenosine triphosphate on brain adenylic deaminase, *J. Biol. Chem.,* **233**: 178–183, 1958.
26. Waitzman, M. B., Adenylic acid deaminase activity at the site of aqueous humor production, *Fed. Proc.,* **18**: 165, 1959.
27. Krebs, E. G., Graves, D. J., and Fischer, E. H.: Control of phosphorylase *b* kinase activity in muscle, *Fed. Proc.,* **18**: 266, 1959.
28. Riley, G. A., and Haynes, R. C., Jr.: Unpublished observations.
29. Smith, L., Reuter, S., Sutherland, E. W., and Rall, T. W.: Unpublished observations.
30. Haynes, R. C., Jr., Koritz, S. B., and Péron, F. G.: Influence of adenosine 3′,5′-monophosphate on corticoid production by rat adrenal glands, *J. Biol. Chem.,* **225**: 115–124, 1957.
31. Rall, T. W., and Sutherland, E. W.: Unpublished observations.
32. Krebs, E. G., Graves, D. J., and Fischer, E. H.: Factors affecting the activity of muscle phosphorylase *b* kinase. *J. Biol. Chem.,* **234**: 2867–2873, 1959.
33. Rall, T. W.: Unpublished observations.

34. Makman, M. H., Makman, R. S., and Sutherland, E. W.: Glucagon in human plasma. In: *Hormones in Human Plasma,* edited by H. W. Antoniades. Boston: Little, Brown and Company. In press, 1960.
35. Sutherland, E. W.: The effect of the hyperglycemic factor and epinephrine on enzyme systems of liver and muscle, *Ann. N. Y. Acad. Sci.,* **54**: 693–706, 1951.
36. Wosilait, W. D., and Sutherland, E. W.: The relationship of epinephrine and glucagon to liver phosphorylase. II. Enzymatic inactivation of liver phosphorylase, *J. biol. Chem.,* **218**: 469–481, 1956.
37. Sutherland, E. W., and Wosilait, W. D.: The relationship of epinephrine and glucagon to liver phosphorylase. I. Liver phosphorylase; preparation and properties, *J. biol. Chem.,* **218**: 459–468, 1956.
38. Sutherland, E. W., and Wosilait, W. D.: Inactivation and activation of liver phosphorylase, *Nature,* **175**: 169–170, 1965.
39. Rall, T. W., Sutherland, E. W., and Wosilait, W. D.: The relationship of epinephrine and glucagon to liver phosphorylase. III. Reactivation of liver phosphorylase in slices and in extracts, *J. biol. Chem.,* **218**: 483–495, 1956.
40. Wosilait, W. D.: Studies on the organic phosphate moiety of liver phosphorylase, *J. biol. Chem.,* **233**: 597–600, 1958.
41. Rall, T. W., and Sutherland, E. W.: Unpublished observations.
42. Berthet, J., Sutherland, E. W., and Rall, T. W.: The assay of glucagon and epinephrine with use of liver homogenates, *J. biol. Chem.,* **229**: 351–361, 1957.
43. Cori, C. F., and Cori, G. T.: The activity and crystallization of phosphorylase *b*. *J. Biol. Chem.,* **158**: 341–345, 1945.
44. Cori, C. F., Cori, G. T., and Green, A. A.: Crystalline muscle phosphorylase. III. Kinetics, *J. Biol. Chem.,* **151**: 39–55, 1943.
45. Cori, G. T.: The effect of stimulation and recovery on the phosphorylase *a* content of muscle, *J. Biol. Chem.,* **158**: 333–339, 1945.
46. Cori, G. T., and Cori, C. F.: Crystalline muscle phosphorylase. IV. Formation of glycogen, *J. Biol. Chem.,* **151**: 57–63, 1943.
47. Cori, G. T., and Cori, C. F.: The enzymatic conversion of phosphorylase *a* to *b*, *J. Biol. Chem.,* **158**: 321–332, 1945.
48. Cori, G. T., and Green, A. A.: Crystalline muscle phosphorylase. II. Prosthetic group, *J. Biol. Chem.,* **151**: 31–38, 1943.
49. Green, A. A., The diffusion constant and electrophoretic mobility of phosphorylases *a* and *b*, *J. Biol. Chem.,* **158**: 315–319, 1945.
50. Green, A. A., and Cori, G. T.: Crystalline muscle phosphorylase I. Preparation, properties, and molecular weight, *J. Biol. Chem.,* **151**: 21–29, 1943.
51. Keller, P. J., and Cori, G. T.: Purification and properties of the phosphorylase-rupturing enzyme. *J. Biol. Chem.,* **214**: 127–134, 1955.
52. Sutherland, E. W.: The effect of the hyperglycemic factor and epinephrine on liver and muscle phosphorylase. In: *Phosphorus Metabolism,* edited by W. D. McElroy and B. Glass, Vol. 1, pp. 53–66. Baltimore: The Johns Hopkins Press (1951).
53. Krebs, E. G., and Fisher, E. H.: Phosphorylase activity of skeletal muscle extracts, *J. biol. Chem.* **216**: 113–120, 1955.
54. Fischer, E. H., and Krebs, E. G.: Conversion of phosphorylase *b* to phosphorylase *a* in muscle extracts, *J. Biol. Chem.,* **216**: 121–132, 1955.
55. Fischer, E. H., Graves, D. J., Crittenden, E. R. S., and Krebs, E. G.: Structure of the site phosphorylated in the phosphorylase *b* to *a* reaction, *J. Biol. Chem.,* **234**: 1698–1704, 1959.
56. Krebs, E. G., Kent, A. B., and Fischer, E. H.: The muscle phosphorylase *b* kinase reaction, *J. Biol. Chem.,* **231**: 73–83, 1958.
57. Graves, D. J., Fischer, E. H., and Krebs, E. G.: Specificity studies on muscle phosphorylase phosphatase, *J. Biol. Chem.,* **235**: 805–809, 1960.

58. Cori, C. F.: Regulation of enzyme activity in muscle during work. In: *Enzymes; Units of Biological Structure and Function,* edited by Gaebler, O. H., New York: Academic Press (1956), pp. 573–83.

59. Cori, G. T., and Illingworth, B.: The effect of epinephrine and other glycogenolytic agents on the phosphorylase *a* content of muscle, *Biochim. Biophys. Acta,* **21**: 105–110, 1956.

60. Ellis, S.: Relation of biochemical effects of epinephrine to its muscular effects, *Pharmacol. Rev.,* **11**: 469–479, 1959.

61. Ellis, S.: Increased hexosemonophosphate, a common factor in muscular contraction potentiated by treppe, a short tetanus, epinephrine, or insulin, *Amer. J. Med. Sci.,* **229**: 218–219, 1955.

62. Ellis, S., Davis, A. H., and Anderson, H. L. Jr.: Effects of epinephrine and related amines on contraction and glycogenolysis of the rat's diaphragm. *J. Pharmacol.,* **115**: 120–125, 1955.

THE EFFECT OF CONTRACTION AND OF EPINEPHRINE ON THE PHOSPHORYLASE ACTIVITY OF FROG SARTORIUS MUSCLE*

William H. Danforth†*, Ernst Helmreich and Carl F. Cori*

Department of Biological Chemistry, Washington University School of Medicine, St. Louis, Missouri

(Communicated May 25, 1962)

Glycogen phosphorylase in skeletal muscle exists as phosphorylase *a*, fully active, and phosphorylase *b*, active only in the presence of 5′-AMP‡[1]. Phosphorylase *b* is phosphorylated at the expense of ATP by phosphorylase *b* kinase[2] and dimerizes to form phosphorylase *a*. The reverse reaction is catalyzed by phosphorylase phosphatase, formerly called PR enzyme[3,4].

In 1956, C. F. Cori[5] emphasized the speed with which glycogen breakdown occurs in muscle during contraction and suggested that the speed of the $b \rightleftharpoons a$ interconversion was of sufficient magnitude to be the pacemaker of the concomitant changes in the rates of glycolytic reactions. Recently, Rulon *et al.*[6] reported a statistically significant increase in phosphorylase *a* content of rat anterior tibial muscle during tetanic contraction. The high and variable baseline levels of phosphorylase *a* in resting muscle obtained with previous methods have precluded a quantitative analysis of changes in the phosphorylase system.

Improvements in techniques for the arrest of enzymatic reactions make it now possible to study the time course of the interconversion of phosphorylase *a* and *b* in the thin frog sartorius muscle during and following contraction. The present studies show that phosphorylase *a* rises from less than 5 per cent

*William H. Danforth, Ernst Helmreich, and Carl F. Cori, "The Effect of Contraction and of Epinephrine on the Phosphorylase Activity of Frog Sartorius Muscle," *Proc. Natl. Acad. Sci.,* **48**, 1191–99 (1962). Reprinted with the permission of the authors and of the publisher.

This work was supported in part by research grants E-3765 and A-1984 from the National Institutes of Health, U.S. Public Health Service.

†Special postdoctoral fellow of the National Heart Institute, U.S. Public Health Service.

‡Abbreviations are as follows: AMP, adenosine monophosphate; EDTA, ethylenediaminetetraacetate.

to nearly 100 per cent of the total phosphorylase (phosphorylase *a* and *b* combined) within three seconds after the onset of tetanic contraction at 30°. On relaxation, phosphorylase *a* activity decreases exponentially to baseline values. Kinetic analysis suggests that changes in the phosphorylase *b* kinase rather than phosphorylase phosphatase activity are responsible for the increase and decrease in phosphorylase *a*.

Variations in muscle phosphorylase *a* in response to epinephrine, as previously studied by G. T. Cori and B. Illingworth(7), have been reinvestigated with the use of the improved fixation technique. The increase and decrease of phosphorylase *a* in muscle as a consequence of the addition and removal of epinephrine is slow when compared with the effect of contraction, requiring minutes rather than seconds. Moreover, the response to epinephrine but not that to muscular work can be inhibited by an inactive analogue of epinephrine.

Experimental Procedure

Female *Rana pipiens* weighing approximately 50 to 70 gm were obtained from Minnesota between November and March. The animals were kept in water at about 3° prior to use. After low spinal transection, the sartorii were carefully dissected and transferred to ice-cold Krebs-Ringer's-bicarbonate solution(8) adapted for frog muscle(9) and equilibrated with 5 per cent CO_2 in argon (pH$\sim$6.8). The muscles were allowed to recover for about 30 min before the experiment was initiated.

Muscle stimulation

Muscles to be stimulated were left attached to the pubic bone. The proximal end of the muscle was mounted in a platinum holder, which served as an electrode. The distal tendon was anchored to a small chain. A second platinum electrode surrounded the distal end of the muscle. Once in place, the muscle was immersed in anaerobic Ringer's solution at the temperature desired for the experiment. Five to ten min before stimulation, the fluid was removed and the muscle surrounded by a constant temperature chamber equilibrated with moist 5 per cent CO_2-95 per cent argon. Stimuli were supplied by a Grass model S-4-B stimulator(10). The muscle contracted isometrically. Details are given in the legend to Figure 1. Tension was monitored by a strain gauge and a Sanborn strain gauge amplifier and recorded by a Sanborn single-channel direct writer(11). Immediately before termination of the experiment, the constant temperature chamber was removed. The muscle was then suddenly immersed in isopentane or dichlorodifluoromethane chilled to near its freezing point (—160°) in liquid nitrogen. The freezing left a mark on the tension record which allowed the duration of the contraction to be estimated to the nearest 0.1 second.

Incubation with epinephrine

Sartorii were incubated in closed 25 ml Erlenmeyer flasks containing 4 ml of Ringer's solution; 0.1 mg of ascorbate per ml was added to prevent the oxidation of epinephrine. The system was made anaerobic by flushing with a stream of 95 per cent argon-5 per cent CO_2. Experiments were terminated by lifting the muscles by the tendon and quickly freezing them in isopentane.

FIXATION

Phosphorylase *b* kinase and phosphorylase phosphatase are inhibited by EDTA* and fluoride respectively. These agents do not affect phosphorylase *a* or *b* activity. The selective action of these inhibitors was the basis for previous techniques for the study of phosphorylase *a* and *b* content of living muscle. Muscles with or without prior freezing were homogenized in an ice-cold solution of these inhibitors[7]. In the present study, it has been established that neither technique is adequate to preserve the true resting baseline, mainly because the kinase is able to act with such great speed.

In preliminary experiments, the phosphorylase *a* content of resting frog sartorius muscle, although lower than previously reported, was still quite variable despite rapid freezing by a method similar to that of Seraydarian *et al.*[12] The frozen muscle was placed in a precooled cartridge in which it was pulverized by shaking with a steel ball. NaF and EDTA were then added in aqueous solution to the cold cartridge, so that ice formed in intimate contact with the frozen muscle powder. On further shaking, heat generated by friction of the steel ball caused thawing. It was soon realized that the rate of penetration of the inactivating solution into the muscle particles was in most cases too slow to prevent significant activity of phosphorylase *b* kinase (Table 1).

In the method finally adopted, the inhibitors were added in a glycerol-water solution which remained liquid at —35°. This mixture allowed the inhibitors to penetrate the muscle powder at a considerably lower temperature and acted as a lubricant to facilitate rapid and smooth mixing of the muscle powder in the inactivating solution. Heat resulting from friction between the ball and the cartridge was also kept at a minimum. Use of glycerol for this purpose was possible only because it did not inhibit phosphorylase at the final concentrations present in the test system.

The procedure was carried out as follows. Each frozen muscle, weighing 100–150 mg, was transferred to liquid nitrogen and then to a stainless steel cartridge of about 20 ml capacity precooled in liquid nitrogen. One stainless steel ball large enough to slide easily but with minimal clearance into the cartridge was added and the cartridge closed. The muscle was pulverized by shaking for 8 sec at about 100 oscillations per sec in a Nossal shaker[13]. The cartridge was then removed and warmed to —35° in an alcohol bath. 2 ml of a precooled 60 per cent glycerol solution containing 2×10^{-2} *M* NaF and 1×10^{-3} *M* EDTA (*p*H 6.7) were added to the powder. The cartridge was again closed and shaken for an additional 6 sec. 8 ml of an aqueous solution of 2×10^{-2} *M* NaF, 1×10^{-3} *M* EDTA, 1.9×10^{-2} *M* Na glycerophosphate, and 1.9×10^{-2} *M* cysteine (*p*H 6.7) at 3° were then added and the cartridge was again shaken for 4 sec. The importance of introducing the inactivating substances at such low temperatures is shown in Table 1. No other method gave such a consistently low baseline of phosphorylase *a* content of resting muscle.

Larger particles were removed by centrifugation at 1,400 × *g* for 10 min at close to 2°. The supernatant fluid was analyzed for phosphorylase activity.

*See the third footnote on p. 708.

PHOSPHORYLASE ASSAY

The reaction mixture for assay of total phosphorylase activity at 30° at *p*H 6.7 contained 7.5×10^{-2} *M* glucose-1-phosphate, 5×10^{-4} *M* 5′-AMP, 1 per cent glycogen, 7.5×10^{-3} *M* Na-glycerophosphate buffer, 7.5×10^{-3} *M* cysteine, 1×10^{-2} *M* NaF and 5×10^{-4} *M* EDTA. The total volume was 1 ml and contained 0.5 ml of the centrifuged muscle extract (final dilution of the muscle approximately 1:200). For measurements of phosphorylase *a* activity, 5′-AMP was omitted from the reaction mixture. At two or more suitably chosen time intervals, 0.2 ml aliquots were removed and inactivated in 8 ml 0.125 *M* Na-acetate buffer *p*H 4. Inorganic phosphate released by phosphorylase was measured according to Lowry and Lopez(14). In this method, hydrolysis of the acid-labile glucose-1-phosphate is held to a minimum. It was thus possible to measure phosphorylase activity in the presence of high concentrations of glucose-1-phosphate (i.e., 15 times the K_m value of glucose-1-phosphate for phosphorylase). At these concentrations, the phosphorylase reaction proceeds essentially linearly until about 12 per cent of the added glucose-1-phosphate have been used up. Phosphoglucomutase activity should not be significant in the presence of EDTA (5×10^{-4} *M*) in the reaction mixture.

Phosphorylase *a* activity is defined as the activity demonstrable in the absence of 5′-AMP and is expressed as per cent of total phosphorylase activity (i.e., phosphorylase $a + b$ combined, measured in the presence of 5′-AMP).

In order to determine whether the muscle extract contributed significant amounts of 5′-AMP to the final reaction mixture, it was treated in a 1:100 dilution with acid-washed Norit-A (4 mg per ml at 0° for 10 min). Such treatment resulted in variable losses of total phosphorylase activity but did not significantly change the per cent of phosphorylase *a* in control muscles or in muscles stimulated electrically or incubated with epinephrine. It was shown in control procedures that 5′-AMP added to the muscle extract was effectively removed by treatment with Norit.

Since phosphorylase had been exposed to 60 per cent glycerol during fixation of muscle and 6 per cent glycerol was present in the final phosphorylase reaction mixture, possible effects of glycerol on phosphorylase *a* and *b* activity were investigated. Table 1 shows that the total phosphorylase activity of frog muscle extracted with water or with glycerol was essentially the same. When 15 per cent glycerol was present in the final reaction mixture, there was an 18 per cent inhibition of both crystalline rabbit muscle phosphorylase *a* and *b*. At the concentration used in the assay, little or no inhibition could be detected.

Materials

EDTA, glucose-1-phosphate, and 5′-AMP were products of the Sigma Chemical Company. Norit-A was purchased from the Pfanstiehl Laboratories, Inc. Epinephrine hydrochloride was obtained from Parke Davis and Company and norepinephrine bitartrate from Winthrop Laboratories. 1-(3,4-dichlorophenyl)-2-isopropylaminoethanol hydrochloride, referred to as dichloroisoproterenol (DCI), was a gift of Dr. I. H. Slater, Eli Lilly Co. Five times recrystallized rabbit muscle phosphorylase *a*, prepared as described by B. Illingworth and G. T. Cori[15], was used to standardize the phosphorylase assay. Three times recrystallized rabbit muscle phosphorylase *b* was prepared by the method of Fischer and Krebs[16].

Results

Phosphorylase *a* content of resting muscle

The average phosphorylase *a* content of resting frog sartorius muscle, determined as described under *Methods,* was 2.8 per cent of total phosphorylase with a range of values from 1 to 5 per cent (Table 1). These values are much

Table 1

The Influence of the Method of Fixation on the Phosphorylase *a* Activity of Resting Frog Muscle

Frog sartorii were frozen in isopentane or dichlorodifluoromethane at −160° and were powdered in the frozen state. In one series of experiments, the muscle powder was extracted at 0° with an aqueous solution containing 2×10^{-2} *M* NaF and 1×10^{-3} *M* EDTA, and in the other series at −35° with a solution containing 60 per cent glycerol in addition to 2×10^{-2} *M* NaF and 1×10^{-3} *M* EDTA.

No. of experiments	Conditions for Fixation: Solution	Volume (ml)	Temperature (°C)	Phosphorylase *a* in per cent of total: Mean	Range	Total phosphorylase activity per gm muscle μmoles Pi/min
25	EDTA-NaF in H_2O	10	0	10.1	(3–32)	71.4
10	EDTA-NaF in glycerol	2	−35	2.8	(1–5)	72.4

lower and more uniform than any previously reported. Extensive experience with various methods of fixation and extraction of muscle leads us to conclude that higher values for the phosphorylase *a* content of resting frog sartorius muscle are artifacts. In this context, it should be emphasized that the phosphorylase activity found in resting muscle is more than enough to account for the rate of anaerobic production of lactate in an isolated frog sartorius at 30°, i.e., approximately 3.4 μmoles lactate per gm fresh weight of muscle per hr.

Effect of stimulation

Although the ratio of phosphorylase *a* to *b* was altered as the result of stimulation (see below), the total phosphorylase activity was not significantly changed. In ten resting sartorii, the phosphorylase activity assayed in the presence of 5′-AMP corresponded to the formation of 7.2 (S.D. ±1.4) $\times 10^{-5}$ moles of inorganic phosphate per gm muscle per min. The mate of each of these muscles was inactivated during tetanic contraction at 10°. The comparable value for the stimulated muscles was 7.7 (S.D. ± 1.3) $\times 10^{-5}$ moles of inorganic phosphate per gm muscle per min.

In Figure 1, curve (A), is shown the effect of a sustained isometric contraction at 30° on the phosphorylase *a* content of muscle; values of 50 per cent *a* are reached after 1 sec and greater than 80 per cent after 2.5 sec of stimulation. Tetanic stimulation at 10°, curve (B), results in a slower rise and lower steady-state level of phosphorylase *a* than stimulation at 30°. With single twitches at 10°, curve (C), the rate of increase and the steady-state level of phosphorylase *a* are still less. Curves (A) and (B) in Figure 2 show the rate of decrease in phosphorylase *a* after a stimulation of sufficient duration to produce maximum phosphorylase *a* activity for the respective temperatures. After about 60 sec, the phosphorylase *a* content has returned to the basal level of five per cent or less of the total phosphorylase content.

The kinetics of the rise and fall of phosphorylase *a* can be treated as follows: phosphorylase $b \underset{k_2}{\overset{k_1}{\rightleftharpoons}}$ phosphorylase *a*. The velocity constants, k_1 and k_2, during stimulation can be calculated by means of the equation:

$$k_1 b_i / a_{\text{st. st.}} = t^{-1} \ln a_{\text{st. st.}} / (a_{\text{st. st.}} - a_t), \tag{1}$$

where $k_1 b_i / a_{\text{st. st.}} = k_1 + k_2$. Here, b_i is the initial activity of phosphorylase *b*, a_t the activity of phosphorylase *a* at time *t*, and $a_{\text{st. st.}}$ the final activity, which was 94, 60, and 30 per cent for curves A, B, and C in Figure 1 respectively. During relaxation $k_2 \gg k_1$, as shown by the almost complete disappearance

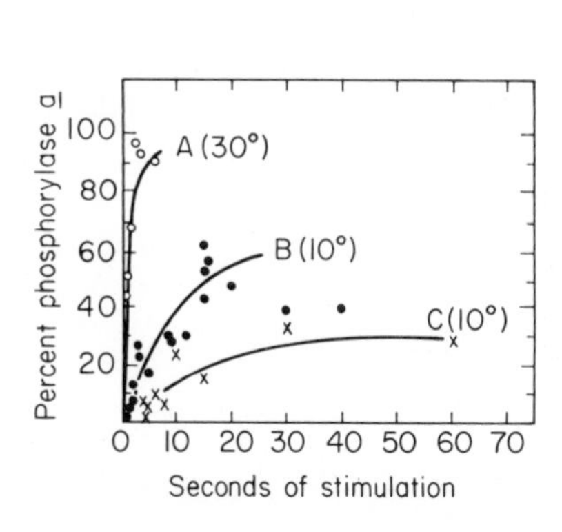

Fig. 1. Increase in phosphorylase *a* content of frog sartorius during isometric contraction. Each point represents a muscle from a different frog. Stimulation was by 12-volt shocks, 1.5 msec in duration. Curve (A), (0), muscles tetanized with 35 shocks per sec at 30°; Curve (B), (0), muscles tetanized with 15 shocks per sec at 10°; Curve (C), (X), single switches at 1 per sec at 10°. Theoretical curves are drawn according to equation (1) (see text).

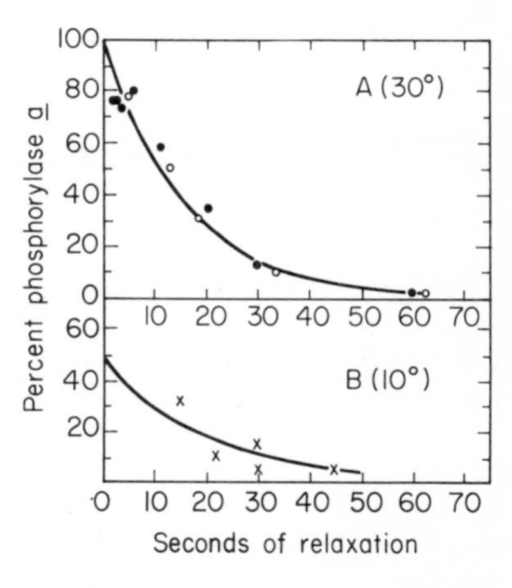

Fig. 2. Decrease in phosphorylase *a* content of frog sartorius following cessation of isometric contraction. The conditions of stimulation were identical to those in Figure 1. (0) preceding tetanus was 2 sec; (0) preceding tetanus was 4 sec; (X) preceding tetanus was 15 sec. Theoretical curves are drawn according to equation (2) (see text).

of phosphorylase *a* (Fig. 2). Therefore, equation (1) becomes:

$$k_2 = t^{-1} \ln a_i/a_t. \tag{2}$$

The kinetic treatment is based on the assumption that the two opposing reactions follow the first-order reaction rate equation. This has been shown to be the case for the phosphorylase phosphatase reaction *in vitro*[3] and is reasonably well supported by the curve in Figure 2*A* for *in vivo* conditions. On the other hand, the assumption that k_1 represents a first-order velocity constant may be an oversimplification. During stimulation, for example, it is not known how fast the activation of the kinase is relative to its rate of action on phosphorylase *b*. Furthermore, fatigue during more prolonged stimulation may result in a slowing down of kinase activity. (See the points at 30 and 40 sec in curve (B), Fig. 1.) The chief factor to be considered, however, is the unavoidable variability of the enzyme content of different muscles. It should be emphasized that each point in the curves in Figures 1, 2, and 3 represents a different muscle so that any variation in the kinase and phosphatase content will be reflected in a scatter of results and in a different steady-state.

All curves in Figures 1, 2, and 3 were drawn according to equation (1) or (2). The apparent steady-state values of the ascending curves were obtained from preliminary plots of the data. The points were then plotted on a semilogarithmic graph. The best fit straight line was determined by the method of least squares. Values for $k_1 + k_2$ and for k_2 were calculated from the slope of this line.

The values of k_1 and k_2 are shown in Table 2. Apart from temperature-dependent changes, the value for k_2 remained relatively constant. Thus, at 30° the decline in phosphorylase *a* activity during rest was the same whether preceded by a two- or a four-sec tetanus. At 10°, k_2 calculated from the rising curve of phosphorylase *a* activity during tetanus was not significantly different

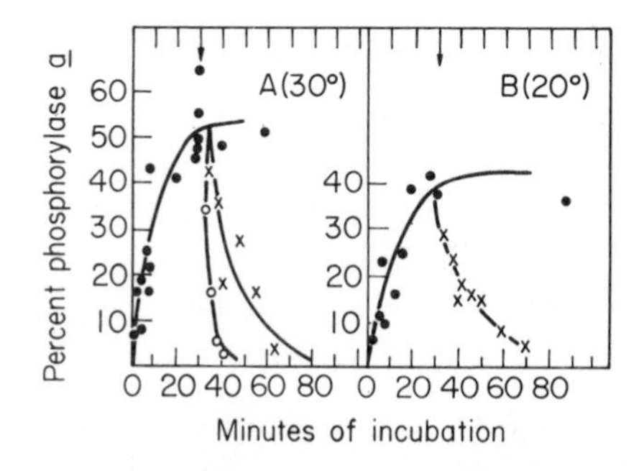

Fig. 3. Effect of epinephrine on phosphorylase *a* content of frog sartorius. Individual sartorii were incubated anaerobically in Ringer's solution containing 1.1 × 10^{-6} *M* epinephrine and 0.1 mg per ml ascorbate. After 30 min incubation (see arrow) some muscles were rinsed and transferred to a similar medium but without epinephrine. (0) Epinephrine present; (X) epinephrine removed; (0) epinephrine removed and 1.1 × 10^{-4} *M* dichloroisoproterenol added. Theoretical curves are drawn according to equations (1) and (2) (see text).

Table 2

EFFECT OF STIMULATION ON THE INTERCONVERSION OF PHOSPHORYLASE *a* AND *b* IN FROG SARTORIUS

First-order velocity constants for the conversion of phosphorylase *b* to *a* (k_1) and *a* to *b* (k_2) have been calculated from the data of Figures 1 and 2. k_2 has been calculated from the data obtained during contraction as well as during relaxation. At 30°, the high ratio of phosphorylase *a* to *b* make estimation of k_2 during stimulation inexact

Temperature (°C)	Type of stimulation	Stimulation (sec^{-1})		Relaxation (sec^{-1})	
		k_1	k_2	k_2	
30	Tetanus	0.97	(0.063)	0.057*	0.058†
10	Tetanus	0.080	0.054	0.050	
10	Single twitches at 1 per sec	0.021	0.048		

*After 2-sec tetanus.
†After 4-sec tetanus.

from k_2 calculated from the decline in phosphorylase *a* activity after cessation of stimulation. Also at 10°, the type of stimulation (single twitches at 1 per sec or a sustained contraction in the form of a tetanus) had no marked effect on k_2.

By contrast, k_1, indicative of the $b \longrightarrow a$ transformation, changed markedly with the type of stimulation and with temperature and fell to low values immediately after cessation of stimulation. No exact estimate can be made for the value of k_1 during relaxation except that since the equilibrium is 95 per cent or more to the side of phosphorylase *b*, k_1 would be maximally 1/20 k_2. The rapidity with which phosphorylase *b* kinase activity decreases during rest would explain why repetitive single twitches give a lower k_1 value than a tetanus.

Temperature coefficients for the two enzymes involved in the $b \rightleftharpoons a$ transformation are remarkably different. From k_1 in Table II, one calculates a Q_{10} of 3.5 for the kinase activity and from k_2 a Q_{10} of 1.1 for the phosphatase activity. As a consequence, the steady-state level of phosphorylase *a* attained during stimulation changes markedly with temperature.

EFFECT OF EPINEPHRINE

Incubation with epinephrine or norepinephrine increased the phosphorylase *a* content of frog sartorius muscle (Table 3). Epinephrine was equally and presumably maximally effective at a concentration of 10^{-5} and 10^{-6} *M*, whereas norepinephrine had a pronounced effect at the higher concentration only. Dichloroisoproterenol(17), an inactive analogue, which probably acts by competitive inhibition, completely blocked epinephrine and norepinephrine action at a concentration 10 times greater than that of the sympathomimetic amines. The effect of stimulation on phosphorylase *a* could not be blocked by this antagonist.

In Figure 3 is shown a time curve for the action of epinephrine on phosphorylase *a*. The time scale is in minutes as compared to seconds in the case of stimulation. In fact, on comparing the rise of phosphorylase *a* in the two cases, there is an approximately 500-fold difference in rate (*t* 1/2 at 30° is 0.7 sec for stimulation and 370 sec for the action of epinephrine, Table 4). The rate-limiting step in the action of epinephrine on the $b \longrightarrow a$ conversion could be

Table 3

DIFFERENTIAL EFFECT OF DICHLOROISOPROTERENOL ON EPINEPHRINE ACTION AND STIMULATION

Paired muscles from a single frog were placed in Ringer's solution containing 0.1 mg per ml ascorbate and gassed with 95 per cent argon-5 per cent CO_2. Both muscles were incubated at 30° for 30 min, one in the presence and one in the absence of 1.1×10^{-4} M dichloroisoproterenol (DCl). The flasks were then opened and epinephrine or norepinephrine was added. The media were again equilibrated with the gas phase and reincubation was carried out for an additional 30 min. The muscles to be stimulated were incubated in DCI alone and then mounted in a moist chamber at 30° and gassed with 95 per cent argon-5 per cent CO_2. They were fixed after a 2- or 4-sec tetanus.

Additions	Phosphorylase *a* in per cent of total: No DCl	+ DCl
Norepinephrine 1.1×10^{-6} M	10	4
Norepinephrine 1.1×10^{-5} M	46	6
Epinephrine 1.1×10^{-6} M	64	3
Epinephrine 1.1×10^{-5} M	62	8
Muscle contraction	84*	72

*See Figure 1.

Table 4

COMPARISON OF EFFECT OF STIMULATION AND EPINEPHRINE ON RATE OF CHANGE IN PHOSPHORYLASE *a* IN FROG SARTORIUS

The half-time of rise and fall in phosphorylase *a* was calculated from the data given in Figures 1, 2, and 3.

	Rise in Phosphorylase *a*		Fall in Phosphorylase *a*	
Temperature (°C)	Experimental conditions	Half-time (sec)	Experimental conditions	Half-time (sec)
30	Stimulation	0.7	Relaxation	12
30	Incubated with epinephrine	370	Incubated with epinephrine, then washed	660
30	—	—	Incubated with epinephrine, then washed with DCI*	150
20	Incubated with epinephrine	520	Incubated with epinephrine, then washed	870

*Dichloroisoproterenol.

a preceding enzymatic step concerned with formation of cyclic 3′,5′-AMP or penetration to and fixation of epinephrine at the binding sites.

Figure 3 shows that the action of epinephrine can be reversed under anaerobic conditions with a return of phosphorylase *a* to the resting level of five per cent or less. The muscles were first incubated with epinephrine until a maximum effect was produced; they were then quickly rinsed with large quantities of Ringer's solution and transferred to fresh medium without epinephrine but containing in some cases 1.1×10^{-4} M dichloroisoproterenol. It is interesting to note that the presence of this antagonist hastened the reversal, presumably by displacing epinephrine from its binding sites. The half-time for reversal at 30° was 660 sec in the absence of dichloroisoproterenol and 150 sec in its presence (Table 4).

During incubation with epinephrine at 20°, the final level of phosphorylase *a* was somewhat lower than at 30°, but as can be seen from Figure 3 and Table 4, the effect of temperature on the rate of either the increase or the decrease in phosphorylase *a* was not great. At 10° (not illustrated), the response to epinephrine was slowed down very markedly.

Discussion

The principal energy-yielding reaction during anaerobic work in an isolated muscle is the conversion of glycogen to lactic acid. During a short tetanus, the rate of glycogenolysis may increase several hundred times and then on relaxation return to the resting rate[5]. These changes in rate are paralleled by a corresponding rise and fall in the phosphorylase *a* content of muscle. From the kinetic analysis presented in this paper, it seems likely that the actual regulatory control over glycogenolysis is exerted by the phosphorylase *b* kinase system which is rapidly activated and deactivated during and following contraction. Calcium ions have been implicated in this mechanism because they are liberated during contraction and activate the kinase *in vitro*[18]. However, it has not so far been possible to deactivate the kinase *in vitro* by removal of calcium ions[18].

The differential inhibition of the action of epinephrine by dichloroisoproterenol suggests that muscle work and epinephrine activate the kinase by different mechanisms. The much slower rise and fall in phosphorylase *a* during and after epinephrine action may be mediated by the enzymatic formation and subsequent removal of cyclic 3',5'-AMP. In fact, it has been shown recently that dichloroisoproterenol inhibits the effect of epinephrine on the formation of cyclic 3',5'-AMP in a particulate enzyme preparation from dog heart[19].

Summary

Phosphorylases *a* and *b* have been determined in frog sartorii fixed by rapid freezing. Extraction was at −35° with a glycerol-fluoride-EDTA solution. In resting muscle, less than five per cent of the total phosphorylase was present as the active or *a* form. Starting with this low and uniform baseline, it was possible to determine the kinetics of the $b \underset{k_2}{\overset{k_1}{\rightleftharpoons}} a$ interconversion, mediated by phosphorylase *b* kinase (k_1) and phosphorylase phosphatase (k_2). It was found that during contraction $k_1 \gg k_2$ and during relaxation $k_2 \gg k_1$ as the result of changes exclusively in k_1.

Increase in phosphorylase *a* during incubation with epinephrine at 30° had a half-time of 370 seconds as compared to 0.7 sec, during an isometric contraction. Dichloroisoproterenol inhibited the effect of epinephrine and norepinephrine and also hastened the reversal of epinephrine action but had no effect on the increase in phosphorylase *a* produced by muscle work. Two different mechanisms appear to be involved in the activation of the phosphorylase kinase by muscle work and by epinephrine.

References

1. Brown, D. H., and C. F. Cori, in *The Enzymes,* ed. P. D. Boyer, H. A. Lardy, and K. Myrbäck (2d ed.; New York: Academic Press, Inc., 1961), vol. 5, pp. 207–228.
2. Krebs, E. G., and E. H. Fischer, *J. Biol. Chem.,* **216**, 113 (1955).
3. Cori, G. T., and C. F. Cori, *J. Biol. Chem.,* **158**, 321 (1945).
4. Keller, P. J., and G. T. Cori, *Biochim. Biophys. Acta,* **12**, 235 (1953); *J. Biol. Chem.,* **214**, 127 (1955).
5. Cori, C. F., in *Enzymes, Units of Biological Structure and Function,* ed. O. H. Gaebler (New York: Academic Press, Inc., 1956), pp. 573–583.
6. Rulon, R. R., D. D. Schottelius, and B. A. Schottelius, *Am. J. Physiol.,* **200**, 1236 (1961).
7. Cori, G. T., and B. Illingworth, *Biochim. Biophys. Acta,* **21**, 105 (1956).
8. Krebs, H. A., and K. Henseleit, *Hoppe Seyler's Z. physiol. Chem.,* **210**, 33 (1932).
9. Narahara, H. T., P. Özand, and C. F. Cori, *J. Biol. Chem.,* **235**, 3370 (1960).
10. Grass Instrument Company, Quincy, Mass.
11. Sanborn Company, Waltham, Mass.
12. Seraydarian, K., W. F. H. M. Mommaerts, A. Wallner, and R. J. Guillory, *J. Biol. Chem.,* **236**, 2071 (1961).
13. McDonald Engineering Company, Bay Village, Ohio.
14. Lowry, O. H., and J. A. Lopez, *J. Biol. Chem.,* **162**, 421 (1946).
15. Illingworth, B., and G. T. Cori, *Biochem. Preparations,* **3** (1953).
16. Fischer, E. H., and E. G. Krebs, *J. Biol. Chem.,* **231**, 65 (1958).
17. Powell, C. E., and I. H. Slater, *J. Pharmacol. and Exper. Therap.,* **122**, 480 (1958).
18. Krebs, E. G., D. J. Graves, and E. H. Fischer, *J. Biol. Chem.,* **234**, 2867 (1959).
19. Murad, F., Y.-M. Chi, T. W. Rall, and E. W. Sutherland, *J. Biol. Chem.,* **237**, 1233 (1962).

INFLUENCE OF ATP CONCENTRATION ON RATES OF SOME PHOSPHORYLATION REACTIONS*

Henry A. Lardy and R. E. *Parks, Jr.*

Department of Biochemistry and Institute for Enzyme Research, University of Wisconsin, Madison

Influence of ATP and Magnesium Concentration

The processes which regulate respiratory activity in cells have been studied in our laboratory for some years[6,10]. Since we have on previous occasions[6,9] discussed the role of phosphate acceptors and inorganic phosphate concentration in the regulation of muscle respiration, this discussion today will be limited to a consideration of the interrelation of Mg^{++} and ATP in controlling some phosphorylation reactions.

Some years ago Professor Cori[1] pointed out that "hexosemonophosphate . . . a normal constituent of muscle . . . can increase considerably under certain experimental conditions without any increase in the formation of lactic acid[2]. This indicates that the reaction between fructose-6-phosphate and adenosinetriphosphate in intact muscle is a limiting factor as regards the rate at which lactic acid is formed and carbohydrate is oxidized."

Studies of several phosphorylation reactions have revealed a phenomenon which may, at least in part, explain the nature of the control mechanism involved. In essence, the rate of the reaction depends on the relative concentration of ATP and Mg^{++}. Where the latter is present in fixed amount, we observe striking inhibition of the reaction by adding ATP in excess of the amount required for the maximum rate of reaction.

*Henry A. Lardy and R. E. Parks, Jr., "Influence of ATP Concentration on Rates of Some Phosphorylation Reactions," in *Enzymes: Units of Biological Structure and Function,* ed. Oliver H. Gaebler, Henry Ford International Symposium, Detroit, 1955. New York: Academic Press, Inc., 1956. Reprinted with the permission of the authors and of Academic Press.

In 1952 Hers[4] reported that the fructokinase reaction proceeded maximally when the concentration of Mg was equivalent to the ATP present. At about the same time[8] we were studying another ATP-requiring system, namely, the conversion of propionate and CO_2 to succinate, and reaffirmed the concept of Hers for this reaction.

In addition, however, we found a striking inhibition of the reaction by increasing concentrations of ATP (Fig. 1)[7]. Since the propionate and CO_2 reaction is a complicated one—it appears that ATP is required for activation of both the propionate and the CO_2[7]—it is difficult to examine the nature of this inhibition. We were able to study the phenomenon with three other pure or highly purified enzyme systems.

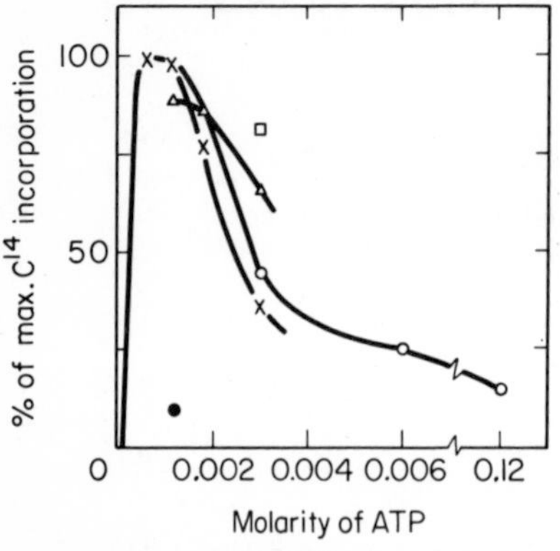

Fig. 1. Effect of ATP concentration on $C^{14}O_2$ fixation into succinate. ○, 1 μM. Mg^{++} per milliliter; X1 μM. Mn^{++} per milliliter; △, 2 μM. Mg^{++} per milliliter; ●, no added divalent cation; □, 3 μM. Mn^{++} per milliliter.

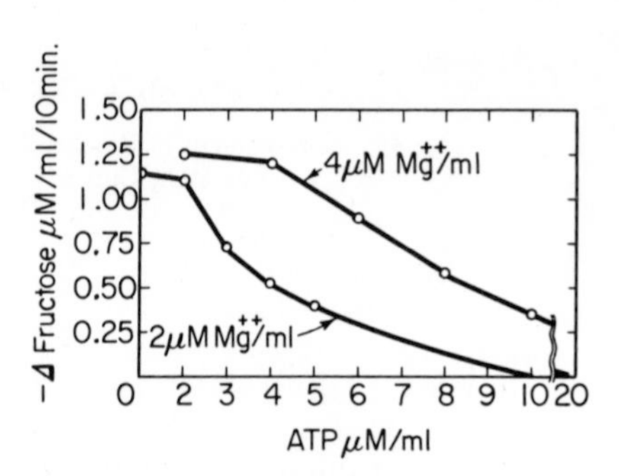

Fig. 2. Inhibition of purified beef liver fructokinase by excess ATP.

The Fructokinase Reaction

The relationship betwen Mg^{++} and ATP observed by Hers[4] has been studied in our laboratory with a 400-fold purified preparation of beef liver fructokinase. When thus purified, the enzyme exhibits maximum activity when the Mg: ATP ratio is 1 or greater. Mg^{++} in concentrations as great as fivefold in excess of the ATP concentration is not inhibitory, but ATP in excess of the molar concentration of Mg^{++} is strongly inhibitory (Fig. 2). In fact, the reaction is brought to zero velocity when the ATP is increased to a concentration fivefold greater than that of Mg^{++}. The inhibition in this case is not due to accumulating ADP, for the reaction is carried out in the presence of a sufficient excess of phosphocreatine and ATP-creatine transphosphorylase to keep all but a trace of the adenosine nucleotide in the triphosphorylated form. The inhibition produced by a given amount of ATP can be reversed by adding a molar equivalent of Mg^{++}.

ATP-Creatine Transphosphorylase

The crystalline enzyme[5] which phosphorylates creatine, using ATP, is most active when Mg^{++} and ATP are present at equimolar concentrations. An excess of either Mg^{++} or ATP relative to the other component results in an appreciable inhibition of the rate of transphosphorylation, but the effect is not so striking as with fructokinase or phosphohexokinase.

Phosphohexokinase

This enzyme catalyzes the reaction

$$\text{Fructose-6-phosphate} + \text{ATP} \longrightarrow \text{Fructose-1,6-diphosphate} + \text{ADP}$$

It has been purified more than 200-fold from an alkaline extract of rabbit muscle[11] As is shown in Fig. 3, Mg^{++} in great excess did not appreciably inhibit the phosphorylation of fructose-6-phosphate. ATP, in excess of the molar equivalent of Mg^{++} present, decreases the rate of the reaction. This enzyme, like fructokinase, is strongly inhibited by ADP, but the extent of the reaction in the experiments summarized in Fig. 3 is so slight as to allow accumulation of only a trace of ADP.

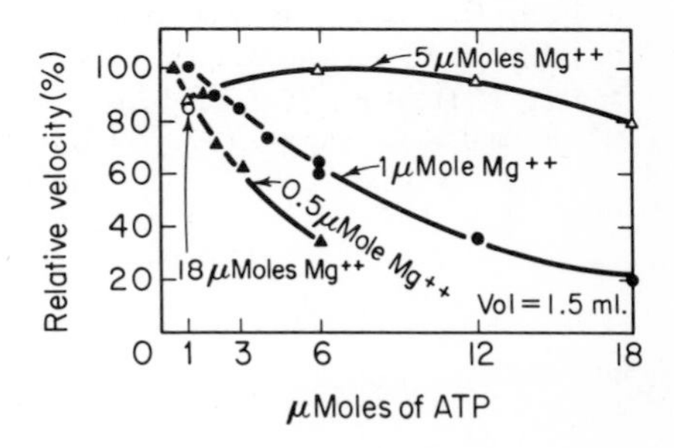

Fig. 3. Inhibition of purified phosphohexokinase by excess ATP.

It seems likely that the strong inhibition of phosphohexokinase by ATP may help explain the lack of hexosediphosphate accumulation in stimulated muscle which was pointed out by Professor Cori. It is known that the molar concentration of magnesium in muscle exceeds that of ATP by several-fold. But it is very likely that much of the magnesium will be bound by proteins and nucleotides other than ATP. The low activity of phosphohexokinase in muscle may be due to the high ATP concentration relative to free Mg^{++}. Only after muscle has been working for some time does the ATP concentration fall[3]. Perhaps at that time enzymes like phosphohexokinase may become more active than in resting muscle.

At the present time it is not possible to offer an explanation for the mechanism by which ATP inhibits each of these enzymes. Perhaps, as Hers[4] first suggested, the actual substrate for the enzyme is a complex of Mg and ATP. Addition of excess ATP might decrease the concentration of active substrate by favoring the formation of a different chelate form containing a greater amount of ATP, for example, Mg-$(ATP)_2$. It is also possible that free Mg^{++} is required for the reaction. Were this the case, excess ATP might inhibit by chelating free Mg^{++}, leaving none for the enzyme. Another possibility is that free ATP is inactive as a phosphate donor but a competitive antagonist of the active ATP-Mg chelate.

Regardless of the mechanism, it is important to keep in mind that the concentration of ATP in living cells may have a direct regulatory influence on several enzyme reactions.

References

1. Cori, C. F., *in* "Respiratory Enzymes, A Symposium," p. 175. University of Wisconsin Press, Madison, 1941.
2. Cori, G. T., and Cori, C. F., *J. Biol. Chem.,* **116**, 119, 129 (1936).
3. Flock, E. V., Ingle, D. J., and Bollman, J. L., *J. Biol. Chem.*, **129**, 99 (1939).
4. Hers, H. G., *Biochim. et Biophys. Act,* **8**, 416, 424 (1952).
5. Kuby, S. A., Noda, L. H., and Lardy, H. A., *J. Biol. Chem.* **209**, 191 (1954).
6. Lardy, H. A., *in* "The Biology of Phosphorus," p. 131. Michigan State College Press, East Lansing, 1952.
7. Lardy, H. A., and Adler, J., *J. Biol. Chem.,* **219**, 933 (1956).
8. Lardy, H. A., and Fischer, J., *Abstr. Am. Chem. Soc., 123rd Meeting*, p. 106 (1953).
9. Lardy, H. A., Rapports 3ème Congrès International de Biochimie. In press (1955).
10. Lardy, H. A., and Wellman, H., *J. Biol. Chem.,* **195**, 215 (1952).
11. Ling, K. H., Byrne, W. L., and Lardy, H. A., *in* "Methods in Enzymology" (S. P. Colowick and N. O. Kaplan, eds.), Vol 1, p. 306. Academic Press, New York, 1955.

PHOSPHOFRUCTOKINASE AND THE PASTEUR EFFECT*

Janet V. Passonneau and Oliver H. Lowry

Department of Pharmacology and the Beaumont-May
Institute of Neurology, Washington University
School of Medicine, St. Louis, Missouri
(Received for publication, December 13, 1961)

Several recent studies, based on measurement of substrate levels, indicate that the enzyme reaction primarily responsible for the Pasteur effect is phosphofructokinase (PFK) in yeast(1), ascites tumor cells(2), heart(3), and diaphragm(4). The last confirms earlier less clearcut evidence for skeletal muscle(5). The rapid increase in fructose diphosphate (EDP) following onset of ischemia in brain(6,7) may be interpreted in the same way, and this is substantiated by *in vitro* studies with supernatant fluid from brain homogenates supplemented with liver mitochondria(8,9).

Bücher(10) concludes that in insect wing muscles PFK is activated when the metabolism is increased during muscular activity. He stated the PFK could be activated *in vitro* by changes in concentrations of ATP, Mg and fructose-6-P (F6P).

Mansour and Menard found that glycolysis in liver flukes is controlled by PFK(11). When glycolysis was stimulated by serotonin or 3′,5′-cycloadenylate (3′,5′-AMP), the concentrations of glucose-6-P (G6P) and F6P fell and FDP role(12). In addition it was shown that partially purified PFK could be activated *in vitro* by ATP, Mg and 3′,5′-AMP, and that activation was characterized by decrease in the concentration of F6P required for activity in the presence of high levels of ATP(13). Dr. Mansour has recently found a similar phenomenon in heart muscle (personal communication).

*Janet V. Passonneau and Oliver H. Lowry, "Phosphofructokinase and the Pasteur Effect," *Biochem. Biophys. Res. Comm.,* 7: No. 1, 10–15 (1962). Reprinted with the permission of the authors and of the publisher.

This work was supported by grants from the American Cancer Society (P-38) and the National Institute of Health (B-434(C8)).

In studies to be reported elsewhere, every member of the glycolytic cycle was measured in mouse brain at short intervals after decapitation. Within 3 seconds (10 day old mice) G6P and F6P fell to half and FDP nearly doubled. Clearly F6P was being phosphorylated at a more rapid rate than before decapitation. All other substrate and cofactor changes were those expected from the increased rate of glycolysis and eventual deficit in ATP. There was no sign that any other step except that of glucose phosphorylation was stimulated by the anoxia.

This communication presents evidence that the peculiar kinetic properties of PFK may explain the Pasteur effect.

Experimental

PFK activity was measured at 26 to 29°. In most experiments the FDP generated was led enzymatically to glycerol-P, and the disappearance of DPNH (initially .01 mM) was followed fluorometrically. The substrate was F6P in studies with purified PFK, or a mixture of G6P and F6P, kept near equilibrium with P-glucoisomerase, in the case of crude enzyme preparations.

In a few experiments ADP, rather than FDP, was measured with added lactic dehydrogenase, DPNH, P-enolpyruvate and its kinase. Again the decrease in DPNH was followed fluorometrically.

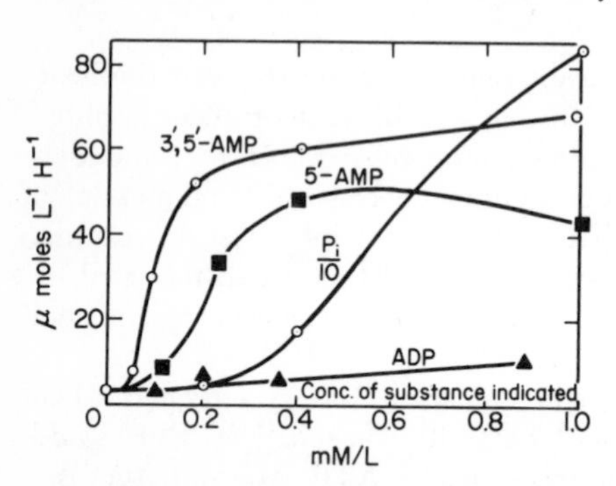

Fig. 1. Effect of P_i and three nucleotides on activity of rabbit muscle PFK. The other concentrations of reactants were ATP, 2.3 mM; Mg, 5 mM; F6P, 0.1 mM. The P_i concentrations, as indicated, are drawn to a 10 fold smaller scale.

In all cases, in an effort to partly simulate conditions *in vivo,* the studies were conducted in 0.02 M imidazole buffer, pH 7.0, with 0.15M potassium acetate, 0.01% bovine plasma albumin, and, unless noted, 5 mM $MgCl_2$. Rabbit muscle PFK was purified 20 fold according to Ling et al.[14].

In the presence of 2.3 mM ATP and 0.1 mM F6P, activity was increased 10 fold by either 0.09 mM 3′,5′-AMP, 0.22 mM AMP or 4.5 mM P_i (estimated from Fig. 1) whereas ADP was comparatively inactive. If, however, all samples contained 2 mM P_i, activity was increased by much lower concentrations of 3′,5′-AMP and AMP; ADP also became quite effective (Table I). To simulate the changes which occur early in anoxia, P_i was raised, ATP lowered, and AMP and ADP added at low levels. Activity increased 30 fold (Table I), which is 4 times the sum of the individual increments when these components were changed one at a time.

The PFK stimulators appear to exert their action chiefly by lowering, as much as 25 fold, the concentration of F6P required for activity (Fig. 2), whereas increasing ATP concentration has the opposite effect.

Table I

ACTIVATION OF RABBIT MUSCLE PHOSPHOFRUCTOKINASE

Except where noted, concentrations were ATP, 2.3 mM; Mg, 5 mM; F6P, 0.1 mM; P_i, 2 mM. The enzyme preparation was used at a concentration of 0.1 μg per ml and pre-incubated 5 minutes with all components present, except F6P, which was added to start the reaction. Velocity is expressed as micromoles per liter per hour.

Addition		Velocity	Addition or change		Velocity
0		2	3′, 5′-AMP	.047 mM	71
ADP	.058 mM	7	ATP	1.8 〃	3
ADP	.144 〃	12	ATP	1.8 〃	
5′-AMP	.02 〃	5	5′-AMP	.02 〃	
5′-AMP	.12 〃	61	ADP	.058 〃	84
3′, 5′-AMP	.007 〃	4	P_i	4 〃	

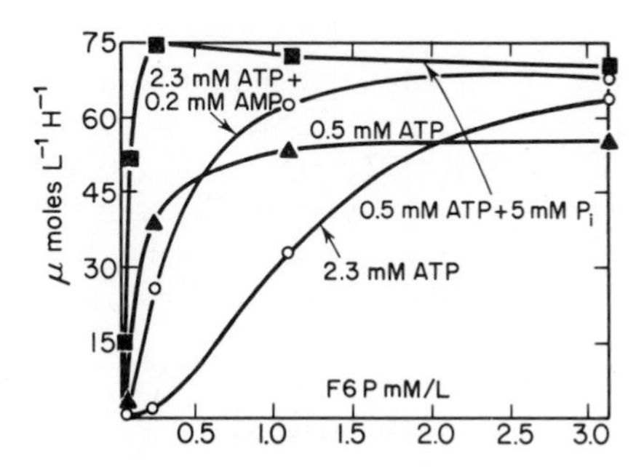

Fig. 2. Effect of F6P concentration of PFK activity under four different conditions.

Mg concentration is not critical. With low levels of F6P and high ATP, both P_i and AMP stimulated markedly whether Mg was 1, 2, 5 or 10 mM. However, absolute activity was decreased at the low Mg levels in agreement with Lardy and Parks[15].

Whole mouse brain homogenate and supernatant fluid were studied extensively and the PFK behavior was very similar to that of purified muscle PFK. Whole homogenates from four rabbit tissues behaved in a comparable manner as far as they were tested (Table II).

The following compounds had little or no stimulating activity for muscle PFK at concentrations equal to or greater than those likely to be found in the cell: glucose, 6-P-gluconate, glucose-1-P, 3-P-glycerate, 2-P-glycerate, P-enolpyruvate, pyruvate, lactate, α-glycerol-P, creatine, P-creatine, 5′-mono-P or di-P derivatives of cytosine, guanosine, uridine or inosine, 5′-TMP, 3′-AMP or adenosine tetraphosphate. G6P was about a fifth to a tenth as effective as F6P as a stimulator.

The most active stimulator of all so far discovered is the product FDP itself. At a concentration of 0.01 mM it was as active as 0.2 mM AMP in the presence of 2.5 mM ATP, and 0.1 mM F6P. As little as 0.001 mM FDP tripled activity with 1 mM ATP. With ATP raised to 4 mM, FDP was ineffective even at 0.1 mM concentration.

Table II

PHOSPHOFRUCTOKINASE ACTIVITY OF WHOLE HOMOGENATES OF RABBIT TISSUES

The concentrations were ATP, 2.5 mM; Mg, 5 mM; F6P, 0.25 mM; (G6P + F6P = 1 mM). The velocity is expressed as millimoles per kilo of tissue per hour.

Tissue	Dilution during assay	Activity		
		No addition	+5 mM P_i	+.2 mM 5′-AMP
Muscle	93,000	304	3580	
Muscle	55,000			2580
Brain	8,600	8	328	
Kidney	8,800	9	107	125
Liver	12,800	12	128	74

In addition to these kinetic effects, it should be noted that PFK is exceedingly unstable at pH 7 to 8 unless some stabilizing substance is present. Under the conditions of assay, ATP (1 to 2.5 mM) or P_i (10 mM or greater) preserve activity unchanged for 10 minutes at least. Without either of these, activity is almost completely lost in 2 minutes. This stability behavior is not involved in the phenomena described, although it may of course be related.

Discussion

There are some striking similarities between the kinetic behavior of PFK from muscle and that from the liver fluke[13]. The activity of each is increased by 3′,5′-AMP, provided that ATP is high and F6P is low. However, Mansour and Mansour report that 5′-AMP is inactive for the fluke enzyme and it also appears that a preliminary activation or solubilization by ATP plus Mg is involved[13]. How closely the kinetic findings of Bücher for insect wing muscles resemble those reported here, is not clear from the preliminary report[10].

Clearly the kinetics of mammalian PFK are very complex, and it may be premature to try to explain them. However, the following hypothesis would account for most of the results to date: (1) that there are two ATP sites, a primary active site and a second inhibitory one, and (2) that P_i, AMP, F6P, FDP and the other stimulators all can compete with ATP for the second site and when these occupy the second site they are not inhibitory.

The results if they can be carried over to the cell, indicate that glycolysis can be turned on if either ATP falls, or P_i, ADP or AMP rises. This would provide multiple protection in the case of emergency. The fact that FDP also stimulates sets the stage for a trigger phenomenon. It seems a truly unique situation to have inhibition by one substrate, and stimulation by the other substrate and both products. Lardy and Parks[15] as well as Mansour and Mansour[13] have already suggested that the increase in PFK activity which occurs when ATP decreases may have significance for control of glycolysis.

The question of whether or not hexokinase is in turn controlled by PFK through variation in G6P levels, as suggested by several authors, is beyond the scope of this note.

References

1. Lynen, F., Hartmann, G., Netter, K. F., and Schuegraf, A., in *Regulation of Cell Metabolism,* p. 256, Eds. Wolstenholme, G. E. W., and O'Connor, C. M., Little Brown and Co., Boston, 1959.
2. Lonberg-Holm, K. K., *Biochim. Biophys. Acta,* **35**, 464 (1959).
3. Park, C. R., Morgan, H. E., Henderson, M. J., Regen, D. M., Cadenas, E., and Post, R. L, in *Recent Progress in Hormone Research,* **7**, 493, Ed. Pincus, G., Academic Press, New York and London, 1961.
4. Newsholme, E. A., and Randle, P. J., *Biochem. J.,* **80**, 655 (1961).
5. Iwakawa, Y., *J. Biochem.* (Japan), **36**, 191 (1944).
6. Thorn, W., Pfleiderer, G. P., Frowein, R. A., and Ross, I., *Pflügers Archiv.,* **261**, 334 (1955).
7. Thorn, W., Scholl, H., Pfleiderer, G., and Mueldener, B., *J. Neurochem.,* **2**, 150 (1958).
8. Aisenberg, A. C., Reinfarge, B., and Potter, V. R., *J. Biol. Chem.,* **224**, 1099 (1957).
9. Aisenberg, A. C., *J. Biol. Chem.,* **234**, 441 (1959).
10. Bücher, T., *Ang. Chem.,* **71**, 744 (1959).
11. Mansour, T. E., and Menard, J. S., *Fed. Proc.,* **19**, 50 (1960).
12. Mansour, T. E., *J. Phar. Exp. Therap.,* (in press) (1961).
13. Mansour, T. E., and Mansour, J. M., *J. Biol. Chem.,* (in press) (1962)
14. Ling, K.-H., Byrne, W. L., and Lardy, H. A., in *Methods in Enzymology,* **1**, 306, Eds. Colowick, S. P., and Kaplan, N. O., Academic Press, New York, 1955.
15. Lardy, H. A., and Parks, R. E., Jr., in *Enzymes: Units of Biological Structure and Function,* 584, Ed. Gaebler, O. H., Academic Press, New York, 1956.

PART X

SOME ASPECTS OF ORGANIC CHEMISTRY AS A BASIS FOR "BIOENGINEERING"

LIPMANN[1] SYNTHESIZED acetyl phosphate by a method stemming from an ancient recipe (1864) of Kämmerer and Carius. Dry trisilver phosphate was permitted to react with acetyl chloride. The resulting acetylphosphate (with heavy admixtures of di- and tri-acetyl phosphates) turned out to be active as a phosphoryl donor of 5-adenylic acid if incubated in the presence of a crude enzyme preparation from a strain of lactic acid bacteria[cf. 1].

In 1940 Lynen[2] synthesized pure monoacetylphosphate by using monosilver dibenzylphosphate. He obtained the crystalline monosilver salt of acetyl phosphate. A few years later Lipmann and Tuttle[3] worked out a particularly simple variation of acetyl phosphate synthesis. They simply added acetylchloride to a dry mixture consisting of 1 mole trisilver phosphate and 2 moles phosphoric acid. A high yield of acetylphosphate was obtained. This "happy" mixture was also used successfully with succinyl dichloride to give succinylphosphate[3].

As early as 1934, H. O. L. Fischer and E. Baer[4] accomplished the chemical synthesis of D-glyceraldehyde-3-phosphate. This was the component of triosephosphate which Embden had singled out in 1933 as the genuine hydrogen donor in the glycolysis of hexoses in muscle. The starting point of the Fischer-Baer synthesis was governed by the chemical synthesis of glycerophosphoric esters by E. Fischer and E. Pfähler and their own synthesis of the methylated cycloacetale of glyceraldehyde[cf. 4].

In 1938, α-glucose-1-phosphate was chemically synthesized by C. F. Cori, S. P. Colowick, and G. T. Cori[5]. As mentioned previously, this was the phosphoglucosyl compound which the Coris had discovered as a product of glycogen phosphorolysis in muscle. The early achievement of obtaining chemical synthesis of the α-glucose-1-phosphate with no admixture of the β-ester can be partly ascribed to the choice of acetobromoglucose and trisilverphosphate as reactants[6].

Alex Todd and his elite team successfully synthesized the polyphosphates of nucleosides between 1947 and 1950[7]. A most significant event in the recent history of bioengineering was marked by Khorana's observation[8] that monoesters of phosphoric acid plus orthophosphoric acid react readily with aliphatic or aromatic carbodiimides ($R'N = C = Nr'$) to give the corresponding pyrophosphate compounds (and the hydration product of carbodiimide). As Hotchkiss phrased it in a personal communication to me, "by this bifunctional agent two high yielding low-key reactions are substituted for the old fire and brimstone." In this way Khorana was able to make large amounts of all the nucleosides or deoxynucleoside triphosphates and also of polynucleotides[6].

Chantrenne, in 1948, found a common key to the initiation of various types of peptide synthesis. He discovered that benzoylphosphate is an almost useless benzoyl donor but that a *di* substituted derivative, dibenzoyl phosphoric ester, is a very effective benzoyl donor in chemical peptide synthesis[9]. If glycine was added, excellent yields of hippuric acid were obtained. This observation remained unnoticed for some time but, as mentioned in the section on protein synthesis, it was a forerunner of the type of compounds which turned out to be crucial in protein synthesis.

References

1. Lipmann, F.: *Cold Spring Harbor Symp.,* **8** (1939), 248.
2. Lynen, F.: *Ber. Chem. Ges.,* **73B** (1940), 367.
3. Lipmann, F., and Tuttle, L. C.: *J. Biol. Chem.,* **153** (1944), 571.
4. Fischer, H. O. L., and Baer, E.: *Ber. Chem. Ges.,* **65** (1932), 337.
5. Cori, C. F., Colowick, S. P., and Cori, G. T.: *J. Biol. Chem.,* **121** (1937), 465.
6. Khorana, H. Gobind: *Some Recent Developments in the Chemistry of Phosphate Ester of Biological Interest.* New York John Wiley and Sons, Inc., 1961.
7. Baddiley, J., and Todd, A. R.: *J. Chem. Soc.,* p. 648 (1947); ———, Michelson, A. M., and Todd, A. R.: *J. Chem. Soc.,* p. 582 (1949).
8. Smith, M., and Khorana, H. G.: *J. Amer. Chem. Soc.,* **80** (1958), 1141. ———, Moffatt, J. G., and Khorana, H. G.: *J. Amer. Chem. Soc.,* **80** (1958), 620.
9. Chantrenne, H., and Carlsberg, C. R., Lab.: *Ser. Chim.,* **26** (1948), 297.

EPILOGUE

Developments in biochemistry, organic chemistry, and biophysics have brought us to a place in the odyssey of the life sciences which one may call "bioengineering." Within the last few years it has been made possible to synthesize entirely new genetic code messages. The "Watson-Crick" type of double helix formation is as crucial in the transcription of such messages as it is in naturally occurring gene replication. Proteins have recently been synthesized chemically, and it will probably become feasible in the near future to synthesize enzymes by the same means. Perhaps we shall witness a rapid advance in biology governed largely by developments in "bioengineering." Such a development will demand as much creativity as did the previous ones. As Albert Szent-Györgyi once remarked, "Research is not a systematic occupation but an intuitive artistic vocation."

Our present development of biology has been greatly influenced by personalities perceptive to philosophical and dialectic issues. One of the least dogmatic and most understanding figures in the field of biochemistry during the first half of our century was Sir Frederick Gowland Hopkins who, in spite of his many successes in biochemistry, intuitively worried much about the direction of his own research. It was indicative of Hopkins' open mind that he recognized the serious challenge which he encountered through his close friend, John Scott Haldane, one of the great physiologists, who at that time seriously questioned the values of the more analytical sciences like biochemistry. Haldane, in his Gifford lectures of 1925 at Yale University, had predicted a very dim future for the biochemical approach to the life sciences. Hopkins even quoted one of Haldane's most forceful statements in a lecture in 1927. Haldane's serious criticism demanded concentrated thinking and a strong, yet carefully formulated reply. Haldane had stated, "The new physiology is biological physiology—not biophysics or biochemistry. The attempt to analyze living organisms into physical and chemical mechanisms is probably the most colossal failure in the whole history of modern science." (J. S. Haldane, 1925). This pronouncement had as its source a strong belief in the "wholistic" character of an organism.

A decade later, in 1937, Haldane's son, J. B. S. Haldane, the illustrious geneticist and a student of Hopkins, although deeply impressed by his father's lifework, expressed almost the opposite view. In a little essay dedicated to Hopkins on his 75th birthday, he wrote a prophetic statement about the future of molecular biology and the life sciences. Young Haldane said, "The geneticist today is in a rather difficult position. He must have at least a bowing acquaintance with anatomy, cytology and mathematics. He must dabble in taxonomy, physics, and even psychology. So unless he is a pupil of Hopkins he may sometimes forget that the only precise account of the most fundamental phenomena which he studies must ultimately be a biochemical account." (J. B. S. Haldane, 1937.)

Biochemical genetics began as "physiological genetics" more than 30 years ago with the transplantation studies of eye colors in *Drosophila* by Beadle and Ephrussi and the studies on flower colors by Scott-Moncrief and J. B. S. Haldane. In the latest development of physiological genetics, that of Nirenberg and Khorana, new genetic code units are made enzymatically or synthetically and are "put to work" through molecular translation units, the polysomes, and new phenotypic products, the polypeptides, have been "deciphered." An ultimate biochemical account has indeed been made. Young Haldane's early prophecies were certainly not far off.

What about the development of other branches of cell biology pertaining more to function, for instance, cell motion? Since this book has described the development of our understanding of muscle contraction, let us briefly recapitulate some of the highlights we have encountered in the study of muscle biochemistry and biophysics. Muscle can be looked upon as an instrument which converts chemical and electrical energy into mechanical energy. Upon arrival of a signal, excitation first propagates along the surface of the muscle fiber and then is carried into the interior, the contractile element is transformed from the inactive to the active state. Activity brings about tension and shortening and is accompanied by release of heat. The source of energy, manifested in the form of work (and heat) can be traced ultimately to the splitting of phosphate from ATP. However, the most immediate source still remains unknown. Entropy changes may occur, and if they do they most likely are of very different nature from those seen in the release of a stretched rubber string. The electron microscope work by H. Huxley and J. Hanson demonstrates the existence of an ultrastructure which suggests a sliding mechanism of rods relative to each other. The rods themselves may not change in length. Rather the thick rods, myosin molecules, seem to slide past the thin rods, actin fibers. The cross bridges between the two types of rods and alterations in these structures are probably crucial for this mechanism. The head part of the myosin molecules that carried the cross bridges possesses the ability to split ATP specifically. Perhaps the free energy of this splitting drives a reaction which brings about a large decrease in entropy. Perhaps we are on the road to understanding the nature of cellular motion.

Biochemical genetics and muscle physiology are not the only disciplines which have led to insight into the nature of the living cell. An example of the creative fusion of several disciplines of the life sciences is that initiated from a special field of sensory physiology, vision. The broad approach of Hecht, Wald and Rushton to the biological basis of vision may have reassured even Haldane Senior that biologists are able to use physiology, molecular biology, and biophysics as well as biochemistry to arrive at a wholistic view of the living organism. The text of the caption from Hopkins' picture in this book may fit well here. It was prompted by the challenge from Alfred North Whitehead's essays on the wholistic approach, an approach related to J. S. Haldane's reflections. Hopkins graciously summed up his thoughts on these major biological problems in his Harvard Tercentenary Address in 1936 as follows.

"It is impossible at Harvard to forget the teaching of that profound philosopher Alfred North Whitehead, who came one day from Cambridge to Cambridge. We have his assurance that the conception of organism must replace in thought the unrelated or accidentally related entities which were the abstract units of Newtonian physics. Reality always involves relations internal and external, while an event and no static entity must be the unit of things real.

Biology from its very nature has never been much tempted to abstraction and for it the organism has always been the only significant unit, while the living organism as it exists in time is essentially a directed event. The question that arises is whether the modern biochemist, in analysing the organism into the parts which he is best able to study, has so departed from reality that his studies have no longer biological meaning. I myself would venture to answer that question, if it troubles the minds of any, by saying that *so long as his analysis involves the isolation of events, and not merely of substances, he is not in danger of such departure.*"

Even this approach could be further questioned. In a decade of successful molecular biology largely sparked by Max Delbrück, the following argument was raised by Delbrück himself.

"It is not to be taken as a foregone conclusion that structure on the molecular scale on the one hand and integrated function on the other hand, are compatible observables." And he meant here to bring up difficulties far beyond the purely technical, related to those encountered in the description of atomic events.

Recent developments of molecular biology do not indicate that contradictions of a more serious nature, like those encountered in quantum mechanics, have not as yet been confronted in biology. Should they arise, we can only hope that we have sufficient ingenuity to tackle them.

Perhaps Niels Bohr has given us the best advice for the future of biology in the little line also quoted recently by O. R. Frisch. "What we try in science–" so Bohr insisted "is to evolve a way of speaking unambiguously about our experiences. How far the word 'reality' is suitable for that purpose is in itself a question to be decided in the light of 'experience.' "

References

Reference to and citation from John Scott Haldane's Gifford Lecture stem from two Hopkins essays, "The Influence of Chemical Thought on Biology" (Harvard, 1936) p. 281, and "A Lecture on Organicism Delivered Upon an Unknown Occasion" (1927), p. 179, in *Hopkins and Biochemistry, 1861–1947*, eds. Joseph Needham and Ernest Baldwin, A Commemorative Volume prepared on the occasion of the First International Congress of Biochemistry, Cambridge, 1949. Cambridge: W. Heffer and Sons Ltd., 1949.

Quotation of John Burdon Sanderson Haldane's words about Hopkins and biochemistry is from "The Biochemistry of the Individual" by J. B. S. Haldane, in *Perspectives in Biochemistry,* eds. Joseph Needham and David E. Green. Thirty-one essays presented to Sir Frederick Gowland Hopkins by past and present members of his Laboratory. Cambridge: Cambridge University Press, 1937, pp. 1 and 10.

Albert Szent-Györgyi's observation is taken from his essay, "Lost in the Twentieth Century," in the *Annual Review of Biochemistry,* **32** (1963), 7.

Quotation of Max Delbrück is from "Atom Physics in 1910 and Molecular Biology in 1957," in Epilogue Seminar at M.I.T. after Bohr's Compton Lectures (an unpublished essay).

O. R. Frisch's article was "Two Biographies of Niels Bohr," in *Scientific American* (June, 1967), p. 146.